Phytohormones and Related Compounds–A Comprehensive Treatise, Volume I

The Biochemistry of Phytohormones and Related Compounds

Phytohormones and Related Compounds - A Comprehensive Treatise

Volume I

The Biochemistry of Phytohormones and Related Compounds

edited by
D.S. LETHAM, P.B. GOODWIN AND T.J.V. HIGGINS

1978

ELSEVIER/NORTH-HOLLAND BIOMEDICAL PRESS
AMSTERDAM · OXFORD · NEW YORK

ISBN 0–444–80053–0

Published by:

Elsevier/North-Holland Biomedical Press
335 Jan van Galenstraat, P.O. Box 211
Amsterdam, The Netherlands

Sole distributors for the U.S.A. and Canada:

Elsevier North-Holland, Inc.
52 Vanderbilt Avenue
New York, N.Y. 10017

Library of Congress Cataloging in Publication Data

Main entry under title:

Phytohormones and related compounds.

Bibliography:
Includes index.
CONTENTS: v. 1. The biochemistry of phytohormones and related compounds.
1. Hormones (Plants) I. Letham, D.S. II. Goodwin, P.B. III. Higgins, T.J.V.
QK731.P58 581.1'927 78–18843
ISBN 0–444–80053–0 (v. 1)

PRINTED IN THE NETHERLANDS

List of Contributors

J.E. GRAEBE
Pflanzenphysiologisches Institut der Universität Göttingen, Untere Karspüle 2, D-3400 Göttingen, G.F.R.

P.B. GOODWIN
Department of Agronomy and Horticultural Science, University of Sydney, Sydney, N.S.W. 2006, Australia.

T.J.V. HIGGINS
Division of Plant Industry, Commonwealth Scientific and Industrial Research Organization, P.O. Box 1600, Canberra City, ACT. 2601, Australia.

J.V. JACOBSEN
Division of Plant Industry, Commonwealth Scientific and Industrial Research Organization, P.O. Box 1600, Canberra City, ACT. 2601, Australia.

D.S. LETHAM
Research School of Biological Sciences, The Australian National University, Canberra City, ACT 2601, Australia.

B.V. MILBORROW
School of Biochemistry, University of New South Wales, Kensington, Sidney, N.S.W., Australia.

D. OSBORNE
Agricultural Research Council Unit of Developmental Botany, 181A Huntingdon Road, Cambridge, England.

H.J. ROPER
Pflanzenphysiologisches Institut der Universität Göttingen, Untere Karspüle 2, D-3400 Göttingen, G.F.R.

E.A. SCHNEIDER
Biology Department, Carleton University, Ottawa, Ontario, K15. 5B6 Canada.

F. WIGHTMAN
Biology Department, Carleton University, Ottawa, Ontario, K15. 5B6 Canada.

Preface

Forty-one years have elapsed since the classic work "Phytohormones" (MacMillan, 1937) was written by the plant hormone pioneers, F.W. Went and K.V. Thimann. The volume produced by Went and Thimann discussed the only type of plant hormone then known, namely, auxin. Today five types of hormones are known to occur in higher plants and considerable progress has been made in elucidating the structure and function of algal and fungal hormones. In spite of the great advances now made in the important field of phytohormones, there has been no attempt in recent years to discuss these compounds collectively in a comprehensive way. It was for this reason that the present volumes were prepared.

Volume I discusses the biochemistry of the hormones of both higher and lower plants and also deals in a chemical way with other naturally occurring compounds which influence plant growth when supplied exogenously. In Volume I, the activities of higher plant hormones at the subcellular level are emphasized (Chapters 8–11). The volume commences with a historical and somewhat critical essay (Chapter 1) which is intended primarily for "beginners". We hope it will enable them to avoid some of the pitfalls encountered by earlier workers. Volume II is intended to be a companion volume and the two volumes are cross-referenced. Volume II discusses the hormonal control of the development and senescence of higher plants.

Over the last 25 years, the emphasis on unitary effects of single hormones and the simplistic interpretation of their actions has been replaced by a recognition of chemically diverse groups of hormones acting synergistically, antagonistically and/or sequentially to regulate and coordinate organ development. Furthermore, we wish to emphasize that there should be an integrated approach to plant biology placing the role of hormones in proper perspective within the overall context of plant metabolism. It is our hope that the present volumes will enable people to adopt a broader approach to the study of hormonal regulation of plant development.

Finally the Editors wish to thank all contributors for their helpful attitude and for their great patience. Thanks are also due to Bettina I. Gollnow and Rosemary Metcalfe for their help with many editorial problems.

Abbreviations used in Volumes I and II

A	Adenosine
ABA	Abscisic acid
ACh	Acetylcholine
ADH	1-Acetoxy-2,4-dihydroxy-*n*-heptadeca-16-ene
AMO-1618	2′-Isopropyl-4′-(trimethylammonium chloride)-5′-methylphenyl piperidine-1-carboxylate
ARF	Adventitious root formation
atm.	Atmosphere
ATP	Adenosine 5′-triphosphate
ATP/O_2 Ratio (P/O ratio)	Moles of ATP formed relative to the number of gram atoms of oxygen consumed
6-BAP (or BAP)	6-Benzylaminopurine
OH-BAP	6-(*o*-Hydroxybenzylamino)purine
BOH	β-Hydroxyethylhydrazine
BTOA	Benzothiazol-2-yloxyacetic acid
C	Cytidine
°C	Degrees Celsius
C-3 Pathway	The Calvin–Bassham cycle of photosynthesis
C-4 Pathway	The Hatch–Slack pathway describing carboxylation of 3-carbon compounds to yield 4-carbon acids which donate the carbon to the C-3 pathway of photosynthesis
C_6/C_1 Ratio	An estimate of the relative contributions of glycolysis and the pentose phosphate pathway in the breakdown of glucose
cAMP	Adenosine 3′:5′ cyclic monophosphate (cyclic-AMP)
CCC	2-Chloroethyltrimethylammonium chloride (chlorocholine chloride)
CCCP	Carbonylcyanide *m*-chlorophenylhydrozone
CD	Circular dichroism
Ci	Curie
mCi	millicurie
μCi	microcurie
CoA	Coenzyme A
conc.	Concentration
C_0t	Concentration of DNA in a hybridization reaction multiplied by the time of incubation (moles of nucleotide × sec/litre)
CPA	4-Chlorophenoxyacetic acid
CPIB	α-(4-Chlorophenoxy)isobutyric acid

cpm	Counts per minute
CPP	Copalyl pyrophosphate
cv., cvs.	Cultivar, cultivars, respectively
2,4-D	2,4-Dichlorophenoxyacetic acid
DBcAMP	6-*N*,2′-*O*-Dibutyryladenosine 3′:5′-cyclic monophosphate
DCCD	Dicyclohexylcarbodiimide
DIHB	3,5-Diiodo-4-hydroxybenzoic acid
DMAPP	γ,γ-Dimethylallylpyrophosphate
DNA-RNA	DNA and RNA hybrid•
DNBP	4,6-Dinitro-2-*sec*-butylphenol
DNP	Day-neutral plant
2,4-DNP	2,4-Dinitrophenol
dpm	Disintegrations per minute
equiv.	Equivalent
ER	Endoplasmic reticulum
FITC	Fluorescein isothiocyanate
FMN	Flavin mononucleotide
FPP	*trans*-Farnesyl pyrophosphate
5-FU	5-Fluorouracil
FUDR	Fluorodeoxyuridine
G	Guanosine
g	Gram
kg	kilogram
mg	milligram
μg	microgram
ng	nanogram
pg	picogram
g	Acceleration due to gravity
GA	Gibberellin
GA_3	Gibberellic acid
GC-MS	Combined gas chromatography-mass spectrometry
GC-RC	Combined gas chromatography-radioisotope counting
GC-SICM	Gas chromatography with selected ion current monitoring
GGPP	*trans*-Geranylgeranyl pyrophosphate
GLC	Gas-liquid chromatography
GPP	*trans*-Geranyl pyrophosphate
HEPES	*N*-2-Hydroxyethylpiperazine-*N*′-2-ethanesulphonic acid
H-ol	Helminthosporol
hr	Hour
IAA	Indol-3-ylacetic acid
IAAld	Indol-3-ylacetaldehyde
IAAsp	Indol-3-ylacetylaspartic acid
IAM	Indol-3-ylacetamide
IAN	Indol-3-ylacetonitrile

IAOx	Indol-3-ylacetaldoxime
IBA	γ-(Indol-3-yl)butyric acid
ICA	Indole-3-carboxylic acid
IMP	Inosine 5′-monophosphate
iP	N^6-($\triangle^2$-isopentenyl)adenine, i.e. 6-(3-methylbut-2-enylamino)purine (in Chapter 4 of Vol. 1, a system of abbreviations for derivatives of iP is used; substituents are indicated in brackets or parentheses before the symbol iP)
[9R]iP	N^6-($\triangle^2$-Isopentenyl)adenosine
IPA	Indol-3-ylpropionic acid
IPP	$\triangle^3$-Isopentenyl pyrophosphate
IPyA	Indol-3-ylpyruvic acid
IR	Infrared
K	Kinetin
KMBA	α-Keto-γ-methylthiobutyric acid
Krebs Cycle	Tricarboxylic acid cycle
l	Litre
μl	microlitre
nl	nanolitre
LD	Long day(s)
LDP	Long-day plant
LSDP	Long-short-day plant
LWD	Leaf water deficit
M	Molar (concentration)
m*M*	millimolar
μM	micromolar
n*M*	nanomolar
m	Metre
mm	millimetre
μm	micron
mμm	millimicron
MAK	Methylated albumin kieselguhr
MCPA	2-Methyl-4-chlorophenoxyacetic acid
mcs	Meter-candle-seconds
MH	Maleic hydrazide
min	Minute
mole	Mole
mmole	millimole
μmole	micromole
mol.wt.	Molecular weight
m.p.	Melting point
MVA	Mevalonic acid
NAA	Naphth-1-ylacetic acid (also α-napthalenacetic acid)
NAAm	Naphth-1-ylacetamide

NAD (NAD^+)	Nicotinamide-adenine dinucleotide
NADH	Nicotinamide-adenine dinucleotide (reduced)
NADPH	Nicotinamide-adenine dinucleotide phosphate (reduced)
NHP	Non-Histone proteins
NMR	Nuclear magnetic resonance
NMSP	α-(Napth-1-ylmethylthio)propionic acid
NOXA	Naphth-2-yloxyacetic acid
NPA	*N*-(Naphth-1-yl)phthalamic acid
ODC	Ornithine decarboxylase
O.P.	Osmotic pressure
ORD	Optical rotary dispersion
P_{fr}	Phytochrome; far-red absorbing form
PAA	Phenylacetic acid
PAL	Phenylalanine ammonia lyase
PEP carboxylase	Phospho*enol*pyruvate carboxylase
PIB	α-Phenoxyisobutyric acid
PMA	Phenylmercuric acetate
Polymerase I II III	DNA-dependent RNA polymerases which transcribe rRNA, mRNA and sRNA respectively
poly(U)	Polyuridylic acid
ppb	Parts per billion
ppm	Parts per million
RH	Relative humidity
RNase	Ribonuclease
RNA	Ribonucleic acid
D-RNA	DNA-like (base composition) RNA
hrRNA	Heavy ribosomal RNA
lrRNA	Light ribosomal RNA
mRNA	Messenger RNA
met-tRNA	Methionyl-tRNA
mRNP	mRNA-protein complexes
poly(A) RNA	RNA molecules with long stretches of A residues attached to the 3′ end
pre-mRNA	Precursor mRNA
pre-rRNA	Precursor rRNA
pre-tRNA	Precursor tRNA
rRNA	Ribosomal RNA
sRNA	Soluble (4 and 5s) RNA
tRNA	Transfer RNA
TB-RNA	Tenaciously-bound RNA
tcRNA	Translational control RNA
RuDP Carboxylase	Ribulose 1,5-bisphosphate carboxylase
SADH	*N*-(Dimethylamino)succinamic acid, also termed B-995, B-9 and Alar
SD	Short day(s)

SDP	Short-day plant
SLDP	Short-long-day plant
SDS	Sodium dodecyl sulphate
SDS-PAGE	Sodium dodecyl sulphate-polyacrylamide gel electrophores
SF	Senescence factor
sp.; spp. (plural)	Species
2,4,5-T	2,4,5-Trichlorophenoxyacetic acid
2,4,6-T	2,4,6-Trichlorophenoxyacetic acid
TIBA	2,3,5-Triiodobenzoic acid
TLC	Thin-layer chromatography
TMSi	Trimethylsilyl
TNH_2	Tryptamine
TOL	Tryptophol
TPP	Tryptophan
Tris	2-Amino-2-hydroxy-methylpropane-1,3-diol
U	Uridine
UV	Ultraviolet
V	Volt
vol.	Volume
v/v	Relative volumes of two components
W	Watt
wt.	Weight
Z	Zeatin (in Chapter 4 of Vol. 1, a system of abbreviations for derivatives of zeatin is used; substituents are indicated in brackets or parentheses before the letter Z)
[9R]Z	Zeatin riboside
[9R-5′P]Z	Zeatin riboside 5′-monophosphate
%	Per cent

Note concerning chemical nomenclature

Auxins

In naming indole auxins, the indole moiety is denoted by the term indol-n-yl, where n is the site of substitution. Thus, IAA is usually named indol-3-ylacetic acid, though in some instances it may be referred to as 3-indoleacetic acid or indole-3-acetic acid. The same system of nomenclature is used for other aromatic auxins; e.g. naphth-1-ylacetic acid, which also appears as α-napthalene acetic acid.

Gibberellins

The complex system of nomenclature for gibberellins and related compounds is clearly defined in Chapter 3 (section 3.2.1) and requires no further comment.

Configuration of optically active compounds

In Chapters 3 and 6 particularly, the Cahn-Ingold-Prelog system, i.e. *(R, S)*-nomenclature, is used to designate absolute configuration of optically active molecules. Plant physiologists naturally find this system confusing and would find the following lucid explanations helpful:

(a) Cahn, R.S., Ingold, C.K. and Prelog, V. (1956) Experientia (Basel), 12, 81–94.
(b) Mislow, K. (1965) Introduction to Stereochemistry, Benjamin, New York.
(c) Eliel, E.L. (1971) J. chem. Educ., 48, 163–167.

Cytokinins

In discussing sites of substitution of the adenine molecule, the exocyclic nitrogen atom, at position 6 of the purine ring, is designated position N_6. When specific compounds are named as derivatives of adenine, adenosine or adenylic acid (an infrequent occurrence), the group attached to the exocyclic nitrogen is prefixed by N^6. Except when excessively cumbersome or otherwise inappropriate, purine cytokinins are always given systematic names as derivatives of purine and are listed thus in the index; thus, 6-benzylaminopurine not N^6-benzyladenine. In Chapter 4, a system of abbreviations for common naturally occurring cytokinins is used and is defined in section 4.2.1.

Contents

Chapter 4
Cytokinins – D.S. Letham

Chapter 6
Abscisic Acid – B.V. Milborrow

Chapter 7

Natural-occurring Plant Growth Regulations other than the Principal Hormones of Higher Plants – D.S. Letham

Chapter 8
Phytohormones and Subcellular Structural Modification – T.J.V. Higgins and J.V. Jacobsen

Chapter 9
The Influence of Plant Hormones on Selected Aspects of Cellular Metabolism – T.J.V. Higgins and J.V. Jacobsen

Chapter 10
The Influence of Phytohormones on Replication and Transcription – J.V. Jacobsen and T.J.V. Higgins

Chapter 11
Posttranscriptional, Translational and Posttranslational Effects of Plant Hormones – J.V. Jacobsen and T.J.V. Higgins

Phytohormones and Related Compounds – A Comprehensive Treatise, Volume I
Editors: Letham, Goodwin and Higgins

1

Phytohormones in Retrospect

D.S. LETHAM, T.J.V. HIGGINS, P.B. GOODWIN and J.V. JACOBSEN

Introduction

For a multicellular organism to exist there must be intercellular coordination of growth and development, and of responses to the environment. Even very simple cell associations may show such integrations of activity. For example, the blue-green alga *Anabaena* consists of vegetative cells aligned in filaments. Differentiated cells (heterocysts) arise at regularly spaced intervals in each filament. In higher plants, cells are able to interact in a considerable number of modes. Physical interactions include those mediated by turgor pressure; for example adequate turgor pressure is necessary for the elongation of stems and roots, the expansion of leaves, and for normal cambial induction and function. Pressure deficits arise in the leaves but are very rapidly transmitted throughout the plant. Certain rapid movements in plants are mediated by electrical signals, passing through electrotonically interconnected cells, as in the sensitive plant *Mimosa,* and in the insectivorous *Drosera* (Pickard 1973).

More potential for flexibility exists in the chemical interchanges between cells. A large number of chemicals can potentially be exchanged, the limitations being only very broadly set by molecular weight, water- and lipid-solubility and binding to other molecules. Of this wide range of potentially and actually exchanged molecules, a small number of organic compounds have now been shown to act as coordinators of the growth and development of the plant and to be active in minute amounts. Such compounds are termed phytohormones, or plant hormones. Phytohormones move from a site of production to a site of action to evoke characteristic responses. However, it should be remembered that higher plants also contain other highly active compounds which regulate growth when supplied exogenously, but their movement to a site of action in the intact plant has not been established (see Chapter 7). Although such compounds may not be chemical "messengers" or coordinators of growth as are the phytohormones, certain of them may still play a significant role in the regulation of plant growth in vivo.

1.1. The chemical diversity of phytohormones and related compounds

The sequence of publications which led to the concept of phytohormones and culminated in the isolation of the first phytohormone from a higher plant can be traced back to 1758. Duhamel du Monceau (1758) observed that swellings and root formation occurred above, but not below, ring wounds on stems. To account for observations of this type, Sachs (1882, 1893) proposed that organ-forming substances were produced in the plant and were distributed polarly; for example, root-forming substances were considered to be produced in the leaves and to move basipetally in the stems. However a half century was to elapse before root formation was associated with phytohormone (auxin) derived from the shoot (Thimann and Went 1934; Thimann and Koepfli 1935) and it was the study of tropisms that led to the first definite knowledge of this type of hormone.

Darwin and Darwin (1880) showed that coleoptiles curve towards a light source, but this phototropic curvature did not occur if the tip of the coleoptile was shaded with a small cap. Although the curvature was distributed over almost the whole coleoptile, the tip must have been the site of perception. Hence the Darwins stated: "we must, therefore, conclude that when seedlings are freely exposed to a lateral light, some influence is transmitted from the upper to the lower part, causing the latter to bend". The phototropic stimulus was found to pass through gelatin and thus was probably chemical in nature (Boysen Jensen 1911, 1913). Paal (1919) established that if the coleoptile tip was excised and replaced asymmetrically, the coleoptile curved even in darkness. He proposed that the correlation carrier was a growth substance secreted by the tip, and that in darkness it diffused symmetrically down the coleoptile, but was distributed asymmetrically when the plant was illuminated unilaterally. The concept that normal growth is controlled by a symmetrically distributed growth substance produced in the coleoptile tip was confirmed by Soding (1923, 1925) in straight growth experiments involving tip excision and replacement.

The above observations constituted strong evidence that the coleoptile tip secreted a phytohormone controlling normal elongation and phototropism; the next step was to extract and isolate the active compound. After a number of unsuccessful attempts to prepare active aqueous extracts, a new approach was adopted by Went (1928) who placed excised coleoptile tips on agar and then positioned the agar asymmetrically on freshly decapitated coleoptiles. The growth hormone was secreted into the agar which then produced coleoptile curvatures proportional to the number of tips or to the time they were on the agar. Went's classical experiments were significant in two respects; firstly, the existence of a hormone produced by the coleoptile tips was confirmed, and secondly, a quantitative bioassay had been developed. The term auxins derived from auxein (Greek, to increase) was proposed by Kogl and Haagen Smit (1931) for the group of compounds active in this assay and these workers also found that human urine was a rich source of activity (see also Chapter 2, section 2.3). From this source a crystalline product was isolated and termed auxin a by Kogl et al. (1933) and by similar purification methods another active product (auxin b) was obtained from malt and corn oil (Kogl et al. 1934a). By what appeared to be a brilliant piece of work, the structural formulae of auxins a and b were determined, both compounds being identified as carboxylic acids with a cyclopentene moiety (Kogl and Erxleben 1935). Subsequently another active compound was isolated from urine, identified as indol-3-ylacetic acid (I) and termed "hetero-auxin", since it differed markedly in structure from auxins a and b (Kogl et al. 1934b). This indole derivative was also isolated from yeast plasmolysate and found to account for a large part of the auxin activity of yeast (Kogl and Kostermans 1934). In the following year indol-3-ylacetic acid was identified as the auxin in *Rhizopus suinus* cultures (Thimann 1935).

After the isolation of the above products, much attention was given to the identity of the auxins in higher plants. Unfortunately unreliable indirect methods of comparison were adopted for convenience, such as stability to acid and base and calculation of molecular weights from diffusion constants. From this type of evidence it was concluded that: "the auxin of coleoptiles is auxin a. It is, indeed, probable that other higher plants contain the same auxin" (Went and Thimann 1937). This view persisted until well after the Second World War when a number of observations indicated that the previous work concerning auxins a and b was misleading (for key references, see Audus 1972). Firstly, auxins a and b could not be isolated from plants or detected in urine by later workers. Secondly, a synthetic compound with the proposed structure of auxin b lacked significant auxin activity. Thirdly, authentic samples of auxins a and b were examined using modern physical methods and the structures assigned by Kogl and coworkers were not confirmed. Fourthly, indol-3-ylacetic acid was found to account for most of the auxin activity of urine and was isolated from immature corn kernels by Haagen Smit et al. (1946). In modern work indol-3-ylacetic acid has become firmly established as an auxin of higher plant tissues including *Avena* shoots. Related indole compounds and phenylacetic acid also occur in higher plants as auxins (see Chapter 2, section 2.3).

Work in the 1930's established the probable regulatory role of auxin in coleoptile growth, tropisms, root formation, apical dominance and cambial growth. During this period auxins were also found to induce callus formation and parthenocarpic fruit development. These multiple effects led to the view that the growth of plants was controlled by a single type of hormone (auxin) and this view persisted until after 1950 (see e.g. Thimann 1954). However in the period 1954–1963, two further major groups of phytohormones emerged, namely, the gibberellins and cytokinins.

In 1956, it was reported that extracts of peas promoted growth of sections of the first leaf of wheat seedlings (Radley 1956). Soon afterwards plant extracts, and particularly extracts of immature runner bean seeds, were found to promote elongation of dwarf pea seedlings (Radley 1958). This activity resembled that of a group of compounds termed gibberellins which had been isolated from cultures of the fungus *Gibberella fujikuroi*. The first of these to be defined structurally was termed gibberellic acid (see Chapter 3, section 3.1). The above responses suggested that gibberellins occur in higher plants and this was established unequivocally when in 1958 workers at Imperial Chemical Industries in England isolated a pure gibberellin from immature bean seeds; this was shown to be identical to the fungal gibberellin GA_1 (II) (MacMillan and Suter 1958; MacMillan et al. 1960). Gibberellins, which are diterpenoid acids with a tetracyclic ring system (III), now appear to be involved in the regulation of almost every aspect of plant development. However their most dramatic effect on growth is the promotion of shoot elongation of certain genetic dwarfs, e.g. dwarf peas and maize. Of the 51 gibberellins with established structures, 29 have been isolated only from higher plants, 13 only from *G. fujikuroi* and 9 from both the fungus and higher plants.

The third group of phytohormones, the cytokinins, was detected as a result of tissue culture studies. Tobacco pith explants cultured in the presence of auxin grew by cell enlargement. However cell division was induced by placing vascular tissue in contact with the explants on the auxin-containing medium, or by adding to the medium certain plant products, notably coconut milk or malt extract (Jablonski and Skoog 1954). The term "kinin", derived from cytokinesis, was first proposed for compounds with this cell-division-inducing activity. Because of the use of this term by animal physiologists for certain substances inducing muscle contraction, it has now been replaced by cytokinin. All known naturally-occurring cytokinins are N_6-substituted adenines (see Chapter 4), the first to be identified being termed zeatin (IV) because of its isolation from *Zea mays* (Letham 1963, 1966). In zeatin and all naturally-occurring cytokinins with high activity in tissue culture bioassays, the N_6-substituent is a C_5 isoprenoid moiety. However it soon became clear that cytokinins are not specifically cell division-inducing hormones, and like gibberellins and auxins they evoke a spectrum of responses in higher plants. The observation that cell size in apple fruits is positively correlated with the incidence of internal breakdown, a physiological storage disorder (Martin et al. 1954; Letham 1961), and the possibility of using exogenous cytokinin to promote cell division and to reduce cell size in apple fruits, provided the stimulus which resulted in the isolation of zeatin. Application of cytokinins, including zeatin, to apple flowers at full bloom and soon after petal fall was later found to induce parthenocarpic fruit set, to promote development around the calyx of seeded fruit, and to alter the shape, reduce the cell size and increase the density of seeded fruit (Williams and Stahly 1968; Letham 1969; Williams and Letham 1969). It is also of interest that, while the growth regulatory activity of auxins and gibberellins appears to be restricted to plant species, cytokinins also evoke responses from animal cells (see Chapter 4, section 4.1).

Unlike other phytohormones, cytokinins are known to occur covalently bound to tRNA. Forty-three minor nucleosides, which are modifications of the four common nucleosides, are now known to occur in tRNA and certain of these are located at the 3′-side of the anticodon (for a review of tRNA structure and minor bases in tRNA, see Letham and Wettenhall 1977). This position is always occupied by a purine nucleoside which is often adenine or a simple methyl purine, but certain tRNA species have complex (hypermodified) nucleosides including cytokinins at this position. Cytokinins occur not only in higher plant tRNA, but also in tRNA from animal tissues, yeast and bacteria (see Chapter 4, section 4.2.2).

The discovery of the fourth type of phytohormone was the result of physiological studies relating to abscission and bud dormancy. Addicott and coworkers extracted a substance from young cotton fruits which accelerated abscission of the cotyledonary petioles of *Phaseolus* in a bioassay using stem sections (Ohkuma et al. 1963). The active compound was originally termed "abscisin II" and was shown to be the sesquiterpene acid (V) now termed abscisic acid or ABA (Ohkuma et al. 1965). Other workers, but notably Wareing and colleagues in Wales, had observed that an inhibitory fraction in leaf and shoot apex extracts of sycamore and birch appeared to be involved in the induction of photoperiodic bud dormancy in these species (Phillips and Wareing 1959; Eagles and Wareing 1963, 1964). In particular, these workers observed that this fraction of birch leaf extracts, when applied to young leaves of actively growing seedlings of this species, induced formation of dormant apical buds. Hence compounds in the extract appeared to duplicate the effect of short days in causing bud dormancy. Subsequently, an active compound termed dormin, which was considered to possess such activity, was isolated from sycamore leaves by Cornforth et al. (1965) and was shown to be identical to abscisin II, that is to ABA. Although ABA was initially implicated in the regulation of onset of bud dormancy and leaf abscission, there is still no unequivocal evidence that these phenomena are directly controlled by ABA in the intact plant. However, work initiated by Wright (1969) has established that ABA plays an important role in the response of the plant to water stress. When plants are subject to water shortage, the ABA content of the leaves increases markedly (about 50-fold) and this induces stomatal closure (see Chapter 11, Volume II). At low concentrations exogenous ABA is also known to inhibit many phases of plant growth.

The fifth and final type of identified higher plant hormone is ethylene. The sequence of observations which led to the recognition of ethylene as a phytohormone began when Neljubow (1901) found that illuminating gas induced a horizontal growth habit in pea plants. He showed that this response could be attributed to ethylene present in the illuminating gas. In the presence of ethylene, etiolated pea seedlings exhibited a triple response – thickening of the subapical portion of the stem, reduced rate of stem elongation, and a horizontal growth habit of the shoots (Neljubow 1901). The effects of ethylene on plant growth were in fact well documented before it was established that ethylene is produced by plants. This was achieved by Gane (1934) who identified ethylene in the volatile compounds produced by apples. With the realization that ethylene is produced by ripe fruit and can induce ripening of other fruit, ethylene was accepted as a fruit-ripening hormone (see Chapter 10, Volume II). However ethylene now appears to have important regulatory roles in the vegetative plant. For example, endogenous ethylene production appears to be responsible for the formation and maintenance of the seedling hook which protects the young leaves during their emergence from the soil (Goeschl and Pratt 1968; Burg 1973). When the seedling shoot reaches light, ethylene production is suppressed and the plumular hook opens. Endogenous ethylene now appears to be involved in the regulation of many aspects of plant development including stem elongation, sex expression and abscission (see the Chapters in Volume II for detailed discussion concerning the role of ethylene and other plant hormones in the development of higher plants).

Many synthetic compounds have now been tested for auxin and cytokinin activity, and an appreciable number of analogues of ABA have also been prepared and bioassayed. The structural requirements for ABA-like activity are very stringent; even small changes in structure result in very weak activity or in complete loss of activity. In contrast, compounds which are quite different in a chemical sense to IAA and to zeatin may exhibit strong auxin or cytokinin activity, respectively. Thus substituted benzoic, naphthoic, phenylacetic, phenoxyacetic and picolinic acids all exhibit strong auxin activity; even certain dithiocarbamate derivatives (e.g. VI) without a ring system show high auxin activity. Similarly a diversity of 6-(substituted amino)purines exhibit cytokinin activity and certain substituted *N,N'*-diphenylureas are also cytokinins. In the case of auxins and cytokinins we are faced with the paradox of biological specificity being associated with an apparent lack of structural specificity. However, in the case of auxins, the critical molecular features may be physical parameters such as charge distribution as proposed by Porter and Thimann (1965) rather than molecular architecture per se.

Auxins, gibberellins, cytokinins, abscisic acid and ethylene are the 5 types of phytohormones of higher plants which have been characterized chemically. However a number of other important hormonal factors, four of which are referred to below, appear to occur in higher plants, but their chemical identity is at present unknown. The first to be mentioned is florigen. In plant species where flowering is under photoperiodic control, an unidentified hormonal signal which has been termed flowering hormone or florigen moves from the leaf, the site of photoperiodic perception, to the shoot apex where it induces flower formation. The concept of florigen, first introduced by Chailakhyan (1937) is discussed in Chapter 6, Volume II. Several unidentified compounds which induce leaf senescence and abscission have been detected in plants (see Chapter 7, Volume II). Perhaps the most significant of these is a substance termed senescence factor (SF), which diffuses from senescent bean and *Coleus* leaves (Osborne et al. 1972). This factor may induce abscission by promoting ethylene production in the petiole. The third unknown factor, first detected over 60 years ago by Ricca (1916), appears to be involved in leaf closure of the sensitive plant *Mimosa.* The stimulatory substance moves through the stem following certain injurious kinds of stimulation. This and similar unknown substances are reviewed by Sibaoka (1969) and by Pickard (1973). Finally, unidentified inhibitors may be involved in maintenance of apical dominance in stems and in other correlative inhibitions (see reviews by Libbert 1964; Phillips 1969, 1975). For example, a neutral inhibitor is present in potato tubers bearing inhibited sprouts and disappears from the tuber tissue within 24 hr after excision of dominant sprouts. It has been suggested this inhibitor plays a key role in apical dominance in potato tubers (Goodwin and Cansfield 1967). Identification of the above unknown growth regulators would be of great significance.

The phytohormones are of course not the only organic compounds present in higher plants which can affect plant growth when supplied exogenously at low concentrations. In addition to the phytohormones and vitamins, a great diversity of such growth regulators has been isolated from plants and characterized chemically (see Chapter 7). However, the responses they evoke are often observed in plant species unrelated to the species from which the compounds were isolated. For example, certain norditerpenoid dilactones isolated from *Podocarpus* spp. are highly active inhibitors of the growth of pea stem and *Avena* coleoptile segments; some are more active than ABA in these systems. Generally, there is little evidence to suggest that the identified growth regulators other than phytohormones contribute to the control of growth in vivo. However there are certain exceptions to this generalization which merit mention below.

Certain bicyclic ketones with a peroxide linkage isolated from *Eucalyptus grandis* (e.g. VII) appear to regulate root development in *Eucalyptus* spp. (see Chapter 7, section 7.10.3). There is strong evidence that batatasin I (formula VIII), a phenanthrene derivative, and batatasin III (formula IX), a bibenzyl derivative, which occur in yam bulbils, regulate the dormancy of these bulbils (see Chapter 7, section 7.13.1). Dihydroconiferyl alcohol appears to be produced by lettuce cotyledons and enhances the ability of gibberellin to promote lettuce hypocotyl elongation (see Chapter 7, section 7.12.5). Perhaps the most interesting of the compounds discussed in this paragraph is strigol (X), an unsaturated dilactone, which is produced by cotton roots and induces germination of seeds of the Angiosperm parasite, *Striga lutea.* It is active at 10^{-11} *M* and similar compounds appear to be produced by other plant species (see Chapter 7, section 7.12.2). The high activity of strigol and the fact that it moves from a host ("victim") plant root to induce germination of the seed of a parasite make it a unique plant growth regulator. Attempts to isolate such compounds began about 1948; the recent elucidation of the structure of strigol and its chemical synthesis were great achievements by the organic chemist.

For many years, coumarins, phenolic acids and flavonoids were regarded as probable important regulators of plant growth (see Chapter 7). Today, however, such compounds have fallen in status. However the possibility remains, and must not be ignored, that they are capable of profoundly modifying plant growth (see Chapter 7). For example, there is no question that exogenous coumarin can markedly promote cell expansion at very low concentrations (10 μM in one instance) and alter cell shape. Phenolic compounds may influence IAA levels by affecting IAA oxidase activity, inhibit gibberellin-induced growth and inhibit ATP production.

Finally many workers tend to restrict their interest to higher plants and forget that fungi and algae have hormonal systems which regulate sexual reproduction (see Chapter 7). A number of highly active hormones have now been identified in these lower plants. The first four compounds of this type were characterized structurally about 1968 – sirenin, a sesquiterpenoid alcohol (Machlis et al. 1968); antheridiol, a steroid with a lactone side chain (Arsenault et al. 1968); and trisporic acids B and C, compounds related to β-carotene (Van den Ende 1968).

The elucidation of the chemical diversity of phytohormones and related compounds has resulted largely from the endeavour of organic chemists, especially their development of modern physical methods for determination of structure. The contribution of organic chemists to the phytohormone field has been characterized by great perseverence and strict adherence to the logic of scientific method; their individual work has often been monumental (e.g. the elucidation of the structure and synthesis of ABA by Ohkuma and collaborators, and Cornforth and coworkers), while their conclusions are invariably unequivocal and unchanging. Unfortunately these attributes are often lacking in many studies in the physiological area of the phytohormonal domain.

1.2. Phytohormones as coordinators of plant growth and morphogenesis

Plants consist of a large number of tissues and organs, and if a plant is to continue to survive and function as a unit, it is essential that the growth and development of its parts be co-ordinated. It was suggested by Sachs in the nineteenth century that this co-ordination of growth and development was performed by chemical messengers. Today it is generally agreed that movement of the known types of phytohormones through the plant plays a critical role in mediating co-ordinated development.

Five routes for exchange of compounds exist in higher plants. Firstly, the intercellular air spaces, present throughout most higher plant tissues, allow rapid long distance movement of substances which are readily volatile, and have limited water solubility.

Secondly, long distance movement of water-soluble compounds can occur in the xylem stream; however, flow is essentially from the root to the shoot. Thirdly, short distance, or very slow long distance transport can occur in cell walls. Both water-soluble compounds and oils are able to move in this way. For the former, the direction of flow may be influenced by the direction of the transpiration stream, and substances interacting with the bound negative charges in the cell walls (for example heavy metals) are likely to move only slowly. Most of the plant is accessible to compounds which move in the fourth pathway, the phloem. In this tissue the major flow is from the leaves to other organs, but under appropriate conditions movement can be to or from most of the organs in the plant. The fifth route for movement is in the symplast, consisting of the multicellular cytoplasmic continuum established by the plasmodesmata. These appear to offer a lower resistance pathway between cells than the plasmalemma plus cell walls, but there is still considerable resistance to flow. Movement confined to the symplast (excluding the phloem) appears likely to be slow, except where elongated cells are involved. Pfeiffer suggested nearly a century ago that compounds moving specifically in the symplast may play a key role in the coordination of plant cells, especially over distances of a few cells. However, no such compounds have as yet been identified.

Each of the 5 channels mentioned above appears to be used by the plant to coordinate activities. The gas ethylene is able to move in the intercellular spaces, to control such processes as hypocotyl hook opening, leaf expansion, petal senescence and leaf and plant abscission. The natural non-volatile phytohormones appear to be able to move in the other four pathways. As a consequence they can be translocated over considerable distances. For example, there is considerable evidence to suggest that the shoot is dependent on cytokinins produced in the root system, and which move to the shoot in the xylem stream. Auxin shows a peculiar style of movement, in that it moves in a polar fashion through the cell, but moves between cells in the cell walls. Gibberellins and abscisic acid show a pattern of distribution consistent with movement in the phloem. However, the evidence would suggest that all of the non-gaseous pathways are accessible to these phytohormones – they can move in xylem or phloem, symplast or apoplast. This perhaps facilitates entry into the plant and enhances activity in certain bioassays, and may be the reason why these hormones were discovered in the first place.

Despite a considerable amount of work which has led to the general acceptance of the concept that phytohormones coordinate plant growth, there are few systems in which we can state unequivocally that a particular hormone is the coordinating agent. In the remainder of this section, we will consider studies which provide substantial evidence that phytohormones move from a site of production to one of action. The limitations of the data relating to source–target relationships will be emphasized.

Identification of the source tissue of a hormone is sometimes straightforward, but difficulties may occur when the source and target tissues are very close, for example, in ethylene-mediated responses. Sources of coordinating hormones are most commonly identified by removal of the tissue followed by replacement with exogenous hormones. Thus many of the inhibitory effects of the expanding leaf blades of the *Phaseolus vulgaris* stem apex on lateral bud development can be duplicated if the leaf blades are removed, and replaced by 1 mg/g of IAA in lanoline (White 1976). This concentration replaces the apex as a source of auxin (Patrick and Woolley 1973), and the endogenous auxin has rigorously been shown to be IAA (White et al. 1975). Similarly, the coleoptile tip has unequivocally been shown to contain IAA (Greenwood et al. 1972), and auxin at the appropriate rate of supply will replace the coleoptile tip in promoting coleoptile cell elongation (Went and Thimann 1937). Without unequivocal identification of the hormone, and/or without precise substitution of the appropriate rate of supply, other studies implicate the embryo as a source of gibberellin for aleurone cell activation in barley (Cohen and Paleg 1967); young developing fruits as a source of abscisic acid for the promotion of flower abscission in yellow lupin (Van Steveninck 1957, 1958, 1959a, b); the leaf blade as a source of auxin for the inhibition of leaf abscission (Jacobs 1968), and the promotion of vascular strand development in *Coleus* (Jacobs 1970); and the roots as sources of cytokinins in the inhibition of leaf senescence (see Chapter 4, Volume II). Other studies of this type are discussed in Chapters 2 and 4, Volume II. Although complete only if the natural hormone has been rigorously identified, and if the artificial source provides a quantitative as well as a qualitative replacement, such work has provided much valuable information. There is little doubt that the responses mentioned above are mediated by a hormone moving from a site of production to one of action.

A number of approaches have been used to study the target tissue. Perhaps the simplest has been localized hormone application to the intact plant. If the phytohormone is rate limiting to the specified process, then exogenous application might promote the process. Negative responses may be due to lack of penetration, or to rapid metabolism of the phytohormone (for example, IAA applied to the roots of *Pisum* seedlings (Kendall et al. 1971)). Positive or negative effects may be spurious in that they reflect an unusual direction or level of supply. For example, auxin applied to *Pisum* roots in the natural direction, via the stem or root base, stimulates root growth, whereas when the roots are immersed in auxin solutions, growth is inhibited (McDavid et al. 1972). On the other hand, such experiments are simple to perform, and have frequently provided useful insights into the system. For example, such trials led to the discovery of the ethylene promotion of leaf abscission (Harvey 1913), the gibberellin promotion of stem elongation (Kurosawa 1926), and the auxin (Gustafson 1936) and gibberellin (Crane et al. 1960) promotion of parthenocarpic fruit set.

Experiments involving replacement of a source organ with exogenous phytohormone are of course relevant to a study of target tissues. However, the situation is different to that in the intact plant, as not only the source of one hormone is removed, but probably a source of other hormones, a sink for some hormones, and possibly a competing sink or a source for assimilate, water and minerals. Furthermore, the overall hormonal circulation of the plant is disturbed. Nevertheless, if the source organ can be replaced by a specific chemical, provided at the same rate of supply as in vivo, then this is strong evidence that the influence of the source on the target tissue is due, directly or indirectly, to the action of the chemical. On the other hand the interpretation of results showing incomplete or no replacement is very uncertain. Problems are particularly evident in dealing with hormones moving appreciable distances in the vascular tissue, for example in replacing root-produced cytokinins and gibberellins in their effects on leaf senescence or shoot development, or in replacing leaf-produced abscisic acid in regard to its effects on stem apex development.

The target tissue is commonly recognised by its response to applied phytohormones, but for stronger evidence of a relationship a correlation between endogenous hormonal levels and developmental responses is sought. A number of factors limit the effectiveness of such efforts. It is very difficult to measure accurately amounts of growth substances in target tissues (see section 1.5). Even where amounts of specific molecules are measured in an organ, it is possible that the target of the hormone is a single tissue, or a single compartment within a cell. Furthermore, what is being measured is an overall balance between hormone import, synthesis, export, storage, metabolism and inactivation. For example, if the rate of auxin immobilisation is correlated with growth rate and appears to be important, then total endogenous concentrations may mean little (Steeves and Briggs 1960). Similarly, in regard to auxin, it is uncertain whether diffusible or extractable auxin is the important parameter (Scott and Jacobs 1964). In addition, phytohormones interact, not only to modify the response of the tissue to one another, but also to affect hormone synthesis, metabolism and transport (see section 1.3). Furthermore, one is dealing not just with a simple reaction, but with a dynamic, integrated system involving a multitude of processes at biochemical, ultrastructural and physiological levels. The response to a given phytohormone concentration is likely to be determined by many factors within this framework.

In spite of the above problems, correlations between extractable phytohormone activity measured by bioassays and physiological behaviour of organs have been reported frequently. Particularly well-reported are correlations between extractable auxin and gibberellin activity, and stem growth rates (see Chapter 2, Volume II). While it is interesting that correlations are often found, these are an uncertain guide as to whether the phytohormone is regulating the system. Quantitation techniques based on bioassays are often unreliable (see section 1.5); the correlations observed could be due simply to covariance.

Another area into which much effort has been directed is the study of responses in isolated target tissues (for example, stem and root segments, leaf discs, root and shoot tips). In such systems, much of the complexity of the original in vivo situation has been removed, so that mechanisms of action may be more accurately delineated. However, not only have complexities been removed, but so also have many elements of the original control systems. Thus, auxin promotes the growth of isolated stem segments, but there is no convincing evidence that auxin limits cell elongation in the intact stem. Similarly, in monocotyledons, cytokinins delay the senescence of detached leaves, but have little effect on leaves in the intact plant. Hence, extrapolation of results obtained with isolated tissues to the intact plant is extremely hazardous. On the other hand, such studies do at least indicate that the tissue response, where it parallels the in situ response, is a direct consequence of the supply of hormone – it is not a secondary effect, consequent on changes in other tissues. Thus the response of barley aleurone cells to the germinating embryo appears to be mediated largely by gibberellins, as these are necessary and sufficient for the full response of the isolated aleurone tissue (Coombe et al. 1966).

In summary, numerous approaches should be used to determine whether a given hormone is involved in a particular developmental co-ordination in a plant. No single approach is conclusive.

1.3. Hormonal interplay in the regulation of plant growth

In Volume II, the hormonal regulation of many aspects of the growth and physiology of higher plants is discussed – cell division and cell enlargement; cytodifferentiation and organogenesis in callus tissue; stem, leaf, root and fruit growth; flower formation; shoot senescence, leaf abscission and fruit ripening; bud and seed dormancy; correlative growth; and response to water stress. When supplied exogenously to plants, each of the established types of phytohormone affects almost every aspect of plant development mentioned above. While each type of plant hormone is distinctive in a chemical sense and in that it can also display unique activity in carefully selected bioassays, the classes of phytohormones exhibit considerable overlap in their activities when considered in the overall context of developmental botany.

When supplied exogenously to isolated plant tissues or to intact plants, phytohormones may act synergistically or antagonistically to regulate growth, or their effects may simply be additive. Such synergism and antagonism is commonly referred to as hormonal interaction. In the growth of the intact plant, interaction between endogenous hormones is undoubtedly of great significance.

The complexity of hormonal interactions is well illustrated by the hormonal regulation of growth of tobacco pith callus tissue in vitro. Auxin and cytokinin act synergistically to induce cell division and growth but the expression of the synergism depends on the cytokinin/auxin ratio. Low, high and intermediate ratios induce roots, shoots and callus (essentially undifferentiated tissue) respectively (Skoog and Miller 1957; Miller 1961). Although exogenous gibberellin enhances tobacco callus growth (Lance et al. 1976), it suppresses the shoot formation induced by auxin-cytokinin interaction; this inhibitory activity of gibberellin is alleviated by exogenous ABA (Thorpe and Meier 1973), which is, however, a weak inhibitor of tobacco callus growth (see references in Milborrow 1974). Furthermore, the form of shoots induced from the callus tissue is determined by the gibberellin/cytokinin ratio in the culture medium; high ratios give tall spindly shoots with narrow leaves, whereas low ratios yield short shoots with rounded leaves (Engelke et al. 1973). Finally, it should be noted that certain phenylpropanoic and benzoic acids interact with the phytohormones to modify the growth of tobacco callus. Thus in media containing kinetin and IAA, tyrosine, phenylalanine, 2-hydroxycinnamic acid, 4-hydroxybenzoic acid and 2-hydroxybenzoic acid (salicyclic acid) all promote bud formation but inhibit callus growth. In contrast, caffeic, ferulic and sinapic acids promote callus growth but not bud development (Lee and Skoog 1965a). The phenolic acids mentioned probably act by affecting enzyme systems which inactivate IAA (Lee and Skoog 1965b).

Many examples of hormonal interaction in the regulation of plant growth are recorded in Volume II. However, two examples which are not discussed are worthy of note. The first concerns the hormonal interactions which appear to regulate lateral bud development (for discussion see Phillips 1975). While auxin inhibits, cytokinin promotes development and antagonizes auxin; correlative inhibition of lateral buds is also influenced by nitrogen supply (see Chapter 4, Volume II), photoperiod and light intensity which may alter hormonal levels. The second is the demonstration that auxin and cytokinin can interact synergistically in artichoke tuber tissue to promote either cell division or cell enlargement depending on the pretreatment of the tissue, the osmotic environment and the availability of divalent cations (Adamson 1962). These two examples emphasize the importance of environmental and nutritional factors in hormonal interaction.

The concept that growth is regulated by hormonal interactions has several corollaries, one of which merits mention. When a particular response is evoked specifically by one exogenous hormone, other endogenous hormones are interacting. The tissue may of course synthesize or receive the endogenous hormones normally, or may accumulate them in response to the applied hormone. In this context, it is relevant to ask the question: would auxin plus cytokinin induce cell division in tobacco callus if this tissue was devoid of endogenous gibberellin? It is gratifying that the negative answer has recently received considerable experimental support. Tobacco callus does produce gibberellin and the level is affected by culture conditions including the exogenous cytokinin concentration. These culture conditions are important in controlling the growth rate and also affect the ability of the tissue to respond to exogenous gibberellin (Lance et al. 1976). These workers also found that substances known to inhibit gibberellin biosynthesis or action reduced the growth rate of the tissue in the presence of auxin and cytokinin. These data support the view that endogenous gibberellins are important in the response of the tissue to exogenous auxin and cytokinin. The above corollary can be extended. Not only is the presence or absence of non-applied hormone important, but its residual effects in newly isolated tissues may be of equal importance. In many instances, organs are isolated, depleted of endogenous hormone and assumed to be in some basic metabolic state. However, there is no reason to believe that a tissue depleted of a hormone always "forgets" its association with the hormone rapidly (Cline 1976).

Hormonal interaction has so far been viewed at the level of growth responses. However in recent years it has become clear that there is another aspect to phytohormonal interaction – each hormone can influence the biosynthesis, degradation, conjugation or translocation of at least one of the other types of hormone. Examples of this are mentioned below. Thus, exogenous gibberellin and cytokinin both elevate endogenous auxin levels, but application of ethylene, the production of which is markedly enhanced by auxin, reduces these levels in plant tissue (see Chapter 2). Cytokinin appears to promote auxin synthesis and also to suppress IAA conjugation, but the mechanism by which gibberellin elevates auxin levels is uncertain. Auxin transport is promoted by cytokinin and also by gibberellin; in contrast ethylene is inhibitory when applied as a pretreatment. Cytokinins have been reported to increase levels of gibberellin-like activity (see Chapter 3 and Chapter 4 in Volume II), to reduce ABA levels and to promote ABA conjugation (Even-Chen and Itai 1975). In several plant tissues, exogenous ABA reduces the level of gibberellin-like substances and ABA may enhance conversion of some gibberellins to conjugates (Chapter 3). However in one tissue, ABA appears to increase the gibberellin content (Railton and Wareing 1973). Exogenous ethylene induces increases in ABA or ABA-like substances (see Volume II).

The ability of one hormone to affect the level of another, is probably of great significance in regulation of plant growth and merits more critical assessment. Known interactions of this type are interrelated for the first time in Figure 1.1. The complexity of these introduces a new dimension to the mechanism of hormone action. One cannot assume that in exogenous hormone acts per se; the response may be mediated by another hormone. The fact that some responses to high auxin are mediated by increased ethylene levels is well known. Ethylene in turn suppresses auxin level (Figure 1.1) and may provide a feedback control mechanism in cases of excessive auxin production.

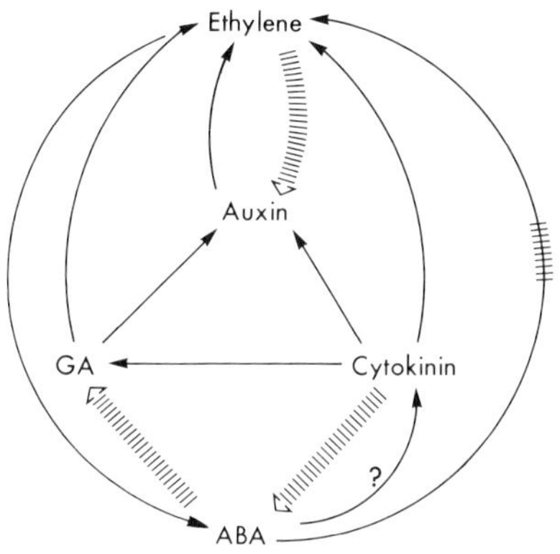

Figure 1.1. Hormonal interactions which affect hormonal levels in plant tissues. A——>B indicates hormone A increases the level of hormone B; A||||||||>B indicates A reduces the level of B. The crossed arrow pointing from ABA to ethylene is intended to indicate that ABA has been reported to both increase and decrease ethylene levels. Very limited evidence suggests that ABA has a cytokinin-sparing effect (see Chapter 4); hence the arrow to cytokinin from ABA bears a question mark. In this figure, GA denotes gibberellin-like substances.

1.4. The action of phytohormones at the subcellular level

If phytohormones are to be used to manipulate the growth and development of plants, hormone action should probably be studied at two levels. The first involves an empirical approach in which all classes of plant hormones, singly and in combination, are tested for their effects on growth at all developmental stages of the important plant species. The second is to describe the more fundamental (biochemical) processes of plant biology and to determine whether plant hormones play a role in regulating those processes. It may then be possible to describe how the hormones act to regulate a particular system and to design a protocol for the manipulation of a particular stage of plant development.

The empirical approach to the study of growth has had a number of successes (see Volume II for examples) but the biochemical one has been much less useful so far (see below) The reasons for the latter probably include the facts that many of the fundamental processes of plant biology have only recently been studied in any great detail and that the techniques available have not had sufficient resolution to pinpoint the site of action of a hormone when it was implicated, e.g. auxin action in wall loosening or gibberellin action in increasing enzyme levels. Two features of recent research on plant biology may increase the rate of advance of our knowledge of how hormones act in plant cells. The first is that modern techniques allow higher resolution of our understanding of the fundamental processes in plant growth and the second is that there is considerable attention being paid to the early biochemical responses to phytohormones within plant cells.

1.4.1. The concept of mechanism and mode of action of phytohormones

To determine how phytohormones act on plant cells especially when there is a great diversity of responses to phytohormones, it is essential to distinguish between the initial trigger reaction(s) (which we will term mechanism) and the series of steps which lead to the physiological response. The latter we denote as the mode of action of the hormone. This concept is essentially that first described by Paleg (1965) who described the mechanism of hormone action as a specific change in metabolism in direct response to the chemical. Recent advances in the knowledge of hormone action suggest that his definition might best be modified to include only the primary reaction of the hormone with some cell component(s). As illustrated below, with the example of barley endosperm and its response to gibberellin, attempts to characterize the specific change in metabolism have revealed a great complexity of events. No matter whether the various events are sequential and interdependent or are largely independent of one another, it has proved difficult to determine what constitutes the specific change in metabolism.

In the gibberellin-promoted increase in storage-reserve (protein, lipid, carbohydrate and mineral) mobilization during the early stages of barley seed germination, the mode of action of gibberellin is thought to include a stimulation of the synthesis of hydrolytic enzymes in the aleurone cells followed by transport of the enzymes to the starchy endosperm cells where stored reserves are hydrolyzed. It was proposed (Varner and Ram Chandra 1964) that the trigger mechanism of gibberellin was to promote the synthesis of mRNAs for the enzymes. Thus, the mechanism of action of gibberellin could be thought to be regulation of gene expression and the production of a series of mRNAs which was followed by a sequence of steps (mode of action), dependent upon the initial event, finally leading to the stimulation of storage-reserve hydrolysis. In the intervening years this picture has become more complex (for reviews see Jones 1973; Varner 1974; Varner and Ho 1976; Jacobsen 1977). Great attention has been paid to the sequence of the responses and their timing in relation to hormone application. For example, in the barley aleurone layer some of the earliest responses (within 90–120 min) to gibberellin include activation of some enzymes involved in membrane metabolism and increased oligosaccharide release; there is also an early ferric iron requirement for gibberellin action. It is only after 2–4 hr after hormone treatment that the mRNA for α-amylase begins to accumulate and this is accompanied by a number of changes such as an apparent rearrangement of intracellular membranes and the attachment of ribosomes to some of these membranes. Wall synthesis is inhibited while wall hydrolysis is promoted. Later, there is a release of ions from the aleurone cells and there are massive ultrastructural changes within the aleurone cells while the hydrolases, ions and other material such as carbohydrate are being released. Gibberellin does not appear to have a direct effect upon the secretion of molecules across the plasmamembrane. Some enzymes are synthesized de novo (e.g. α-amylase and protease), others are activated (e.g. β-amylase in the starchy endosperm), while yet others are released early from the aleurone cell walls (e.g. phosphatase) along with other material. If hormone is applied for a short period of time, then removed and readded some time later, the lag-period-determining events set in motion by the hormone do not seem to decay, in that when the hormone is added back there is an immediate resumption in α-amylase synthesis without an apparent lag period. However, there seems to be a continuous requirement for the hormone showing that an initial triggering event is not sufficient to sustain the responses. There is also a complex interplay between gibberellin and abscisic acid, and possibly with ethylene, in bringing about the hydrolysis of reserves in barley. Auxin and cytokinin have also been implicated in reserve hydrolysis in wheat endosperm (Laidman et al. 1974). This brief account of the spectrum of effects of gibberellin on the biochemical events in aleurone cells, which seem to be directly involved in mediating endosperm storage-reserve hydrolysis, give some indication of the difficulty in assigning a mechanism of action (i.e. a specific change in metabolism) to gibberellin. However, by studying the kinetics of the responses, it would seem that differential gene expression, if involved at all, is not the only factor concerned in the production and release of hydrolytic enzymes. Furthermore, mRNA accumulation is clearly not the earliest detectable response to the hormone and is thus not necessarily closely related to the mechanism of action. Some other aspect of differential gene expression such as the gibberellin-promoted production of regulatory RNAs or proteins cannot be ruled out, however.

The multiplicity of responses to gibberellin detected so far in the barley aleurone constitute the mode of action of the hormone in promoting storage-reserve hydrolysis, but it has not been resolved whether all of these responses are the expression of one mechanism of action or of several independent mechanisms. It should be pointed out that the aleurone layer system is one of the simplest which is currently under investigation with respect to phytohormone action. It represents a single tissue type with fairly uniform cells preprogrammed with a limited number of functions in the early germination of the seed. It will undoubtedly be more difficult to determine mechanism of action in systems as complex as the gibberellin-promoted cell elongation in pea epicotyl or oat coleoptile. Further complexity is introduced by considering the diverse responses of a number of species of gibberellin (see Jones 1973) and this is greatly compounded if one accepts that growth and morphogenesis in plants are regulated by interplay between several classes of hormones (Cleland 1974 and section 1.3). Clearly, it is a very difficult task to attempt elucidation of the mechanism of hormone action in the control of a process as complicated as differentiation or even cell expansion, and it may be wise to concentrate on somewhat simpler systems such as those involving increased enzyme levels, rapid stomatal response or wall loosening.

1.4.2. Towards elucidation of mechanism

Lack of progress in elucidating mechanisms of hormone action is due partly to the fact that we have attempted to study at the biochemical level, complex processes often involving growth. It may be appropriate to reflect on the various approaches made to elucidate the mechanism of auxin action. Thimann (1974) defined four major approaches to which can be added a further two of recent origin. These included early (1930s) studies on some physical characteristics (plasticity and elasticity) of cell walls. A good correlation was found between plasticity changes and cell elongation especially in oat coleoptile. In the 1940s it was popular to consider that auxin promoted growth by regulating oxidative phosphorylation. The next major aspect of cell metabolism to be studied in relation to auxin action was RNA synthesis and function. This has been eclipsed by studies on the acidification of the medium bathing excised tissue sections which are being treated with auxin.

The fifth approach has been to study accurately the kinetics of the many growth responses to hormones, especially auxin, and to relate these to biochemical and physiological data. By studying the kinetics of the responses, a great deal of data has been collected and it may now be possible to place in sequence some of the steps in a particular mode of action. Similarly, probable cause and effect relationships have sometimes been disclosed. However, it may be misleading to overemphasize the importance of timing of responses unless it is known that a single sequence of events is being studied. If, for example, there are several independent sequences set in motion by a hormone, then it may be misleading to study and relate individual steps although they are separated temporally. Auxin-activated gene expression has been discounted as mediating the elongation response in oat coleoptiles because some features of elongation are measurable before new protein synthesis could occur (See Volume II for detailed discussion). However, until more is known about the rates of turnover, synthesis and accumulation of proteins in plant cells it is premature to select 1 hr as the limit to define whether a response to a stimulus involves RNA and protein synthesis or not. In other eukaryotic cells, it has been shown that estradiol stimulates the synthesis of a specific rat uterine protein within 20 min by promoting both transcription and translation (Katzenellenbogen and Gorksi 1972), while dibutyryl adenosine 3′:5′-cyclic monophosphate stimulates (5-fold) the production of phosphoenolpyruvate carboxykinase in rat liver also within 20 min (Iynedjian and Hanson 1977). Furthermore, current techniques might not detect changes in protein synthesis if the protein was turning over rapidly; it is known that, in oviduct for example, mRNA synthesis can occur at the rate of 25 molecules/min and that protein synthesis may proceed at a rate of 10 molecules/molecule of mRNA/min (Palmiter 1975). Thus, because a growth response occurs in less than 1 hr it might not mean that protein synthesis is not involved.

The sixth approach to the study of mechanism of action has been to search for auxin perceptors (receptors or recognition factors*) in responsive tissues. This approach is conceptually the most satisfactory, since if the perceptor(s) could be identified, it should then be possible to define the primary biochemical response to the hormone and possibly the sequence of events leading to the physiological response. So far, the results have been disappointing. The ambitious attempts of the late sixties and early seventies, to link the perception event to a major change in gene expression have either proved irreproducible, or seem to have been neglected, and considering the poor understanding of the control of

* We propose to distinguish between two possible types of perceptors. Receptors: high affinity, low capacity molecules which bind the hormone and transport it to a new site where the complex acts (definition based on that used in mammalian systems). Recognition factors: molecules which specifically recognize the hormone and usually undergo conformational or structural changes after binding the hormone. Binding time could be short allowing the hormone to recycle almost immediately.

gene expression in eukaryotes, this is not too surprising. The more modest approach of first defining perceptors is to be recommended and this is now underway in a number of laboratories (see e.g. Batt and Venis 1975).

A similar diverse array of biochemical processes have been studied in relation to each of the hormones (see Chapters 8–11). Auxin is unique in that it has been studied for a longer time spanning the period when many of the major biochemical pathways and processes were being elucidated in plants. Thus, it tended to be associated with the popular process being studied at particular times, e.g. oxidative phosphorylation in the 1940s. Only general conclusions can be drawn from the collected data concerning responses to phytohormones at the biochemical level (Chapters 8–11). Mechanisms of action are not apparent and any speculation is often a reflection of the current fashions in the study of subcellular biology. Examples include, (a) the proposed role of cytokinins as effectors of translational control by binding to ribosomes but not because of their occurrence in tRNA although the latter was popular for many years and (b) auxin-mediated control of a membrane-bound enzyme such as ATPase because of acidification of the incubation medium around elongating coleoptile segments and because auxin may bind to a component in plant membranes. Proposals such as these, which are in vogue at present, may be abandoned if critical experimental proof of their validity becomes difficult to collect.

In attempting to elucidate mechanism of action we need to solve two basic questions. The first is: does the great diversity of responses reflect one or multiple mechanism(s) of action and what is the nature of the cellular component(s) which perceive the hormone? The second major question to be answered is: what is the role of the physiological state of the cells being exposed to the hormone? The genes being expressed at the time of exposure could clearly play an important role in how the cells respond to a hormone.

It is difficult to formulate a single mechanism of action for each hormone applicable to all situations. For example, it is hard to envisage that the auxin-promoted increase in wall plasticity is mediated by the same mechanism as auxin-promoted accumulation of mRNA for cellulase in expanding or elongating epicotyls. Nevertheless, it is possible that a single master-reaction (mechanism) is responsible for triggering the several modes of action which lead to all of the observed and varied responses to auxin. If we employ the concept of a perceptor, then the single master-reaction system would require that the hormone be perceived always by the same factor. All of the observed responses could be the result of differing expressions of the master reaction depending on the state of genetic expression, ionic condition and metabolic status of the responding cell. The importance of the physiological status of the cell in determining the type of response to auxin and kinetin was well-illustrated by Adamson (1962). He reviewed extensive studies with the hormones on cell division and cell expansion in tuber tissue of *Helianthus tuberosus* and concluded that the process (division or expansion) which was stimulated, clearly depended upon the status of the cells at the time of treatment. The physiological status of the responding cell is also emphasized in studies on the responses of mammalian cells to their hormones. Receptors for mammalian hormones are better-documented than is the corresponding case in plants and in those cases studied it appears as if the receptor molecule is the same in all tissues of the animal while the hormone-induced responses are quite diverse (see Wicks 1974 for an account of glucocorticoid action). This supports the view that a single master-reaction involving the hormone and a perceptor (receptor in the case of steroids) can mediate a diversity of responses.

The alternative view on mechanism of action is that there are multiple trigger events (i.e. mechanisms of action). In this case diverse responses are envisaged as arising from independent primary interactions of the hormone with the cell. Multiple perceptors allow considerably more flexibility in the ways in which hormones might act but probably introduce more complexity than is currently required to explain hormone action. However, there are some instances where it seems essential to invoke more than one mechanism of hormone action. An examples of this we have the differential activity of the optically active isomers of cytokinins and abscisic acid. Both isomers of two optically active N_6-substituted adenines were active in promoting tobacco callus growth while only the levorotatory compounds were active in inducing shoot formation (Yamada et al. 1972). A probable interpretation of these data is that different functional perceptors exist for cytokinins within the same tissue. Similarly, the optical isomers of abscisic acid are equally active in inhibiting germination processes but only the (+) enantiomer is active in controlling stomatal aperture in barley leaves (Cummins and Sondheimer 1973; see also Chapter 6, section 6.4.2).

The study of perceptors for phytohormones is quite new and a lot of effort was probably wasted in the early stages because of the unfortunate choice of differential gene transcription as an assay. This led Kende and Gardner (1976), having critically reviewed the existing literature, to conclude that the evidence for receptors was almost non-existent. Some of the newer techniques such as photo-affinity labelling (Leonard et al. 1975) and group-modifying reagents (Venis 1977) may allow a more fruitful search for factors in plant cells which perceive phytohormones and then initiate a mode or modes of action which climax in physiological, growth or developmental phenomena.

1.5. The problem of the quantitation of phytohormones

Although several, and possibly all, types of phytohormones may be involved in a particular aspect of plant development, often one hormone appears to limit development and so assumes a dominant regulatory role. For example, the sequential response of excised wheat coleoptiles and leaf sections during growth to exogenous gibberellin, cytokinin and IAA, suggests that these hormones limit growth in sequence in these organs of the intact wheat seedling (Wright 1968). In an endeavour to assign regulatory roles to phytohormones, plant physiologists have attempted to quantitate phytohormones in plant tissues and to correlate changes in hormonal levels with onset or cessation of developmental phenomena. Indeed, a large proportion of the hormonal literature concerns investigations of this type. Unfortunately this is a most unsatisfactory area of endeavour.

Frequently no consistent correlations have been found, e.g. in relation to bud dormancy (see Chapter 9, Volume II). This could be due to one of two reasons. Firstly, there may in fact be no correlation between the level of the hormone in the tissue or organ and the developmental phenomenon. However changes in hormonal gradients across a tissue, or in intra-cellular distribution may be of great regulatory significance but would not be detected in tissue extracts. Secondly, because of non-quantitative extraction methods and/or the presence of compounds which interfere in the bioassays used, the correlation may not be detected. Conversely, a correlation found may be spurious because of these factors. The problem of the determination of phytohormone levels in plant tissues is very great and is of such importance that the discussion which follows seems warranted.

Most published determinations of phytohormones (ethylene excepted) in plant tissues involve the following steps: (1) extraction by a solvent such as aqueous methanol; (2) fractionation of the extract by solvent partition and/or chromatography; (3) estimation of the hormone activity in the fractions by a bioassay method. In some cases step 2 is omitted. There are serious problems associated with such methods. Firstly, there is no assessment of the recovery of the hormone from the plant tissue. When extracting a trace compound from a plant tissue, the loss due to incomplete extraction could be very great indeed and might vary with the type and maturity of the tissue. Thus, recovery of radioactive IAA added to corn shoots was found to be only about 20% (Hamilton et al. 1961). Even more disturbing is the report that isolation of chloroplasts from wheat leaves followed by treatment of the preparations with a non-ionic detergent yields a thousand times more gibberellin (GA_4 + GA_9) than is extracted directly from the leaves with 80% methanol (Browning and Saunders 1977)! The adequacy of extraction procedures for all phytohormones requires careful assessment. Secondly, there is no measure of the losses which occur during fractionation of the extracts. However, when certain ion-exchange resins are used, it is known that cytokinin losses can be considerable. Thirdly, the specificity of bioassays is far from absolute. For example, the soybean callus bioassay is widely used by cytokinin workers, but it is known to be markedly affected by abscisic acid, another phytohormone. High concentrations of abscisic acid inhibit the callus growth while low concentrations actually act synergistically with cytokinin to promote growth (Blumenfeld and Gazit 1970). Some common plant metabolites, e.g. glutathione, are now known to interfere in the soybean assay (Bergmann and Rennenberg 1975). Gibberellins and ABA interact in most bioassays for these hormones and many gibberellins and ABA have similar R_f-values on paper chromatograms.

These observations lead to one certain conclusion – quantitation of cytokinins, auxins, gibberellins and ABA by the methods outlined above, must often have yielded meaningless results. For example, results based on the bioassay of paper chromatogram zones led to the conclusion that a change from long to short days caused a marked increase in the level of inhibitor activity extractable from sycamore apices and leaves. This activity was considered to be very largely due to ABA (see references in Wareing and Saunders 1971). However in a later study, quantitation of ABA by GLC using 2-*trans*-ABA as an internal standard (Lenton et al. 1972), established that there was no difference whatsoever induced by transfer from long to short days. In spite of the above observations, the publication of papers dealing with quantitation of hormones by bioassay using either relatively unpurified extracts, or extracts chromatographed on paper, continues unabated. Little confidence can, for example, be placed in determinations of a growth promoter separated by paper chromatography when the adjacent zone yields 50% inhibition in the bioassay (see Fig. 3.2 of Chapter 3, Volume II). Similarly, assignment of significance to observed changes in gibberellin-like activity in a paper chromato-

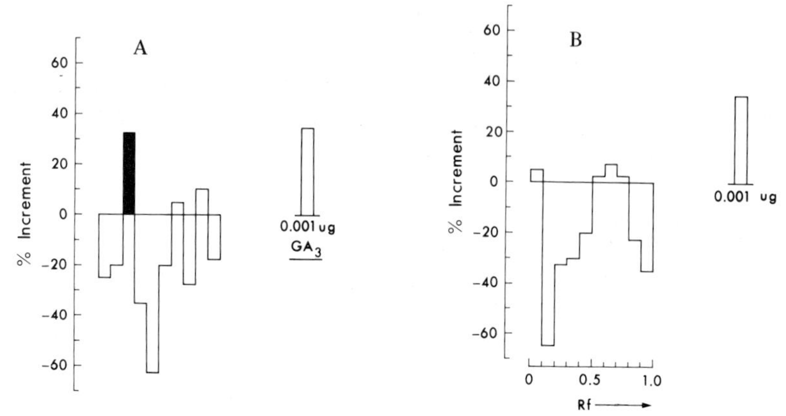

Figure 1.2. Gibberellin activity of zones of paper chromatograms. A, chromatogram of extract from control plants; B, chromatogram of extract from treated plants. The workers who published this figure, which appeared in a well-known journal, concluded that the treatment caused a very marked reduction in the level of gibberellin at R_f 0.25. The gibberellin activity was assessed by bioassay using dwarf pea plants.

gram zone, when this is flanked on both sides by zones which are strongly inhibitory (see Figure 1.2), is illogical.

Bioassay results are particularly unreliable when the identities of the active compounds differ in a series of samples being compared. This is conveniently illustrated by reference to the cytokinin literature. Response in a bioassay usually depends on uptake of an active compound by plant tissue from a culture solution. Compounds differ in their degree of uptake and in the case of cytokinins and related purines these differences can be very large indeed (Terrine et al. 1972). Such differences will be reflected in differing responses in the bioassay. For this reason alone, if a series of extracts contains different cytokinins, bioassays for total cytokinin activity yield no definite information regarding cytokinin activity levels in the original tissue. The situation becomes particularly complex when, as in the case of poplar leaves, extracts from young and mature leaves contain at least 7 cytokinins and extracts from senescing leaves only one (Hewett and Wareing 1973). Furthermore, the identity of most of these compounds is unknown. Reported values for total cytokinin activity per unit weight of tissue in such cases are likely to be misleading.

Development of reliable chemical methods for quantitation of phytohormones in plant tissues is essential. Attempts have been made to quantitate ABA, gibberellins, IAA and cytokinins extracted from plants using GLC. However all reported determinations based on this approach have been unsatisfactory for one or both of the following reasons. Firstly, mass spectrometry was not used to assess the GLC peak attributed to the hormone being determined. This is absolutely essential when quantitating a trace compound by GLC, since mass spectrometry is the only reliable indication of the identity and purity of peaks. Without mass spectral monitoring, GLC procedures are no more reliable than bioassays. Secondly, a suitable internal standard was not used to correct for losses which must occur during extraction and purification. Bridges et al. (1973) using GC-MS claim, for example, that *Zea mays* root steles contain 53.33 μg of IAA/kg fresh weight, but no internal standard whatsoever was used to correct for losses in extraction and in the 6 purification steps used. Although these workers claim "to determine quantitatively the levels of IAA" in root tissues, the recovery of IAA may have been very low, and the error, therefore, very high. Certainly if one multiplies their values by 10, they are similar to those of Rivier and Pilet (1974) who did use the best available internal standard.

For quantitation of phytohormones by methods based on GLC, three internal standards have been used. The first to be introduced was 2-*trans*-ABA (Lenton et al. 1971) for use as an internal standard for ABA determinations. This is separated from ABA during the GLC. However it is not a completely satisfactory internal standard for this purpose. There is no certainty that its recovery will be identical to that of ABA especially when the ratio of ABA to 2-*trans*-ABA differs appreciably from 1, as it normally will in an actual estimation. Furthermore, ABA and 2-*trans*-ABA are interconvertable by light to a 1:1 mixture and hence extreme care must be taken during the extraction procedures. Finally, 2-*trans*-ABA has been reported to occur naturally. Nevertheless, the use of 2-*trans*-ABA as an internal reference by Saunders and coworkers (Lenton et al. 1971, 1972) was a significant contribution to ABA quantitation and a vast advance compared to GLC without any standard. However it is surprising that these workers added the 2-*trans*-ABA to the extract and not the tissue; hence the internal standard did not correct for losses which must inevitably occur during extraction. Other internal standards which have been employed by plant hormone workers for GC-MS quantitation are 5-methyl-IAA in studies of IAA levels (Rivier and Pilet 1974), and 6-(4-hydroxybenzylamino)purine riboside for determination of the cytokinin 6-(2-hydroxybenzylamino)purine riboside in leaf extracts (Thompson et al. 1975). These two studies probably constitute the most reliable GC-MS quantitations of phytohormones. However the use of internal standards which are only structurally similar to the phytohormone being determined is not satisfactory and the peaks revealed by selected ion current monitoring (SICM) and used to quantitate the hormone were not characterized by complete mass spectral scans in even one sample. Furthermore it is disturbing to note that the relative intensities of the monitored ions derived from the extract do not correspond with those for IAA (cf. Fig. 2A and 2B, Rivier and Pilet 1974).

Methods other than the GLC-based procedures outlined above have been of limited value for phytohormone quantitation in extracts. Thus reliable determinations of ABA levels have been achieved by the racemate dilution method using spectropolarimetry (Milborrow 1968) and possibly by an isotopic dilution method (Harrison and Saunders 1975). IAA can probably be determined reliably by spectrofluorimetry using [^{14}C]IAA as an internal standard to correct for losses (Knegt and Bruinsma 1973).

Having reviewed briefly the previous endeavours to quantitate phytohormones, it is appropriate to consider how chemical methods might be developed further. The extremely low levels of phytohormones, especially cytokinins, in plant tissue render conventional methods of analysis useless. One appears to be forced to exploit two types of procedure: sophisticated forms of GC-MS; or radiochemical methods involving specific derivatization of the partially purified hormone with a radioactive reagent of high specific activity. However since this second approach has not been attempted experimentally it is not discussed further here. High pressure liquid chromatography coupled with field desorption mass spectrometry may have some potential value, but there are at present considerable technical problems in this area.

GC-MS methods if properly executed hold great promise for the plant hormone worker. However any such method is valueless unless an appropriate internal standard is added to the plant tissue at the moment of extraction to correct for losses in extraction and purification. The only acceptable standard is isotopically labelled hormone; the labelling must be at specific sites and the replacement of the normal isotope at these sites must be complete. The convenient but inappropriate practice of using compounds which are chemically similar to the hormone (see above) can only lead to more meaningless results. Deuterium-labelled hormones would be ideal GC-MS internal standards. The preparation of hexadeutero-abscisic acid from ABA has been reported by Milborrow (1971) but it has never been added to plant tissue as an internal standard to quantitate

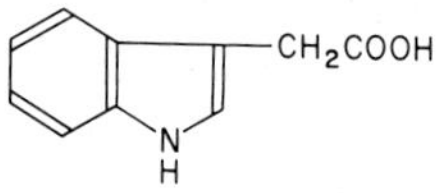

I

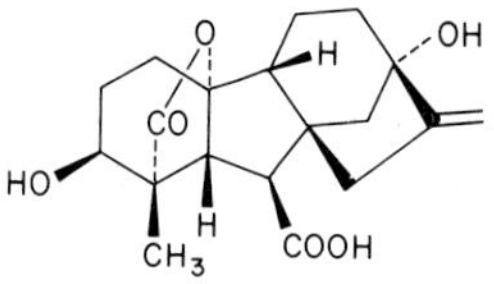

II

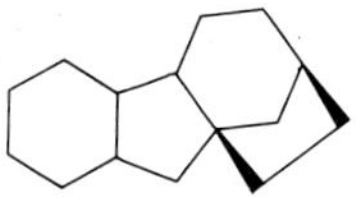

III

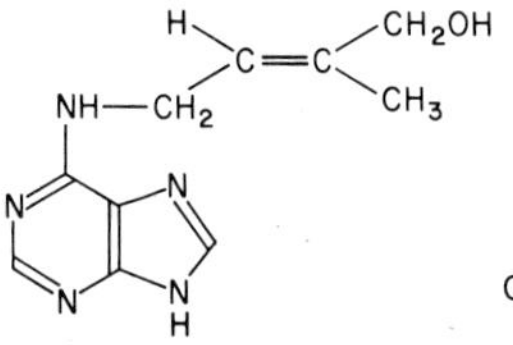

IV

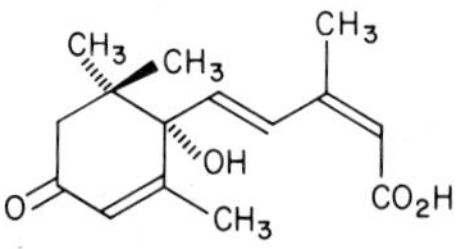

V

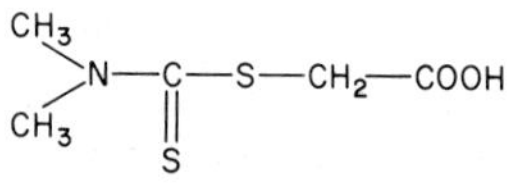

VI

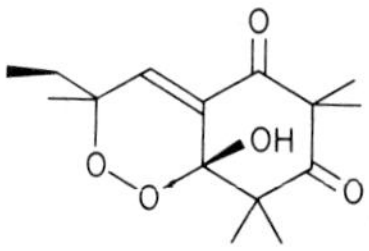

VII

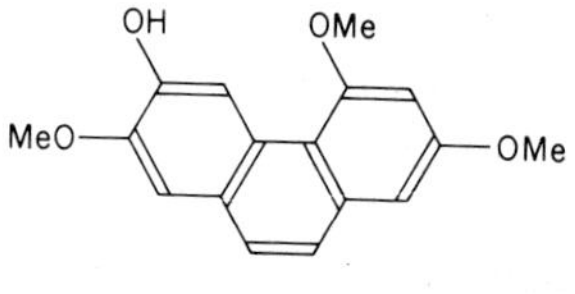

VIII

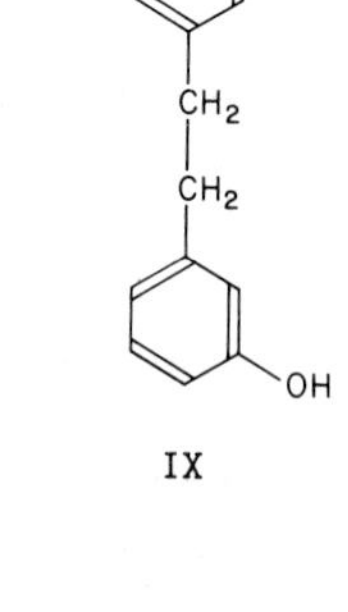

IX

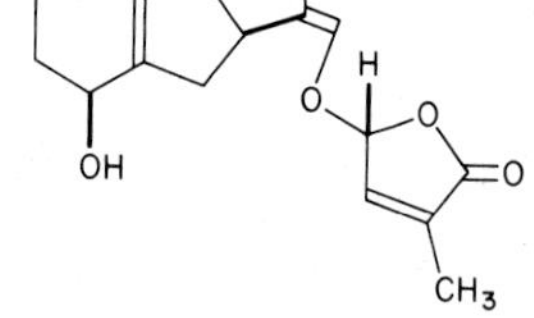

X

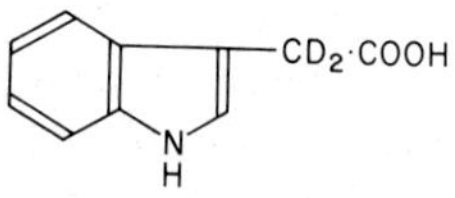

XI

ABA using mass spectrometry. Preparation of deuterium-labelled cytokinins and IAA has not been reported and, in order to stimulate interest in this area, syntheses are outlined below. [Methylene-$^2H_2$]IAA (XI) could be prepared readily from the commercially available indole derivative gramine by reacting it with sodium cyanide in D_2O and EtOD under reflux (for an analogous synthesis, see Beck and Sedvall 1975). The preparation of deuterated cytokinins is more difficult but a synthesis of [2′-$^2H_2$]zeatin (XII) is outlined below (Scheme 1.1) as typical of syntheses for this type of compound. XII and the 7-β-D-glucopyranoside of XII have recently been prepared by this type of synthesis and used to quantitate zeatin and 7-β-D-glucopyranosylzeatin levels in germinating radish seed by GC-MS (Summons, MacLeod, Parker and Letham, unpublished results).

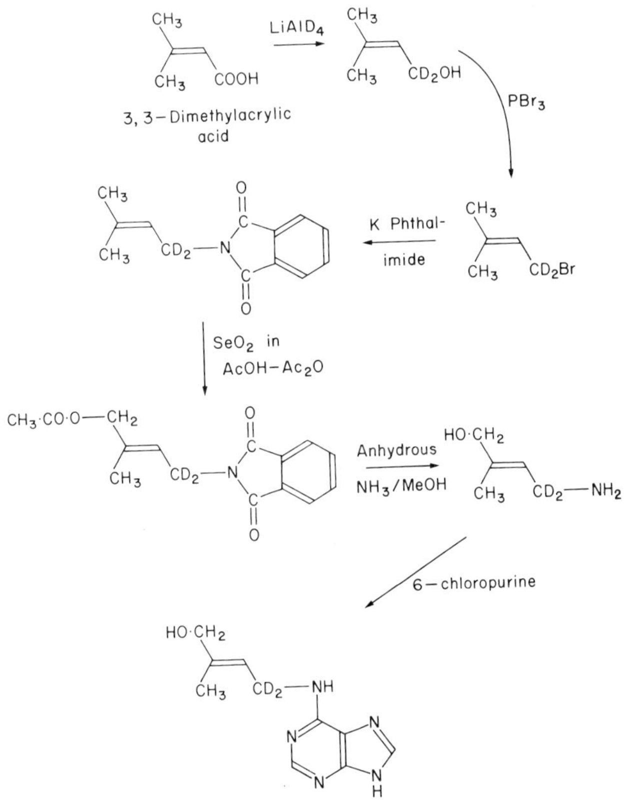

XII

Scheme 1.1. A synthesis for deuterium-labelled zeatin. Deuterium atoms are denoted by the symbol D.

Deuterium-labelled hormone would constitute an ideal internal standard as it will have physical properties almost identical to those of the endogenous hormone and the two will eventually leave the gas chromatograph and enter the mass spectrometer simultaneously. For mass spectral assessment a GC-MS unit with a data system (computer) which would record and store a mass spectrum every 3–4 sec during a GC-MS run would be desirable. The data system would facilitate extraction of meaningful data from the inherently high background due to column bleed and coeluting impurities. Alternatively, a GC-MS unit with the capability of multiple ion detection (i.e. selected ion current monitoring) could be used, provided at least four selected ions could be monitored simultaneously, namely, two ions in the mass spectrum of the endogenous hormone and the two analogous ions in the mass spectrum of the deuterated compound. From the mass spectral assessment outlined above, the ratio of deuterated compound to endogenous hormone could be calculated, and, knowing the precise amount of deuterated compound added to the tissue, the amount of endogenous hormone orginally present could be calculated. The accuracy of the method does not depend on the degree or reproducibility of the recovery of the endogenous hormone. The mass spectral techniques mentioned above can easily cope with a 10- to 20-fold excess of the deuterated compound used as calibrant, which can therefore act as a "carrier" and thus will reduce the absolute detection limit which might be as low as 10^{-11} g using a SICM method.

The bioassay techniques to which the plant hormone worker has become so addicted may provide useful clues in studies of hormonal levels in relation to growth responses. However, such results should be substantiated by methods which are both more specific and more accurate. It is hoped that the above discussion will promote interest in such methods. The new quantitation methods which are now emerging may eventually cause us to abandon some long established concepts, but can only lead us one step nearer to our objective – a complete understanding of the hormonal control of the integrated development of the diverse organs of the flowering plant.

References

Adamson, D. (1962) Canad. J. Bot., 40, 719–744.

Arsenault, G.P., Biemann, K., Barksdale, A.W. and McMorris, T.C. (1968) J. Amer. chem. Soc., 90, 5635–5636.

Audus, L.J. (1972) Plant Growth Substances, Vol. 1, Leonard Hill, London.

Batt, S. and Venis, M.A. (1976) Planta, 130, 15–21.

Beck, O. and Sedvall, G. (1975) J. Labelled Compd., 11, 57–61.

Bergmann, L. and Rennenberg, H. (1975) Naturwissenschaften, 62, 349–350.

Blumenfeld, A. and Gazit, S. (1970) Plant Physiol., 45, 535–536.

Boysen Jensen, P. (1911) Bull. Acad. Roy. (Denmark), 1, 3–24.

Boysen Jensen, P. (1913) Ber. Bot. Ges., 31, 559–566.

Bridges, I.G., Hillman, J.R. and Wilkins, M.B. (1973) Planta, 115, 189–192.

Browning, G. and Saunders, P.F. (1977). Nature (Lond.), 265, 375–377.

Burg, S.P. (1973). Proc. nat. Acad. Sci. (Wash.), 70, 591–597.

Burg, S.P. and Burg, E.A. (1966) Proc. nat. Acad. Sci. (Wash.), 55, 262–269.

Chailakhyan, M.Kh. (1937) C.R. (Dokl.) Acad. Sci. USSR, 16, 227–230. English translation in: "Papers on Plant Growth and Development". W.M. Laetsch and R.E. Cleland, (eds.), Little, Brown and Co., Boston, pp. 365–369.
Cleland, R. (1974) in: Biochemistry of Hormones, H. Rickenberg, (ed.), Butterworths, London, pp. 283–303.
Cline, M.G. (1976), Plant Sci. Lett., 6, 181–184.
Cohen, D. and Paleg, L.G. (1967). Plant Physiol., 42, 1288–1296.
Coombe, B.G., Cohen, D. and Paleg, L.G. (1966) Plant Physiol., 42, 105–112.
Cornforth, J.W., Milborrow, B.V., Ryback, G. and Wareing, P.F. (1965) Nature (Lond.), 205, 1269–1270.
Crane, J.C., Primer, P.E. and Campbell, R.C. (1960) Proc. Amer. Soc. hort. Sci., 75, 129–137.
Cummins, W.R. and Sondheimer, C. (1973) Planta, 111, 365–369.
Darwin, C. and Darwin, F. (1880) The power of movement in plants, John Murray, London.
Duhamel du Monceau (1758) La Physique des Arbres, Paris.
Eagles, C.F. and Wareing, P.F. (1963) Nature (Lond.), 199, 874.
Eagles, C.F. and Wareing, P.F. (1964) Physiol. Plant, 17, 697–709.
Engelke, A.L., Hamzi, H.Q. and Skoog, F. (1973) Amer. J. Bot., 60, 491–495.
Even-Chen, Z. and Itai, C. (1975) Physiol. Plant, 34, 97–100.
Gane, R. (1934) Nature (Lond.), 134, 1008.
Goeschl, J.C. and Pratt, H.K. (1968) in: Biochemistry and Physiology of Plant Growth Substances, F. Wightman and G. Setterfield (eds.), Runge Press, Ottawa, pp. 1229–1242.
Goodwin, P.B. and Cansfield, P.E. (1967) J. exp. Bot., 18, 297–307.
Greenwood, M.S., Shaw, S., Hillman, J.R., Ritchie, A. and Wilkins, M.B. (1972) Planta, 108, 179–183.
Gustafson, F.G. (1936) Proc. nat. Acad. Sci. (Wash.), 22, 628–636.
Haagen Smit, A.J., Dandliker, W.B., Wittwer, S.H. and Murneek, A.E. (1946) Amer. J. Bot., 33, 118–120.
Hamilton, R.H., Bandurski, R.S. and Grigsby, B.H. (1961) Plant Physiol., 36, 354–359.
Harrison, M.A. and Saunders, P.F. (1975) Planta, 123, 291–298.
Harvey, E.M. (1913) Bot. Gaz. (Chic.), 56, 439–442.
Hewett, E.W. and Wareing, P.F. (1973) Physiol. Plant, 29, 386–389.
Iynedjian, P.B. and Hanson, R.W. (1977) J. biol. Chem., 252, 655–662.
Jablonski, J.R. and Skoog, F. (1954) Physiol. Plant, 7, 16–24.
Jacobs, W.P. (1968) Plant Physiol., 43, 1480–1495.
Jacobs, W.P. (1970) Int. Rev. Cytol., 28, 239–275.
Jacobsen, J.V. (1977) Ann. Rev. Plant Physiol., 28, 537–564.
Jones, R.E. (1973) Ann. Rev. Plant Physiol., 24, 571–598.
Katzenellenbogen, B.S. and Gorski, J. (1972) J. biol. Chem., 247, 1299–1305.
Kendall, F.H., Park, C.K. and Mer, C.L. (1971) Ann. Bot. (Lond.), 35, 565–579.
Kende, H. and Gardner, G. (1976) Ann. Rev. Plant Physiol., 27, 267–290.
Knegt, E. and Bruinsma, J. (1973) Phytochemistry, 12, 753–756.
Kogl, F. and Erxleben, H. (1935) Z. Physiol. Chem., 235, 181–200.
Kogl, F. and Haagen Smit, A.J. (1931) Proc. Kon. ned. Akad. Wet., 34, 1411–1416.
Kogl, F. and Kostermans, D.G.F.R. (1934) Z. Physiol. Chem., 228, 113–121.
Kogl, F., Haagen Smit, A.J. and Erxleben, H. (1933) Z. Physiol. Chem., 214, 241–261.
Kogl, F., Erxleben, H. and Haagen Smit, A.J. (1934a) Z. Physiol. Chem., 225, 215–229.
Kogl, F., Haagen Smit, A.J. and Erxleben, H. (1934b) Z. Physiol. Chem., 228, 90–103.
Kurosawa, E. (1926) J. Nat. Hist. Soc. (Formosa), 16, 213–227. in: Source Book on Gibberellins, F.H. Stodola (ed.), Agr. Res. Serv. U.S.D.A. Peoria, Illinois, 1958.
1958.
Laidman, D.L., Colborne, A.J., Doig, R.I. and Varty, K. (1974) in: Mechanisms of Regulation of Plant Growth, R.L. Bieleski, A.R. Ferguson and M.M. Cresswell (eds.), The Royal Society of New Zealand, Wellington, pp. 581–590.

Lance, B., Reid, D.M. and Thorpe, T.A. (1976) Physiol. Plant, 36, 287–292.
Lee, T.T. and Skoog, F. (1965a) Physiol. Plant, 18, 386–402.
Lee, T.T. and Skoog, F. (1965b) Physiol. Plant., 18., 577–585.
Lenton, J.R., Perry, V.M. and Saunders, P.F. (1971) Planta, 96, 271–280.
Lenton, J.R., Perry, V.M. and Saunders, P.F. (1972) Planta, 106, 13–22.
Leonard, N.J., Greenfield, J.C., Schmitz, R.Y. and Skoog, F. (1975). Plant Physiol., 55, 1057–1061.
Letham, D.S. (1961) Aust. J. agric. Res., 12, 600–611.
Letham, D.S. (1963) Life Sci., 2, 569–573.
Letham, D.S. (1966) Phytochemistry, 5, 269–286.
Letham, D.S. (1969) N. Z. J. agric. Res., 12, 1–20.
Letham, D.S. and Wettenhall, R.E.H. (1977) in: The Ribonucleic Acids, P.R. Stewart and D.S. Letham (eds.), Springer, New York, pp. 129–193.
Libbert, E. (1964) in: Regulateurs Naturels de la Croissance Végétale. C.N.R.S., Paris, pp. 387–405.
Machlis, L., Nutting, W.H. and Rapoport, H. (1968) J. Amer. chem. Soc., 90, 1674–1676.
MacMillan, J. and Suter, P.J. (1958) Naturwissenschaften, 45, 46–48.
MacMillan, J., Seaton, J.C. and Suter, P.J. (1960) Tetrahedron, 11, 60–66.
Martin, D., Lewis, T.L. and Cerny, J. (1954) Aust. J. biol. Sci., 7, 211–220.
McDavid, C.R., Sagar, G.R. and Marshall, C. (1972) New Phytol., 71, 1027–1032.
Milborrow, B.V. (1968) in: Biochemistry and Physiology of Plant Growth Substances. F. Wightman and G. Setterfield (eds.) Runge Press, Ottawa, pp. 1531–1545.
Milborrow, B.V. (1971) in: Aspects of Terpenoid Chemistry and Biochemistry, T.W. Goodwin (ed.), Academic Press, London, pp. 137–151.
Milborrow, B.V. (1974) Ann. Rev. Plant Physiol., 25, 259–307.
Miller, C.O. (1961) Ann. Rev. Plant Physiol., 12, 395–408.
Neljubow, D.N. (1901) Beit. Bot. Zentralbl., 10, 128–138.
Ohkuma, K., Lyon, J.L., Addicott, F.T. and Smith, O.E. (1963) Science, 162, 1592–1593.
Ohkuma, K., Addicott, F.T., Smith, O.E. and Thiessen, W.E. (1965) Tetrahedron Lett., pp. 2529–2535.
Osborne, D.J., Jackson, M. and Milborrow, B.V. (1972). Nature New Biology, 240, 98–101.
Paal, A. (1919) Jber. Wiss. Bot., 58, 406–458.
Paleg, L.G. (1965) Ann. Rev. Plant Physiol., 16, 291–322.
Palmiter, R.D. (1975) Cell, 4, 189–197.
Patrick, J.W. and Woolley, D.J. (1973) J. exp. Bot., 24, 949–957.
Phillips, I.D.J. (1969) in: Physiology of Plant Growth and Development, M.B. Wilkins (ed.), McGraw-Hill, New York, pp. 165–202.
Phillips, I.D.J. (1975) Ann. Rev. Plant Physiol., 26, 341–367.
Phillips, I.D.J. and Wareing, P.F. (1959). J. exp. Bot., 10, 504–514.
Pickard, B.G. (1973) Bot. Rev., 39, 172–201.
Porter, W.L. and Thimann, K.V. (1965) Phytochemistry, 4, 229–243.
Radley, M. (1956) Nature (Lond.), 178, 1070–1071.
Radley, M. (1958) Ann. Bot. (Lond.), 22, 297–307.
Railton, I.D. and Wareing, P.F. (1973), Planta, 112, 65–69.
Ricca, U. (1916). Arch. ital. Biol. (Pisa), 65, 219–232.
Rivier, L. and Pilet, P.E. (1974) Planta, 120, 107–112.
Sachs, J. (1882) Arb. Bot.Inst. (Wurzburg), 2, 689–718.
Sachs, J. (1893) Ueber Wachstumsperioden und Bildungsreize, Physiologische Notizen VI, Marburg.
Scott, T.K. and Jacobs, W.P. (1964) in: Regulateurs Naturels de la Croissance Végétale, C.N.R.S., Paris, pp. 457–474.
Sibaoka, T. (1969) Ann. Rev. Plant Physiol., 20, 165–184.
Skoog, F. and Miller, C.O. (1957) Symp. Soc. exp. Biol., 11, 118–131.
Soding, H. (1923). Ber. Deut. Bot. Ges., 41, 396–400.
Soding, H. (1925). Jber. Wiss. Bot., 64, 587–603.
Steeves, T.A. and Briggs, W.R. (1960). J. exp. Bot., 11, 45–67.

Terrine, C., Doree, M., Guern, J. and Hall, R.H. (1972) in: Plant Growth Substances 1970, D.J. Carr (ed.), Springer Berlin, pp. 467–475.
Thimann, K.V. (1935) J. biol. Chem., 109, 279–291.
Thimann, K.V. (1954) Amer. Sci., 42, 589–606.
Thimann, K.V. (1974) Plant Physiol., 54, 450–453.
Thimann, K.V. and Koepfli, J.B. (1935) Nature (Lond.), 135, 101.
Thimann, K.V. and Went, F.W. (1934) Proc. Kon. ned. Akad. Wet., 37, 456–459.
Thompson, A.G., Horgan, R. and Heald, J.K. (1975) Planta, 124, 207–210.
Thorpe, T.A. and Meier, D.D. (1973) Physiol. Plant., 29, 121–124.
Van den Ende, H. (1968) J. Bacteriol., 96, 1298–1303.
Van Steveninck, R.F.M. (1957) Bot. Gaz., 119, 63–70.
Van Steveninck, R.F.M. (1958) J. exp. Bot., 9, 372–383.
Van Steveninck, R.F.M. (1959a) Nature (Lond.), 183, 1246–1248.
Van Steveninck, R.F.M. (1959b) J. exp. Bot., 10, 367–376.
Varner, J.E. (1974). Recent Advan. Phytochem., 7, 123–130.
Varner, J.E. and Ho, D.T.-H. (1976) in: The Molecular Biology of Hormone Action, J. Papaconstantinou (ed.), Academic Press, New York, pp. 173–194.
Varner, J.E. and Ram Chandra, G. (1964) Proc. nat. Acad. Sci. (Wash.), 52, 100–106.
Venis, M.A. (1977) Planta, 134, 145–149.
Wareing, P.F. and Saunders, P.F. (1971) Ann. Rev. Plant. Physiol., 22, 261–288.
Went, F. (1928) Rec. Trav. Bot. Neer., 25, 1–116; 25a, 483–489.
Went, F.W. and Thimann, K.V. (1937) Phytohormones, MacMillan, New York.
White, J.C. (1976) Ann. Bot. (Lond.), 40, 521–529.
White, J.C., Medlow, G.C., Hillman, J.R. and Wilkins, M.B. (1975) J. exp. Bot., 26, 419–424.
Wicks, W.D. (1974) in: Biochemistry of Hormones, H. Rickenberg (ed.), Butterworths, London, pp. 211–241.
Williams, M.W. and Letham, D.S. (1969) HortScience, 4, 215–216.
Williams, M.W. and Stahly, E.A. (1968) Proc. Amer. Soc. hort. Sci., 94, 17–18.
Wright, S.T.C. (1968). In: "Biochemistry and Physiology of Plant Growth Substances." F. Wightman and G. Setterfield (eds.), Runge Press, Ottawa, pp. 521–542.
Wright, S.T.C. (1969) Planta, 86, 10–20.
Yamada, Y., Sekiya, J. and Koshimizu, K. (1972) Phytochemistry, 11, 1019–1021.

Phytohormones and Related Compounds – A Comprehensive Treatise, Volume I
Editors: Letham, Goodwin and Higgins

2

Auxins

ELNORA A. SCHNEIDER and FRANK WIGHTMAN

2.1. Introduction

The existence of auxins was first inferred from observation of their physiological action. Charles and Francis Darwin (1880) noticed that grass coleoptiles tended to curve towards the light, and found that this response could be abolished by shading the coleoptile tip, suggesting the formation of a diffusible substance in the tip which promoted growth of the coleoptile tissue. These observations led eventually to the demonstration by Went in 1928 that decapitated coleoptiles could be induced to curve in the dark by the application of a substance obtained from coleoptile tips by diffusion. This substance was shown by Kögl and coworkers (Kögl et al. 1934) to have identical diffusion and physiological properties to the compound, indol-3-ylacetic acid, which they had recently isolated from human urine, but it was not until several years later that this first growth promoting hormone was isolated from higher plant tissues (Haagen-Smit et al. 1946). Research on indol-3-ylacetic acid (IAA) and related auxins has tended to fall into two main areas; physiological studies on the growth-regulating effects and mechanism of action of this type of plant hormone, and biochemical studies focussed on the hormone itself, its occurrence, biosynthesis, structure – activity relationships, metabolism and transport. These latter aspects of auxin research will be the topic of the present chapter. Many interesting questions may be asked in this area; for example, is IAA the only substance with auxin activity which occurs naturally in plants? What are the pathways by which auxins are synthesised and degraded? What levels of these compounds are present in plants at different stages of their growth and development, and how are these levels controlled? Do the auxins, like animal hormones, move from a site of synthesis to a site of action? What are the sites of auxin synthesis in a plant and what are the characteristics of the auxin transport system? How do environmental factors and other types of plant hormones affect the metabolism and transport of auxins? In the following chapter we shall summarise some of the progress which has been made towards answering these questions.

2.2. Assay methods for auxins

2.2.1. Methods of extraction and separation of auxins

In early work on the extraction of auxins from plants, the solvent employed was usually ether. However, this solvent was later found to be unsatisfactory, since the enzymes of the plant are not inactivated sufficiently rapidly, and auxin production may continue for several hours. More recent workers have usually ground the plant material rapidly in methanol or acetone. After removal of the plant residue by filtration, the solvent is evaporated off under reduced pressure, and the aqueous concentrate is again filtered to remove precipitated chlorophyll. If necessary, neutral indoles can be removed at this stage by partitioning with ether or ethyl acetate at pH 8. The aqueous solution is then acidified to pH 3, and the acidic auxins are removed by partitioning again with peroxide-free ether or ethyl acetate. Partitioning with methylene chloride has also been employed (Powell 1964), and partitioning with this solvent appears to be especially useful when indol-3-ylpyruvic acid is present, since this solvent does not extract indol-3-ylpyruvic acid but will extract IAA (Atsumi et al. 1976).

Until recently, most extracts were purified by paper chromatography, often using the solvent system isopropanol – ammonia – water (8:1:1, by vol.) (Stowe and Thimann 1954). Systems have also been developed for the separation of indoles by thin-layer chromatography (Ballin 1964; Grunwald et al. 1967; Raj and Hutzinger 1970) and by column chromatography (Powell 1964; Raj and Hutzinger 1970). However, it is now becoming clear that plants contain other auxins besides IAA, such as indol-3-ylpropionic acid (IPA) and phenylacetic acid (PAA), and these methods of separation have not proved to have the resolving power and sensitivity required for analysis of nanogram amounts of these closely related compounds. Recently, therefore, attention has been focussed on gas chromatography, which can be used for both separation and quantitative analysis. Pioneer work on the gas chromatography of the indole auxins was carried out in the laboratory of Stowe in the 1960's (Stowe and Schilke 1964; Powell 1964; Grunwald et al. 1967, 1968; Stowe et al. 1968). In most cases the methyl ester derivatives of the acid auxins have been chromatographed. Methylation with diazomethane requires careful procedures, but is more specific than silylation when crude extracts are used. Silylation has also been used and is a simple rapid procedure, but since silylation can take place on the ring N and on phenolic OH groups, as well as on the carboxyl group, care must be taken that the reaction goes to completion or multiple peaks will result. Extracts from some plants may require prepurification by other chromatographic methods (Bandurski and Schulze 1974) but extracts of other plants may be chromatographed directly after derivatisation (Shindy and Smith 1975; Wightman 1977). Linear temperature programming has proved to be very useful when several auxins are present. Wightman (1977) chromatographed the methylated acidic ether fraction of tobacco leaves directly, using dual glass columns packed with 3% OV-17 and a linear temperature program starting at 95 °C and rising to 310 °C at 3 °C per min. By this method, the auxins IAA, IPA and PAA could readily be separated.

It is important to note that considerable losses of IAA can occur during extraction and chromatography. For example, Hamilton et al. (1961) found that the recovery of 100–200 μg amounts of radioactive IAA added to 1 kg of corn shoots was about 25–30%, while recovery of 1–3 μg aliquots was only about 13%. Methods for minimising loss were discussed by Mann and Jaworski (1970) who suggested that maximum recoveries could be obtained by grinding the tissue under nitrogen, by avoiding vacuum evaporation which led to losses by sublimation, and by adding an antioxidant to the origin of paper chromatograms before applying the sample. Using these methods, recovery of 10–50 μg amounts of IAA was about 60%. A somewhat different approach was taken by Bandurski and Schulze (1974) who employed an isotope dilution method in the gas chromatographic separation and estimation of the IAA of *Zea* and *Avena* shoots. In this method, x μg of radioactive IAA of specific activity C_o are added to the shoots at the beginning of the experiment. If the specific activity of the IAA extracted from the shoots is C, then the amount of non-radioactive IAA in the shoots, y, is given by the expression $y = (\frac{C_o}{C} - 1)x$.

This method has the advantage that the error of the determination is independent of the yield of IAA. Problems associated with the determination of auxin levels in plant tissues are also discussed in Volume II, Chapter 2, section 2.3.2.3.1.

2.2.2. Methods for the determination of auxin concentration

2.2.2.1. Bioassays. Because of the low concentration of auxin in plant tissues, bioassay has been until recently the only satisfactory method for the quantitative analysis of auxins. Bioassays generally show good sensitivity; Nitsch and Nitsch (1956) present a table showing the minimal amounts of IAA detectable by various bioassays, and these vary from 7.5×10^{-7} μg (pea root section test) to 5×10^{-4} μg (*Avena* first internode straight growth test). On the other hand, bioassays show relatively low accuracy and are non-specific.

Many comprehensive accounts of bioassay methods have appeared (Larsen 1955; Mitchell and Livingston 1968; Thimann 1972) and this subject will be treated here only briefly. The most venerable bioassay is Went's *Avena* curvature test (Went and Thimann 1937). In this test a block of agar containing auxin is placed asymmetrically on the top of a decapitated coleoptile. The degree of curvature of the coleoptile is measured, and this is proportional to the concentration of auxin in the block up to a maximum value. This test is more specific for IAA than many other bioassays, but has the disadvantage of requiring highly standardised conditions for satisfactory results. A curvature test of a different kind is the pea slit stem test (Van Overbeek and Went 1937). In this test a longitudinal slit is made in a pea stem segment, leaving the two half-segments joined at the base. If the slit segments are incubated in an auxin solution, an inward curvature develops which is proportional to the logarithm of the concentration of auxin in the solution. This test is responsive to a wide range of auxins and has been extensively used in the study of synthetic auxins. Perhaps the most useful bioassays, especially when paper chromatography is used, are the straight growth tests. These include the pea stem straight growth test (Christiansen and Thimann 1950), the *Avena* first internode test (Nitsch and Nitsch 1956) the wheat coleoptile segment test (Wain and Wightman 1954; Mitchell and Livingston 1968) and the *Avena* coleoptile straight growth test (Sirois 1966). The response in these straight growth tests is generally proportional to the log of the auxin concentration, although careful selection of the conditions may result in a linear response (Sirois 1966). Assays may also be based on the elongation of root segments (Pilet et al. 1960). Many other bioassays have been developed, based on such phenomena as parthenocarpic fruit development in tomatoes (Zimmerman and Hitchcock 1944), water uptake by slices of artichoke tissue (Hackett and Thimann 1952) and lateral root induction (Mitchell and Livingston 1968), but these methods have been used to only a limited extent. Mitchell and Livingston (1968) have published an extensive collection of these more unusual bioassays in addition to providing details of the standard auxin test procedures.

2.2.2.2. Physical and chemical methods for assay of auxins. Two colorimetric methods based on the Salkowski and Ehrlich reactions have been used to some extent for the measurement of indole auxins, particularly IAA. The Salkowski method of Gordon and Weber (1951) can be used to measure amounts of IAA as low as 5 μg/ml. It gives no colour with other auxins such as phenylacetic acid, and cannot be used with crude extracts which contain reducing agents. The Ehrlich reaction in presence of trichloracetic acid can be used to measure IPA (1–15 μg) as well as IAA (2–50 μg) (Anthony and Street 1970). The most sensitive colourimetric method for identifying indole auxins is a modification of the standard Ehrlich procedure and uses *p*-dimethylaminocinnamaldehyde as the chromogenic reagent (Wightman 1964).

Spectrofluorimetric methods can have sensitivities comparable to those of bioassays, but are much more accurate. Fluorescent impurities in solvents and plant extracts may interfere, especially when purification is by paper chromatography (Burnett and Audus 1964). To avoid these difficulties, spectrofluorimetry may be combined with gas chromatography (Stowe and Schilke 1964). Stoessl and Venis (1970) have described a method in which IAA is converted to the indolo-α-pyrone and the concentration of this compound is measured spectrofluorimetrically. Amounts of IAA from 10^{-3} to 100 μg can be measured and the method can be used on crude extracts since it is specific to IAA. Knegt and Bruinsma (1973) combined this method with an internal standard of [^{14}C]IAA to correct for losses of IAA during extraction and were able to measure the IAA contents of various plant tissues using 0.05–5 g fresh wt. of starting material (25–125 ng IAA).

The most sensitive and specific auxin analyses are those based on gas chromatography. The separated auxins may be measured using the peak heights obtained with the flame ionisation detector of the gas chromatograph (Bayer 1969; Bandurski and Schulze 1974; Shindy and Smith 1975; Wightman 1977) or by mass fragmentometry. In the latter technique, the auxin separated by gas chromatography is then subjected to mass spectrometry and quantitation is dependent on the peak heights of characteristic fragments of the auxin molecule (Bridges et al. 1973; Rivier and Pilet 1974). This method requires costly investment in equipment, but gives complete assurance of the identity of the compound being measured.

2.3. Occurrence and identity of auxins

At 3 a.m. on April 17, 1926, the young Fritz Went found that the application of a diffusate from coleoptile tips to decapitated coleoptiles resulted in a growth response (Went 1974). Thus it was demonstrated that coleoptile tips contain a diffusible growth-promoting substance, which was later to be named "auxin" (Kögl and Haagen-Smit 1931). At about the same time, Nielson (1930) found that the culture filtrate from the fungus *Rhizopus* was highly active in the *Avena* test, and the active fraction of this "rhizopin" was found to be a weak acid (Dolk and Thimann 1932). Then, in 1934, Kögl and coworkers isolated IAA from human urine and showed that it had a striking promotive effect on coleoptile growth (Kögl et al. 1934). Went (pers. comm.) describes the events which led up to this discovery: "When Kögl started to work on the extraction of growth hormones in plants, he looked for starting materials which were high in the substance. Therefore, he started to look for natural products which might contain the hormone. There was a slight possibility that the female sex hormone would have plant growth activity. When he received from Germany an impure preparation of female sex hormones, this turned out to be highly active, while the pure product was inactive. Therefore, the material from which the female sex hormone was extracted apparently contained the plant hormone. This was urine of pregnant women. Afterwards it was found that urine from all persons contained auxin, and since urine turned out to be very rich in it, they continued to extract auxin from it."

Twelve years later, IAA was isolated from immature corn kernels, which are a relatively rich source of auxin (Haagen-Smit et al. 1946), but the low concentration of IAA in vegetative organs made chemical identification of the compound much more difficult. Using the conventional methods of solvent extraction, paper chromatography and bioassay, IAA has been identified in a liverwort, *Marchantia* (Schneider et al. 1967), two gymnosperms, *Picea* (Steen 1972) and *Pinus* (Alden 1971), and at least 18 angiosperms (see Schneider and Wightman 1974a). Recently, IAA has been conclusively identified in methanol extracts of sterile *Zea* roots. This identification is based on bioassay (Greenwood et al. 1973), thin-layer chromatography in six solvent systems, three chromogenic tests and gas chromatography-mass spectrometry (GC-MS) of the methylated derivative (Elliott and Greenwood 1974). GC-MS has also recently been used to identify IAA in *Avena* shoots (Bandurski and Schulze 1974), cotton ovules (Shindy and Smith 1975) and shoots of Douglas fir (DeVoe and Zaerr 1976).

When the amount of IAA in vegetative tissues of higher plants is measured by bioassay of chromatographed extracts (Figure 2.1), the values usually fall in the range 1–100 μg per kg fresh wt. (see Table 4 in Schneider and Wightman 1974a). Bandurski and Schulze (1974) have measured the IAA content of *Zea* and *Avena* shoot extracts using gas chromatography and an isotope dilution technique (see 2.2.2.2. above). The IAA content of *Zea* shoot tissue was found to be about 24 μg per kg fresh wt., and that of *Avena* shoots about 16 μg/kg fresh wt. In addition the *Zea* tissue was found to contain 352 μg per kg of bound IAA which could be liberated by hydrolysis with 1 N NaOH. These values were quite comparable with earlier bioassay measurements (Van Overbeek 1941). Other GLC measurements of IAA have shown that various segments of *Zea* roots contain from 77–357 μg/kg (Rivier and Pilet 1974) (see Table 2.1), cotton ovules 9.5 μg/kg (Shindy and Smith 1975) and tobacco leaves of various ages 40–300 μg/kg (Wightman 1977), all on a fresh wt. basis.

IAA, however, is not the only indole auxin occurring in higher plants. Gas chromatography of methylated acidic extracts from *Nicotiana* plants using 3% SE-30 columns at 190 °C revealed peaks having the retention times of IPA and indol-3-ylbutyric acid (Bayer 1969). Wightman (1977) also found a peak corresponding to IPA in tobacco leaf ether

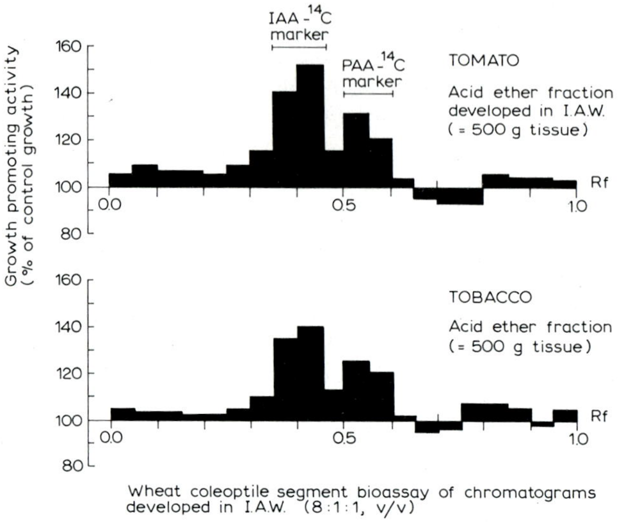

Figure 2.1. Typical bioassay results obtained when paper chromatograms containing acidic ether fractions from tomato and tobacco shoots are assayed for auxin activity with the wheat coleoptile segment test.

Table 2.1.

Average values for the IAA content of different regions of the maize root as determined by mass fragmentography (from Rivier and Pilet 1974)

Root sections from the tip (mm)	Number of sections in each extract	No. of replicate extracts	Fresh wt. (mg/section)	Dry wt. (mg/section)	IAA (μg)	
					per kg fresh wt.	per g dry wt.
I: 0–0.5 (root cap)	1000	2	0.263	0.035	356.6	2.67
II: 0.5–1.0 (meristem)	1000	2	0.370	0.058	179.9	1.16
III: 1.0–4.0 (region of cell elongation)	500	4	4.122	0.561	76.5	0.56
IV: 4.0–10.0 (region of cell maturation)	250	5	8.858	0.695	114.8	1.46

extracts after methylation and gas chromatography on 3% OV-17 columns using linear temperature programming procedures (Figure 2.2). The peak of the suspected IPA was shown to cochromatograph identically with authentic IPA methyl ester. It is hoped that the identification of these new indole auxins in tobacco plants will soon be confirmed by mass spectroscopy.

Gandar and Nitsch (1964) isolated a neutral growth substance from immature pea seeds which showed auxin activity in the *Avena* mesocotyl, *Avena* curvature and artichoke bioassays. This compound was later identified as the methyl ester of a chloro-substituted IAA (Gandar and Nitsch 1967), and further work showed that the chlorine atom was in the 4-position (Marumo et al. 1968a). The free acid was present in the same tissue at a concentration of about 14 μg/kg of seeds (Marumo et al. 1968b) and three other related compounds, monomethyl-4-chloroindol-3-ylacetyl-L-aspartate (Hattori and Marumo 1972), α-*N*-carboethoxyacetyl-D-4-chlorotryptophan and α-*N*-methoxyacetyl-D-4-chlorotryptophan (Marumo and Hattori 1970) were also found. 4-chloro-IAA is an extremely active auxin; in a study of the activities of a large range of halogenated indoles, Porter and Thimann (1965) showed that 4-chloro-IAA had 700% greater activity than IAA in the pea stem curvature test, which was the highest level of activity shown by any of the halogenated indoles tested. 4-chloro-IAA shows an R_f in paper chromatography very similar to that of IAA when the standard solvent system, isopropanol–ammonia–water, is used to analyse auxin extracts (Marumo et al. 1971) and it is possible therefore that with further investigation this highly active auxin will be found in a wider variety of tissues.

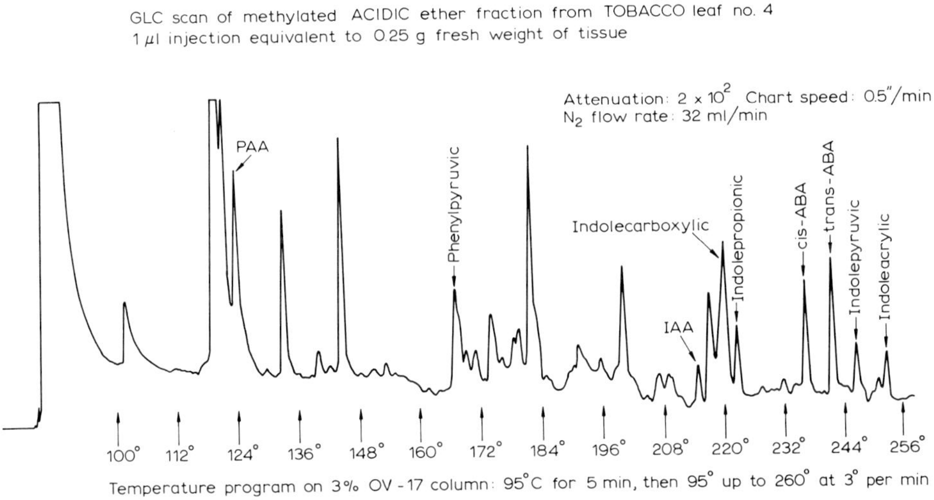

Figure 2.2. GLC scan of 1 μl sample of methylated acidic ether fraction from typical tobacco leaf (no. 4 down stem) assayed on a 3% OV-17 column using temperature programming procedures. Peaks of suspected methylated derivatives of phenyl- and indolyl-acids are indicated on the scan. The identity of each peak was confirmed by co-injecting further 1 μl samples of the ether fraction with samples of each authentic methylated phenyl or indole acid.

Lastly, evidence is accumulating for the presence in higher plants of auxins which are non-indolic in nature. Khalifah et al. (1963, 1965) reported the isolation of a substance from developing citrus fruits which showed appreciable activity in the oat coleoptile curvature test but did not exhibit the properties of an indole compound. Since they could not detect free IAA in fruit extracts, these workers considered that the unknown active substance might be the natural auxin of the tissue and accordingly referred to the substance as "citrus auxin". In later work, Goren and Goldschmidt (1970) observed several zones of growth-promoting activity on chromatograms of extracts from young citrus fruits none of which coincided with the position of IAA and some were also distinct from Khalifah's citrus auxin. Further investigation showed, however, that the auxin present in citrus shoots appears to be IAA (Goldschmidt et al. 1971) and so it seems that citrus plants may contain both IAA and several non-indolic auxins. In recent studies concerned with the metabolism of phenylalanine in higher plants, Wightman and co-workers have demonstrated the natural occurrence of the auxin, phenylacetic acid (PAA), in a range of shoot tissues (Wightman 1973; Wightman and Rauthan 1974). The identification of PAA was based on bioassay and gas chromatography of ether extracts of tomato, tobacco, sunflower, pea, barley and corn shoots and quantitative analysis showed that this auxin occurs in young shoots in amounts ranging from 70–200 μg/kg fresh wt. (Wightman 1977). PAA has also been shown to occur in the marine alga, *Undaria pinnatifida,* from which it was crystallised and identified by IR, UV and NMR spectroscopy (Abe et al. 1974). In a gas chromatographic study of the auxin substances in developing tobacco leaves, Wightman (1977) found that PAA is the main auxin present at all stages of growth and occurs at the highest level in leaves which subsequently show the greatest rate of expansion (Figure 2.15). Like 4-chloro-IAA, PAA runs rather close to the R_f of IAA in paper chromatography using the solvent system, isopropanol – ammonia – water, and thus its distribution in plants may be far more widespread than has been suspected.

In Volume II, auxin levels are discussed in relation to root, stem, leaf, bulb and tuber development (Chapter 2), fruit development (Chapter 3), flower formation (Chapter 6), leaf senescence (Chapter 7), bud dormancy (Chapter 9), and fruit ripening (Chapter 10). It has been appropriate to present some data concerning the occurrence and level of auxins in plant tissues in section 2.5.3 which deals with sites of biosynthesis.

2.4. Structure and activity relationships

After the discovery by Kögl and coworkers in 1934 that the simple indole compound IAA is strikingly active in promoting cell elongation, plant physiologists began to examine other compounds of similar chemical structure to determine whether they also possessed growth-regulating properties. In 1935, further indole compounds were found to be active, such as IPA and γ-(indol-3-yl)butyric acid (I, IBA) (Hitchcock 1935; Kögl and Kostermans 1935) and a number of related arylacetic acids, such as PAA (II), naphth-1-ylacetic acid (III, NAA) and anthrylacetic acids (V), were also reported to show auxin activity (Zimmerman and Wilcoxon 1935; Thimann and Schneider 1939). Then in 1938, naphth-2-yloxyacetic acid (VI) was shown to be an active growth substance (Irvine 1938) and this finding soon stimulated interest in the phenoxyacetic acids which led eventually to the important demonstration by Zimmerman and Hitchcock in 1942 that some of these compounds with ring substituents, such as 2,4-dichlorophenoxyacetic acid (2,4-D, VII), show extremely high activity. It is from among the series of substituted phenoxyacetic acids that most of the auxin-type weedkillers, such as 2,4-D, MCPA (2-methyl-4-chlorophenoxyacetic acid, VIII) and 2,4,5-T(2,4,5-trichlorophenoxyacetic acid, IX), have been discovered (Templeman and Halliday 1951; Crafts 1953).

These early investigations on the auxin activity of a limited range of indole, naphthyl and phenyl compounds gave rise to the fascinating study of the relationships between molecular structure and plant growth-regulating activity. Based on these early findings, Koepfli et al. (1938) proposed that the following structural features were necessary for auxin activity: 1, a ring system as a nucleus; 2, at least one double bond in this ring; 3, a side-chain possessing a carboxyl group, or a group easily converted to a carboxyl group; 4, at least one carbon atom between the ring and the carboxyl group in the side-chain; 5, a particular spatial relationship between the ring-system and the carboxyl group.

These early investigations on the auxin activity of a limited range of indolyl, naphthyl they realized that the steric relationships in a molecule played an important role in determining auxin activity. For several years, these basic structural requirements proved a very useful guide to further research in the chemical structure – activity field, but as many new compounds were tested it became evident that the initial proposals of Koepfli et al. (1938) needed to be modified to account for the high activity shown by some compounds whose molecules do not possess all the basic structural features listed above. In particular, a number of substituted benzoic and naphthoic acids which do not possess a side-chain with a terminal carboxyl group, were found to be highly active compounds (Bentley 1950; Jones et al. 1951; Muir and Hansch 1951; Veldstra 1952) and certain dithiocarbamate derivatives not possessing a ring system (e.g., X and XI) have also been reported to show high activity (Kerk et al. 1955; Garraway and Wain 1962). After investigations comparing the polarographic properties of substituted benzoic and naphthoic acids, as well as several phenoxy and naphthyl acids, Veldstra and Booij (1949) stated that the auxin activity of these compounds can be better explained if the five structural requirements of Koepfli et al. are condensed to two basic requirements, namely: (1) a basal unsaturated ring system (nonpolar part) with high interfacial activity and (2) a carboxyl group or other acidic grouping (polar part) in a definite spatial relationship to the ring system.

In a later study attempting to correlate physiological activity with the molecular structure and steric relationships of a wide range of compounds, Jönsson (1955) further defined this second structural requirement for activity by proposing that "the side-chain must be able to assume a spatial orientation such that it is almost in the same plane as the ring and so allowed the formation of a pseudo-ring with the carboxyl group close to the main ring and near to the centre of the extended ring system so formed". This proposal is simply an elaboration of Veldstra's views in which his "definite spatial relationship" between the ring and the carboxyl group now appears as an "extended ring system".

Other interesting proposals to explain structure – activity relationships in synthetic auxins have been made over the last 20 years (see Muir and Hansch 1953; Wain 1953; Linser 1956; Porter and Thimann 1965; Muir et al. 1967) but to date no comprehensive list of the structural and steric requirements necessary for auxin activity has been formulated which satisfactorily covers all the molecules that are known to possess growth-promoting properties. However, many detailed studies have been made on the different molecular features that are important in determining auxin activity (see reviews by Thimann 1951; Wain 1953; 1958, 1968; Veldstra 1953, 1956; Muir and Hansch 1955; Fawcett 1961; Jönsson 1961; Fredga and Aberg 1965; Wain and Fawcett 1969; Andus 1972) and the relative effects of these structural and steric requirements will now be considered.

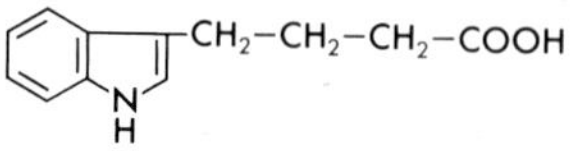

I.

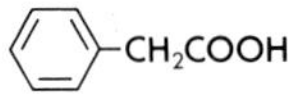

II

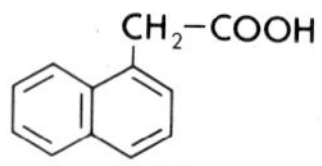

III

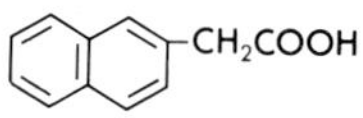

IV

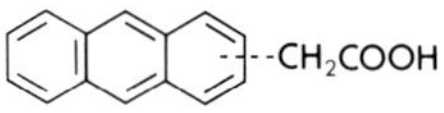

V

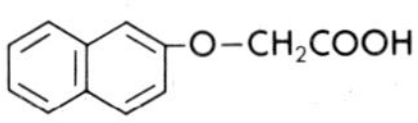

VI

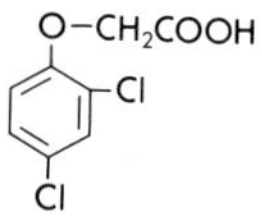

VII

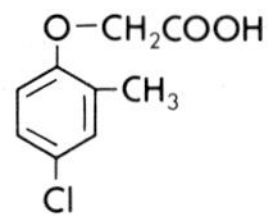

VIII

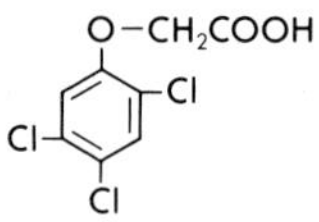

IX

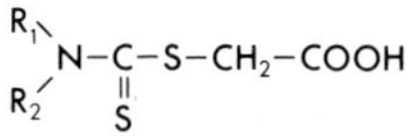

X

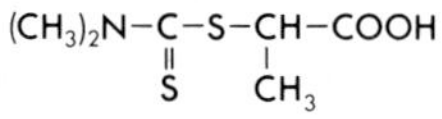

XI

R_1 and R_2 = alkyl

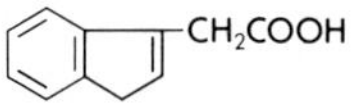

XII

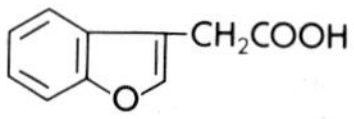

XIII

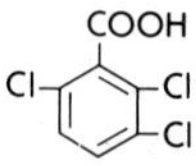

XIV

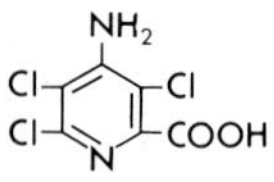

XV

2.4.1. Structure of the ring system

The nature of the ring system has a major effect on auxin activity. Early studies by Thimann (1935) on the importance of the indole ring showed that replacement of the nitrogen in the pyrrole ring with carbon to produce inden-3-ylacetic acid (XII), or with oxygen to give benzofuran-3-ylacetic acid (XIII), leads to a considerable reduction in activity. On the other hand, acetic acids attached to larger ring systems such as naphthalene or anthracene rings, show high auxin activity and a number of polycyclic carboxylic acids are also active substances (Heacock et al. 1958). Compounds with unsaturated rings smaller than indole, such as benzoic and phenylacetic acids, are generally less active than IAA, and substances with five-membered ring systems show no auxin activity.

A second important feature of the ring is that it must be unsaturated and should contain at least one double bond located adjacent to the side-chain (Went 1949). No compound with a completely saturated ring system has yet been found to show auxin activity. That the number of double bonds in the ring does influence activity is well demonstrated by the properties of hydrogenated derivatives of phenylacetic and naphthylacetic acids. Progressive hydrogenation of the double bonds in these two ring systems produces a series of compounds which show a progressive decrease in auxin activity, a property which is entirely lost in the fully saturated compounds, cyclohexylacetic acid and decahydronaphthylacetic acid. The degree of unsaturation in the ring considerably influences the surface activity of the molecule, which Veldstra (1944, 1953) considers to be one of the key factors determining the growth-regulating activity of a compound.

Substitution of halogen atoms or certain alkyl groups into the ring system may also profoundly affect auxin activity. The nature of the substituent and its position on the ring both influence activity. Early studies on the effect of different substituents in the pyrrole ring of indolylacetic acid showed a reduction in activity in all derivatives (Kögl and Kostermans 1935). Introduction of a halogen atom or methyl group into the phenylene ring, however, results in increased activity over that of IAA, the greatest effect being obtained by chloro-substitution in the 4- or 6-position (Hoffman et al. 1952; Sell et al. 1953; Porter and Thimann 1965). The most striking effects of ring substitution on auxin activity is seen in the series of monochloro-, dichloro- and trichlorosubstituted compounds derived from phenoxyacetic, phenylacetic or benzoic acids. With all these compounds, substitution of halogen atoms or lower alkyl groups into the ring may greatly increase activity. depending on the position of the substituents. So far, it has not been possible to formulate any clear rules relating level of auxin activity to the nature of the ring substituent, but generally the most effective substituent is the chlorine atom with bromine being only slightly less effective, followed by a methyl group, iodine and fluorine atoms in decreasing order of effectiveness.

Although the influence of nuclear substitution on activity in synthetic auxins was first studied in benzoic acids (Zimmerman and Hitchcock 1942; Synerholm and Zimmerman 1945), the demonstration by Zimmerman and Hitchcock (1942) of the extremely high growth-promoting activity of the substituted phenoxy acid, 2,4-dichlorophenoxyacetic acid (2,4-D), stimulated much interest in this class of compounds. When complete series

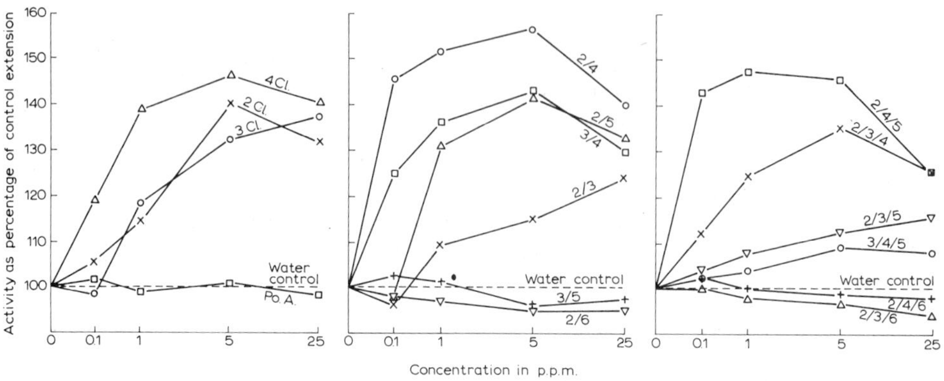

Figure 2.3. Graphs showing activity of phenoxyacetic acid (PoA) and the different monochloro- (left), dichloro- (centre) and trichloro-phenoxyacetic acids (right) in the wheat coleoptile segment test.

of monochloro-, dichloro- and trichloro-phenoxyacetic acids were examined in a range of tests (Muir et al 1949; Leaper and Bishop 1951; Wain and Wightman 1953), the results showed clearly that substitution in the *para*-position of the ring is more effective than substitution in the *ortho* or *meta*-positions, although substitution in either of these latter two positions also leads to a considerable increase in activity over that shown by the parent acid (Figure 2.3). Optimum activity in this series of compounds is shown by the disubstituted isomers, 2,4-, 2,5- and 3,4-dichlorophenoxyacetic acids, whereas disubstitution in both *ortho* positions or in both *meta*-positions, as in the 2,6- and 3,5-dichloro-isomers, results in complete loss of activity by the molecule. With the tri-substituted isomers, only 2,3,4- and 2,4,5-trichlorophenoxyacetic acids show high activity; all the other isomers exhibit low activity or are completely inactive, which is consistent with the fact that these four isomers all have di-*ortho*- or di-*meta*-substituents as part of their trichloro-substitution pattern (Figure 2.3).

Several theories have been put forward to account for the negligible activity shown by chlorophenoxyacetic acids with substituents in both *ortho*- or both *meta*-positions in the ring. One proposal that received much attention was suggested by Muir and coworkers (Muir et al. 1949; Hansch and Muir 1950; Hansch et al. 1951; Muir and Hansch 1951, 1953) who pointed out that the ability of chlorophenoxyacetic acids to induce a growth response is associated with the presence of at least one free *ortho*-position in the ring and they suggested that this *ortho*-position might be involved in some essential reaction in plant cells whereby the growth response is induced. These workers envisaged a two-point attachment between the synthetic auxin molecule and its receptor site in the cell, which they suggested might be the cysteinyl group of a protein since this amino acid moiety could conceivably react with the carboxyl group of the phenoxyacetic side-chain and with any free *ortho*-position in the ring. While most workers in the field soon accepted the possibility that auxin molecules must react with some cellular constituent in order to initiate the growth response, Thimann (1952) proposed that ring positions other than the *ortho*-position might be involved in this initial reaction, as is suggested by the complete inactivity shown by chlorophenoxy acids with substituents in both *meta*-positions of the ring (Wain and Wightman 1953; Toothill et al. 1956). Wain and coworkers also provided evidence (Wain 1953; Toothill et al. 1956) that the size of the substituent atoms or groupings is an important factor in determining activity in di-*ortho*-substituted phenoxyacetic acids, since large atoms such as chlorine or bromine can restrict the free rotation of the side-chain which may prevent the carboxyl group from achieving the right spatial relationship to the ring essential for a two-point reaction with its site of action.

The influence of ring substitution on auxin activity has also been extensively investigated with benzoic, phenylacetic and indolylacetic acids, and most recently with phenols. With each of these series, the pattern of activity shown by the different substituted isomers is quite different from that observed with phenoxyacetic acids. Benzoic acid itself is not an auxin and mono-substitution in the *meta*- or *para*-positions with a halogen or methyl group does not confer any activity to the molecule. *Ortho*-substitution, however, does confer slight activity to the molecule (Muir and Hansch 1951; Veldstra and Westeringh 1952) and this activation is increased with further substitution in the 3-, 5- or 6-positions of the ring and reaches a maximum in the tri-substituted isomer, 2,3,6-trichlorobenzoic acid (XIV) (Jones et al. 1951; Veldstra 1952; Thimann 1952; Zimmerman et al. 1952; Pybus et al. 1959a). This striking activation of benzoic acid by di-*ortho*-substitution is the reverse of the situation observed with substituted phenoxyacetic acids where the presence of 2,6-substituents greatly reduces or destroys the molecule's potential to show activity. Also in marked contrast with the phenoxyacetic acids, *para*-substitution has no activating effect in benzoic acids whereas in phenoxy acids it confers high activity to the molecule. The size of the substituent atom also appears to be important in the *ortho*-activation of benzoic acids; *ortho*-iodo substitution for example greatly reduces activity compared with that shown by the corresponding chloro- and bromo-derivatives (James and Wain 1969). The activating influence of such *ortho*-substituents appears to be related to their ability to displace the carboxyl group from the plane of the ring by steric influence; this effect might permit the molecule to attach itself initially through its carboxyl group to its receptor site in the cell (Muir and Hansch 1953; Fukui et al. 1958). It also appears that the *ortho*-substituent must be large enough to cause this steric displacement, but not so large as to prevent the reaction between the carboxyl group and its receptor site (James and Wain 1969).

In the phenylacetic acid series, the influence of ring substitution on activity follows a similar pattern to that observed with benzoic acids. Phenylacetic acid itself has moderate auxin activity and halogen substitution in the 2-, 3- and 6-position greatly enhances this activity (Weintraub and Norman 1949; Mel'nikov et al. 1953). The 2,3-dichloro-, 2,6-dichloro- and 2,3,6-trichloro-derivatives show activity equal to and even greater than that of IAA or 2,4-D, depending on the test method employed (Pybus et al. 1959b; Chamberlain and Wain 1971). With indolylacetic acid, substitution of chlorine or bromine atoms into any ring position except the imino group in the pyrrole ring results in greater auxin activity than that of the parent molecule (Hoffman et al. 1952; Porter and Thimann 1965). The greatest increase in activity is obtained with monochloro-substitution in the 4- and 6-positions of the indole ring (Porter and Thimann 1965). Methyl substituents on the other hand, are relatively ineffective in enhancing the activity of IAA. Another heterocyclic compound showing high auxin activity which appears to be determined by the nature and position of the nuclear substituents is 4-amino-3,5,6-trichloropicolinic acid (XV) (Hamaker et al. 1963; Kefford and Kaso 1966).

With regard to phenolic substances, the first clear evidence of auxin activity in this class of weakly acidic compounds was provided by Wain and Taylor (1965). They reported that phenols with dichloro-or dibromo-substituents in the 2- and 6-positions of the ring showed appreciable activity in three auxin tests. Further studies by Wain and coworkers revealed that the development of auxin activity in this class of compounds requires the presence of electron-attracting substituents in the 2- and 6-positions and that at least one of these *ortho*-substituents must be capable of intramolecular bonding with the hydrogen of the phenolic hydroxyl group. Substitution in the *para*-position to the hydroxyl group results in complete loss of activity and any substituent in the *meta*-position causes a major reduction in growth-promoting properties (Harper and Wain 1969).

2.4.2. *Importance of the carboxyl group*

With few exceptions, all compounds showing appreciable auxin activity possess a carboxyl group or grouping which may be easily converted in plants to a carboxyl group, such as esters, amides and nitriles. These derivatives are generally less active than the parent acid, but this is not always the case. For example, with esters of IAA, the methyl and ethyl derivatives show activity at least equal to that of the free acid (Zimmerman and Hitchcock 1937; Seeley et al. 1956), but the activity of other esters declines with increasing size of the ester grouping, which probably relates to the decreasing ease of hydrolysis of these higher esters in the test plant (Kögl and Kostermans 1935). The methyl esters of other indole acids and of a series of 2,4-dichlorophenoxy acids have also been reported to show the same activity as the parent compound (Fawcett et al. 1958, 1960). Amide derivatives of indole and chlorophenoxy acids, however, are generally much less active than the free acid, but the corresponding nitrile derivatives may be more active than the parent acid depending on the type of test tissue employed (Seeley et al. 1956; Fawcett et al. 1958, 1960). With indolylacetonitrile and 2,4-dichlorophenoxyacetonitrile, for example, these compounds show lower activity in pea curvature or segment tests, whereas in the *Avena* or *Triticum* coleoptile tests, both compounds show much higher activity than the parent acids (Bentley and Housley 1952; Fawcett et al. 1956, 1958, 1960). The higher activity of these acetonitrile derivatives in coleoptile tissue may be due to the exceptional activity per se of this type of molecule, but it may also be due to better penetration of the compound before it is hydrolysed in the cell.

The effect on activity of replacing the carboxyl group with other acidic groupings such as sulphonic, phosphonic or phosphonous groups, has been examined in a number of indolyl and naphthyl compounds. Indol-3-ylmethanesulphonic (XVI) and naphth-1-ylmethanesulphonic acids have been tested in a wide range of auxin bioassays and in all cases were found to have good activity, comparable to that of IAA and naphth-1-ylacetic acid when high concentrations were used (Veldstra 1956; Vlitos and Hitchcock 1958). The corresponding phosphonic (XVII) and phosphonous acids have also been examined and shown to be active compounds but again only when tested at high concentrations (Veldstra 1956).

There are, however, several compounds which do not possess a carboxyl group yet show moderate or even high auxin activity in certain tests. The activity of some of these compounds, such as indol-3-ylacetaldehyde and naphth-l-ylacetaldehyde, can be readily explained by their conversion in test plants to the corresponding carboxylic acid (Larsen 1951). Similarly, the high activity of the alcohols, β-(indol-3-yl)ethanol (tryptophol) and β-(naphth-l-yl)ethanol is also probably due to their conversion to the corresponding acids (Rayle and Purves 1967b; Veldstra 1953). A number of chloro-substituted phenylethylamines have been shown to have appreciable auxin activity and again it is reasonable to assume that such activity arises from their conversion to the corresponding active chlorophenylacetic acid by the amine oxidase system in the test plants (Albores and Wain 1974). On the other hand, several monophenols have been shown to possess high auxin activity when the ring is substituted with halogen atoms in both *ortho*-positions (Wain and Taylor 1965) and the activity of these compounds cannot be explained in terms of their conversion to a carboxylic acid. However, the electron-withdrawing halogen subsituents in the two *ortho*-positions of the ring will markedly increase the ionization of the phenolic hydroxyl group and the weak acidic property so induced may well be an essential factor influencing the auxin activity of these molecules (Harper and Wain 1969).

All the above observations indicate that while other polar groups can confer auxin activity on many compounds, their effect is generally inferior to that of the carboxyl group. It is clear that for a compound to show high auxin activity, its molecule must possess a carboxyl group or grouping readily converted to a carboxyl group, as was originally postulated by Koepfli et al. (1938). This essential requirement of an acidic grouping for auxin activity indicates that the carboxyl group is one of reactive groupings involved in the attachment of the auxin molecule to its site of action in the cell.

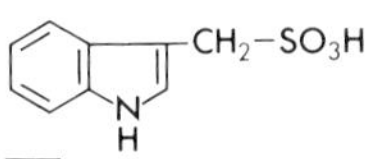

XVI

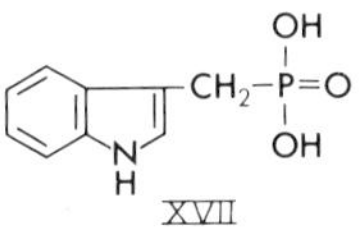

XVII

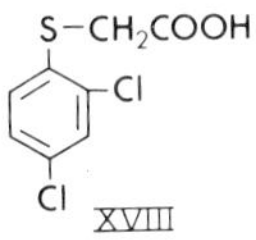

XVIII

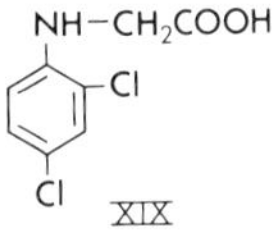

XIX

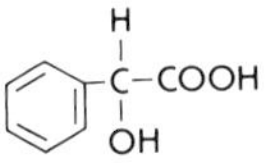

XX

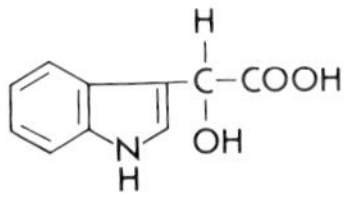

XXI

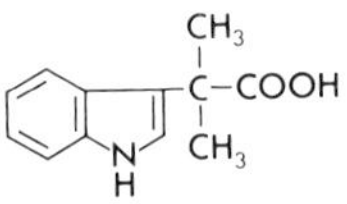

XXII

2.4.3. Structure of the side chain. According to the fourth structural requirement for auxin activity proposed by Koepfli et al. (1938), the carboxyl group must be separated from the ring by at least one carbon atom. Later work has shown that this structural feature is not essential for activity, since a number of highly active benzoic, naphthoic and picolinic acids have been found in which the carboxyl group is attached directly to the ring. Nevertheless, among the many indolyl-, naphthyl- and phenyl-aliphatic acids that show auxin activity, those compounds with one bridging carbon atom between the ring and the carboxyl group usually show the highest activity. When the number of these bridging atoms is increased, as for example in compounds possessing propionic, butyric or valeric acid side-chains, there is generally a decrease in activity, but this loss of activity does not parallel the increase in number of carbon atoms in the side-chain. This situation was first observed by Grace (1939) who investigated the root-inducing capacity of the higher homologues of naphthylacetic acid and found that high activity is shown by those acids possessing an odd number of bridging methylene (-CH_2-) groups in the side-chain, whereas those acids with an even number showed only low activity. In a later study using a series of seven ω-(2,4-dichlorophenoxy)-aliphatic acids and employing the tomato-leaf epinasty test, Synerholm and Zimmerman (1947) found a similar alternation in activity to occur with increasing length of side-chain. They suggested that such alternation in activity might be due to β-oxidation degradation of the side-chain resulting in the progressive removal of two-carbon units to yield either the inactive parent phenol or the highly active acetic homologue, depending on whether there is an even number or an odd number of methylene groups in the side-chain. Chemical evidence that the side-chains of such ω-phenoxy-aliphatic acids are indeed susceptible to β-oxidation in plants was provided by Fawcett et al. (1954) who fed flax plants with the different homologues of the ω-phenoxyalkanecarboxylic acid series and showed that considerable quantities of phenol was formed only in those plants that had been supplied with acids containing an even number of methylene groups in the side-chain (i.e., the propionic valeric and heptanoic homologues).

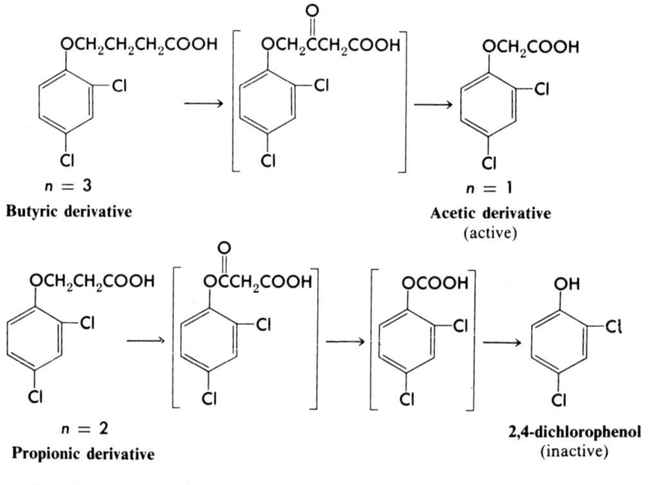

Figure 2.4. Formulae showing β-oxidative degradation of the side-chain of ω-(2,4-dichlorophenoxy)-aliphatic acids.

Subsequent physiological and biochemical studies by Wain and Wightman (1954) and coworkers using several homologous series of ω-chlorophenoxy acids with different ring substituents established that the alternation in activity observed in most series (see Figures 2.5 and 2.6) can be fully explained in terms of β-oxidative metabolism of the aliphatic side-chains in the test plants to yield the active acetic derivative or the corresponding inactive phenol (Fawcett et al. 1956, 1958) (Figure 2.4). The same conclusion was also reached by Luckwill and Woodcock (1956) who examined the homologous series of ω-(naphth-2-yloxy)-aliphatic acids and found clear alternation in growth-

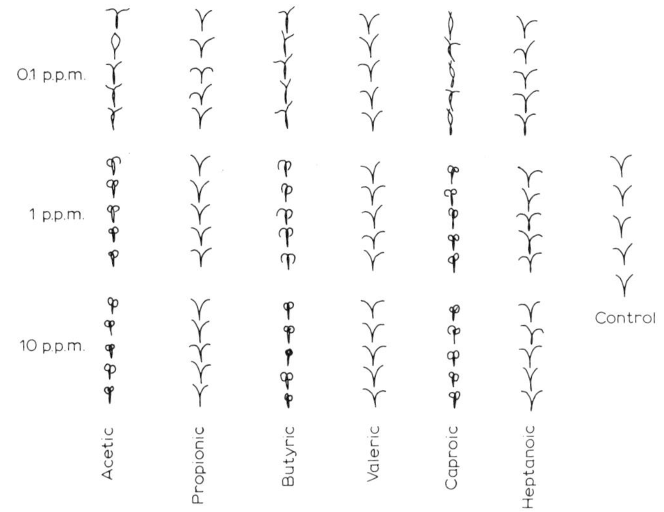

Figure 2.5. Alternation in growth-promoting activity shown by ω-(2,4-dichlorophenoxy)-aliphatic acids in the pea stem curvature test. Water control on right.

promoting activity when the first eight homologues were tested in the tomato parthenocarpy test. Further investigations by the Wye College group using 18 homologous series of ring-substituted ω-phenoxy-aliphatic acids, showed that the ability of the β-oxidase system in pea and tomato tissues to degrade these acids depends on the number and

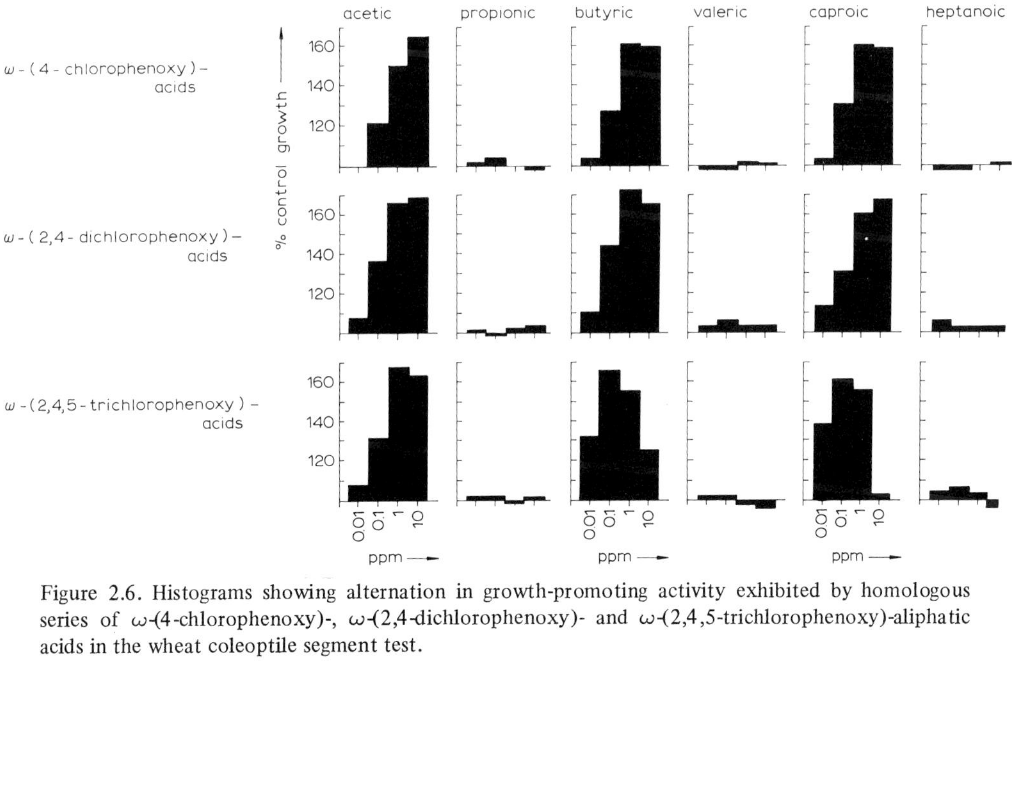

Figure 2.6. Histograms showing alternation in growth-promoting activity exhibited by homologous series of ω-(4-chlorophenoxy)-, ω-(2,4-dichlorophenoxy)- and ω-(2,4,5-trichlorophenoxy)-aliphatic acids in the wheat coleoptile segment test.

position of the ring substituents (Fawcett et al. 1959). This finding indicated the high substrate specificity of the β-oxidase enzymes in some dicotyledonous plants and led to the development of a new method of weed control based on the ability of many dicotyledonous weeds to degrade certain chloro-substituted phenoxybutyric acids to the highly active acetic acid, and thus succumb to herbicidal action. Other dicotyledonous species, such as certain crop plants possessing β-oxidase enzymes showing much higher substrate specificity, would not degrade these chlorophenoxybutyric acids to the active end product, and so would remain unharmed (Wain 1955, 1958). This metabolic approach to weed control resulted in the development of the herbicides, MCPB (γ-(2-methyl-4-chlorophenoxy)butyric acid) and 2,4-DB (γ-(2,4-dichlorophenoxy)butyric acid), which have proved very successful in the control of many dicotyledonous weeds in several dicotyledonous crops.

In further studies on the metabolism of side-chain in synthetic auxins, Fawcett et al. (1960, 1961) examined the growth-regulating activity and metabolism in pea and wheat tissues of homologous series of ω-(indol-3-yl)-aliphatic acids and their corresponding amide, nitrile and methyl ester derivatives. Although hydrolysis of all the derivatives to the corresponding acids was shown to be an essential first step in the metabolism of these compounds, the pattern of activity observed in all series except the nitrile series, can be fully explained in terms of β-oxidative degradation of the higher homologues.

Substitution of other elements linking the side-chain to the ring system can also greatly modify auxin activity. The introduction of a bridging oxygen atom in phenylacetic acid to produce phenoxyacetic acid, completely eliminates the moderate activity of the phenyl compound. Similarly and even more strikingly, introduction of an oxygen bridging atom in naphth-l-ylacetic acid to form naphth-l-yloxyacetic acid causes a great reduction in activity, whereas the reverse effect is found with the weakly active naphth-2-ylacetic acid (IV) since naphth-2-yloxyacetic acid (VI) is a much more active compound. Thus it appears that the effect on activity of introducing an oxygen bridging atom between the ring and the side-chain is influenced by the type of ring system and the position of the side-chain. An explanation of these effects may possibly be found in the stereochemistry of the molecule or in its electronic configuration.

Although oxygen as a linkage element in the side-chain can bring about a major reduction in growth-promoting properties, substitution of halogen atoms or alkyl groups in the ring can reverse this effect, as is seen by the high activity shown by such compounds as 2,4-dichlorophenoxyacetic or 2-methyl-4-chlorophenoxyacetic acids. With such highly active phenoxyacetic acids, however, replacement of the bridging oxygen atom with sulphur (as in XVIII) considerably lowers the activity of the resulting phenylthioacetic analogues (Wilske and Burstrom 1950; Jönsson et al. 1952; Fawcett et al. 1955) and activity is reduced to a very low level or completely lost when the bridging oxygen atom is replaced by a -SO- or $-SO_2-$ group (Wilske and Burstrom 1950). The effect on activity caused by substitution of a linking imino (-NH-) group to form the phenylglycines (e.g. XIX) varies with the nature of the ring substituents. Comparison of the activity shown in three auxin bioassays by 10 chloromethyl-phenylglycines with that given by their corresponding phenoxyacetic acid analogues shows that phenylglycines are generally less active than the corresponding phenoxyacetic acids, although the effects of ring-

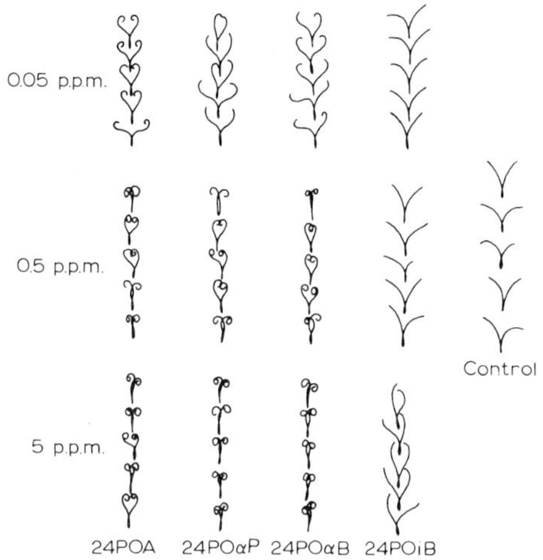

Figure 2.7. Growth curvatures induced in the pea stem curvature test during treatment with solutions of 2,4-dichlorophenoxyacetic acid (24POA), α-(2,4-dichlorophenoxy)-propionic (24POαP), -butyric (24POαB) and -isobutyric acids (24POiB). Water control on right. The structural formulae for these acids are presented below.

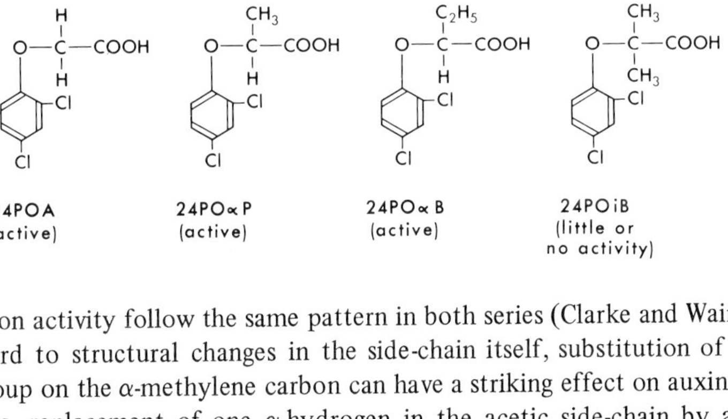

24POA (active) 24POαP (active) 24POαB (active) 24POiB (little or no activity)

substitution on activity follow the same pattern in both series (Clarke and Wain 1963).

With regard to structural changes in the side-chain itself, substitution of an alkyl or hydroxyl group on the α-methylene carbon can have a striking effect on auxin activity. In aryloxy acids, replacement of one α-hydrogen in the acetic side-chain by a methyl or ethyl group, to give the propionic or butyric derivative, does not appreciably change auxin activity, but replacement of both α-hydrogen atoms by methyl groups to give the isobutyric derivative produces a molecule that is usually completely inactive as an auxin. These effects of alkyl substitution in the side-chain were first studied by Osborne and Wain (1950, 1951) using derivatives of naphth-2-yloxyacetic acid, and in later investigations some 13 series of substituted phenoxyacetic, α-propionic, α-butyric and α-isobutyric acids and their corresponding phenylthio- analogues were examined in a range of auxin tests (Fawcett et al. 1953, 1955). In most series, only the acetic, propionic and butyric derivatives showed auxin activity whereas, with few exceptions, all the isobutyric acids were inactive (Figure 2.7). These extensive results indicate that the structure of the isobutyric acid side-chain is not suitable for the initiation of a growth response and such considerations led to the suggestion that the presence of at least one hydrogen atom on the carbon alpha to the carboxyl group is necessary for auxin activity (Osborne and Wain 1951; Wain 1953). Although a few exceptions to this so-called α-hydrogen rule have been found, notably among aryl compounds such as α-(indol-3-yl)isobutyric acid (XXII) which has some auxin activity (Fawcett et al. 1955; Aberg 1958; Thimann 1958), in general, the presence of at least one hydrogen on the α-carbon of the side-chain appears to be an important, if not an absolute structural requirement for activity in all synthetic auxins of the aryloxy-aliphatic acid type. It seems probable that the inactivating influence of di-alkyl substitution on the α-carbon atom in such compounds is due to the interference caused by these substituents on the ability of the molecule to attain the right spatial configuration necessary for activity (Veldstra 1953).

Although mono-substitution of the acetic side-chain with small alkyl groups (e.g. methyl, ethyl or propyl) does not change the activity of the molecule from that of the parent compound, substitution of larger, straight-chain or branched alkyl groups does greatly reduce or eliminate activity completely, as seen with several substituted indole (Schlender et al. 1966) and phenyl derivatives (Fujita et al. 1966). In contrast to the minimal effect observed with mono-alkyl substitution, introduction of a single hydroxyl group on the α-carbon of the side-chain results in a major loss in auxin activity, so that indol-3-ylglycollic acid (XXI) (Thimann 1958) and phenylglycollic acid (XX) (Takeda 1959) show very low activity compared with their acetic homologues. Such side-chain hydroxylation will undoubtedly change the hydrophilic: lipophilic balance of the molecule, which Veldstra (1953) considers of vital importance in determining the activity of a synthetic auxin.

2.4.4. Importance of spatial configuration

There is now abundant evidence indicating that in addition to the required presence in most auxin molecules of an unsaturated ring system and a carboxyl group, the spatial configuration of these groups is also of vital importance in determining the primary action of the molecule at its receptor site in the cell. The importance of stereochemical considerations in determining auxin activity was first observed by Kögl (1937) who resolved α-(indol-3-yl)propionic acid (XXIII) into its two stereoisomers and found that the dextro-rotatory isomers is about 30 times more active in the *Avena* curvature test than the laevo-rotatory form. Although later work indicated that both isomers appear to have equal activity (Kögl and Verkaaik 1944; Aberg 1958), other early investigations on the structure – activity relationships of a range of compounds led Veldstra (1944) to conclude that for a molecule to exhibit auxin activity, its steric configuration must allow for the carboxyl group to lie out of the plane of the ring, and that maximum activity will be seen when the carboxyl dipole is perpendiçular to that plane. The key observation that led to this conclusion was the high activity shown by *cis*-cinnamic acid (XXV) whereas the *trans*-isomer (XXIV) was found to be completely inactive. Veldstra's examination of Stuart models of these two isomers showed that while the side-chain in both forms is free to rotate around the C – C bond attaching it to the ring system, in the case of the *trans*-isomer this rotation would take place through 360° and therefore the average rotation of the carboxyl group would be in the plane of the ring. In the *cis*-isomer, however, complete rotation through 360° is not possible and movement of the side-chain is restricted to one side of the linkage C – C bond. This steric hinderance to the movement of the side-chain will result in the average orientation of the carboxyl dipole being at a definite angle to the plane of the ring and will have a component which is perpendicular to the ring. It is this latter orientation which Veldstra considers to be the spatial configuration responsible for the high activity shown by *cis*-cinnamic acid.

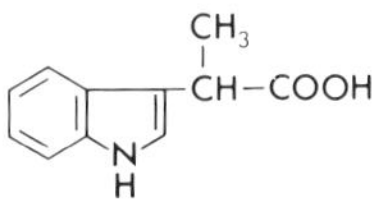

XXIII

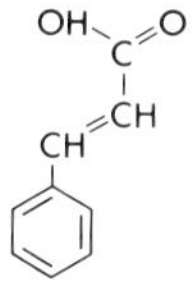

XXIV

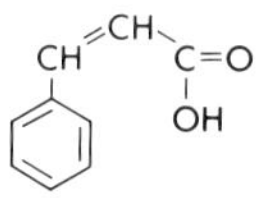

XXV

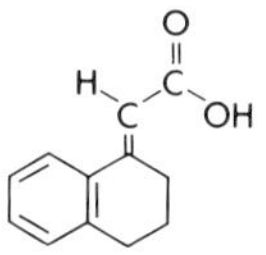

XXVI

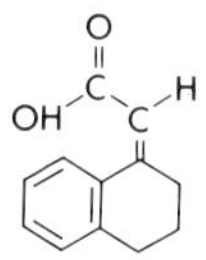

XXVII

Strong support for this view came from later studies on the auxin activity and comparative sterochemistry of the two geometric isomers of tetrahydronaphthylidene-l-acetic acid. Stuart models show that the physiologically inactive *trans*-isomer (XXVII) has the carboxyl group in the plane of the ring, whereas the active *cis*-isomer (XXVI) cannot be constructed without some strain in the hydrogenated part of the nucleus, and this means that the side-chain cannot lie in the plane of the ring. There will therefore be a component of the carboxyl dipole perpendicular to the plane of the ring, which Veldstra (1953) considers is the decisive steric factor accounting for the activity of the *cis*-isomer. Independent evidence for the non-flat configuration of *cis*-cinnamic acid and other *cis*-isomers has come from analysis of UV absorption spectra which show that hindrance of conjugation occurs with these isomers, indicating that the carboxyl group (or its dipole) is out of the plane of the ring (Havinga and Nivard 1948). Such stereochemical considerations have also been applied, with some success, to explaining the different activities observed with naphthyl- and naphthylxy-acetic acid isomers, and also to account for the unexpected activity of naphthoic acids and certain substituted benzoic acids (Veldstra 1953, 1956).

Another stereochemical relationship that can strongly influence auxin activity is the phenomenon of optical isomerism. The early work of Kögl (1937) with α-(indol-3-yl) propionic acid was the first to suggest that the two optical isomers in a racemic mixture may have very different auxin activity. This possibility was further investigated by a number of workers in the 1950's who successfully resolved several highly active α-aryl- and α-aryloxy-propionic acids and then demonstrated that most, if not all of the auxin activity of each compound resided in only one of its two isomeric forms. Thus, Smith and Wain (1951) showed that the (+)-form of α-(naphth-2-yloxy)propionic acid is highly active in six bioassays whereas the (–)-form shows negligible activity. In the same year Veldstra and Westeringh (1951) reported that the (–)-isomer of tetrahydro-l-naphthoic acid and the (+)-isomer of α-allylphenylacetic acid are highly active in the pea curvature test while the corresponding enantiomorphs show only low activity. In further studies, Smith et al. (1952) isolated the isomeric forms of α-(2,4-dichlorophenoxy)propionic and

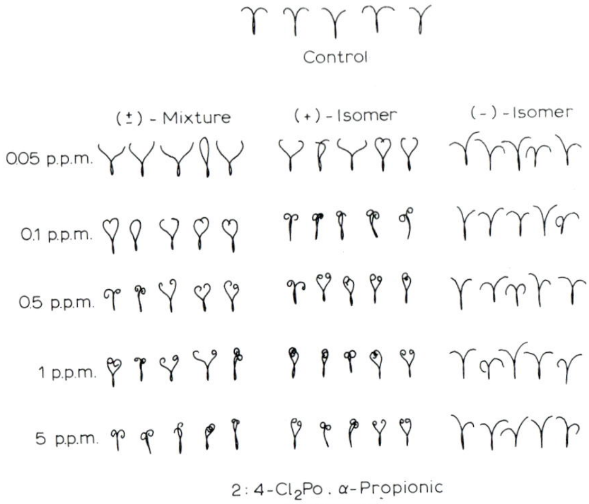

Figure 2.8. Growth-promoting activity of the racemic (±) mixture and (+)- and (–)-forms of α-(2,4-dichlorophenoxy)propionic acid in the pea stem curvature test.

α-(2,4,5-trichlorophenoxy)-propionic acids and showed that all the activity of the racemic mixture is due to the (+)-form of each compound (Figure 2.8). Similar experiments with the optical isomers of several α-phenoxy- and α-naphthoxy-propionic acids using root growth inhibition tests and the tomato parthenocarpy test for assessing activity has shown that while both forms of some compounds are active, only one isomer in every case showed high activity (Aberg 1953; Luckwill and Woodcock 1956). From all these and other investigations we now have much information on the comparative activities of the optical isomers of many synthetic auxins and in most instances, the evidence indicates that the ability to induce a growth response resides only in one configurational form of each isomeric pair (see reviews by Jönsson 1961; Fredga and Aberg 1965; Wain and Fawcett 1969). Experiments to determine the absolute configuration of many of these stereoisomers have shown that the active isomers belong to the D-configuration series; they are thus related sterically to D(-)lactic acid and are mostly dextrorotatory (+) (Matell 1953; Fredga 1955; Fredga and Aberg 1965).

The finding that the growth-promoting property of an optically active auxin resides mainly or entirely in only one stereoisomer suggests that this active isomer is capable of specific interaction with some asymmetric receptor site in the cell which is concerned with the initiation of the growth-response. This view led to a proposal by Wain and coworkers on the primary mode of action of auxins which explains the specific differences in activity of stereoisomers and also accounts for the low activity usually shown by α-aryloxyisobutyric acids (Smith and Wain 1951; Smith et al. 1952; Wain 1953). These workers suggested that activity in aryloxy acids depends on the presence in the molecule of three essential structural units, namely, an unsaturated ring system, a carboxyl group, and at least one α-hydrogen atom, all having a definite spatial relationship with each other

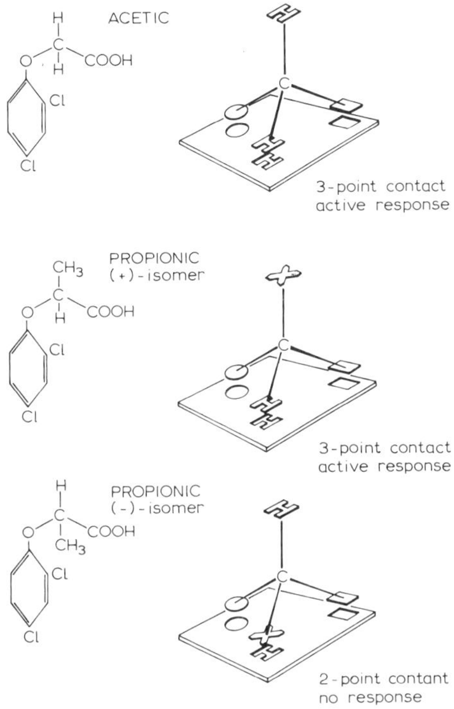

Figure 2.9. Diagram illustrating the spatial presentation of three essential groupings in an auxin molecule to cellular receptor sites (from Wain 1953).

so that the molecule can "fit on" to its receptor site in the plant cell. This attachment to the receptor site was visualized as a three-point contact involving the three essential structural units, each with its own specific reaction point in the receptor site (Figure 2.9). The absence of any one of the three structural units, as seen for example in α-aryloxy-isobutyric acids, or an unsuitable configuration of the units, as would be the case in one stereoisomer of an optically active α-aryloxypropionic acid, would result in no induction of the growth response. Moreover, it could be expected that inactive compounds possessing structural units that would fit at two points only on the receptor site, might compete with active molecules in the cell and so reduce their growth-promoting activity. Such competitive inhibition of auxin action was demonstrated by Smith et al. (1952) who showed that each (−)-isomer of three α-aryloxypropionic acids when present in coleoptile cells as the dominant molecular species, could effectively antagonize the growth-promoting action of the corresponding (+)-isomer (Figure 2.10). This type of anti-auxin action in wheat coleoptile and pea stem cells was further demonstrated by Wain and Wightman (1957) using several α-aryloxyisobutyric acids and a number of other inactive compounds possessing only two of the essential structural units. Their results confirmed the findings of earlier workers that inactive α-aryloxyisobutyric acids, and even the weakly active α-(indol-3-yl)-isobutyric acid (XXII) can effectively antagonise the inhibitive action of

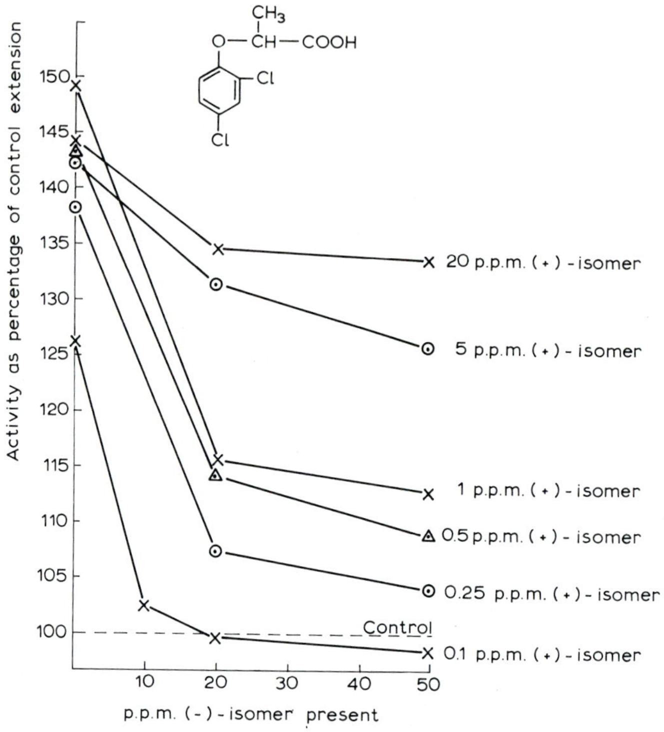

Figure 2.10. Decreasing activity shown by (+)-α-(2,4-dichlorophenoxy)propionic acid in presence of increasing amounts of the (−)-isomer in the *Avena* coleoptile segment test.

auxins on root growth (Burstrom 1950; Aberg 1950; Erdtman and Jönsson 1954). All these investigations show that many molecules inactive as auxins, but possessing a substituted ring system and a carboxyl group, can bind to the auxin receptor site in plant cells and so competitively antagonise the binding of auxin molecules, the net result being the reduction or elimination of any auxin-induced growth response.

If we accept the view that an auxin must become attached to a stereospecific receptor site in a cell in order to initiate a growth response, then we should consider other aspects of its molecular structure and configuration which might influence this attachment. One such aspect that has received attention arose from consideration of electron distributions in auxin molecules, which led Porter and Thimann (1959) to conclude that all active compounds possess a fractional positive charge at some point in the molecule at a distance of about 0.55 nm from the negative charge of the ionized carboxyl group. This distance of 0.55 nm between the positive and negative charge is consistently obtained with molecules in each of the major classes of auxins, including the indole acids, phenoxyacetic, benzoic and picolinic acids, as shown in Figure 2.11. Also shown in this figure is the charge distribution on the non-ring auxin, carboxymethyldimethyl-dithiocarbamate, which can assume a flat spatial orientation which resembles a ring (Veldstra 1956) and in this form the nitrogen atom takes a slight positive charge at the usual 0.55 nm distance from

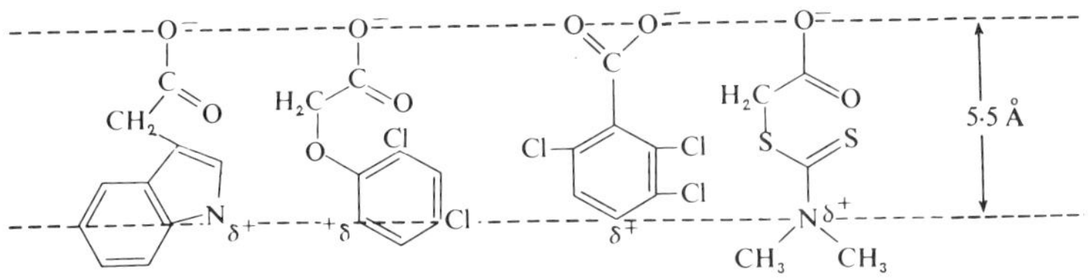

Figure 2.11. Diagram illustrating the theory of Porter and Thimann (1959) showing the 0.55 nm spatial separation of the positive and negative charges exhibited by the molecules of four active auxins, indoleacetic acid (IAA), 2,4-dichlorophenoxyacetic acid (2,4-D), 2,3,6-trichlorobenzoic acid (2,3,6-TBA) and carboxymethyl-*N*,*N*-dimethyldithiocarbamate.

the carboxyl group (Thimann 1963). Porter and Thimann (1965) feel that this specific charge separation of the dipolar auxins allows them to be ionically bound to the receptor site in the plant cell which has a pattern of similar spaced complementary charges. Such electrostatic association of an auxin molecule with its site of action does not involve a chemical reaction of the type envisaged in the "two-point attachment" theory earlier proposed by Muir and Hansch (1955).

In summary, we see that the structural and spatial requirements for auxin activity seem to relate to the need for attachment of the molecule to some receptor site in the cell, which itself has characteristic charge requirements and steric limitations. The recent experiments of Hertel et al (1972, 1976) and of Venis and coworkers using isolated membrane fragments from corn coleoptiles show clearly that the binding of ^{14}C-labelled auxin molecules to cellular membranes does indeed occur and that the presence of two, and possibly three kinetically different receptors can be demonstrated (Batt and Venis 1976; Batt et al. 1976). Similarly, the presence of IAA-binding proteins in the cytosol fraction of soybean cotyledons has also been reported by Ihl (1976) who found that these proteins exhibited binding specificity for other compounds with molecular structures appropriate for auxin activity.

2.5. Pathways and sites of biosynthesis

2.5.1. Pathways of biosynthesis of indole auxins

2.5.1.1. Tryptophan as the precursor of indole auxins. The amino acid tryptophan (TPP) is a logical precursor for IAA because of the close chemical similarity of the two compounds, and since tryptophan is an essential constituent of proteins, it is presumably present in all cells. Free tryptophan levels in leaves are much higher than those of IAA; for example, tomato shoots contain 50 μg per kg fresh wt. of IAA and 12,000 μg per kg fresh wt. of tryptophan (Schneider et al. 1972). However, free tryptophan may show compartmentation in tissues. Widholm (1974) has shown that in tissue cultures of carrot, soybean and rice the free tryptophan concentrations were greater than those required to completely inhibit anthranilate synthetase in vitro. This suggests that a large fraction of the free tryptophan is normally separate from the anthranilate synthetase in vivo. By analogy, it is possible that much of the free tryptophan of the cell is normally not available to the enzymes of IAA biosynthesis. [^{14}C] IAA can be formed from [^{14}C] TPP by at least 13 tissues (see references in Schneider and Wightman 1974a) and this conversion can also proceed in cell-free extracts of tobacco (Phelps and Sequeira 1967), mung bean (Wightman and Cohen 1968) and pea (Moore 1969). Moreover, when sterile pea plants were fed with either [^{14}C]indole plus [^{3}H]serine, or [^{14}C, ^{3}H]TPP, the ratios (^{3}H/^{14}C)IAA: (^{3}H/^{14}C) TPP observed when tryptophan and IAA were extracted from the plants were the same in each case, showing that there had been no bypass from indole to IAA (Erdmann and Schiewer 1971). Similar experiments with sterile grown maize coleoptiles using double labelling procedures combined with isotope competition methods, have further confirmed the central role of tryptophan in IAA biosynthesis (Heerkloss and Libbert 1976). No studies have yet been made of the biosynthesis of 4-chloro-IAA, but since the immature peas in which this compound occurs also contain D-4-chlorotryptophan derivatives (Marumo and Hattori 1970), it is reasonable to suppose that the precursor of 4-chloro-IAA is 4-chloro-TPP.

Our present knowledge of the pathways of IAA biosynthesis rests on several kinds of evidence; the presence of intermediates as native compounds, the biological activity of intermediates, in vivo interconversions of the intermediates and isolation of the neccessary enzyme systems. Using these methods, four pathways of IAA biosynthesis have been distinguished in various species of higher plants. The pathways have been named after their key intermediates, indol-3-ylpyruvic acid (IPyA), tryptamine (TNH_2), indol-3-ylacetaldoxime (IAOx) and tryptophol (TOL).

2.5.1.2. The indolepyruvic acid pathway. This pathway is outlined in Scheme 2.1. The instability of IPyA has so far precluded its isolation as a native compound from plant tissues. However, there is good evidence for the occurrence of indol-3-ylacetaldehyde (IAAld) in sterile pea shoots (Rajagopal 1968) and in cucumber seedlings (W.K. Purves, pers. comm.). The biological activity of IPyA is difficult to test, since one of its decomposition products is IAA. IAAld also forms some IAA spontaneously in solution, but the amount of IAA produced in this way is insufficient to account for the considerable activity which IAAld shows in several test systems (Larsen and Rajagopal 1964). When the conversion of [^{14}C] TPP to [^{14}C] IPyA is studied in vivo, the IPyA produced must be stabilised by the formation of a derivative such as the 2,4-dinitrophenylhydrazone to prevent its decomposition during extraction, and Gibson and coworkers (1972a) used this method to demonstrate the formation of [^{14}C] IPyA from labelled TPP by excised

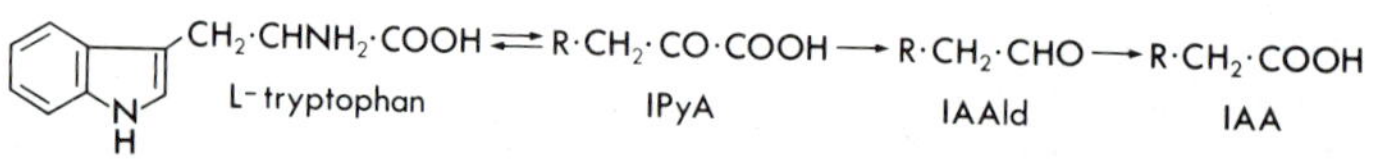

Scheme 2.1. The indolepyruvic acid pathway. IPyA and IAAld denote indol-3-ylpyruvic acid and indol-3-ylacetaldehyde respectively. R denotes an indol-3-yl moiety in all schemes.

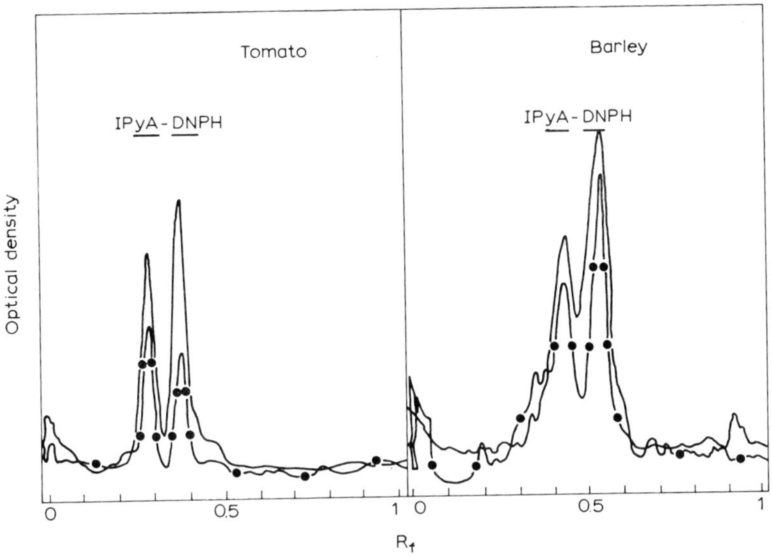

Figure 2.12. Densitometer and radioactivity scans of thin-layer chromatograms of suspected indol-3-ylpyruvic acid-2,4-dinitrophenylhydrazones (IPyA-DNPH) isolated from acidic ether fractions of tomato and barley shoots fed with [^{14}C] tryptophan. Continuous line: densitometer scan at 410 nm. Line with black circles: radioactivity scan. Solvent system: benzene – acetic acid – ethyl acetate (9:5:5, by vol.).

tomato and barley shoots. The shoots were allowed to metabolise [2-^{14}C] TPP for 24 hr and the acid – ether soluble metabolites were then extracted. Unlabelled carrier IPyA was added to the acid fraction and the 2,4-dinitrophenylhydrazone was then prepared to stabilise the IPyA. The IPyA-dinitrophenylhydrazone was identified by thin-layer chromatography and shown to be radioactive (Figure 2.12). [^{14}C] TPP may also be converted to [^{14}C] IAAld in vivo (Khalifah 1967; Gibson et al. 1972a), but since IAAld may also arise by the tryptamine pathway, this is not conclusive evidence for the operation of the IPyA route. The conversion of [^{14}C] TPP to [^{14}C] IPyA, [^{14}C] IAAld and [^{14}C] IAA has also been demonstrated in cell-free extracts of mungbean seedlings (Wightman and Cohen 1968) and tomato shoots (Wightman 1973). The enzyme catalysing the first step of the IPyA pathway has been found to be an unspecific aminotransferase which can utilise nonaromatic amino acids, such as aspartic and glutamic acids, as well as aromatic amino acids as amino group donors (Forest and Wightman 1972). The presence of this aminotransferase has been demonstrated in 30 species from 16 families (Truelsen 1973), and the enzyme has been purified 60-fold from mungbean seedlings (Gamborg 1965) and 600-fold to a single protein from bushbean seedlings (Forest and Wightman 1972). IPyA decarboxylase, the second enzyme of the pathway, has been purified 234-fold from yeast (Sukanya et al. 1971) and was also present in a cytosol preparation from tomato shoots (Gibson et al. 1972a). The nature of the last enzyme in the pathway, which converts IAAld to IAA, appears to vary with the species. In mungbean seedlings, this reaction is catalysed by an NAD-dependent IAAld dehydrogenase (Wightman and Cohen 1968), while the *Avena* coleoptile enzyme is an oxidase which appears to be quite different from the mungbean enzyme in its pH optimum, lack of NAD dependence and insensitivity to ionising radiation (Rajagopal 1971). Thus, although the instability of IPyA has made its identification as a native compound difficult, several other lines of evidence are consistent with the operation of the IPyA pathway in vivo. In fact, since many plants do not contain the more stable key intermediates of the other biosynthesis pathways to be discussed below, the IPyA pathway is therefore probably the most important route of IAA formation in plants. Where two pathways do appear to operate in the synthesis of IAA, as for example in tomato plants, the amounts of radioactivity found in IAA and other indole metabolites from shoots fed with [3-^{14}C]tryptophan or [2-^{14}C]tryptamine have confirmed that the IPyA pathway is the predominant route of IAA formation (Gibson et al. 1972a) (see Table 2.2).

2.5.1.3. The tryptamine pathway. This pathway is shown in Scheme 2.2. Tryptamine is of sporadic occurrence in higher plants. It was first isolated from *Acacia* (White 1944), and has since been found in several other species, although it is not detectable in such

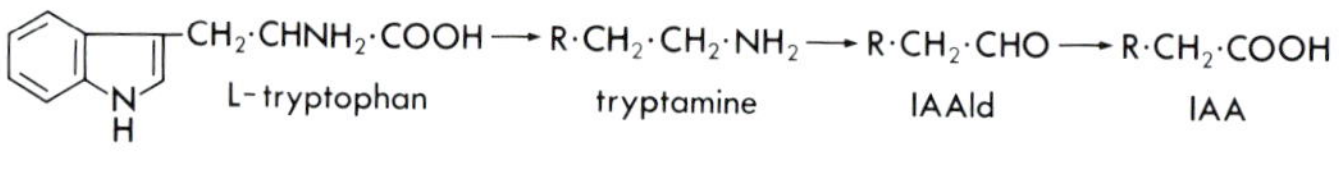

Scheme 2.2. The tryptamine pathway.

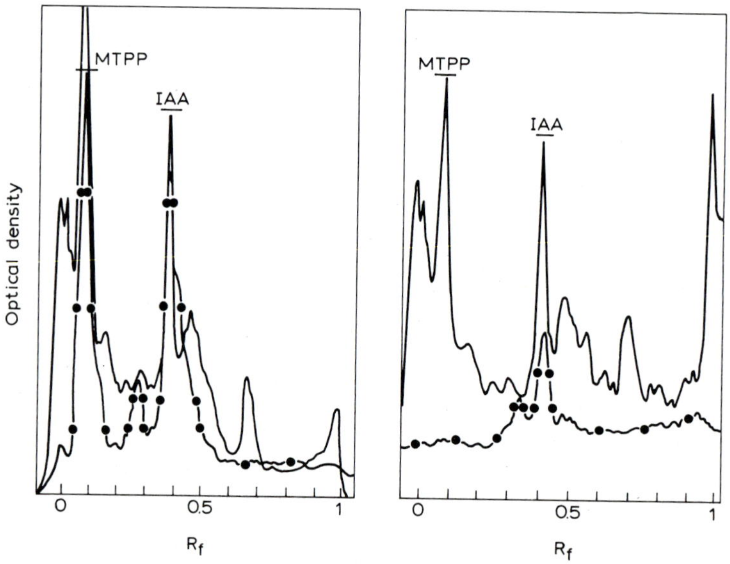

Figure 2.13. Comparative amounts of [^{14}C]IAA present in thin-layer chromatograms of acidic ether fractions obtained from barley shoots fed with either [^{14}C]tryptophan (left) or [^{14}C]tryptamine (right). Continuous line: densitometer scan obtained after treatment of chromatogram with Ehrlich reagent. Line with black circles: radioactivity scan. Solvent system: ethyl acetate – isopropanol – ammonia (9:7:1, by vol.). MTPP: malonyltryptophan.

Table 2.2.

Distribution of radioactivity in the indole compounds isolated from tomato shoots 24 hr after feeding with either [3-^{14}C] tryptophan or [2-^{14}C] tryptamine

	Unfed (0 time)	Unfed (24 hr)	Feeding solution	Shoots fed [3-^{14}C] tryptophan			Shoots fed [2-^{14}C] tryptamine		
Compound	μg/ g.f.wt.+	μg/ g.f.wt.	dpm/ μg	μg/ g.f.wt.	cpm*/ g.f.wt.	Dilution**	μg/ g.f.wt.	cpm*/ g.f.wt.	Dilution**
Tryptophan	12.0	12.0	2,750	313.0	497,000	1.5	14.0	600	52.2
Tryptamine	1.0	1.0	2,600	1.6	710	5.4	160.0	28,000	1.3
Indol-3-ylactic acid	0.005	0.005	–	0.102	127	2.0	0.005	–	–
Tryptophol	0.005	0.005	–	0.014	185	1.9	0.066	154	1.0
IAA	0.05	0.05	–	0.06	53	2.8	0.056	19	6.6

* Corrected for counting efficiency of 89%

$$** \text{ Dilution} = \frac{\text{Sp. act. of feeding solution, dpm}/\mu g}{\text{Sp. act. of compound, cpm}/\mu g}$$

F.wt. = fresh weight

common plants as pea, squash and cabbage (Schneider et al. 1972). Tryptamine is active, after a lag period, in the *Avena* curvature test (Skoog 1937), and also promotes elongation growth in excised segments of several species (Ferenczy and Matkovics 1961; Thimann and Grochowska 1968). Intact shoots of tobacco (Phelps and Sequeira 1967), tomato and barley (Gibson et al. 1972a) are capable of converting [^{14}C]TPP to [^{14}C]TNH_2, and [^{14}C]TNH_2 to [^{14}C]IAA (see Figure 2.13). In tomato and barley shoots [^{14}C]IAAld was also isolated, as its 2,4-dinitrophenylhydrazone derivative (see Fig. 2.14) after feeding with either [^{14}C]TPP or [^{14}C]TNH_2 (Gibson et al. 1972a). The enzyme which catalyses the first step in the tryptamine pathway, tryptophan decarboxylase, has been demonstrated in tobacco buds (Phelps and Sequeira 1968), cucumber hypocotyls (Sherwin 1970) and barley and tomato shoots (Gibson et al. 1972b). The tomato shoot enzyme has been purified 86-fold, and shown to be separable from tryptophan aminotransferase. It appears to be a more specific enzyme than the aminotransferase, since it can decarboxylate L-TPP and 5-hydroxy-L-TPP but not D-TPP, L-tyrosine or L-phenylalanine (Gibson et al. 1972b). In barley, both tryptophan decarboxylase and tryptamine are found in the shoot but not in the roots, and the levels of both tryptamine and the decarboxylase begin to rise only after the first week of seedling growth (Schneider and Wightman 1974b). Homogenates of the cotyledons, hypocotyls and roots of cucumber seedlings are able to catalyse the release of $^{14}CO_2$ from [^{14}C]DL-TPP labelled in the carboxyl group (Elkinaway and Raa 1974); however, the formation of tryptamine was not demonstrated in this system. Such a demonstration is essential, since plants may contain a tryptophan decarboxylase which does not yield tryptamine as a product (Reed

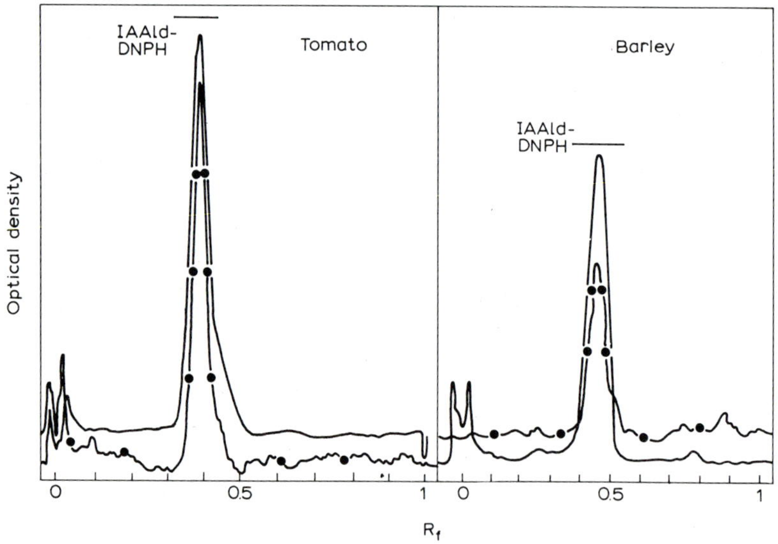

Figure 2.14. Densitometer and radioactivity scans of thin-layer chromatograms of suspected indol-3-ylacetaldehyde-2,4-dinitrophenylhydrazones (IAA1d-DNPH) isolated from neutral ether fractions of tomato and barley shoots fed with [^{14}C] tryptophan. Continuous line: densitometer scan at 410 nm. Line with black circles: radioactivity scan. Solvent system: benzene.

1968). Pea seedlings contain an amine oxidase which catalyses the oxidative deamination of tryptamine, as well as several other amines (Hill and Mann 1964) and at least three amine oxidases have been isolated from cucumber seedlings (Percival and Purves 1974). The cucumber enzymes differ in substrate specificity, but like the pea enzyme, none is specific for tryptamine. This broad specificity is also characteristic of mammalian amine oxidases.

2.5.1.4. The indoleacetaldoxime pathway. The indoleacetaldoxime pathway (Scheme 2.3) is characteristic of the family Brassicaceae (Kutacek and Kefeli 1968). *Brassica* plants probably contain small amounts of indol-3-ylacetonitrile (IAN) as a native constituent, although greater amounts of this compound are often formed during tissue extrac-

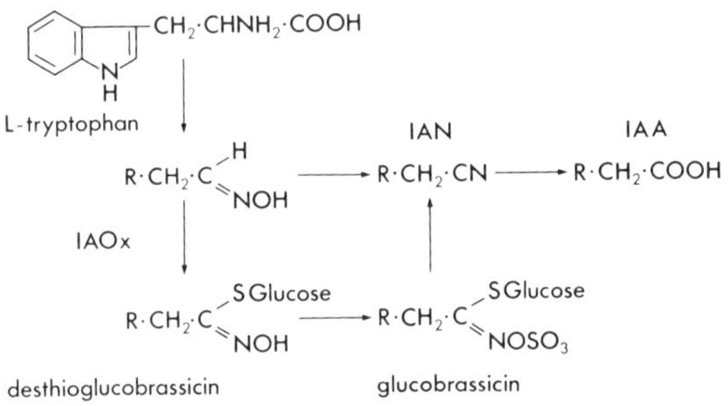

Scheme 2.3. The indoleacetaldoxime pathway. IAOx and IAN denote indol-3-ylacetaldoxime and indol-3-ylacetonitrile respectively.

tion by myrosinase-catalysed breakdown of glucobrassicin (Kutacek and Prochazka 1964). IAOx is also found naturally in cabbages (Kindl 1968) as is glucobrassicin (Gmelin and Virtanen 1961). IAOx promotes growth in wheat coleoptile and pea stem segments (Fawcett 1964) and IAN is an active growth promoter in many species (Bentley and Housley 1952; Seeley et al. 1956; Michel 1957; Ballin 1962). Other species show little growth response to IAN, apparently due to an inability to convert this compound to IAA (Thimann 1953; Seeley et al. 1956; Wightman 1962). Glucobrassicin is active in the oat and wheat coleoptile tests (Kutacek et al. 1966; Skytt-Andersen and Muir 1966) although it has little promotive effect on cabbage hypocotyls (Skytt-Andersen and Muir 1966). In vivo metabolism experiments have shown that cabbages can convert [^{14}C]TPP to IAOx (Kindl 1968), to IAN (Wightman 1962; Kutacek and Kefeli 1970) and to glucobrassicin (Kutacek et al. 1962; Kutacek and Kefeli 1968). This species can also convert [^{14}C]IAN to IAA (Wightman 1962). Rajagopal and Larsen (1972) fed unlabelled IAOx to a wide variety of plants and found that IAA was formed in all cases. Several species accumulated IAN and in one species, *Amaranthus*, an increase in IAAld was found. [^{3}H]IAOx can also be converted to desthioglucobrassicin and glucobrassicin by Woad leaves (Mahadevan and Stowe 1972). The enzyme responsible for converting TPP to IAOx in vivo has not yet been isolated, although this reaction can be catalysed in vitro by horseradish peroxidase (Kindl 1968). Indol-3-ylacetaldoxime hydrolyase, which converts IAOx to IAN, is present in banana leaves (Mahadevan 1963) and has been partially purified from *Gibberella* (Kumar and

Mahadevan 1963; Skukla and Mahadevan 1968, 1970). Nitrilase, which converts IAN to IAA, is limited to the Gramineae, Cruciferae and Musaceae, and does not appear to be a highly specific enzyme (Mahadevan and Thimann 1964; Thimann and Mahadevan 1964). Glucobrassicin can be converted to IAA by myrosinase, but since this enzyme is active below pH 5.2, it may not be important in the synthesis of IAA in vivo (Schraudolf and Weber 1969).

2.5.1.5. The tryptophol pathway. This pathway (Scheme 2.4) is really a modification of the indolepyruvic acid pathway for IAA synthesis. It appears to involve the two main intermediates required for the IPyA pathway and additionally involves the formation of tryptophol (indol-3-ylethanol, TOL) as a transitory side-reaction product. Few species, however, appear to synthesise IAA by this route.

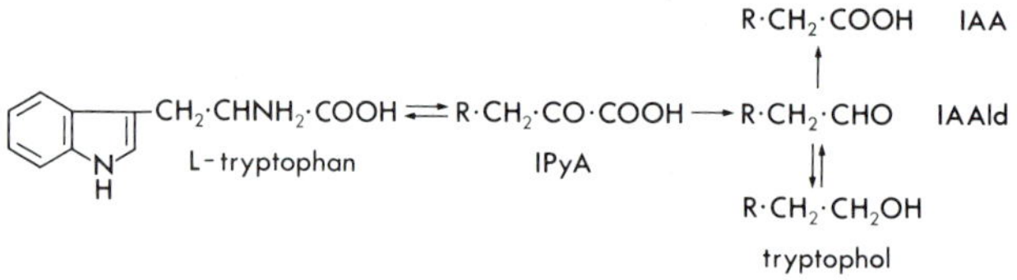

Scheme 2.4. The tryptophol pathway.

Native tryptophol has been extracted from cucumber shoots and identified by gas chromatography and mass spectroscopy (Rayle and Purves 1967a), and is also present in cucumber seedlings (Rajagopal 1967) and tomato shoots (Schneider et al. 1972). However, it is active as a growth promoter in several species in which it does not appear to occur naturally (Rayle and Purves 1968). [^{14}C]TPP is a precursor of [^{14}C]TOL in tomato shoots (Gibson et al. 1972a) and [^{14}C]TOL can be converted to [^{14}C]IAA by cucumber hypocotyls (Rayle and Purves 1967b). Wightman and Cohen (1968) partially purified an alcohol dehydrogenase from mung bean seedlings which catalysed the reversible reduction of IAAld to TOL in the presence of NADH. In cucumber shoots, however, a tryptophol oxidase catalyses the irreversible aerobic oxidation of TOL to IAAld and hydrogen peroxide. This enzyme has been purified 3,000-fold and was found to have no cofactor requirements other than the presence of an electron acceptor (Vickery and Purves 1972). The oxidase was inhibited by its product, IAAld, and the inhibition was overcome by oxygen. IAA and synthetic auxins were also found to be inhibitory, but gibberellins had no effect (Percival et al. 1973).

2.5.2. Pathways of biosynthesis of non-indole auxins

The only non-indole auxin whose biosynthesis has been investigated is phenylacetic acid (Wightman 1973). When excised tomato shoots were fed with [3-^{14}C]DL-phenylalanine, radioactive phenylpyruvic acid, phenylethylamine, phenylacetaldehyde and PAA were formed. Similar conversion of [3-^{14}C]DL-phenylalanine to [^{14}C]PAA was also found to occur in excised shoots of tobacco, sunflower, pea, corn and barley plants (Wightman and Rauthan 1974). As mentioned above, the aromatic aminotransferase which catalyses the conversion of tryptophan to IPyA can also use phenylalanine as a substrate. Thus it seems probable that the pathways for the biosynthesis of PAA in tomato shoots are analogous to the indolepyruvate and tryptamine routes of IAA synthesis in the same tissue (see Scheme 2.5).

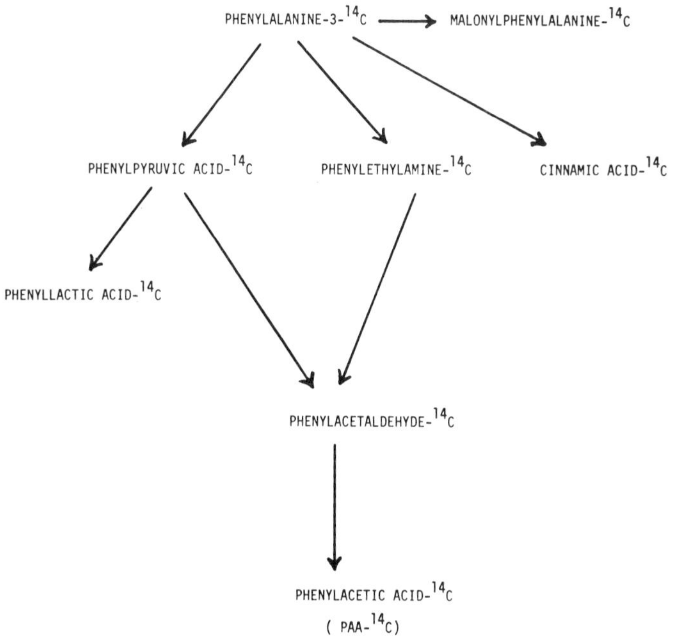

Scheme 2.5. Biochemical pathways for the metabolism of [3-^{14}C]DL-phenylalanine to [^{14}C]phenylacetic acid and other ^{14}C-labelled phenyl compounds in tomato shoots.

2.5.3. Sites of biosynthesis of indole auxins

It has proved very difficult to define exactly the sites at which auxin is synthesised under natural conditions. There have been several studies of auxin distribution within the plant, but it must be remembered that the level of auxin in a given organ reflects a balance between synthesis, metabolism and transport into and out of the organ. Again, the ability of certain organs to synthesise IAA from tryptophan has been measured, but this only describes the capacity of the system to form IAA under certain experimental conditions which may not obtain in vivo. We have already noted that it may be impossible to come to a definite conclusion about the pathway of IAA synthesis in a given tissue, and that we can only suggest the most likely route by a consideration of several lines of evidence. The same situation exists when we try to define the site of IAA synthesis within the plant.

2.5.3.1. Leaves and shoots IAA has been identified in leaves and shoots of many species, including *Coleus* (Jacobs and Morrow 1957), *Phaseolus* (Wheeler 1968) barley and tomato (Schneider et al. 1972) and tobacco (Wightman 1977). The concentration of IAA tends to be high in the youngest leaves and to decline with age, although there may

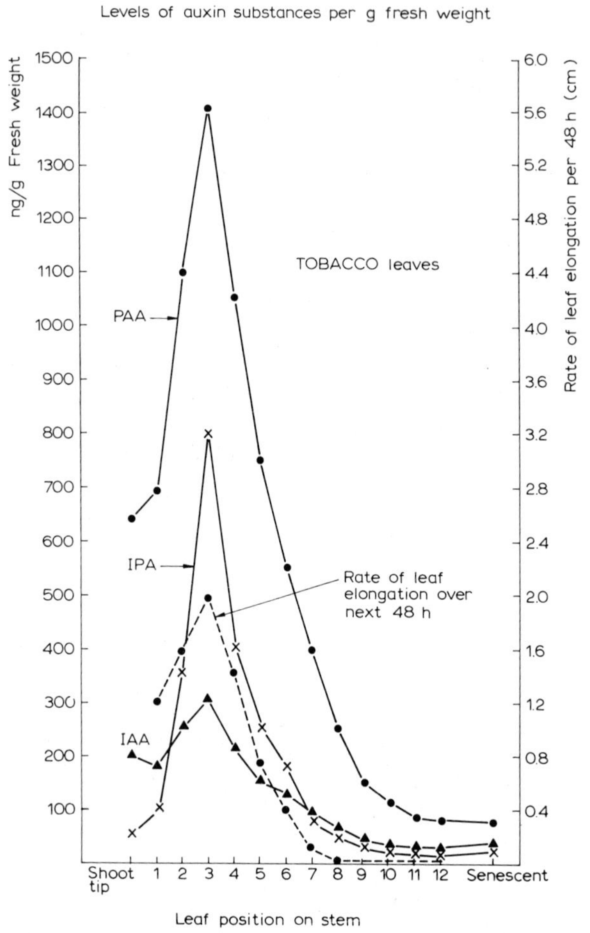

Figure 2.15. Endogenous levels of indol-3-ylacetic acid (IAA), indol-3-ylpropionic acid (IPA) and phenylacetic acid (PAA) in tobacco leaves at different stages of development.

be a secondary increase in IAA content in senescent tissue. For example, in *Phaseolus* leaves the free IAA content at 3 days was 143 μg/kg, at 7 days 15 μg/kg, at 19 days 9 μg/kg and at 40 days 196 μg/kg (Wheeler 1968). Recently, Wightman (1977) has reported gas chromatographic analysis of the auxins present in developing tobacco leaves. His results show that IAA, IPA and PAA are at their highest level in the youngest, most rapidly expanding leaves and decline as the leaf ages (see Fig. 2.15). Leaves of wheat (Kim and Rohringer 1969), barley, tomato (Gibson et al. 1972a) and cabbage (Wightman 1962) are capable of converting labelled tryptophan to IAA. When the shoot tips and first 4 leaves of tomato plants were fed individually with [^{14}C] TPP, it was found that all the tissues formed radioactive IAA and that the greatest amount of conversion occurred in leaf 1, the young, expanding leaf immediately below the tip (Wightman 1973). Similar results were obtained when cell-free systems from the tips and individual leaves were used. Thus it seems that leaves are able to synthesise IAA throughout a great part of their life. The highest rate of IAA production is probably found in young expanding leaves, and a second peak may occur in senescent leaves where the level of free tryptophan rises during proteolysis (Sheldrake and Northcote 1968a). The elevated levels of IAA in young tissues may also be related to the occurrence of compounds in these tissues which protect the IAA from oxidation (Phipps 1966; Yoneda and Stonier 1967).

2.5.3.2. Stems. IAA has been isolated from the stems of tobacco (Bayer and Ahuja 1968), cotton (Wiese and deVay 1970) and *Solanum* (Wakhloo 1965), and also in the cambium and differentiating xylem and phloem of the trunks of *Acer, Fraxinus* and *Populus* (Sheldrake 1971). Presumably, some of the IAA present in stems is in transit from other sites of synthesis, but some may also be made in situ. Cucumber hypocotyls (Sherwin and Purves 1969) and tomato stems (Wightman 1973) can convert [^{14}C]TPP to [^{14}C]IAA, as can enzyme preparations from tomato (Wightman 1973) and pea stems (Moore 1969). Young pea stems converted about twice as much tryptophan to IAA per unit time as did older stems (Moore 1969). Tobacco internode sections cultured on a basal medium retained the ability to produce a callus at the basal end for several months, indicating a continued ability to produce auxin (Sheldrake and Northcote 1968b). Auxin production was related to the continuing activity of the vascular cambium and the authors suggest that autolysis of the cell contents of the differentiating xylem releases tryptophan, which is then converted to auxin.

2.5.3.3. Coleoptiles. Coleoptile tips were the source of the first auxin isolated by Went (1928) and the classical view of the coleoptile has been that auxin from the tip is transported basipetally to supply the rest of the coleoptile, which is relatively auxin deficient. The tips of *Zea* coleoptiles contain a diffusible auxin which has recently been identified as IAA by mass spectrometry (Greenwood et al. 1972) and a gradient of auxin concentration, with the highest levels occurring in the tip was demonstrated by Thimann (1934). Diffusates from *Avena* coleoptiles tips also contain a second auxin, "P", which can be converted to IAA by heating (Shen-Miller and Gordon 1966). Sheldrake (1973a) suggests that the IAA and related indole compounds which are present in the tip are not synthesised in situ, but rather originate in the seed and are transported via the xylem to the tip where they accumulate. This hypothesis is supported by the observation that these compounds are present in the guttation fluid from decapitated coleoptiles. However, Black and Hamilton (1971) found that sterile *Avena* coleoptiles were able to incorporate label from [^{14}C]TPP into IAA, so it appears that coleoptiles do have the ability to synthesise IAA independently.

2.5.3.4. Flowers, fruits and seeds. A comprehensive account of earlier work on the auxin relations of reproductive tissues may be found in the review by Nitsch (1965). Relatively little work has been done on the auxin metabolism of flowers. Kaldewey (1957) followed the production of diffusible auxin during the development of the flower in *Fritillaria* and found a peak of auxin production in the immature bud stage which could be prevented by removal of the developing carpels. At anthesis, production of auxin was minimal. Developing *Chrysanthemum* flowers showed two peaks of auxin activity. One peak corresponded to a stage of development where the flower bud was greater than 10 mm in diameter but the florets were not yet visible on the capitulum, and the second peak coincided with the beginning of floret expansion (Jeffcoat and Cockshull 1972). The unfertilised styles and ovaries of *Lilium regale* are able to convert tryptophan to IAA (Nitsch 1952), and pollination and pollen tube growth stimulate a wave of auxin production in the style and ovary of *Nicotiana* (Lund 1956). Developing seeds often produce large amounts of auxin. In corn, a high content of free IAA is found at the "milk" stage (Stehsel 1950), while in the mature seed much of the IAA is in the form of IAA-*myo*-inositol esters (see section 2.6.2 below). Most of the auxin in the rye seed is in the endosperm and aleurone layer (Hatcher 1945). In these early papers, IAA was identified by bioassay, but the auxin of cotton ovules has recently been identified as IAA by gas chromatography (Shindy and Smith 1975). In some fruits, a peak in auxin production is found several days after pollination. In the strawberry, for example, IAA rose from 3.6 to 127.0 ng per berry during the first 12 days after pollination (Nitsch 1950), and similar peaks have been found in apple (Luckwill 1948), grape (Coombe 1960), tomato (Iwahori 1967) and blueberry (Kender and Desrochers 1970). The first peak of auxin production in apple seeds was found to coincide with the development of the endosperm from a coenocytic to a cellular tissue. The peak in grapes coincided with the peak of mitotic activity in the seeds, and the seeds are also the principal source of auxin in the strawberry, where the receptacle tissue is devoid of auxin and does not develop normally if the seeds are removed. Very few studies of the biosynthesis of IAA in fruit have been made. Dannenberg and Liverman (1957) demonstrated the conversion of [^{14}C]TPP to [^{14}C] IAA in watermelon slices, and this conversion can also be carried out by lime fruits (Khalifah 1967) and immature pea seeds (Moore 1969).

2.5.3.5. Roots. IAA has been identified by gas chromatography and mass spectrometry in extracts of *Zea* roots (Elliot and Greenwood 1974) and is also present in the roots of *Alnus* (Dullaart 1970). Most of the auxin in *Zea* roots was found to be in the stele, which contained 142 μg/kg, while the concentration in the cortex was only 1 μg/kg (Greenwood et al. 1973). The root cap also seems to have a high auxin content. Rivier and Pilet (1974) showed by quantitative mass fragmentography that the IAA content of *Zea* root caps was 356.6 μg/kg fresh wt., while that of the 10 mm apical section of the root was 109.6 μg/kg fresh wt. The question of the source of auxin in the root has not yet been satisfactorily resolved (Scott 1972). In some plants, such as peas, it has been shown that exogenous auxin applied to the apical bud can be translocated to the roots, suggesting that the roots are normally dependent on the shoot for their auxin supply (Morris et al. 1969). It is also possible that root auxin may be synthesised in situ. Excised sterile root tips of *Phaseolus coccineus* can convert [^{3}H]TPP to [^{3}H]IAA (Mitchell and Davies 1972) and the first enzyme of the IPyA pathway for IAA synthesis, tryptophan aminotransferase, is present in bush bean roots (Forest and Wightman 1972). Scott (1972) also suggests that auxin may be obtained by the roots as a metabolic product of the microorganisms of the rhizosphere. While some auxin from this source may enter the root, it seems inherently unlikely that plants should routinely depend on an exogenous source for one of their most important hormones.

Thus, in summary, it seems that the most important sites of IAA synthesis in plants are the young expanding leaves, the cambial tissues, immature ovaries and developing seeds. Other tissues, such as mature and senescent leaves, stems and roots also have the capacity to synthesise IAA, although the extent of their contribution in vivo may be less. Sheldrake (1973b) points out that most of the active sites of auxin synthesis are associated with the death of cells, either during vascular differentiation, the digestion of endosperm or the senescence of leaves. He therefore hypothesises that tryptophan is normally the limiting factor for auxin synthesis and that the level of tryptophan in living cells is normally too low for synthesis to occur. In tissues where cell death is occurring, tryptophan is released by the autolysis of proteins, allowing the concentration of tryptophan to reach a level at which the synthesis of IAA can proceed.

2.6. Metabolism of indole auxins

2.6.1. Oxidation of indol-3-ylacetic acid

It has been known for 25 years that IAA can be oxidised by plant peroxidases, and during this time there have been continued attempts to settle the question of the molecular site of IAA-oxidase activity. Are peroxidase and IAA-oxidase activities functions of the same molecule, or do molecules exist which have IAA-oxidase activity alone? It has been found that both peroxidase and IAA-oxidase can exist in numerous isoenzymatic forms (Shannon 1968; Schneider and Wightman 1974a) and these isoenzymes can be classified into at least four groups on the basis of their relative peroxidase and IAA-oxidase activities: (1) typical peroxidases, such as horseradish peroxidase, which catalyse peroxidation reactions at high rates while oxidising IAA much more slowly; (2) enzymes which have both high IAA-oxidase and high peroxidase activities, e.g. turnip peroxidase P_7 (Ricard et al. 1972) (3) peroxidases devoid of IAA-oxidase activity, e.g. several isoenzymes of spinach root peroxidase (Endo 1968); (4) IAA-oxidases without peroxidase activity; these appear to be quite rare. Sequeira and Mineo (1966) found an IAA-oxidase in tobacco roots and in a crystalline horseradish peroxidase preparation which had no peroxidative activity towards guaiacol, although this report could not be confirmed by Hoyle (1972). The only other report of a peroxidase-free IAA-oxidase appears to be that of Van der Mast (1969), who found two IAA-oxidase fractions in pea roots, one of which had no peroxidase activity.

The products of the oxidation of IAA by horseradish peroxidase were studied by Hinman and Lang (1965), who found that the main product at IAA concentrations below 2×10^{-4} *M*, was 3-methyleneoxindole. Intact pea seedlings and enzyme extracts from these plants can convert IAA to 3-hydroxymethyloxindole (Tuli and Moyed 1967) and this compound can be dehydrated to 3-methyleneoxindole either spontaneously or enzymatically (Basu and Tuli 1972). The 3-methyleneoxindole reductase of peas, which reduces 3-methyleneoxindole to 3-methyloxindole, has been highly purified and shown to occur in several isoenzymatic forms (Moyed and Williamson 1967a, b).

As indicated in Hinman and Lang's reaction scheme (Scheme 2.6), an alternate end

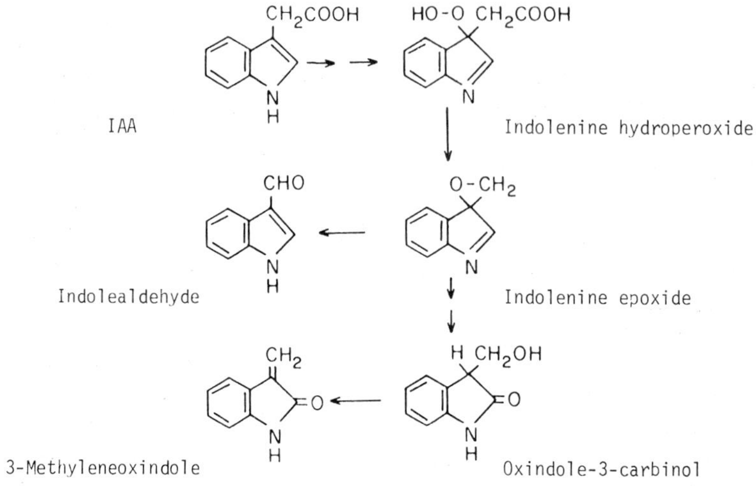

Scheme 2.6. Pathways for IAA oxidative metabolism.

product of IAA oxidation is indolealdehyde, which would be obtained from the epoxide intermediate shown if ring opening occurred before rather than after hydration. This compound is much more stable than the oxindoles and has been demonstrated as a native compound in *Zea* seed (Stowe et al. 1968), in shoots of *Equisetum* (Bourdoux et al. 1971), in bark of *Murrya exotica* (Chowdhury and Chakraborty 1971) and in shoots of barley and tomato (Schneider et al. 1972). Recently, indolealdehyde has also been isolated from cotton ovules and identified by gas chromatography and mass spectroscopy (Shindy and Smith 1975). Exogenous IAA can be converted to indolealdehyde by several species (see references in Schneider and Wightman 1974a) and indolealdehyde has also been detected after oxidation of IAA by plant enzyme preparations (Stutz 1958; BeMiller and Colilla 1972). In pea stem sections, indolealdehyde can be further oxidised to indole-carboxylic acid and also undergoes a reversible reduction to indol-3-ylmethanol, in a reaction analogous to the formation of tryptophol from indol-3-ylacetaldehyde (Magnus et al. 1971). Indolecarboxylic acid has been found as a native compound in *Zea* seeds (Stowe et al. 1968) and *Nicotiana* stems (Bayer 1969).

Chemical oxidation of IAA results in the cleavage of the pyrrole ring with the formation of 2-formamidoacetophenone and 2-aminoacetophenone (Abramovitch and Ahmed 1961). Incubation of pea root tips with [^{14}C] IAA results in the formation of radioactive 2-formamidoacetophenone (Collet 1968), and a pyrrolooxegenase which can catalyse the oxidation of IAA to 2-formamidoacetophenone has been isolated from wheat germ, spinach and Swiss chard (Frydman et al. 1971, 1972). The 2-formamidoacetophenone was further hydrolysed by a formylase present in the wheat germ preparation to yield 2-aminoacetophenone.

2.6.2. Conjugation of indol-3-ylacetic acid

In many important metabolites of IAA, the auxin molecule is conjugated rather than being oxidised, and when exogenous IAA is supplied to plants it is usually converted either to 1-(indol-3 ylacetyl)-β D-glucose or to indol-3-ylacetyl-L-aspartic acid. The aspartate was first reported as a metabolite of exogenous IAA in pea stem sections by Andreae and Good (1955) and has since been demonstrated in many species, both after feeding IAA and as a native compound (see references in Schneider and Wightman 1974a). Monomethyl 4-chloroindol-3-ylacetyl-L-aspartate has also been isolated as a native compound from immature peas (Hattori and Marumo 1972). Indol-3-ylacetyl-glucose was isolated by Zenk (1961) after feeding [^{14}C]IAA to *Colchicum* leaves and is probably also formed from exogenous IAA in barley leaves (Schneider 1965) and pea and bean stems (Davies 1972). *Zea* seeds, which have an unusually high IAA content, have been found to contain an interesting group of IAA conjugates; these include IAA-*myo*-inositols, IAA-*myo*-inositol-arabinosides and IAA-*myo*-inositol-galactosides (Ueda and Bandurski 1969; Ueda et al. 1970; Piskornik and Bandurski 1972). An enzyme preparation from immature *Zea* kernels has been shown to catalyse the formation of several *myo*-inositol and glucose esters of IAA in presence of CoA and ATP (Kopcewicz et al. 1974).

Some aspects of auxin metabolism are discussed in relation to root, stem and leaf growth in Volume II Chapter 2 (sections 2.1.3.3, 2.3.3.5 and 2.4.3.3 respectively).

2.7. Transport of indol-3-ylacetic acid and synthetic auxins

The definition of a hormone includes the stipulation that the hormone moves through the organism from its site of synthesis to its site of action. This characteristic of hormones is not so marked in plants as it is in animals, and we have seen that auxin may be synthesised at several different sites. However, plants do possess a well-defined system by which auxin is moved throughout the organism, and this movement is generally called "transport", although the use of this term does not imply that movement necessarily proceeds against a potential gradient. Valuable discussions of earlier work on this topic may be found in the reviews by McCready (1966) and Goldsmith (1968b, 1969).

2.7.1. Methods of studying auxin transport

Studies of auxin transport almost always depend upon the application of exogenous hormone, and the observed characteristics of the transport system may vary considerably depending upon the degree of intactness of the plant and the site of application of the hormone. Many workers have used isolated segements of coleoptiles, stems and roots, and the agar block method usually used with these tissues derives from the classical paper of Went (1928). A small block of agar containing auxin, the donor, is placed on one end of the section, and a small block of plain agar, the receiver, on the other. The transport of auxin may then be measured either by analysis of the receiver at various time intervals after application of the donor, or by analysis of succesive parts of the segment. The amount of auxin transported is small, about 2×10^{-10} g/hr/mm^2 in corn and oat coleoptiles (Goldsmith 1969), and therefore the most sensitive methods must be used to analyse the moving auxin. In early work, or for measurement of endogenous auxin, the *Avena* curvature test has generally been used. More recently, ^{14}C- and ^{3}H-labelled auxins have allowed greater speed and sensitivity of measurement. [^{3}H]IAA has been used relatively rarely; it has the advantage of high specific activity, so that IAA concentrations as low as 10^{-9} *M* may be used. This is advantageous in studies on roots, the growth of which may be inhibited by the 10^{-7} to 10^{-5} *M* solutions often used when [^{14}C] IAA is employed (Davies and Mitchell 1972). The position of the label must be considered, since it is vital to know that the radioactivity being measured is actually attached to IAA. Carboxyl labelling is often used, since if decarboxylation occurs, much of the $^{14}CO_2$ produced may diffuse out of the section. Fixation of this $^{14}CO_2$ is still possible, however, and most workers have found it necessary to verify that the transported label is in fact in the form of IAA. Chromatographic analysis of receiver blocks generally shows that at least 90% of the auxin is transported unchanged: this has been demonstrated for IAA (Hertel and Leopold 1963), 2,4-D (McCready 1963) and NAA (Gorter and Veen 1966), and in both shoot (Gorter and Veen 1966) and root (Scott and Wilkins 1968) tissue. However, analysis of the segments may show the presence of several metabolites (Table 2.3). Thus, measurements of label in the segment may not be related to the fraction of the applied auxin which is actually moving.

The amount of auxin which is observed in the receiver block in a classical transport experiment depends on several processes; uptake from the donor, transport itself, immobilisation and metabolism of the auxin, and movement into the receiver. Some auxin may be lost during uptake from the donor, as decarboxylation may take place at the cut surface (Iversen and Aasheim 1970). During transport of auxin through *Avena* coleoptiles, as much as 68% of the radioactivity may be immobilised (Goldsmith and Thimann 1962). The immobilised auxin can be extracted from the tissue with ether, and when extracted can be shown to be chromatographically identical to IAA, Centrifugation of a tissue homogenate from oat coleoptiles containing the immobilised auxin showed that the immobilised activity was sedimentable at 240 X *g* and appeared to be associated with cell walls. Label could be extracted from the sediment by urea, trypsin or chymotrypsin, but not by water, phosphate buffer or ribonuclease. It was therefore concluded that the immobilised IAA was attached to protein by hydrogen bonds (Winter and Thimann 1966). In addition to bonding to protein, IAA may be metabolised to low molecular weight compounds, principally indolylacetylaspartic acid, indolylacetylglucose and indolealdehyde (Table 2.3). The identification of these compounds is often based solely on chromatography in a limited number of solvent systems and it is quite probable that other metabolites remain to be identified. Some decarboxylation may take place at the cut surface during passage of the auxin into the receiver block; however Goldsmith and Thimann (1962) showed that identical results were obtained when the receiver was a further length of the section itself instead of a receiver block, so no large artefacts appear to occur during this step. In experiments on *Zea* root segments, Wilkins et al. (1972) found that after transport for 8 hr at 25 °C the amount of radioactivity in the receiver block began to decline due to a metabolically dependent resorbtion of radioactivity into the section; thus it seems that transport experiments should not be unduly prolonged.

In addition to the many experiments which have been carried out on isolated segments, some studies have been made on the transport of auxins in intact plants. Auxins may be applied to the surface in aqueous solution, often in the presence of wetting agents (Morris et al. 1969) or injected into the plant (Bourbouloux and Bonnemain 1973). As will be discussed below, the pathway of auxin translocation in intact plants is strongly influenced by the site of application.

Table 2.3.

Metabolic products of indol-3-ylacetic acid detected during transport

Species	Compound	Method of identification	Reference
Gossypium	Indolylacetyl-aspartic acid	TLC, 2 solvents, Prochazka reagent	Robinson et al. 1968
Vicia	Indolylacetyl-aspartic acid	TLC, 3 solvents	Eschrich 1968
Vicia	Indolylacetyl-aspartic acid	PC*, 1 solvent	Bourbouloux and Bonnemain 1974
Pisum	Indolylacetyl-aspartic acid	PC, 4 solvents, Ehrlich reagent, UV fluorescence	Morris et al. 1969
Pisum	Indolylacetyl-aspartic acid	PC, 1 solvent	Morris and Kadir 1972
Pisum	Indolylacetyl-aspartic acid	PC, 2 solvents	Patrick and Wooley 1973
Phaseolus	Indolylacetyl-aspartic acid	TLC, 1 solvent	Craker et al. 1970
Phaseolus	Indolylacetyl-aspartic acid	TLC, 5 solvents	Iversen et al. 1971
Salix	Indolylacetyl-aspartic acid	PC, 1 solvent	Lepp and Peel 1971
Nicotiana	Indole aldehyde	GLC	Bayer 1972
Pisum	Indole aldehyde	PC, 4 solvents, 4 colour reagents, UV fluorescence	Morris et al. 1969
Pisum	Indolylacetyl-glucose	PC, 1 solvent; TLC, 2 solvents; hydrolysis yields IAA	Patrick and Wooley 1973

* PC: paper chromatography

2.7.2. Velocity and flux of auxin transport

The amount of auxin entering an agar receiver block may be called "transport intensity", I (Van der Weij 1932; Kaldewey 1966) or "flux", F (McCready 1966). Then:

F (ng/hr) = V (mm/hr) × C (ng/mm)

where V = velocity and C = capacity.

The velocity of auxin transport is usually measured by the intercept method of Van der Weij (1932). In this method, the amount of auxin arriving in the receiver is plotted against time. If the increase in auxin in the receiver follows a linear time course, indicating a steady state of flux, then the intercept on the time axis will indicate the time, T, needed for the auxin to travel through the segment. Then V = L/T, where L = the length of the segment. Difficulties may arise with this method if the arrival curves are not linear, as the estimation of T then becomes difficult (McCready 1963; De La Fuente and Leopold 1966). The velocity of auxin movement can also be measured by applying the IAA in the form of a short pulse, then following the movement of this pulse down the section. The values obtained by this method for the velocity of auxin movement in corn coleoptiles (12–15 mm/hr, Goldsmith 1967) were comparable to those obtained earlier by the intercept method (Hertel and Leopold 1963). Goldsmith (1969) lists the velocities of auxin transport in 6 root and shoot tissues, and some more recent determinations are collected in Table 2.4. In general, it has been found that the velocity of auxin transport has the following characteristics:

(1) Velocity is independent of the length of the segment in both shoots (McCready 1963) and roots (Wilkins and Cane 1970).
(2) Velocity is independent of the auxin concentration in the donor (McCready 1963).
(3) The acropetal movement of auxin in roots appears to be slower than the basipetal movement of auxin in shoot tissues. For example, for *Zea* at 25 °C, the velocity of IAA transport in root segments is 5 mm/hr (Wilkins and Cane 1970) and the velocity in coleoptile segments is 15 mm/hr (Goldsmith 1967).
(4) Velocity varies with the age of the tissue. In coleoptiles, velocity varies considerably from the tip to the base of the organ, and the lowest velocities are associated with the regions which show the most rapid rates of growth (Shen-Miller 1973). In fruit stems of *Fritillaria,* V = 7 mm/hr near the tip, declining to 5 mm/hr near the base (Kaldewey 1966).
(5) Velocity varies with temperature. This has been clearly shown in *Zea* root segments, where the velocity of IAA transport rises from 0.2 mm/hr at 1 °C to an optimum value of 8 mm/hr at 31 °C. At higher temperatures the rate decreases, falling to 1 mm/hr at 50 °C (Wilkins and Cane 1970). Very similar results were obtained by Gregory and Hancock (1955) in apple shoots, where the maximum velocity was 10 mm/hr at 27 °C.
(6) The velocity of transport varies with the nature of the auxin applied. Generally IAA shows the most rapid rate, NAA moves slightly more slowly (Hertel and Flory 1968) and 2,4-D is transported at a significantly lower rate (McCready and Jacobs 1963a).
(7) When auxins are applied to intact plants, the velocity of auxin movement depends on the site of application of the auxin. Application to the apical bud or to young leaves or fruits results in rates similar to those observed in isolated segments. However,

Table 2.4.

The velocity of transport of indol-3-ylacetic acid

Species	Tissue		Velocity (mm per hr)	Reference
A.	Isolated shoot segments (basipetal movement)			
Fritillaria	Fruit stem		6.5	Kaldewey 1963
Phaseolus	Petiole		5.7	McCready and Jacobs 1963a
Lens	Stem		6.5	Pilet 1965a
Coleus	Stem		2.1–3.3	Jacobs and McCready 1967
Phaseolus	Hypocotyl		6.7	Smith and Jacobs 1969
Nicotiana	Stem		8.0	Bayer 1972
Zea	Coleoptile	apex	41	Shen-Miller 1973
		middle	13	
		base	34	
Avena	Coleoptile	apex	28	
		remainder	14	
B.	Isolated root segments (acropetal movement)			
Lens	Root		2.2	Kirk and Jacobs 1968
Zea	Root		4–6	Scott and Wilkins 1968
Zea	Root		8.0	Wilkins and Cane 1970
C.	Intact plants (basipetal movement in stem unless otherwise noted)			
Vicia	Applied to mature leaf		150–200	Eschrich 1968
Pisum	Applied to apex		11	Morris et al. 1969
Fraxinus	Applied to apex.	active	13	Hollis and Tepper 1971
		dormant	3	
Vicia	Applied to mature leaf		160	Bonnemain 1971
	Applied to young leaf		10	
Populus	Applied to leaf		70	Eliasson 1972
Phaseolus	Applied to decapitated apex		28	Patrick and Wooley 1973
Vicia	Applied to young pod		20	Bourbouloux and Bonnemain 1973
Vicia	Applied to apex measured in root		9	Bourbouloux and Bonnemain 1974

auxin applied to mature leaves appears to enter the phloem and is carried at much higher rates (Table 2.4).

Many studies have also been made of the flux of auxin into receivers, and this parameter has been shown to have the following characteristics:

(1) Unlike velocity, flux does increase with increased auxin concentration in the donor. The relationship is not a linear one, and saturation appears to occur at higher concentrations (Goldsmith and Thimann 1962).

(2) Flux is often strongly affected by the degree of maturity of the tissue. This topic is discussed in section 2.7.8 below.

(3) In *Zea* roots, flux showed a complex relationship to temperature, rising to a maximum at 10–15 °C, declining again to 31 °C and rising again to 50 °C (Wilkins and Cane 1970). The authors suggest that this complex curve is the resultant of two processes with different temperature optima; these might be, (a) the true flux of IAA through the segment, and (b) immobilisation and metabolism of IAA.

(4) Flux may show rhythmic fluctuations. Hertel and Flory (1968) found that the addition of IAA to auxin-depleted corn coleoptiles resulted in oscillations in the rate of flux of IAA into receivers with a period of about 25 min. These oscillations may be associated with the oscillations of electric potential of similar period found during IAA transport in *Avena* coleoptiles by Newman (1963). Similar fluctuations with a period of about 20 min were found in *Avena* and *Zea* coleoptiles by Shen-Miller (1973).

2.7.3. Polarity of auxin transport

Perhaps the most interesting characteristic of the auxin transport system is its polarity. The direction in which auxin moves in the plant depends upon the organ to which the auxin is applied and, when intact plants are used, on the site of application of the hormone.

In isolated segments of the shoot, auxin moves predominantly in a basipetal direction, that is towards the root. This basipetal polarity has been demonstrated in many tissues including *Avena* coleoptiles (Went 1928), *Phaseolus* hypocotyls (Jacobs 1950), *Arachis* gynophores (Jacobs 1951), apple shoots (Gregory and Hancock 1955), *Coleus* internodes (Jacobs 1961), sunflower epicotyls (Leopold and Lam 1961), *Zea* coleoptiles (Hertel and Leopold 1963), *Lens* stems (Pilet 1965b), *Fritillaria* fruit stems (Kaldewey 1966), *Pinus* seedling stems (Zaerr 1968), young *Nicotiana* stems (Bayer 1972) and sugar cane internodes (Scott and Most 1972). However, movement in stems is not exclusively basipetal; the ratio of basipetal to acropetal movement varies greatly with tissue, segment length, temperature and auxin concentration in the donor (McCready 1968). The ratio may be as high as 170:1 (sugar cane internodes, Scott and Most 1972) or as low as 3:1 (*Coleus* internodes, Jacobs 1961).

Much recent work shows that auxin transport in roots is also polar. The direction of movement in roots, however, is predominantly acropetal, that is towards the root tip. Conflicting results were obtained by earlier workers (see discussion in Kirk and Jacobs 1968), but Pilet (1964) was able to show a definite acropetal polarity in segments of *Lens* roots, with a ratio of acropetal:basipetal movement of 2:1. In 1968, Kirk and Jacobs reinvestigated the *Lens* root system using [^{14}C]IAA of high specific activity and found a strong acropetal polarity with a polar ratio of 33:1 (Figure 2.16). At about the same time, Wilkins and Scott (1968a) demonstrated acropetal auxin transport in root segments of *Zea,* and this system has subsequently been investigated in detail by Wilkins and coworkers. Using 6-mm segments and times greater than 1 hr, polar ratios of from 4 to 30 were found (Scott and Wilkins 1968). Other root tissues in which acropetal polar transport has been demonstrated include peas (Hillman and Phillipps 1970), sunflower and cabbage (Iversen and Aasheim 1970). In fact, the idea that acropetal transport is characteristic of roots has become accepted to such a degree that the presence of acropetal transport in the rhizophore of *Selaginella* has been used as evidence that this organ is morphologically a root (Wochok and Sussex 1974).

The direction of movement of exogenous auxin in intact plants appears to depend mainly on the site of application of the hormone. When low concentrations of labelled IAA are applied to the apices of intact pea plants, the auxin is transported to the root system, where it tends to accumulate in the developing root primordia (Morris et al. 1969). However, when auxin is applied to adult leaves transport appears to be in the phloem (Little and Blackman 1963; Eschrich 1968) and the direction of transport may be affected by the presence of sinks, such as the roots (Hoad et al. 1971). Very little work has been done on the polarity of movement of exogenous auxin in intact roots. Davies and Mitchell (1972) applied [^{3}H]IAA of high specific activity to intact primary roots of *Phaseolus* at the low, non-inhibitory concentration of 2×10^{-9} *M*. The auxin was

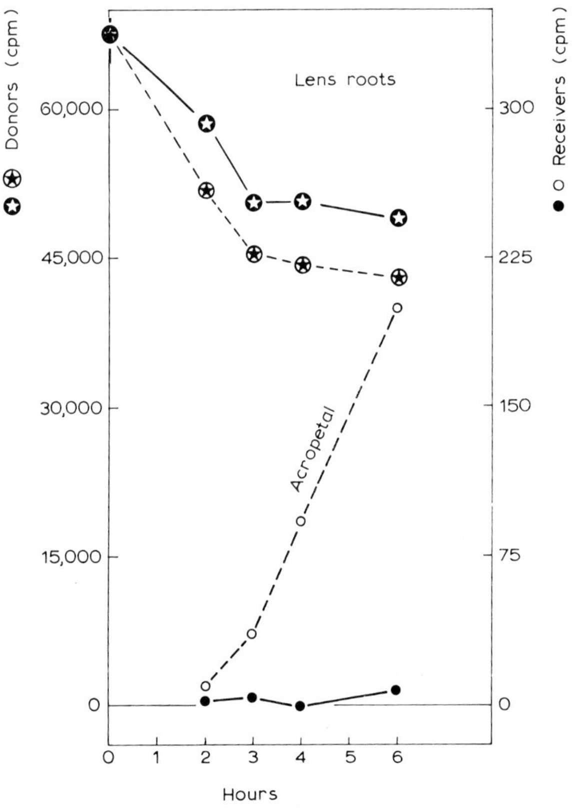

Figure 2.16. Time course of movement of radioactivity in *Lens* root segments supplied with donor blocks containing [2-^{14}C]IAA at an initial concentration of 2.8 mg/l. Radioactivity levels in donor blocks are shown at top of figure (legend on left), counts in receiver blocks are shown at bottom of figure (legend on right). Solid symbols with continuous lines refer to basipetal movement; hollow symbols with dashed lines refer to acropetal movement (from Kirk and Jacobs 1968).

supplied either to the root tip or 2 cm above the tip. Short-distance basipetal transport took place from both locations, and there was also acropetal movement of smaller amounts of auxin from the upper application point to the root tip, where the auxin tended to accumulate. Thus, the results differed somewhat from those obtained using intact segments; however, the authors felt that long-distance acropetal movement from the shoot is the usual means by which the root receives its auxin supply.

2.7.4. The pathway of auxin transport

It has not proved an easy matter to locate the physical pathway taken by auxin during polar transport. Investigators of this problem have mainly used three techniques, autoradiography, surgery and the aphid-honeydew method, and each of these techniques is subject to artefacts. Since IAA is a soluble compound, frozen sections must be used for autoradiography, and it is difficult to obtain good resolution and freedom from artefacts, such as bleaching and spurious fogging of the film (Wangerman 1970). Perhaps more importantly, autoradiography cannot distinguish between IAA which is in transport and IAA which has been immobilised or metabolised to other compounds. Furthermore, autoradiography does not give any information about transport velocity, and a tissue which showed a low level of labelling on an autoradiograph could still be transporting IAA rapidly. Difficulties with surgical techniques are mainly those of working on the small scale required. Here careful selection of a tissue with a suitable anatomy is helpful (e.g., *Helianthus,* Wangerman 1974). The aphid honeydew method can be used only to investigate the presence or absence of IAA in the phloem. Possible metabolism of IAA complexes to IAA by the aphid should be considered, and small colonies of aphids should be used so that movement in the phloem is not affected by the insects acting as sinks.

Earlier work on the pathway of auxin transport had suggested that polar movement of auxin could occur in the absence of vascular tissue in *Zea* coleoptiles (Hertel and Leopold 1963) and *Coleus* stems (Jacobs and McCready 1967). More recently, Wangerman (1974), using *Helianthus* stem segments, showed by surgical techniques that interruption of the pith only, or of the xylem only, did not affect the movement of basipetally moving IAA or 2,4-D. However, cutting of the region containing the young phloem fibres, sieve tubes, phloem parenchyma and procambium caused a large accumulation of label above the wound. Autoradiographs showed the presence of label mainly in the phloem and phloem fibres. Transport in a woody tissue has been studied by Sheldrake (1973c), who carried out surgical experiments on stem segments excised from 3–5 month old *Nicotiana* plants. In this system most of the auxin was transported through cells associated with the internal phloem and in cells close to the cambium. Xylem parenchyma and mature xylem cells did not translocate any auxin, and since auxin transport was not reduced by the application of osmotic gradients in a direction which would reverse an osmotically active mass flow, Sheldrake concluded that mature sieve tubes are unable to transport auxin. The low velocity of auxin transport in this tissue (about 5 mm/hr) is also inconsistent with movement through the phloem. The pathway of movement of auxin in root segments of *Zea* has been studied by autoradiography and girdling techniques (Bowen et al. 1972) and by microapplication of IAA to particular tissues at the cut end of the segment (Shaw and Wilkins 1974). Girdling experiments, which removed about 80% of the cross-sectional area of the root, did not greatly affect the uptake of IAA from the donor but led to a 50% reduction in the acropetal movement of IAA into the receiver. Thus, although movement could occur in both tissues, the flux per unit area was much greater in the stele than in the cortex. Autoradiographic data also supported this view, since the heaviest labelling was found in non-xylem areas of the stele and in the inner cortex. Basal movement of radioactivity was slight, and occurred with equal efficiency in the cortex and the stele. Microapplication of droplets of IAA to the cut surface of intact and desteled roots (Shaw and Wilkins 1974) led to the conclusion that acropetal transport occurs only when stelar tissues are present. When all stelar tissue was removed from sections taken 6 mm behind the apex, acropetal and basipetal movement became equal and polarity was abolished.

Thus, in isolated segments in both roots and shoots polar transport appears to be associated with non-xylem areas of the vascular tissue. In shoots, there is evidence that mature sieve tubes are not involved, and in woody stems movement may take place in the cambium.

As we have mentioned above, the pathway of transport of exogenous auxin in intact plants is strongly dependent on the site of application. When labelled auxin was applied to the apical bud of pea seedlings, the auxin moved through the stem to the roots within 2.75 hr (Morris et al. 1969). However, no radioactivity was found in aphids feeding on the stem, although the internode upon which the aphids were feeding contained considerable amounts of radioactivity (Morris and Kadir 1972). This finding, together with the low velocity of auxin transport (11 mm/hr, Morris et al. 1969) showed that transport was not occurring in the phloem sieve tubes. Low rates of transport, similar to those found in isolated stem segments, were also found when auxin was applied to the apical buds of intact white ash shoots (Hollis and Tepper 1971), suggesting again that transport in the sieve tubes was not involved. Auxin applied to young (2 cm) leaves of bean moved at about 10 mm/hr, and autoradiography showed a concentration of label in the cambium and differentiating xylem, and little radioactivity in the sieve tubes (Bonnemain 1971). When auxin was applied to young fruits of tomato (Bonnemain 1970) and bean (Bourbouloux and Bonnemain 1973) the rates of transport were less than 20 mm/hr, and radioactivity was concentrated in the internal phloem (tomato) and the cambium, protoxylem and developing phloem regions (bean). The resolution of the autoradiographs was not sufficient to decide exactly which of the phloem cells was responsible for transport. Thus, when auxin is applied to apical buds, young leaves or young fruits of intact plants, the characteristics of transport are similar to those found in excised stem segments. Transport velocities and aphid experiments show that mature sieve tubes are unlikely to be involved, but autoradiography suggests that the pathway of transport is located somewhere in the phloem-cambium region.

However, when auxin is applied to mature leaves, transport does probably take place in the sieve tubes. When labelled IAA was applied to mature leaves of *Vicia* (Eschrich 1968) or pea seedlings (Morris and Kadir 1972), radioactivity was recovered in aphid honeydew. Bonnemain (1971) found that movement of exogenous IAA out of mature leaves of *Vicia* occurred at rates up to 160 mm/hr, and autoradiography showed that most of the radioactivity was in the phloem, although some was present in the cambium and the peripheral parts of the xylem. In the intact pea plant, there is very little lateral transfer between the non-phloem pathway followed by auxin applied to the apical bud, and the phloem route used by auxin applied to the leaves (Morris and Kadir 1972). Recently, endogenous IAA has been identified in both phloem sap and root-pressure (xylem) sap of *Ricinus* by gas chromatography (Hall and Medlow 1974). Phloem transport in *Ricinus* is non-polar, and phloem space may be freely available to IAA in this species. The origin of the IAA in the xylem sap is not known.

2.7.5. Specificity of the auxin transport system

It is clear from the above discussion that the system responsible for polar transport of auxin in plants is distinct from that responsible for movement of assimilates in the phloem. It has therefore been of interest to determine the degree of specificity of the auxin transport system. Can it carry only IAA, or are other auxins or even other hormones moved also? Implicit in many studies of specificity has also been the idea that the primary recognition site for auxin transport might be identical to the primary site for auxin action.

Leopold and Lam (1961) measured the rate of polar movement of three auxins, IAA, NAA and IBA in sunflower epicotyls. All three compounds showed basipetal polarity of transport, and movement was inhibited in each case by the auxin transport inhibitor, triiodobenzoic acid. The velocity of movement of NAA (6.7 mm/hr) was almost as fast as that of IAA (7.5 mm/hr), while IBA moved much more slowly (3.2 mm/hr). Thus it appears that the auxin transport system is not exclusively specific for IAA itself. For some time it was thought that the synthetic auxin 2,4-D did not move in the polar auxin transport system, but McCready and Jacobs (1963a) demonstrated basipetal polar movement in *Phaseolus* petioles, although the velocity was very low (1 mm/hr). 2,4,5-Trichlorophenoxyacetic acid also showed basipetal polar movement in bean petioles, and its rate of transport was much greater than that of 2,4,6-trichlorophenoxyacetic acid, which has little auxin activity (Jacobs 1968). Another synthetic auxin with a completely different chemical structure, 4-amino-3,5,6-trichloropicolinic acid (Picloram) showed basipetal polar transport in *Phaseolus* petioles, although as was the case for 2,4-D, the velocity of transport was very low (Horton and Fletcher 1968). The IAA metabolite, indolealdehyde, appears in receiver blocks after application of IAA to the apices of intact pea plants (Morris et al. 1969) and may be mobile. In a comprehensive experiment, Hertel et al. (1969) studied the movement of several active auxins and inactive auxin analogues in *Zea* coleoptile sections. In this system IAA and 2,4-D showed basipetal transport. The active auxin α-NAA was also transported, but the inactive β-NAA was not. Both the dextro- and the less active laevo-isomers of indol-3-yl-2-methylacetic acid were transported, but the flux of the laevo form into the receivers was much less than that of the dextro-isomer. Benzoic acid and gibberellin A_1 did not show any polar movement. Thus, basipetal polar transport in *Zea* coleoptiles is limited to active auxins, and there is some degree of correlation between the degree of activity of the auxin and the amount of the auxin which is moved in a given time. The authors suggested that it was possible that the lack of transport of inactive analogues was due to rapid immobilisation; however, Veen (1972) showed that this was not so, at least in the case of the α- and β-isomers of NAA. Three other possibilites suggested by Hertel et al. (1969) to account for the correlation between growth activity and transportability remain to be tested: (1) transport may be required for the auxin to reach the site of action; (2) the specificity of the sites of transport and action may be similar and (3) a specific step in transport may be identical to the primary site of action.

The specificity of auxin transport has also been tested in intact plants. Morris and Thomas (1974) applied eight compounds, three of which were auxins, to the apical buds of intact pea plants. Only IAA and NAA showed basipetal transport during the first 24 hr. After 72 hr, a considerable amount of 2,4-D had moved downwards at a very low rate (1.4–2.2 mm/hr), but the other, non-auxin compounds (kinetin, adenine, sorbose, sucrose and leucine) did not move out of the bud. Thus in intact plants, as in segments, the auxin transport system is a specific one.

2.7.6. Effects of light on auxin transport

Since the early experiments of Cholodny (1927) and Went (1928), it has been realised that unilateral illumination of coleoptiles results in a greater flow of auxin down the shaded side, ascribed by these workers to a lateral transport of auxin. In 1964, this phenomenon was reinvestigated by Pickard and Thimann, using [^{14}COOH] IAA. Agar blocks containing the labelled auxin were applied to the intact tips of apical 6.5 mm *Zea* coleoptile sections. Each coleoptile was impaled on a razor blade which bisected the base to a depth of 1 mm, and an agar receiver block was set on each side of the base. The coleoptiles were then illuminated unilaterally with 1,000 meter-candle-seconds (m.c.s.) of white light, which is in the region of the dose response curve where "first positive curvature" occurs and curvature is a product of the intensity and time of illumination (Thimann and Curry 1960). Immediately afterwards, agar donor blocks containing [^{14}C] IAA were placed in contact with the intact apices. After illumination, 25% of the total radioactivity in the receivers was found on the lighted side and 75% on the shaded side, and when the two halves of the tissue were assayed, the ratio was 35:65. Coleoptiles were then exposed to 3 hr of light at 500 mcs, a dosage which leads to "second positive curvature", a response which depends on the length of illumination rather than the total dosage. Again the distribution of auxin was unequal, with 34% of the radioactivity being found on the light side and 66% on the dark side. Since the region of sensitivity to second positive curvature dosages extends several mm below the tip, the experiment was repeated with subapical sections, and asymmetry was again found (46% light: 54% dark). All the radioactivity applied to the sections in this experiment was recovered, showing that the lower auxin content on the lighted side was not due to photolytic destruction of the hormone, as had been suggested earlier (Galston 1962). Very early experiments by Boysen-Jensen and Nielsen (1926) showed that the insertion of a piece of mica into the tip of a coleoptile parallel to the plane of the unilateral illumination abolished the phototropic curvature, indicating that an actual lateral transfer of a growth substance was required. A similar experiment was performed by Briggs (1963), using *Zea* coleoptile tips and analysing the auxin content of the receivers by the *Avena* curvature test. Briggs found that coleoptile tips partially split at the base yielded more diffusible auxin from the shaded than from the illuminated side, but if the two halves of the coleoptile were completely separated by the insertion of a coverslip, the amount of auxin diffusing from each half was identical.

Thus, physical contact between the two halves is neccessary for the development of an auxin gradient between the illuminated and dark sides, and actual lateral transfer of hormone must take place.

It has also been suggested (Naqvi and Gordon 1967) that lateral transport may be the result of a photoinduced decrease in the flux of auxin. When *Zea* coleoptiles were illuminated bilaterally with white light in the first positive curvature range, there was a decrease in the amount of endogenous auxin diffusing from the tip, and also a decrease in the amount of exogenous [^{14}C] IAA transported to the base of the coleoptile. Naqvi and Gordon therefore suggested that in unilaterally illuminated coleoptiles, lateral transport may be a secondary effect of a concentration gradient established by the retardation of auxin transport on the lighted side. This possibility was further explored by Naqvi (1972a). In the first experiment reported in this paper, two agar blocks were placed on the top of a 12 mm *Zea* coleoptile segment, cut 1–2 mm below the tip. One block contained 0.2 mg/l of [2-^{14}C] IAA and the other an equal concentration of unlabelled IAA. Two receiver blocks were placed at the bottom of the coleoptile, and when these blocks and the tissue were analysed after a 90 min transport period, it was found that 13% of the absorbed radioactivity had moved towards the unlabelled half of the segment. Thus lateral transport can occur in the absence of a concentration gradient. In the second experiment, the concentration of IAA in the labelled donor block was 0.3 mg/l and in the unlabelled donor 0.2 mg/l. Under these conditions 19% of the radioactivity moved laterally, showing that lateral movement is increased by the introduction of a concentration gradient. Transport could also take place against a concentration gradient; when the concentration of labelled IAA was 0.2 mg/l and that of the unlabelled IAA 0.3 mg/l, 7% of the radioactivity moved laterally against the gradient. These results are partially consistent with the hypothesis that lateral transport is the result of the formation of an IAA concentration gradient in that lateral movement was increased when a lateral gradient was imposed. However, since lateral movement also took place in the absence of, and even against a gradient, it would seem that lateral redistribution can be independent of the existence of such gradients.

Bilateral illumination may increase or decrease the transport of auxin, depending on the tissue. As mentioned above, illumination of etiolated coleoptiles is generally inhibitory to auxin transport (Naqvi and Gordon 1967). Thornton and Thimann (1967) found that illumination of *Avena* coleoptile segments imposed a transient oscillation on the transport of IAA. Exposure to 3×10^5 ergs/cm^2 of blue light for 25 min led to a decline in the flux of IAA into the receivers. After 25 min the rate rose again to a maximum at 35 min, then again declined. The net result was a 12–17% decrease in the flux over an 80-min transport period. The red – far red wavelengths did not appear to have any effect on this phenomenon. Light also inhibits auxin transport in *Zea* coleoptiles. Hagar and Schmidt (1968) found that after a 1-hr exposure to 13,000 lux of white light there was a 24% reduction in auxin movement during a subsequent 1-hr transport period in the dark. Again, in this coleoptile tissue, only blue light was effective. When transport through the node is studied, however, the red – far red system is found to have a considerable effect (Yamaki and Fujii 1968). Irradiation with 1 min of red light resulted in a considerable decrease in the flux of IAA into receivers, and a corresponding increase in the "trapping" of IAA in the node tissue. The authors suggest that this trapping effect explains the inhibitory effect of red light on the mesocotyl tissue below the node.

In contrast to the inhibitory effect of light on auxin transport in coleoptiles, light appears to have a promotive effect on this process in roots and green shoots. In *Zea* root segments, white light promoted acropetal movement and both red and blue wavelengths were as effective as the white light. Promotion occurred only when the segments were irradiated during the transport period (Scott and Wilkins 1969). A similar effect was found with green stem segments of *Coleus*, where irradiation with blue, red or far red light more than doubled the flux of [^{14}COOH] IAA into basal receivers (Koevenig and Jacobs 1972).

2.7.7. Effects of gravity on auxin transport

Inversion of coleoptiles, so that longitudinal basipetal movement has to take place against gravity, generally reduces the rate of basipetal transport. Hertel and Leopold (1963) found that inversion of *Zea* coleoptile segments caused a 10–20% reduction in the amount of IAA arriving in receiver blocks during a 90-min transport period. Ouitrakul and Hertel (1969) confirmed this result, and found that transport of the synthetic auxin NAA showed an even more pronounced inhibition (20–50%) after inversion of the segments. This inhibitory effect on NAA transport appeared within 30 min. Centrifugation at $10 \times g$ caused basipetal transport of NAA against the centrifugal force to be inhibited more severely. Little and Goldsmith (1967) studied the effect of inversion on basipetal and acropetal transport in *Zea* and *Avena* coleoptiles under aerobic and anaerobic conditions. It was found that under aerobic conditions, inversion inhibited basipetal movement and promoted acropetal movement. The promotion of acropetal movement was caused by the reduction in "recycling" of acropetally moving auxin towards the base by the reduced level of basipetal transport. Under anaerobic conditions, inversion had no effect on movement in either direction. This indicates that the effect of inversion is not simply an effect of gravity on a passive process such as the rate of flow of extracellular water, but rather acts on the metabolically dependent process, basipetal polar transport. In some tissues, inversion may be without effect; for example, Pilet (1965b) found no effect of inversion on transport in either direction in *Lens* stem segments.

The effect of transverse gravity stimulation on longitudinal auxin transport appears to differ with the conditions of the experiment. In *Zea* coleoptiles (Gillespie and Thimann 1963) and *Lens* stem segments (Pilet 1965b), horizontal placement did not affect the total amount of radioactivity transported basipetally, and rotation on a horizontal clinostat did not affect the movement of auxin through *Avena* coleoptiles (Dedolph et al. 1966). On the other hand, Goldsmith and Wilkins (1964) found that horizontal orientation of *Zea* coleoptile sections slightly reduced flux into basal receivers at times between 2 and 4 hr. Transverse stimulation of intact plants may affect longitudinal auxin movement quite markedly. Lyon (1963) grew *Phaseolus* or *Brassica* plants either in the normal erect fashion or horizontally on a clinostat. When radioactive IAA was applied to the leaf blades, it was found that less label reached the roots in the horizontally-grown plants than in the controls, suggesting that the ability of the horizontal plants to transport auxin was impaired. The appearance of horizontally-grown sunflower plants also suggests that auxin is accumulating in the shoots. The stems show an inhibition of elongation, together with increased thickening and petiole epinasty (Palmer and Halsall 1969). When the sunflower plants were laid horizontally and rotated once every 48 hr, stem segments excised from the plants showed a marked decrease in basipetal transport of [^{14}C] IAA. After 60 hr of horizontal pretreatment, flux into receivers was reduced by 69%. Segments were then cut from plants which had been grown in a normal vertical position. These segments were placed either horizontally or vertically and basipetal auxin transport was assayed. No difference was found between the two treatments, and the authors concluded that at least 12 hr of horizontal pretreatment were required in order to show inhibition of transport. Reduced transport may be caused by some factor which develops in the shoots as a result of prolonged horizontal orientation, and Palmer and Halsall suggest that this factor may be ethylene. Pretreatment of erect plants with ethylene caused a 62% reduction in the transport of IAA in stem segments.

Horizontal orientation of plant tissues also results in the lateral migration of auxin, and this lateral movement has been presumed to underlie the geotropic movements of roots and shoots since the days of Cholodny and Went. Evidence for a difference in the flow of auxin through the upper and lower halves of horizontal coleoptiles was obtained by Dolk (1936) who found that 60% more endogenous auxin diffused out of the lower than the upper half. Dolk also showed that asymmetric distribution occurred when exogenous auxin was applied to horizontal decapitated coleoptiles. Dolk's experiments were repeated in 1963 by Gillespie and Thimann, using [^{14}COOH] IAA. In 6.5-mm horizontal corn coleoptile segments, the ratio of the radioactivities appearing in the upper and lower receiver blocks was 30:70, while in the tissue the ratio was 40:60. An actual lateral migration of radioactive IAA was demonstrated by Goldsmith and Wilkins (1964). The labelled auxin was applied in an agar block to one side of the apical end of a *Zea* coleoptile segment. When the coleoptile remained erect, 10% of the radioactivity was collected in a receiver placed on the opposite side of the base. When the coleoptile was placed in a horizontal position with the donor applied to the upper half, 25% of the radioactivity moved to the receiver on the lower half, clearly showing a promotion of lateral transport by the gravitational stimulus. Shaw et al. (1973) found a net downward movement of [^{3}H] IAA of 30% in horizontal shoot tissues of intact *Zea* plants, and lateral transport was also found in the shoots of intact *Avena* plants although its magnitude was much less. Lateral movement can occur at any point along the length of a 25-mm corn coleoptile segment, but responsiveness to gravity is much greater near the tip (De La Fuente and Leopold 1968). Shaw et al. (1973) found that a net downwards movement of label was detectable in excised *Zea* coleoptiles after 10 min of horizontal treatment. This result suggested that lateral auxin transport preceded the geoelectric effect (Grahm and Herz 1962) and the negative geotropic curvature. However, a somewhat different time course of lateral movement was described for *Zea* coleoptile sections by Filner et al. (1970). Within 5 min a significant flow of auxin to the upper side was measured, which reversed within 15 min to the normal downward flow. This time course of auxin movement closely parallels the geotropic bending of the coleoptile, which shows a transient downwards and then a normal upwards curvature. Shaw et al. (1973) suggest that the presence of endogenous regulators in their intact coleoptiles may have modified the transport of endogenous auxin, and also point out that Filner et al. did not demonstrate an actual upward movement in the tissue but only an increased export into the upper receiver blocks.

Lateral transport under the action of gravity has several features in common with normal basipetal transport. Firstly, lateral transport is not confined to the natural auxin, IAA. Ouitrakul and Hertel (1969) measured the transport of radioactive NAA through half-sections of *Zea* coleoptiles by applying the synthetic auxin in an apical donor block and measuring its appearance in a lateral receiver. It was found that placing the half sections horizontally with the receiver downwards increased the flux of auxin into the receiver within 10 min. Gravity dependent lateral movement of another synthetic auxin, 2,4-D, has been demonstrated in sunflower hypocotyl segments (Abrol and Audus 1973). In these experiments, a donor block containing [^{14}C] 2,4-D was placed asymmetrically on the apical end of the segment and the radioactivity in split receiver blocks at the base of the segment was measured. It was found that gravity induced a polar migration to the lower side involving about 10% of the total 2,4-D in transit. Both gravity-induced lateral transport and longitudinal basipetal transport tended to show saturation as the concentration of 2,4-D in the donor was increased to 5 mg/ml. The response of both processes to temperature was also similar. Like basipetal movement, gravity-induced lateral movement can occur against a concentration gradient (De La Fuente and Leopold 1968) and is inhibited by triiodobenzoic acid (De La Fuente and Leopold 1968), *N*-(naphth-l-yl)phthalamic acid (Steyer 1969), morphactins (Bridges and Wilkins 1973) and DPX 1840 (Beyer 1972); the action of these synthetic inhibitors is further discussed in section 2.7.10. However, ethylene completely inhibits lateral transport in pea stem segments while basipetal transport is little affected (Burg and Burg 1966).

Since gravity-induced lateral transport has many features in common with basipetal transport, it has been suggested that the two processes might be interdependent. Naqvi and Gordon (1966) found that in horizontally oriented *Zea* coleoptiles, flux through the lower half was 3.4 times greater than that through the upper half. Even when the coleoptile was completely split longitudinally, flux through the lower half was still greater. They therefore suggested that the lateral downward movement might be caused by this difference in the flux through the two halves of the segment. However, Cane and Wilkins (1969) demonstrated by an ingenious method that polarised lateral transport can occur independently of any differential basipetal movement. In this experiment, *Zea* coleoptiles were opened out into flat sheets so that all the cells in the coleoptile could be oriented with respect to gravity as if they were on the top, bottom or side of a horizontally oriented coleoptile. When all the cells were placed so that they were equivalent to those on the side of a horizontal coleoptile, there was a lateral movement of IAA from the upper to the lower half of the flat sheet of cells, but there was no difference between the basipetal longitudinal transport of auxin in the upper and lower halves of the tissue. Thus it was concluded that lateral transport can occur independently of any difference in the ability of the upper and lower halves of the coleoptile to transport auxin basipetally.

2.7.8. Effects of plant age on auxin transport

One of the earliest studies on the effect of age of the plant on auxin transport is that of

Jacobs (1950) on the movement of IAA through *Phaseolus* hypocotyls. No acropetal transport was detected in this tissue. The capacity for basipetal transport appeared between the 3rd and 5th days after germination, and sections from 5- to 8-day-old plants showed a steady increase in the amount of auxin transported into basal receivers. When the ability of various regions of the 8-day-old hypocotyl to transport auxin was studied, it was found that the greatest amount of movement was present in the apical centimeter, with a steep decline towards the base where very little transport occurred. A more extensive study of the same tissue (Smith and Jacobs 1969) showed that in young hypocotyls (4–8 days) maximal flux occurred 25–30 mm below the cotyledons. At 10 days, transport was greatest in the segment nearest the apex and declined towards the base, and by 14 days, very little transport could be detected. This general pattern of decline in basipetal transport with age has also been found in several other tissues. In *Phaseolus* petioles there was a steady decline in polarity with age (4–28 days); in this tissue acropetal transport was present and the decline in polarity seemed to be due to an increase in acropetal movement (McCready and Jacobs 1963b). In *Coleus* petioles the greatest amount of movement, expressed as a percentage of uptake, occurred in petiole 2 (9.6%) and there was a steady decline to 2.5% in petiole 8, which was ready to abscise (Veen and Jacobs 1969). When the abilities of different segments of the same petiole to transport auxin applied to the leaf blade were measured, it was found that the amount of transportable auxin per unit length did not vary greatly with position, although there was some increase at each end of the petiole. The abscission layer did not constitute a barrier to the movement of auxin (Kaldewey and Jacobs 1974). Similarly, in *Phaseolus* petioles, the abscission layer did not block transport. However, marked differences existed in the velocity and flux of auxin through the petiole and through the pulvinus. In both old and young plants the velocity and flux were both considerably less in the pulvinus than in the petiole. The velocity of transport and the flux through petiole segments showed little change with age, but the flux through pulvinar tissue in old plants was less than half of that found in young plants (Jacobs et al. 1966). Cane and Wilkins (1970) measured the flux and velocity of IAA movement through 6-mm *Zea* root segments cut 1, 7 and 13 mm from the apex. At 10 °C, the velocity of transport was essentially the same in all the segments, but the acropetal flux was 50 cpm/hr in the apical section, 20 cpm/hr in the middle section and 10 cpm/hr in the basal section. In a few cases, older tissues have been shown to transport more auxin than younger. For example, in fruit stems of *Fritillaria,* although the velocity of transport declined towards the base, the capacity increased, so that the flux through older tissues was greater (Kaldewey 1966). In *Pinus* seedlings, the flux appeared to be greatest near the point of insertion of the cotyledons (Zaerr 1968).

The ability to transport auxin in a polar manner may appear very early during germination; Greenwood and Goldsmith (1970) found that hypocotyl segments excised from *Pinus* embryos were capable of polar transport after 2 hr of hydration. However, Jacobs (1950) found little or no basipetal transport in *Phaseolus* hypocotyls until after the third day of growth.

Ageing of the segments after removal from the plant may also affect the ability of the tissue to transport auxin. De La Fuente and Leopold (1970) tested the ability of sunflower stem sections to transport auxin after ageing the sections for periods up to 25 hr after cutting. In sections taken from both old and young tissues there was a 50% decline in flux into receiver blocks after 10 hr. In a second experiment, a pulse of auxin was applied to the sections, and transport into receivers was measured after periods up to 6 hr. In this case, flux into the receivers declined very rapidly, reaching 50% of the original value after only 1 hr. De La Fuente and Leopold suggest that the decline of transport capacity with ageing is due either to a decline in the numbers of transport sites, or to an increase in the capacity of the tissue to immobilise and metabolise auxin; the latter factor was probably responsible for the rapid decline found in the second experiment. The ability of stem segments to transport synthetic auxins such as NAA (Veen 1969) and 2,4,5-T (Osborne et al. 1968) also declines with ageing, and *Zea* root segments lose their capacity to transport IAA, either acropetally or basipetally, within 6 hr after excision (Wilkins et al. 1972).

2.7.9. Effects of hormones on auxin transport

As discussed above, the ability of stem explants to transport auxin declines with ageing. This decline can be prevented in some tissues by treatment of the tissue with unlabelled auxin during the ageing process. For example, treatment with unlabelled NAA prevented the loss of the ability of *Coleus* explants to transport radioactive NAA after ageing (Veen 1969).

Hertel and Flory (1968) found that the addition of 2 μM IAA to donor blocks containing radioactive 2,4-D greatly enhanced the movement of the 2,4-D in *Zea* coleoptiles. Treatment with unlabelled IAA can also stimulate the movement of radioactive IAA. A 30-min IAA pretreatment was found to increase the rate of movement of a pulse of labelled IAA through *Zea* coleoptiles, and increased flux could be shown with a pretreatment as short as 20 min. IAA concentrations as low as 3×10^{-8} *M* were stimulatory, but when high concentrations (10^{-4} *M*) of IAA were used, subsequent IAA transport was inhibited. Rayle et al. (1969) found that transport of radioactive α-NAA was increased by treatment with IAA or unlabelled α-NAA; the inactive analogue β-NAA did not promote transport and was in fact inhibitory.

Gibberellin may also promote the transport of auxin in both segments and whole plants. Pilet (1965c) found that a 2-hr treatment with GA_3 increased the velocity of IAA transport in *Lens* stem segments from 5.69 mm/hr to 9.83 mm/hr. Pretreatment of the apices of sunflower plants with GA_3 for 24 hr greatly enhances the flux of [^{14}C]IAA through stem segments cut from the plants. More than twice as much IAA passed through segments from treated plants after 24 hr, but the effect fell to zero about 150 hr after treatment. GA_3 also reduced or prevented the inhibition of IAA transport in sunflower plants under the influence of a transverse gravitational stimulus (Palmer and Halsall 1969). Jacobs and Case (1965) treated the apices of decapitated pea plants with lanolin paste containing either IAA alone or IAA + GA_3. In presence of gibberellin, more IAA was found to be translocated to the lower parts of the plant. Somewhat different results were found by Basler (1974) who injected [^{14}C]2,4,5-T into the cotyledonary nodes of intact *Phaseolus* plants with or without GA_3. GA_3 treatment resulted in increased accumulation of auxin in the shoots and decreased basipetal movement into the roots. It is likely that the differences between these two experiments can be explained by the different methods of application of the auxin; the apically applied auxin in the experiment of Jacobs and Case was probably moving basipetally in the endogenous transport system, while in Basler's work movement was probably both basipetal and acropetal in the phloem stream.

Pretreatment of *Lens* stem segments with abscisic acid (ABA) caused a 22% decrease in the uptake of IAA from donor blocks, and the velocity of transport was reduced from 5.71 mm/hr to 3.36 mm/hr. In the ABA-treated segments the percentage of the total radioactivity present in the basal part of the segment was greater than that of the control, suggesting a possible "damming" effect or inhibition of transfer into the receiver (Pilet 1971). In these experiments, when the ABA was applied together with IAA in the donor blocks instead of as a pretreatment, uptake of IAA was also inhibited. However, Philipson et al. (1973) did not find any affect of ABA on the uptake of IAA from solution by *Avena* coleoptile segments even after 100 min of pretreatment, and Osborne and Mullins (1969) did not find any effect of ABA on the transport of [^{14}C] IAA in bean petioles. Injection of ABA together with [^{14}C] 2,4,5-T into intact bean plants appeared to enhance basipetal transport of the auxin, which tended to accumulate in the roots and nutrient solution (Basler 1974). Much of this effect may have been due to the closure of stomata by the ABA, since high humidity greatly reduced the inhibitory effect of ABA on transport into the primary leaves.

The role of ethylene in auxin transport has recently been summarised by Abeles (1973). Ethylene inhibits the transport of auxin, but only when applied to the tissues as a pretreatment. Beyer and Morgan (1969) studied the effect of ethylene pretreatment on the movement of a pulse of auxin down cotton stem segments. As the pretreatment time increased from 1.5 to 3 hr, a progressively larger fraction of the auxin was removed from the pulse, and after 3 hr of exposure to ethylene, 26% of the auxin in the pulse was no longer transported. The effect of ethylene on the velocity of movement of the pulse was very slight. Ethylene can also prevent the lateral movement of auxin in horizontal pea stem segments (Burg and Burg 1966), although lateral transport in corn coleoptiles was not affected (Burg and Burg 1967). The inhibitory effect of ethylene on auxin transport does not seem to be associated with decreased IAA uptake or with increased decarboxylation (Burg and Burg 1967; Ernest and Valdovinos 1971). Ethylene pretreatment may increase the formation of indoleacetylaspartic acid; however, since the synthetic auxin NAA is metabolised more slowly in ethylene-treated tissues, it seems unlikely that inhibition of auxin transport can be accounted for by an increase in auxin conjugation. Rather, it appears that the increase in indoleacetylaspartic acid is the result of a local increase in auxin concentration caused by the lowered transport capacity (Beyer and Morgan 1970). Osborne and Mullins (1969) found that pretreatment of *Phaseolus* petiole segments with ethylene for 6 hr caused a significant reduction in the flux of IAA into receivers in a subsequent transport experiment. However, if the sections were supplied with non-radioactive IAA during the ethylene pretreatment, no inhibition of transport was found. Treatment with auxin or kinetin also protected the segments against the loss of transport capacity which occurs during ageing and maintained the ability of the segments to incorporate [^{14}C] leucine into protein. Osborne and Mullins propose that transport of auxin requires binding of the auxin to a specific protein located at the plasmalemma. This protein would possess an auxin receptor site and an ethylene receptor site, and the filling of one of these sites would prevent binding of the other compound to the second site. Thus, when the auxin concentration in the cell is high the auxin receptor sites are filled, but when auxin is depleted by ageing the tissue, ethylene can bind to its acceptor site and prevent the binding and subsequent transport of auxin.

As mentioned above (Osborne and Mullins 1969), in bean petioles kinetin may have an enhancing effect on auxin transport, especially after a period of ageing, possibly due to its ability to maintain protein synthesis and thus the supply of auxin transport proteins. In *Lens* stem segments, however, pretreatment for 2 hr with 100 μg/ml of kinetin reduced uptake from the donors to 37% of the controls. The velocity of transport was also reduced from 6.48 to 2.99 mm/hr and the flux fell from 8.1 to 5.2% of uptake (Pilet 1965a).

Thus all the natural hormones may affect the transport of auxin. Generally the effect is greater if the hormone is applied as a pretreatment, although auxins and ABA may have some effect if applied together with the auxin whose transport is being measured. Within a certain concentration range, auxins and gibberellin tend to stimulate auxin transport. Cytokinins are stimulatory in some tissues but inhibit in others, and ethylene and ABA are generally inhibitory.

2.7.10. Effects of synthetic inhibitors on auxin transport

The transport of auxins may be inhibited, more or less specifically, by several synthetic compounds, notably 2,3,5-triiodobenzoic acid (TIBA), *N*-(naphth-l-yl)phthalamic acid (NPA), the morphactins (derviatives of fluorene-9-carboxylic acid) and 3,3a-dihydro-2-(*p*-methoxyphenyl)-8*H*-pyrazolo[5,1a]isoindol-8-one (DPX 1840).

TIBA can cause complete inhibition of basipetal polar auxin transport. For example, when stem segments were cut from sunflower shoots which had been pretreated with 10^{-4} *M* TIBA, there was 100% inhibition of the flux of IAA into basal receivers in a subsequent transport period (Niedergang-Kamien and Leopold 1957). TIBA can also inhibit auxin transport in intact plants; TIBA applied to the stem of intact pea plants inhibited the movement of [^{14}C]IAA when the auxin was applied to the apical bud (Morris et al. 1973). TIBA can also inhibit geotropic and phototropic responses, suggesting that it may inhibit lateral as well as longitudinal transport (Van der Beek 1959; Keitt 1960). TIBA moves polarly through *Zea* coleoptile segments, possibly following the same route as auxin. About 20 times more TIBA moves basipetally than acropetally, and the velocity of transport (2.5 mm/hr) is about 20% that of IAA (Thomson et al. 1973).

The effect of TIBA on the exit of auxin from segments is much more marked than its effect on uptake. Christie and Leopold (1965) found that exit was inhibited by 50% in corn coleoptiles treated with 3 μ*M* TIBA, while uptake was little affected. This suggested that the limiting step in transport is part of the process governing exit from the cell. Inhibition of transport by TIBA might also be caused by increased auxin immobilisation, resulting in a decrease in auxin in the transportable pool. Immobilisation of IAA in TIBA-treated *Avena* coleoptiles has in fact been demonstrated by Winter (1968). Hertel and Flory (1968) point out that TIBA could either have a direct stimulatory effect on immobilisation, or an indirect effect by which TIBA-induced inhibition of export caused an increased auxin concentration in the tissue and this increase in substrate concentration stimulated immobilisation. They predicted that if the effect of TIBA on immobilisation were direct, then at a given time less auxin would be available for transport than in controls. If the TIBA were then removed, the treated tissue should show a lower auxin exit rate than the controls. However, if the effect of TIBA on immobilisation was indirect, via an inhibition of export, auxin should accumulate in the tissue and when the TIBA was removed more auxin should exit from the treated tissue than from the control. Experiments with *Zea* coleoptile sections showed that more auxin was exported from TIBA-treated tissues than from controls after removal of the inhibitor, supporting the hypothesis that TIBA acts by inhibiting exit. TIBA can cause a 50% decrease in the flux of [^{3}H]IAA into basal receivers within 1 min, and this rapid response suggests that the action of the inhibitor is a direct one (Thomson et al. 1973). TIBA may act at the auxin-binding site on the plasma membrane, since it can inhibit the binding of [^{14}C]NAA to plasma membrane vesicles (Hertel et al. 1972). Inhibition of transport by TIBA does not seem to be caused by an effect on protein synthesis or respiration, since protein synthesis as measured by the incorporation of [^{14}C]leucine is not inhibited by TIBA (Osborne and Mullins 1969) and a concentration of TIBA (10^{-4} *M*) which completely abolishes transport in sunflower stem segments was found to be without effect on respiration (Niedergang-Kamien and Leopold 1957).

NPA was shown to be an inhibitor of basipetal auxin transport in *Avena* coleoptiles by Morgan and Söding in 1958. Evidence that NPA also blocks IAA transport in intact plants was obtained by Morgan (1964) who showed that when the apex of decapitated bean seedlings was replaced by a [^{14}C]IAA solution, the presence of a ring of lanolin paste containing NPA below the apex reduced transport of radioactivity to the lower parts of the plant. If NPA is injected into *Avena* coleoptiles immediately before photo- or geo-stimulation, the geotropic and first and second positive phototropic curvatures are inhibited, suggesting that NPA can inhibit lateral auxin transport (Steyer 1969). Unlike TIBA, NPA does not move polarly in corn coleoptiles (Thomson et al. 1973).

A very rapid effect of NPA has been observed in *Zea* coleoptiles, where the efflux of [^{3}H]IAA into basal receivers was reduced to 60% of that of controls within 1 min of NPA application (Thomson et al. 1973). As was the case with TIBA, such rapid effects suggest an interaction with the plasma membrane, and recent experiments have provided some direct evidence for this. Fractions containing plasma membrane vesicles were prepared from *Zea* coleoptiles by sucrose density gradient centrifugation, and the percentage of plasma membrane in each fraction was estimated by phosphotungstic acid – chromic acid staining. The binding of [^{3}H]NPA to each fraction was then tested, and a highly significant correlation was found between binding capacity and plasma membrane content. The K_m of binding was about 10^{-7} *M*, and the interaction appeared to be non-covalent. Unlabelled IAA did not compete with [^{3}H]NPA binding, indicating that NPA and IAA bind at different sites (Lembi et al. 1971).

A detailed review of the physiology of the morphactins has been published by Schneider (1970). These compounds act on a very wide variety of processes within the plant; germination, seedling growth, stem elongation and cell division in the meristem are inhibited and the polarity of cell division is disturbed. Apical dominance is weakened and the emergence of lateral buds is favoured. One of the actions of morphactins which seems to underlie these varied phenomena is their inhibitory effect on the transport of auxins. Krelle and Libbert (1968) found that the morphactin CFM (methyl-2-chloro-2-hydroxy-fluorene-9-carboxylate) inhibited both IAA-induced and gravity-induced curvatures in wheat coleoptiles, and that a 30 min pretreatment with CFM inhibited basipetal IAA transport through sunflower hypocotyl segments. Krelle and Libbert also found that CFM can inhibit auxin transport through whole plants. Pea seedlings were decapitated, and the outgrowth of lateral buds which would normally follow decapitation was prevented by the application of IAA in lanolin paste to the apex. This action of IAA could be inhibited by applying CFM between the site of IAA application and the lateral bud. The action of CFM on the basipetal transport of $[^{14}C]$IAA through *Zea* coleoptile segments is rapid; a 5 min pretreatment with 10^{-6} *M* morphactin reduced auxin transport by 27%. Uptake and polarity were not affected but velocity and flux were reduced (Naqvi 1972b). Besides their inhibitory action on basipetal auxin transport, morphactins also suppress photo- and geo-tropisms which are dependent on lateral transport (Jones et al. 1954; Khan 1967; Krelle and Libbert 1968). Direct evidence that CFM inhibits downwards lateral transport of auxin in horizontal *Zea* coleoptile segments has been obtained by Bridges and Wilkins (1973). Concentrations of CFM greater than 10^{-9} *M* inhibited both the geotropic curvature of the coleoptiles and the downwards lateral movement of $[^{3}H]$IAA supplied to the upper half of the segment. The pattern of IAA metabolism in the coleoptiles was unchanged by the CFM. Like NPA, morphactins show apolar movement, and the direction of movement may depend on the stage of development of the plant (Schneider 1969).

The inhibition of auxin transport by DPX 1840 has been discussed in a comprehensive paper by Beyer (1972). DPX 1840 at a concentration of 0.5 m*M* inhibited the flux of $[^{14}C]$NAA into basal receivers in stem sections of *Vigna, Pisum, Phaseolus, Glycine, Helianthus* and *Zea* by 89–98%, while uptake and recovery were inhibited by only 2–9%. The inhibitory effect was fairly rapid and the transport of auxin was reduced by 80% in *Gossypium* stem sections after 0.5 hr of treatment. DPX 1840 was also effective when applied to intact plants; when $[^{14}C]$auxin was applied to the apex of intact bean seedlings, very little radioactivity moved past a ring of lanolin containing DPX 1840 which had been applied to the stem. Besides its inhibitory effect on basipetal transport, DPX 1840 also inhibited the lateral movement of auxin in geotropically stimulated *Helianthus* hypocotyls. $[^{14}C]$DPX 1840 moved both basipetally and acropetally in excised stem segments at a velocity of 1–2 mm/hr.

Thus, the actions of the four synthetic inhibitors are similar in many ways. All inhibit basipetal transport in both stem segments and whole plants, and all also inhibit the lateral transport of auxin during tropistic responses. All act more or less rapidly, suggesting a direct action at the membrane, and specific binding at the plasma membrane has been demonstrated for NPA. TIBA shows some qualitative differences from the other compounds in that it is an auxin antagonist, it moves with basipetal polarity and it has a very marked effect on auxin exit from the cell. Quantitatively, the order of activities of the four inhibitors on basipetal transport is NPA $>$ morphactin $\gtrsim$ DPX 1840 $>$ TIBA (Beyer 1972).

movement in *Zea* roots could be supported by anaerobic metabolism, but when sodium fluoride was added to the anaerobic segments, acropetal movement was inhibited (Wilkins and Scott 1968b). The velocity of acropetal movement through *Zea* roots increased with temperature over the physiological range and declined at 40–50 °C (Wilkins and Cane 1970).

As we have discussed above (section 2.7.7), gravity dependent lateral transfer of auxin has many characteristics in common with basipetal longitudinal transport, including temperature relations, saturation at higher donor concentrations and inhibition by synthetic inhibitors. Phototropic responses can also be inhibited by TIBA and morphactins. Thus it seems likely that the mechanisms underlying basipetal and lateral auxin movements will prove to be similar or identical.

Evidence from the rapid action of synthetic auxin transport inhibitors (section 2.7.10) suggests that the active step of auxin transport is situated at the plasma membrane, and binding of auxin to particulate fractions probably containing plasma membrane vesicles has in fact been demonstrated (Hertel et al. 1972). Goldsmith and Ray (1973) have provided good evidence that the active transport process is located at the basal plasmalemma in *Zea* coleoptiles. In these experiments, the coleoptiles were treated with the fungal metabolite cytochalasin B, which has been shown to completely inhibit cytoplasmic streaming, but which has little effect on auxin transport (Cande et al. 1973). The cytoplasm in the treated coleoptiles was then displaced either to the apex or the base of the cells by a brief centrifugation, and the transport of [^{3}H]IAA into basal receivers was measured at various time intervals after centrifugation. The presence of the cytochalsin B prevented the redistribution of the cytoplasm during the transport period. Two hr after centrifugation it was found that the ^{3}H present in the receivers in the apically centrifuged coleoptiles was only 10% of the uncentrifuged controls, while in those coleoptiles where the cytoplasm had been centrifuged towards the base, the ^{3}H present in the receivers was 160% of the controls. Thus, removal of cytoplasm from the base of the cell was strongly inhibitory to basipetal transport. This inhibition was not due to the absence of amyloplasts, since a short reverse centrifugation restored the amyloplasts of apically-centrifuged coleoptiles to the base, but relieved the inhibition only slightly. Goldsmith and Ray therefore suggested that the inhibitory effect of centrifugation was caused by removal of the mitochondria which supply energy for transport. Indeed, it was found that the inhibitory effect of displacement of the cytoplasm to the apex could be partially overcome by supplying ATP or phosphoenolpyruvate to the coleoptiles. Acropetal transport was affected very little by centrifugation, and basipetal polarity was retained even after apical centrifugation.

Goldsmith and Ray then considered these results in relation to five possible models of auxin transport which are outlined in Table 2.5. Since only model 5 accorded with the experimental data, it was concluded that "transport of auxin is due to polar metabolic secretion of auxin localised at the basal plasma membrane of each cell." This conclusion is also consistent with the relatively greater effect of TIBA on basal secretion than uptake (Christie and Leopold 1965). Intercellular movement cannot be due to cytoplasmic streaming, since transport can occur when streaming is stopped by cytochalasin B (Cande et al. 1973). Movement can occur through the vacuole or along the cell membranes since

Table 2.5.

Effect of position of cytoplasm in the cell after centrifugation on basipetal auxin transport (after Goldsmith and Ray 1973)

Model		Effect of position of cytoplasm on basipetal transport	
		Cytoplasm apical	Cytoplasm basal
1.	Passive movement through apical and basal membranes, active intracellular transport.	–	–
2a.	Active uptake and secretion; no passive permeability of end membranes.	–	–
2b.	Active uptake and secretion; some passive permeability of end membranes.	+	+
3.	Active secretion towards the base slightly greater than active secretion towards the apex (model of Leopold and Hall 1966).	– (polarity reversed by apical centrifugation)	+
4.	Active uptake, passive movement through basal membrane.	+	–
5.	Passive uptake, active secretion.	– (polarity not reversed by apical centrifugation)	+

basipetal transport continues when the bulk of the cytoplasm is displaced to the basal end of the cell. It has been calculated that the concentration difference required to transport IAA within the cell is less than half of that present in the transport stream (Cande et al. 1973). Thus, basipetal auxin transport probably involves uptake through the apical plasmalemma, diffusion downwards through the cell, and active secretion through the basal plasmalemma at specific transfer sites. Interaction of auxin with the transport site probably involves non-covalent binding, since when NAA was labelled with ^{18}O in the carboxyl group, no ^{18}O was lost during transport (Hertel and Flory 1968). This finding eliminates several types of covalent binding such as peptide bonding or the formation of *S*-acyl bonds to compounds such as CoA.

Auxin transport is also discussed in relation to tropisms in Chapter 5 and to root and stem growth in Chapter 2, from Volume II.

2.8. Effects of environmental and hormonal stimuli on auxin levels

2.8.1. Effects of environmental factors on auxin levels

Earlier literature on the effects of environmental factors on auxin levels has been summarised by Hillman and Galston (1961). As discussed above (sections 2.7.6 and 2.7.7), both light and gravity have profound effects on the basipetal and lateral transport of IAA, and therefore on the distribution of the hormone within the plant. Other environmental factors which may affect auxin levels are temperature and water stress. Exposure to low temperatures for 40 days increased the activity of IAA-oxidase in winter wheat seedlings 10-fold (Bolduc et al. 1970). The authors concluded that this increased IAA-oxidase level should result in low auxin levels during vernalisation, but no measurements of IAA content were made. Conflicting results have been obtained regarding the effect of water stress on IAA levels. Darbyshire (1971) found that the IAA-oxidase activities of pea and tomato plants increased with increasing plant water deficit. However, the IAA-oxidase levels in wheat leaves were found to decrease under drought conditions in both normal and drought-resistant cultivars (Mills and Todd 1973). The effect of water stress on auxin levels is discussed further in Chapter 11, Volume II.

In Volume II, Chapter 2, the effects of light (sections 2.3.2.3.2 and 2.3.7.2), photoperiod (2.3.6, 2.4.2.3 and 2.6.1) and mineral nutrition (2.3.2.3.5) on auxin levels are discussed in relation to regulation of growth. The effects of photoperiod and temperature on auxin levels in relation to induction of flower formation are reviewed in Chapter 6, section 6.2.2, also in Volume II.

2.8.2. Effects of hormones on auxin levels

In contrast to the relatively small amount of work which has been done on the effect of environmental factors on auxin levels, the effect of hormones on auxin concentration has received considerable attention. It should be noted, however, that the actual amounts of auxin found after hormone treatment have seldom been measured, and inferences on auxin level are often drawn from the effects of hormones on such factors as IAA-oxidase activity.

Hormones may affect auxins in the plant by their action on transport (see section 2.7.2), synthesis, conjugation or oxidation of the auxin. The application of IAA itself to the plant may affect auxin levels in each of these ways. In pith explants of hybrid tobacco, IAA may induce the formation of the enzymes required for its own synthesis (Cheng 1972), and auxin may also affect enzymes responsible for its conjugation and oxidation. Pretreatment of pea tissue with any active auxin will induce the formation of the system synthesizing indolylacetylaspartic acid (Venis 1972), and treatment with IAA can also induce or repress specific isoenzymes of peroxidase or IAA-oxidase (see references in Schneider and Wightman 1974a). For example, in tobacco callus, low, growth-promoting concentrations of IAA increased the synthesis of two fast-moving IAA-oxidase isoenzymes, while high, inhibitory concentrations of IAA caused an increase in isoenzymes with low and moderate mobility. The promotive effect of IAA on the fast-moving isoenzymes was also dependent on the kinetin concentration (Lee 1971a). The complexity of these effects, and the lack of actual measurements of IAA concentration following the various treatments, makes specific suggestions for the mechanism of action of IAA in regulating its own biosynthesis difficult. However, it does appear possible that IAA exerts feedback control via an effect on synthesis, conjugation or oxidation.

In many cases, the application of GA_3 to plants results in an increased auxin content. For example, Kuraishi and Muir (1964) found that treatment of dwarf pea seedlings with GA_3 resulted in a several-fold increase in diffusible auxin in the apices 24 hr after treatment. Paleg (1965) cites at least 16 species in which the auxin level was found to increase after gibberellin treatment. It is not yet clear whether GA_3 acts to promote auxin synthesis, or whether it tends to decrease the conjugation and oxidation of the hormone. The effect of GA_3 on auxin oxidation varies widely with the species. Gibberellins have been found to promote IAA-oxidases in dwarf pea buds (Ockerse and Waber 1970), wheat seedlings (Bolduc et al. 1970) and tobacco callus (Lee 1971b) and to inhibit IAA-oxidase in cucumber seedlings (Katsumi and Sano 1968). In sunflower (Valdovinos and Ernest 1967) and *Coleus* (Valdovinos et al. 1967), GA_3 had no effect on IAA decarboxylation. These diverse effects make it unlikely that a general explanation for the effect of GA_3 on auxin levels will be found at the level of auxin oxidation. There is some evidence that gibberellin may enhance auxin synthesis. The growth of *Avena* coleoptile tips is promoted by GA_3 in presence of tryptophan (Sastry and Muir 1965). However, Ockerse and Galston (1967) failed to find any GA-tryptophan synergism in cuttings and sections of dwarf pea. In sunflower, dwarf pea and *Coleus,* pretreatment with GA_3 enhanced the decarboxylation of $[1-^{14}C]$TPP by enzyme breis; this might indicate an increase in the rate of conversion of tryptophan to tryptamine, but as no tryptamine was isolated the significance of the decarboxylation remains uncertain (Valdovinos and Ernest 1966). The only direct effect of GA_3 on an IAA-forming enzyme system is that reported by Lantican and Muir (1967). An enzyme system from *Avena* coleoptiles which converted tryptophan to IAA was found to lose activity on dialysis, and activity could be restored by adding GA_3 to the reaction mixture. GA_3 was not found to have any direct effect on purified tryptophan aminotransferase or decarboxylase from tomato shoots (Wightman and Gibson, unpublished). Thus, although an effect of GA_3 on auxin synthesis remains possible, the site and mechanism of such an effect remains unknown.

Cytokinins, like gibberellins, have a promotive effect on auxin levels, and a supply of cytokinin may be important in maintaining the endogenous auxin level. The growth rate of the coleoptiles of derooted *Avena* seedlings can be restored to normal by cytokinin treatment and application of cytokinin increased both the rate and duration of auxin secretion from excised coleoptile tips (Jordan and Skoog 1971). The authors noted that since this latter effect could be observed when tips were excised 5 min after cytokinin treatment, it was unlikely that a cytokinin-directed increase in supply of auxin precursors from other parts of the seedling was involved, and that the cytokinin probably increased auxin synthesis in the tip. Rajagopal and Larsen (1974) found that application of a cytokinin to coleoptiles of derooted seedlings did not affect the activity of the enzyme which converts indolylacetaldehyde to IAA, suggesting that cytokinins may stimulate an earlier step in IAA biosynthesis. Treatment of excised *Coleus* shoots with kinetin results in a considerable increase in methanol-extractable auxin (Hemberg 1972) and tobacco callus cultures which normally require auxin may grow in an auxin-free medium in the presence of high kinetin concentrations (Syono and Furuya 1972). IAA was identified in these calluses after three years of subculture in a high-kinetin medium without added IAA. Thus it seems that the high kinetin level allowed the calluses to synthesise sufficient IAA to maintain growth in the absence of an exogenous supply. Cytokinins may also affect auxin conjugation and decarboxylation. Kinetin treatment reduced the rate of conjugation of exogenous IAA by mungbean hypocotyl segments, thus maintaining a higher level of the free acid in the tissue (Lau and Yang 1973). Specific isoenzymes of IAA-oxidase and peroxidase may also be affected by cytokinins (see references in Schneider and Wightman 1974a, Table I; note that references 56 et seq. refer to cytokinins and not gibberellins). Perhaps the most comprehensive study of this aspect of the interaction between cytokinins and auxins is that of Lee (1971c, 1972, 1974). In tobacco callus cultures, low concentrations (0.2 μM) of kinetin induced the formation of fast-moving anionic isoenzymes of IAA-oxidase and peroxidase. This increase was prevented by actinomycin D and cycloheximide, suggesting a requirement for RNA and protein synthesis. Higher concentrations of kinetin (2 μM) repressed the formation of these isoenzymes. The fast-moving isoenzymes promoted by kinetin were found predominantly in the cytoplasm, but also in the ribosome and plasma-membrane fractions. The plasma-membrane fraction was particularly interesting since the IAA-oxidase/peroxidase ratio was several times higher than in other parts of the cell, and was increased more than five times at high kinetin concentrations. The effect of cytokinins on IAA-oxidising enzymes is evidently complex, and will have to be correlated with actual measurements of IAA before their significance can be assessed.

Application of ethylene reduces auxin levels in oats (Botjes 1942), bean (Von Guttenberg and Steinmetz 1947), pea (Burg et al. 1971), *Coleus* (Ernest and Valdovinos 1971) and cucumber (Rudich et al. 1972). In seeds of *Prunus salicina,* ethylene treatment led to an increase in auxin content, which was attributed to inhibition of the transport of auxin out of the seeds (Domoto and Hewitt 1974). In *Coleus,* pretreatment with ethylene reduced the auxin content of the apical part of the stem from 8.3 to 4.3 μg/kg fresh wt. Enzyme breis prepared from the pretreated plants showed a 36% reduction in their capacity to form auxin from tryptophan and the amount of indolylacetyl-aspartic acid formed from exogenous IAA was twice as great in the treated tissues (Ernest and Valdovinos 1971). In some tissues, ethylene may cause an increase in peroxidase (Imaseki 1970; Ridge and Osborne 1970; Birecka and Miller 1974). In most cases the IAA-oxidase activity of these peroxidases was not investigated, but ethylene has been shown to increase IAA-oxidase activity in cotton (Hall and Morgan 1964). Ethylene pretreatment of *Coleus* plants did not affect the decarboxylation of $[1-^{14}C]$IAA by enzyme preparations from the shoots (Ernest and Valdovinos 1971).

Thus, it seems that auxins, gibberellins, cytokinins and ethylene can each modify the transport, synthesis and metabolism of IAA. Combinations of hormones may show different effects from those obtained when a single hormone is applied, and IAA may itself alter the level of other hormones, such as ethylene. These complex inter-relationships should provide a fruitful field for future research.

References

Abe, H., Uchiyama, M. and Sato, R. (1974) Agr. Biol. Chem., 38, 897–898.
Abeles, F.B. (1973) Ethylene in Plant Biology, Academic Press, New York.
Åberg, B. (1950) Physiol. Plant., 3, 447–461.
Åberg, B. (1953) Kungl. Lantbrukshögsk. Ann., 20, 241–295.
Åberg, B. (1958) Kungl. Lantbrukshögsk. Ann., 24, 375–395.
Åberg, B. (1970) Kungl. Lantbrukshögsk. Ann., 36, 235–253.
Abramovitch, R.A. and Ahmed, K.S. (1961) Nature (Lond.), 192, 259–260.
Abrol, B.K. and Audus, L.J. (1973) J. exp. Bot. 24, 1209–1223.
Albores, M. and Wain, R.L. (1974) Meded. Fakul. Landbouw-Weten. (Ghent) 39, 617–624.
Alden, T. (1971), Physiol. Plant., 25, 54–57.
Andreae, W.A. and Good, N.E. (1955) Plant Physiol., 30, 380–382.
Anthony, A. and Street, H.E. (1970) New Phytol., 69, 47–50.
Atsumi, S., Kuraishi, S. and Hayashi, T. (1976) Planta, 129, 245–247.
Audus, L.J. (1972) Plant Growth Substances, Vol. 1, Leonard Hill, London, pp. 66–103.
Audus, L.J. and Thresh, R. (1956) Ann. Bot., 20, 439–459.
Ballin, G. (1962) Planta, 58, 261–282.
Ballin, G. (1964) J. Chromatogr., 16, 152–156.
Bandurski, R.S. and Schulze, A. (1974) Plant Physiol., 54, 257–262.
Basler, E. (1974) Plant Cell Physiol., 15, 351–361.
Basu, P.S. and Tuli, V. (1972) Plant Physiol., 50, 503–506.
Batt, S. and Venis, M.A. (1976) Planta, 130, 15–21.
Batt, S., Wilkins, M.B. and Venis, M.A. (1976) Planta, 130, 7–14.
Bayer, M.H. (1969) Plant Physiol., 44, 267–271.
Bayer, M.H. (1972) J. exp. Bot., 23, 801–812.
Bayer, M.H. and Ahuja, M.R. (1968) Planta, 79, 292–298.
BeMiller, J.N. and Colilla, W. (1972) Phytochemistry, 11, 3393–3402.
Bentley, J.A. (1950) Nature (Lond.), 165, 449.
Bentley, J.A. and Housley, S. (1952) J. exp. Bot., 3, 393–405.
Beyer, E.M. (1972) Plant Physiol., 50, 322–327.
Beyer, E.M. and Morgan, P.W. (1969) Plant Physiol., 44, 1690–1694.
Beyer, E.M. and Morgan, P.W. (1970) Plant Physiol., 46, 157–162.
Birecka, H. and Miller, A. (1974) Plant Physiol., 53, 569–574.
Black, R.C. and Hamilton, R.H. (1971) Plant Physiol., 48, 603–606.
Bolduc, R.J., Cherry, J.H. and Blair, B.O. (1970) Plant Physiol., 45, 461–464.
Bonnemain, J.-L. (1970) C.R. Acad. Sci. (Paris), 270, 1326–1329.
Bonnemain, J.-L. (1971) C.R. Acad. Sci. (Paris), 273, 1699–1702.
Botjes, J.O. (1942) Proc. kon. ned. Akad. Wet., 45, 999.
Bourbouloux, A. and Bonnemain, J.-L. (1973) Planta, 115, 161–172.
Bourbouloux, A. and Bonnemain, J.-L. (1974) Planta, 119, 169–182.
Bourdoux, P., Vandervorst, D. and Hootele, C. (1971) Phytochemistry, 10, 1934–1935.
Bowen, M.R., Wilkins, M.B. and McCorquodale, I. (1972) Planta, 105, 273–292.
Boysen-Jensen, P. and Nielsen, N. (1926) Planta, 1, 321–331.
Bridges, I.G. and Wilkins, M.B. (1973) J. exp. Bot., 24, 711–723.
Bridges, I.G., Hillman, J.R. and Wilkins, M.B. (1973) Planta, 115, 189–192.
Briggs, W.R. (1963) Plant Physiol., 38, 237–247.
Burg, S.P. and Burg, E.A. (1966) Proc. nat. Acad. Sci. (Wash.), 55, 262–269.
Burg, S.P. and Burg, E.A. (1967) Plant Physiol., 42, 1224–1228.
Burg, S.P., Apelbaum, A., Eisinger, W. and Kang, B.G. (1971) HortScience, 6, 359.
Burnett, D. and Audus, L.J. (1964) Phytochemistry, 3, 395–415.
Burstrom, H. (1950) Physiol. Plant., 3, 277–292.

Cande, W.Z., Goldsmith, M.H.M. and Ray, P.M. (1973) Planta, 111, 279–296.
Cane, A.R. and Wilkins, M.B. (1969) Plant Physiol., 44, 1481–1487.
Cane, A.R. and Wilkins, M.B. (1970) J. exp. Bot., 21, 212–218.
Chamberlain, V.K. and Wain, R.L. (1971) Ann. appl. Biol., 69, 65–72.
Cheng, T.Y. (1972) Plant Physiol., 50, 723–727.
Cholodny, N. (1927) Biol. Zentralbl., 47, 604–626.
Chowdhury, B.K. and Chakraborty, D.P. (1971) Phytochemistry, 10, 481–483.
Christiansen, G.S. and Thimann, K.V. (1950) Arch. Biochem., 26, 230–247.
Christie, A.E. and Leopold, A.C. (1965) Plant Cell Physiol., 6, 337–345.
Clarke, G.C. and Wain, R.L. (1963) Ann. appl. Biol., 51, 453–458.
Collet, G.F. (1968) Canad. J. Bot., 46, 969–978.
Coombe, B.G. (1960) Plant Physiol., 35, 241–250.
Crafts, A.S. (1953) Ann. Rev. Plant Physiol., 4, 253–280.
Craker, L.E., Chadwick, A.V. and Leather, G.R. (1970) Plant Physiol., 46, 790–793.
Dannenburg, W.N. and Liverman, J.L. (1957) Plant Physiol., 32, 263–269.
Darbyshire, B. (1971) Physiol. Plant., 25, 80–84.
Darwin, C. and Darwin, F. (1880) "The Power of Movement in Plants", London.
Davies, P.J. (1972) Physiol. Plant., 27, 262–270.
Davies, P.J. and Mitchell, E.K. (1972) Planta, 105, 139–154.
Dedolph, R.R., Naqvi, S.M. and Gordon, S.A. (1966) Plant Physiol., 41, 897–902.
De La Fuente, R.K. and Leopold, A.C. (1966) Plant Physiol., 41, 1481–1484.
De La Fuente, R.K. and Leopold, A.C. (1968) in: Biochemistry and Physiology of Plant Growth Substances, F. Wightman and G. Setterfield (eds.), Runge Press, Ottawa, pp. 1129–1143.
De La Fuente, R.K. and Leopold, A.C. (1970) Plant Physiol., 45, 19–24.
DeYoe, D.R. and Zaerr, J.B. (1976) Plant Physiol., 58, 299–303.
Dolk, H.E. (1936) Rec. Trav. Bot. Neer., 33, 509–585.
Dolk, H.E. and Thimann, K.V. (1932) Proc. nat. Acad. Sci. (Wash.), 18, 30–46.
Domoto, P.A. and Hewitt, A.A. (1974) HortScience, 8, 503–504.
Dullaart, J. (1970) J. Exp. Bot., 21, 975–984.
Eliasson, L. (1972) Physiol. Plant., 27, 412–416.
Elkinaway, M. and Raa, J. (1974) Physiol. Plant., 30, 172–175.
Elliott, M.C. and Greenwood, M.S. (1974) Phytochemistry, 13, 239–241.
Endo, T. (1968) Plant Cell Physiol., 9, 333–341.
Erdmann, N. and Schiewer, U. (1971) Planta, 97, 135–141.
Erdtman, H. and Jönsson, A. (1954) Acta chem. scand., 8, 119–126.
Ernest, L.C. and Valdovinos, J.G. (1971) Plant Physiol., 48, 402–406.
Eschrich, W. (1968) Planta, 78, 144–157.
Fawcett, C.H. (1961) Ann. Rev. Plant Physiol., 12, 345–368.
Fawcett, C.H. (1964) Nature (Lond.), 204, 1200–1201.
Fawcett, C.H., Osborne, D.J., Wain, R.L. and Walker, R.D. (1953) Ann. appl. Biol., 40, 231–243.
Fawcett, C.H., Ingram, J.M.A. and Wain, R.L. (1954) Proc. roy. Soc. B, 142, 60–72.
Fawcett, C.H., Wain, R.L. and Wightman, F. (1955) Ann. appl. Biol., 43, 342–354.
Fawcett, C.H., Taylor, H.F., Wain, R.L. and Wightman, F. (1958) Proc. roy. Soc. B, 148, 543–570.
Fawcett, C.H., Taylor, H.F., Wain, R.L. and Wightman, F. (1958) Proc. roy. Soc. B, 148, 543–570.
Fawcett, C.H., Pascal, R.M., Pybus, M.B., Taylor, H.F., Wain, R.L. and Wightman, F. (1959) Proc. roy. Soc. B, 150, 95–119.
Fawcett, C.H., Wain, R.L. and Wightman, F. (1960) Proc. roy. Soc. B, 152, 231–254.
Fawcett, C.H., Wain, R.L. and Wightman, F. (1961) in: Plant Growth Regulation. R.M. Klein (ed.), Iowa State Univ. Press, Ames, pp. 71–91.
Ferenczy, L. and Matkovics, B. (1961) Acta Biol. (Budapest), 12, 107–118.
Filner, B., Hertel, R., Steele, C. and Fan, V. (1970) Planta, 94, 333–354.
Forest, J.C. and Wightman, F. (1972) Canad. J. Biochem., 50, 813–829.

Fredga, A. (1955) Arkiv. Kemi., 8, 463–468.
Fredga, A. and Åberg, B. (1965) Ann. Rev. Plant Physiol., 16, 53–72.
Fredga, A. and Matell, M. (1952) Arkiv. Kemi, 4, 325–330.
Frydman, R.B., Tomaro, M.L. and Frydman, B. (1971) FEBS Lett., 17, 273–276.
Frydman, R., Tomaro, M.L. and Frydman, B. (1972) Biochim. biophys. Acta (Amst.), 284, 63–79.
Fujita, T., Kawazu, K., Mitsui, T. and Kato, J. (1966) Agr. Biol. Chem., 30, 1277–1279.
Fukui, K., Nagata, N. and Yonezawa, T. (1958) J. Amer. Chem. Soc., 80, 2267.
Galston, A.W. (1962) In: Encyclopedia of Plant Physiology, Vol. XVII, W. Ruhland (ed.), Springer, Berlin, pp. 492–529.
Gandar, J.C. and Nitsch, C. (1964) Colloq. Int. Cent. nat. Rech. Sci., 123, 169–178.
Gandar, J.-C. and Nitsch, C. (1967) C.R. Acad. Sci. (Paris), 265, 1795–1798.
Gamborg, O. (1965) Canad. J. Biochem., 43, 723–730.
Garraway, J.L. and Wain, R.L. (1962) Ann. appl. Biol., 50, 11–20.
Gibson, R.A., Schneider, E.A. and Wightman, F. (1972a) J. exp. Bot., 23, 381–399.
Gibson, R.A., Barett, G. and Wightman, F. (1972b) J. exp. Bot., 23, 775–786.
Gillespie, B. and Thimann, K.V. (1963) Plant Physiol., 38, 214–225.
Gmelin, R. and Virtanen, A.I. (1961) Ann. Acad. Sci., Fennicae Ser. A II Chemica 1–23.
Golschmidt, E.E., Monselise, S.P. and Goren, R. (1971) Canad. J. Bot., 49, 241–245.
Goldsmith, M.H.M. (1966a) Plant Physiol., 41, 15–27.
Goldsmith, M.H.M. (1966b) Plant Physiol., 41, 749–754.
Goldsmith, M.H.M. (1967) Plant Physiol., 42, 258–263.
Goldsmith, M.H.M. (1968a) In: Biochemistry and Physiology of Plant Growth Substances, F. Wightman and G. Setterfield (eds.), Runge Press, Ottawa, pp. 1037–1050.
Goldsmith, M.H.M. (1968b) Ann. Rev. Plant Physiol., 19, 347–360.
Goldsmith, M.H.M. (1969) In: Physiology of Plant Growth and Development. M.B. Wilkins, (ed.), McGraw-Hill, London, pp. 127–162.
Goldsmith, M.H.M. and Ray, P.M. (1973) Planta, 111, 297–314.
Goldsmith, M.H.M. and Thimann, K.V. (1962) Plant Physiol., 37, 492–505.
Goldsmith, M.H.M. and Wilkins, M.B. (1964) Plant Physiol., 39, 151–162.
Goldsmith, M.H.M., Cataldo, D.A., Karn, J., Brenneman, T. and Trip, P. (1974) Planta, 116, 301–317.
Gordon, S.A. and Weber, R.P. (1951) Plant Physiol., 26, 192–195.
Goren, R. and Goldschmidt, E.E. (1970) Physiol. Plant., 23, 937–947.
Gorter, C.J. and Veen, H. (1966) Plant Physiol., 41, 83–86.
Grace, N.H. (1939) Canad. J. Res., C. 17, 247–255.
Grahm, L. and Herz, C.H. (1962) Physiol. Plant., 15, 96–114.
Greenwood, M.S. and Goldsmith, M.H.M. (1970) Planta, 95, 297–313.
Greenwood, M.S., Shaw, S., Hillman, J.R., Ritchie, A. and Wilkins, M.B. (1972) Planta, 108, 179–183.
Greenwood, M.S., Hillman, J.R., Shaw, S. and Wilkins, M.B. (1973) Planta, 109, 369–374.
Gregory, F.C. and Hancock, C.K. (1955) Ann. Bot. (Lond.), 19, 451–465.
Griffiths, J., Intarakosit, P., Taylor, H.F. and Wain, R.L. (1966) Ann. appl. Biol., 58, 183–191.
Grunwald, C., Vendrell, M. and Stowe, B.B. (1967) Analyt. Biochem., 20, 484–494.
Grunwald, C., Mendez, J. and Stowe, B.B. (1968) in: Biochemistry and Physiology of Plant Growth Substances, F. Wightman and G. Setterfield (eds.), Runge Press, Ottawa, pp. 163–171.
Haagen-Smit, A.J., Leech, W.D. and Bergren, W.R. (1942) Amer. J. Bot., 29, 500–505.
Haagen-Smit, A.J., Dandliker, W.B., Wittwer, S.H. and Murneek, A.E. (1946) Amer. J. Bot., 33, 118–120.
Hackett, D.P. and Thimann, K.V. (1952) Proc. nat. Acad. Sci. (Wash.), 38, 770–775.
Hagar, A. and Schmidt, R. (1968) Planta, 83, 347–371.
Hall, S.M. and Medlow, G.C. (1974) Planta, 119, 257–261.
Hall, W.C. and Morgan, P.W. (1964) Colloq. Int. Cent. Rech. Sci., 123, 727–745.
Hamaker, J.W., Johnston, H., Martin, R.T. and Redeman, C.T. (1963), Science, 141, 363.
Hamilton, R.H., Bandurski, R.S. and Grigsby, B.H. (1961) Plant Physiol., 36, 354–359.

Hansch, C. and Muir, R.M. (1950) Plant Physiol., 25, 389–393.
Hansch, C., Muir, R.M. and Metzenberg, R. (1951) Plant Physiol. 26, 812–821.
Harper, D.B. and Wain, R.L. (1969) Ann. appl. Biol., 64, 395–407.
Hatcher, E.S.J. (1945) Ann. Bot. (Lond.), 9, 236–266.
Hattori, H. and Marumo, S. (1972) Planta, 102, 85–90.
Havinga, E. and Nivard, R.J. F. (1948) Rech. Trav. chim. Pays Belg., 67, 84.
Heacock, R.A., Wain, R.L. and Wightman, F. (1958) Ann. appl. Biol., 46, 352–365.
Heerkloss, R. and Libbert, E. (1976) Planta, 131, 299–302.
Hemberg, T. (1972) Physiol. Plant., 26, 98–103.
Hertel, R. and Flory, R. (1968) Planta, 82, 123–144.
Hertel, R. and Leopold, A.C. (1963) Planta, 59, 535–562.
Hertel, R., Evans, M.L., Leopold, A.C. and Sell, H.M. (1969) Planta, 85, 238–249.
Hertel, R., Thomson, K.-S. and Russo, V.E.A. (1972) Planta, 107, 325–340.
Hertel, R., Dohrmann, U., Jesaitis, A.J. and Petersen, W. (1976) in: Collected Abstracts of 9th International Conference on Plant Growth Substances. Lausanne, Switzerland, p. 140.
Hill, J.M. and Mann, P.J.G. (1964) Biochem. J., 91, 171–182.
Hillman, S.K. and Phillips, I.D.J. (1970) J. exp. Bot., 21, 959–967.
Hillman, W.S. and Galston, A.W. (1961) in: Encyclopedia of Plant Physiology, Vol. XIV W. Ruhland, (ed.), Springer-Verlag, Berlin, pp. 683–702.
Hinman, R.L. and Lang, J. (1965) Biochemistry, 4, 144–158.
Hitchcock, A.E. (1935) Contrib. Boyce Thomspon Inst., 7, 87–95.
Hoad, G.V., Hillman, S.K. and Wareing, P.F. (1971) Planta, 99, 73–88.
Hoffman, O.L., Fox, S.W. and Bullock, M.W. (1952) J. biol. Chem., 196, 437–441.
Hollis, C.A. and Tepper, H.B. (1971) Plant Physiol., 48, 146–149.
Horton, R.F. and Fletcher, R.A. (1968) Plant Physiol., 43, 2045–2048.
Hoyle, M.C. (1972) Plant Physiol., 50, 15–18.
Ihl, M. (1976) Planta, 131, 223–228.
Imaseki, H. (1970) Plant Physiol., 46, 172–174.
Irvine, V.C. (1938) Univ. Colorado Studies, 26, 69–70.
Iversen, T.-H. and Aasheim, T. (1970) Planta, 93, 354–362.
Iversen, T.-H., Aasheim, T. and Pedersen, K. (1971) Physiol. Plant., 25, 417–424.
Iwahori, S. (1967) Plant Cell Physiol., 8, 15–22.
Jacobs, W.P. (1950) Amer. J. Bot., 37, 248–254.
Jacobs, W.P. (1951) Amer. J. Bot. 38, 307–310.
Jacobs, W.P. (1961) In: Plant Growth Regulation, R.M. Klein (ed.), Iowa State Univ. Press, Ames, pp. 397–409.
Jacobs, W.P. (1968) Plant Physiol., 43, 1480–1495.
Jacobs, W.P. and Case, D.B. (1965) Science, 148, 1729–1731.
Jacobs, W.P. and McCready, C.C. (1967) Amer. J. Bot., 54, 1035–1040.
Jacobs, W.P. and Morrow, I.B. (1957) Amer. J. Bot., 44, 823–842.
Jacobs, W.P., McCready, C.C. and Osborne, D.J. (1966) Plant Physiol., 41, 725–730.
James, C.S. and Wain, R.L. (1969) Ann. appl. Biol., 63, 205–210.
Jeffcoat, B. and Cockshull, K.E. (1972) J. exp. Bot., 23, 722–732.
Jones, R.L., Metcalfe, T.P. and Sexton, W.A. (1951) Biochem. J., 48, 422–425.
Jones, R.L., Metcalfe, T.P. and Sexton, W.A. (1954) J. Sci. Food. Agric., 5, 32–47.
Jönsson, A. (1955) Svensk kem. Tidskr., 67, 166–187.
Jönsson, A. (1961) in: Encyclopedia of Plant Physiology, Vol. 14, W. Ruhland (ed.), Springer, Berlin, pp. 959–1006.
Jönsson, A., Nilsson, G. and Burström, H. (1952) Acta chem. scand., 6, 993.
Jordan, W.R. and Skoog, F. (1971) Plant Physiol., 48, 97–99.
Kaldewey, H. (1957) Planta, 49, 300–344.
Kaldewey, H. (1963) Planta, 60, 178–204.

Kaldewey, H. (1966) Proc. Intl. Symp. Plant Stimulation, Sofia, Bulgaria, pp. 211–221.
Kaldewey, H. and Jacobs, W.P. (1974) Amer. J. Bot., 61, 699–706.
Katsumi, M. and Sano, H. (1968) Physiol. Plant, 21, 1348–1355.
Kefford, N.P. and Kaso, O.H. (1966) Bot. Gaz. (Chic.), 127, 159–163.
Keitt, G.W. (1960) Bot. Gaz. (Chic.), 122, 51–62.
Kender, W.J. and Desrochers, J.C. (1970) J. Amer. Soc. Hort. Sci., 95, 699–702.
Kerk, G.J.M. van der, Rallte, M.H. van, Sijpersteijn, A.K. and Veen, R. van der (1955) Nature (Lond.), 176, 308–310.
Khalifah, R.A. (1967) Physiol. Plant., 20, 355–360.
Khalifah, R.A., Lewis, L.N. and Coggins, C.W. (1963) Science, 142, 399–400.
Khalifah, R.A., Lewis, L.N., Coggins, C.W. and Radlick, P.C. (1965) J. exp. Bot., 16, 511–517.
Khan, A.A. (1967) Physiol. Plant., 20, 306–313.
Kindl, H. (1968) Hoppe-Seyler's Z. Physiol. Chem. , 349, 519–520.
Kim, W.K. and Rohringer, R. (1969) Canad. J. Bot., 47, 1425–1433.
Kirk, S.C. and Jacobs, W.P. (1968). in: Biochemistry and Physiology of Plant Growth Substances, F. Wightman and G. Setterfield (eds.), Runge Press, Ottawa, pp. 1077–1094.
Knegt, E. and Bruinsma, J. (1973) Phytochemistry, 12, 753–756.
Koepfli, J.B., Thimann, K.V. and Went, F.W. (1938) J. biol. Chem., 122, 763–780.
Koevenig, J.L. and Jacobs, W.P. (1972) Plant Physiol., 49, 866–867.
Kögl, F. (1937) Naturwissenschaften, 25, 465.
Kögl, F. and Haagen-Smit, A.J. (1931) Proc. kon. ned. Akad. Wet. 34, 1411–1416.
Kögl, F. and Kostermans, D.G.F.R. (1935) Hoppe-Seyler's Z. Physiol. Chem., 235, 201–206.
Kögl, F. and Verkaaik, B. (1944) Hoppe-Seyler's Z. Physiol. Chem., 280, 167–176.
Kögl, F., Haagen-Smit, A.J. and Erxleben, H. (1934) Hoppe-Seyler's Z. Physiol. Chem., 228, 90–206.
Kopcewicz, J., Ehmann, A. and Bandurski, R.S. (1974) Plant Physiol., 54, 846–851.
Krelle, E. and Libbert, E. (1968) Planta, 80, 317–320.
Kumar, S.A. and Mahadevan, S. (1963). Arch. Biochem. Biophys., 103, 516–518.
Kuraishi, S. and Muir, R.M. (1964) Plant Cell Physiol., 5, 61–69.
Kutacek, M. and Kefeli, V.I. (1968) in: Biochemistry and Physiology of Plant Growth Substances, F. Wightman and G. Setterfield, (eds.), Runge Press, Ottawa, pp. 127–152.
Kutacek, M. and Kefeli, V.I. (1970) Biol. Plant., 12, 145–158.
Kutacek, M. and Prochazka, Z. (1964) Colloq. Int. Cent. Rech. Sci., 123, 445–456.
Kutacek, M., Prochazka, Z., Grünberger, D. and Stajkova, R. (1962) Collect. Czech. Chem. Commun., 27, 1278–1283.
Kutacek, M., Bulgakov, R. and Oplistilova, K. (1966) Biol. Plant., 8, 252–255.
Lantican, B.P. and Muir, R.M. (1967) Plant Physiol., 42, 1158–1160.
Larsen, P. (1951) Plant Physiol., 26, 697–707.
Larsen, P. (1955) in: Modern Methods of Plant Analysis, Vol. III, K. Paech and M. Tracy (eds.), Springer, Berlin, pp. 565–625.
Larsen, P. and Rajagopal, R. (1964) Colloq. Int. Cent. Rech. Sci., 123, 221–233.
Lau, O.-L. and Yang, S.F. (1973) Plant Physiol., 51, 1011–1014.
Leaper, J.M.P. and Bishop, J.R. (1951). Bot. Gaz., (Chic.), 112, 250–258.
Lee, T.T. (1971a) Plant Physiol., 48, 56–59.
Lee, T.T. (1971b) Canad. J. Bot., 49, 687–693.
Lee, T.T. (1971c) Plant Physiol., 47, 181–185.
Lee, T.T. (1972) Canad. J. Bot., 50, 2471–2477.
Lee, T.T. (1974) Phytochemistry, 13, 2445–2453.
Lembi, C.A., Morre, D.J., Thomson, K.-S. and Hertel, R. (1971) Planta, 99, 37–45.
Leopold, A.C. and Hall, O.F. (1966) Plant Physiol., 41, 1476–1480.
Leopold, A.C. and Lam, S.C. (1961) in: Plant Growth Regulation, R.M. Klein (ed.), Iowa State Univ. Press, Ames, pp. 411–418.
Lepp. N.W. and Peel, A.J. (1971) Planta 96, 62–73.

Linser, H. (1956) in: Chemistry and Mode of Action of Plant Growth Substances, R.L. Wain and F. Wightman (eds.), Butterworths, London, pp. 148–158.
Little, C.H.A. and Goldsmith, M.H.M. (1967) Plant Physiol., 42, 1239–1245.
Little, E.C.S. and Blackman, G.E. (1963) New Phytol., 62, 173–197.
Luckwill, L.C. (1948) J. Hort. Sci., 24, 32–44.
Luckwill, L.C. and Woodcock, D. (1956) in: Chemistry and Mode of Action of Plant Growth Substances, R.L. Wain and F. Wightman, (eds.), Butterworths, London, pp. 195–201.
Lund, H.A. (1956) Amer. J. Bot., 43, 502–568.
Lyon, C.J. (1963) Plant Physiol., 38, 567–574.
Magnus, V., Iskric, S. and Kveder, S. (1971) Planta, 97, 116–125.
Mahadevan, S. (1963) Arch. Biochem. Biophys., 100, 557–558.
Mahadevan, S. and Stowe, B.B. (1972) Plant Physiol., 50, 43–50.
Mahadevan, S. and Thimann, K.V. (1964) Arch. Biochem. Biophys., 107, 62–68.
Mann, J.D. and Jaworski, E.G. (1970) Planta, 92, 285–291.
Marumo, S. and Hattori, H. (1970) Planta, 90, 208–211.
Marumo, S., Hattori, H. and Abe, H. (1971) Analyt. Biochem., 40, 488–490.
Marumo, S., Abe, H., Hattori, H. and Munakata, K. (1968a) Agr. Biol. Chem., 32, 117–118.
Marumo, S., Hattori, M., Abe, H. and Munakata, K. (1968b) Nature (Lond.), 219, 959–960.
Matell, M. (1953) Kungl. Lantbrukshögsk. Ann., 20, 205–240.
McCready, C.C. (1963) New Phytol., 62, 3–18.
McCready, C.C. (1966) Ann. Rev., Plant Physiol. 17, 283–294.
McCready, C.C. (1968) in: Biochemistry and Physiology of Plant Growth Substances, F. Wightman and G. Setterfield (eds.), Runge Press, Ottawa, pp. 1005–1023.
McCready, C.C. and Jacobs, W.P. (1963a) New Phytol., 62, 19–34.
McCready, C.C. and Jacobs, W.P. (1963b) New Phytol., 62, 360–366.
McRae, D.H. and Bonner, J. (1952) Plant Physiol., 27, 834–838.
McRae, D.H. and Bonner, J. (1953) Physiol. Plant., 6, 485–510.
Mel'nikov, N.N., Turetskaya, R.K., Baskakov, Y.A., Boyarkin, A.N. and Kuznetsova, M.S. (1953) Dokl. Akad. Nauk (SSSR), 89, 953–955.
Michel, B. (1957) Plant Physiol., 32, 632–639.
Mills, V.M. and Todd, G.W. (1973) Plant Physiol., 51, 1145–1146.
Mitchell, E.K. and Davies, P.J. (1972) Plant Cell Physiol., 13, 1135–1138.
Mitchell, J.W. and Livingston, G.A. (1968) Methods of Studying Plant Hormones and Growth-regulating Substances. Agriculture Handbook No. 336, U.S. Dept. Agr., Beltsville.
Moore. T.C. (1969) Phytochemistry, 8, 1109–1120.
Morgan, D.G. (1964) Nature (Lond.), 201, 476–477.
Morgan, D.G. and Söding, H. (1958), Planta, 52, 235–249.
Morris, D.A. and Kadir, G.O. (1972) Planta, 107, 171–182.
Morris, D.A. and Thomas, A.G. (1974) Planta, 118, 225–234.
Morris, D.A., Briant, R.E. and Thomson, P.G. (1969) Planta, 89, 178–197.
Morris, D.A., Kadir, G.O. and Barry, A.J. (1973) Planta, 110, 173–182.
Moyed, H.S. and Williamson, V. (1967a) Plant Physiol., 42, 510–514.
Moyed, H.S. and Williamson, V. (1967b) J. biol. Chem., 242, 1075–1077.
Muir, R.M. and Hansch, C. (1951) Plant Physiol., 26, 369–374.
Muir, R.M. and Hansch, C. (1953) Plant Physiol., 28, 218–232.
Muir, R.M. and Hansch, C. (1955) Ann. Rev. Plant Physiol., 6, 157–176.
Muir, R.M., Hansch, C. and Gallup, A.H. (1949) Plant Physiol., 24, 359–366.
Muir, R.M., Fujita, T. and Hansch, C. (1967) Plant Physiol., 42, 1519–1526.
Naqvi, S.M. (1972a) Physiol. Veg., 10, 495–501.
Naqvi, S.M. (1972b) J. exp. Bot. 23, 763–767.
Naqvi, S.M. and Gordon, S.A. (1966) Plant Physiol., 41, 1113–1118.
Naqvi, S.M. and Gordon, S.A. (1967) Plant Physiol., 42, 138–143.

Newman, I.A. (1963) Aust. J. biol. Sci., 16, 629–646.
Niedergang-Kamien, E. and Leopold, A.C. (1957), Physiol. Plant., 10, 29–38.
Nielson, N. (1930) Jb. Wiss. Bot., 73, 125–191.
Nitsch, J.P. (1950) Amer. J. Bot., 37, 211–215.
Nitsch, J.P. (1952) Quart. Rev. Biol., 27, 33–57.
Nitsch, J.P. (1965) in: Encyclopedia of Plant Physiology, Vol. XV pt. 1, W. Ruhland (ed.), Springer, Berlin, pp. 1537–1647.
Nitsch, J.P. and Nitsch, C. (1956) Plant Physiol., 31, 94–111.
Ockerse, R. and Galston, A.W. (1967) Plant Physiol., 42, 47–54.
Ockerse, R. and Waber, J. (1970) Plant Physiol., 46, 821–824.
Osborne, D.J. and Mullins, M.G. (1969) New Phytol., 68, 977–991.
Osborne, D.J. and Wain, R.L. (1950) J. Hort. Sci., 26, 60–70.
Osborne, D.J. and Wain, R.L. (1951) Science, 114, 92.
Osborne, D.J., Horton, R.F. and Black, M.F. (1968) in: The Transport of Plant Hormones, Y. Varder (ed.), North-Holland, Amsterdam, pp. 79–96.
Ouitrakul, R. and Hertel, R. (1969) Planta, 88, 233–243.
Paleg, L.G. (1965) Ann. Rev. Plant Physiol., 16, 291–322.
Palmer, J.H. and Halsall, D.M. (1969) Physiol. Plant., 22, 59–67.
Patrick, J.W. and Wooley, D.J. (1973) J. exp. Bot., 24, 949–957.
Percival, F.W. and Purves, W.K. (1974) Plant Physiol., 54, 601–607.
Percival, F.W., Purves, W.K. and Vickery, L.B. (1973) Plant Physiol., 51, 739–743.
Phelps, R.H. and Sequeira, L. (1967) Plant Physiol., 42, 1161–1163.
Phelps, R.H. and Sequeira, L. (1968) in: Biochemistry and Physiology of Plant Growth Substances, F. Wightman and G. Setterfield, (Eds.), Runge Press, Ottawa, pp. 197–212.
Philipson, J.J., Hillman, J.R. and Wilkins, M.B. (1973), Planta, 114, 87–93.
Phipps, J. (1966) Le Catabolisme Auxinique chez le Tabac: ses Modalites dans le Plante Saine et Parasitee par le Virus de la Mosaique, Toulouse, Imprimerie du Commerce.
Pickard, B.G. and Thimann, K.V. (1964) Plant Physiol., 39, 341–350.
Pilet, P.E. (1964) Nature (Lond.), 204, 561–562.
Pilet, P.E. (1965a) C.R. Acad. Sci. (Paris), 260, 4053–4056.
Pilet, P.E. (1965b) Physiol. Plant., 18, 687–702.
Pilet, P.E. (1965c) Nature (Lond.), 208, 1344–1345.
Pilet, P.E. (1971) Physiol. Plant., 25, 28–31.
Pilet, P.E., Kobr, M. and Siegenthaler, P.A. (1960) Rev. Gen. Bot., 67, 573–588.
Piskornik, Z. and Bandurski, R.S. (1972) Plant Physiol., 50, 176–182.
Porter, W.L. and Thimann, K.V. (1959) Abstr. 9th Int. Bot. Congr., Univ. Toronto Press, p. 305.
Porter, W.L. and Thimann, K.V. (1965). Phytochemistry, 4, 229–243.
Powell, L.E. (1964) Plant Physiol., 39, 836–842.
Pybus, M.B., Smith, M.S., Wain, R.L. and Wightman, F. (1959a) Ann. appl. Biol., 47, 173–181.
Pybus, M.B., Wain, R.L. and Wightman, F. (1959b) Ann. appl. Biol., 47, 593–600.
Raj, R.K. and Hutzinger, O. (1970) Anal. Biochem., 46, 471–474.
Rajagopal, R. (1967) Physiol. Plant., 20, 655–660.
Rajagopal, R. (1968) Physiol. Plant., 21, 378–385.
Rajagopal, R. (1971) Physiol. Plant., 24, 272–281.
Rajagopal, R. and Larsen, P. (1972) Planta, 103, 45–54.
Rajagopal, R. and Larsen, P. (1974) Physiol. Plant., 31, 119–124.
Rayle, D.L. and Purves, W.K. (1967a) Plant Physiol., 42, 520–524.
Rayle, D.L. and Purves, W.K. (1967b) Plant Physiol., 42, 1091–1093.
Rayle, D.L. and Purves, W.K. (1968) in: Biochemistry and Physiology of Plant Growth Substances, F. Wightman and G. Setterfield, (eds.), Runge Press, Ottawa, pp. 153–161.
Rayle, D.L., Ouitrakul, R. and Hertel, R. (1969) Planta, 87, 49–53.
Reed, D.J. (1968) in: Biochemistry and Physiology of Plant Growth Substances, F. Wightman and

G. Setterfield (eds.), Runge Press, Ottawa, pp. 243–258.
Ricard, J., Mazza, G. and Williams, R.J.P. (1972) Europ. J. Biochem., 28, 566–578.
Ridge, I. and Osborne, D.J. (1970) J. exp. Bot., 21, 843–856.
Rivier, L. and Pilet, P.E. (1974) Planta, 120, 107–112.
Robinson, B.J., Forman, M. and Addicott, F.T. (1968) Plant Physiol., 43, 1321–1323.
Rudich, J., Halevy, A.H. and Kedar, N. (1972) Plant Physiol., 50, 585–590.
Sastry, K.S.K. and Muir, R.M. (1965) Plant Physiol., 40, 294–298.
Schlender, K.K., Bukovac, M.J. and Sell, H.M. (1966) Phytochemistry, 5, 133–139.
Schneider, E.A. (1965) Studies on the Biosynthesis and Degradation of 3-Indoleacetic acid and Gramine in Barley Shoots, PhD Thesis, Carleton University, Ottawa.
Schneider, E.A. and Wightman, F. (1974a) Ann. Rev. Plant. Physiol., 25, 487–513.
Schneider, E.A. and Wightman, F. (1974b) Canad. J. Biochem., 52, 698–705.
Schneider, E.A., Gibson, R.A. and Wightman, F. (1972) J. exp. Bot., 23, 152–170.
Schneider, G. (1969) Ber. Dtsch. Bot. Ges., Votr. Gesamtgeb. Bot. N.F. 3, 19–42.
Schneider, G. (1970) Ann. Rev. Plant Physiol., 21 499–536.
Schneider, M.J., Troxler, R.F. and Voth, P.D. (1967) Bot. Gaz. (Chic.), 128, 174–179.
Schraudolf, H. and Weber, H. (1969) Planta, 88, 136–143.
Scott, T.K. (1972) Ann. Rev. Plant Physiol., 23, 235–258.
Scott, T.K. and Most, B.H. (1972) in: Hormonal Regulation in Plant Growth and Development, H. Kaldewey and Y. Vardar (eds.), Chimie, Weinheim, pp. 58–67.
Scott, T.K. and Wilkins, M.B. (1968), Planta, 83, 323–334.
Scott, T.K. and Wilkins, M.B. (1969) Planta, 87, 249–258.
Seeley, R.C., Fawcett, C.H., Wain, R.L. and Wightman, F. (1956) in: Chemistry and Mode of Action of Plant Growth Substances, R.L. Wain and F. Wightman (eds.), Butterworths, London, pp. 234–247.
Sell, M.H., Wittwer, S.H., Rebstock, T.K. and Redeman, C.T. (1953) Plant Physiol., 28, 481–487.
Sequeira, L. and Mineo, L. (1966) Plant Physiol., 41, 1200–1208.
Shannon, L. (1968) Ann. Rev. Plant Physiol., 19, 187–210.
Shaw, S., Gardner, G. and Wilkins, M.B. (1973) Planta, 115, 97–111.
Shaw, S. and Wilkins, M.B. (1974) J. exp. Bot., 25, 199–207.
Sheldrake, A.R. (1971) J. exp. Bot., 22, 735–740.
Sheldrake, A.R. (1973a) New Phytol., 72, 433–447.
Sheldrake, A.R. (1973b) Biol. Rev., 48, 509–559.
Sheldrake, A.R. (1973c) J. exp. Bot., 24, 87–96.
Sheldrake, A.R. and Northcote, D.H. (1968a) Nature (Lond.), 217, 195.
Sheldrake, A.R. and Northcote, D.H. (1968b) New Phytol., 67, 1–13.
Shen-Miller, J. (1973) Plant Physiol., 51, 615–619.
Shen-Miller, J. and Gordon, S.A. (1966) Plant Physiol., 41, 831–841.
Sherwin, J.E. (1970) Plant Cell Physiol., 11, 865–872.
Sherwin, J.E. and Purves, W.K. (1969) Plant Physiol., 44, 1303–1309.
Shindy, W.W. and Smith, O.E. (1975) Plant Physiol., 55, 550–554.
Shukla, P.S. and Mahadevan, S. (1968) Arch. Biochem. Biophys., 125, 873–883.
Shukla, P.S. and Mahadevan, S. (1970) Arch. Biochem. Biophys., 137, 166–174.
Sirois, J. (1966) Plant Physiol., 41, 1308–1312.
Skoog, F. (1937) J. gen. Physiol., 20, 311–334.
Skytt-Andersen, A. and Muir, R.M. (1966) Physiol. Plant., 19, 1038–1048.
Smith, C.W. and Jacobs, W.P. (1969) Amer. J. Bot., 56, 492–497.
Smith, M.S. and Wain, R.L. (1951) Proc. roy. Soc. B, 139, 118–127.
Smith, M.S., Wain, R.L. and Wightman, F. (1952) Ann. appl. Biol., 39, 295–307.
Steen, I. (1972) Physiol. Plant., 26, 92–97.
Stehsel, M.L. (1950) I. Interrelations between Tryptophane, Auxin and Nicotinic Acid during Development of the Corn Kernel. II. Studies on the Nature of the Auxin Complex in the Corn Kernel, PhD Thesis, Univ. Calif. Berkeley. U.S.A.

Steyer, B. (1969) Flora, 159, 484–493.
Stoessl, A. and Venis, M.A. (1970) Anal. Biochem., 34, 344–351.
Stowe, B.B. and Schilke, J.F. (1964) Colloq. Int. Cent. Nat. Rech. Sci., 123, 409–419.
Stowe, B.B. and Thimann, K.V. (1954) Arch. Biochem. Biophys., 51, 501–516.
Stowe, B.B., Vendrell, M. and Epstein, E. (1968) in: Biochemistry and Physiology of Plant Growth Substances, F. Wightman and G. Setterfield (eds.), Runge Press, Ottawa, pp. 173–182.
Stutz, R.L. (1958) Plant Physiol., 33, 207–212.
Sukanya, N.K., Vaidyanathan, C.S. and Mahadevan, S. (1971) Indian J. Biochem. Biophys., 8, 235–238.
Synerholm, M.E. and Zimmerman, P.W. (1945) Contrib. Boyce Thompson Inst., 14, 91.
Synerholm, M.E. and Zimmerman, P.W. (1947) Contrib. Boyce Thompson Inst., 14, 369–382.
Syono, K. and Furuya, T. (1972) Plant Cell Physiol., 13, 843–856.
Takeda, A. (1959) Contrib. Boyce Thompson Inst., 20, 197–204.
Templeman, W.G. and Halliday, D.J. (1951) Ann. Rev. Plant Physiol., 2, 210–230.
Thimann, K.V. (1934) J. gen. Physiol., 18, 23–24.
Thimann, K.V. (1935) Proc. kon. ned. Acad. Wet., 38, 896–912.
Thimann, K.V. (1951) in: Plant Growth Substances, F. Skoog (ed.), University of Wisconsin Press, Madison, pp. 21–36.
Thimann, K.V. (1952) Plant Physiol., 27, 392–404.
Thimann, K.V. (1953) Arch. Biochem., 44, 242–243.
Thimann, K.V. (1958) Plant Physiol., 33, 311–321.
Thimann, K.V. (1963) Annu. Rev. Plant Physiol., 14, 1–18.
Thimann, K.V. (1972) in: Plant Physiology, a Treatise, Vol. VIB, F.C. Steward (ed.), Academic Press, New York, pp. 25–40.
Thimann, K.V. and Curry, G.M. (1960) in: Comparative Biochemistry, Vol 1, M. Florkin and H. S. Mason (eds.), Academic Press, New York, pp. 243–310.
Thimann, K.V. and Grochowska, M. (1968) in: Biochemistry and Physiology of Plant Growth Substances, F. Wightman and G. Setterfield (eds.), Runge Press, Ottawa, pp. 231–242.
Thimann, K.V. and Mahadevan, S. (1964) Arch. Biochem. Biophys., 105, 133–141.
Thimann, K.V. and Schneider, C.L. (1939) Amer. J. Bot., 26, 328–333.
Thomson, K.S., Hertel, R., Muller, S. and Tavares, J.E. (1973) Planta, 109, 337–352.
Thornton, R.M. and Thimann, K.V. (1967) Plant Physiol., 42, 247–257.
Toothill, J., Wain, R.L. and Wightman, F. (1956) Ann. appl. Biol., 44, 547–560.
Truelsen, T.A. (1973) Physiol. Plant., 28, 67–70.
Tuli, V. and Moyed, H.S. (1967) Plant Physiol., 42, 425–430.
Ueda, M. and Bandurski, R.S. (1969) Plant Physiol., 44, 1175–1181.
Ueda, M., Ehmann, A. and Bandurski, R.S. (1970) Plant Physiol., 46, 715–719.
Valdovinos, J.G. and Ernest, L.C. (1966) Plant Physiol., 41, 1551–1552.
Valdovinos, J.G. and Ernest, L.C. (1967) Physiol. Plant., 20, 682–687.
Valdovinos, J.G., Ernest, L.C. and Perley, J.E. (1967) Physiol. Plant., 20, 600–607.
Van der Beek, L.C. (1959) Plant Physiol., 34, 61–65.
Van der Mast, C.A. (1969) Acta bot. neerl., 18, 620–626.
Van der Weij, H.G. (1932) Rec. Trav. Bot. Neerl., 29, 380–496.
Van Overbeek, J. (1941) Amer. J. Bot., 28, 1–10.
Van Overbeek, J. and Went, F.W. (1937) Bot. Gaz. (Chic.), 99, 22–41.
Veen, H. (1969) Acta bot. neerl., 18, 447–454.
Veen, H. (1972) Planta, 103, 35–44.
Veen, H. and Jacobs, W.P. (1969) Plant Physiol., 44, 1157–1162.
Veldstra, H. (1944) Enzymologia, 11, 97–163.
Veldstra, H. (1952) Rec. Trav. chim. Pays-Bas, 71, 15–32.
Veldstra, H. (1953) Ann. Rev. Plant Physiol., 4, 151–198.
Veldstra, H. (1956) in: The Chemistry and Mode of Action of Plant Growth Substances, R.L. Wain and F. Wightman (eds.), Butterworths, London, pp. 117–133.

Veldstra, H. and Booij, H.L. (1949). Biochim. biophys. Acta (Amst.), 3, 278–432.
Veldstra, H. and Westeringh, C. van der (1951) Rec. Trav. chim. Pays-Bas, 70, 1113–1126.
Veldstra, H. and Westeringh, C. van der (1952) Rec. Trav. chim. Pays-Bas, 71, 318–320.
Venis, M.A. (1972) Plant Physiol., 49, 24–27.
Vickery, L.E. and Purves, W.K. (1972) Plant Physiol., 49, 716–721.
Vlitos, A.J. and Hitchcock, T.S. (1958) Contrib. Boyce Thompson Inst., 19, 255–266.
Von Guttenberg, H. and Steinmetz, E. (1947) Pharmazie, 2, 17–21.
Wain, R.L. (1953) Monograph No. 2, Royal Inst. Chem. London.
Wain, R.L. (1955) Ann. appl. Biol., 42, 151–157.
Wain, R.L. (1958) Adv. Pest Control Res., 2, 263–305.
Wain, R.L. (1968) Monograph No. 31, Soc. Chem. Ind., pp. 3–20.
Wain, R.L. and Fawcett, C.H. (1969) In: Plant Physiology, a Treatise. Vol. VA, F.C. Steward (ed.), Academic Press, New York, pp. 229–296.
Wain, R.L. and Taylor, H.F. (1965) Nature (Lond.), 207, 167–169.
Wain, R.L. and Wightman, F. (1953) Ann. appl. Biol., 40, 244–249.
Wain, R.L. and Wightman, F. (1954) Proc. roy. Soc. B., 142, 525–536.
Wain, R.L. and Wightman, F. (1957) Ann. appl. Biol., 45, 140–157.
Wakhloo, J.L. (1965) Planta, 65, 301–314.
Wangerman, E. (1970) New Phytol., 69, 919–927.
Wangerman, E. (1974) New Phytol., 73, 623–636.
Weintraub, R.L. and Norman, A.C. (1949), Econ. Bot., 3, 289.
Went, F.W. (1928) Rec. Trav. Bot. Neerl., 25, 1–116.
Went, F.W. (1949) Arch. Biochem., 20, 131–136.
Went, F.W. (1974) Ann. Rev. Plant Physiol., 25, 1–26.
Went, F.W. and Thimann, K.V. (1937) Phytohormones. Macmillan, New York.
Wheeler, A.W. (1968) J. exp. Bot., 19, 102–107.
White, E.P. (1944) N.Z. J. Sci. Technol. B., 25, 137–162.
Widholm, J.M. (1974) Physiol. Plant., 30, 323–326.
Wiese, M.V. and deVay, J.E. (1970) Plant Physiol., 45, 304–309.
Wightman, F. (1962) Canad. J. Bot., 40, 689–718.
Wightman, F. (1964) in: Regulateurs Naturels de la Croissance Vegetale. Colloq. Int. No. 123, C.N.R.S. Paris, pp. 191–212.
Wightman, F. (1973) in: Nitrogen metabolism in plants, Biochem. Soc. Symp., 38, 247–275.
Wightman, F. (1977) in: Plant Growth Regulation, P.E. Pilet (ed.), Springer, Heidelberg, pp. 77–90.
Wightman, F. and Cohen, D. (1968) in: Biochemistry and Physiology of Plant Growth Substances. F. Wightman and G. Setterfield, (eds.), Runge Press, Ottawa, pp. 273–288b.
Wightman, F. and Rauthan, B.S. (1974) in: Plant Growth Substances, 1973, Hirokawa, Tokyo. pp. 15–27.
Wilkins, M.B. and Cane, A.R. (1970) J. exp. Bot., 21, 195–211.
Wilkins, M.B. and Scott, T.K. (1968a) Nature (Lond.), 219, 1388–1389.
Wilkins, M.B. and Scott, T.K. (1968b) Planta, 83, 335–346.
Wilkins, M.B. and Whyte, P. (1968) in: Biochemistry and Physiology of Plant Growth Substances. F. Wightman and G. Setterfield (eds.), Runge Press, Ottawa, pp. 1051–1062.
Wilkins, M.B., Cane, A.R. and McCorquodale, I. (1972) Planta, 106, 291–310.
Wilske, C. and Burstrom, H. (1950) Physiol. Plant., 3, 58–67.
Winter, A. (1968) in: Biochemistry and Physiology of Plant Growth Substances, F. Wightman and G. Setterfield (eds.), Runge Press, Ottawa, pp. 1063–1076.
Winter, A. and Thimann, K.V. (1966) Plant Physiol., 41, 335–342.
Wochok, Z.S. and Sussex, I.M. (1974) Plant Physiol., 53, 738–741.
Yamaki, T. and Fujii, T. (1968) in: Biochemistry and Physiology of Plant Growth Substances F. Wightman and G. Setterfield (eds.), Runge Press, Ottawa, pp. 1025–1036.
Yoneda, Y. and Stonier, T. (1967) Plant Physiol., 42, 1017–1020.

Zaerr, J.B. (1968) Physiol. Plant., 21, 1265–1269.
Zenk, M.H. (1961) Nature (Lond.), 191, 493–494.
Zimmerman, P.W. and Hitchcock, A.E. (1937) Contrib. Boyce Thompson Inst., 8, 337–350.
Zimmerman, P.W. and Hitchcock, A.E. (1942) Contrib. Boyce Thompson Inst., 12, 321–343.
Zimmerman, P.W. and Hitchcock, A.E. (1944) Proc. Amer. Soc. Hort. Sci., 45, 353–361.
Zimmerman, P.W. and Wilcoxon, F. (1935). Contrib. Boyce Thompson Inst., 7, 209–229.
Zimmerman, P.W., Hitchcock, A.E. and Prill, A.E. (1952) Contrib. Boyce Thompson Inst., 16, 419–000.

Phytohormones and Related Compounds – A Comprehensive Treatise, Volume I
Editors: Letham, Goodwin and Higgins

3

Gibberellins

J.E. GRAEBE and H.J. ROPERS

3.1. Introduction

The gibberellins (GAs) are present in minute amounts in higher plants. Their most obvious function within the hormonal system is to regulate the rate of growth and final length of internodes. This is best seen in certain "dwarf" cultivars which due to disturbed GA relations have very short internodes but upon treatment with GA grow to normal or intermediate heights depending on the dose of GA supplied. The bolting of cold-requiring and long-day rosette plants in response to induction, the extension of floral organs, the development of fruits, seeds and bulbs, the germination of seeds and the breaking of dormancy also depend on the action of endogenous GA.

The GAs were not discovered in higher plants but first became known as metabolites of the ascomycete *Gibberella fujikuroi* (Saw) Wr (imperfect state: *Fusarium moniliforme* Sheld). Early in this century, this fungus was recognized by Japanese plant pathologists as the cause of the "bakanae" disease of rice, which causes overgrowth symptoms now known to be typical effects of GA. The Japanese scientist Kurosawa (1926) showed that a sterilized filtrate of the medium in which *G. fujikuroi* had been growing could cause the same overgrowth symptoms in rice as the disease and thus that the increased growth was due to some substance produced by the fungus. Pursuing the problem of the identity of this substance, Yabuta and Sumiki (1938) and Yabuta and Hayashi (1939) at the University of Tokyo succeeded in obtaining crystals of the active material, which they had named "gibberellin". These crystals later turned out to be a mixture of compounds. The first pure GA was isolated by a British team (Cross 1954; Borrow et al. 1955) closely followed by research teams in U.S.A. (Stodola et al. 1955) and Japan (Takahashi et al. 1955). The British product, "gibberellic acid" (now GA_3), was the first one to be structurally fully elucidated (Cross et al. 1959) and the structures of several other GAs were determined by relating them to gibberellic acid.

The work with the fungus led to the discovery of GAs in higher plants. Bioassays, which had been developed on the basis of the action of fungal GAs on higher plants, indicated that substances with similar effects were present in higher plants also and particularly in immature seeds (Radley 1956; Brian 1957; Phinney et al. 1957; Lona 1957). The first GA of a higher plant was isolated from immature bean seeds and found to be identical with fungal GA_1 (MacMillan et al. 1960).

This chapter is an account of our present knowledge of the kinds of GAs occurring in

G. fujikuroi and in higher plants, their activities, biosynthesis, metabolism and dependence on environmental factors. This information is fundamental to further investigation of the precise regulation of GA activity in higher plants, which is the main difference between GAs as hormones and GAs as products of a pathogenic fungus.

The history of GA discovery has been treated extensively in the early reviews by Stowe and Yamaki (1957, 1959), Stodola (1958), Brian (1959) and Brian et al. (1960). The "Source Book on Gibberellin, 1828–1957" by Stodola (1958) is a valuable reference to the earlier work containing many abstracts of publications otherwise only available in Japanese. Several reviews on the general subject of GAs have appeared (Brian 1959, 1966; Brian et al. 1960; Phinney and West 1960, 1961; Paleg 1965; Lang 1970 and Jones 1973). A review in Japanese has been published by Tamura et al. (1969). Reviews specializing on the chemistry or biosynthesis of GAs are mentioned in sections 3.4 and 3.7, respectively.

3.2. Occurrence and identity

3.2.1. Nomenclature

The GAs (Figure 3.1) are diterpenoid acids biogenetically derived from the tetracyclic diterpenoid hydrocarbon *ent*-kaur-16-ene or (–)-kaurene (I). Somewhat less than half of the known GAs retain the full 20 carbon atoms of this precursor; they are referred to as C_{20}-GAs and have the carbon skeleton II. The others have lost one carbon atom (number 20) in the course of their biosynthesis; they are referred to as C_{19}-GAs and have the skeleton III. The numbering of the carbon atoms is in accord with accepted rules for

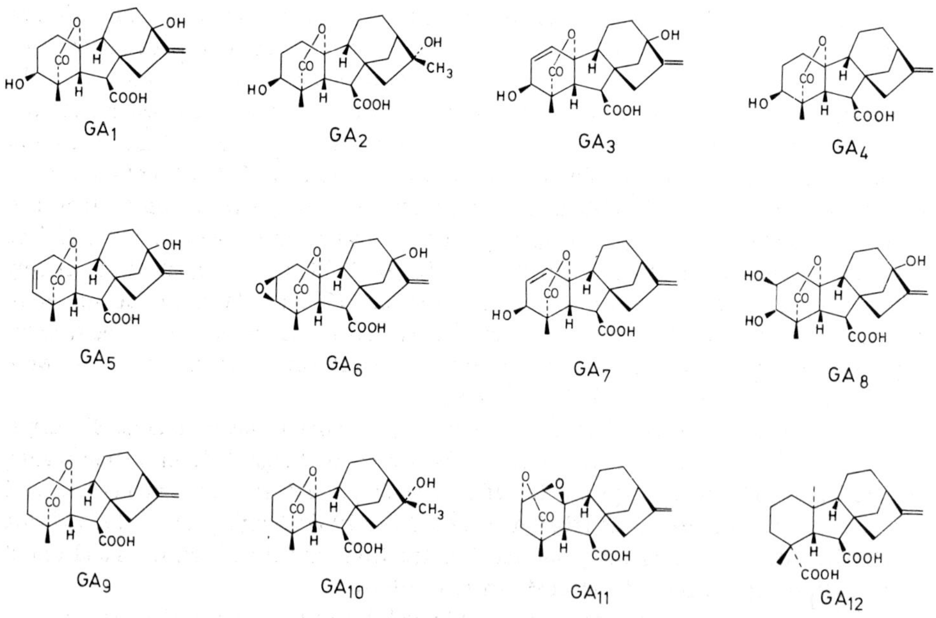

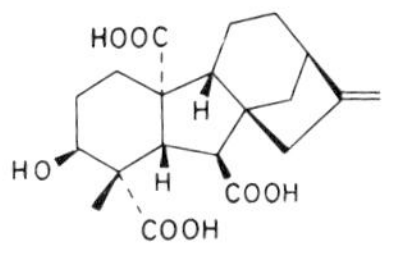
GA$_{13}$

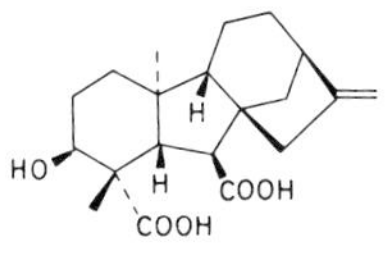
GA$_{14}$

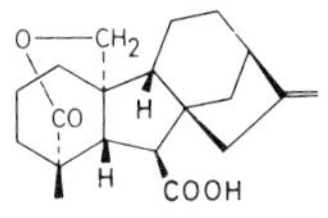
GA$_{15}$

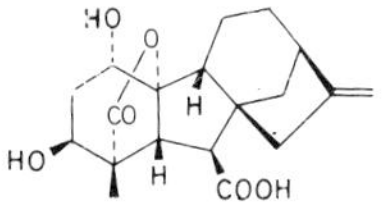
GA$_{16}$

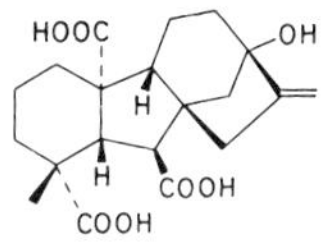
GA$_{17}$

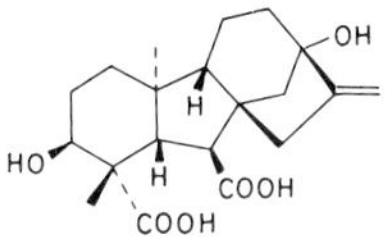
GA$_{18}$

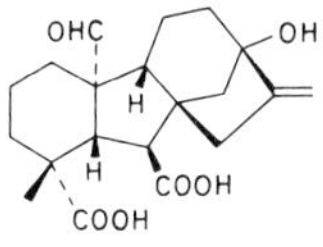
GA$_{19}$

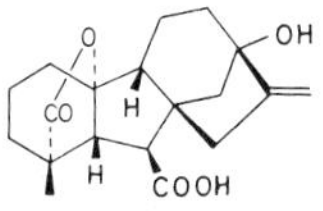
GA$_{20}$

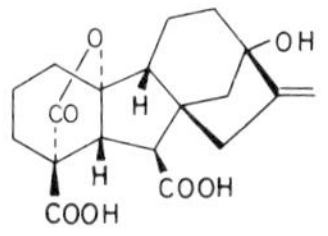
GA$_{21}$

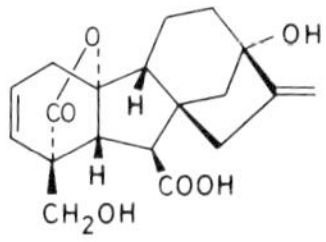
GA$_{22}$

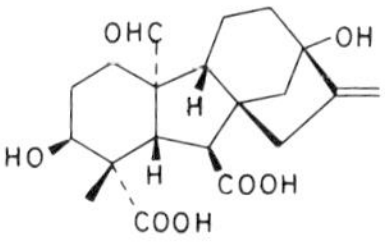
GA$_{23}$

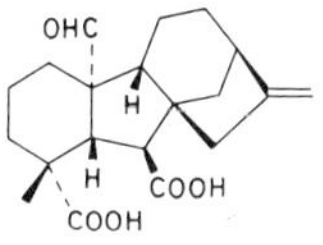
GA$_{24}$

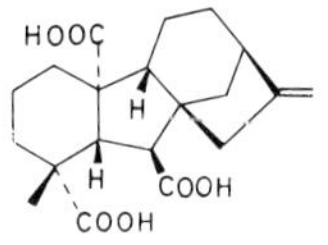
GA$_{25}$

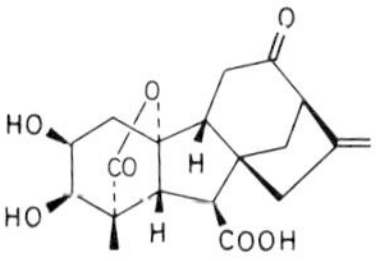
GA$_{26}$

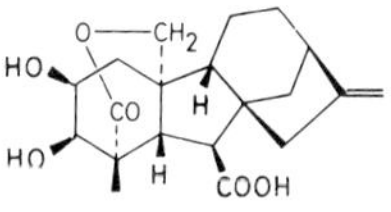
GA$_{27}$

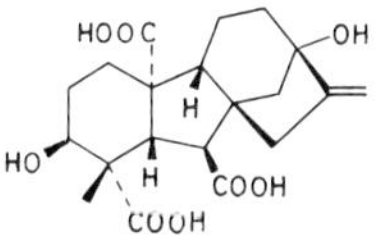
GA$_{28}$

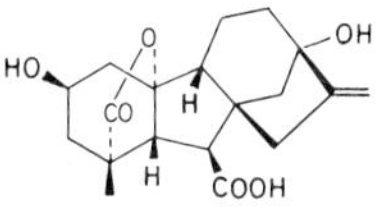
GA$_{29}$

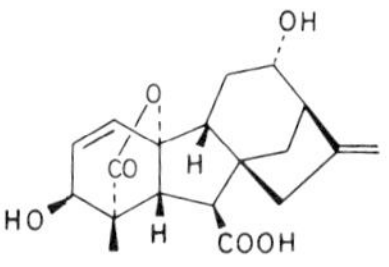
GA$_{30}$

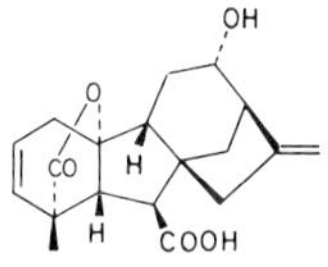
GA$_{31}$

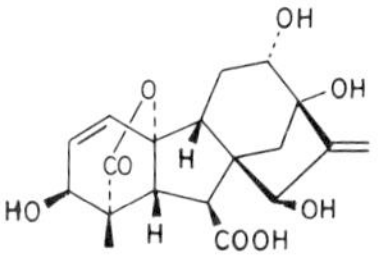
GA$_{32}$

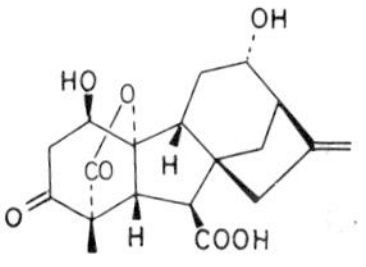
GA$_{33}$

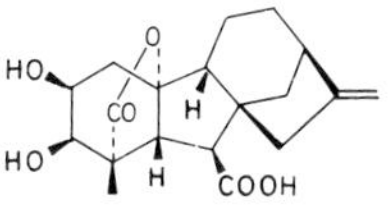
GA$_{34}$

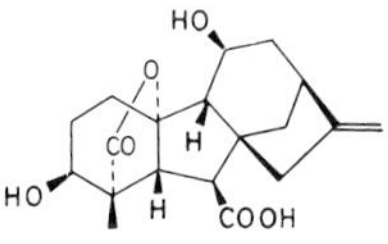
GA$_{35}$

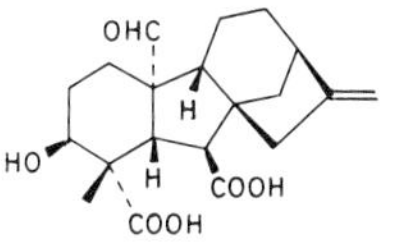
GA$_{36}$

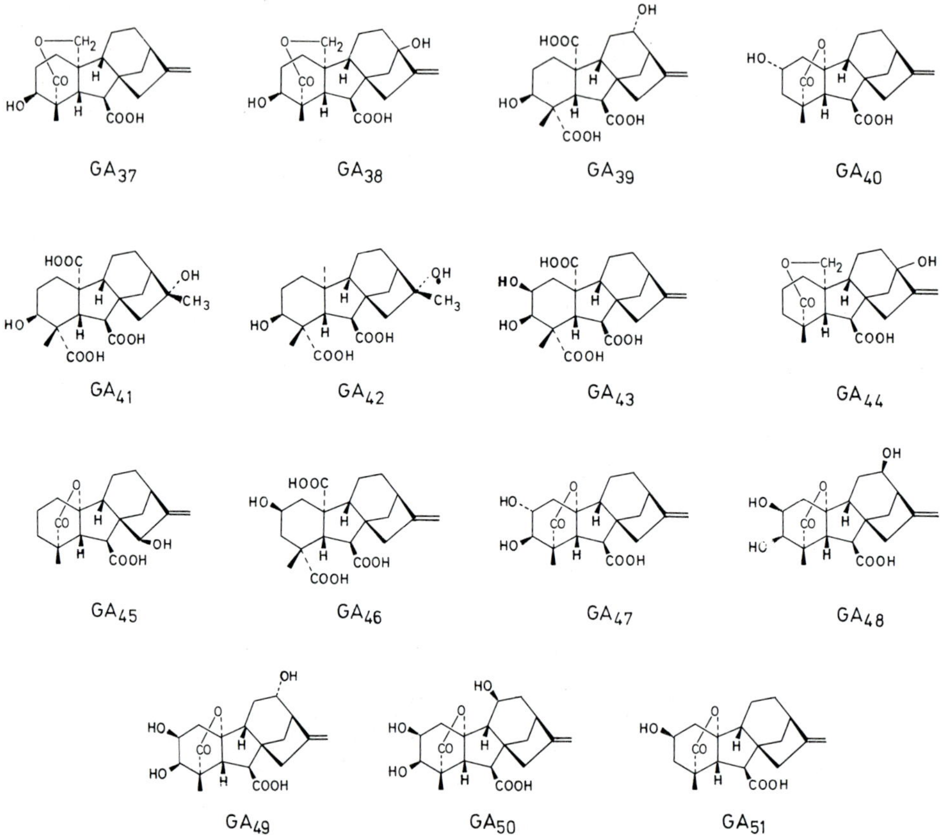

Figure 3.1. Known gibberellin structures. References are given in Table 3.1.

tetracyclic diterpenes (McCrindle and Overton 1965). The systematic nomenclature is based on the name "kaurane" for the structure IV and "gibberellane" for the structure V (Rowe 1968). Since the GAs and their precursors are enantiomeric to these structures, they are systematically denoted *ent*-gibberellanes (C_{20}-GAs, II), *ent*-20-norgibberellanes (C_{19}-GAs, III) and *ent*-kauranes (I), respectively. Naturally occurring compounds with the stereochemistry of kaurane (IV) are known but no representative of the gibberellanes (V) has been found.

An earlier used nomenclature was based on the carbon skeleton VI, named "gibbane" (Grove and Mulholland 1960). However, this system, which came into being before the stereochemistry and biogenesis of the GAs were known, is unrelated to other diterpenoids, notably the GA precursors, and is therefore no longer used. Note that in the gibbane system the positions 2 and 3 are exactly reversed from those of the gibberellane system; this is particularly treacherous in reading older literature.

Since the systematic nomenclature is cumbersome, the individual GAs are usually referred to by their A-numbers as gibberellin $A_1 \ldots A_n$ or abbreviated $GA_1 \ldots GA_n$.

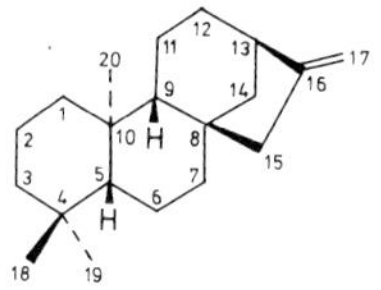

I

ent-Kaurene-*ent*-kaur-16-ene. (–)-kaurene

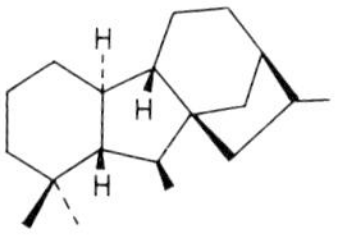

III

ent-20-Norgibberellane (C_{19}-GAs)

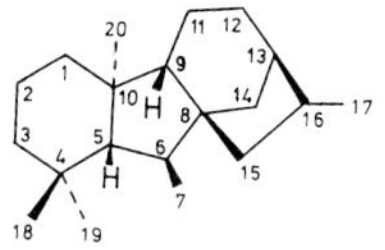

II

ent-Gibberellane (C_{20}-GAs)

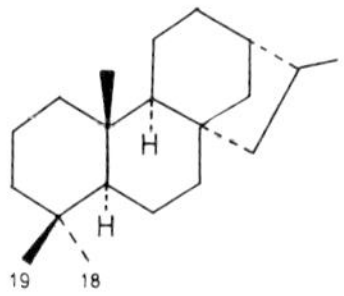

IV

Kaurane

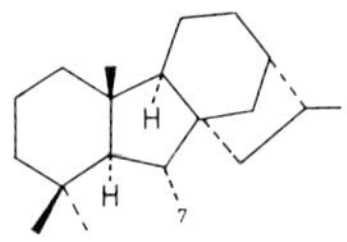

V

Gibberellane

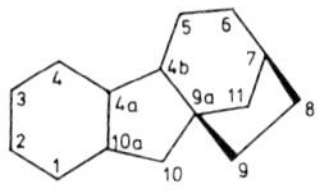

VI

Gibbane

According to rules proposed by MacMillan and Takahashi (1968), the A-numbers should be used only for naturally occurring, fully characterized compounds which have a gibbane skeleton and show activity in the specific GA bioassays. Common practice has come to deviate from these rules insofar that the gibbane system is no longer being used and many of the numbered GAs have little or no biological activity. At present, Dr. MacMillan at the University of Bristol and Dr. Takahashi of the University of Tokyo handle the allocation of A-numbers to avoid duplications. The A-numbers of GAs, which can be regarded as trivial names for *ent*-gibberellanes and *ent*-20-norgibberellanes, are used in this chapter.

In contrast to the GAs, their kauranoid precursors are often called by their systematic or semi-systematic names, which has caused some confusion particularly in describing GA biosynthesis. The *ent*-system is useful in distinguishing between enantiomeric carbon skeletons but it is not practical in the description of substituents. In this system, α and β refer to the projection of the standard skeletons (here IV and V) and are retained (with the prefix "*ent*") in the names of the enantiomers although the projections of the substituents are reversed. Thus the structure VII, an intermediate in GA biosynthesis, is

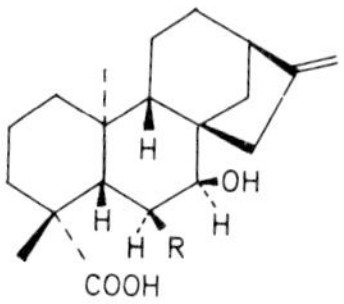

VII R = H

VIII R = T

ent-7α-Hydroxykaurenoic acid

ent-7α-Hydroxy-[6α-$^3H_1$]kaurenoic acid

named *ent*-7α-hydroxykaurenoic acid although the hydroxyl group projects out of the plane of the paper and must be referred to as a β-hydroxyl group. This is particularly unfortunate in the description of stereospecific labelling, as for example in *ent*-7α-hydroxy-[6α-$^3H_1$]kaurenoic acid (VIII), where the tritium atom must be referred to as a 6β- or *ent*-6α-tritium.

In naming kaurenolides, we shall be using the trivial names, designating substituents the way they are actually drawn, e.g. 7β-hydroxykaurenolide (XLI). The conventional way of designating the substituents is also used with the A-numbers of the GAs.

3.2.2. Structures

Fifty-one GAs have been fully characterized to date, just one short of a full deck of cards. Their structures are shown in Figure 3.1 and references to the species in which they have been discovered or conclusively identified are given in Table 3.1. Common features of the structures are the *ent*-gibberellane (II) or *ent*-20-norgibberellane (III) skeletons, respectively, and the carboxyl group at C-6. Basically, all the GAs also have a 19-oic acid group, which is involved in the typical γ-lactone ring of the C_{19}-GAs and in the δ-lactone ring of those C_{20}-GAs which have C-20 at the oxidation state of an alcohol (e.g. GA_{15}). GAs with C-20 at the oxidation state of an aldehyde (e.g. GA_{24}) form δ-lactols in pyridine and probably occur as equilibrium mixtures of lactols and free acids in water solution.

The numerical order of the GAs is not systematic but approximately represents the order of discovery. Many of the GAs fit into a simple scheme consisting of four major and two minor series characterized by the hydroxylation patterns at C-3, C-12 and C-13 (Table 3.2). Within each series, the first members are C_{20}-GAs arranged in order of increasing oxidation levels of C-20 which is present as a methyl, δ-lactone, formyl (in equilibrium with the δ-lactol) or carboxyl group. The next member of each series is the corresponding C_{19}-GA with the 19 → 10 lactone bridge. The second C_{19}-GA, where present, has a Δ-1 double bond. Two of the series also have members with Δ-2 double bonds. This scheme, which is related to the biosynthesis of GAs, accommodates 26 of the known structures and several others are related to them in a simple manner. Thus GA_8, GA_{27}, GA_{29}, GA_{34}, GA_{43}, GA_{46} and GA_{51} are 2β-hydroxy derivatives; GA_{40} and GA_{47} are 2α-hydroxy derivatives, and GA_2, GA_{10}, GA_{41} and GA_{42} are 16,17-hydrates of GAs in the scheme. The remaining GA_6, GA_{11}, GA_{16}, GA_{21}, GA_{22}, GA_{26}, GA_{33}, GA_{35}, GA_{45}, GA_{48}, GA_{49} and GA_{50} show individual features.

A number of naturally occurring GA conjugates have been identified; they are listed in Table 3.3. Most prominent in higher plants are the β-D-glucosyl ethers, usually referred to as GA glucosides. In these, the glucosyl moiety is often attached to the 2β-hydroxyl group as, for example, in GA_8-glucoside (IX), which is the single glucosyl ether most frequently found. Another kind of glucosidation is present in the glucosyl esters, for example GA_1-glucosyl ester (X) in which the C-7 carboxyl group is esterified with glucose. Both the glucosyl ethers and the glucosyl esters are very polar compounds. The glucosyl esters are neutral (cf. section 3.3.2). Both glucoside ethers and glucoside esters have been prepared synthetically (Schneider et al. 1974; Hiraga et al. 1974c). The glucosyl derivatives have metabolic significance and will be discussed extensively in section 3.8. An acetate of GA_3, which is produced by *G. fujikuroi*, was the first GA derivative to be

Table 3.1.

Sources of conclusively identified GAs

GA	Source	Reference
GA_1	*Gibberella fujikuroi*	Stodola et al. (1955)
		Grove et al. (1958)
	Phaseolus vulgaris	West and Phinney (1959)
	Phaseolus coccineus	MacMillan et al. (1960)
	Citrus reticulata	Kawarada and Sumiki (1959)
	Althaea rosea	Harada and Nitsch (1967)
	Cucumis sativus and *C. melo*	Hemphill et al. (1972)
	Corylus avellana	Williams et al. (1974)
	Sonneratia apetala	Ganguly and Sircar (1974)
GA_2	*G. fujikuroi*	Takahashi et al. (1955)
		Grove (1961)
GA_3	*G. fujikuroi*	Cross (1954)
		Stodola et al. (1955)
	A. rosea	Harada and Nitsch (1967)
	Pharbitis nil	Murofushi et al. (1968)
	C. sativus and *C. melo*	Hemphill et al. (1972)
	S. apetala	Ganguly and Sircar (1974)
	Pinus attenuata	Kamienska et al. (1976a)
GA_4	*G. fujikuroi*	Takahashi et al. (1957)
	Pyrus malus	Sińska et al. (1973)
	P. coccineus	Bowen et al. (1973)
	C. sativus	Hemphill et al. (1972)
	P. vulgaris	Hiraga et al. (1974b)
	Echinocystis macrocarpa	Beeley et al. (1975)
	P. attenuata	Kamienska et al. (1976a)
GA_5	*P. vulgaris*	West and Phinney (1959)
	P. coccineus	MacMillan et al. (1960)
	P. nil	Murofushi et al. (1968)
	Rhizophora mucranata	Ganguly and Sircar (1974)
	Prunus persica	Yamaguchi et al. (1975c, d)
GA_6	*P. coccineus*	MacMillan et al. (1962)
	P. vulgaris	Durley et al. (1971)
GA_7	*G. fujikuroi*	Cross et al. (1962)
	P. malus	Sińska et al. (1973)
	P. attenuata	Kamienska et al. (1976a)
	E. macrocarpa	Beeley et al. (1975)
GA_8	*P. coccineus*	MacMillan et al. (1962)
	A. rosea	Harada and Yokota (1970)
	Calonyction aculeatum	Murofushi et al. (1970)
	P. vulgaris	Durley et al. (1971)
	P. nil	Yokota et al. (1971c)
GA_9	*G. fujikuroi*	Cross et al. (1962)
	A. rosea	Harada and Nitsch (1967)
	P. malus	Sińska et al. (1973)
	C. avellana	Williams et al. (1974)
	Pisum sativum	Frydman et al. (1974)

Table 3.1. cont.

	R. mucranata	Ganguly and Sircar (1974)
GA_{10}	*G. fujikuroi*	Hanson (1966)
GA_{11}	*G. fujikuroi*	Brown et al. (1967)
GA_{12}	*G. fujikuroi*	Cross and Norton (1965)
GA_{13}	*G. fujikuroi*	Galt (1965)
GA_{14}	*G. fujikuroi*	Cross (1966)
GA_{15}	*G. fujikuroi*	Hanson (1967)
GA_{16}	*G. fujikuroi*	Galt (1968)
		Bearder and MacMillan (1973)
GA_{17}	*P. coccineus*	Pryce and MacMillan (1967)
	C. aculeatum	Murofushi et al. (1970)
	P. nil	Takahashi et al. (unpublished, cited by Murofushi et al. 1973)
	P. sativum	Frydman et al. (1974)
GA_{18}	*Lupinus luteus*	Koshimizu et al. (1966)
	Wistaria floribunda	Koshimizu et al. (1972)
GA_{19}	*Phyllostachys edulis*	Murofushi et al. (1966)
	P. coccineus	Pryce et al. (1967)
	C. aculeatum	Murofushi et al. (1970)
	L. luteus	Fukui et al. (1972)
GA_{20}	*P. nil*	Murofushi et al. (1968)
	P. coccineus	MacMillan and Pryce (1968)
	P. sativum	Frydman and MacMillan (1973)
	Bryophyllum daigremontianum	Gaskin et al. (1973)
GA_{21}	*Canavalia gladiata*	Murofushi et al. (1969a, b)
GA_{22}	*C. gladiata*	Murofushi et al. (1969a, b)
GA_{23}	*L. luteus*	Koshimizu et al. (1968)
	W. floribunda	Koshimizu et al. (1972)
GA_{24}	*G. fujikuroi*	Harrison and MacMillan (1971)
	E. macrocarpa	Beeley et al. (1975)
GA_{25}	*G. fujikuroi*	Harrison and MacMillan (1971)
	E. macrócarpa	Beeley et al. (1975) (see footnote a)
GA_{26}	*P. nil*	Yokota et al. (1971b)
GA_{27}	*P. nil*	Yokota et al. (1971b)
	C. aculeatum	Murofushi et al. (1970)
GA_{28}	*L. luteus*	Fukui et al. (1971)
	P. coccineus	Gaskin and MacMillan (1975)
GA_{29}	*C. aculeatum*	Murofushi et al. (1970)
	P. nil	Yokota et al. (1970)
	P. sativum	Frydman and MacMillan (1973)
GA_{30}	*C. aculeatum*	Murofushi et al. (1970)
GA_{31}	*C. aculeatum*	Murofushi et al. (1970)
GA_{32}	*P. persica*	Yamaguchi et al. (1975c, d)
	Prunus armeniaca	Coombe and Tate (1972)
GA_{33}	*C. aculeatum*	Murofushi et al. (1971)
GA_{34}	*C. aculeatum*	Murofushi et al. (1971)

Table 3.1. cont.

GA_{35}	*Cytisus scoparius*	Yamane et al. (1974)
GA_{36}	*G. fujikuroi*	Bearder and MacMillan (1972)
GA_{37}	*G. fujikuroi*	Bearder and MacMillan (1973)
	P. vulgaris	Hiraga et al. (1972, 1974b)
GA_{38}	*P. vulgaris*	Hiraga et al. (1972, 1974b)
	P. sativum	Frydman et al. (1974)
GA_{39}	*Cucurbita pepo*	Koshimizu (unpublished, cited by Takahashi 1974)
GA_{40}	*G. fujikuroi*	Yamaguchi et al. (1975b)
GA_{41}	*G. fujikuroi*	Bearder and MacMillan (1973)
GA_{42}	*G. fujikuroi*	Bearder and MacMillan (1973)
GA_{43}	*E. macrocarpa*	Beeley et al. (1975)
	Cucurbita maxima	Graebe et al. (1974c)
GA_{44}	*P. sativum*	Frydman et al. (1974)
GA_{45}	*Pyrus communis*	Bearder et al. (1975c)
GA_{46}	*E. macrocarpa*	Beeley and MacMillan (1976)
GA_{47}	*G. fujikuroi*	Beeley and MacMillan (1976)
GA_{48}	*C. pepo*	Koshimizu et al[b].
GA_{49}	*C. pepo*	Koshimizu et al[b].
GA_{50}	*Lagenaria leucantha*	Koshimizu et al[b].
GA_{51}	*P. sativum*	Sponsed and MacMillan (1977)

[a] The name GA_{25} was first given to a GA which was isolated from *Sonneratia apetala* (Ganguly et al. 1970). The A_{25}-allocation was, however, withdrawn after the structure was conclusively shown to be that of tetrahydro-GA_3 (Gaskin et al. 1972).

[b] Unpublished information from Dr. J. MacMillan.

characterized and propyl esters of GA_1 and GA_3 have been found in seeds of *Cucumis sativus* (Table 3.3).

Related to the GAs are gibberethione (formerly pharbitic acid) (XI), which is a sulphur-containing derivative of GA_3 from immature seeds of *Pharbitis nil* (Yokota et al. 1974) and antheridiogen (XII), the antheridium-inducing factor of a fern, *Anemia phyllitidis* (Nakanishi et al. 1971; Endo et al. 1972). Antheridiogen may be seen as a rearranged GA_4. Antheridium-inducing factors are discussed further in Chapter 7.

3.2.3. Occurrence

The species in which the GAs have been conclusively identified are listed in Table 3.1. Of the 51 structures, 13 have been found only in *G. fujikuroi*, 29 only in higher plants and nine have been found both in *G. fujikuroi* and in higher plants. There is no basic difference between the GAs of the fungus and those of the higher plants but the latter have a more varied pattern of hydroxylation. A certain degree of species specificity is evident in this pattern, which is brought out in Table 3.4 showing GAs in the immature seeds of some thoroughly investigated species. Apparently *Pisum sativum* lacks a 3β-

Table 3.2.

Systematic arrangement of GAs

	Hydroxylation pattern					
A-Ring structure	no OH	13-OH	3β-OH	3β,13-$(OH)_2$	3β,12α-$(OH)_2$	3β,12α,13,15β $(OH)_4$
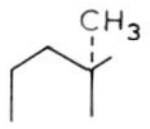	GA_{12}		GA_{14}	GA_{18}		
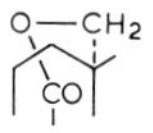	GA_{15}	GA_{44}	GA_{37}	GA_{38}		
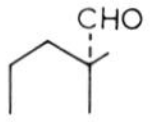	GA_{24}	GA_{19}	GA_{36}	GA_{23}		
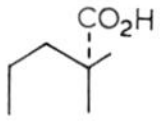	GA_{25}	GA_{17}	GA_{13}	GA_{28}	GA_{39}	
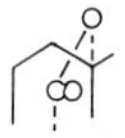	GA_9	GA_{20}	GA_4	GA_1		
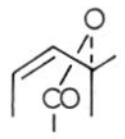			GA_7	GA_3	GA_{30}	GA_{32}
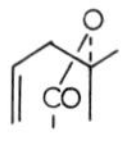				GA_5 *	GA_{31} *	

* GA_5 and GA_{31} are here viewed as dehydration products of 3-hydroxy GAs, with which they occur (Table 3.1).

hydroxylating system (only GA_{38}, which occurs in minute amounts, disturbs the pattern). *Echinocystis macrocarpa* lacks a 13-hydroxylating system, which may be common in Cucurbitaceae. The two *Phaseolus* species and the two representatives of Convulvulaceae have both 3β- and 13-hydroxylating systems although in *Calonyction,* 12α-hydroxylation apparently competes with 13-hydroxylation (cf. Figure 3.1).

Twenty-six species of higher plants representing 11 families are listed in Table 3.1. With one exception (Pinaceae) they are dicotyledonous. In addition, a much wider distribution of GAs is indicated by the presence of so-called "gibberellin-like activity" in numerous plant extracts. This term is used for the activity of unidentified substances ("GA-like substances") in bioassays that respond very specifically to GAs (Phinney et al. 1957). Tamura et al. (1969) in their review recorded the occurrence of GA-like activity in 9 gymnosperms and approximately 130 species of angiosperms and more recently Murakami (1970b) found GA-like activity in 75 of 86 mono- and dicotyledonous species from 82 families investigated. The conclusive identification of GA in higher plants together

Table 3.3.

Naturally occurring gibberellin conjugates

Conjugate	Source	Reference
GA_3-*O*(3)-β-D-glucoside	*P. nil*	Tamura et al. (1968)
GA_8-*O*(2)-β-D-glucoside	*P. coccineus*	Schreiber et al. (1970)
	A. rosea	Harada and Yokota (1970)
	P. nil	Yokota et al. (1971c)
	P. vulgaris	Hiraga et al. (1974a, b)
GA_{26}-*O*(2)-β-D-glucoside	*P. nil*	Yokota et al. (1971c)
GA_{27}-*O*(2)-β-D-glucoside	*P. nil*	Yokota et al. (1971c)
GA_{29}-*O*(2)-β-D-glucoside	*P. nil*	Yokota et al. (1971c)
GA_{35}		
GA_{35}-*O*(11)-β-D-glucoside	*C. scoparius*	Yamane et al. (1974)
GA_1 β-D-glucosyl ester	*P. vulgaris*	Hiraga et al. (1974a, b)
GA_4 β-D-glucosyl ester	*P. vulgaris*	Hiraga et al. (1974a, b)
GA_{37} β-D-glucosyl ester	*P. vulgaris*	Hiraga et al. (1974a, b)
GA_{38} β D-glucosyl ester	*P. vulgaris*	Hiraga et al. (1974a, b)
3-*O* β-Acetyl GA_3	*G. fujikuroi*	Schreiber et al. (1966)
GA_1-*n*-propyl ester	*C. sativus*	Hemphill et al. (1973)
GA_3-*n*-propyl ester	*C. sativus*	Hemphill et al. (1973)

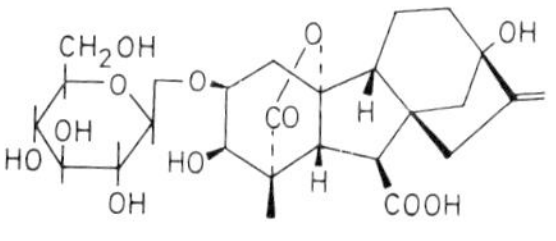

IX

GA_8-glucoside

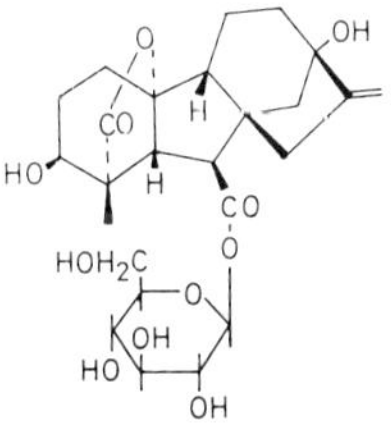

X

GA_1-glucosyl ester

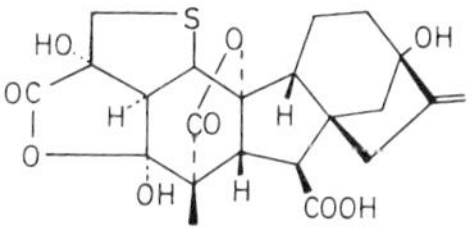

XI

Gibberethione (earlier name: pharbitic acid)

with the wide-spread occurrence of GA-like activity suggest that GAs are of universal occurrence in higher plants.

Already at an early stage, GA-like activity was recorded from all parts of higher plants including shoots, roots, leaves, flower buds, fruits and seeds (see reviews by Phinney and West 1960; Brian 1966; Tamura et al. 1969; and, specifically for seeds, Bhalla and Gairola 1972). Subsequently GA-like activity has also been demonstrated in chloroplasts (Stoddart 1968; Railton and Reid 1974a), flower petals and anthers (Murakami 1973, 1975) and GAs have been conclusively identified in pollen (Kamienska et al. 1976a). The GA-like activity may be qualitatively different in different parts of the same plant as shown for mature and immature leaves and runner buds of strawberry plants (Goodwin and Gordon 1972). In general, vegetative parts of the plant contain less GA-like activity (a few μg or less per kg fresh weight) than reproductive parts, particularly stamens and

Table 3.4.

GAs in some thoroughly investigated immature seeds

Family/species	3-Deoxy		3β-Hydroxy	
	13-Deoxy	13-Hydroxy	13-Deoxy	13-Hydroxy
Leguminosae				
Phaseolus coccineus		A_5 A_{17} A_{19} A_{20}	A_4	A_1 A_6 A_8 A_{28}
Phaseolus vulgaris		A_5	A_4 A_{37}	A_1 A_6 A_8 A_{38}
Pisum sativum	A_9 A_{51}	A_{17} A_{20} A_{29} A_{44}		A_{38}
Convolvulaceae				
Pharbitis nil		A_5 A_{17} A_{20} A_{29}	A_{26} A_{27}	A_3 A_8
Calonyction aculeatum	A_{31}	A_{17} A_{19} A_{29}	A_{27} A_{30} A_{33} A_{34}	A_8
Cucurbitaceae				
Echinocystis macrocarpa	A_{24} A_{25} A_{46}		A_4 A_7 A_{43}	

immature seeds (often 10–100 mg per kg fresh weight). Since immature seeds are both rich in GA and often large in size, they have become the favourite starting material for the isolation of GAs. They were also the sources of most GAs listed in Table 3.1, although other sources are represented: GA_1 and GA_9 of *C. avellana* and GA_4, GA_7 and GA_9 of *P. malus* were obtained from mature seeds (cf. 3.11.4); GA_1, GA_3, GA_8 and GA_9 of *A. rosea*, GA_{19} of *P. edulis* and GA_{20} of *B. daigremontianum* were obtained from shoot apices, in the latter case including flower buds; GA_1, GA_3, GA_5 and GA_9 of *S. apetala* and *R. mucranata* were found in leaves; GA_1 of *C. reticulata* was found in water sprouts; and GA_3, GA_4 and GA_7 of *P. attenuata* were obtained from pollen. The GAs of seeds to a great degree become conjugated as the seeds mature. Hence maturing seeds were the sources of most of the glucosides in Table 3.3. Only GA_8-glucoside was also isolated from shoot apices of *A. rosea.*

Changes in levels of GA-like activity in relation to growth of the vegetative plant (Chapter 2), fruit development (Chapter 3), flower formation (Chapter 6), seed dormancy and germination (Chapter 8), and bud dormancy (Chapter 9) are discussed in Volume II.

Tamura et al. (1969) as quoted by Lang (1970) also recorded the occurrence of GA-like activity in 28 species of bacteria, 7 algae, 8 fungi (in addition to *G. fujikuroi*), 1 moss and 7 ferns. Since then, activity has been reported in further bacteria (Brown 1972; Gairola et al. 1972; Kucherenko 1973), algae (Abe et al. 1974) and fungi (Gogola 1972; Pegg 1973a, b). The occurrence of compounds like antheridiogen (XII) in a fern and helminthosporol (XXX) in a fungus, compounds that are not typical GAs but nevertheless have some GA-like activity (sections 3.6.2, 3.6.3), warns against equating GA-like activity with the presence of GA. Indeed, GAs have not been conclusively identified in any lower plant other than *G. fujikuroi.*

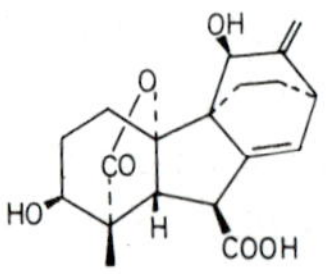

XII

Antheridiogen

3.3. Isolation and identification

3.3.1. Recovery

The recovery of GAs from plant material is variable and much attention must be paid to the isolation procedures, if the composition of the final extract is to bear any resemblance to that of the intact plant. Faull et al. (1974a) after preliminary purification through solvent partition and a single TLC step recovered 58–59% of GA_3, which had been added to the plant material during extraction. It is our non-committal estimate that some 40–80% of endogenous GA may be recovered by the standard methods. Even if the methods are rigorously controlled, variability cannot be excluded because GAs of different polarities may be recovered to different extents and because of other variables which are discussed in the following sections. An appreciation of these difficulties is important for proper experimental design and for a critical evaluation of results claimed and cited in the literature. This point will be elaborated upon in section 3.5.

3.3.2. Extraction

For extraction of GAs, the plant material is usually macerated with cold methanol or acetone containing up to 50% water. After evaporation of the organic solvent and centrifugation, a preliminary purification is achieved by partitioning the aqueous residue against ethyl acetate at different pH values (Curtis and Cross 1954). The GAs are extracted into ethyl acetate at pH 2.5–3.0, re-extracted into buffer at pH 8.0 (Curtis and Cross used pH 6.5) and again extracted into ethyl acetate after adjusting pH to 2.5–3.0. Almost all free GAs may be obtained in the final (acid) ethyl acetate fraction by this method. Non-polar GA glucosyl esters (those of GA_4 and GA_{37}) also become extracted by ethyl acetate (Takahashi 1974) but remain in the first (neutral) ethyl acetate fraction. Not extracted by ethyl acetate are GA_{32}, part of GA_{28}, probably GA_{41} and GA_{43}, all the GA glucosides and the most polar glucosyl esters (those of GA_1 and GA_{38}). These compounds can be extracted with butanol from the acidified water phase and the butanol extract may be separated into a neutral fraction containing the esters and an acidic fraction containing the other components (Takahashi 1974). An example of a separation of these fractions, although in a slightly different sequence, is found in Hiraga et al. (1974a).

Durley and Pharis (1972) have investigated the effects of pH and buffer concentration on the partition coefficients for 27 GAs between phosphate buffer and ethyl acetate. Most GAs distribute well into ethyl acetate at pH 2.5 and into phosphate buffer at pH 8.0, but quite substantial losses may occur in both cases. A high concentration of the phosphate buffer favours distribution into the organic phase, which is an advantage is the extraction of polar GAs into ethyl acetate at low pH, but may lead to incomplete recovery of apolar GAs in the buffer at pH 8.0. Thus, at pH 2.5 GA_3 would be almost completely extracted by three equal volumes of ethyl acetate from 1.5 *M* buffer, but only to 82% from 0.05 *M* buffer. At pH 8.0, however, GA_9 would be recovered from the EtOAc phase to 99% by three volumes of 0.05 *M* buffer but only to 58% by the same amount of 1.5 *M* buffer. From these results it appears that complete extraction would only be ensured by using high concentration of buffer for the extraction at low pH and vice versa. The study also showed that apolar GAs, e.g. GA_9, do not distribute well into the buffer phase at pH 6.5

although GA_3 does. Extraction of acids at pH 2.5 rather than at pH 3.5 is of advantage mainly for the extraction of dicarboxylic GAs. Extreme pH values should be avoided because they bring about degradation of GAs (section 3.4.1). Petroleum ether and ethyl ether have been used to remove lipids, which otherwise would have interfered with subsequent separation and bioassay, at neutral or slightly alkaline pH before the extraction of GAs. Whereas petroleum ether was not found to extract GAs from the buffer phase at pH 8.0, ethyl ether removed significant amounts of the less polar GAs (Durley and Pharis 1972). It should be kept in mind that the extraction experiments of Durley and Pharis (1972) were done with pure solutions. Plant extracts contain a multitude of surface active, buffering, adsorbing and other substances that are likely to change the partition properties considerably and unpredictably. Many constituents become very difficult or impossible to recover after a plant extract has been taken to complete dryness.

The GA moieties of GA conjugates may be liberated by hydrolysis and extracted with ethyl acetate. Since acid- or base-catalyzed hydrolysis results in destruction of most GAs, enzymatic hydrolysis is preferred (e.g. Yokota et al. 1971c; Hiraga et al. 1974a). The efficiencies of β-glucosidases, cellulases and other enzymes from different sources in hydrolyzing GA glucosides have been investigated by Knöfel et al. (1974). The most rapid and complete hydrolysis without changes in the GA molecule was obtained with a cellulase from *Aspergillus niger* and β-glucosidases from some sources but not from others.

Instead of organic solvents, buffers of neutral or slightly basic pH can be used for the extraction of GAs from plant material (McComb 1961). This method has the advantage that less lipid impurities are obtained in the extract and, on the basis of bioassay measurements, it has been claimed to give better results than extraction with organic solvents (R.L. Jones 1968; Dale 1969). In spite of this, extraction with organic solvents is more commonly used.

An early method of removing GAs from the medium of *G. fujikuroi* (Yabuta and Hayashi 1939; Curtis and Cross 1954) and from extracts of higher plants (West and Phinney 1959; MacMillan et al. 1960) consisted of adsorbing them onto charcoal and subsequently eluting the GAs with acetone. The advantage was rapid reduction of volume. Although still in use, the method has been largely abandoned owing to variable activity of different lots of charcoal and owing to destruction of GAs by acid sometimes present in the charcoal.

Experiments have been done on the isolation of GA_3 on columns packed with Sepharose to which GA_3-specific antibodies were bound (Fuchs and Gertman 1974). Attempts to quantitatively determine GA_3 in an extract by pre-loading the column with radioactive GA_3 and measuring the amount displaced by the extract yielded no reproducible results because different GAs cross-reacted with the antibodies to different degrees which resulted in unpredictable displacement. However, it seems as if the method could be useful for bulk isolation of GAs if antibodies to all the known GAs or at least to certain typical ones could be prepared for the column.

3.3.3. Purification and separation

Crude plant extracts usually contain impurities that interfere with GA bioassays and make identification by physical methods inconclusive. However, experience has shown that unless special measures are taken to avoid contamination with solvent residues and, above all, plasticizers from plastic vessels, tubing or bottle screw caps, the purification may introduce more impurities than it removes and make analysis by GLC and GLC-mass spectrometry impossible (Binks et al. 1970; Gaskin et al. 1971).

Paper chromatography was used extensively to purify the extracts during early stages of GA research (see review by Phinney and West 1961). It is now greatly replaced by TLC but is still used for the purification of extracts to be bioassayed by the lettuce hypocotyl and cereal endosperm assays, which can be carried out directly on pieces of the chromatograms. Care must be taken when solvent systems containing NH_4OH are used, since they may cause degradation (Cross et al. 1961).

TLC is preferred by many for the purification of small amounts of GA-containing extracts for bioassay. Solvent systems for both adsorption and partition chromatography devised at an early stage (Sembdner et al. 1962; MacMillan and Suter 1963; Kagawa et al. 1963; Cavell et al. 1967) are still being used with success although the multitude of GAs which have become known in the meantime have rendered the tables of R_f values insufficient. A useful system for the removal of bioassay inhibitors has been used by Neskovic and Konjevic (1974).

Two problems deserve special attention in the TLC of GAs: the elution of GAs from the adsorbent and the removal of solvent residues. Water-saturated ethyl acetate is one of the most commonly used solvents for elution of GAs. However, this elutes the more polar GAs poorly, for which combinations of acetone and MeOH or even pure MeOH may be necessary. Such polar solvents dissolve considerable amounts of silicic acid which then contaminates the eluate. In our experience with radioactive GAs, even very apolar GAs such as GA_{12} and GA_{15} may become almost insoluble in acetone (but not in MeOH) after evaporation of the solvent from eluates containing silica gel. This may have led to losses in many cases, since acetone is a preferred solvent for GAs. Solubility can often be restored by partitioning between ethyl acetate and weak aqueous HCl (pH 3). Perhaps the problem arises through complex formation between GAs and metal ions present in some silica gels (Donike et al. 1973). As for the second problem, if the extracts are to be analyzed by bioassay, it is absolutely necessary to remove solvent residues, particularly acetic acid, which is contained in some of the most popular TLC solvent systems but is highly toxic to the barley endosperm and lettuce hypocotyl bioassays even in trace amounts. It can only be removed by prolonged evaporation in vacuo.

A number of systems have been developed for the purification of GAs on a preparative scale and they are also being used increasingly to purify extracts for bioassay. A good preliminary purification may be obtained by counter-current distribution (West and Phinney 1959; Crozier et al. 1969; Murofushi et al. 1973). Column chromatography is more common. Preliminary purification without much separation of the individual GAs has been obtained on columns of charcoal-Celite (West and Phinney 1959; MacMillan et al. 1962; Durley et al. 1971), silica gel for adsorption chromatography (West and Phinney 1959), insoluble polyvinylpyrrolidone (Glenn et al. 1972) and Sephadex G-10 (Crozier et al. 1969). Separation of individual GAs is best accomplished by partition chromatography. Thus Stodola et al. (1957) separated GA_1 and GA_3 on a phosphate-buffered Celite

3.4.1. Decomposition and degradation

GAs are moderately stable in water solution and GA_3 is one of the least stable ones, forming the products XIII to XVII to a significant degree at room temperature (Cross et al. 1961; Pryce 1973). The major decomposition product in unbuffered solution at room temperature is gibberellenic acid (XIII) (Moffatt 1960). Autoclaving GA_3 in aqueous solution results in 98–99% destruction with the formation of the products XIII to XVIII (Pryce 1973, 1974).

The decomposition of GAs is greatly catalyzed by acids and bases. Aqueous mineral acid at room temperature causes the cônversion of GA_3 to allogibberic acid (XIV) and epiallogibberic acid (XV) (Grove and Mulholland 1960). At 100 °C a Wagner-Merwein

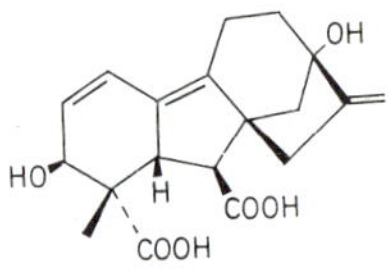

XIII
Gibberellenic acid

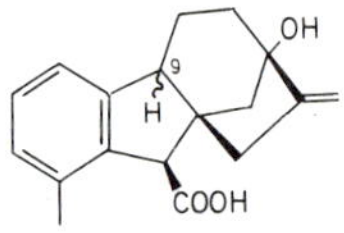

XIV 9α- H
XV 9β- H
allo Gibberic acid
epiallo Gibberic acid

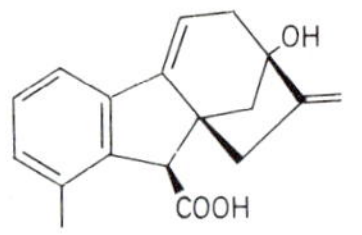

XVI
Dehydro*allo*gibberic acid

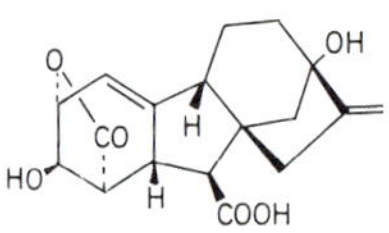

XVII
iso-GA_3

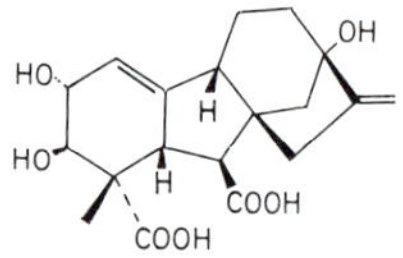

XVIII
iso-GA_3 hydroxy acid

rearrangement of the C/D rings occurs with formation of gibberic acid (XIX) and *epi* gibberic acid (XX) (Cross 1954; Grove et al. 1960). The aromatization of ring A is typical

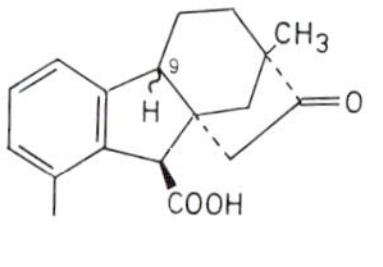

XIX 9α- H
XX 9β - H
Gibberic acid
epi Gibberic acid

for GAs with 1,2-unsaturation (e.g. GA_3, GA_7) and does not occur in GAs with saturated A–rings (e.g. GA_1, GA_4). The rearrangement of the C/D rings is common for GAs with a 13-hydroxyl group; GA_1 therefore from the product XXI with hot aqueous mineral acid.

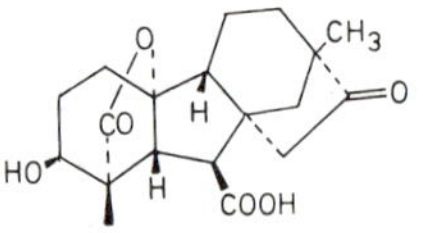

XXI
Rearranged GA_1

In GAs that have no 13-hydroxyl group, treatment with aqueous mineral acid may result in hydration of the 16,17-double bond as in the conversion of GA_4 to GA_2 (Grove 1961) and GA_9 to GA_{10} (Hanson 1966). It may also result in the migration of the 16,17-double bond to the 15,16-position as shown for the conversion of GA_4 to the isomer XXII and

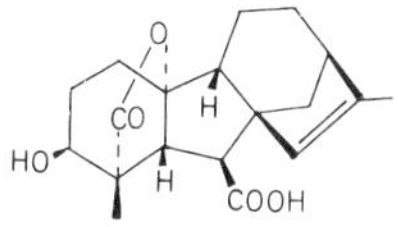

XXII

iso-GA_4

of GA_{13} to the corresponding isomer (Aldridge et al. 1965; Galt 1965).

Treatment of GA_3 with weak aqueous alkali (0.1 N NH_4OH at room temperature) causes the isomerization of the lactone ring to the isomer XVII (Cross et al. 1961). Treatment with strong aqueous alkali yields the hydroxyacid XVIII. The alkali-catalyzed

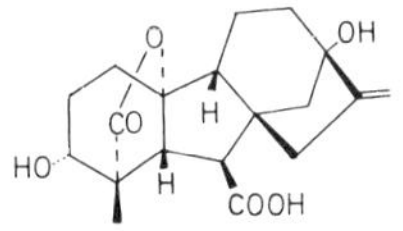

XXIII

3-*epi* GA_1 (pseudo-GA_1)

isomerization of the lactone ring is common for 1,2-unsaturated GAs and does not occur with saturated A-rings, which instead may undergo epimerization at C-3. Thus GA_1 and GA_4 upon treatment with dilute aqueous alkali very easily yield mixtures of epimers (e.g. XXIII) in which the 3-equatorial hydroxy epimers predominate (Cross et al. 1961; MacMillan and Pryce 1967). The 3-epimerisation does not occur in the presence of a 1,2-double bond or a 2-hydroxyl group (e.g. in GA_8).

Thus it is important to keep decomposition to a minimum during extraction and isolation by careful control of pH and temperature. It is even more important to keep the possibility of decomposition in mind when interpreting experimental results. The decomposition products have little or no activity in GA bioassays (Brian et al. 1967; Pryce 1973; Hoad et al. 1976). Their formation therefore represents a loss of unknown magnitude where bioassay is used for measurement. It is sometimes argued that an underestimation results in less pretentious results and is therefore permissible. But a great deal of our knowledge of GA relations in plants is based on the presence of *less* GA-like activity in one extract relative to another and many authors draw conclusions on the basis of small variations in activity between extracts. Furthermore, if metabolism of radioactive GAs is investigated without identification of the products, the formation of decomposition products may be mistaken for metabolic conversion. This is particularly a danger in feeding radioactive compounds to intact plants, since no proper control on the decomposition of the precursor within the plant can be obtained. For example, Durley et al. (1975) in feeding $[^3H]GA_{20}$ to *Bryophyllum* obtained as products 3-*epi*-GA_1 and C/D-ring-rearranged GA_{20} in addition to GA_{29}. On the other hand, identified degradation products may also aid in the identification of the parent GA as in the identification of degradation products of $[^3H]GA_1$ after acid hydrolysis of an unknown conjugate

(Nadeau and Rappaport 1974b). Because of the ready hydration of the 16,17-double bond in some GAs, the occurrence of 16-hydroxy-GAs must be interpreted with particular caution. For example, GA_2 results from the acid hydrolysis of GA_4 glucosyl ester (Hiraga et al. 1974a).

3.4.2. Preparation of labelled GAs and GA precursors

Isotopically labelled GAs for use in metabolic studies have been prepared by a number of methods including direct chemical labelling of GAs, microbiological conversion of labelled precursors and combinations of both. An important and difficult problem is the removal of radioactive by-products of the labelling reaction. They may otherwise turn up in subsequent metabolic studies and become mistaken for metabolic products or they may truly become metabolized to products, which then erroneously appear to originate from the main starting material

A useful reaction for the labelling of GAs with ^3H is the selective hydrogenation of the 1,2-double bond in GA_3 in the presence of Pd catalyst, which must be partially poisoned with pyridine to suppress hydrogenation of the exocyclic 16,17-double bond. Main products are the dicarboxylic acid (XXIV), which arises through hydrogenolysis of the

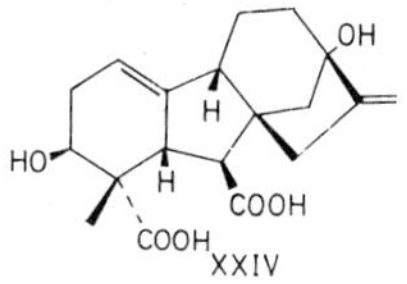

XXIV

Hydrogenolysis product from GA_3

lactone bridge, and GA_1 (Jones and McCloskey 1963). Using this method and a mixture of $^3H_2$ and $^1H_2$, Kende (1967) prepared [1,2-^{3}H]GA_1, separating it from labelled XXIV by TLC. Pitel and Vining (1970) by a modification of Kende's method obtained [1,2-^{3}H]GA_1 (487 mCi/mmole) in 20% yield. Separating the reaction products by column partition chromatography they discovered several by-products that apparently had been overlooked in the earlier work. The difficult separation of GA_1 and dihydro-GA_1, which appears as a product of over-hydrogenation, was only achieved by argentation partition chromatography (Vining 1971). Nadeau and Rappaport (1974a) using carrier-free $^3H_2$ and further modifications of the hydrogenation method obtained [1,2-^{3}H]GA_1 of very high specific activity (43 Ci/mmole). They uncovered even more by-products, among them most disconcertingly [^{3}H]GA_3 (9.4% 13 Ci/mmole) which indicated a considerable amount of non-specific exchange of alkane hydrogen with tritium. In fact, [^{3}H]GA_6 and [^{3}H]GA_3 have been prepared by direct tritiation according to the Wilzbach method (Sembdner et al. 1968; Asakawa et al. 1974a). However, Bearder et al. (1973b) examining a sample of [1,2-^{3}H]GA_1 (5 Ci/mmole) by partition chromatography could not find any contaminating [^{3}H]GA_3. Selective hydrogenation of the 1,2-double bond in GA_7 yields [^{3}H]GA_4 (Durley and Pharis 1973).

[1-^{3}H]GA_5 may be obtained by dehydration of [1,2-^{3}H]GA_1 (Musgrave and Kende 1970). Since the yield is very low by this method, Murofushi et al. (1974) preferred to use a selective reduction of GA_3 methyl ester monomesyl ether with $^3H_2/^1H_2$, which yields [1-^{3}H]GA_5 methyl ester directly by migration of the 1,2-double bond and elimination of the 3-*O*-mesyl group. [1-^{3}H]GA_5 (5.3 Ci/mmole) was obtained after hydrolysis and purification. [1-^{3}H]GA_8 was obtained from [1-^{3}H]GA_5 by oxidation with OsO_4 followed by treatment with hydrogen sulphide. Murofushi et al. (1974) also prepared [2,3-^{3}H]GA_{20} (3.3 Ci/mmole) from GA_5 via the GA_5 methyl ester 16,17-epoxide, which was hydrogenated with $^3H_2/^1H_2$.

The ring-D exocyclic methylene group is the preferred site for labelling with ^{14}C. The compound is first converted to the corresponding 17-*nor*-16-ketone, e.g. XXV, by treat-

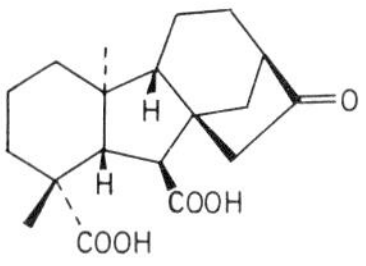

XXV

GA_{12}-norketone (GA_{12}-17-nor-16-ketone)

ment with OsO_4 and metaperiodate or by ozonolysis; then it is regenerated by a Wittig reaction using [^{14}C]-methyltriphenylphosphonium halide. This method, which was first used by Cross et al. (1964) to prepare *ent*-[17-^{14}C]kaurene and [17-^{14}C]GA_9, is best suited for labelling apolar GAs, e.g. GA_{12} and GA_{14} (Cross et al. 1968b). It has also been used extensively for ^{14}C-labelling of *ent*-kaurenoid precursors of GA, e.g. *ent*-kaurenoic acid (XL) (Geissman et al. 1966). The same method, but with tritiated Wittig reagent, introduces ^{3}H in the 17-position as, for example, in the preparation of [17-^{3}H]GA_{14} (Durley et al. 1974) and [17-^{3}H]GA_9 (Bearder et al. 1976b).

More highly hydroxylated GAs cannot be labelled by the Wittig method because a large excess of reagent is necessary, which may give rise to base catalyzed isomerization (Aldridge et al. 1965). Such GAs may be prepared by microbiological conversion of labelled precursors. Thus Hanson and Hawker (1973) prepared GA_{12}-alcohol (LXIX), which was readily labelled with ^{14}C via the *nor*-ketone, and was then transformed by cultures of *G. fujikuroi* to [17-^{14}C]GA_3 (5.58 mCi/mmole; 31.4% yield). GA_{12}-alcohol is not a natural precursor of GAs in *G. fujikuroi* (section 3.7.3), but it is converted more rapidly than the endogenous intermediates, which results in less dilution of label. Another method for obtaining labelled GAs of high specific activity is the use of the mutant B1-41a of *G. fujikuroi* to transform labelled GA precursors which occur after *ent*-kaurenoic acid in the pathway. B1-41a is blocked in the biosynthetic pathway between *ent*-kaurenal and *ent*-kaurenoic acid (section 3.7.2.3) and therefore produces minimal amounts of GA from endogenous precursors. Normally the main product is GA_3 (Bearder et al. 1975a), but Bearder et al. (1976a) obtained good yields of [^{3}H]GA_{20} by feeding [^{3}H]-steviol acetate as a precursor (section 3.7.3). Feeding normal radioactive precursors such as [2-^{14}C]mevalonate to wild type *G. fujikuroi* yields GA_3 with too low a specific radioactivity (a few μCi/mmole) for metabolic studies (Zweig and DeVay 1959; McComb 1964a; Pitel et al. 1971a).

The *ent*-kaurenoid precursors of GA (section 3.7; Scheme 3.3) do not accumulate in cultures of *G. fujikuroi* but may be obtained by partial synthesis starting with 7β-hydroxykaurenolide (XLI), which is a major metabolite of the fungus (Cross et al. 1963b).

7β-hydroxykaurenolide (XLI) may be converted to *ent*-7-oxokaurenoic acid (LXII) (Hanson and Hawker 1972a), which on Wolff-Kishner reduction yields *ent*-kaurenoic acid (XL) (Galt and Hanson 1966; Bearder et al. 1973a). *ent*-Kaurenol (XXXVIII) may then be obtained by reduction of methyl *ent*-kaurenoate with $LiAlH_4$ and can in turn be oxidized with CrO_3 to *ent*-kaurenal (XXXIX) (e.g. Murphy and West 1969; Bearder et al. 1974). *ent*-7-Oxokaurenoic acid (LXII) can also be converted to *ent*-7α-hydroxykaurenoic acid (VII Scheme 3.3), another GA precursor, whereby a tritium atom may be conveniently introduced specifically in the 6β-position (Hanson and Hawker 1972a; Hanson et al. 1972). *ent*-7α-Hydroxykaurenoic acid (VII) is also obtained by microbiological hydroxylation of *ent*-kaurenoic acid (XL) (Beilby et al. 1973; Croft et al. 1974). GA_{12}-alcohol (LXIX) (Hanson and Hawker 1973), GA_{12}-aldehyde (XLV) and GA_{12} (Cross et al. 1968b; Bearder et al. 1973a) are obtained from 7β-hydroxykaurenolide (XLI) by a ring-contraction reaction. If 3β, 7β-dihydroxykaurenolide is used as a starting material, the same method yields GA_{14}-aldehyde (XLVI), which can be reduced to GA_{14}-alcohol (LXX) (Hedden et al. 1974). All these compounds may be labelled by the Wittig method. $[6\text{-}^3H_2]GA_{14}$-aldehyde and $[6\text{-}^3H_2]GA_{14}$-alcohol have also been prepared by treatment of the corresponding unlabelled substances with ^3H_2O and sodium methoxide in tetrahydrofuran (Hedden et al. 1974).

ent-Kaurenoic acid (XL) may also be obtained from a number of higher plants. Thus, for example, it is contained in large amounts in the resin of the leaves of Australian *Beyeria* species (Jefferies and Payne 1965) and also in the dry ray florets of *Helianthus annuus* (Pyrek 1970; Kasprzyk et al. 1974). The dried sunflower florets are available for medical purposes. *ent*-$[17\text{-}^{14}C]$kaurenol has been prepared via the *nor*-ketone and Wittig reaction from *ent*-16β, 17,19-trihydroxykaurene (Verbiscar et al. 1967), another Australian plant terpene. *Espeletia grandiflora* may be a good but exotic source of *ent*-kaurenol (XXXVIII) and *ent*-kaurenal (XXXIX) (Piozzi et al. 1971). *ent*-Kaurene (I) is most conveniently obtained from *Cryptomeria japonica*, a widely grown ornamental tree. Different varieties yield either *ent*-kaurene (I) or 13β-kaurene ((+)-phyllocladene) (XXVI) (Appleton et al. 1968; MacMillan and Walker 1972) with at most 1% contamination of the other isomer (Appleton et al. 1970). The stereoisomers can easily be distinguished by their m.p. or optical rotation. *ent*-$[15,17\text{-}^3H]$kaurene and *ent*-$[15,17\text{-}^2H]$kaurene have been prepared by treating *ent*-kaurene with $[^3H]CF_3CO_2H$ and $[^2H]CF_3CO_2H$, respectively, in benzene (Bearder et al. 1974).

GA precursors and several C_{20}-GAs of high specific activity can be prepared in μg amounts from labelled mevalonate by the use of cell-free systems from higher plants (West and Upper 1969; Graebe 1969; Graebe et al. 1974a, c) (section 3.7).

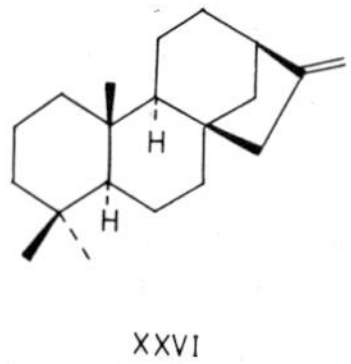

XXVI

13β-Kaurene ((+)-phyllocladene)

3.5. Bioassays

3.5.1. Introduction

Bioassays have been instrumental in advancing GA research. In the original purification and isolation of GA from fungal cultures, the fractions containing GA were followed by means of their activity in a rice seedling bioassay. The indications that GA is present in higher plants, that immature seeds contain enough of it to be isolated, and that conjugated GAs occur were first obtained by bioassay. Our knowledge of possible sites of biosynthesis, of changes in GA content during the life of a plant, and of occurrence beyond the species shown in Table 3.1, still depends heavily on bioassay work. Bioassays probably will continue to play a role in GA research, particularly as a means of discovering areas of interest for further investigation by other methods. The determination of biological activity of new GAs and research into structure-activity relationships can of course only be done by bioassay.

Unfortunately, work with bioassays has fallen into disrepute because their precision has been overestimated and they have been used in experiments of inferior design. However, this should not be held against the bioassays. Properly used and interpreted, they do detect GA and, within certain limitations, also give an estimate of the relative amounts. This section gives a general description of the most commonly used GA bioassays and a discussion of such aspects as pertain to their use and the interpretation of published data. For a complete description of GA bioassays, the reader is referred to the review by Bailiss and Hill (1971) which is devoted entirely to this subject.

3.5.2. The bioassays

Practically every physiological effect of GA has been put to use as a bioassay at some time or other. However, only some of the bioassays have become widely used due to their ease of performance, reliability, sensitivity and range of response. These are of three types: the dwarf seedling, the hypocotyl and the cereal endosperm bioassays.

3.5.2.1. Dwarf seedling bioassays. The dwarf seedling bioassays depend on the increased growth of certain dwarf cultivars in response to applied GA. A small amount (5–100 μl) of test solution is applied to the seedlings grown in soil, vermiculite or water culture. After a period of continued growth, the length of certain internodes or leaf sheaths is measured and used as the response parameter. The dwarf seedling assays are easy to perform, reliable and specific. However, they are of longer duration than the other bioassays and, with the exception of the rice seedling bioassay, require much space.

In the *dwarf pea bioassay* the test solution (usually 5 μl) is applied to the apical buds of pea seedlings. In its original version, this test was carried out over a period of 6 weeks and the log dose-log response curve for GA_3 was linear from 10^{-1} to 10 μg per plant (Brian and Hemming 1955; Brian et al. 1964). McComb and Carr (1958) using closely controlled conditions shortened the assay time to 16 days and obtained linear response with the cv. Meteor and GA_3 for 10^{-3}–10 μg per plant (later 10^{-3}–5 μg per plant; Skene and Carr 1961a). Further shortening of the duration of the assay was obtained by Köhler and Lang

(1963) using cv. Progress No. 9, which was germinated in the dark (4 days) and grown under red light during the treatment (5 days). The dose – response curve for GA_3 was linear for $5 \times 10^{-3} - 5 \times 10^{-1}$ µg per plant.

Except for some response to α-tocopherol (Bruinsma 1963), steroids (Kopcewicz 1969) and in the earliest study occasionally to IAA (Brian and Hemming 1955), the dwarf pea assay as far as is known is specific to GAs. The assay was used to demonstrate the inhibitory action of unknown components in plant extracts (Köhler and Lang 1963) and of tannins (Green and Corcoran 1975) but there is no reason to believe that it is more sensitive to such inhibitors than other assays.

The *dwarf maize bioassay* may be carried out with one of the single gene mutants d-1, d-2, d-3, d-5 and an_1 of *Zea mays* (Phinney 1961; Phinney and West 1961). The most commonly used are d-5 and d-1. The test solution (100 µl) is applied in the cone formed by the first leaf of young (7-day-old) seedlings. The length of the first or preferably the first plus the second leaf sheaths is measured after 7–10 days of treatment. The log dose – log response curve for GA_3 is linear in the range 10^{-2}–1 µg per plant for most of the mutants and 10^{-3}–1 µg per plant for d-1. The most complete account of the variables and statistics of this test is contained in the thesis by Neely (1959).

α-Tocopherol is the only non-GA or -GA precursor known to have a slight activity on d-5 (Bruinsma 1963). The test is particularly easy to perform but a draw-back is that only segregating seeds are available (tall/dwarf = 3:1) so that only 1/4th of the germinated seedlings can be used.

The *rice seedling bioassay* is very similar to the dwarf maize bioassay. The growth of a leaf sheath, usually the second, serves as a measure of response. The plants may be grown in test tubes, which is a space saving feature. Originally the test solutions were added to the tubes and thus became available to the plants through the roots. In this version, the test was thoroughly investigated by Ogawa (1963c). Soaking the seeds for 2 days and growing the plants for 7 days in continuous light resulted in significant response to 10^{-3} µg/ml of GA_3 and the log dose – log response curve for GA_3 was linear in the range 10^{-3}–1 µg/ml (1.5–3 ml of test solution). In the now more commonly used variant called the *dwarf rice microdrop.assay* (Murakami 1968a, 1970c), the test solution (1 µl) is applied to the coleoptiles of 4-day-old dwarf seedlings and the second leaf sheath is measured after another 3 days of growth in continuous light. This way the test is sensitive to GA_3 in the range 10^{-4}–1 µg per plant, the linear range being somewhat more narrow. The dwarf cultivar Tan-ginbozu is widely used because it is particularly sensitive, responds to a wider spectrum of GAs than other varieties (section 3.6.1.2) and has no endogenous GA-like activity (Suge and Murakami 1968; Murakami 1972). Helminthosporol and some of its derivatives are the only non-GAs or -GA precursors known to be slightly active in the rice seedling assays (Kato et al. 1968; section 3.6.3).

3.5.2.2. Hypocotyl assays. The hypocotyl assays depend on the increased elongation of the hypocotyls of certain seedlings in response to GA. The assays are easy to perform, rapid and require less space than the dwarf seedling bioassays, but they are also less sensitive.

The *lettuce hypocotyl assay* first described by Frankland and Wareing (1960) and further developed by Brian et al. (1964) is by far the most used test of this type. Lettuce seeds are pregerminated in the dark (36 hr) and then placed in small Petri dishes on filter paper moistened with the test solution (often 3 ml). The hypocotyls are measured after 48–72 hr in continuous light. Since the total response falls within about 15 mm, measurements should be made to the nearest 0.5 mm. The test solutions are best made up with weak mineral solution, since mineral nutrients otherwise may have a promoting effect. The test may conveniently be performed directly on cut out sections of paper chromatograms (cf. section 3.3.3). Since most lettuce cultivars do not give a linear response, the choice of assay material is very important. Arctic King and Arctic have been found most suitable (Brian et al. 1964; K.C. Jones 1968). Under optimal conditions, the log dose–log response curve for Arctic King and GA_3 is linear for 10^{-2}–10 μg/ml.

The lettuce hypocotyl bioassay is very specific for GAs. Known inhibitors are 2,4-D and kinetin in concentrations above 10^{-2} μg/ml and IAA above 10 μg/ml (Brian et al. 1964). The assay is very sensitive to inhibition by acetic acid residues, which are often present after TLC (cf. section 3.3.3).

The *cucumber hypocotyl bioassay* is similar to the lettuce hypocotyl assay but utilizes cucumber seedlings (Halvey and Cathey 1960). In the modification by Brian et al. (1964), the linear response is only in the range 10^{-1}–10 μg/plant for GA_3. Furthermore, the cucumber hypocotyl assay is the least specific of the bioassays described here. IAA in concentrations up to 1 μg/plant may have the same activity as GA_3 (Katsumi et al. 1965), although this was not observed by Brian et al. (1964). Also helminthosporol and its derivatives and dihydroconiferyl alcohol are active in the assay (Kato et al. 1968; Sakurai et al. 1974). Kinetin in higher concentrations inhibits the response (Brian et al. 1964). In spite of these disadvantages, the assay is of interest because of its specificity for 13-deoxy GAs (section 3.6.1.2). Only intact seedlings can be used for the assay since the combined effects of GA and IAA on isolated sections are complex (Kazama and Katsumi 1973, 1974; Katsumi and Kazama 1974).

3.5.2.3. Cereal endosperm bioassays. In germinating cereal seeds, GA produced by the embryo stimulates the formation of hydrolytic enzymes in the aleurone layer. The enzymes are secreted into the endosperm, where they hydrolyze reserve carbohydrates and other polymers. For bioassay purposes, the embryos are removed by cutting the seeds transversly and the embryo-less half-seeds are incubated with the test solutions. The hydrolytic enzymes formed in response to GA may be measured by the reducing sugars released by their action on endogenous starch, or an enzymatic activity, usually α-amylase, is measured with added substrate after the half-seeds have been removed. These assays are the most sensitive, rapid and space-saving ones but also the most difficult to perform in a reproducible manner. If α-amylase is measured, the assay is highly specific for GA.

In the *barley endosperm or barley aleurone bioassay,* which was originally described by Nicholls and Paleg (1963) and later thoroughly studies by Coombe et al. (1967a, b), the formation of reducing sugars is the response measured. For this purpose, the seeds are soaked for 24 hr, then cut and placed in the sterile test solution (1 ml) contained in vials for another 24 hr. Reducing sugars are assayed by the colorimetric method of Somogyi-

Nelson. The log dose – log response curve for GA_3 is linear from 10^{-5}–10^{-2} $\mu g/ml$. In a recent simplification of the assay the test for reducing sugars is omitted and the refraction of the solution after incubation is measured instead (Coombe 1971).

A blank (without endosperm) must be included for each unknown extract tested, since it may contain reducing or refracting substances. It has been pointed out that the colorimetric determination of reducing sugars is unreliable in the presence of Mn^{2+}, NH_4^+, amino acids and substances like citrate which form chelates with Cu^{2+} (Briggs 1966b). The test is supposed to be less specific than the following one (Briggs 1973). However, in routine work these disadvantages may be outweighed by the ease of performance: more replicas can be included and the risk for errors due to manipulative difficulties seems minimal.

In the other version, the *barley endosperm α-amylase bioassay,* the activity of starch degrading enzymes is determined directly (Jones and Varner 1967). The half seeds are allowed to imbibe on sterile moist sand for 2 days and are then incubated for 24–48 hr in the sterile test solution (2 ml). GA_3 gives good response after 24 hr but other GAs may need longer incubation (Jones, K.C. 1968; Jackson 1971; Ogawa 1975). After incubation, amylase degrading activity in the medium is measured by the colorimetric assay with iodine. Significant response is obtained with 5×10^{-5} $\mu g/ml$ of GA_3 and the response – log concentration curve for GA_3 is linear in the range $5 \times 10^{-4} - 5 \times 10^{-2}$ $\mu g/ml$.

This test is highly specific and specificity can be improved even more, if the enzyme solution is heated to 70 °C to degrade enzymes other than α-amylase before the activity is measured (Briggs 1967). Performed in this manner, the assay only responds to helminthosporol and helminthosporic acid in addition to GAs (Briggs 1973) (cf. 3.6.3). The formation of α-amylase is inhibited by ABA (Chrispeels and Varner 1966). Organic solvent residues were found to have no effect (Jones and Varner 1967) but the assay is highly sensitive to acetic acid residues. Different varieties of barley have very different activity and even the same variety from different years or grown in different localities may have different activities (Jackson 1971).

The *rice endosperm α-amylase assay* (Murakami 1966; Ogawa 1966) is similar to the barley endosperm assay and is supposed to have a particularly low background.

The *barley endosperm phosphatase assay* (Jones 1969) measures the formation of acid phosphatase instead of α-amylase. It is little used but may deserve more attention as it is very easy to perform, rapid and reliable.

3.5.2.4. Miscellaneous assays. Many bioassays that never reached wide usage have been designed. Many of them utilize isolated parts of plants, e.g. cereal leaf segments (Radley 1958; Hayashi and Murakami 1954; Skene and Carr 1961b), leaf discs of bean or apple (Wheeler 1960), epicotyls of pea seedlings (Hayashi and Murakami 1953; Phillips and Jones 1964) and senescing leaves of *Rumex* and *Taraxacum* (Whyte and Luckwill 1966; Fletcher and Osborne 1966). These assays are sometimes both rapid and convenient but mostly fall short of the standard assays in specificity and sensitivity.

3.5.3. The use of GA bioassays

The absolute amounts of GAs in mixtures of unknown composition cannot be measured by bioassay because the dose–response relationships are different for different GAs (section 3.6). Only relative amounts can be estimated. Bioassays are therefore used to compare the amounts of GA in extracts. The activities of the extracts are either expressed directly in terms of the response parameter of the bioassay (e.g. percent growth of control, units of α-amylase activity), or they are first related to a standard curve obtained with GA_3 and then expressed as GA_3-equivalents. The variability of plant bioassays in general is great and the multitude of GAs gives rise to some special problems. Although this has been pointed out repeatedly, the magnitude of the possible errors is still being underestimated in much GA research. Such underestimation often results in physiological significance being assigned to observed differences in GA-like activity that could in fact be due to the variability of the methods alone.

The variance in the response of bioassays to pure GA can be accurately estimated by the use of many replications and statistical methods including linear regression, test of linearity, homogeneity of variance and confidence limits (see Coombe et al. 1967b; Hill and Wimble 1969; Bailiss and Hill 1971). Hill and Wimble (1969) assaying by the Meteor dwarf pea bioassay a pure GA_3 solution, which they applied at a rate of 5 ng per plant, obtained a mean response corresponding on the standard curve to 5.9 ng per plant with the 95% confidence limits at 3.5 and 9.7 ng per plant. Assaying a solution which contained 0.5 ppm GA_3 by the lettuce hypocotyl assay, they obtained a mean response corresponding to 0.87 ppm with the 95% confidence limits at 0.58 and 1.34 ppm. Thus, in both cases the confidence interval was greater than the content of GA in the solution. In the latter case, the confidence interval did not even include the true value.

The variance of the response of the assay to GA in plant extracts is bound to be even greater because of the mixture of compounds present. But the greatest variance probably lies in the recovery of GAs by extraction and after chromatography (section 3.3.1). This recovery will be different for GAs of different polarity. The confidence limits of the bioassay alone therefore are inadequate for assessing a difference between the original plant materials from which the extracts were made. Furthermore, the slope, linearity and range of linearity for an extract are likely to be different from those of the standard curve. These parameters primarily depend on which GAs are present in the extracts (section 3.6.1) but also on interfering contaminants. They are going to be different in different zones of a chromatogram. But a comparison of activities is meaningless, unless the dose – response curves are linear and parallel and unless the measured response falls within the linear range. Measurements at the lower and upper ends of the linear range may give severely distorted results both because of the inaccurracy of the standard curves and because solvent residues may induce small significant responses (Hartley et al. 1969). It is therefore essential to determine the dose–response curve for the extracts being compared.

To obtain valid statistical parameters, it would be necessary to make several extractions of each material, to assay each chromatographic fraction at several dilutions and to use at least ten replications for each bioassay point. A single value for the GA content of a plant would require some five hundred assay plants and a comparison between two values would require about eight hundred. We do not want to recommend this, but rather draw the conclusion that small to medium differences in the response of bioassays to plant extracts cannot be accepted as evidence for differences in GA content of the original plant materials even if they are significant according to the confidence limits or least significant difference of the bioassay. Many workers in the field have realized this for a long time (Lang 1970). More recently, Crozier et al. (1973) after having purified the same extract by three different methods and assayed it by two different assays concluded that it would be " . . . an advisable precaution to regard apparent differences in GA levels of less than three to four orders of magnitude* as nothing more than a consequence of inaccurate estimation." However, new papers are still appearing and old ones are being quoted in which these principles are completely disregarded.

Finally, statistical analysis of the bioassays is not a substitute for proper experimental design. This is not a problem specific to GA research and it is the same whether GA is measured by bioassay or some other method. Necessary information concerning the reproducibility of results is lacking in numerous papers concerned both with measurements of GA levels by bioassay and with the metabolism of radioactive GAs or GA precursors. We have taken this into account in reviewing the literature for the following sections.

3.6. Structure–activity relationships

This section reviews the relative activities of different GAs, their derivatives, degradation products and precursors in the standard bioassays. Such comparisons are often imperfect both because of the variability inherent in the bioassays (3.5.3) and because the data originate from different laboratories where the bioassays may be performed under very different conditions. A critical variable is also the purity of the tested samples; if an inactive GA is contaminated with a small amount of a highly active one, the results are misleading. Finally, the assignment of a GA to a certain activity group requires a good portion of subjective judgement taking both the minimum detection level and the slope of the dose – activity curve into account (Crozier et al. 1970). It is therefore not surprising that data from different laboratories vary considerably and it is clear that only very marked trends in the structure–activity relationship can be considered.

3.6.1. Naturally occurring GAs

The naturally occurring GAs lend themselves to structure–activity studies because of the systematic differences between their structures. Most thoroughly investigated have been GA_1–GA_9, which were supplied as research samples for several years following their discovery. They were therefore not only compared in the standard bioassays (Brian et al. 1964; Paleg et al. 1964; Brian et al. 1967), but also in more unusual ones, e.g. the *Avena* first leaf assay (Bentley-Mowat 1966). They were also compared with respect to other effects such as induction of bolting in rosette plants (Michniewicz and Lang 1962) and induction of parthenocarpic fruit set (Bukovac and Nakagawa 1967). The results of the early comparisons are tabulated in the review by Brian (1966).

The GAs that were discovered later have been investigated more sporadically and primarily in connection with their discovery. The study by Crozier et al. (1970) is there-

* They meant 3- to 4-fold

fore very valuable; the activities of GA_1–GA_{27} (except GA_{16}) in several bioassays were investigated or re-investigated under a single set of conditions. In a later paper, which also contains the complete references to other structure–activity studies, the data was further complemented to include the relative activities of GA_1–GA_{38} in five bioassays (Reeve and Crozier 1974). Sections 3.6.1.1 and 3.6.1.2 are largely based on these two publications.

3.6.1.1. General influence of structures. The activity of a GA is to a certain degree dependent on which bioassay is used, but certain structures are more generally active than others. Thus the C_{19}-GAs are more active than the C_{20}-GAs (Tables 3.5–3.8). Within the group of the C_{19}-GAs, 3β-hydroxylation, 3β, 13-dihydroxylation and 1,2-unsaturation result in higher activity, if the responses in the different bioassays are summarized (Tables 3.5, 3.6). The highest activities are found among the 1,2-unsaturated GAs. Since they are all 3β-hydroxylated, it is unknown whether the double bond alone would

Table 3.5.

Relative activities of GAs. I. C_{20}- and ring-A saturated C_{19}-GAs. The influence of 3β- and 13-hydroxylation[a]

Relative activities: ++++ very high, +++ high, ++ moderate, + low, 0 very low to inactive

Structure	GA	Barley α-amylase	Dwarf pea	Dwarf rice	Dwarf-5 maize	Lettuce hypocotyl	Cucumber hypocotyl
C_{20}-GAs, C-20 = CH_3							
no OH	GA_{12}	0	0	+	+	0	+
13-OH				0[b]		0[b]	
3β-OH	GA_{14}	0	0	+	0	0	0
3β, 13-$(OH)_2$	GA_{18}	0	++	+[c]/+++	+[c]/+++	+	0
C_{20}-GAs, C-20 = CH_2-O-							
no OH	GA_{15}	0	+	++	+++	0[d]/++	0[e]/++
13-OH	GA_{44}			++[c]			
3β-OH	GA_{37}	++	++	+++		++	+++
3β, 13-$(OH)_2$	GA_{38}	+	+++	+++[c]/+	+++	0	+
C_{20}-GAs, C-20 = CHO							
no OH	GA_{24}	0	+	+++	+	0	+++
13-OH	GA_{19}	0	0	+++	++	+	0
3β-OH	GA_{36}	++	++	+++		+	+++
3β, 13-$(OH)_2$	GA_{23}	++	++	+++	+++	+	0
C_{20}-GAs, C-20 = CO_2H							
no OH	GA_{25}	0	0	0	0	0	+
13-OH	GA_{17}	0	0	+	0	0	0
3β-OH	GA_{13}	+	0	+	0	+	0
3β, 13-$(OH)_2$	GA_{28}	0	0	0[c]/+	0	0	0
C_{19}-GAs. A-ring saturated							
no OH	GA_9	+	++	++	++	+++	+++
13-OH	GA_{20}	+	+	+++	+++	+++	0
3β-OH	GA_4	+++	+++	++	++	++	+++
3β, 13 $(OH)_2$	GA_1	++++	+++	+++	++	+++	++

[a] The basic data is after Crozier et al. (1970) and Reeve and Crozier (1974). Data for GA_{29} - GA_{38} in the d-5 assay were supplemented from Yokota et al. (1971a), Fukui et al. (1972) and Yamane et al. (1973). Significant deviations of other authors are noted: [b] Bearder et al. (1975b), 13-hydroxy GA_{12} has no number of its own; [c] Fukui et al. (1972); [d] Brian et al. (1967); [e] Isogai et al. (1974).

Table 3.6.

Relative activities of GAs. II. Unsaturated C_{19}-GAs[a]

Structure	GA	Barley α-amylase	Dwarf pea	Dwarf rice	Dwarf-5 maize	Lettuce hypocotyl	Cucumber hypocotyl
C_{19}-GAs, 1,2-unsaturated							
3β-OH	GA_7	+++	+++	+++	++++	++++	++++
3β,13-$(OH)_2$	GA_3	++++	++++	++++	+++	+++	++
3β,12α-$(OH)_2$	GA_{30}	+++	+++	++	++	+	+
3β,12α,13,15β-$(OH)_4$	GA_{32}	+++	+++	++++	++++	++	+++
C_{19}-GAs, 2,3-unsaturation							
13-OH	GA_5	++	+++	+++	++++	++	+
13,18-$(OH)_2$	GA_{22}	+++	+++	+++	++	++	0
12α-OH	GA_{31}	+	++	++	++	0	0

[a] Relative activities and basic sources of data as in Table 3.5.

Table 3.7.

Relative activities of GAs. III. The effect of 2β-hydroxylation[a]

Structure	GA	Barley α-amylase	Dwarf pea	Dwarf rice	Dwarf-5 maize	Lettuce hypocotyl	Cucumber hypocotyl
2-Deoxy-GAs							
C_{20}, 3β-OH	GA_{37}	++	++	+++		++	+++
C_{19}, 13-OH	GA_{20}	+	+	+++	+++	+++	0
C_{19}, 3β-OH	GA_4	+++	+++	++	++	++	+++
C_{19}, 3β,13-$(OH)_2$	GA_1	++++	+++	+++	++	+++	++
Corresponding 2β-OH GAs							
C_{20}, 2β,3β-$(OH)_2$	GA_{27}	0	+	+	0	0	0
C_{19}, 2β,13-$(OH)_2$	GA_{29}	+	0	+	+	0	0
C_{19}, 2β,3β-$(OH)_2$	GA_{34}	0	+	+	0	0	0
C_{19}, 2β,3β,13-$(OH)_3$	GA_8	+	+	+	0	+	0
C_{19}, 2β,3β-$(OH)_2$-12-oxo	GA_{26}	0	0	0	0	0	0

[a] Relative activities and basic sources of data as in Table 3.5.

Table 3.8.

Relative activities of GAs. IV. Miscellaneous GAs[a]

Structure	GA	Barley α-amylase	Dwarf pea	Dwarf rice	Dwarf-5 maize	Lettuce hypocotyl	Cucumber hypocotyl
All C_{19}-GAs							
16α-OH	GA_{10}	+	0	+++	++	++[b]/0	++
1α,3β-$(OH)_2$	GA_{16}	+	++	+		+	+
3β,11β-$(OH)_2$	GA_{35}	+++	++	++	+	+	+++
3β,16α-$(OH)_2$	GA_2	++++	++	+++	+	++	++
13-OH, 2,3-epoxy	GA_6	++	++	+++	++	++	+
1β,12α-$(OH)_2$, 3-oxo	GA_{33}	+	+	+	+	+	0
13-OH, 18-CO_2H	GA_{21}	0	0	0	+	+	0
1, 10-epoxy, isolactone	GA_{11}	+	0	+	0	0	+

[a] Relative activities and basic sources of data as in Table 3.5.
[b] Brian et al. (1967).

also result in high activity. The effect of a 2,3-double bond cannot be ascertained because only GA_{20} (Table 3.5) and GA_5 (Table 3.6) are available for comparison.

One of the most striking effects of structure is the loss of activity on 2β-hydroxylation (Table 3.7). Hydroxylation in this position occurs during metabolism and may be regarded as an inactivation mechanism (section 3.8).

The C_{20}-GAs with the exception of GA_{27} are fully represented in Table 3.5. Within this group, the GAs with δ-lactone or aldehyde functions in the C-20 position have moderate to high activity, whereas those with methyl or carboxyl groups in this position are almost inactive. This may be due to a certain tendency of the δ-lactones or C-20 aldehydes to partially fit a hypothetical receptor site for the γ-lactones of the C_{19}–GAs. This explanation is particularly attractive since the C-20 aldehydes probably form δ-lactols in solution (section 3.2.2). An alternative explanation would be that the δ-lactones or aldehydes are converted by the assay plants to C_{19}-GAs whereas the C-20 carboxylic acids are not (Fukui et al. 1972); Reeve and Crozier 1974). This hypothesis remains a possibility, but no evidence exists for the conversion of the δ-lactones or C-20 aldehydes to C_{19}-GAs whereas the C-20-methyl GAs GA_{12} and GA_{14} are known to be precursors of C_{19}-GAs but have low activity in the bioassays. Furthermore, the activities of the *ent*-kaurenoid GA precursors are not correlated with their roles as GA precursors (section 3.6.4).

3.6.1.2. Bioassay specificity. Superimposed on the differences in activities of the GAs are the specificities of the bioassays. Thus the dwarf rice bioassay (cv. Tan-ginbozu) responds better than the others to C_{20}-GAs, whereas the barley endosperm and lettuce hypocotyl assays respond particularly poorly to them (Table 3.5). The barley aleurone and dwarf pea assays are more dependent on 3β-hydroxylation than the others (Table 3.5). The cucumber hypocotyl assay has a specificity which is unique: it responds poorly or not at all to 13-hydroxylated GAs (Tables 3.5, 3.6) except GA_{32}. The response to GA_{32} is not typical either, since it is high only with the highest concentration tested (10 μg per plant). At lower concentrations, GA_{32} has only a fraction of the activity of GA_4 and GA_7 in the cucumber assay (Yamane et al. 1973).

3.6.2. Conjugates, derivatives and degradation products

No derivatives have been found which have higher biological activity than the natural GAs with γ-lactones, 3β-hydroxylation and 1,2-unsaturation. These structures therefore are probably optimal. Furthermore, practically all parts of these structures must be necessary for full activity, since most changes lead to loss of activity. The following discussion is mainly based on the monumental work by Brian et al. (1967) in which 134 compounds related to GAs were tested in several assays.

Derivatives involving the hydroxyl group. The activity of GAs with conjugated hydroxyl groups is difficult to ascertain because the conjugates may become hydrolyzed in the assay plants giving rise to free GAs. Thus acylation of the 3β- and 13-hydroxyl groups in GA_3 and of the 3β-hydroxyl group in GA_7 changes the activity only little, which Brian et al. (1967) attributed to hydrolytic fission of the acyl bonds in vivo. Glucosidation of the hydroxyl groups, on the other hand, results in considerable loss of activity as shown for the natural glucosides by Yokota et al. (1971a) and Yamane et al. (1973) and for synthetic glucosides by Sembdner et al. (1974). The glucosides may be hydrolyzed to some degree in vivo, particularly if they are fed via the roots under non-sterile conditions, in which case they are cleaved by microorganisms (Sembdner et al. 1974). Methyl ethers are less easily hydrolyzed and methylation of the hydroxyl groups in GA_3 accordingly results in much lower activity than that of GA_3 (Agnistikova et al. 1974).

Epimerization of the 3β-hydroxyl group results in great loss of activity. The 3α-epimers of GA_1 (XXIII), GA_3 and GA_4 have only 10^{-3}–10^{-2} the activity of the corresponding 3β-hydroxy GAs in most assays (Brian et al. 1967; Murakami 1970a; Agnistikova et al. 1974). An inhibitory action of 3α-*epi*-GA_1 in one assay but not in others needs confirmation before its significance can be evaluated (Kato and Katsumi 1967).

The ring-B carboxyl group. GA methyl esters are considerably less active than the corresponding GAs with free carboxyl groups (Brian et al. 1967), although this effect is less noticeable in the cucumber hypocotyl assay. A free carboxyl group is therefore a likely requirement for high activity. In spite of this, the naturally-occurring glucosyl esters have biological activities approaching those of the free GAs. Like the glucosyl ethers, they are cleaved by the assay plants and microorganisms, which may be the reason for their activity (Hiraga et al. 1974c; Liebisch 1974). GA_{12}-aldehyde (XLV, Scheme 3.4). GA_{14}-aldehyde (XLVI, Scheme 3.4), GA_{12}-alcohol (LXIX) and GA_{14}-alcohol (LXX), which differ from GA_{12} and GA_{14}, respectively, in the oxidation stage of C-7, have been tested in several bioassays (Hoad et al. 1976). Except for marginal activity in the Tanginbozu and d-5 assays, the GA_{12}-derivatives were inactive, which approximates earlier results with GA_{12}. The derivatives of GA_{14}, on the other hand, had low to intermediate activity in several assays, notably in the dwarf pea, Tan-ginbozu and d-5 assays. The conclusion of the authors that the derivatives are more active than GA_{14} in these assays would, however, have to be more rigorously tested to be accepted. The differences in the low activities were very small and the data for GA_{14} was from previous work. A further claim that both the GA_{12}- and GA_{14}-derivatives were more active than GA_{12} and GA_{14}, respectively, in the α-amylase assay was also based on inconclusive data. A clear and consistent difference between the C-7 alcohols, aldehydes and carboxylic acids does not follow from this work.

The lactone ring. Hydrolytic or reductive fission of the lactone ring results in complete or almost complete loss of biological activity (Sembdner et al. 1965; Brian et al. 1967). On the other hand, isomerization of the (4→10) lactone ring in GA_3 to the (4→2) position of isoGA_3 (XVII) affects the activity surprisingly little. Thus the activity of isoGA_3 in most assays is the same as, or one step lower than, that of GA_3; only in the dwarf maize assays is the activity drastically reduced (Brian et al. 1967; Hoad et al. 1976). The very low activity of GA_{11} (Table 3.8), therefore, is probably due mainly to the epoxy function rather than to the isomeric lactone. The very low activity of GA_{15} in the dwarf rice bioassay is completely absent in the isomer with the lactone reversed from the (4→20) to the (20→4) position (Isogai et al. 1974), and the corresponding isomer of GA_{37} also has less activity in almost all assays (Hoad et al. 1976). Because of the relatively high activity of isoGA_3 (Hoad et al. 1976), the role of the γ-lactone of GAs for activity is unclear.

The exocyclic double bond. Hydrogenation of the 16,17-double bond or its isomerization to the endocyclic 15,16-position are also changes that have relatively little effect on activity. Thus tetrahydro-GA_3 has approximately the same activity as GA_1. Dihydro-GA_4 and the 15,16-ene isomer of GA_4 have approximately the same activity as GA_4 (Brian et al. 1967). In contrast, 16,17-dihydro-GA_5 and dihydro-GA_9 have considerably less activity than GA_5 and GA_9, respectively. Apparently the double bond is not an absolute requirement for activity. Indeed, the entire methylene group at C-16 may not be required for activity, since the *nor*-ketone XXVII is almost as active as GA_7 in both the lettuce hypocotyl and barley endosperm assays (Jones et al. 1971). Hydration of the double bond usually decreases activity somewhat, which may be due to the introduction of a hydroxyl group.

Fluorination. The biological activity of the monofluoro-analogues of GA_9 and GA_{12} (XXVIII and XXIX respectively) is about the same or less than that of the non-substituted GAs (Stoddart 1972b). Only in one case (F-GA_9 in combination with the lettuce hypocotyl assay) did the fluorination lead to a clear (5-fold) increase in activity. Replacement of a hydrogen atom with fluorine results in much greater electronegativity and hydrogen bonding potential without great changes in size. The lack of effect of fluorination must therefore be considered when possible models for GA receptors are constructed.

Changes in the carbon skeleton. All changes in the carbon skeleton, such as fission of rings A and D, aromatization of ring A, Wagner – Merwein rearrangement of the C/D-rings or epimerization of the B/D-ring junction, severely decrease, and in most cases completely destroy, activity (Brian et al. 1967; Hoad et al. 1976). Antheridiogen (XII), which can be regarded as a ring-C/D rearranged GA_4, has activity in the dwarf maize (d-5) and barley endosperm α-amylase assays. The activity seemed to be high on d-5 (1/20–1/2 that of GA_3) and in the α-amylase test medium (1/250 of GA_3) but the shapes of the response curves were very different from that of GA_3 and a comparison is therefore difficult (Sharp et al. 1975).

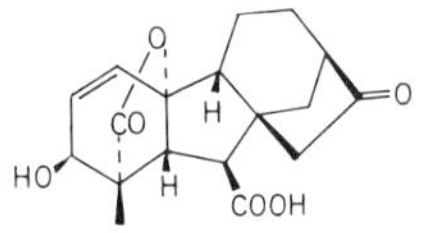

XXVII

GA_7-norketone (GA_7-17-nor-16-ketone)

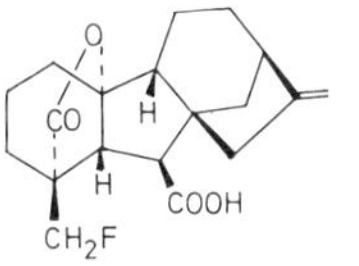

XXVIII

18-Fluoro-GA_9

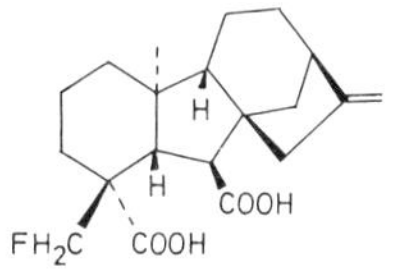

XXIX

18-Fluoro-GA_{12}

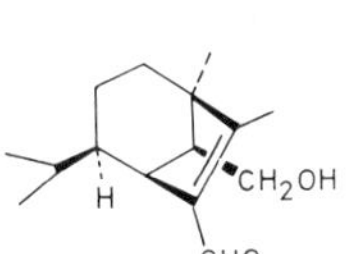

XXX

Helminthosporol

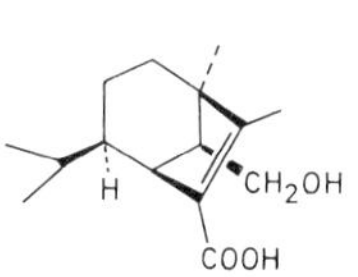

XXXI

Helminthosporic acid

3.6.3. *Helminthosporol and β-ecdyson*

From cultures of the fungus *Helminthosporium sativum,* Tamura et al. (1965) isolated a compound, helminthosporol (XXX), with activity in the rice seedling bioassay. The activity in this assay of XXX and its autoxidation product, helminthosporic acid (XXXI), has been confirmed (Sakurai and Tamura 1965; Kato et al. 1966) and both compounds are also active in releasing reducing sugars or α-amylase in the rice and barley endosperm assays (Mori et al. 1965; Ogawa 1966; Briggs 1966a; Okuda et al. 1967; Coombe et al. 1974). Specific molecular requirements must also be fulfilled for activity. Thus oxidation of the hydroxymethyl group of XXX to an aldehyde or carboxyl group or reduction of the aldehyde group of XXX to a hydroxymethyl or methyl group removes all activity in the rice seedling test (Sakurai and Tamura 1965). Reduction of the double bond in helminthosporic acid (XXXI) does not lead to loss of activity, provided that the resulting dihydrohelminthosporic acid has the functional group in the *cis*-position. However, the corresponding *trans*-structure is completely inactive (Tamura and Sakurai 1965). Removal of the hydroxymethyl group in XXXI, on the other hand, does not decrease the activity (Coombe et al. 1974). In the light of this specificity, the suggestion ascribed to Briggs (1966a) that helminthosporol and helminthosporic acid fit a part of the hypothetical GA receptor and constitute a sort of minimal requirement for GA activity, seemed reasonable. A comparison of spatial structures (Figure 3.2) shows both similarities and differences.

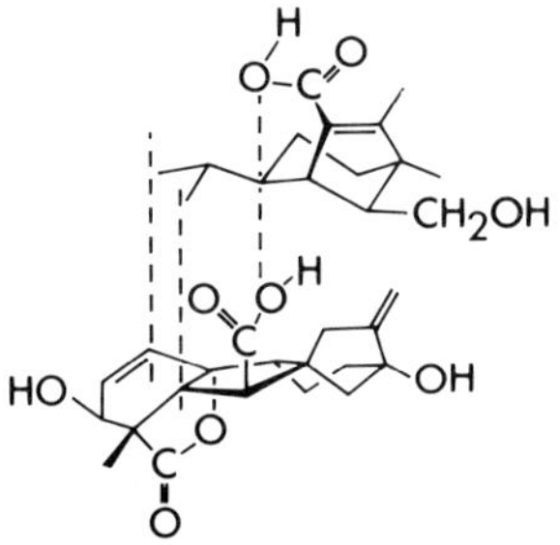

Figure 3.2. Spatial formulae of helminthosporic acid and GA_3 (Coombe et al. 1974).

However, the activity of XXX and XXXI does not show a very broad spectrum. Thus helminthosporol is not active in the lettuce hypocotyl assay and helminthosporic acid is only active at high concentrations (10–100 μg/ml) (Sakurai and Tamura 1965). No true activity has been found in the pea segment, dwarf pea and dwarf maize assays (Kato et al. 1966), although Hashimoto et al. (1967) and Katsumi et al. (1967) have recorded the effects of very high concentrations (50–500 mg per plant) in the dwarf maize assay as positive response. Even in the rice seedling and endosperm biosassays, the activity is only 10^{-3}–10^{-2} times that of GA_3. Furthermore, Kato et al. (1968) on the basis of competition studies concluded that helminthosporol and GA_3 act at different sites. This conclusion is further supported by the findings that the distribution of elongated cells in the leaf sheaths of rice seedlings treated with helminthosporol is different from that of cells in seedlings treated with GA_3 (Okuda et al. 1966) and that different isoenzymes of α-amylase are formed in response to the two compounds (Momotani and Kato 1967). On the basis of this and other evidence reviewed by Kato et al. (1968) it does not seem very likely that helminthosporol (XXX) and its analogues act by partially fitting a GA receptor site.

The insect hormone ecdyson was once reported to have considerable activity in the dwarf pea bioassay (Carlisle et al. 1963). Careful reinvestigation, however, showed that β-ecdyson has no activity in three GA bioassays (Hendrix and Jones 1972).

3.6.4. GA precursors

Compounds known to be precursors of GAs in higher plants might be expected to show activity in GA bioassays because of their potential conversion to GAs within the assay plants. However, no conversion of *ent*-kaurenoid precursors to GAs in intact plants have been demonstrated (section 3.7.5). Table 3.9 shows that GA precursors at the most have moderate activity in the bioassays and that many of them are inactive. Whereas the activity in some cases could conceivably be due to conversion, the activity of *ent*-6α, 7α-dihydroxy-kaurenoic acid, which is not a precursor of GAs (3.7.2.4), cannot be explained in this way. The same is true for the activity of steviol (*ent*-13-hydroxy-kaurenoic acid, section 3.7.3) in the reducing sugar assay, since it is unlikely that barley endosperm or aleurone have the necessary enzymes for GA biosynthesis. Most likely, the activity of the precursors, where it is present, results from their resemblance to GA molecules.

Table 3.9.

Relative activities of some GA precursors and related compounds

Relative activities: ++ moderate, + low, (+) very low, 0 inactive

Compound	Barley α-amylase	Dwarf pea	Dwarf rice	Dwarf maize d-1	Dwarf maize d-5	Lettuce hypocotyl	Cucumber hypocotyl
ent-Kaurene	0[a]/(+)[b]/0[c]	0[d]	(+)[e]	0[d]	0[c]/0[d]	0[c]/0[d]	0[d]
ent-Kaurenol	(+)[b]/0[c]	0[f]	+[e]	0[f]	+[f]/0[c]	0[c]	0[f]
ent-Kaurenoic acid	+[b]/0[c]	0[f]/0[d]	++[e]	0[f]/0[d]	(+)[d]/+[f]/(+)[c]	0[d]/(+)[c]	0[f]/0[d]
ent-7α-Hydroxy-kaurenoic acid					+[g]		
GA_{12}-aldehyde	0[h]/(+)[c]	0[h]	+[h]		0[c]	0[h]/0[c]	0[h]
ent-6α,7α-dihydroxy-kaurenoic acid	(+)[c]				+[c]	+[c]	
Steviol	++[b]	0[d]	++[e]	0[d]	(+)[d]/+[i]	0[d]	0[d]

[a] Murphy and Briggs (1973b); [b] Ogawa (1975) (reducing sugar test); [c] Cross et al. (1970); [d] Brian et al. (1967); [e] Murakami (1972); [f] Katsumi et al. (1964); [g] Lew and West (1971); [h] Reeve and Crozier (1974); [i] Ruddat et al. (1963).

3.7. Pathway of biosynthesis

A biogenetic relationship of gibberellic acid (GA_3) to the diterpenes was proposed by Cross et al. (1956) as soon as the main features of its structure emerged. A few years later, Birch et al. (1959) demonstrated the incorporation of acetate and the then newly discovered mevalonic acid (MVA) into GA_3 by cultures of *G. fujikuroi,* thus proving that GA biosynthesis is indeed a variant of the terpenoid pathway. Although this early work by Birch et al. outlined the main features of the pathway, it has taken many years to find the actual intermediates and to demonstrate the step-wise conversions involved. This is understandable, for the biosynthesis of GA_3 from MVA proceeds via 18 or more true intermediates and with the formation of at least 15 related products. Some points are still unclear but most of the steps are known. Knowledge about the enzymes involved is still restricted to the beginning of the pathway. GA biosynthesis has been reviewed by Cross (1968), Hanson (1971), MacMillan (1971, 1974a) and West (1973).

3.7.1. The systems

Much of the progress with the GA pathway has been made with cultures of *G. fujikuroi* which incorporate acetate, MVA and other precursors into GA_3, thus performing the entire pathway. For biosynthetic studies, the fungus is usually grown on a defined medium, e.g. the "ICI-medium" devised by Borrow et al. (1964), which is often modified by reducing NH_4NO_3 to 1/10–1/2 of the specified amount in order to hasten the onset of GA production. This occurs after a few days when the nitrogen source has become exhausted and coincides with the onset of secondary metabolism in general, which in many strains is marked by pigmentation of the culture (Borrow et al. 1961, 1964). Labelled precursors may be added at this point or at the beginning of fermentation. A more sophisticated method is the use of resuspension cultures: the mycelium is harvested after a few days of growth, washed and reincubated with added precursor in a medium without a nitrogen source (e.g. Bearder et al. 1974; Hedden et al. 1974). Phinney (unpublished) by UV irradiation has produced a valuable mutant, B1–41a, which is blocked in the GA pathway between *ent*-kaurenal and *ent*-kaurenoic acid (Scheme 3.3) but readily converts *ent*-kaurenoic acid and later intermediates of the pathway to GA_3 (Bearder et al. 1974). The advantage of this is that when these intermediates are radioactively labelled and fed to this mutant, they are incorporated with minimal dilution of the label by endogenous compounds. With this mutant it has even been possible to survey the entire pathway by feeding unlabelled substrates and analyzing the products by GC-MS (Bearder et al. 1975a).

Which compounds accumulate in the fungal cultures depends to a certain degree on the cultural conditions and the strain of fungus. However, the most common strains under normal (acidic) conditions yield GA_3, 7β-hydroxykaurenolide (XLI), 7β,18-dihydroxykaurenolide (XLII) and fujenal (XLIV) (Scheme 3.3) as main end products. Some other GAs may also accumulate to a smaller extent, notably GA_{13}. MacMillan and Wels (1974) made a complete survey by GC-MS of compounds formed from [2-^{3}H]MVA by a wild type strain, GF-1a, and found 15 GAs and 10 diterpenes of known structure, 7 compounds which were assigned tentative structures and 40 compounds of unknown structure.

Cell-free systems have been used for study of the early part of the pathway, where the intermediates do not accumulate in cultures of *G. fujikuroi,* and for the study of GA biosynthesis in higher plants. Frequent mention will be made of such systems prepared from immature seeds of *Echinocystis macrocarpa* Greene (Graebe et al. 1965) and *Cucurbita maxima* (Graebe 1969). *E. macrocarpa* has been renamed *Marah macrocarpus* (Greene) Greene but the system shall be referred to by the old name. The *Echinocystis* system has been useful in studying the steps from MVA to *ent*-7α-hydroxykaurenoic acid (see section 3.7.2.3) and their enzymology. The *Cucurbita* system was only discovered after these steps had been investigated and has mainly been used to study the formation of C_{20}-GAs. So far, it is the only cell-free system which is active from MVA to C_{19}-GAs. The preparation of active cell-free systems from *G. fujikuroi* is problematic and therefore was achieved relatively late (Shechter and West 1969; Evans and Hanson 1972). Nevertheless, these systems have contributed important information on the formation of *ent*-kaurene.

3.7.2. The pathway

3.7.2.1. Formation of geranylgeranyl pyrophosphate. The GA pathway proper begins with the cyclization of *trans*-geranylgeranyl pyrophosphate (GGPP) (XXXVI, Scheme 3.2) to *ent*-kaurene (I) but MVA is such a widely used precursor that a few comments on its conversion to GGPP are pertinent. This follows the normal isoprenoid pathway (Scheme 3.1) in which MVA is first activated and converted to the key intermediate isopentenyl

Scheme 3.1. Conversion of mevalonic acid to geranylgeranyl pyrophosphate.

pyrophosphate (IPP), which in turn is converted to *trans*-geranyl pyrophosphate (GPP), *trans*-farnesyl pyrophosphate (FPP) and GGPP by step-wise condensations. The formation of IPP requires ATP and Mn^{2+} or Mg^{2+}. Recent work on MVA-activating enzymes in plants has been reviewed by Hill and Rogers (1974).

Plant extracts often catalyze the conversion of MVA to GGPP (Nandi and Porter 1964; Jungalwala and Porter 1967; Oster and West 1968; Ogura et al. 1972; Davies et al. 1975) or its alcohol moiety, *trans*-geranylgeraniol (Graebe et al. 1965; Graebe 1968; Staby et al. 1973; Buggy et al. 1974). Geranylgeraniol probably arises through the action of phosphatases on GGPP in the crude preparations. Often FPP and *trans*-farnesol are also obtained. Although they are part of the same pathway, FPP and GGPP are synthesized by separate enzymes, at least in pumpkin fruits (Ogura et al. 1968; 1972). Thus purified GGPP synthetase catalyzes the formation of GGPP from IPP and its isomer, dimethylallyl pyrophosphate, without accumulation of FPP (Scheme 3.1). This enzyme is stimulated by Mn^{2+}. Purified FPP synthetase, on the other hand, converts IPP to FPP and no further and is stimulated by Mg^{2+}. The same may be true in cell-free systems from pea seedlings and pea fruits, since the formation of GGPP or geranylgeraniol is favoured over that of FPP or farnesol by a high Mn^{2+}/Mg^{2+} ratio (Graebe 1967, 1968). FPP synthetase has also been obtained from seedlings of *Ricinus communis* (Green and West 1974).

The enzymes that catalyze the formation of GGPP from MVA are usually contained in the soluble fraction of cell extracts but they partly originate from cell organelles. Thus soluble MVA-activating enzymes, prenyl transferase and geranylgeraniol biosynthesis have been obtained associated with plastid preparations (Rogers et al. 1966; Buggy et al. 1974; Green et al. 1975). Some of the enzyme activity must be directly associated with the chloroplast structure, since washed chloroplast fragments stimulate the formation of geranylgeraniol from MVA as much as fifty-fold in the presence of supernatant enzymes (Graebe 1968; Buggy et al. 1974). Preparations which contain chloroplast fragments do not yield GGPP but free geranylgeraniol, presumably due to strong phosphatase activity associated with the plastid membranes.

Several other neutral compounds are usually obtained in addition to farnesol and geranylgeraniol when MVA is incubated with plant extracts. The tertiary alcohols geranyllinalool and nerolidol arise as acid hydrolysis products of GGPP and FPP, respectively. Squalene, phytoene and unidentified lipids of the steroid and carotenoid pathways may also be found (e.g. Graebe 1968; Buggy et al. 1974). Some of these products migrate like *ent*-kaurenoid intermediates of GA biosynthesis in several TLC systems; hence identification of the products by TLC only is inconclusive in tracer studies (cf. 3.3.4).

3.7.2.2. Formation of ent-kaurene. The discovery by Cross et al. (1963a) that *ent*-kaurene (I) is a metabolite of *G. fujikuroi* was of fundamental significance to further pathway studies. Soon afterwards, *ent*-[17-^{14}C]kaurene was shown to become incorporated into [17-^{14}C]GA_3 to the extent of 5.7% by cultures of the same fungus, which established its role as an intermediate in the pathway (Cross et al. 1964). *ent*-Kaurene is not excreted into the medium like GAs but remains associated with the mycelium of the fungus (Bearder et al. 1974).

The biosynthesis of *ent*-[^{14}C]kaurene from [2-^{14}C]MVA was first shown in the *Echinocystis* cell-free system (Graebe et al. 1965). It has since been studied in cell-free systems from immature seeds of *Cucurbita maxima* (erroneously identified as *C. pepo*) (Graebe 1969), immature pea seeds (Anderson and Moore 1967; Graebe 1968; Coolbaugh and Moore 1969), pea shoots (Coolbaugh et al. 1973), etiolated seedlings and seeds of *Ricinus communis* (Robinson and West 1969a; West 1973) and germinating tomato seeds (Yafin and Shechter 1975). The *Echinocystis* and *Cucurbita* systems are the most active ones, incorporating 50% or more of the available *(R)*-MVA. The system from pea seeds is surprisingly inactive although peas contain large amounts of GAs. The cause for this low activity could be an important factor in the regulation of *ent*-kaurene biosynthesis. The system from pea shoots is even less active. In contrast to the other systems, including the one from *R. communis* seeds, the system from *R. communis* seedlings not only yields *ent*-kaurene (I) but at least four other diterpene hydrocarbons of which *ent*-beyrene (XXXII) and *ent*-isopimaradiene ((+)-sandaracopimaradiene) (XXXIII) have been fully identified, *ent*-13,16-cycloatisane (trachylobane) (XXXIV) has been partly characterized, and a hydrocarbon, which was named casbene, has been assigned the structure (XXXV) (Robinson and West 1969a).

The conversion of [^{14}C]GGPP to *ent*-[^{14}C]kaurene was shown by Upper and West (1967) using the *Echinosystis* system and by Robinson and West (1969b) using the *Ricinus* seedling system. In the latter, [^{14}C]GGPP was converted to all five hydrocarbons mentioned above. Shechter and West (1969) working with the cell-free system from *G. fujikuroi* discovered that the conversion of GGPP to *ent*-kaurene proceeds in two steps via *ent*-labda-8(17),13-dien-15-yl pyrophosphate (XXXVII), which they named copalyl

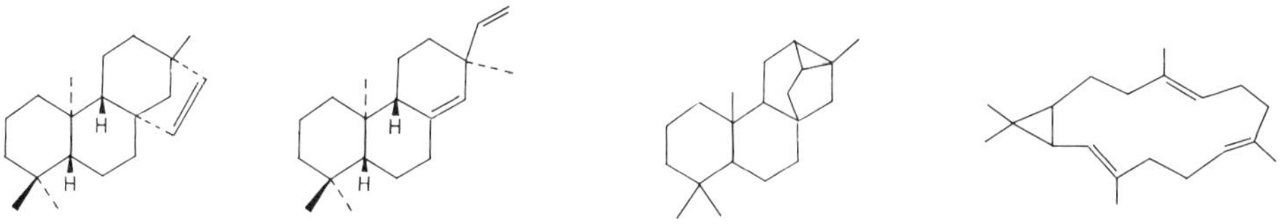

XXXII *ent*-Beyrene

XXXIII *ent*-Isopimaradiene ((+)-sandaracopimaradiene)

XXXIV *ent*-13,16-Cycloatisane (trachylobane)

XXXV Casbene

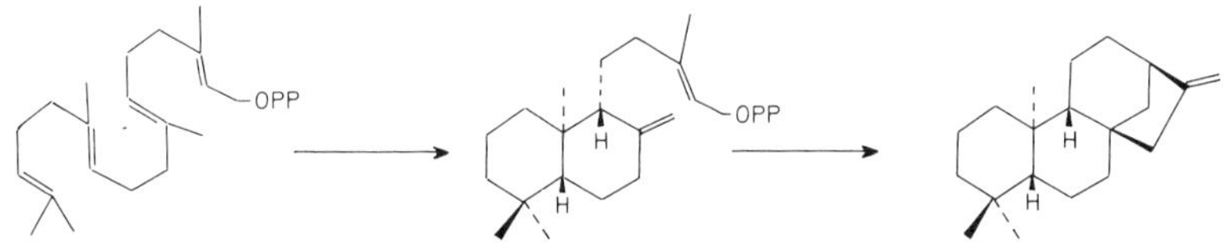

XXXVI trans.-Geranylgeranyl pyrophosphate

XXXVII Copalyl pyrophosphate

I ent-Kaurene

Scheme 3.2. Conversion of geranylgeranyl pyrophosphate to *ent*-kaurene.

pyrophosphate (CPP), as a stable intermediate (Scheme 3.2). They also showed that this is the case in the cell-free systems from *E. macrocarpa* and *R. communis*. In the latter system, CPP is an intermediate in the formation of XXXII to XXXIV in addition to *ent*-kaurene (I) but not in the formation of casbene (XXXV) (Robinson and West 1969b). In the *Echinocystis* system, CPP is converted very rapidly to *ent*-kaurene and therefore does not accumulate (West 1973). Hanson and White (1969) independently showed the incorporation of CPP into *ent*-kaurene, kaurenolides and GA_3 in fungal cultures. The stereochemistry of the cyclization of GGPP to *ent*-kaurene is discussed in section 3.7.4.2.

Fall and West (1971) purified kaurene synthetase from *C. fujikuroi* about 170-fold and found that its two activities, cyclization of GGPP to CPP and of the latter to *ent*-kaurene, retained a constant ratio throughout purification. The two steps may therefore be catalyzed by a single enzyme or enzyme complex, the mol. wt. of which was estimated to be about 4.6×10^5 daltons. Although both activities require Mg^{2+} for full activity, they differ in their response to other metal ions, pH and inhibitors, which suggests that they are located at different sites. CPP shows some product inhibition. Attempts to purify the enzyme from *E. macrocarpa* were less successful, since it is less stable, but some purification of the enzyme from *Ricinus* seedlings was achieved (West 1973). Preliminary results suggest that in this system, the formation of the four hydrocarbons from CPP is catalyzed by different enzymes, which all have mol. wt. of about 2×10^5 daltons are stimulated by Mg^{2+} or Mn^{2+} and are inhibited by sulfhydryl binding agents. These enzymes have a definitive peak of activity at a certain age of the seedlings (2.5–3 days), which indicates a possible role in development.

The biosynthesis of *ent*-kaurene is inhibited by a number of synthetic compounds which are classified as plant growth retardants because they cause a dwarf growth habit in several plants. Ninnemann et al. (1964) showed that two of these compounds, AMO-1618 and CCC (2-chloroethyltrimethylammonium chloride), inhibit the formation of GA_3 in cultures of *G. fujikuroi.* Subsequently, Dennis et al. (1965) showed that AMO-1618, phosphon D (tributyl-2,4-dichlorobenzyl-phosphonium chloride) and some other related compounds inhibit the biosynthesis of *ent*-kaurene in the *Echinocystis* cell-free system. They also obtained evidence that the conversion of GGPP to *ent*-kaurene is the actual step inhibited. This has since been confirmed for several plant growth retardants in both cell-free preparations and fungal cultures (see review by Lang 1970).

Further characterization of the effects of plant growth retardants and other synthetic compounds on purified *ent*-kaurene synthetase from *G. fujikuroi* and partly purified preparations from *E. macrocarpa* has shown that it is primarily the first activity, the conversion of GGPP to CPP (Scheme 3.2) which is sensitive to inhibition. In the *G. fujikuroi* system, the most effective inhibitors of this activity are AMO-1618, its isomer Carvadan and phosphon D, all inhibiting to an extent of 50% or more at 5×10^{-7} *M.* Several other growth retardants (phosphon S, Q-64, Q-58 and Q-53) and inhibitors of steroid biosynthesis (SKF 3301A and SKF 525A) were effective at 5×10^{-6} to 5×10^{-5} *M,* whereas CCC was inhibitory only at the high level of 10^{-2} *M* (Fall and West 1971). Similar results were obtained when some of these compounds were tested with the *Echinocystis* system, SKF 525A being the most effective, followed by phosphon and AMO-1618 (West 1973).

The second activity, conversion of CP to *ent*-kaurene, is much less sensitive to inhibition by most of the growth retardants, although Q-64, Q-58, Q-53 and the steroid inhibitors inhibit both activities equally much in the *G. fujikuroi* system (Fall and West 1971). AMO-1618 had very little activity at the highest concentration tested in both cell-free systems and a 5×10^3 times higher concentration of phosphon D was needed to obtain the same inhibition as with the first activity.

The high concentrations of CCC needed to inhibit *ent*-kaurene biosynthesis in the cell-free systems are interesting because much less (6.3×10^{-7} *M*) is needed to inhibit GA biosynthesis in the intact fungus (Ninnemann et al. 1964; Barnes et al. 1969; Cross and Myers 1969). This discrepancy is unexplained. Recently, the growth retardant ancymidol [α-cyclopropyl-α-(*p*-methoxyphenyl)–5-pyrimidine methyl alcohol] has been shown to act on the further conversion of *ent*-kaurene (Coolbaugh and Hamilton 1976). The well known growth retardant *N*-(dimethylamino)succinamic acid (termed B-995 or SADH), on the other hand, has no effect on the biosynthesis of *ent*-kaurene or GAs, either in cell-free systems or in cultures of *G. fujikuroi.* The dwarfing effect of growth retardants on higher plants may be partly explained by their inhibition of GA biosynthesis, but it has become clear that they have other effects as well in intact plants (section 3.11.5).

3.7.2.3. Sequence from ent-kaurene to ent-7α-hydroxykaurenoic acid. In the further pathway (Scheme 3.3) *ent*-kaurene (I) is oxidized step-wise at C-19 to form *ent*-kaurenol (XXXVIII), *ent*-kaurenal (XXXIX) and *ent*-kaurenoic acid (XL). The latter becomes hydroxylated to *ent*-7α-hydroxykaurenoic acid (VII). This sequence was first established in the *Echinocystis* cell-free system although most of the steps were found more or less simultaneously in cultures of *G. fujikuroi*. The latter will be considered first.

The incorporation of partially synthesized *ent*-[17-^{14}C]kaurenol into GA_3 and GA_{13}

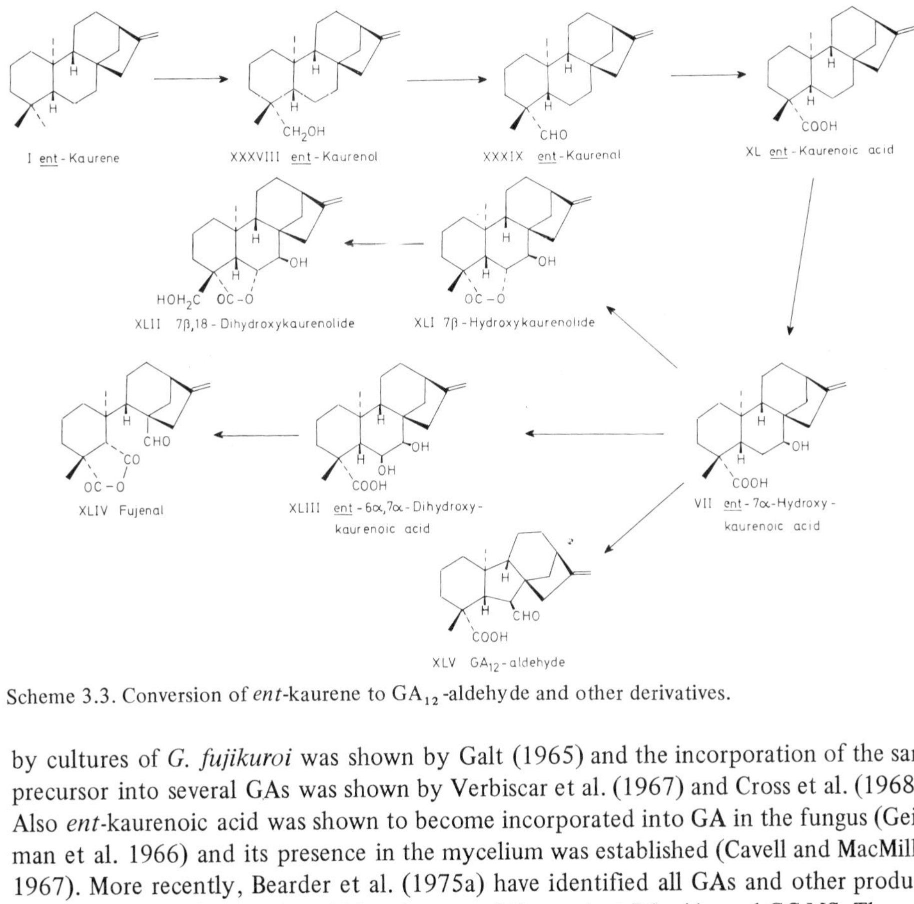

Scheme 3.3. Conversion of *ent*-kaurene to GA_{12}-aldehyde and other derivatives.

by cultures of *G. fujikuroi* was shown by Galt (1965) and the incorporation of the same precursor into several GAs was shown by Verbiscar et al. (1967) and Cross et al. (1968a). Also *ent*-kaurenoic acid was shown to become incorporated into GA in the fungus (Geissman et al. 1966) and its presence in the mycelium was established (Cavell and MacMillan 1967). More recently, Bearder et al. (1975a) have identified all GAs and other products formed from *ent*-kaurenoic acid by the use of the mutant B1–41a and GC-MS. The conversion of *ent*-7α-hydroxy[17-^{14}C]kaurenoic acid specifically into [17-^{14}C]GA_3 and other products in cultures of *G. fujikuroi* was shown by Hanson et al. (1972). The yield of *ent*-7α-hydroxykaurenoic acid from *ent*-kaurene was, however, so low (0.006%) that it did not allow the conclusion to be drawn that this compound is a normal intermediate. Nevertheless, the formation of *ent*-7α-hydroxykaurenoic acid from *ent*-kaurenoic acid in the fungus has been verified by Bearder et al. (1975a), who found that it is very rapidly converted further and therefore does not accumulate. In spite of this rapid conversion of *ent*-7α-hydroxykaurenoic acid formed in situ, it is less readily converted to GAs than is *ent*-kaurenoic acid when added exogenously (Lew and West 1971; Bearder et al. 1975a). This may be due to poor translocation within the mycelium. Conversion of *ent*-kaurenoic acid to *ent*-7α-hydroxykaurenoic acid in high yield has also been shown in *Rhizopus nigricans* (Beïlby et al. 1973).

In the *Echinocystis* system, *ent*-[^{14}C]kaurenol (Graebe et al. 1965), *ent*-[^{14}C]kaurenal and *ent*-[^{14}C]kaurenoic acid (Dennis and West 1967) were identified as products from [2-^{14}C]MVA and their sequence of formation was demonstrated by isolation and re-incubation in the same system. *ent*-7α-Hydroxy-[^{14}C]kaurenoic acid was obtained in mg amounts and identified by a monumental incubation of *ent*-[^{14}C]kaurenoic acid with 1440 separate 1-ml portions of *Echinocystis* endosperm (Lew and West 1971). Since the *Echinocystis* system does not produce GAs, the intermediacy of all these products in GA biosynthesis was proven by incubation with cultures of *G. fujikuroi.*

The conversion of *ent*-kaurene to *ent*-7α-hydroxykaurenoic acid in the *Echinocystis* system is catalyzed by the microsomal fraction and requires oxygen and NADPH (Dennis and West 1967; Lew and West 1971). Furthermore, as shown by incubation with $^{18}O_2$, the oxygen in *ent*-kaurenol originates from atmospheric oxygen. At least two of the reactions in the sequence are inhibited by CO and the inhibition can be reversed by light (Murphy and West 1969). These properties suggest that the sequence is catalyzed by mixed function oxidases. The action spectrum of the light effect suggests the participation of cytochrome P_{450}. An ATP-activated transhydrogenase which catalyzes the reduction of $NADP^+$ by NADH is associated with the microsomal fraction of the *Echinocystis* endosperm. $NADP^+$ added together with NADH and ATP therefore supports the reactions from *ent*-kaurene to *ent*-7α-hydroxykaurenoic acid as efficiently as NADPH (Hasson and West 1973).

The sequence from *ent*-kaurene to *ent*-7α-hydroxykaurenoic acid has also been shown to occur in the *Cucurbita* system (section 3.7.1) by the use of GC-MS for identification of the products (Graebe et al. 1972, 1974a). In contrast, the cell-free system from *G. fujikuroi* does not convert *ent*-kaurene, although it converts *ent*-kaurenoic acid, to further products (section 3.7.2.4) (West 1973). Similarly, no metabolic conversion of *ent*-[17-^{14}C]kaurene was obtained with cell-free systems from germinating seeds of barley although *ent*-kaurene is present naturally and decreases during the first 10–12 hr of germination indicating that it is being metabolized in vivo (Murphy and Briggs 1973b). This latter work is valuable because it identifies artefacts which give the impression that metabolism has occurred. Products were found on TLC which migrated like *ent*-[^{14}C]-kaurenol and *ent*-[^{14}C]kaurenal and a third product was also detected. However, on careful identification they turned out to be *ent*-[^{14}C]kauran-17-ol, *ent*-[^{14}C]kauran-16,17-epoxide and *ent*-[^{14}C]-kauran-16,17-diol, which are all not part of the pathway. Microsomal preparations from barley grains do perform later steps in the sequence, although the conversion is very low: *ent*-kaurenol, *ent*-kaurenal and *ent*-kaurenoic acid is each converted one step further (Murphy and Briggs 1975). Cell-free systems from immature pea seeds do not convert *ent*-[^{14}C]kaurene when it is added exogenously but there may be some conversion of [2-^{14}C]MVA to *ent*-[^{14}C]kaurenol, which must be assumed to proceed via *ent*-[^{14}C]kaurene formed in situ (Coolbaugh and Moore 1971b).

Moore et al. (1972) reported that *ent*-kaurene becomes bound to a large proteinaceous aggregate in the $10^5 \times g$ supernatant fraction of the pea system and concluded that the protein is a soluble carrier protein necessary for further metabolism of *ent*-kaurene. Although this would be logical for the water-insoluble substrate, the evidence could also be interpreted as a binding of *ent*-[^{14}C]kaurene to microsomal fragments that often do not sediment completely by this force. This view is supported by the 10^5 supernatant forming a pellet when centrifuged a second time, by the very large mol. wt. of the presumptive carrier protein (15×10^6 daltons) and by the demonstrated association of *ent*-[^{14}C]kaurene with the regular microsomal pellet. Furthermore, conversion of the *ent*-[^{14}C]kaurene-protein complex to products was never demonstrated but only deduced from small differences in the recovery of *ent*-[^{14}C]kaurene after incubation. The highest "conversion" according to this assay was obtained with a boiled protein complex, which makes its significance doubtful.

3.7.2.4. Formation of GA_{12}-aldehyde, ent-6α, 7α-dihydroxykaurenoic acid and kaurenolides: a triple branch-point *ent*-7α-Hydroxykaurenoic acid (VII) can be metabolized in three ways (Scheme 3.3): by an oxidative ring contraction which yields GA_{12}-aldehyde (XLV), by 6β-hydroxylation which yields *ent*-6α,7α-dihydroxykaurenoic acid (XLIII) and by 6α-hydroxylation which yields the lactonic 7β-hydroxykaurenolide (XLI). Only GA_{12}-aldehyde is a precursor of GAs.

In line with predictions made by Birch et al. (1959), the conversion of *ent*-7α-hydroxykaurenoic acid to GAs occurs by contraction of ring-B with extrusion of C-7 (cf. 3.7.4). On the basis of high incorporation (15.4%) of [17-^{14}C]GA_{12}-aldehyde specifically into [17-^{14}C]GA_3 by *G. fujikuroi*, Cross et al. (1968b) suggested that GA_{12}-aldehyde is a natural intermediate in GA biosynthesis and the first product formed after ring contraction. Its formation in fungal cultures was subsequently established by Hanson et al. (1972) who showed incorporation of ent-7α-hydroxy-[17-^{14}C]kaurenoic acid specifically into [17-^{14}C]GA_{12}-aldehyde. Furthermore, Bearder et al. (1973a) using GC-RS and GC-MS have demonstrated the incorporation of [6-^{3}H]GA_{12}-aldehyde into all the major fungal GAs by a slow-growing wild type strain and Bearder et al. (1975a) by GC-MS determined the complete spectrum of GAs formed from GA_{12}-aldehyde in the mutant B1–41a.

In the *Cucurbita* cell-free system, [^{14}C]GA_{12}-aldehyde as identified by GC-MS is a product of incubations with [2-^{14}C]MVA, *ent*-[^{14}C]kaurene, *ent*-[^{14}C]kaurenol, *ent*-[^{14}C]kaurenoic acid or *ent*-7α-hydroxy[^{14}C]kaurenoic acid (Graebe et al. 1972, 1974a). GA_{12}-aldehyde is a major product in the presence of Mn^{2+} in this system because its further conversion is strongly suppressed. In the absence of Mn^{2+}, it is completely converted further to a number of GAs (section 3.7.2.5). The cell-free systems from *E. macrocarpa* and *G. fujikuroi* according to preliminary identifications also produce GA_{12}-aldehyde (West and Fall 1972; West 1973).

ent-6α-7α-Dihydroxykaurenoic acid (XLIII) is a metabolite of *G. fujikuroi* which does not accumulate in the cultures but becomes further converted to fujenal (XLIV) by an oxidative ring cleavage (Cross et al. 1970). Fujenal and its derivatives are major metabolites of normal cultures (Cross et al. 1963d; Bearder et al. 1975a), one metabolite being the corresponding diacid (MacMillan and Wels 1974). In the *Cucurbita* cell-free system, *ent*-6α,7α-dihydroxykaurenoic acid is formed simultaneously with and at the same rate as GA_{12}-aldehyde. It is therefore a major product of incubations with *ent*-7α-hydroxykaurenoic acid or any earlier precursor in the pathway (Graebe et al. 1974a, 1975; cf. section 3.7.4.4). Also in the *Cucurbita* system *ent*-6α,7α-dihydroxykaurenoic acid is oxidized further to an *ent*-trihydroxykaurenoic acid and other products (Graebe and Hedden 1974). According to preliminary identification, *ent*-6α,7α-dihydroxykaurenoic acid is also formed in the cell-free systems from *E. macrocarpa* and *G. fujikuroi* (West 1973). The formation of *ent*-6α,7α-dihydroxykaurenoic acid represents a major branch-point of unknown significance in the pathway, since it is not a precursor of GAs in the fungus or any of the cell-free systems. It may be a true by-product of the reaction that results in the formation of GA_{12}-aldehyde (section 3.7.4.4).

7β-Hydroxykaurenolide (XLI), 7β,18-dihydroxykaurenolide (XLII) and other kaurenolides are also major metabolites of the fungus (Cross et al. 1963b, c; MacMillan and Wels 1974). The formation of 7β-hydroxy-[17-^{14}C]kaurenolide from *ent*-[17-^{14}C]kaurene in cultures of *G. fujikuroi* and its further conversion to 7β,18-dihydroxy-[17-^{14}C]kaurenolide (Scheme 3.3) was shown by Cross et al. (1964, 1968a). They also showed that it was not converted to GA_3. The formation of 7β-hydroxykaurenolide from *ent*-kaurenoic acid or *ent*-7α-hydroxykaurenoic acid has also been shown in the mutant B1–41a (Bearder et al. 1975a) and in the cell-free system from *G. fujikuroi* (West 1973). Kaurenolides are also obtained in the *Cucurbita* system but in much lower amounts than e.g. *ent*-6α-7α-dihydroxykaurenoic acid (Graebe et al. 1974a). Formation of kaurenolides in higher plants is also implied by the presence of 7β,13-dihydroxykaurenolide in bean seeds (Gaskin and MacMillan 1975). The kaurenolides are useful as starting material for partial synthesis of several intermediates of GA biosynthesis (3.4.2).

The enzymes catalyzing the conversions of *ent*-7α-hydroxykaurenoic acid across the triple branch point are microsomal in the *G. fujikuroi* as well as in the *Cucurbita* cell-free system (West 1973; Graebe and Hedden 1974).

3.7.2.5. Conversion of GA_{12}-aldehyde to C_{20}- and C_{19}-GAs. GA_{12}-aldehyde (XLV) is converted to GA_{14}-aldehyde (XLVI) in *G. fujikuroi* and both are converted to both C_{20}- and C_{19}-GAs giving rise to a 3-deoxy and a 3β-hydroxy series, respectively (Scheme 3.4). In the *Cucurbita* cell-free system, GA_{12}-aldehyde is converted to C_{20}- and C_{19}-GAs via a slightly different route.

Cross et al. (1968b) found that *G. fujikuroi* (strain ACC.917) converted [17-^{14}C]-GA_{12}-aldehyde very well and [17-^{14}C]GA_{14} reasonably well to [17-^{14}C]GA_3, whereas it converted [17-^{14}C]GA_{12} mainly to three unidentified acids and only a little to [17-^{14}C]GA_3. They concluded that GA_{12}-aldehyde is converted mainly via GA_{14}-aldehyde to GA_3 in the fungus and that GA_{14} and GA_{12} are not on the main pathway (cf. Scheme 3.4). Similar results were obtained by Bearder et al. (1973a) using a slow-growing strain, REC-193A. Direct evidence for the intermediacy of GA_{14}-aldehyde was obtained by Hedden et al. (1974), who identified remnants of [^{3}H]GA_{14}-aldehyde (0.5%) after almost complete conversion of [6-^{3}H]GA_{12}-aldehyde to [^{3}H]GA_{14} during 2 hr of incubation. They also showed efficient conversion of [6-^{3}H]GA_{14}-aldehyde and [17-^{14}C]GA_{14}

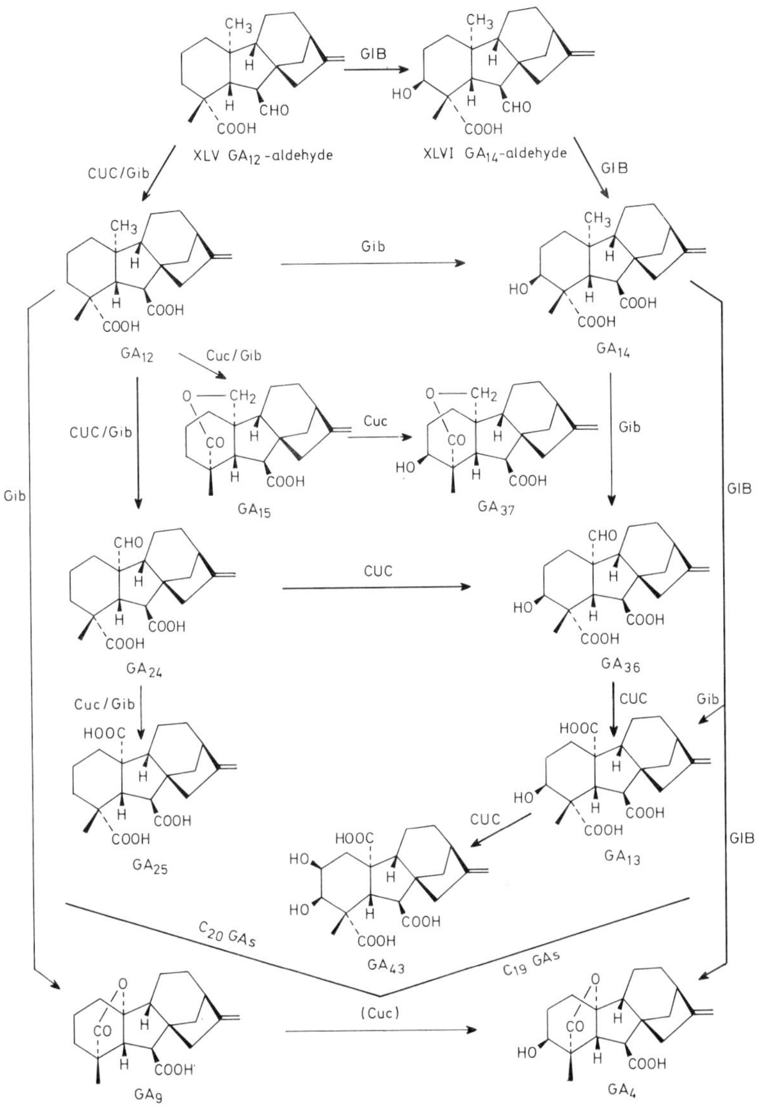

Scheme 3.4. Conversion of GA_{12}-aldehyde to C_{20}- and initial C_{19}-GAs by cultures of *G. fujikuroi* (Gib) and cell-free extracts of *C. maxima* (Cuc). Capital letters indicate the most prominent pathways. The production of GA_{37} in cultures of *G. fujikuroi* and the conversion of GA_{12} to GA_4 by *C. maxima* are not accounted for by the scheme (cf. text). GA_{43} is not produced by the fungus but is the main terminal GA of the *C. maxima* pathway. GA_4 is an intermediate of the pathway leading to GA_3 in the fungus (see Scheme 3.5).

to all the major GAs of the fungus and concluded that the pathway goes via GA_{14}-aldehyde and GA_{14}. The study by Bearder et al. (1975a) with the mutant B1–41a completed the picture. GA_{12}-aldehyde is converted mainly to 3β-hydroxy-GAs including the C_{20}-GAs GA_{14}, GA_{42} (GA_{14}-hydrate), GA_{36} and GA_{13} (Scheme 3.4), and several C_{19}-GAs with GA_3 as the main terminal product of the pathway. Thus in the fungus, 3β-hydroxylation is the first reaction of this sequence, taking place immediately after ring contraction. GA_{12}-aldehyde to a less extent is also a precursor of 3-deoxy GAs, presumably via GA_{12}. GA_{12} mainly becomes converted to the 3-deoxy GAs GA_9, GA_{15}, GA_{24} and GA_{25} but some 3β-hydroxy GAs are formed in addition, indicating that some 3β-hydroxylation takes place at this stage also. This secondary 3β-hydroxylation is completely suppressed at pH 7 (vs. normal pH 3–4). GA_9 is not converted to GA_4 in cultures of *G. fujikuroi*, so there is no 3β-hydroxylation at the stage of C_{19}-GAs (Cross et al. 1964; Bearder et al. 1976b).

GA_{24} becomes converted to GA_{25} and GA_{14} becomes converted to GA_{13} but GA_{15}, GA_{37} and GA_{36} are not converted at all in mutant B1–41a (Bearder et al. 1975a). Since GA_{36} would be expected to be the precursor of at least GA_{13}, it was suggested that GA_{36} may not penetrate to the site of conversion or the equilibrium between free aldehyde and lactol may be unfavourable. The presence of GA_{37} in cultures of *G. fujikuroi* remained unexplained by the feeding studies.

The *Cucurbita* cell-free system also produces 3-deoxy and 3-hydroxy GAs but the preferred sequence is different from that of the fungus. Thus [^{14}C]GA_{12} – aldehyde is converted to [^{14}C]GA_{12}, -GA_{15}, -GA_{37}, -GA_{24}, -GA_{36}, -GA_{13}, -GA_{43} and -GA_4 (Scheme 3.4). [^{14}C]GA_{12} is converted to the same series (Graebe et al. 1974a, c). Reincubation of the intermediates proved the main sequence to be GA_{12}, GA_{24}, GA_{36}, GA_{13} to GA_{43} (Graebe et al. 1974b, c). Thus, in the *Cucurbita* system, 3β-hydroxylation occurs at the level of GA_{24} resulting in the formation of GA_{36}. Unlike the situation in *G. fujikuroi,* 3β-hydroxylation can also take place at the level of C_{19}-GAs in this system: GA_9 is converted very efficiently to GA_4 (Graebe and Hedden 1974). However, since GA_9 has not been identified in the system, the significance of this reaction is unknown. GA_{43}, which so far has only been found in *Echinocystis* and *Cucurbita* endosperms, is the main product of the cell-free system and the main endogenous GA of the endosperm. The C_{20}-pathway therefore is not an artefact of the cell-free system.

GA_{15}, which would be expected to be an intermediate between GA_{12} and GA_{24}, was not incorporated into anything but GA_{37}, which was not converted further. However, in the conversion of GA_{12}-aldehyde to GA_{13} evidence was obtained by GC for the intermittent appearance of a compound yielding GA_{15} on extraction (Graebe et al. 1974b). Thus a compound related to GA_{15}, e.g. the corresponding non-lactone hydroxyacid, may be the true intermediate. GA_{14}-aldehyde and GA_{14} can also be obtained in the *Cucurbita* system, but only by loading it with high concentrations of GA_{12}-aldehyde. Some GA_{25} was formed from GA_{24} but it was converted to GA_{13} much less efficiently than was GA_{36} and was therefore not considered to be part of the main pathway.

The conversion of GA_{12}-aldehyde to GA_{12} in the *Cucurbita* system represents the end of the part of the pathway which, beginning with the oxidation of *ent*-kaurene to *ent*-kaurenol, is catalyzed by microsomal enzymes. This last step is catalyzed by Fe-requiring supernatant enzymes as are all steps occurring after GA_{12} (Graebe and Hedden 1974). In the *G. fujikuroi* cell-free system, $[^{14}C]GA_{14}$ has been identified as a product obtained from *ent*-$[17\text{-}^{14}C]$kaurenoic acid with microsomal preparations (West 1973). Thus the 3β-hydroxylation is catalyzed by supernatant enzymes in *Cucurbita* but by microsomal enzymes in *G. fujikuroi,* which may account for the difference in the pathways.

Durley et al. (1974) fed $[17\text{-}^{3}H]GA_{14}$ to etiolated pea plants and identified labelled GA_{18}, GA_{38}, GA_{23}, GA_{28}, GA_1 and GA_8 as products by GC-RC using three different columns. The first five of these GAs form a series of 3β,13-dihydroxylated GAs corresponding to the series of 3-deoxy- and 3β-hydroxy-GAs obtained in *G. fujikuroi* and the *Cucurbita* cell-free system (cf. Table 3.2). With the exception of GA_{38}, the 3β,13-dihydroxy-GAs obtained in the feeding experiment of Durley et al. are not known to be endogenous to pea plants, but the corresponding 13-hydroxy-GAs, GA_{44}, GA_{17}, GA_{20} and GA_{29}, have been conclusively identified in immature seeds of the same variety (Table 3.4). Thus the pea plants have oxidized the supplied 3β-hydroxy-GA in the 13-position giving rise to a series which differs from their natural GAs by the presence of the 3β-hydroxyl group. Had GA_{12} been fed instead, the plants might have produced the natural GAs. A lack of specificity with regard to the substrate is common in the GA pathway as will be seen in section 3.7.3.

One of the questions still remaining to be answered is the exact stage at which the C-20 group is removed to form the first C_{19}-GA (cf. Scheme 3.4). GA_{12} is converted to GA_9 in the fungus and to GA_4 in the *Cucurbita* system. GA_{14} is a precursor of GA_4 in the fungus. The C_{20}-GAs arising from GA_{12} and GA_{14} represent successive oxidative stages of the methyl group that becomes removed and it was only logical to suggest that C-20 becomes removed by an oxidative decarboxylation at the stage of a carboxylic acid. However, Cross et al. (1968b) did not get conversion of GA_{13} to C_{19}-GAs in cultures of *G. fujikuroi* and Bearder et al. (1975a) did not get conversion of GA_{13} or GA_{25} to C_{19}-GAs (cf. Scheme 3.4). They did not get conversion of GA_{13}-anhydride (XLVII) or GA_{25}-anhydride either, although Hanson and Hawker (1972b) reported incorporation of $[17\text{-}^{14}C]GA_{13}$-anhydride to the extent of 0.07% into $[17\text{-}^{14}C]GA_4/GA_7$ and to the extent of 0.14% into $[17\text{-}^{14}C]GA_3$. However, as was pointed out by the authors, this very small incorporation does not permit the conclusion that GA_{13}-anhydride is a normal intermediate in the pathway to C_{19}-GAs.

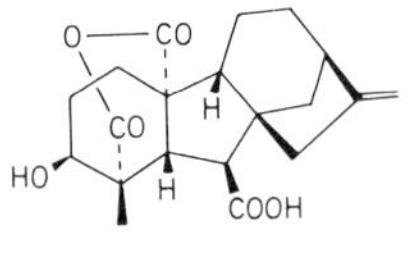

XLVII
GA_{13}-anhydride

To explain the facts that GA_{13} is not incorporated into C_{19}-GAs and that all of the H-atoms at the positions 1, 5 and 9 are retained in the transformation of C_{20}-GAs to C_{19}-GAs (section 3.7.4), which precludes the possibility of decarboxylation via a β,γ-unsaturated acid, Durley (1968) and Hanson and White (1969) suggested that the C-20 group might be lost at the oxidation state of an aldehyde, for example by a Baeyer – Villiger type oxidation. Whereas this remains a distinct possibility, feeding studies have supported it no more than the decarboxylation theory. Thus Bearder et al. (1975a) got no conversion of the aldehydic GA_{24} and GA_{36} to C_{19}-GAs (Scheme 3.4). The argument by Fukui et al. (1972) that the higher activities in GA bioassays of C-20 aldehydes than of C-20 carboxylic acids would indicate that the former but not the latter are precursors of C_{19}-GAs is not valid as long as the higher activity may be due to a better fit at the receptor site (cf. 3.6.1.1). The conclusion by Durley et al. (1974) that their experimental results on the transformations of $[17\text{-}^{14}C]GA_{14}$ in pea plants (see above) indicate a distinct pathway in which the conversion to C_{19}-GAs occurs via the C-20 aldehyde, GA_{23}, seems little related to their data. Thus there is no experimental evidence now for a conversion to C_{19} at any stage later than GA_{12} and GA_{14}, in which compounds C-20 is present as a methyl group. However, all the C_{20}-GAs are converted poorly or not at all in *G. fujikuroi* even to products of which they would be expected to be the precursors; the negative results with respect to their conversion to C_{19}-GAs prove nothing and the question about the oxidation state at which C-20 is lost remains open.

3.7.2.6. Metabolism of C_{19}-GAs in G. fujikuroi. This section is concerned with the metabolism of C_{19}-GAs in *G. fujikuroi* only, while the corresponding events in higher plants will be discussed in section 3.8. In *G. fujikuroi,* the two C_{19}-GAs first formed are GA_9 and GA_4 (Scheme 3.4). GA_9 becomes metabolized mainly to GA_{40} and GA_{20} with small amounts of GA_{10} and GA_{11} also forming (Bearder et al. 1976b) (Scheme 3.5).

Evidence that GA_4 is a precursor of GA_3 was presented by Geissman et al. (1966), who demonstrated by TLC the formation of a $[^{14}C]GA_4/GA_7$ fraction from $[17\text{-}^{14}C]$-kaurenoic acid, and Verbiscar et al. (1967), who showed that this fraction was incorporated into $[^{14}C]GA_1$ and $[^{14}C]GA_3$. The full course of the pathway (Scheme 3.5) was conclusively determined by Pitel et al. (1971b) and McInnes et al. (1973) by incubation with $[^{14}C]$-GA_4, -GA_7, -GA_3, -GA_{16} and $[1,2\text{-}^3H]GA_1$. GA_3 is terminal, it only gave small amounts of isoGA_3 and allogibberic acid, which were probably artefacts (3.4.1). Bearder et al. (1975a) determined the same sequence as part of their feeding studies with the mutant B1–41a. Both working groups agreed that GA_3 is formed mainly via GA_7 and only to a very small extent via GA_1. This conversion has been elaborated upon particularly because Geissman et al. (1966) had suggested that GA_3 is formed from GA_1 also. However, Bearder et al. (1973b) using the mutant R9 also showed that the conversion of GA_1 to GA_3 is very low. Most convincingly Bearder et al. (1975b) found that GA_1 rather than GA_3 was the main product when the mutant B1–41a was incubated with steviol (*ent*-13-hydroxykaurenoic acid) (XLVIII). The fungus carried this substrate through the same reactions as it did *ent*-kaurenoic acid, but since steviol has a 13-hydroxyl group, the first C_{19}-GA was GA_1 instead of GA_4 which therefore was by-passed (cf. Scheme 3.5). The accumulation of GA_1 and the absence of substantial amounts of GA_3 in this conversion proves that the formation of GA_3 is dependent on the presence of GA_4. The formation of GA_3 via GA_4 and GA_7 is therefore well established at least for some strains.

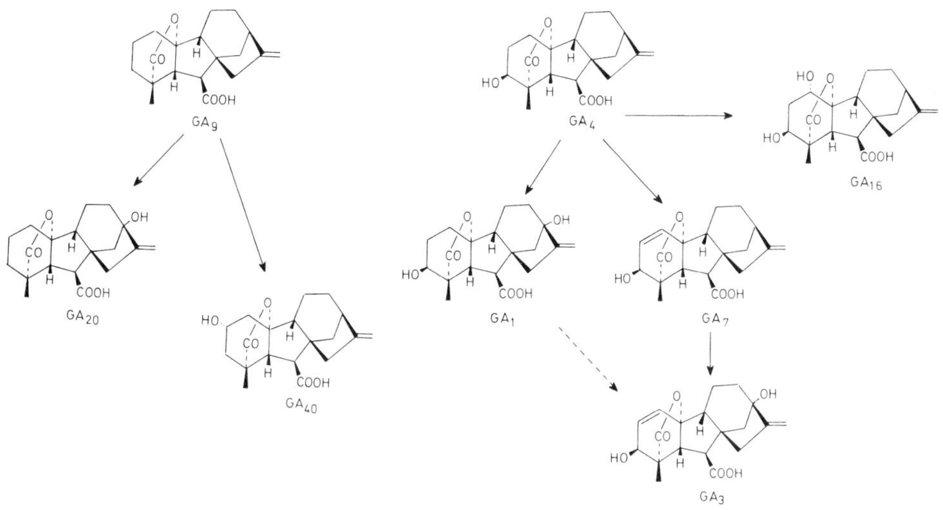

Scheme 3.5. The pathway of C_{19}-gibberellins in *Gibberella fujikuroi.*

3.7.3. Conversion of steviol and other substrates by G. fujikuroi

The fates of several naturally occurring and synthetic derivatives of *ent*-kaurenol, *ent*-kaurenoic acid, GA_{12} and GA_{14} in cultures of *G. fujikuroi* have been investigated. Some of these investigations have provided further insight into the sequences in GA biosynthesis; others have shown the specificity, or lack of specificity, of the enzymes, and yet others have provided new routes to compounds that are otherwise difficult to obtain.

Steviol (*ent*-13-hydroxykaurenoic acid) (XLVIII) occurs naturally as a glycoside in the leaves of *Stevia rebaudiana* (Ruddat et al. 1965) where it is biosynthesized via acetate, MVA, *ent*-kaurene and *ent*-kaurenoic acid (Ruddat et al. 1965; Bennett et al. 1967; Hanson and White 1968). Since many higher plants have 13-hydroxylated C_{20}-GAs, 13-hydroxylation must take place at an earlier stage in them than in *G. fujikuroi* where it only occurs in the hydroxylation of GA_4 and GA_7 (cf. Scheme 3.5). It has therefore been suggested (see Lang 1970) that steviol may be the precursor of 13-hydroxylated GAs in higher plants, i.e. that 13-hydroxylation might take place already before the contraction of ring-B (cf. Scheme 3.3). There is no direct evidence for or against this theory but when fed to *G. fujikuroi,* steviol (XLVIII) is converted to C_{19}-GAs along much the same pathway as *ent*-kaurenoic acid, only with 13-hydroxylated intermediates. Thus Bearder et al. (1975b), feeding steviol to the mutant B1–41a, found the 7β-hydroxylated and 6β,7β-dihydroxylated derivatives (XLIX and L), 7β,13-dihydroxykaurenolide (LII), 13-hydroxy-GA_{12} (LIII) and GA_1 as main products identified by GC-MS (cf. normal pathway in Schemes 3.3–3.5). Smaller amounts of GA_{20} and other GAs were also found. The by-passing of GA_4 and GA_7 in this incubation was discussed in section 3.7.2.6. Steviol acetate (LI) also is converted by the fungus to 13-acetoxy-derivatives but in this case 3β-hydroxylation is suppressed, so that the main product is GA_{20}-acetate rather than GA_1-acetate (cf. Scheme 3.5) (Bearder et al. 1976a). Hydrolysis yields GA_{20}, which is

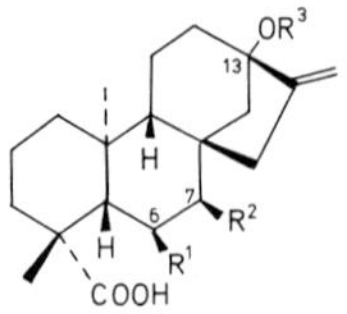

	R^1	R^2	R^3
XLVIII	H	H	H
XLIX	H	OH	H
L	OH	OH	H
LI	H	H	$COCH_3$

Steviol (*ent*-13-hydroxykaurenoic acid)
7β-Hydroxysteviol (*ent*-7α,13-dihydroxykaurenoic acid)
6β,7β-Dihydroxysteviol (*ent*-6α,7α,13-tri-hydroxykaurenoic acid)
Steviol acetate

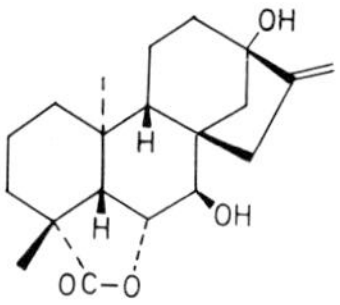

LII

7β,13-Dihydroxykaurenolide

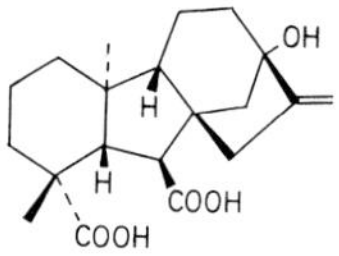

LIII

13-Hydroxy GA_{12}

otherwise difficult to obtain. The conversion of *ent*-kaurenoic acid (XL), steviol (XLVIII) and steviol acetate along the same pathway already indicates a certain lack of specificity by the enzymes. Particularly surprising is the conversion in good yield of the rearrangement product isosteviol (LIV) to the corresponding analogues of C_{20}-GAs and GA_9

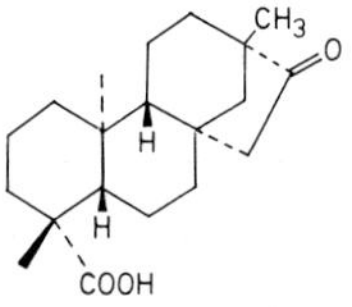

LIV

Isosteviol (*ent*-16-oxobeyran-19-oic acid)

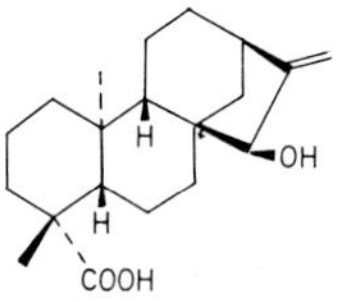

LV

Desacetoxyxylopic acid (*ent*-15α-hydroxykaurenoic acid)

(Bearder et al. 1976a). Also desacetoxyxylopic acid (*ent*-15α-hydroxykaurenoic acid) (LV) is transformed in good yield to 15α-hydroxylated products (MacMillan 1974b). Xylopic acid, the acetate of (LV) occurs naturally in fruits of *Xylopia aethiopica.*

The *ent*-kaurenyl succinate (LVI) is converted to its 7β-hydroxy- and 6β,7β-dihydroxy-derivatives and to GA_3 by *G. fujikuroi* (Jefferies et al. 1974). Apparently this conversion takes place via *ent*-7α-hydroxykaurenoic acid (VII), since the alcohol (LVII) is only poorly converted to GA_3. Thus 7β-hydroxylation before oxidation of C-19 to carboxyl does not permit ready conversion to GA_3 (cf. Scheme 3.3). The 3α-hydroxy-, 3β-hydroxy- and 3-oxo-derivatives (LVIII, LIX and LX, respectively) become 7β-hydroxylat-

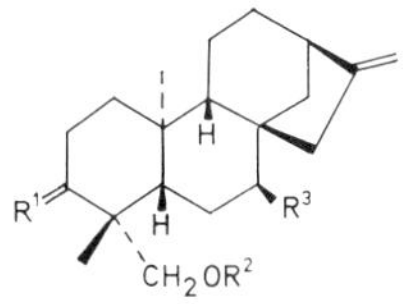

	R^1	R^2	R^3
LVI	H, H	$CO(CH_2)_2COOH$	H
LVII	H, H	H	OH
LVIII	H,α-OH	$CO(CH_2)_2COOH$	H
LIX	H, β-OH	"	H
LX	O	"	H

ent-Kaurenyl succinate
ent-Kaur-16-ene-7α,19-diol, *ent*-7α-hydroxykaurenol
ent-3β-Hydroxykaurenyl succinate
ent-3α-Hydroxykaurenyl succinate
ent-3-Oxokaurenyl succinate

ed, but in contrast to LVI, no ring contraction takes place and no GA_3 is formed. Apparently the oxygen functions at C-3 prevent ring contraction, either directly or by interfering with the oxidation of C-19. This is interesting because in the natural pathway 3β-hydroxylation is the very first step after ring contraction (section 3.7.2.5, Scheme 3.4). Thus, in spite of the lack of specificity for the C/D-ring system discussed above, the specificity for the A- and B-rings is high and the sequence of oxidative steps rigidly prescribed. Also the dienol (LXI) is a very poor precursor of GA_3 (0.48% incorporation) (Cross et al. 1968a) and *ent*-7-oxokaurenoic acid (LXII) does not undergo ring contraction, neither in the *G. fujikuroi* cell-free system (West 1973) nor in cultures of the mutant B1–41a (Bearder et al. 1975a). *ent*-7-Oxokaurenoic acid could conceivably have been a

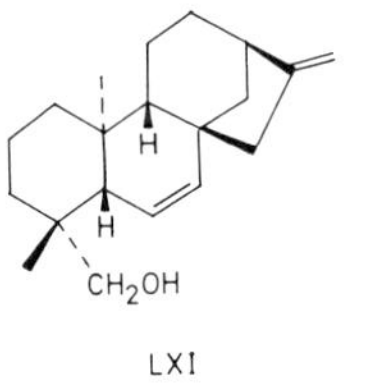

LXI

ent-Kaur-6,16-dien-19-ol

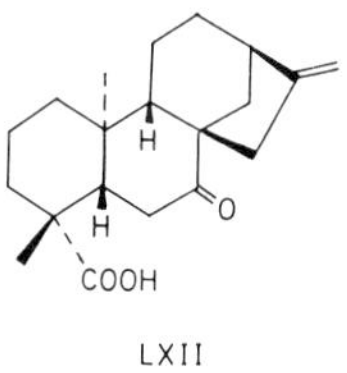

LXII

ent-7-Oxokaurenoic acid

natural intermediate between *ent*-7α-hydroxykaurenoic acid and GA_{12}-aldehyde, since this reaction involves the transfer of two electron pairs. No derivatives of *ent*-kaurenol or *ent*-kaurenoic acid with both the 6- and 7-positions oxidized are converted by the fungus (Cross et al. 1968a; Jefferies et al. 1974).

The dienol succinate (LXIII) is converted by *G. fujikuroi* along the normal pathway yielding several C_{20}- and C_{19}-GAs with 2,3-unsaturation or 2,3-epoxides, e.g. 13-deoxy-GA_5 (LXIV) and 13-deoxy-GA_6 (LXV) (Bakker et al. 1974). The diene-diol (LXVI) was also converted in good yield to similar products, in this case including GA_5 (20% yield). In contrast, the diol (LXVII) is only incorporated to 7.5% and the triol (LXVIII) to 0.23% into GA_3 (Cross et al. 1968b). However, comparison is difficult, since most feeding data depend on single experiments and different strains were used in the experiments of Bakker et al. and Cross et al.

Wild type *G. fujikuroi* and the mutant B1–41a convert GA_{12}-alcohol (LXIX) very

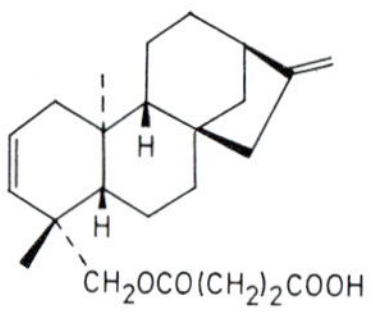

LXIII

ent-Kaura-2,16-dien-19-yl succinate

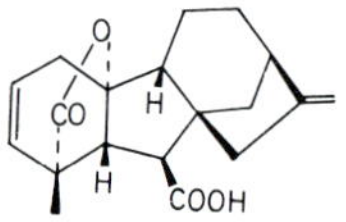

LXIV

13-Deoxy GA_5

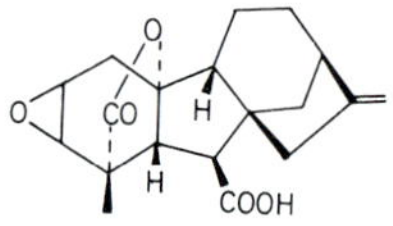

LXV

13-Deoxy GA_6

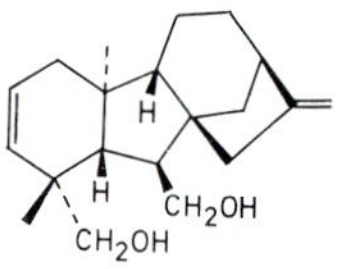

LXVI

ent-7,19-Dihydroxygibberella-2,16-diene

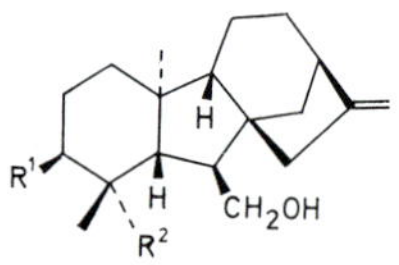

LXVII R^1= H R^2= CH_2OH

LXVIII R^1= OH R^2= CH_2OH

LXIX R^1= H R^2= COOH

LXX R^1= OH R^2= COOH

ent-7,19-Dihydroxygibberell-16-ene
ent-3α,7,19-Trihydroxygibberell-16-ene
GA_{12}-alcohol
GA_{14}-alcohol

efficiently along the normal pathway yielding GA_3 as the main product (Hanson and Hawker 1973; Hedden et al. 1974; Bearder et al. 1975a) whereas mutant R-9, which does not perform 13-hydroxylation, yields mainly GA_4 and GA_7 (Bearder et al. 1973b). The conversion of GA_{12}-alcohol probably takes place via GA_{14}-alcohol (LXX), since it gives rise to 3β-hydroxylated GAs only (cf. Scheme 3.4) and since GA_{14}-alcohol is also incorporated very efficiently into GA_3 (Hedden et al. 1974). In spite of its good conversion, GA_{12}-alcohol is not a natural intermediate of GA biosynthesis in *G. fujikuroi* (Hanson and Hawker 1973).

Fluorination of C-18 does not interfere with conversion. Thus 18-fluoro-GA_{12}-aldehyde (LXXI) is converted along the usual pathway giving rise to fluoro-GA_9 and fluoro-GA_3 (Bateson and Cross 1974) (cf. Schemes 3.4 and 3.5).

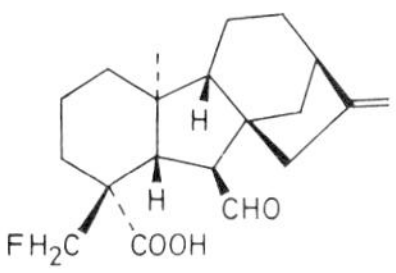

LXXI

18-Fluoro-GA_{12}-aldehyde

3.7.4. *Stereochemistry and mechanisms*

3.7.4.1. Over-all incorporation of MVA and acetate. Mevalonic acid (MVA) has an asymmetric carbon atom (C-3) and therefore exists in two enantiomeric forms. However, the biosynthesis of MVA from three molecules of acetyl-CoA is stereospecific and gives rise only to (3*R*)-MVA (Figure 3.4), which is also the sole form utilized by biological systems for the biosynthesis of isoprenoid compounds (Cornforth et al. 1960). In the conversion of MVA to isoprenoids, C-1 and the 3-hydroxyl group are lost but the asymmetry is retained and the mevalonoid C-2 always appears in the isoprenoid carbon chain rather than in the typical methyl groups which originate from the mevalonoid methyl groups. Figure 3.3 shows the labelling patterns of GA_3 biosynthesized by *G. fujikuroi* from [2-^{14}C]MVA and [1-^{14}C]acetate (Birch et al. 1959).

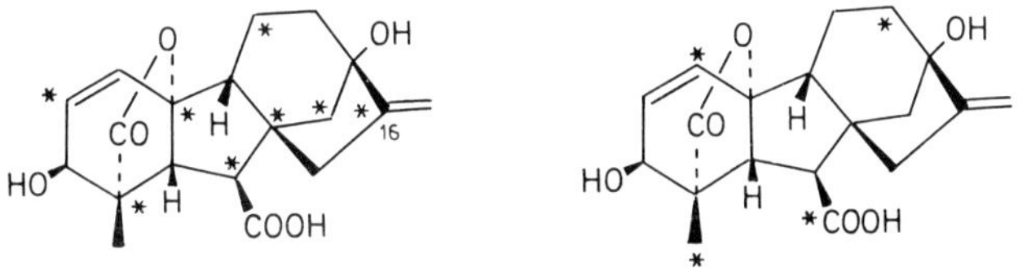

Figure 3.3. Labelling pattern in GA_3 biosynthesized from [1-^{14}C]acetate (left) and [2-^{14}C]mevalonate (right) (Birch et al. 1959).

The H-atoms attached to C-2, C-4 and C-5 of MVA can be labelled stereospecifically, which introduces additional chirality (Figure 3.4). These H-atoms retain their positions during isoprenoid biosynthesis (Popjak and Cornforth 1966). Their fates during bio-

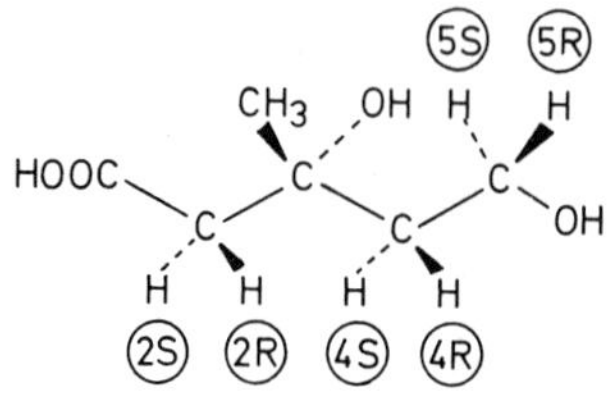

Figure 3.4. (3*R*)-Mevalonic acid and the positions of the label in [2-^{3}H]-, [4-^{3}H]- and [5-^{3}H]-labelled MVAs.

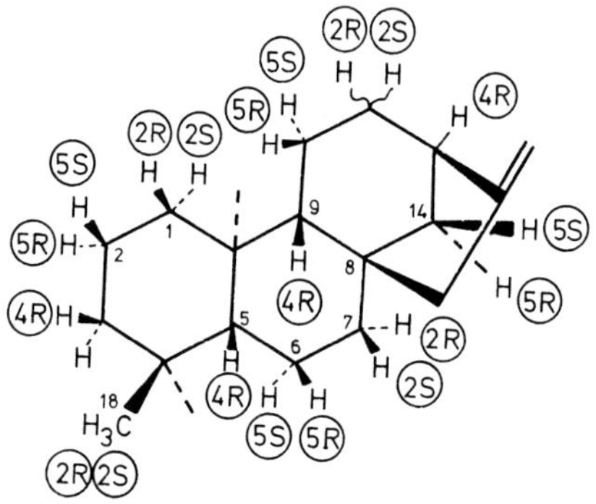

Figure 3.5. Origin of the H-atoms of the *ent*-kaurane skeleton from specifically labelled mevalonic acids.

synthesis of *ent*-kaurene, kaurenolides and GAs have been determined by Hanson and coworkers who fed specifically labelled MVAs to *G. fujikuroi* and located the labels in the products by degradation. The results are summarized in Figure 3.5 showing the positions of the *pro*-chiral mevalonoid H-atoms in the *ent*-kaurenoid structure, and in Table 3.10

Table 3.10.

Number of ^{3}H-atoms retained in different products after feeding specifically labelled [^{3}H,2-^{14}C]MVAs to cultures of *G. fujikuroi*

Position of ^{3}H label in precursor	*ent*-Kaurene	7β-OH-Kaurenolide	7β,18-$(OH)_2$-Kaurenolide	GA_{13}	GA_4	GA_3
[2-$^3H_2$][a]	8	7	6	6	6	5
(2R)-[2-$^3H_1$][b]	4	4	3.6	3	3	3
(2S)-[2-$^3H_1$][b]	4	3	2.6			2
(4R)-[4-$^3H_1$][a]		4	4	3		2
(5R)-[5-$^3H_1$][b]	4	4	4	3		2
(5S)-[5-$^3H_1$][c]						4

[a] Hanson and White (1969); [b] Evans et al. (1970); [c] Evans et al. (1973).

showing the total numbers of labelled H-atoms retained in the different products. These results will be discussed in the following sub-sections.

3.7.4.2. Cyclization of geranylgeranyl pyrophosphate. Scheme 3.6 shows the generally accepted mechanism for the cyclization of *trans*-geranylgeranyl pyrophosphate (GGPP)

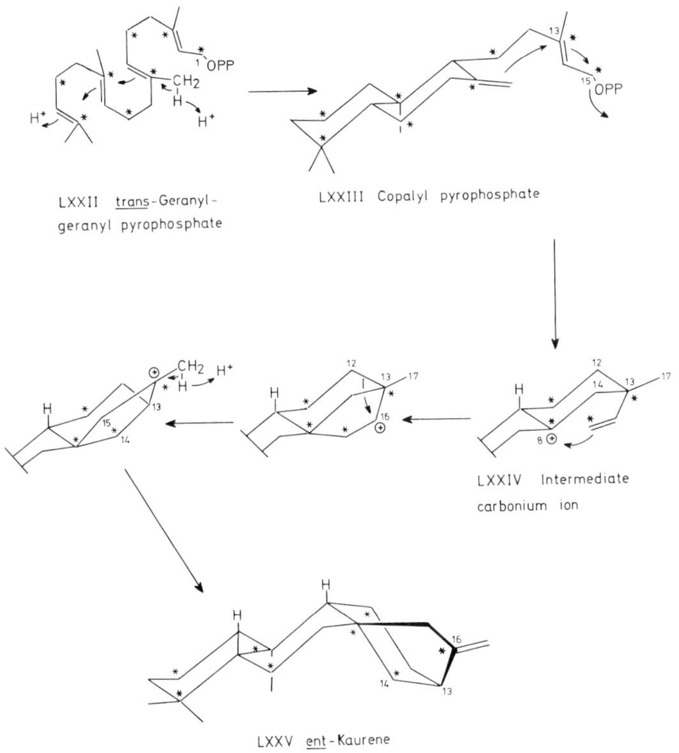

Scheme 3.6. Cyclization of geranylgeranyl pyrophosphate to *ent*-kaurene. Label as from [1-^{14}C] acetate.

(LXXII) to *ent*-kaurene (LXXV) with the labelling pattern expected from [1-^{14}C]acetate. The reaction is initiated by proton attack on the terminal double bond of GGPP followed by electron shifts and proton elimination from the methyl group at C-7 to form the bicyclic copalyl pyrophosphate (CPP) (LXXIII) (cf. section 3.7.2.2). An initial C-ring is formed by loss of the pyrophosphate group, a shift of the allylic double bond and ring closure resulting in the tricyclic carbonium ion (LXXIV), in which the previous methylene carbon of CPP occupies the 14-position. An attack by the vinyl group at C-8 leads to formation of an initial D-ring. Rearrangement of the 12,13-carbon bond to the 12,16-position then gives rise to the final ring system in which the previous methylene group of CPP occupies position 15 and the previous vinyl group positions 13 and 14. Elimination of a proton results in the final exocyclic double bond and stabilizes the molecule. This mechanism for the formation of the C/D ring system, which was originally suggested by Wenkert (1955) for tetracyclic diterpenes in general, receives strong support from three lines of evidence. Firstly, C-16 in GA_3 becomes labelled from [1-^{14}C]acetate (Fig. 3.3). Secondly, the label from either [1-^{3}H]geranylgeraniol or [15-^{3}H]CPP (these positions are analogous) becomes incorporated specifically into the 14-position of *ent*-kaurene (Scheme 3.6) (Hanson and White 1969). Thirdly, the simultaneous formation of *ent*-beyrene, *ent*-13,16-cycloatisane and *ent*-kaurene in the cell-free system from *Ricinus* seedlings (section 3.7.2.2) lends support to the mechanism because they can all be formally derived by stabilization of the intermediate steps in the formation of *ent*-kaurene from GGPP through loss of single protons (Shechter and West 1969; Robinson and West 1969b).

The participation of any of the three possible isomers of *ent*-pimara-8,15-dienes (LXXVI) as an intermediate in the cyclization is ruled out by the labelling patterns

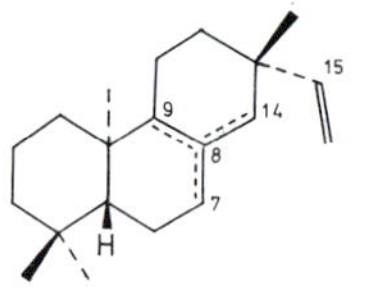

LXXVI

ent-Pimara-8,15-diene isomers

obtained with [^{3}H]MVAs (Table 3.10, Figure 3.5). Thus *ent*-pimara-7(8),15-diene is precluded because *ent*-kaurene contains eight ^{3}H-atoms from [2-^{3}H,2-^{14}C]MVA, two of which are specifically located at C-7 (Hanson and White 1969). This is confirmed by the retention of all four ^{3}H-atoms from either (2*R*)- or (2*S*)-[2-$^3H_1$,2-^{14}C]MVA in *ent*-kaurene but the loss of one mevalonoid (2*S*)-^{3}H in the formation of 7β-hydroxykaurenolide (XLI, Scheme 3.3). The intermediacy of *ent*-pimara-8(9),15-diene is precluded because two ^{3}H-atoms from (4*R*)-[4-$^3H_1$,2-^{14}C]MVA located specifically at positions 5 and 9 are retained in GA_3. Finally, *ent*-pimara-8(14),15-diene is ruled out by the incorporation of ten out of twelve ^{2}H-atoms from Me-[^{2}H]MVA into *ent*-kaurene (Evans and Hanson 1972) (not shown in Table 3:10). This agrees with the loss of two ^{2}H-atoms in the formation of the Δ8-double bond of CPP (LXXIII, Scheme 3.6) and the Δ16-double bond of *ent*-kaurene. An 8(14)-unsaturated *ent*-pimaradiene as intermediate would have resulted in loss of a third ^{2}H-atom at the stage of LXXIV (Scheme 3.6). As further evidence, both the (5*R*)- and the (5*S*)-mevalonoid ^{3}H-atoms are retained specifically in position 14 of GA_3 (Evans et al. 1970, 1973).

Earlier results by the same authors had shown that *ent*-[16-^{3}H,16-^{14}C]pimara-8(14),15-diene becomes incorporated (0.021%) into GA_3 with location of the ^{3}H-label specifically in position 14 (Hanson and White 1969). This labelling is still not questioned after the new results discussed above (Evans and Hanson 1972) were obtained, and is another reminder that incorporation even if specific does not prove the intermediacy of a compound in the normal pathway.

3.7.4.3. Formation of ent-7α-hydroxykaurenoic acid and the kaurenolides One ^3H-atom originating from (2*S*)-[2-$^3H_1$,2-^{14}C]MVA is lost when *ent*-kaurene is converted to 7β-hydroxykaurenolide (XLI, Scheme 3.3), whereas all the ^3H-atoms from (2*R*)-[2-$^3H_1$, 2-^{14}C]MVA are retained (Table 3.10, Figure 3.5). Thus the introduction of the 7β-hydroxyl group in the formation of *ent*-7α-hydroxykaurenoic acid occurs with retention of configuration. Also the 6α-hydroxylation that results in the kaurenolide lactone takes place without inversion, since all four (5*R*)-[5-$^3H_1$,2-^{14}C]MVA ^3H-atoms are retained. Both hydrogen atoms that are removed in the oxidation of ring-B are in axial positions. Biosynthesis of 7β,18-dihydroxykaurenolide proceeds with partial loss of both (2*S*)- and (2*R*)-mevalonoid H-atoms from the ring-A methyl group.

3.7.4.4. Ring contraction and 3β-hydroxylation. The conversion of *ent*-7α-hydroxykaurenoic acid to GA_{12}-aldehyde involves loss of the 6β-hydrogen, a shift of the 7,8-bond to the 6,8-position and loss of a proton from the extruded C-7 (Scheme 3.7). That

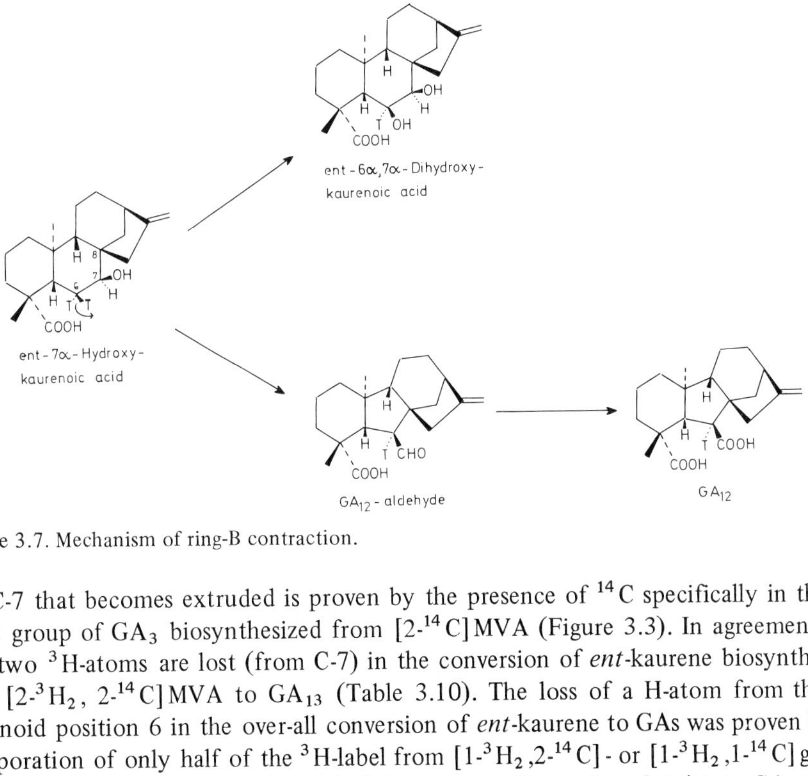

Scheme 3.7. Mechanism of ring-B contraction.

it is C-7 that becomes extruded is proven by the presence of ^{14}C specifically in the carboxyl group of GA_3 biosynthesized from [2-^{14}C]MVA (Figure 3.3). In agreement with this, two ^3H-atoms are lost (from C-7) in the conversion of *ent*-kaurene biosynthesized from [2-$^3H_2$, 2-^{14}C]MVA to GA_{13} (Table 3.10). The loss of a H-atom from the *ent*-kauranoid position 6 in the over-all conversion of *ent*-kaurene to GAs was proven by the incorporation of only half of the ^3H-label from [1-$^3H_2$,2-^{14}C]- or [1-$^3H_2$,1-^{14}C] geranyl pyrophosphate (condensed with unlabelled isopentenyl pyrophosphate) into GA_3 (Evans et al. 1970; Hanson et al. 1972). These substrates introduce ^3H specifically into the two 6-positions of the kauranoid nucleus. That it is the 6β-hydrogen that is lost is shown by the difference of one ^3H-atom between *ent*-kaurene and GA_{13} biosynthesized from (5*R*)-[5-$^3H_1$,2-^{14}C]MVA but retention of all four ^3H-atoms from (5*S*)-[5-$^3H_1$,2-^{14}C]-

MVA in GA_3 (Table 3.10). Direct loss of the 6β-H was demonstrated by using *ent*-7α-hydroxy-[6α-$^3H_1$,-17-^{14}C] kaurenoic acid of which 4.5% of the ^{14}C-label but none of the ^3H-label was incorporated into GA_3 (Hanson et al. 1972).

Hanson et al. (1972) on the basis of incorporation studies with [1-$^3H_2$,1-^{14}C]-geranyl pyrophosphate (see above) in *G. fujikuroi* concluded that both H-atoms at C-6 were retained in the cyclization of *ent*-7α-hydroxykaurenoic acid to GA_{12}-aldehyde and that the 6β-H was lost only in the subsequent oxidation to GA_{12} (cf. Scheme 3.7). This would require a specific transfer of the 6β-H to C-7 during cyclization. However, Graebe et al. (1975) by incubation of *ent*-7α-hydroxy-[6-$^3H_2$,^{14}C] kaurenoic acid with the *Cucurbita* system clearly showed that one hydrogen atom is already lost from C-6 in the formation of GA_{12}-aldehyde (Scheme 3.7). Furthermore, the formation of *ent*-6α,7α-dihydroxy-kaurenoic acid, which in this system occurs simultaneously with the formation of GA_{12}-aldehyde (section 3.7.2.4), also proceeds with loss of a 6-H. It is therefore likely that the cyclization is initiated by the removal of the 6β-H to form an unstable intermediate, which is then converted to either GA_{12}-aldehyde or *ent*-6α-7α-dihydroxykaurenoic acid.

The 3β-hydroxylation by which GA_{12}-aldehyde is converted to GA_{14}-aldehyde in *G. fujikuroi* (Scheme 3.4) occurs with retention of configuration, since GA_{13} has one ^3H-atom less than the kaurenolides when both are formed from (4*R*)-[4-$^3H_1$,2-^{14}C] MVA (Table 3.10) and since the 3α-hydrogen in GA_3 formed from the same precursor is unlabelled (Hanson and White 1969). Full retention of the 3α-H in the conversion of *ent*-[3α-$^3H_1$,17-^{14}C] kaurene to GAs has been shown by Dawson et al. (1975). The loss of a second (4*R*)-^3H-atom in GA_3 is due to the 13-hydroxylation.

3.7.4.5. Loss of C-20, formation of the lactone ring and the double bond of GA_3. Loss of the C-20 group must occur without the formation of a double bond involving any of the positions 1, 5 or 9 (Table 3.10, Figure 3.5). Position 1 is ruled out because six ^3H-atoms from [2-$^3H_2$,2-^{14}C] MVA are retained in GA_{13} and in GA_4 (two such ^3H-atoms are lost in the formation of the C-7 carboxyl). Positions 5 and 9 are excluded because GA_3 contains the two ^3H-atoms from (4*R*)-[4-$^3H_1$,2-^{14}C]MVA, which are specifically located in these positions. The lactone carbon is derived from the mevalonoid methyl group, since it is unlabelled in GA_3 coming from [2-^{14}C] MVA (Figure 3.3). Instead the C-18 methyl group becomes labelled from this precursor and is therefore derived from C-2 of MVA. Formation of the 1,2-double bond of GA_3 involves the loss of a *pro*-(2*S*)-mevalonoid hydrogen atom, since one ^3H-atom is lost in the conversion of GA_4 to GA_3 when [2-$^3H_2$,2-^{14}C] MVA is the precursor, but not when (2*R*)-[2-$^3H_1$,2-^{14}C] MVA is used (cf. Scheme 3.5). Thus it is the 1α-H that is removed in the formation of the double bond. The H-atom removed from C-2 is also α, since GA_3 has one (5*R*)-^3H-atom less than GA_{13} but retains all possible (5*S*)-^3H-atoms. Thus the removal of the two H-atoms to form the double bond is *cis* and from the α-face of the A-ring (Evans et al. 1970).

3.7.5. Feeding of GA precursors to intact higher plants

GA precursors have been fed to intact plants and in some cases the results have been encouraging. However, lack of analytical facilities has rendered most of the experiments inconclusive. The incorporation of [2-^{14}C] MVA into *ent*-[^{14}C] kaurene (Hanson and

White 1968) and of [2-^{14}C]acetate, [2-^{14}C]MVA, *ent*-[17-^{14}C]kaurene and *ent*-[7-^{3}H]-kaurenoic acid into steviol (XLVIII) has been conclusively demonstrated in shoots of *Stevia rebaudiana* (Ruddat et al. 1965; Bennett et al. 1967; Hanson et al. 1968) (cf. section 3.7.3). However, the identification by TLC of *ent*-[^{14}C]kaurenol as a product of [2-^{14}C]MVA in root apices of *Helianthus annuus* (Sitton et al. 1967) is inconclusive. A peak with the retention time of *ent*-kaurenol on GLC was shown, but there was no indication whether it was radioactive or otherwise related to the precursor and it was not determined whether it was present before incubation. The identification by TLC of *ent*-kaurenol (XXXVIII) and *ent*-kaurenal (XXXIX) as products of *ent*-kaurene (I) in seedlings of *Pharbitis nil* (Bennett et al. 1966; Takeba and Takimoto 1971) is also inconclusive, particularly in the light of the results obtained by Murphy and Briggs (1973b) (cf. section 3.7.2.2). The conversion of *ent*-[^{14}C]kaurene to unidentified polar products in seedlings of *P. nil* (Barendse and Kok 1971) and of *ent*-[^{14}C]kaurenoic acid to unidentified polar material in clover leaves, cabbage chloroplasts and barley and rice seedlings has been reported (Stoddart and Lang 1968; Stoddart 1969; Faull et al. 1974b). Whereas some of these products could be GAs, they could also be something else. Stoddart (1972a) has reported incorporation of [2-^{14}C]acetic acid into [^{14}C]GA_3 in developing barley grains. [^{14}C]GA_3 was identified by recrystallization with GA_3 to constant specific activity. However, the accurate determination of specific activity of the very low value obtained is difficult and the location of label in the molecule is essential when working with acetate, which becomes incorporated into many primary metabolites. To our knowledge, incorporation of MVA or *ent*-kaurenoid precursors into GA in an intact higher plant has not yet been conclusively demonstrated. However, the results obtained by feeding GAs to higher plants (sections 3.7.2.5 and 3.8) leave little doubt that such conversions are possible.

3.8. Metabolism

This section is concerned with the metabolism of endogenous GAs in higher plants, primarily in seeds, and with the fates of exogenously supplied GAs. We have cited here all conversions of C_{19}-GAs in higher plants although the corresponding events in *G. fujikuroi* were treated as part of the biosynthetic pathway. This separation is logical for several reasons. In the fungus GAs have no known function and our interest in their biosynthesis continues through to the terminal GA_3. The last part of the pathway was needed as a reference in sections 3.7.3 and 3.7.4. In higher plants, the first formed C_{19}-GAs may be regarded as the first functional hormones (although this may not be entirely true since several C_{20}-GAs have activity, see section 3.6.1). The further metabolism of the C_{19}-GAs may serve to regulate their activity or it may be the result of their action. Furthermore, in the fungus this part of the metabolic sequence is relatively simple and has been studied in a single organism under defined conditions. In higher plants, the conversions of C_{19}-GAs have been investigated in different species at different stages of development and by methods ranging from GC-MS to intuition. It is complicated by the formation of conjugates as main end-products. The metabolism of C_{19}-GAs in higher plants is best understood on the basis of our knowledge of the pathway in *G. fujikuroi*.

3.8.1. Endogenous metabolism and development of the field

A large body of evidence indicates that GAs are synthesized in immature seeds and become converted to polar conjugates as the seeds mature. Radley (1958) noticed that mature beans contain less ethyl acetate-extractable GA-like activity than immature ones, and many later studies have established a typical rise in free endogenous GAs during seed or fruit development which is followed by a decrease as the seeds approach maturity*. The peaks of GA-like activity are often of very short duration. At first it was believed that the GAs became bound in some way to cellular structures and therefore could not be extracted. This opinion was strengthened by the discovery of McComb (1961) that the crude enzyme preparation ficin, which has proteolytic and other activities, released GA in an ethyl acetate-extractable form from a bean seed homogenate. Also Lazer et al. (1961) found that more GA could be extracted from sprouting barley after treatment of the homogenate with dilute acid. However, the formation of a GA-glucoside in leaf discs after feeding GA_3 (Murakami 1961b) and the demonstration that GA-like activity in seeds could be recovered by butanol extraction after the normal ethyl acetate extraction lead to the suspicion that GAs do not become bound to macromolecules but rather are converted into very polar glucosides as the seeds mature. This was confirmed when Sembdner et al. (1964) isolated and partially characterized a sugar-containing GA conjugate from seeds of *P. coccineus* and Schreiber et al. (1970) conclusively identified GA_8-*O*(2)-β-D-glucopyranoside (IX) from the same source. It is now known that GAs in mature seeds primarily occur as glucoside ethers and glucoside esters, GA_8-glucoside being the most frequently found representative (section 3.2.2, Table 3.3). Sembdner and Schreiber (1965) also isolated GA conjugates from shoot apices and flower buds of *Nicotiana tabacum,* one of which contained both sugars and amino acids, but this conjugate has never been fully identified. The suggestion by Sembdner et al. (1968) that the term "conjugated" be used instead of "bound" to describe these derivatives has become generally accepted. For the interpretation of the many published experiments in which butanol-extractable GA-like activity is referred to as "bound GA", it is important to remember that the more polar free GAs may also occur in this fraction (section 3.3.2). The ficin treatment of McComb (1961) probably released GAs from GA glucosides and GA glucosyl esters but the possibility that structurally bound GAs exist is not ruled out. Indeed, in some experiments GA was really released from a tissue residue rather than extracted from the water phase with butanol (e.g. Pegg 1966).

---

* For example in *P. vulgaris* (Corcoran and Phinney 1962; Hashimoto and Rappaport 1966a), *E. macrocarpa* (*M. macrocarpus,* cf. 3.7.1), *Lupinus succulentus* (Corcoran and Phinney 1962), *L. luteus* (Ogawa 1963a), *P. nil* (Murakami 1961a; Ogawa 1963b; Zeevaart 1966; Barendse 1971), *P. purpurea* (Reinhard and Sacher 1967), peach (Ogawa 1965), apricot (Jackson and Coombe 1966), grass seeds (Stoddart 1965), water melon (Bhalla 1971), avocado which shows a somewhat abnormal pattern (Blumenfeld and Gazit 1972), apple (Sinska et al. 1973), barley (Stoddart 1972a, Mounla and Michael 1973), *P. sativum* (Frydman et al. 1974). See also the review by Bhalla and Gairola (1972).

Some evidence for the actual conversion of free GAs to conjugated ones was obtained by Hashimoto and Rappaport (1966a, b). But the conclusive demonstration of such conversions had to await the availability of radioactive GAs. Thus Sembdner et al. (1968) showed that [^{14}C]GA_3 and [^{3}H]GA_6 became converted to [^{14}C]GA_3-glucoside and [^{3}H]GA_8-glucoside (preliminary identification), respectively, during maturation of bean fruits. Later, they took advantage of this reaction for the large-scale metabolic synthesis of GA_3-glucoside, injecting GA_3 into 33 kg of maturing bean fruits (Sembdner et al. 1972). Similarly, Barendse et al. (1968) found that [^{3}H]GA_1, which had been injected into maturing fruits of pea and *P. nil,* became increasingly converted to polar, ethyl acetate-inextractable compounds during seed maturation. Musgrave and Kende (1970) found the same fate for [^{3}H]GA_5 in pea fruits. Although the lack of analytical facilities left most of the polar products unidentified and the lack of knowledge about endogenous GAs resulted in the use of GAs that were not endogenous to the plants (except for GA_3 in *P. nil*), these pioneering feeding experiments added to the evidence that free GAs of immature seeds become converted to polar conjugates as the seeds mature.

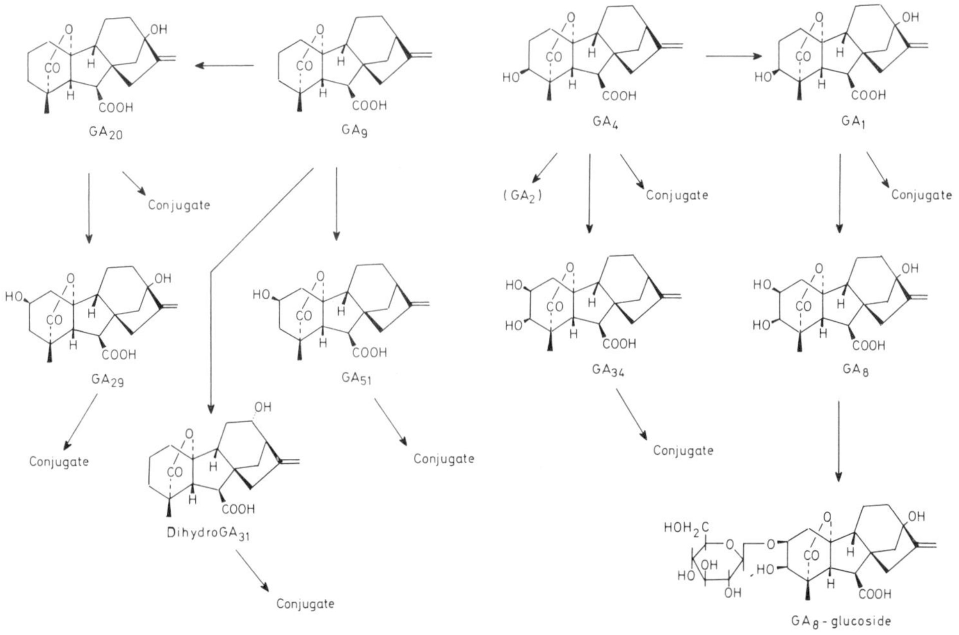

Scheme 3.8. Metabolism of C_{19}-GAs in higher plants.

The following two recent studies of seed metabolism feature conclusive identification of endogenous GAs in combination with the feeding of the same GAs radioactively labelled to arrive at a picture of natural metabolism. At the same time, they conveniently illustrate the metabolic pathways of the 3-deoxy- and 3β-hydroxy-GAs (Scheme 3.8). It can be seen from Scheme 3.8 that the reactions found to date, with one exception, are limited to 2β-hydroxylations, 13-hydroxylations and conjugations.

3.8.2. Metabolism in pea seeds and seedlings. The 3-deoxy pathway

Frydman et al. (1974) determined the native GAs throughout seed maturation of *Pisum sativum* (cv. Progress No. 9) using GC-MS for identification and GC-SICM (section 3.3.4) for the quantitative determinations at each stage. Three C_{20}-GAs, GA_{17}, GA_{38} and

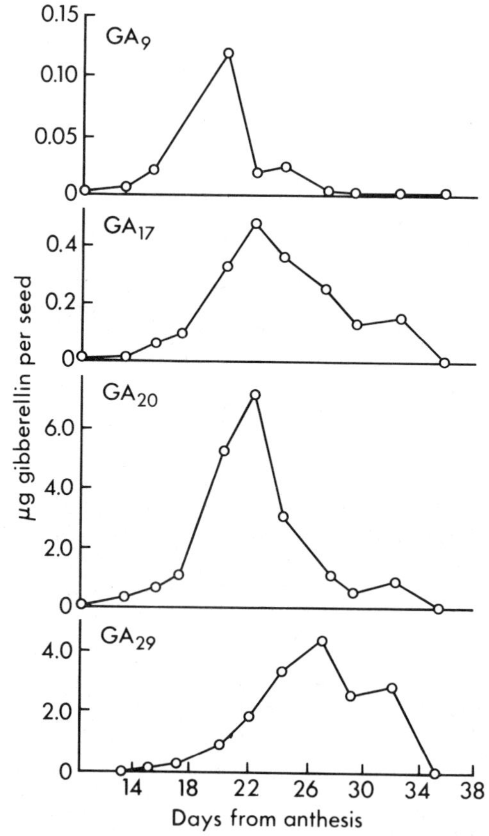

Figure 3.6. The levels of GA_9, GA_{17}, GA_{20} and GA_{29} in pea seeds throughout maturation. (Note differing scales on ordinates) (Frydman et al. 1974).

GA_{44}, were found as were three C_{19}-GAs, GA_9, GA_{20} and GA_{29}. Later GA_{51} was also found (Sponsel and MacMillan 1977). The levels of GA_{38} and GA_{44} were too low for quantitative analysis and GA_{51} had not yet been discovered, but the others showed the typical rise and fall during seed development that had also been found by bioassay in many species. The greater resolution of GC-SICM revealed a sequence of the maxima for the individual GAs (Figure 3.6): that of GA_9 (no hydroxyl) appeared first, then GA_{20} and GA_{17} (both 13-hydroxylated) together, and lastly GA_{29} (2β,13-dihydroxylated). This indicated a possible natural route from GA_9 via GA_{20} to GA_{29} (cf. Scheme 3.8).

In feeding studies with the same cultivar of peas, using GC-MS for identification of all products and GC-RC to relate them with the label, Frydman and MacMillan (1975) found that $[^3H]GA_9$ was converted to $[^3H]GA_{20}$, dihydro-$[^3H]GA_{31}$ and a conjugate of the latter (Scheme 3.8). $[^3H]GA_{20}$ was converted to $[^3H]GA_{29}$ and also to conjugates of both $[^3H]GA_{20}$ and $[^3H]GA_{29}$. Conversion of $[^3H]GA_9$ to H_2-$[^3H]GA_{31}$ and $[^3H]GA_{20}$ as well as the conversion of the latter to $[^3H]GA_{29}$ in pea seedlings and germinating pea seeds was shown independently by Railton et al. (1974a, b). This all seemed to agree with the route mentioned above. However, Frydman and MacMillan (1975) found that the conversion of GA_9 to GA_{20} was rather low (8% of the absorbed radioactivity) and sometimes they got $[^3H]GA_{51}$ instead of $[^3H]GA_{20}$ as a product from $[^3H]GA_9$. The authors therefore doubted that GA_9 is the natural precursor of GA_{20}. Seen in the light of the biosynthetic scheme (Scheme 3.4), this could mean that 13-hydroxylation mainly takes place before the formation of C_{19}-GAs, which is supported by the appearance of endogenous GA_{17}, GA_{38} and GA_{44} in the immature seeds and by the fact that only very little GA_9 was found (Figure 3.6). The occurrence sometimes of GA_{20}, and sometimes of GA_{51}, as products of GA_9 indicates that the 13- and 2β-hydroxylations are activated sequentially as the pea fruit develops and that the stage of 13-hydroxylation had been "missed" in the feeds where GA_{51} was formed. This is also in accord with the sequence of maxima found in endogenous GAs (Figure 3.6). The major product in all $[^3H]GA_9$ feedings was free and conjugated H_2-$[^3H]GA_{31}$, which is not known to be native to pea or any other plant. This may be another indication that the main path does not go via GA_9 or, alternatively, H_2-GA_{31} may be conjugated so rapidly that it has escaped attention. The conversion of $[^3H]GA_{20}$ to $[^3H]GA_{29}$ was very efficient (50–90%) and may well be part of the normal pathway.

3.8.3. The 3β-hydroxy pathway in bean seeds

Conclusive identification has shown that the main free GAs in immature seeds of *Phaseolus vulgaris* are GA_1 and GA_8, but some GA_4, GA_5, GA_6 and the C_{20}-GAs GA_{37} and GA_{38} are also present (Hiraga et al. 1974b). Mature seeds contain mainly GA_8-glucoside but also substantial amounts of GA_1-, GA_4-, GA_{37}- and GA_{38}-glucosyl esters (Hiraga et al. 1974a). $[^3H]GA_1$, $[^3H]GA_4$, $[^3H]GA_5$ and $[^3H]GA_8$ injected into maturing seeds in the pods were all transformed mainly to $[^3H]GA_8$-glucoside, $[^3H]GA_4$ being converted via $[^3H]GA_1$ and $[^3H]GA_8$ (Yamane et al. 1975) (Scheme 3.8). Sembdner et al. (1968) have shown the conversion of $[^3H]GA_6$ to $[^3H]GA_8$ in the closely related *P. coccineus* (section 3.8.1). Thus GA_8 is the major end-product of all the endogenous C_{19}-GAs in *Phaseolus.* As will be seen below, this is a general feature of 3β-hydroxy-GA metabolism in higher plants.

Glucosides of $[^3H]GA_1$ and $[^3H]GA_5$, the glucosyl ester of $[^3H]GA_1$ and several unidentified products were also obtained in the feeding studies of Yamane et al. (1975). $[^3H]GA_{20}$ (Scheme 3.8) was also fed although it is not endogenous to *P. vulgaris* (but is prominent in *P. coccineus*) and was converted to $[^3H]GA_1$, $[^3H]GA_8$ and $[^3H]GA_8$-glucoside like other GAs. This confirms that *P. vulgaris* has a strong 3β-hydroxylating system, which might have been anticipated from the absence of 3-deoxy-GAs in this species (Table 3.4). Whether the natural 3β-hydroxylation takes place at the level of C_{20}-GAs as in *G. fujikuroi*, at the level of C_{19}-GAs or at both is probably a matter of enzyme specificity and relative reaction rates.

3.8.4. Metabolism of [³H] GA₁ in barley aleurone

The study of GA_1 metabolism in barley aleurone was expected to give some clue as to its mode of action in this tissue. $[^3H]GA_1$ is converted to $[^3H]GA_8$, $[^3H]GA_8$-glucoside and a very polar conjugate of $[^3H]GA_1$ in barley aleurones and barley half-seeds (Nadeau et al. 1972; Stolp et al. 1973). The polar conjugate is a major metabolite and therefore deserves some attention. It is amphoteric, has a mol. wt. of 700–800 by Sephadex and may contain a short peptide chain as a non-GA moiety (Nadeau and Rappaport 1974b). In contrast to the glucoside, this conjugate is not cleaved by mild acid or basic hydrolysis but only by treatment with strong acid at 100 °C for 12 hr, which releases rearrangement products of GA_1 (cf. 3.4.1). It is not known whether this conjugate exists naturally. According to preliminary data, it has no biological activity.

Attempts have been made to relate the metabolism of $[^3H]GA_1$ in barley aleurone to its hormonal action by the use of ABA. ABA cancels the action of GA_1, so there is no α-amylase produced in its presence. Nadeau et al. (1972) and Stolp et al. (1973) found that ABA increases the conversion of $[^3H]GA_1$ to the metabolites mentioned above, apparently by increasing GA_1-uptake. In half-seeds this effect was particularly obvious as there was only a little metabolism of GA_1 unless ABA was added. Thus it is unlikely that the hormonal effect of GA_1 is dependent on its metabolism. On the other hand, it is also unlikely that the increased metabolism of GA_1 is the reason for less activity in presence of ABA, since there was always plenty of residual GA_1 even when the metabolism was strong (e.g. 57% in half seeds). The increased metabolism and inhibition of GA action may therefore have a common cause rather than one phenomenon affecting the other. For example, if both the metabolism of GA_1 and the formation of α-amylase are compartmentalized, an effect of ABA on the membrane system might effect both independently. A compartmentalization of GA_1-metabolism is indicated by the work of Musgrave et al. (1972) who showed that $[^3H]GA_1$ which had been fed to barley aleurones could exchange freely between cells and medium whereas the (unidentified) polar products of the feeding could not. Only after 25–30 hr did the cells release the polar products, apparently by lysis.

3.8.5. Metabolism of GAs in seedlings

Feeding studies have shown that GAs are metabolized in seedlings along much the same pathway as in seeds. $[^3H]GA_4$ is converted by dwarf rice seedlings to GA_1, GA_{34}, GA_2 and much polar material (Durley and Pharis 1973) (Scheme 3.8). The formation of

[^{3}H]GA_1 agrees with the pathway in bean seeds; the formation of [^{3}H]GA_{34} completes the possible 2β-hydroxylations and the formation of [^{3}H]GA_2 may be an artefact (section 3.4.1). Essentially the same pattern of [^{3}H]GA_4 metabolism was found in seedlings of *Pseudotsuga menziesii* and in germinating pollen of *Pinus attenuata* (Kamienska et al. 1974, 1976b).

[^{3}H]GA_1 is converted by seedlings and leaf discs of dwarf and normal *Zea mays* to [^{3}H]GA_8, [^{3}H]GA_8-glucoside and a very polar conjugate of [^{3}H]GA_1, perhaps the same as in barley aleurone (Davies and Rappaport 1975a, b). Efficient (60%) conversion of [^{3}H]GA_1 to [^{3}H]GA_8 has also been shown in pea epicotyls (Stoddart et al. 1974) and minor conversions have been shown in seedlings of dwarf rice (1.2%) and *P. coccineus* (0.2%) (Railton et al. 1973; Reeve et al. 1975). Conversion of [^{3}H]GA_1 to an unidentified glucoside in seedlings of *P. nil* (Barendse 1971) and to unidentified products in *P. sativum* and *Solanum andigena* have been reported (Kende 1967; Railton and Wareing 1973b).

The conversions show that the 2β-hydroxylation of GA_1 to form GA_8 is an important step in GA metabolism. This reaction is also carried out by germinating bean seeds and has been obtained in cell-free systems from this material (Nadeau and Rappaport 1972; Patterson and Rappaport 1974; Rappaport et al. 1974). The reaction proceeds with loss of ^{3}H from the 2-position of [1,2-^{3}H]GA_1, which results in stoichiometric formation of ^3H_2O. NADPH and Fe^{3+} are required for full activity and the activity is located in the 95,000 × *g* supernatant fraction. Not only in the metabolism of GA_1 but also in the metabolism of the other GAs discussed in this section, 2β-hydroxylation is a common feature (Scheme 3.8). One might therefore think that 2β-hydroxylation of GAs is a rather unspecific reaction. However, neither [^{3}H]GA_1 methyl ester, 3-*epi*-[^{3}H]GA_1 (XXIII), the Wagner – Merwein rearranged [^{3}H]GA_1 (XXI) nor dihydro-[^{3}H]GA_1 becomes 2β-hydroxylated by barley half-seeds although they are all absorbed (Rappaport et al. 1974). The cell-free system also shows this specificity.

GA_3 is rapidly converted to more polar products in bean shoots (Nitsch and Nitsch 1963; Asakawa et al. 1974b), *P. nil* seedlings (Barendse 1974) and dwarf maize leaves (Davies and Rappaport 1975b). The only conclusively identified product of GA_3 is GA_3 glucoside (see 3.8.1) and this is probably also the sole product in seedlings of *P. nil* (Barendse and Klerk 1975). The formation of less polar products from GA_3, e.g. in tillers of red clover and in willow (Stoddart 1966; Bowen and Wareing 1969), is probably an artefact and may indicate the formation of allo- and epiallo-gibberic acid (section 3.4.1, XIV and XV). Asakawa et al. (1974b) identified [^{3}H]gibberellenic acid glucoside and [^{3}H]isoGA_3 glucoside (cf. XIII and XVII), which are also considered to be artefacts, as products after feeding [^{3}H]GA_3 to dwarf beans.

Conversion of [^{3}H]GA_5 to unidentified polar products has been reported for shoots of dwarf pea and *Silene* and for barley aleurones (Musgrave and Kende 1970; Ende and Zeevaart 1971; Musgrave et al. 1972). A small (0.8%) conversion of [^{3}H]GA_5 to [^{3}H]GA_3 in etiolated dwarf pea seedlings seems to require more rigid identification before its significance can be evaluated (Durley et al. 1973). The results of feeding [^{3}H]GA_9 to shoots was discussed in section 3.8.2. [^{3}H]GA_9 has also been fed to homogenates (Reid et al. 1972) and chloroplast suspensions (Railton and Reid 1974b) but the nature of the data does not permit conclusions as to whether they metabolized it or not. These feedings will be discussed in sections 3.11 and 3.9, respectively.

3.8.6. Physiological significance of GA metabolism

End products of GA metabolism are 2β-hydroxylated GAs and their conjugates. Since both are inactive or have very low biological activity (3.6.1, 3.6.2), the most likely function of this metabolism is to remove active GAs from the system. In developing seeds, this may happen at a certain stage of maturity, at which both C_{19}- and C_{20}-GAs become 2β-hydroxylated (cf. sections 3.8.2 and 3.8.3). In shoots, the GA level may be regulated by a balance between biosynthesis and constant removal through 2β-hydroxylation. This is shown by the incorporation experiments and also by the fact that GA-like activity in shoots rapidly disappears when GA biosynthesis is inhibited by AMO-1618 (Zeevaart 1971; Ende and Zeevaart 1971). Furthermore GA applied exogenously has to be renewed periodically if a persistent effect is desired (e.g. Phinney 1956; Lang and Reinhard 1961). A very great turn-over of GAs resulting in GA conjugates must also take place in tulip bulblets during their development (Aung and Rees 1974).

Murakami (1961b), Hashimoto and Rappaport (1966a) and Sembdner et al. (1968) suggested that GA conjugates might be storage forms of GAs in dormant tissues which could be converted back to free GAs through hydrolysis when they are needed for resumption of growth. There is no good evidence for or against this hypothesis although experiments have been published in which the fate of radioactive GA applied to immature seeds was followed through maturation and subsequent germination of the treated seeds (Barendse et al. 1968; Sembdner et al. 1972; Yamane et al. 1975). Some of these experiments are supposed to show that the proportion of ethyl acetate-extractable GA to "bound" GA (released by acid hydrolysis or extractable with butanol) increases when the seeds germinate, but the significance of the data is difficult to assess. Because of very low and erratic recovery of the radioactive GAs at the time of germination, it has become common practice to interpret these experiments on the basis of relative amounts of free and conjugated GAs at each stage. It is then important to realize that if for some reason one component is lost from the system faster than the other, or if it is recovered less efficiently, the proportion of the other automatically increases. This may give the illusion of a conversion. If, for example, a preferential leakage of polar compounds from the seeds occurs during imbibition, it would result in increased proportions of "free" Gas. Indeed, Sembdner et al. (1972) reported that polar metabolites, which had arisen from application of $[^{14}C]GA_3$ to the cotyledons of cucumber seedlings, spread to all seedling organs and also leaked into the nutrient medium around the roots. Many authors also report very low recovery of applied radioactive GAs, indicating they are lost from the plants. The evidence obtained by Aung et al. (1969), based on bioassay, that "free" GA is converted to "bound" GA in tulip bulbs grown at 13 °C and reconverted to "free" GA when the bulbs are transferred to 18 °C is inconclusive because the loss of GA at 13 °C is based on a single point and the quantitation of "bound" GA after release by acid hydrolysis is inaccurate.

Thus the evidence is insufficient to show that endogenous conjugated GA becomes converted to free GA. However, conjugated GAs are hydrolyzed by several enzymes from microorganisms and higher plants (Sembdner et al. 1974; Knöfel et al. 1974). It has also been shown that microorganisms associated with the roots of rice plants produce exoenzymes that quantitatively hydrolyze GA_8-glucoside (Sembdner et al. 1974). Thus it is important to show that any reconversion of conjugated GAs in higher plants is not due to microbial activity. The storage-reconversion hypothesis has become less attractive with respect to the GA-glucosides after it became known that the GA-moieties are mostly 2β-hydroxylated and therefore probably inactive. Then the glucosyl esters became more attractive candidates for this role, since two of the glucosyl esters found in mature bean seeds contain the highly active GA_1 and GA_4 (section 3.8.3). Direct cleavage of the glucosyl esters of $[^3H]GA_1$ and $[^3H]GA_4$ by rice seedlings has been shown (Hiraga et al. 1974c). It might seem the rapid conversion of free GAs to conjugates in germinating seeds and seedlings (section 3.8.5) is evidence against the reconversion hypothesis. However, in none of the reported cases was the conversion of added GA to conjugates complete. Therefore the presence of an equilibrium which maintains a certain proportion of unconjugated GA cannot be excluded.

Sembdner et al. (1968) also suggested that GA-glucosides might be transport forms, since they are present in the bleeding sap of trees. This hypothesis also is neither proven nor disproven. An obvious need for such a transport system is not indicated, since unconjugated GAs are readily translocated in plants to act in the growing parts (section 3.10).

3.8.7. Interaction with macromolecules

It is only logical to assume that GAs have to interact with macromolecules to express their activity. Several attempts have therefore been made to demonstrate the binding of radioactive GAs to macromolecular fractions. In the last of these attempts, Stoddart et al. (1974) found that $[^3H]GA_1$ when fed to pea epicotyls became associated to some extent with the high and intermediate molecular weight protein fractions as fractionated by Sephadex G-200. The amount of association was very low, but this was expected on the basis of similar experiments with animal tissues. The binding was very specific for GA_1. Only the future can show whether this binding is significant. Previous attempts were unsuccessful.

A different approach was used by Kessler and Snir (1969) and Kessler (1971a, b). They concluded from the formation of a new fraction in CsCl equilibrium density centrifugation and from changes in several physical constants that DNA from plant sources enters into a complex with GA_4/GA_7. A direct binding has not yet been demonstrated.

The binding of GA to macromolecules is discussed further in Chapter 8 (section 8.2.1.1).

3.9. Sites of formation

3.9.1. Immature and germinating seeds and embryos

There is conclusive evidence that the rich supply of GA in immature seeds is biosynthesized by the seeds themselves rather than imported from the mother tissue. The first evidence that GA is biosynthesized in fruits (including seeds) was obtained by Baldev et al. (1965) who showed by bioassay that the GA content of detached pea fruits maturing on a sterile agar medium increased approximately 300-fold and that AMO-1618 in the medium reduced both GA content and the development of the peas. Since the development of the seeds was only influenced by concentrations of AMO-1618 causing

60% or more reduction of GA production, the pea fruits probably produce a slight reserve above their needs. Direct evidence for GA synthesis in seeds is the incorporation of [2-^{14}C]MVA into GAs in *Echinocystis* and *Cucurbita* endosperm preparations (section 3.7). The latter contains all the enzymes needed to convert MVA to GA_4. Coolbaugh and Moore (1971a) have shown that the biosynthesis of *ent*-kaurene in immature pea seeds is mainly located in the cotyledons and not in the embryo or seed coat. The endosperm was not investigated.

Paleg (1960) and Yomo (1960a) discovered that GA could substitute for the embryo in the initiation of α-amylase formation in germinating barley grains. They put forward the hypothesis that GA is normally produced or released by the embryo and transported to the target tissue, the aleurone, where it is active. Subsequent work has confirmed this. The production of GA-like activity has been shown in isolated embryos of barley, *Avena fatua* and corn (Yomo 1960b; Naylor and Simpson 1961; Simpson 1966; Yomo and Iinuma 1966; Cohen and Paleg 1967; MacLeod and Palmer 1967; Radley 1967, 1968, 1969; Murphy and Briggs 1973a) as well as in the embryos of some dicotyledonous plants (Srejovic and Neskovic 1973). As in immature seeds, the time-course of GA-like activity in germinating grains exhibits a maximum which appears 2–3 days after the seeds have been soaked (Fleming and Johnson 1964; Groat and Briggs 1969; Murphy and Briggs 1973a). An extensive review on hormone relations in germinating cereal grains has been presented by Briggs (1973).

Since the production of GA-like activity in isolated embryos is inhibited by CCC, the GAs are most likely synthesized de novo rather than released from a "bound" or compartmented form (Yomo and Iinuma 1966; Simpson 1966; Radley 1967, 1968, 1969). However, direct evidence for GA biosynthesis in germinating embryos is still lacking. Very extensive and careful attempts to show incorporation of [^{14}C]MVA and [^{14}C]isopentenyl pyrophosphate into *ent*-kaurene and of *ent*-[^{14}C]kaurene into further products using both intact barley grains and cell-free systems have failed, although indications of the presence of some later steps in the biosynthetic sequence were obtained (section 3.7.2.3). However, lack of incorporation is not exceptional in attempts to show GA biosynthesis and does not exclude the possibility that the pathway is operating in vivo. In view of the lack of incorporation of *ent*-kaurene, it is difficult to interpret the results of Simpson (1966) and Radley (1968, 1969) which purport to show that (unlabelled) *ent*-kaurene overcomes the inhibition of GA production caused by CCC or AMO-1618. Perhaps in these cases *ent*-kaurene had an effect on the bioassay.

3.9.2. Seedling shoots and flowers

The GA relations of seedlings are not clear. Dale and Felippe (1968) and Dale (1969) found an increase in GA-like activity in seedlings of *P. vulgaris* on the fourth day of germination, whereas Goto (1970) and Goto and Esashi (1973, 1975) found that essentially all fractions of GA-like activity decreased after 24–48 hr of germination. The latter authors used the more reliable methods. On the basis of experiments by Jones and Phillips (1966, 1967) with sunflower seedlings, the site of GA production in seedlings is usually considered to be the young leaves of the apical bud. Excising different parts of the seedlings, Jones and Phillips found that only the removal of the youngest leaves strongly reduced stem growth, whereas removal of the apical dome or older leaf pairs had no such effect. Where the youngest leaves had been removed, maximal growth rate could be fully restored by the application of GA. Moreover, when apical buds were incubated on agar, more GA-like activity diffused into the agar over a 20-hr period than could be extracted from the tissue. The amount extractable from the tissue did not decrease during the diffusion period, so there was a net increase in GA-like activity corresponding to the diffused material. The simplest and perhaps most likely explanation of these data is that the young leaves of the apical buds continued to produce GA on the agar and that this GA represents the GA biosynthesized in the intact seedlings.

The data by Jones and Phillips do not exclude other possibilities. The GA diffusing out of the buds could have been produced somewhere else, e.g. in the roots, in leaves or in small amounts all along the cambium, from where it would be translocated to the apical bud exerting its action on the way (cf. section 3.10). The youngest leaves would retain the GA in situ but it would diffuse out of them after excision. If the "sink" effect of the youngest leaves is removed by excising them, the GA of the plant might no longer be transported through the internode above the youngest remaining leaves and this would result in growth retardation. Thus the proof that the young leaves are the site of synthesis rests on the supposition that the increased amount of GA-like activity detected by diffusion represents biosynthesis. However this could also be due to more efficient determination by bioassay of GA-like material in diffusates; Bailiss and Wilson (1967) found considerably less inhibitory material in diffusates than in extracts. Attempts to show a decrease in diffusible GA-like activity after adding CCC were inconclusive (Jones and Phillips 1967). If CCC was added on the fourth day after sowing and the treatment continued for eight days, the internodes were shortened, but the diffusion of GA-like activity from excised apices was significantly decreased in only one of four experiments. Only by adding CCC on the second day of sowing and continuing the treatment for fourteen days was a significant decrease in diffusible GA obtained, but this would probably have occurred no matter where GA was biosynthesized. Reinvestigating the problem, Phillips (1971) found that incorporation of MVA into the agar blocks on which the sunflower buds were placed increased the amount of diffusible GA-like activity markedly. To the author, this suggested biosynthesis, but no evidence has been presented to establish this. The problem merits further investigation, since 50 apical buds diffused material representing μg-amounts of GA-like activity.

Thus it is not clear where GA is produced in the seedling. Neither is it clear what the diffusing GA-like activity represents. The technique (Jones and Phillips 1964) has been very useful and was employed in several experiments (Bailiss and Wilson 1967; Atsmon et al. 1968; Jones and Lang 1968; Zeevaart 1968; Cleland and Zeevaart 1970; Stoddart and Lang 1968). In these cases, the tissue was not extracted both before and after the diffusion. Where it was extracted at all, it either yielded more GA-like activity than was obtained through diffusion, or the same activity. The possibility that the diffusing GA represents synthesis induced by wounding as has been shown for storage tissue (Kamisaka and Masuda 1968; Rappaport and Sachs 1967) was considered by Phillips (1971). Bradshaw and Edelman (1969) have called attention to the fact that wounded tissue can be regarded as equivalant to explants for tissue culture and that the changes taking place are likely to be precursors of renewed growth (regeneration). Thus there are several objections to regarding the GA diffusing out of excised organs as being equivalent to the GA produced in that organ in the intact plant.

Lockhart (1957) on the basis of excision experiments also found that the apical bud of the pea plant was essential for full growth and that GA would restore growth to excised plants. However, Vlitos and Meudt (1957) and Kuraishi and Muir (1964) did not find that removal of the apical bud decreased growth. Shininger (1972) also by excision experiments concluded that the cotyledons of young pea seedlings are an effective source of GA which is necessary for normal growth. However, GA did not restore the growth of plants without cotyledons to the level of intact GA-treated plants, which shows that the seedlings are dependent on other factors as well from the cotyledons. Furthermore, AMO-1618 inhibited growth of plants lacking cotyledons, which shows that they probably depend on a supply of GA from the shoot as well.

Evidence is gathering that GA is present in and partly biosynthesized by plastids. Thus GA-like activity is associated with the chloroplast or etioplast fraction of leaves, from which it becomes released into the medium by sonication or osmotic disruption (Stoddart 1968; Frydman and Wareing 1973a; Railton and Wareing 1973a; Railton and Reid 1974a, Cooke et al. 1975; Evans and Smith 1976). Direct evidence for biosynthesis has not yet been obtained but it has been shown that about 20% of the *ent*-kaurene synthetase activity in the *Echinocystis* endosperm (cf. section 3.7.2.2) is associated with the etioplast fraction (Simcox et al. 1975). With preparations from etiolated pea seedlings and *Ricinus* endosperm, the results were less clear. Stoddart (1969) showed an unmistakable increase in GA-like activity when sonicated chloroplast preparations were incubated with *ent*-kaurenoic acid. This is suggestive of GA biosynthesis, but a direct incorporation of *ent*-kaurenoic acid into GAs in such preparations could not be demonstrated. Railton and Reid (1974b) concluded from the results of incubating $[^3H]GA_9$ and $[^3H]GA_{20}$ with chloroplast suspensions of *P. sativum* that chloroplasts are a major site of GA metabolism. However, no control incubations were reported, the identification of products was inconclusive and the radioactivity did not correlate with biological activity or reference compounds on TLC. Evidence for the release of GA from chloroplasts in response to red light will be discussed in section 3.11.1.

Considerable amounts of GA-like activity is often found in floral parts, e.g. 2–5 mg GA_3-equiv./kg in *Cassia fistula* (Sircar et al. 1970). Murakami (1975), measuring by TLC and bioassay the GA-like activity of floral organs of *Mirabilis jalapa,* found that the amount of GA in the floral tubes reached a maximum during floral tube development. However, more than 90% of the total GA-like activity of the developing flowers was found in the anthers. This together with the early observation that emasculation decreases the growth of the flowers and that GA increases the growth of petals (Jeffcoat et al. 1969) is evidence that GA of the flower is synthesized in the flower and mainly in the anthers. Furthermore, GA_3, GA_4 and GA_7 have been conclusively identified in the pollen of a number of *Pinus* species (Kamienska et al. 1976a) and the considerable GA-like activity found in dormant pollen increases during the first 15 hr of pollen germination (Kamienska and Pharis 1975). Changes in the levels of GA-like activity have also been shown in developing flowers of *Chrysanthemum* and *P. nil* (Jeffcoat and Cockshull 1972; Murakami 1973).

3.9.3. Seedling roots

Roots probably contain GA-like substances and they are present in xylem sap (section 3.10.1). This has been regarded as circumstantial evidence that they are produced in the roots, although this does not necessarily follow. One group of workers (Sitton et al. 1967) is often quoted as having shown that exudate from decapitated *Helianthus* seedlings continued to deliver a high amount of GA-like substances for several days. However, except for a single high value in one of two fractions on the fourth day, the data in reality shows a steady decline in activity (cf. Crozier and Reid 1971). The conversion of [^{14}C] MVA to unidentified products by root tips (Sitton et al. 1967) is not good evidence for GA biosynthesis either (cf. section 3.7.5). Reid and Carr (1967) tried to study the effect of CCC on the bleeding sap of pea seedlings but the data is inconclusive because of the disappearance of some fractions, appearance of other fractions and the obvious presence of material inhibitory to the bioassay. Experiments like the ones conducted with shoot tips have shown that root tips from both *Helianthus* and *Hedera* diffuse GA-like substances into agar without apparent reduction of the level of extractable GA-like activity (Jones and Phillips 1967; Frydman and Wareing 1973b). The amount of diffusible GA-like substances is strongly reduced by CCC but this has only been shown by applying the inhibitor at the time of sowing, so all GA biosynthesis would be inhibited regardless of where it takes place (Jones and Phillips 1967).

Reid and Crozier (1971) purported to show that GA-like activity in the bleeding sap of tomato seedlings is reduced almost to zero within one day of waterlogging, which has been quoted as circumstantial evidence for GA production in roots. They also concluded that the supply of GA from the root to the shoot is essential for shoot growth. However, the reproducibility of the data is doubtful since the rapid decrease in GA-like activity on which the conclusions were based did not occur in the three experiments shown in the preliminary communication (Reid et al. 1969). In this communication the authors accordingly concluded that there was no rapid decrease in GA-like activity. Furthermore, the decrease in GA-like activity (Crozier and Reid 1971) may not have been due to the water-logging but to the same unaccountable factor which decreased the growth of the control plants between days four and seven (the GA levels of the controls plants were not shown). A hypothesis of considerable entertainment value was presented by Crozier and Reid (1971) suggesting that inactive GAs may be produced in the shoot, translocated to the roots where they are converted to active GAs and exported back as such to the shoot. The hypothesis was based on the inconclusive identification by column chromatography of GA_1 as the main GA in shoots and roots of control plants but GA_{19} as the main GA in shoots and roots of seedlings, the roots of which had been decapitated and the developing lateral roots of which were shaved off every day. Such damaged seedlings show stunted growth and the data were supposed to show that growth was restored by GA. In reality it shows that GA had no more effect on the treated plants than on the control plants and thus was not the factor limiting growth in the damaged plants. Frydman and Wareing (1973b) applying the same technique to *Hedera helix* plants found no accumulation of other GAs in the stem when the roots were damaged and found that an apparent accumulation of GA in the cotyledons in one experiment was not reproducible. A rigid control of reproducibility is absolutely essential before results of extraction experiments can be used as a basis for conclusions with physiological implications.

Additional evidence for GA production in roots is the demonstration of GA-like substances in root tips obtained from a clone of excised tomato roots maintained in isolated culture for 5 years (Butcher 1963).

3.9.4. Conclusions

GA must be produced somewhere in the plant and although some of the evidence is weak it remains likely that GA is indeed produced in shoot and root tips, young leaves, floral parts, immature seeds and germinating embryos. This does not exclude the possibility that more or less all tissues have some ability to produce GA, which is supported by the fact that tissue cultures from many different organs contain GA-like substances (Nickell 1958; Carr and Reid 1968). However, the typical increases in GA-like activity in developing flowers and in developing and germinating seeds point to an increased ability to produce GA when needed at the site of utilization. The decline in GA after these "spurts" probably indicates inactivation after the high levels of hormone are no longer needed (cf. section 3.8.6), which implies there is precise localized metabolic control of GA levels rather than a general supply of GA throughout the whole plant.

Chloroplasts are probably a sub-cellular site of GA production. No efforts seem to have been made to investigate whether GA is produced in specialized cells, e.g. similar to the glandular hairs which produce diterpene resins in *Bayeria viscosa* (Dell and McComb 1974); these resins are produced in young leaves only. We cannot share the view of Sheldrake (1973) that GA production takes place in dying tissue (root caps, endosperm as a regressive nutrient tissue, differentiating vascular tissue, and at or near wound tissue) and perhaps is a consequence of cytolysis. On the contrary, GA levels decrease in senescing tissue (Fletcher et al. 1969; Chin and Beevers 1970; Goldschmidt et al. 1972), which Sheldrake feels may indicate that senescing tissue lacks GA precursors.

3.10. Translocation

Early work on the translocation of growth hormones in plants is described in a review by McCready (1966), which also contains pertinent view-points of general significance.

3.10.1. Whole plants

Zweig et al. (1961) showed by radioautography that $[^{14}C]GA_3$ applied to the cotyledonary node of young bean seedlings becomes distributed to all parts of the plants within a few hours, particularly to the young growing parts. In this case the movement takes place both in the phloem as indicated by accumulation of radioactivity in shoot and root tips, and in the xylem as indicated by transport into mature green leaves. When $[^{14}C]GA_3$ is applied to the leaf, on the other hand, a translocation pattern like that of photosynthetic products is obtained, which indicates movement primarily in the phloem. Thus GA applied to a mature leaf is translocated to the growing region including the immature expanding leaves but not to other mature leaves, nor does it spread very much within the donor leaf itself (Zweig et al. 1961; McComb 1964b). In pea plants, it could also be demonstrated that $[^{14}C]GA_3$ applied to a young expanding leaf did not move out of this leaf until it was almost fully expanded. This was not confirmed in tomato seedlings, where Couillerot and Bonnemain (1975) also by autoradiography observed movement of $[^{14}C]GA_3$ out of young leaves. Further evidence for movement with the photosynthetic products was obtained by Chin and Lockhart (1965) using unlabelled GA_3 and growth response as a sign that the applied GA had reached the growing region of young bean seedlings. GA was readily translocated from the trifoliate leaves to the growing region but it did not reach the latter from a primary leaf, unless the other leaves were removed. Furthermore, they showed that in light enough GA to give a growth response moved out of the leaf within an hour, whereas in the dark no GA moved out of the leaf within 6 hr. The rate of movement of GA_3 from a mature leaf of a pea seedling to the growing point was estimated by McComb (1964b) to be about 5 cm/hr. In tomato seedlings the rate is about 4 cm/hr (Couillerot and Bonnemain 1975).

Thus GA after foliar application primarily moves in the phloem. However, considerable amounts of it may also reach the xylem. Thus Chailakhyan et al. (1974) showed that GA_3 applied to a basal leaf of an otherwise deleafed shoot of *Perilla nankinensis* reached the top of the plant even after girdling of the stem. This may be explained by a radial translocation of the hormone from the phloem to the xylem, which was actually shown for $[^{14}C]GA_3$ in stems of young willow (Bowen and Wareing 1969). The opposite process, translocation from the transpiration stream to the sieve tube sap, was also shown, so that $[^{14}C]GA_3$ practically moved parallel in the two channels. Endogenous GA-like substances are found both in the sieve tube sap (Kluge et al. 1964; Hoad and Bowen 1968) and in the bleeding (xylem) sap of many species (Carr et al. 1964; Phillips and Jones 1964; Reid and Carr 1967; Sitton et al. 1967; Skene 1967; Atsmon et al. 1968; Jones and Lacey 1968; Luckwill and Whyte 1968; Murakami 1968b; Reid and Burrows 1968; Sembdner et al. 1968; Reid and Crozier 1971; Sweet et al. 1974). The presence of GA-like substances in the xylem sap may indicate a purposeful transport or just a general circulation of GA in the whole plant. Further identification has shown that the bleeding sap of *Acer platanoides* and *Ulmus glabra* mainly contained GA-like substances in the form of butanol-extractable conjugates (Sembdner et al. 1968). If such conjugates are indeed inactivated GAs (section 3.8.6), the GAs in the bleeding sap may be GA no longer required. Some evidence for this view may be the appearance of GA-like activity in guttation fluid of barley and wheat seedlings and of radioactivity in the guttation fluid of tomato seedlings to which $[^{14}C]GA_3$ had been applied (Couillerot and Bonnemain 1975).

$[^{3}H]GA_1$ applied to maize roots is readily taken up and distributed throughout the whole plant (Davies and Rappaport 1975a). Using a point-source application of $[^{14}C]GA_3$ to coleoptiles and root tips of *Z. mays*, Webster and Wilkins (1974) found very little longitudinal movement. Apparently the hormone does not reach the vascular system under these circumstances or the transport in the intact plant is strictly acropetal (cf. section 3.10.3).

3.10.2. Excised shoots

If excised shoots are stood in radioactive GA-solution, the radioactivity is rapidly translocated to, and accumulated in, the growing region (Asakawa et al. 1974a; Barendse

1974). In this process, GA is often metabolized to more polar derivatives. Galston and Warburg (1959) and Purves and Hillman (1959) found that unlabelled GA_3 moved to the uppermost section of 10-cm pea epicotyls at a rate of 20–33 mm/min. This very high rate indicates movement with the transpiration stream and was probably an artefact of the direct contact of the open xylem with the solution at the cut surface. A functional transpiration stream is, however, not essential for GA uptake and accumulation, since $[^3H]GA_1$ and $[^3H]GA_5$ are accumulated in the growing region of pea epicotyls floating on solution (Musgrave et al. 1969). This accumulation is supposed to be specific; the biologically inactive derivatives methyl $[^3H]GA_5$ and 16-keto-$[^3H]GA_5$ were not accumulated. This is in contrast to all the other characteristics of GA translocation, which do not differ from those of the translocation of other metabolites (McComb 1964b).

3.10.3. Segments

The movement of a substance in segments is usually explored by placing an agar block containing the substance in contact with the cut surface at one end of the segment and collecting translocated material at the other end in a receiver block. The amount of substance in the receiver block after a specified time as measured by bioassay or radioactivity is a measure of movement. In thorough studies, the amount of GA remaining in the tissue is also determined which is very important. McCready (1966) has pointed out that this situation is totally unnatural. The whole cut surface with its variety of tissues becomes exposed to the substance and neither phloem nor xylem transport functions normally. GA moves very slowly through such segments; it takes hours to cover a few mm in the acropetal as well as in the basipetal direction (Kato 1958; Clor 1967; Hertel et al. 1969; Scott and Most 1972; Wilkins and Nash 1974; Phillips and Hartung 1974). It is questionable whether it is justifiable to speak of "transport" at all, since the rate of movement lies well within the range for diffusion. It is true that Kentzer and Libbert (1961) and Libbert and Gerdes (1964) observed an apparent effect of the transport inhibitor 2,3,5-triiodobenzoic acid (TIBA) on both longitudinal and lateral movement of GA through hypocotyl sections. However the small effect measured by bioassay of the receiver blocks is not convincing and could easily be due to an effect of TIBA on the assay plants rather than on the sections.

Phillips and Hartung (1974) thoroughly investigated the movement of $[^3H]GA_1$ through internode sections of *P. coccineus* by the agar block technique. In agreement with the slow movement found by others, they detected very little radioactivity in the receiver blocks after 6 hr diffusion through 6-mm sections, or after 16 hr diffusion through 80-mm sections. There was no evidence for polarity of movement. However, less radioactivity diffused from young, elongating sections than from older sections of the same internodes. In longer sections also, $[^3H]GA_1$ was retarded in the younger parts of the tissue and since this was at the beginning of the path of movement in the basipetal direction but near the end of the path in the acropetal direction, very different profiles of radioactivity within the tissue resulted after a period of diffusion in the two directions. This was interpreted as being due to a possible binding of the hormone in the younger tissue. However, it can be determined from the data that growth of the internodes in the zone of maximal elongation (4 mm/cm) was of the same magnitude as the rate of movement of [^{3}H] GA_1. The young tissue may therefore simply have outgrown the slow movement of the hormone which was thus only apparently retarded.

Predominantly basipetal transport of GA has been reported to occur in young *Coleus* petioles (Jacobs and Kaldewey 1970; Jacobs 1972; Jacobs and Pruett 1972), *Z. mays* roots (Jacobs and Pruett 1973) and *P. coccineus* roots (Hartung and Phillips 1974). However, the interpretation of the data obtained in diffusion experiments is doubtful. Where it was measured, the amount of hormone in the tissue was about the same whether acropetal or basipetal transport had taken place. Small differences in the relatively small amounts diffusing further into the receiver agar can then give the illusion of a large effect. Thus careful measurement by Hartung and Phillips (1974) of the [^{3}H]GA_1 moving through *Phaseolus* root segments yielded 135 dpm/mm^2 of uptake area in the receiver block by basipetal transport, and 15 dpm/mm^2 by acropetal transport, which is a ninefold difference. However, an average of 1792 dpm remained in the tissue after basipetal transport and 1946 dpm after acropetal transport, which shows that there was no significant difference in the amounts taken up. The difference between the receiver blocks reflects a difference in the rates of diffusion out of the cut surfaces at the two ends rather than in the rate of transport through the sections. Also in the *Coleus* petioles, it seems too optimistic to conclude that at least ten times as much GA_3 moved basipetally as moved acropetally on the basis of an average of 0.7 ng in the basipetal receiver blocks and 0.07 ng in the acropetal ones. The *Coleus* work (references above) is already inconclusive because the receiver blocks were autoclaved for 18 min at 120 °C before being bioassayed (cf. section 3.4.1). Šebánek and Hink (1967) showed a possible polarization of GA movement, which became directed towards added IAA in pea epicotyls whether this was acropetal or basipetal movement. If these results could be confirmed by more reliable methods (unphysiologically high concentrations of GA were added, only the end pieces of the segments were analyzed), they would have interesting implications with respect to interaction of IAA and GA, e.g. in tropisms (section 3.10.5).

3.10.4. Immobile GAs

The presence of GAs in sieve-tube sap and in bleeding sap shows that at least some native GA moves within the plants (section 3.10.1). The diffusion of GA-like activity from excised plant organs also shows mobility (section 3.9.2). However, Jones and Lang (1968) discovered that not all GAs move equally freely. In analyzing by TLC and bioassay the GAs contained in pea seedlings, they found that two components could be extracted by organic solvents whereas only one component was found in the diffusates. Similarly, Cleland and Zeevaart (1970) found a discrepancy between extractable and diffusible GAs in *Silene armeria.* This indicates that some GAs may be immobilized in the tissue, perhaps through compartmentation in plastids (see sections 3.9.2 and 3.11.1).

3.10.5. Tropisms

Lateral redistribution of GA on geotropic and phototropic stimulation has been studied by different methods in different organs, but these studies have not yet given a unified picture. Phillips (1972a) found a marked redistribution of GA-like activity diffusing into agar blocks over a 20-hr period from geotropically stimulated sunflower buds. The activity obtained from the lower halves of horizontally oriented buds was almost ten times that from the upper halves. The data is somewhat marred by a thousand-fold variation in yield of GA-like activity, which ranges from about 6 ng–9 μg per shoot. The latter value seems unreasonably high. A claim that biosynthesis of GA was enhanced three-fold in the geotropically stimulated buds is not justified both because of the variation and because it is not known whether the diffusing GA-like substances represent biosynthesis. Railton and Phillips (1973) claimed a similar distribution of GA-like activity in horizontally oriented maize coleoptiles. The significance of their data cannot be judged from the single experiment shown, in which the GA extract had been exposed to pH 9 and pH 2.5 and the separated GA had been left on the TLC plate for 2–3 hr. The evident presence of large amounts of inhibitory material in the bioassay makes the estimation very unreliable.

Webster and Wilkins (1974) found a slight upwards distribution of $[^{14}C]GA_3$ in roots and coleoptiles of horizontally positioned seedlings of *Z. mays.* $[^{14}C]GA_3$ was applied at one point by puncturing the tissue. During 6 hr 13.8% more $[^{14}C]GA_3$ diffused from a puncture on the lower side to the upper side than in the reverse direction. Apparently an entirely different system is measured by this method than by diffusion into agar blocks, since there was practically no longitudinal movement of $[^{14}C]GA_3$. El Antably and Larsen (1974a, b) showed that extracts from the upper halves of horizontally positioned *V. faba* roots contained 1.8–3.7 times more native GA-like activity than did the lower halves. Native GA_3 as identified by GLC also becomes preferentially distributed into the upper halves. This agrees with the experiments of Webster and Wilkins (1974) and disagrees with those of Phillips (1972a); however, the differences in techniques and experimental systems make comparisons difficult. Phillips (1972b), describing a single experiment, reported that about eight times as much GA-like material diffuses from the shaded side as from the lighted side of unilaterally lighted sunflower buds.

The distribution of GA in relation to geotropism and phototropism is discussed further in Chapter 5, Volume II.

3.11. Effects of environmental stimuli on GA levels and metabolism

3.11.1. Light

The interest in the effects of light on GA content of plants dates back to the findings of Lockhart (1956, 1958a, b) and Lockhart and Deal (1960) that GA treatment prevented the inhibition of stem growth by light in seedlings of a number of species including pea, bean, cucumber and pumpkin. In the case of pea seedlings, the problem of light inhibition became linked with the problem of dwarfism when Sale and Vince (1960) and Gorter (1961) found that dwarfism in this species is determined by light: in darkness all varieties reach the same height but in light the dwarf varieties become more inhibited than the tall ones. That GA could counteract dwarfism in light-grown pea seedlings was already known from the work of Brian and Hemming (1955) and Brian (1957). After extensive experimentation, Lockhart (1959a, b, 1961, 1964) concluded that stem elongation, in species where GA reverses light inhibition, is regulated by light through some effect on GA metabolism, most likely involving decreased biosynthesis but possibly also increased destruction. However, in his earliest work, Lockhart realized that the effect of light on seedling morphology is complex and pointed out that GA acted only on stem elongation without reversing other effects of light, e.g. the greater number of nodes in light-grown plants. The complexity of the problem has been brought out further by later work (Roesel and Haber 1964; Vince 1967; Russell and Galston 1968), which has revealed that not even the reversal of light inhibition of stem growth in pea seedlings by GA is complete. In view of the complexity of the relation between GA effects and light inhibition of growth, it is not surprising that a simple, consistent relationship between light and GA content has not been found.

Kende and Lang (1964) and Jones and Lang (1968) confirming and extending early work by Radley (1958) found no difference in GA-like activity from light- or dark-grown tall and dwarf pea seedlings. Furthermore they found that dwarf peas growing in light did not respond to one of their own native GAs and suggested that the action of light in this case is to lower the responsiveness of the tissue to a particular GA or, alternatively, to prevent the conversion of inactive GA to one that is active in the tissue. In conflict with these results, Köhler (1965, 1970, 1971) found that light-grown pea seedlings contained substantially more GA than dark-grown ones, and that dwarf seedlings contained more than tall seedlings. It is unrewarding to speculate on the reasons for the discrepancy in results, since the experiments were carried out using different methods and with different varieties. Köhler interpreted the inverse relationship between growth and GA content in his experiments as being due to an accumulation of non-utilized GA in the cases of decreased growth. Some evidence for more rapid metabolism in the faster growing seedlings was also found (Köhler 1971). Other workers have also found signs of faster metabolism of GAs in dark-grown than in light-grown seedlings. Thus Musgrave and Kende (1970) found that $[^3H]GA_5$ was metabolized faster in dark-grown seedlings, which also responded more to GA_5 than did light-grown seedlings. Railton (1974a), without documenting the reproducibility, presented some evidence that $[^3H]GA_9$ and $[^3H]GA_{20}$ are metabolized faster in dark-grown than in light-grown shoots of dwarf pea. Stoddart et al. (1974) possibly found that more $[^3H]GA_1$ was obtained bound to a protein fraction isolated from dwarf pea seedlings if the labelled hormone was applied to the shoots in the dark than if it was applied in the light. Also in *P. nil,* more GA has been found in light-grown than in dark-grown plants (Ogawa 1964; Barendse and Lang 1972). However, the differences are mostly small. In the work with radioactive GA, the products were not conclusively identified and in all the studies, the reproducibility is doubtful.

In contrast to the apparent situation in pea and *P. nil,* shoots of *Phaseolus coccineus* seedlings grown in a 16 hr photoperiod contain much less GA-like activity than the corresponding dark-grown controls (Crozier and Audus 1968; Bown et al. 1975). Also the metabolism of GA is different in *P. coccineus,* $[^3H]GA_4$ being metabolized faster in light-grown than in dark-grown plants (Bown et al. 1975). This may in some unknown manner be related to the difference in response of pea and bean seedlings to GA. Whereas both dark- and light-grown pea seedlings respond to GA, bean seedlings respond in light only (Lockhart 1958a, b; Bown et al. 1975).

Somewhat different from the problem of GA content in light- and dark-grown plants and at first sight giving more unified results are the effects of short-term irradiation with red light. Thus brief exposure of lettuce seeds (Köhler 1966), apple seed embryos (Smolenska and Lewak 1971), etiolated barley leaf sections (Reid et al. 1968), etiolated wheat leaf sections (Beevers et al. 1970) and etiolated pea seedlings (Neskovic and Konjevic 1974) to red light causes a transient increase in extractable GA-like activity. In the case of the leaves and shoots, light was given for 5–30 min and the increase in GA-like activity was maximal 10–20 min after termination of the treatment. Thereafter it declined to, or below the levels, of the dark controls within 1–2 hr. A connection has been proposed between the increased GA-levels and unrolling of the tightly rolled leaves of etiolated wheat seedlings, which can be effected either by red light or by GA application (Poulson and Beevers 1970; Beevers et al. 1970).

The red-light-induced increase in GA-like activity is probably due to a phytochrome-mediated release of GA from the etioplasts. The increase is obtained when homogenates of etiolated barley or wheat leaves are irradiated in vitro (Reid et al. 1972; Cooke and Saunders 1975a). In barley leaves it is associated with a defined etioplast fraction which contains phytochrome (Evans and Smith 1976) and in wheat leaves it is also associated with an etioplast-containing fraction (Cooke et al. 1975). In both instances, the increase is prevented by far-red light immediately after red irradiation. The origin of the new GA-like material is unclear. Evans and Smith (1976) concluded that it was released from the etioplasts rather than being newly synthesized, since as much GA-like activity could also be obtained by sonication of the etioplasts in the dark. Stoddart (1968) also reported that more GA-like material could be extracted from chloroplast suspensions after sonication. However, since the phytochrome of these chloroplasts would be in the P_{fr} form already, they should not need sonication, if the situations were comparable. Also in the wheat leaves, the GA-like activity is in reality the same before and after irradiation, but it appears in the neutral fraction of the extracts before, and in the acid fraction after, irradiation (Loveys and Wareing 1971a; Cooke et al. 1975). Therefore the situation is not the same as in the experiments of Evans and Smith (1976), where the GA-like material was not extractable with organic solvents at all, unless the sample had been irradiated or sonicated. The complete inhibition by ABA of the light-stimulated increase in GA-like material which was shown in the original work on wheat leaves (Beevers et al. 1970) remains unexplained in the light of later work, but supports the hypothesis (Evans and Smith 1976) that membrane permeability is involved.

Other conclusions reached in the work with wheat leaves are based on differences in GA-like activity that to our minds are too subtle to justify general conclusions. An unexplained decrease in the effect of red light on wheat leaves seems to have occurred over the years. Thus the original work by Beevers et al. (1970) shows effects corresponding to 4–6 mm elongation above the 1% confidence limit in the lettuce hypocotyl test when extracts from 8 g of irradiated wheat leaves were tested. Loveys and Wareing (1971a) observed a response corresponding to only 1.2 mm in the same bioassay and Cooke and Saunders (1975a, b) and Cooke et al. (1975) based their conclusions on a maximal response of 1 mm or less and on differences between treatments of 0.3–0.5 mm. However this is also the approximate range of the 0.1% confidence limits given. A comparison of controls and equivalent treatments in all these works shows that the confidence limits bear no relation to the variance introduced by extraction and chromatography, not to mention variations in the extracted material itself. It is therefore questionable whether the results are significant at all.

The conclusion by Reid et al. (1972) that [^{3}H]GA_9 is metabolized to other GAs to a greater extent in red light than in darkness is based on slight differences in the distribution of radioactivity which was streaked over whole TLC plates without forming definite peaks and making it necessary to compare at least 70% of the plates to find a difference between the 10341 cpm from the red-light irradiated extract and the 7662 cpm from the extract kept in the dark. The data do not even prove that there was metabolism, let alone light-induced metabolism. Ecklund and Moore (1974) found that cell-free extracts prepared from light-grown and tall pea seedlings had a higher capacity for biosynthesis of *ent*-[^{14}C]kaurene from [2-^{14}C]MVA than did extracts from dark-grown and dwarf seedlings. The general validity of this difference apparently is limited, since it was found in only two cultivars while others yielded inconclusive results.

3.11.2. Dwarfism

Shortly after the discovery by Brian et al. (1954) that GA would prevent the dwarf habit of certain varieties of pea (3.11.1), Phinney (1956) found that several single-gene dwarf mutants of corn respond with normal growth to the application of GA and suggested that the mutant alleles in such plants affected different steps in GA biosynthesis. Effects of this type have been found with *G. fujikuroi* (Spector and Phinney 1966, 1968). However, in higher plants no consistent relationship has been found between dwarfism and GA content. Thus Phinney (1961) found no GA-like activity in three dwarf maize mutants and reduced levels in two, while Suge and Murakami (1968) found great variations in GA contents of different rice cultivars. Although most of the dwarf cultivars had less GA-like material than the tall ones, some of them had as much or even more than the tall ones. Only one rice cultivar, Tan ginbozu, had no detectable GA at all. This cultivar has been most useful in bioassays (section 3.5.2). Less GA-like substances have been found in dwarf than in tall cultivars of *P. nil* (Ogawa 1962; Barendse and Lang 1972), sweet pea (Margara 1963) and *P. vulgaris* (Proano and Greene 1968; Goto 1970; Goto and Esashi 1973, 1975), but no difference was found between tall and dwarf varieties of peach (Wylie and Ryugo 1971) and more GA-like activity was found in dwarf than in normal wheat (Radley 1970). A single gene mutant of tomato, "yellow-green 6", may have more GA than normal strains as indicated by bioassay and GLC analysis (Perez et al. 1974). In most of these studies, the reproducibility was not ascertained.

3.11.3. Photoperiodism

The relationships between GA content and daylength have been studied in connection with flower induction and the regulation of dormancy. Many plants are induced to flower under specific daylength regimes only, including long-day conditions (LD), short-day conditions (SD) and combinations of both. Early work established that GA could substitute for LD, which gave rise to the hypothesis that one effect of LD was to induce increased GA formation. Many experiments have since shown that plants contain more GA by bioassay when grown in LD than in SD (e.g. Lang 1960; Chailakhyan and Lozhnikova 1961, 1966; Harada 1962; Lozhnikova 1967; Lozhnikova and Chailakhyan 1969; Zeevaart 1969b, c; Cleland and Zeevaart 1970; Grigorieva et al. 1971; see also the review by Chailakhyan 1968). Only in one plant, *Bryophyllum daigremontianum,* has the GA (GA_{20}) which increases in LD been conclusively identified (Gaskin et al. 1973).

It is generally believed that the GA formed in LD is instrumental in regulating stem elongation, which particularly in rosette plants is often a prerequisite for flower formation. On the basis of several experiments, it is believed that GA does not have a more direct effect than this on flower induction. It is also clear that increased GA production is not the only effect of LD on growth, e.g. the response of the tissue to GA is also increased (Zeevaart 1971).

In most of the studies cited above, GA was not assayed until flower primordia or flower buds had already been formed. This leaves undecided the question of whether the higher levels of GA are due to a direct effect of the daylength or to some secondary effect, e.g. a higher production of GA in the flower buds (cf. section 3.9.2). In some cases this possibility is ruled out, most obviously in the long-short-day plant *B. daigremontianum.* This plant must first be kept under LD, presumably to induce the formation of GA, then it must be kept in SD for the final induction, which occurs only in the presence of GA. Such plants have higher levels of GA by bioassay and GLC when growing in continuous LD than in continuous SD although neither condition alone induces flower formation (Zeevaart 1969a, c, 1973).

In other plants no significant difference in GA content has been found whether the plants are growing in LD or SD. This may be due to increased metabolism of GA in the induced plant returning the levels to normal values while maintaining a higher rate of turnover. Transient increases in GA-like substances were found in the long-day plant *Silene armeria* after 4–6 LDs (Cleland and Zeevaart 1970) and perhaps in spinach after only 1 LD (Radley 1963). *S. armeria* requires 4–6 LDs for full induction of flowering and spinach needs 7 LDs, but it shows a growth response after only 1 LD. Although the data obtained solely by extraction was variable and erratic, an increase for a single day by a factor of 2.2–3.5 was reproducible in the work with *S. armeria,* if the values for diffusible and extractable GA-like materials are considered together. A more rapid decrease in extractable GA-like activity after treatment with AMO-1618 was found in LD than in SD for both *S. armeria* and spinach (Ende and Zeevaart 1971; Zeevaart 1971). This is evidence for a higher turnover in LD, which had been suggested by Radley (1963). The level of GA at any one time in this dynamic system would be very sensitive to small changes in the conditions, which might explain the erratic results of the extraction experiments (cf. Radley 1970). All this evidence is of a preliminary nature, since the GAs involved have not been identified, their absolute amounts are unknown and the origin of diffusible GAs is unknown.

In contrast to the results obtained with *Silene* and spinach, Stoddart (1966) found that $[^3H]GA_3$ disappeared more rapidly under SD than under LD in leaves of red clover (also a LD-plant). Although the relation of this disappearance to natural metabolism cannot be evaluated, since GA_3 is not known to be endogenous in red clover and since artefacts were obviously formed (cf. 3.8.5), the results are supported by more extractable and diffusible GA-like material being found in the leaves of the same variety when growing in SD than in LD (Stoddart and Lang 1968). The discrepancy in results between different species calls for comparison by conclusive methods. The conversions of *ent*-[^{3}H]kaurene in shoots of *P. nil* (Bennett et al. 1966) and of *ent*-[^{14}C]kaurenoic acid in leaves of red clover (Stoddart and Lang 1968) have been compared under inductive and non-inductive conditions. However, the small variations in small amounts of unidentified products cannot be interpreted, particularly since the reproducibility was not shown. Indeed, the differences in *ent*-kaurene metabolism between LD and SD claimed by Bennett et al. (1966) could not be reproduced by Takeba and Takimoto (1971).

Dormancy in several species is believed to be induced by decreased levels of GA in response to the change from LD to SD (Eagles and Wareing 1964; Digby and Wareing 1966). In line with this hypothesis, tuberization in *Solanum andigena* is induced by SD and the effect is counteracted by GA. Furthermore, application of CCC to the leaves in LD causes formation of tubers in spite of the unfavourable photoperiod. Also in accord with the hypothesis, the contents of GA-like substances in the leaves decrease after only two SDs upon transfer from LD to SD (Railton and Wareing 1973a). The levels of GA-like substances in leaves growing under SD increase within 35 min of interruption of the long night by red light (Railton and Wareing 1973c). Differences in the conversion of [^{3}H]GA_1 to unknown products were not observed when the plants were kept in LD or SD (Railton and Wareing 1973b). The metabolism of native GA-like substances was not investigated.

3.11.4. Temperature

The effects of temperature on GA content of plants have been studied mainly in connection with breaking of dormancy in seeds (stratification) and induction of flowering (vernalization). GA often replaces the cold treatment in both cases (see review by Jones 1973) and it might be anticipated that one important effect of the cold treatment is to increase the production of GA. This seems to be the case when apple seeds are stratified. Dormant apple seeds have low GA-like activity, mainly due to GA_4, GA_7 and traces of GA_9. At mid-stratification, GA_4 increases some 4×10^4-fold within a week* and then returns to the same low level as before stratification. GA_7 and GA_9 levels remain constant throughout stratification. All GAs were conclusively identified by GC-MS at mid-stratification and quantitated by GLC and TLC (Sinska and Lewak 1970; Sinska et al. 1973). It is difficult to imagine that such an increase in GA-level during stratification should not have a role in dormancy breaking.

A different relation has been found with hazel seeds. Dormancy in hazel embryos is induced by dry storage at 15 °C for 28 days during which period the GA content falls to a very low level (Ross and Bradbeer 1971a). Dormancy is broken by stratification, which involves imbibition and moist storage at 5 °C for 42 days. The seeds then can germinate, if they are transferred to 20 °C. Breaking of dormancy in this species does not seem to involve the production of GA, since there is little increase in GA during stratification (Frankland and Wareing 1966) and since inhibitors of GA biosynthesis given during the cold treatment do not inhibit subsequent germination at 20 °C. During the germination period, on the other hand, there is a substantial increase in GAs, mainly GA_1 and GA_9 identifed by GC-MS (Williams et al. 1974). Inhibitors of GA biosynthesis given during

* We have not been able to reproduce these results.

germination inhibit both the formation of GAs (Ross and Bradbeer 1971b) and germination (Bradbeer 1968). These results support the hypothesis by Bradbeer (1968) that dormancy-breaking in this species does not involve GA-production during the cold period but rather the activation of a GA-producing system which then operates during germination at 20 °C.

Earlier work done on GA-levels in connection with vernalization is considered inconclusive and will not be discussed here (see review by Jones 1973). Zeevaart (1968) working with *Lunaria* and Reid et al. (1974) working with winter wheat did not detect increased levels of GA-like substances during vernalization. In the latter case, there was a decrease in GA-like activity during subsequent growth at higher temperature. Also Einert et al. (1972) found no increase during cold treatment of tulip bulbs, but in this case GA-like activity increased during subsequent development at 18 °C. Most of the activity was in the well developed flower buds.

3.11.5. Hormones and growth retardants

Few experiments have been done on the effect of other growth hormones on GA contents. This is understandable, since the primary effect of the hormones is not known and since the situation is apparently very complex; an effect on GA content may be the result of many intervening events.

The well-known ability of cytokinins to retard senescence may be the reason why kinetin prevents the disappearance of GA in pea roots after removal of the epicotyl (Sebanek 1966) and in *Nasturtium* leaves during senescence (Chin and Beevers 1970). Kinetin-treated wheat leaf sections also retain more GA-like material than the control sections (Loveys and Wareing 1971b) and 6-benzylaminopurine is believed to preserve GA-like substances in tomato plants when the roots are flooded (Reid and Railton 1974). On the other hand, kinetin is supposed to decrease the level of GA in germinating seeds (Michniewicz and Kamieňska 1967). In some of these cases it is doubtful whether the effects are really due to the cytokinin treatment. A claim (Railton 1974b) that 6-benzylaminopurine affects the turnover rate of $[^3H]GA_{20}$ in dwarf pea seedlings is based on data which prove neither that the disappearance of radioactivity is due to metabolism nor that the observed differences in the chromatograms are due to the cytokinin.

Wareing et al. (1968) suggested that ABA might inhibit GA biosynthesis, since maize leaves standing in ABA solution for 24 hr had less GA-like substances than the control leaves, and since ABA inhibited the increase in GA levels otherwise found when spinach is exposed to one LD. Direct proof for the hypothesis is still missing. ABA also completely inhibits the increase in GA-like substances occurring on irradiation of etiolated wheat leaves (Beevers et al. 1970) (cf. section 3.11.1). If the light-induced increase of GA is due to a release from the plastids (section 3.11.1), the mode of action of ABA may be on membrane permeability rather than on GA biosynthesis.

The plant growth retardants are a group of synthetic compounds causing a dwarf habit in several species, while not affecting others, and even causing increased growth in a few (see review by Cathey 1964). The most important growth retardants are AMO-1618, CCC, phosphon D and B-995 (SADH), which were mentioned in section 3.7.2.2 because the first three of them inhibit GA biosynthesis at a specific step in *G. fujikuroi* and in cell-free systems of higher plants, while B-995 does not have this effect (Fall and West 1971). A few years ago it seemed almost certain that at least AMO-1618 and CCC acted in higher plants also by inhibiting GA biosynthesis and that the induced dwarfism was a consequence of lowered GA-levels, a hypothesis which stimulated much research (see review by Lang 1970). However, in the earliest work it was pointed out that the growth retardants might have other effects as well. This has been well confirmed by later research and although several cases have been found in which the retardants certainly lower the level of GA and where full growth can be restored by the application of GA, the mechanism of action of the retardants is by no means clear.

A lowered level of GA-like substances was found in pea seeds maturing in excised pods placed on agar containing AMO-1618; they had only 9–40% as much GA-like activity as the controls (Baldev et al. 1965). In *Silene armeria,* AMO-1618 completely inhibits both stem elongation and the increase in GA-levels normally brought about by LD induction (Cleland and Zeevaart 1970). The same is the case in spinach, where the level of GA-like substances falls to a barely detectable level within a few days of application of AMO-1618 (Zeevaart 1971). AMO-1618 also reduced GA-levels in *Pisum sativum* (Reid and Crozier 1970), *Phaseolus coccineus* (Crozier et al. 1973) and *Cupressus arizonica* (Kuo and Pharis 1975), in all cases also retarding growth. CCC fed to the roots of the mother plant very effectively reduces the GA-like substances of maturing seeds in *Pharbitis nil* without affecting the development of the seeds (Zeevaart 1966). Seedlings grown from such seeds still contain CCC, are dwarfed and contain only 21% of the GA-like substances of the controls. In radish, CCC inhibits stem elongation after vernalization and reduces GA content to about 30% of the control (Suge and Rappaport 1968).

In other cases, the relations are less clear. Low concentrations (1 mg/l) of CCC actually appear to increase the content of GA-like substances in pea seedlings although the increase is very variable (Reid and Crozier 1970). Growth is not affected by this low concentration of CCC. AMO-1618 at a low concentration (1 mg/l) has also been supposed to increase GA levels in peas and barley leaves (Reid and Crozier 1972) but the significance of the data is doubtful. CCC at a high concentration (100 mg/pot) has been claimed to raise the level of GA-like substances in tomato seedlings, but not unexpectedly the data shows no significant increase 5 and 29 days after a single application (Bragt 1969); full growth also resumed. Also in cucumber seedlings where the seeds had been treated with CCC 5 days earlier, the same GA-level was found as in the controls (Snir and Kessler 1975). The effect of the treatment on growth was not mentioned in this publication. A different situation was found in *Gladiolus* where application of CCC results in increased stem length, increased number of flowers per spike and also increased GA-levels, particularly associated with the more polar (ethyl acetate-inextractable) fraction (Halvey and Shilo 1970).

Clear increases in GA-like substances, particularly those of a more polar nature have been obtained with B-995. Thus in *Brassica oleracea* GA-levels were increased ten-fold in cold conditions, whereas floral induction was inhibited (Fontes et al. 1970). In *Cupressus arizonica* 1000 mg/l of B-995 increased GA-like activity from 225 to 2580 ng GA_3-equiv. per plant while strongly inhibiting growth (Kuo and Pharis 1975). A relative newcomer among the plant growth retardants, ancymidol, also believed to inhibit GA biosynthesis but at a different site (see section 3.7.2.2), reduces GA-like activity in bean seedlings to virtually nothing. GA_7 overcomes the dwarfing effect of ancymidol in the dark but not in the light (Shive and Sisler 1976). The fungicide triarimol (α-(2,4-dichlorophenyl)-α-phenyl-5-pyrimidinemethanol) has similar effects.

While AMO-1618 and CCC thus reduce the levels of GA-like substances in many plants, they do not do so in others and they also have other effects. Furthermore, although Lockhart (1962) showed that the effects of CCC and phosphon D on growth in *Phaseolus vulgaris* could be completely reversed by GA, this has not been found in many other cases. Thus Crozier et al. (1973) showed that the effect of AMO-1618 on stem elongation in the closely related species *Phaseolus coccineus* could not be completely reversed by the application of GA_3. Several other effects of the retardant, namely reduced cell length, leaf area and root growth, were not affected by GA application at all. Douglas and Paleg (1974) showed that phosphon D, AMO-1618 and CCC inhibit sterol biosynthesis in *Nicotiana tabacum* at steps remote from the part of the pathway which is common to sterol and GA biosynthesis. In this species, the effects of the retardants on growth could be prevented by application of either sterols or GA. Unless the primary action of GA is to activate steroid biosynthesis, this proves that the growth retardants also inhibit biosynthetic steps other than the ones in the GA pathway.

Since there are cases in which growth is very much more affected by the growth retardants than are the GA levels, it must be assumed that factors other than decreased GA biosynthesis can cause dwarfism in response to the retardants. This has also been assumed for those species in which GA does not completely relieve the inhibition (Lang 1970). Assuming such factors exist, there is no way of knowing to what extent they are responsible for the effect in species where the levels of GA are lowered. It is therefore no longer justified to base conclusions on the assumption that GA biosynthesis alone is inhibitied by treatment with growth retardants.

Acknowledgement

The authors wish to thank Dr. J. MacMillan for reading this Chapter and for his suggestions.

References

Abe, H., Uchiyama, M., Sato, R. and Muto, S. (1974) in: Plant Growth Substances 1973, Hirokawa, Tokyo, pp. 201–206.

Agnistikova, V.N., Kobrina, N.S., Kucherov, V.F. and Serebryakov, E.P. (1974) in: Biochemistry and Chemistry of Plant Growth Regulators. K. Schreiber, H.R. Schütte, G. Sembdner (eds.), Academy of Science of the German Democratic Republic, Institute of Plant Biochemistry, Halle (Saale), pp. 63–77.

Aldridge, D.C., Hanson, J.R. and Mulholland, T.P.C. (1965) J. Chem. Soc., 3539–3549.

Anderson, J.D. and Moore, T.C. (1967) Plant Physiol., 42, 1527–1534.

Appleton, R.A., McCrindle, R. and Overton, K.H. (1968) Phytochemistry, 7, 135–137.

Appleton, R.A., McCrindle, R. and Overton, K.H. (1970) Phytochemistry, 9, 581–583.

Asakawa, Y., Tamari, K., Inoue, K. and Kaji, J. (1974a) Agr. Biol. Chem., 38, 713–717.

Asakawa, Y., Tamari, K., Shoji, A. and Kaji, J. (1974b) Agr. Biol. Chem., 38, 719–725.

Atsmon, D., Lang, A. and Light, E.N. (1968) Plant Physiol., 43, 806–810.
Aung, L.H. and Rees, A.R. (1974) J. exp. Bot., 25, 745–751.
Aung, L.H., Hertogh, A.A., de and Staby, G. (1969) Plant Physiol., 44, 403–406.
Bailiss, K.W. and Hill, T.A. (1971) Bot. Rev., 37, 437–479.
Bailiss, K.W. and Wilson, I.M. (1967) Ann. Bot. (Lond.) 31, 195–211.
Bakker, H.J., Cook, I.F., Jefferies, P.R. and Knox, J.R. (1974) Tetrahedron, 30, 3631–3640.
Baldev, B., Lang, A. and Agatep, A.O. (1965) Science, 147, 155–157.
Barendse, G.W.M. (1971) Planta, 99, 290–301.
Barendse, G.W.M. (1974) in: Plant Growth Substances 1973, Hirokawa, Tokyo, pp. 332–341.
Barendse, G.W.M. and Klerk, G.J.M., de (1975) Planta, 126, 25–35.
Barendse, G.W.M. and Kok, N.J.J. (1971) Plant Physiol., 48, 476–479.
Barendse, G.W.M. and Lang, A. (1972) Plant Physiol., 49, 836–841.
Barendse, G.W.M., Kende, H. and Lang, A. (1968) Plant Physiol., 43, 815–822.
Barnes, M.F., Light, E.N. and Lang, A. (1969) Planta, 88, 172–182.
Bartlet, P.A. and Johnson, W.S. (1970) Tetrahedron Lett., 4459–4462.
Bateson, J.H. and Cross, B.E. (1974) J. Chem. Soc. Perkin I, 1131–1136.
Bearder, J.R. and MacMillan, J. (1972) Agr. Biol. Chem., 36, 342–344.
Bearder, J.R. and MacMillan, J. (1973) J. Chem. Soc. Perkin I, 2824–2830.
Bearder, J.R., MacMillan, J. and Phinney, B.O. (1973a) Phytochemistry, 12, 2173–2179.
Bearder, J.R., MacMillan, J. and Phinney, B.O. (1973b) Phytochemistry, 12, 2655–2659.
Bearder, J.R., MacMillan, J., Wels, C.M., Chaffey, M.B. and Phinney, B.O. (1974) Phytochemistry, 13, 911–917.
Bearder, J.R., MacMillan, J. and Phinney, B.O. (1975a) J. Chem. Soc. Perkin I, 721–726.
Bearder, J.R., MacMillan, J., Wels, C.M. and Phinney, B.O. (1975b) Phytochemistry, 14, 1741–1748.
Bearder, J.R., Dennis, F.G., MacMillan, J., Martin, G.C. and Phinney, B.O. (1975c) Tetrahedron Lett., 669–670.
Bearder, J.R., Frydman, V.M., Gaskin, P., MacMillan, J., Wels, C.M and Phinney, B.O. (1976a) J. Chem. Soc. Perkin I, 173–178.
Bearder, J.R., Frydman, V.M., Gaskin, P., Hatton, I.K., Harvey, W.E., MacMillan, J. and Phinney, B.O. (1976b) J. Chem. Soc. Perkin I, 178–183.
Beeley, L.J. and MacMillan, J. (1976) J. Chem. Soc. Perkin I, 1022–1028.
Beeley, L.J., Gaskin, P. and MacMillan, J. (1975) Phytochemistry, 14, 779–783.
Beevers, L., Loveys, B., Pearson, J.A., Wareing, P.F. (1970) Planta, 90, 286–294.
Beilby, J.P., Ghisalberti, E.L., Jefferies, P.R., Sefton, M.A. and Sheppard, P.N. (1973) Tetrahedron Lett., 2589–2592.
Bennett, R.D., Ko, S.T. and Heftmann, E. (1966) Plant Physiol., 41, 1360–1363.
Bennett, R.D., Lieber, E.R. and Heftmann, E. (1967) Phytochemistry, 6, 1107–1110.
Bentley-Mowat, J.A. (1966) Ann. Bot. (Lond.), 30, 167–171.
Bhalla, P.R. (1971) Physiol. Plant., 24, 106–111.
Bhalla, P.R. and Gairola, C. (1972) Indian J. Plant Physiol., 15, 148–170.
Binks, R., MacMillan, J. and Pryce, R.J. (1969) Phytochemistry, 8, 271–284.
Binks, R., Goodfellow, R.J., MacMillan, J. and Pryce, R.J. (1970) Chem. Ind. (Lond.), 565–566.
Binks, R., Cleaver, R.L., Littler, J.S. and MacMillan, J. (1971) Chem. Brit., 7, 8–12.
Birch, A.J., Richards, R.W., Smith, H., Harris, A. and Whalley, W.B. (1959) Tetrahedron, 7, 241–251.
Blumenfeld, A. and Gazit, S. (1972) Physiol. Plant., 27, 116–120.
Borrow, A., Brian, P.W., Chester, V.E., Curtis, P.J., Hemming, H.G., Henehan, C., Jefferys, E.G., Lloyd, P.B., Nixon, I.S., Norris, G.L.F. and Radley, M. (1955) J. Sci. Food Agric., 6, 340–348.
Borrow, A., Jefferys, E.G., Kessell, R.H.J., Lloyd, E.C., Lloyd, P.B. and Nixon, I.S. (1961) Canad. J. Microbiol., 7, 227–276.
Borrow, A., Brown, S., Jefferys, E.G., Kessell, R.H.J., Lloyd, E.C., Lloyd, P.B., Rothwell, A., Rothwell, B. and Swait, J.C. (1964) Canad. J. Microbiol., 10, 407–444.
Bowen, M.R. and Wareing, P.F. (1969) Planta, 89, 108–125.

Bowen, D.H., MacMillan, J. and Graebe, J.E. (1972) Phytochemistry, 11, 2253–2257.
Bowen, D.H., Crozier, A., MacMillan, J. and Reid, D.M. (1973) Phytochemistry, 12, 2935–2941.
Bown, A.W., Reeve, D.R. and Crozier, A. (1975) Planta, 126, 83–91.
Bradbeer, J.W. (1968) Planta, 78, 266–276.
Bradshaw, M.J. and Edelman, J. (1969) J. exp. Bot., 20, 87–93.
Bragt, J., van (1969) Neth. J. Agr. Sci., 17, 183–188.
Brian, P.W. (1957) Symp. Soc. exp. Biol., 11, 166–182.
Brian, P.W. (1959) Biol. Rev., 34, 37–84.
Brian, P.W. (1966) Int. Rev. Cytol., 19, 229–266.
Brian, P.W. and Hemming, H.G. (1955) Physiol. Plant., 8, 669–681.
Brian, P.W., Elson, G.W., Hemming, H.G. and Radley, M. (1954) J. Sci. Food Agric., 5, 602–612.
Brian, P.W., Grove, J.F. and MacMillan, J. (1960) Fortschr. Chem. Org. Naturstoffe, 18, 350–433.
Brian, P.W., Hemming, H.G. and Lowe, D. (1964) Ann. Bot. (Lond.), 28, 369–389.
Brian, P.W., Grove, J.F. and Mulholland, T.P.C. (1967) Phytochemistry, 6, 1475–1499.
Briggs, D.E. (1966a) Nature (Lond.), 210, 418–419.
Briggs, D.E. (1966b) Nature (Lond.), 210, 419–421.
Briggs, D.E. (1967) J. Inst. Brew., 73, 361–370.
Briggs, D.E. (1973) in: Biosynthesis and its Control in Plants, B.V. Milborrow (ed.), Academic Press, New York, pp. 219–277.
Brown, M.E. (1972) J. appl. Bacteriol., 35, 443–451.
Brown, J.C., Cross, B.E. and Hanson, J.R. (1967) Tetrahedron, 23, 4095–4103.
Bruinsma, J. (1963) Chem Weekblad, 59, 599.
Buggy, M.J., Britton, G. and Goodwin, T.W. (1974) Phytochemistry, 13, 125–129.
Bukovac, M.J. and Nakagawa, S. (1967) Experientia (Basel) 23, 865.
Butcher, D.N. (1963) J. exp. Bot., 14, 272–280.
Carlisle, D.B., Osborne, D.J., Ellis, P.E. and Morehouse, J.E. (1963) Nature (Lond.), 200, 1230.
Carr, D.J. and Reid, D.M. (1968) in: Biochemistry and Physiology of Plant Growth Substances, F. Wightman and G. Setterfield (eds.), Runge Press, Ottawa, pp. 1169–1185.
Carr, D.J., Reid, D.M. and Skeene, K.G.M. (1964) Planta, 63, 382–392.
Cathey, H.M. (1964) Ann. Rev. Plant Physiol., 15, 271–302.
Cavell, B.D. and MacMillan, J. (1967) Phytochemistry, 6, 1151–1154.
Cavell, B.D., MacMillan, J., Pryce, R.J. and Sheppard, A.C. (1967) Phytochemistry, 6, 867–874.
Chailakhyan, M.Kh. (1968) Ann. Rev. Plant Physiol., 19, 1–36.
Chailakhyan, M.Kh. and Lozhnikova, V.N. (1961) Soviet Plant Physiol., 7, 432–438.
Chailakhyan, M.Kh. and Lozhnikova, V.N. (1966) Soviet Plant Physiol., 13, 734–741.
Chailakhyan, M.Kh., Khlopenkova, L.P. and Khazhakyan, Kh.K. (1974) Dokl. Botan. Sci., 215, 19–22.
Chin T.Y. and Beevers, L. (1970) Planta, 92, 178–188.
Chin, T.Y. and Lockhart, J.A. (1965) Amer. J. Bot., 52, 828–833.
Chrispeels, M.J. and Varner, J.E. (1966) Nature (Lond.), 212, 1066–1067.
Cleland, C.F. and Zeevaart, J.A.D. (1970) Plant Physiol., 46, 392–400.
Clor, M.A. (1967) Nature (Lond.), 214, 1263–1264.
Cohen, D. and Paleg, L.G. (1967) Plant Physiol., 42, 1288–1296.
Cooke, R.J. and Saunders, P.F. (1975a) Planta, 123, 299–302.
Cooke, R.J. and Saunders, P.F. (1975b) Planta, 126, 151–160.
Cooke, R.J., Saunders, P.F. and Kendrick, R.E. (1975) Planta, 124, 319–328.
Coolbaugh, R.C. and Hamilton, R. (1976) Plant Physiol., 57, 245–248.
Coolbaugh, R.C. and Moore, T.C. (1969) Plant Physiol., 44, 1364–1367.
Coolbaugh, R.C. and Moore, T.C. (1971a) Phytochemistry, 10, 2395–2400.
Coolbaugh, R.C. and Moore, T.C. (1971b) Phytochemistry, 10, 2401–2412.
Coolbaugh, R.C., Moore, T.C., Barlow, S.A. and Ecklund, P.R. (1973) Phytochemistry, 12, 1613–1618.

Coombe, B.G. (1971) Science, 172, 856–857.
Coombe, B.G. and Tate, M.E. (1972) in: Plant Growth Substances 1970, D.J. Carr (ed.), Springer, Heidelberg, pp. 158–165.
Coombe, B.G., Cohen, D. and Paleg, L.G. (1967a) Plant Physiol., 42, 105–112.
Coombe, B.G., Cohen, D. and Paleg, L.G. (1967b) Plant Physiol., 42, 113–119.
Coombe, B.G., Mander, L.N., Paleg, L.G. and Turner, J.V. (1974) Aust. J. Plant Physiol., 1, 473–481.
Corcoran, M.R. and Phinney, B.O. (1962) Physiol. Plant., 15, 252–262.
Corey, E.J., Brennan, T.M. and Carney, R.L. (1971) J. Amer. chem. Soc., 93, 7316–7317.
Cornforth, R.H., Fletcher, K., Hellig, H. and Popják, G. (1960) Nature (Lond.), 185, 923–924.
Couillerot, J.-P. and Bonnemain, J.-L. (1975) C.R. Acad. Sci. (Paris) Ser. D, 280, 1453–1454.
Croft, K.D., Ghisalberti, E.L., Jefferies, P.R., Knox, J.R., Mahoney, T.J. and Sheppard, P.N. (1974) Tetrahedron, 30, 3663–3667.
Cross, B.E. (1954) J. Chem. Soc., 4670–4676.
Cross, B.E. (1966) J. Chem. Soc., 501–504.
Cross, B.E. (1968) in: Progress in Phytochemistry, Vol. 1, L. Reinhold and Y. Liwschitz (eds.), Interscience Publ., New York, pp. 195–222.
Cross, B.E. and Myers, P.L. (1969) Phytochemistry, 8, 79–83.
Cross, B.E. and Norton, K. (1965) J. Chem. Soc., 1570–1572.
Cross, B.E., Grove, J.F., MacMillan, J. and Mulholland, T.P.C. (1956) Chem. Ind. (Lond.), 954–955.
Cross, B.E., Grove, J.F., MacMillan, J., Moffatt, J.S., Mulholland, T.P.C., Seaton, J.C. and Sheppard, N. (1959) Proc. Chem. Soc., 302–303.
Cross, B.E., Grove, J.F. and Morrison, A. (1961) J. Chem. Soc., 2498–2515.
Cross, B.E., Galt, R.H.B. and Hanson, J.R. (1962) Tetrahedron, 18, 451–459.
Cross, B.E., Galt, R.H.B., Hanson, J.R., Curtis, P.J., Grove, J.F. and Morrison, A. (1963a) J. Chem. Soc., 2937–2943.
Cross, B.E., Galt, R.H.B. and Hanson, J.R. (1963b) J. Chem. Soc., 2944–2961.
Cross, B.E., Galt, R.H.B. and Hanson, J.R. (1963c) J. Chem. Soc., 3783–3790.
Cross, B.E., Galt, R.H.B. and Hanson, J.R. (1963d) J. Chem. Soc., 5052–5061.
Cross, B.E., Galt, R.H.B. and Hanson, J.R. (1964) J. Chem. Soc., 295–300.
Cross, B.E., Galt, R.H.B. and Norton, K. (1968a) Tetrahedron, 24, 231–237.
Cross, B.E., Norton, K. and Stewart, J.C. (1968b) J. Chem. Soc., C 1054–1063.
Cross, B.E., Stewart, J.C. and Stoddart, J.L. (1970) Phytochemistry, 9, 1065–1071.
Crozier, A. and Audus, L.J. (1968) Planta, 83, 207–217.
Crozier, A. and Reid, D.M. (1971) Canad. J. Bot., 49, 967–976.
Crozier, A., Aoki, H. and Pharis, R.P. (1969) J. exp. Bot., 20, 786–795.
Crozier, A., Kuo, C.C., Durley, R.C. and Pharis, R.P. (1970) Canad. J. Bot., 48, 867–877.
Crozier, A., Reid, D.M. and Reeve, D.R. (1973). J. exp. Bot., 24, 923–934.
Curtis, P.J. and Cross, B.E. (1954) Chem. Ind. (Lond.), 1066.
Dale, J.E. (1969) Planta, 89, 155–164.
Dale, J.E. and Felippe, G.M. (1968) Planta, 80, 288–298.
Davies, L.J. and Rappaport, L. (1975a) Plant Physiol., 55, 620–625.
Davies, L.J. and Rappaport, L. (1975b) Plant Physiol., 56, 60–66.
Davies, B.H., Rees, A.F. and Taylor, R.F. (1975) Phytochemistry, 14, 717–722.
Dawson, R.M., Jefferies, P.R. and Knox, J.R. (1975) Phytochemistry, 14, 2593–2597.
Dell, B. and McComb, A.J. (1974) Aust. J. Bot., 25, 195–210.
Dennis, D.T. and West, C.A. (1967) J. biol. Chem., 242, 3293–3300.
Dennis, D.T., Upper, C.D. and West, C.A. (1965) Plant Physiol., 40, 948–952.
Digby, J. and Wareing, P.F. (1966) Ann. Bot. (Lond.), 30, 607–622.
Donike, M., Suberg, H. and Jaenicke, L. (1973). J. Chromatogr., 85, 9–18.
Douglas, T.J. and Paleg, L.G. (1974) Plant Physiol., 54, 238–245.
Durley, R.C. (1968) Ph.D. Thesis, University of Bristol.
Durley, R.C. and Pharis, R.P. (1972) Phytochemistry, 11, 317–326.

Durley, R.C. and Pharis, R.P. (1973) Planta, 109, 357–361.
Durley, R.C., MacMillan, J. and Pryce, R.J. (1971) Phytochemistry, 10, 1891–1908.
Durley, R.C., Crozier, A., Pharis, R.P. and McLaughlin, G.E. (1972) Phytochemistry, 11, 3029–3033.
Durley, R.C., Railton, I.D. and Pharis, R.P. (1973) Phytochemistry, 12, 1609–1612.
Durley, R.C., Railton, I.D. and Pharis, R.P. (1974) Phytochemistry, 13, 547–551.
Durley, R.C., Pharis, R.P. and Zeevaart, J.A.D. (1975) Planta, 126, 139–149.
Eagles, C.F. and Wareing, P.F. (1964) Physiol. Plant., 17, 697–709.
Ecklund, P.R. and Moore, T.C. (1974) Plant Physiol., 53, 5–10.
Einert, A.E., Staby, G.L. and DeHertogh, A.A. (1972) Canad. J. Bot., 50, 909–914.
El-Antably, H.M.M. and Larsen, P. (1974a) Nature (Lond.), 250, 76–77.
El-Antably, H.M.M. and Larsen, P. (1974b) Physiol. Plant. 32, 322–329.
Ende, H., Van den and Zeevaart, J.A.D. (1971) Planta, 98, 164–176.
Endo, M., Nakanishi, K., Näf, U., McKeon, W. and Walker, R. (1972) Physiol. Plant., 26, 183–185.
Evans, A. and Smith, H. (1976) Proc. nat. Acad. Sci. (Wash.), 73, 138–142.
Evans, R. and Hanson, J.R. (1972) J. Chem. Soc. Perkin I, 2382–2385.
Evans, R., Hanson, J.R. and White, A.F. (1970) J. Chem. Soc., C 2601–2603.
Evans, R., Hanson, J.R. and Mulheirn, L.J. (1973) J. Chem. Soc., Perkin I, 753–756.
Evans, R., Hanson, J.R. and Siverns, M. (1975). J. Chem. Soc., Perkin I, 1514–1517.
Fall, R.R. and West, C.A. (1971) J. biol. Chem., 246, 6913–6928.
Faull, K.F., Coombe, B.G. and Paleg, L.G. (1974a) Aust. J. Plant Physiol., 1, 183–198.
Faull, K.F., Coombe, B.G. and Paleg, L.G. (1974b) Aust. J. Plant Physiol., 1, 199–210.
Fleming, J.R. and Johnson, J.A. (1964) Science, 144, 1021–1022.
Fletcher, R.A. and Osborne, D.J. (1966) Nature (Lond.), 211, 743–744.
Fletcher, R.A., Oegema, T. and Horton, R.F. (1969) Planta, 86, 98–102.
Fontes, M.R., Ozbun, J.L. and Powell, L.E. (1970) Nature (Lond.), 228, 82–83.
Frankland, B. and Wareing, P.F. (1960) Nature (Lond.), 185, 255–256.
Frankland, B. and Wareing, P.F. (1966) J. exp. Bot., 17, 596–611.
Frydman, V.M. and MacMillan, J. (1973) Planta, 115, 11–15.
Frydman, V.M. and MacMillan, J. (1975) Planta, 125, 181–195.
Frydman, V.M. and Wareing, P.F. (1973a) J. exp. Bot., 24, 1131–1138.
Frydman, V.M. and Wareing, P.F. (1973b) J. exp. Bot., 24, 1139–1148.
Frydman, V.M., Gaskin, P. and MacMillan, J. (1974) Planta, 118, 123–132.
Fuchs, Y. and Gertman, E. (1974) Plant Cell Physiol., 15, 629–633.
Fukui, H., Koshimizu, K. and Mitsui, T. (1971) Phytochemistry, 10, 671–673.
Fukui, H., Ishii, H., Koshimizu, K., Katsumi, M., Ogawa, Y. and Mitsui, T. (1972) Agr. Biol. Chem., 36, 1003–1012.
Gairola, C., Bhalla, P.R., Sabharwal, P.S. and Aleem, M.I.H. (1972) Planta, 106, 177–180.
Galston, A.W. and Warburg, H. (1959) Plant Physiol., 34, 16–22.
Galt, R.H.B. (1965) J. Chem. Soc., 3143–3151.
Galt, R.H.B. (1968) Tetrahedron, 24, 1337–1339.
Galt, R.H.B. and Hanson, J.R. (1966) Tetrahedron, 22, 3185–3187.
Ganguly, S.N. and Sircar, S.M. (1974) Phytochemistry, 13, 1911–1913.
Ganguly, S.N., Sanyal, T., Sircar, P.K. and Sircar, S.M. (1970) Chem. Ind. (Lond.), 832–833.
Gaskin, P. and MacMillan, J. (1975) Phytochemistry, 14, 1575–1578.
Gaskin, P., MacMillan, J., Firn, R.D. and Pryce, R.J. (1971) Phytochemistry, 10, 1155–1157.
Gaskin, P., MacMillan, J., Ganguly, S.N., Sanyal, T., Sircar, P.K. and Sircar, S.M. (1972) Chem. Ind. (Lond.), 424–425.
Gaskin, P., MacMillan, J. and Zeevaart, J.A.D. (1973) Planta, 111, 347–352.
Geissman, T.A., Verbiscar, A.J., Phinney, B.O. and Cragg, G. (1966) Phytochemistry, 5, 933–947.
Glenn, J.L., Kuo, C.C., Durley, R.C. and Pharis, R.P. (1972) Phytochemistry, 11, 345–351.
Gogola, N. (1972) Biol. Vestn., 20, 55–63.
Goldschmidt, E.E., Eilati, S.K. and Goren, R. (1972) in: Plant Growth Substances 1970, D.J. Carr, (ed.), Springer, Heidelberg, pp. 611–617.

Goodwin, P.B. and Gordon, A. (1972) J. exp. Bot. 23, 970–979.
Gorter, C.J. (1961) Physiol. Plant., 14, 332–343.
Goto, N. (1970) Plant Cell Physiol., 11, 355–359.
Goto, N. and Esashi, Y. (1973), Physiol. Plant., 28, 480–489.
Goto, N. and Esashi, Y. (1975) Plant Cell Physiol., 16, 759–766.
Graebe, J.E. (1967) Science, 157, 73–75.
Graebe, J.E. (1968) Phytochemistry, 7, 2003–2020.
Graebe, J.E. (1969) Planta, 85, 171–174.
Graebe, J.E. and Hedden, P. (1974) in: Biochemistry and Chemistry of Plant Growth Regulators, K. Schreiber, H.R. Schütte, G. Sembdner (eds.), Academy of Science of the German Democratic Republic, Institute of Plant Biochemistry, Halle (Saale), pp. 1–16.
Graebe, J.E., Dennis, D.T., Upper, C.D. and West, C.A. (1965) J. biol. Chem., 240, 1847–1854.
Graebe, J.E., Bowen, D.H. and MacMillan, J. (1972) Planta, 102, 261–271.
Graebe, J.E., Hedden, P., Gaskin, P. and MacMillan, J. (1974a) Phytochemistry, 13, 1433–1440.
Graebe, J.E., Hedden, P. and MacMillan, J. (1974b) in: Plant Growth Substances 1973, Hirokawa, Tokyo, pp. 260–266.
Graebe, J.E., Hedden, P., Gaskin, P. and MacMillan, J. (1974c) Planta, 120, 307–309.
Graebe, J.E., Hedden, P. and MacMillan, J. (1975) J. Chem. Soc. Chem. Commun., 161 162.
Green, F.B. and Corcoran, M.R. (1975) Plant Physiol., 56, 801–806.
Green, T.R. and West, C.A. (1974) Biochemistry, 13, 4720–4729.
Green, T.R., Dennis, D.T. and West, C.A. (1975) Biochem. biophys. Res. Commun., 64, 976–982.
Grigorieva, N.Y., Kucherov, V.F., Lozhnikova, V.N. and Chailakhian, M.K. (1971) Phytochemistry, 10, 509–517.
Groat, J.I. and Briggs, D.E. (1969) Phytochemistry, 8, 1615–1627.
Grove, J.F. (1961) J. Chem. Soc., 3345–3547.
Grove, J.F. and Mulholland, T.P.C. (1960) J. Chem. Soc., 3007–3022.
Grove, J.F., Jeffs, P.W. and Mulholland, T.P.C. (1958) J. Chem. Soc., 1236–1240.
Grove, J.F., MacMillan, J., Mulholland, T.P.C. and Turner, W.B. (1960) J. Chem. Soc., 3049–3057.
Halevy, A.H. and Cathey, H.C. (1960) Bot. Gaz. (Chic.), 122, 63–67.
Halevy, A.H. and Shilo, R. (1970) Physiol. Plant., 23, 820–827.
Hanson, J.R. (1965) J. Chem. Soc., 5036–5040.
Hanson, J.R. (1966) Tetrahedron, 22, 701–703.
Hanson, J.R. (1967) Tetrahedron, 23, 733–735.
Hanson, J.R. (1968) in: Progress in Phytochemistry, I. L. Reinhold and G. Lievschitz (ed.), Interscience, London, pp. 161.
Hanson, J.R. (1971) Progr. Chem. Org. Natur. Comp., 29, 395–416.
Hanson, J.R. (1972) in: Chemistry of Terpenes and Terpenoids, A.A. Newman (ed.), Academic Press, New York, pp. 155–199.
Hanson, J.R. and Hawker, J. (1972a) Tetrahedron, 28, 2521–2526.
Hanson, J.R. and Hawker, J. (1972b) Tetrahedron Lett., 4299–4302.
Hanson, J.R. and Hawker, J. (1973) Phytochemistry, 12, 1073–1075.
Hanson, J.R. and White, A.F. (1968) Phytochemistry, 7, 595–597.
Hanson, J.R. and White, A.F. (1969) J. Chem. Soc. C, 981–985.
Hanson, J.R., Hawker, J. and White, A.F. (1972) J. Chem. Soc. Perkin I, 1892–1895.
Hanson, J.R., Savona, G. and Siverns, M. (1974) J. Chem. Soc. Perkin I, 2001–2005.
Hanson, J.R., Siverns, M. Piozzi, F. and Savona, G. (1976) J. Chem. Soc. Perkin I, 114–117.
Harada, H. (1962) in: Eigenschaften und Wirkungen der Gibberelline, R. Knapp (ed.), Springer-Verlag, Berlin, pp. 45–50.
Harada, H. and Nitsch, J.P. (1967) Phytochemistry, 6, 1695–1703.
Harada, H. and Yokota, T. (1970) Planta, 92, 100–104.
Harrison, D.M. and MacMillan, J. (1971) J. Chem. Soc. C, 631–636.
Hartley, R.D., Hill, T.A., Pegg, G.F. and Thomas, G.G. (1969) J. exp. Bot., 20, 276–287.

Lockhart, J.A. (1959b) Plant Physiol., 34, 460–465.
Lockhart, J.A. (1961) Amer. J. Bot., 48, 516–525.
Lockhart, J.A. (1962) Plant Physiol., 37, 759–764.
Lockhart, J.A. (1964) Planta, 62, 97–115.
Lockhart, J.A. and Deal, P.H. (1960) Naturwissenschaften, 47, 141–142.
Lona, F. (1957) Ateneo Parmense, 28, 111–115.
Lorenzi, R., Horgan, R. and Heald, J.K. (1975) Planta, 126, 75–82.
Loveys, B.R. and Wareing, P.F. (1971a) Planta, 98, 109–116.
Loveys, B.R. and Wareing, P.F. (1971b) Planta, 98, 117–127.
Lozhnikova, V.N. (1967) Dokl. Bot. Sci., 168, 92–94.
Lozhnikova, V.N. and Chailakhyan, M.Kh. (1969) Dokl. Akad. Nauk SSSR, 188, 715.
Luckwill, L.C. and Whyte, P. (1968) Soc. Chem. Ind., Monograph No. 31, 87–101.
MacLeod, A.M. and Palmer, G.H. (1967) Nature (Lond.), 216, 1342–1343.
MacMillan, J. (1971) in: Aspects of Terpenoid Chemistry and Biochemistry, T.W. Goodwin (ed.), Academic Press, New York, pp.153–180.
MacMillan, J. (1972) in: Plant Growth Substances 1970, D.J. Carr (ed.), Springer, Heidelberg, New York, pp. 790–797.
MacMillan, J. (1974a) in: The Chemistry and Biochemistry of Plant Hormones, V.C. Runeckles, E. Sondheimer and D.C. Walton (eds.), Recent Advances in Phytochemistry, Vol. 7. Academic Press, New York, pp. 1–19.
MacMillan, J. (1974b) in: Biochemistry and Chemistry of Plant Growth Regulators, K. Schreiber, H.R. Schütte and G. Sembdner (eds.), Acad. Sci. G.D.R., Halle (Saale), pp. 33–49.
MacMillan, J. and Pryce, R.J. (1967) J. Chem. Soc. C, 740–742.
MacMillan, J. and Pryce, R.J. (1968) Tetrahedron Lett., 1537–1542.
MacMillan, J. and Pryce, R.J. (1973) in: Phytochemistry Vol. III, L.P. Miller (ed.), Van Nostrand-Reinhold, New York, pp. 283–326.
MacMillan, J. and Suter, P.J. (1963) Nature (Lond.), 197, 190.
MacMillan, J. and Takahashi, N. (1968) Nature (Lond.), 217, 170–171.
MacMillan, J. and Walker, E.R.H. (1972) J. Chem. Soc. Perkin I, 981–985.
MacMillan, J. and Wels, C.M. (1973) J. Chromatogr., 87, 271–276.
MacMillan, J. and Wels, C.M. (1974) Phytochemistry, 13, 1413–1417.
MacMillan, J., Seaton, J.C. and Suter, P.J. (1960) Tetrahedron, 11, 60–66.
MacMillan, J., Seaton, J.C. and Suter, P.J. (1962) Tetrahedron, 18, 349–355.
MacMillan, J., Pryce, R.J., Eglington, G. and McCormick, A. (1967) Tetrahedron Lett., 2241–2243.
Margara, J. (1963) Ann. Physiol. Vég., 5, 249.
McComb, A.J. (1961) Nature (Lond.), 192, 575–576.
McComb, A.J. (1964a) J. gen. Microbiol., 34, 401–411.
McComb, A.J. (1964b) Ann. Bot. (Lond.), 28, 669–687.
McComb, A.J. and Carr, D.J. (1958) Nature (Lond.), 181, 1548–1549.
McCready, C.C. (1966) Plant Physiol., 17, 283–294.
McCrindle, R. and Overton, K.H. (1965) Advanc. Org. Chem. 5, 47–113.
McInnes, A.G., Smith, D.G., Arsenault, G.P. and Vining, L.C. (1973) Canad. J. Biochem., 51, 1470–1474.
Michniewicz, M. and Kamienska, A. (1967) Naturwissenschaften, 54, 372.
Michniewicz, M. and Lang, A. (1962) Planta, 58, 549–563.
Moffat, J.S. (1960) J. Chem. Soc., 3045–3049.
Momotani, Y. and Kato, J. (1967) Plant Cell Physiol., 8, 439–445.
Moore, T.C., Barlow, S.A. and Coolbaugh, R.C. (1972) Phytochemistry, 11, 3225–3233.
Mori, S., Kumazawa, K. and Mitsui, S. (1965) Plant Cell Physiol., 6, 571–574.
Mounla, M.A.Kh. and Michael, G. (1973) Physiol. Plant., 29, 274–276.
Murakami, Y. (1961a) Bot. Mag. (Tokyo), 74, 241–247.
Murakami, Y. (1961b) Bot. Mag. (Tokyo), 74, 424–425.

Railton, I.D., Murofushi, N., D
Rao, T.R. and Nagar, P.K. (19
Rappaport, E. and Sachs, M. (1
Rappaport, L., Davies, L., Lave
Growth Substances 197
Reeve, D.R. and Crozier, A. (1
Reeve, D.R., Crozier, A., Durl
Reid, D.M. and Burrows, W.J. (
Reid, D.M. and Carr, D.J. (196
Reid, D.M. and Crozier, A. (19
Reid, D.M. and Crozier, A. (19
Reid, D.M. and Crozier, A. (19
berg, pp. 420–427.
Reid, D.M. and Railton, I.D. (1
Reid, D.M., Clements, J.B. and
Reid, D.M., Crozier, A. and Ha
Reid, D.M., Tuing, M.S., Durley
Reid, D.M., Pharis, R.P. and R
Reinhard, E. and Sacher, R. (1
Robinson, D.R. and West, C.A.
Robinson, D.R. and West, C.A.
Roesel, H.A. and Haber, A.H. (
Rogers, L.J., Shah, S.P.J. and G
Ross, J.D. and Bradbeer, J.W. (
Ross, J.D. and Bradbeer, J.W. (
Rowe, J.W. (1968) The Comm
October 1968. U.S. Fore
Ruddat, M., Lang, A. and Mose
Ruddat, M., Heftmann, E. and I
Russell, D.W. and Galston, A.W.
Sakurai, A. and Tamura, S. (19
Sakurai, N., Shibata, K. and Ka
Sale, P.J.M. and Vince, D. (196
Schneider, G., Sembdner, G. an
Schneider, G., Jänicke, S. and S
Schreiber, K., Schneider, G., Se
Schreiber, K., Weiland, J. and S
Scott, T.K. and Most, B.H. (
H. Kaldewey and Y. Var
Šebánek, J. (1966) Naturwissen
Šebánek, J. and Hink, J. (1967)
Seeley, S.D. and Powell, L.E. (1
Sembdner, G. and Schreiber, K.
Sembdner, G., Gross, R. and Sc
Sembdner, G., Schneider, G., W
Sembdner, G., Schneider, G. an
Sembdner, G., Weiland, J., Auri
70–86.
Sembdner, G., Weiland, J., Sch
stances 1970, D.J. Carr (e
Sembdner, G., Adam, G., Lische
Liebisch, H. W. and Schr
pp. 349–355.

Murakami, Y. (1966) Bot. Mag. (Tokyo), 79, 315–327.
Murakami, Y. (1968a) Bot. Mag. (Tokyo), 81, 33–43.
Murakami, Y. (1968b) Bot. Mag. (Tokyo), 81, 334–343.
Murakami, Y. (1970a) Bot. Mag. (Tokyo), 83, 211–213.
Murakami, Y. (1970b) Bot. Mag. (Tokyo), 83, 312–324.
Murakami, Y. (1970c) Jap. Agric. Res. Q., 5, 5–9.
Murakami, Y. (1972) in: Plant Growth Substances 1970, D.J. Carr (ed.), Springer, Heidelberg, pp. 166–174.
Murakami, Y. (1973) Plant Cell Physiol., 14, 91–102.
Murakami, Y. (1975) Plant Cell Physiol., 16, 337–345.
Murofushi, N., Iriuchijima, S., Takahashi, N., Tamura, S., Kato, J., Wada, Y., Watanabe, E. and Aoyama, T. (1966) Agr. Biol. Chem., 30, 917–924.
Murofushi, N., Takahashi, N., Yokota, T. and Tamura, S. (1968) Agr. Biol. Chem., 32, 1239–1245.
Murofushi, N., Takahashi, N., Yokota, T., Kato, J., Shiotani, Y. and Tamura, S. (1969a) Agr. Biol. Chem., 33, 592–597.
Murofushi, N., Takahashi, N., Yokota, T. and Tamura, S. (1969b) Agr. Biol. Chem., 33, 598–609.
Murofushi, N., Yokota, T. and Takahashi, N. (1970) Agr. Biol. Chem., 34, 1436–1438.
Murofushi, N., Yokota, T. and Takahashi, N. (1971) Agr. Biol. Chem., 35, 441–443.
Murofushi, N., Yokota, T., Watanabe, A. and Takahashi, N. (1973) Agr. Biol. Chem., 37, 1101–1113.
Murofushi, N., Durley, R.C. and Pharis, R.P. (1974) Agr. Biol. Chem., 38, 475–476.
Murphy, G.J.P. and Briggs, D.E. (1973a) Phytochemistry, 12, 1299–1308.
Murphy, G.J.P. and Briggs, D.E. (1973b) Phytochemistry, 12, 2597–2605.
Murphy, G.J.P. and Briggs, D.E. (1975) Phytochemistry, 14, 429–433.
Murphy, P.J. and West, C.A. (1969) Arch. Biochem., 133, 395–407.
Musgrave, A. and Kende, H. (1970) Plant Physiol., 45, 56–61.
Musgrave, A., Kays, S.E. and Kende, H. (1969) Planta, 89, 165–177.
Musgrave, A., Kays, S.E. and Kende, H. (1972) Planta, 102, 1–10.
Nadeau, R. and Rappaport, L. (1972) Phytochemistry, 11, 1611–1616.
Nadeau, R. and Rappaport, L. (1974a) Phytochemistry, 13, 1537–1545.
Nadeau, R. and Rappaport, L. (1974b) Plant Physiol., 54, 809–813.
Nadeau, R., Rappaport, L. and Stolp, C.F. (1972) Planta, 107, 315–324.
Nakanishi, K., Endo, M., Näf, U. and Johnson, L.F. (1971) J. Amer. chem. Soc., 93, 5579–5581.
Nandi, D.L. and Porter, J.W. (1964) Arch. Biochem., 105, 7–19.
Naylor, J.M. and Simpson, G.M. (1961) Canad. J. Bot., 39, 281–295.
Neely, P.M. (1959) Ph.D. Thesis, Univ. Calif., Los Angeles.
Nešković, M. and Konjević, R. (1974) J. exp. Bot., 25, 733–739.
Nicholls, P.B.N. and Paleg, L.G. (1963) Nature (Lond.), 199, 823–824.
Nickell, L.G. (1958) Science, 128, 88–89.
Ninnemann, H., Zeevaart, J.A.D., Kende, H. and Lang, A. (1964) Planta, 61, 229–235.
Nitsch, C. and Nitsch, J.P. (1963) Bull. Soc. Bot. Fr., 110, 7–17.
Ogawa, Y. (1962) Bot. Mag. (Tokyo), 75, 449–450.
Ogawa, Y. (1963a) Plant Cell Physiol., 4, 85–94.
Ogawa, Y. (1963b) Plant Cell Physiol., 4, 217–225.
Ogawa, Y. (1963c) Plant Cell Physiol., 4, 227–237.
Ogawa, Y. (1964) Plant Cell Physiol., 5, 11–20.
Ogawa, Y. (1965) Bot. Mag. (Tokyo), 78, 412–416.
Ogawa, Y. (1966) Plant Cell Physiol., 7, 509–517.
Ogawa, Y. (1975) Plant Cell Physiol., 16, 265–270.
Ogura, K., Nishino, T. and Seto, S. (1968) J. Biochem. (Tokyo) 64, 197–203.
Ogura, K., Shinka, T. and Seto, S. (1972) J. Biochem. (Tokyo), 72, 1101–1108.
Okuda, M., Kato, J. and Tamura, S. (1966) Physiol. Plant., 19, 152–158.
Okuda, M., Kato, J. and Tamura, S. (1967) Planta, 72, 289–291.

Sharp, P.B., Keitt, G.W.Jr., Clum, H.H. and Näf, U. (1975) Physiol. Plant., 34, 101–105.

Shechter, I. and West, C.A. (1969) J. biol. Chem., 244, 3200–3209.

Sheldrake, A.R. (1973) Biol. Rev., 48, 509–559.

Shininger, T.L. (1972) Plant Physiol., 49, 341–344.

Shive, J.B., Jr. and Sisler, H.D. (1976) Plant Physiol., 57, 640–644.

Simcox, P.D., Dennis, D.T. and West, C.A. (1975) Biochem. biophys. Res. Commun., 66, 166–172.

Simpson, G.M. (1966) Canad. J. Bot., 44, 115–116.

Sińska, I. and Lewak, S. (1970) Physiol. Veg., 8, 661–667.

Sińska, I., Lewak, S., Gaskin, P. and MacMillan, J. (1973a) Planta, 114, 359–364.

Sińska, I., Grochowska, M.J. and Lewak, S. (1973b) Bull. Acad. Pol. Sci. Ser. Sci. Biol., 21, 291–295.

Sircar, P.K., Dey, B., Sanyal, T., Ganguly, S.N. and Sircar, S.M. (1970) Phytochemistry, 9, 735–736.

Sitton, D., Richmond, A. and Vaadia, Y. (1967) Phytochemistry, 6, 1101–1105.

Skene, K.G.M. (1967) Planta, 74, 250–262.

Skene, K G.M. and Carr, D.J. (1961a) Aust. J. biol. Sci., 14, 13–25.

Skene, K.G.M. and Carr, D.J. (1961b) Phyton, 16, 97–115.

Smoleńska, G. and Lewak, S. (1971) Planta, 99, 144–153.

Snir, I. and Kessler, B. (1975) Planta, 125, 73–75.

Spector, C. and Phinney, B.O. (1966) Science, 153, 1397.

Spector, C. and Phinney, B.O. (1968) Physiol. Plant., 21, 127–136.

Sponsel-Frydman, V.M. and MacMillan, J. (1977) Planta, 135, 129–136.

Srejović, V. and Nešković, M. (1973) Arch. Biol. Nauka (Beograd), 25, 107–117.

Staby, G.L., Hacket, W.P. and De Hertogh, A.A. (1973) Plant Physiol., 52, 416–421.

Stoddart, J.L. (1965) Ann. Bot. (Lond.), 29, 741–749.

Stoddart, J.L. (1966) J. exp. Bot., 17, 96–107.

Stoddart, J.L. (1968) Planta, 81, 106–112.

Stoddart, J.L. (1969) Phytochemistry, 8, 831–837.

Stoddart, J.L. (1972a) in: Hormonal Regulation in Plant Growth and Development, H. Kaldewey and Y. Vardar (eds.), Verlag Chemie, Weinheim, pp. 189–205.

Stoddart, J.L. (1972b) Planta, 107, 81–88.

Stoddart, J.L. and Lang, A. (1968) in: Biochemistry and Physiology of Plant Growth Substances, F. Wightman and G. Setterfield (eds.), Runge Press, Ottawa, pp. 1371–1383.

Stoddart, J., Breidenbach, W., Nadeau, R. and Rappaport, L. (1974) Proc. nat. Acad. Sci. (Wash.), 71, 3255–3259.

Stodola, F.H. (1958) Source Book on Gibberellin, 1828–1957, Agr. Res. Serv. U.S. Dept. Agr., Peoria, Ill.

Stodola, F.H., Raper, K.B., Fennell, D.I., Conway, H.F., Sohns, V.E., Langford, C.T. and Jackson, R.W. (1955) Arch. Biochem., 54, 240–245.

Stodola, F.H., Nelson, G.E.N. and Spence, D.J. (1957) Arch. Biochem., 66, 438–443.

Stolp, C.F., Nadeau, R. and Rappaport, L. (1973) Plant Physiol., 52, 546–548.

Stowe, B.B. and Yamaki, T. (1957) Ann. Rev. Plant Physiol., 8, 181–216.

Stowe, B.B. and Yamaki, T. (1959) Science, 129, 807–816.

Suge, H. and Murakami, Y. (1968) Plant Cell Physiol., 9, 411–414.

Suge, H. and Rappaport, L. (1968) Plant Physiol., 43, 1208–1214.

Sweet, G.B., Zaerr, J.B. and Lavender, D.P. (1974) in: Mechanisms of Regulation of Plant Growth, R.L. Bieleski, A.R. Ferguson and M.M. Cresswell (eds.), Bull. 12, Roy. Soc. New Zealand, Wellington, pp. 709–713.

Takahashi, N. (1974) in: Plant Growth Substances 1973, Hirokawa, Tokyo, pp. 228–240.

Takahashi, N., Kitamura, H., Kawarada, A., Seta, Y., Takai, M., Tamura, S. and Sumiki, Y. (1955) Bull. Agr. Chem. Soc. (Japan), 19, 267–277.

Takahashi, N., Seta, Y., Kitamura, H. and Sumiki, Y. (1957) Bull. Agr. Chem. Soc. (Japan), 21, 396–398.

Takahashi, N., Murofushi, N., Tamura, S., Wasada, N., Hoshino, H., Tsuchiya, T., Aoyama, T. and Morita, H. (1967) Tetrahedron Lett., 895–899.

Takeba, G. and Takimoto, A. (1971) Plant Cell Physiol., 12, 81–88.
Tamura, S. and Sakurai, A. (1965) Agr. Biol. Chem., 29, 595–596.
Tamura, S., Sakurai, A., Kainuma, K. and Takai, M. (1965) Agr. Biol. Chem., 29, 216–221.
Tamura, S., Takahashi, N., Yokota, T., Murofushi, N. and Ogawa, Y. (1968) Planta, 78, 208–212.
Tamura, S., Takahashi, N., Katsumi, M., Phinney, B.O. and Kato, J. (1969) Gibberellins: Biochemistry and Physiology. Univ. Tokyo Press.
Upper, C.D. and West, C.A. (1967) J. biol. Chem., 242, 3285–3292.
Verbiscar, A.J., Cragg, G., Geissman, T.A. and Phinney, B.O. (1967) Phytochemistry, 6, 807–814.
Vince, D. (1967) Planta, 75, 291–308.
Vining, L.C. (1971) J. Chromatogr., 60, 141–143.
Vlitos, A.J. and Meudt, W. (1957) Contrib. Boyce Thompson Inst., 19, 55–62.
Wareing P.F., Good, J. and Manuel, J. (1968) in: Biochemistry and Physiology of Plant Growth Substances, F. Wightman and G. Setterfield (eds.), Runge Press, Ottawa, pp. 1561–1579.
Webster, J.H. and Wilkins, M.B. (1974) Planta, 121, 303–308.
Wenkert, E. (1955) Chem. Ind. (Lond.), 282–284.
West, C.A. (1973) in: Biosynthesis and its Control in Plants, B.V. Milborrow (ed.), Academic Press, New York, pp. 143–169.
West, C.A. and Fall, R.R. (1972) in: Plant Growth Substances 1970, D.J. Carr (ed.), Springer, Heidelberg, pp. 133–142.
West, C A. and Phinney, B.O. (1959) J. Amer. chem. Soc., 81, 2424–2427.
West, C.A. and Upper, C.D. (1969) Methods in Enzymology 15, 481–490.
Wheeler, A.W. (1960) J. exp. Bot., 11, 217–226.
Whyte, P. and Luckwill, L.C. (1966) Nature (Lond.), 210, 1360.
Wilkins, M.B. and Nash, L.J. (1974) Planta, 115, 245–251.
Williams, P.M., Bradbeer, J.W., Gaskin, P. and MacMillan, J. (1974) Planta, 117, 101–108.
Wulfson, N.S., Zaretskii, I.V., Papernaja, I.B., Serebryakov, E.P. and Kucherov, V.F. (1965) Tetrahedron Lett., 4209–4216.
Wylie, A. and Ryugo, K. (1971) Plant Physiol., 48, 91–93.
Yabuta, T. and Hayashi, T. (1939) J. Agric. Chem. Soc. (Japan), 15, 257–266.
Yabuta, T. and Sumiki, Y. (1938) J. Agric. Chem. Soc. (Japan), 14, 1526.
Yafin, Y. and Shechter, I. (1975) Plant Physiol., 56, 671–675.
Yamaguchi, I., Takahashi, N. and Fujita, K. (1975a) J. Chem. Soc. Perkin I, 992–996.
Yamaguchi, I., Miyamoto, M., Yamane, H., Murofushi, N., Takahashi, N. and Fujita, K. (1975b) J. Chem. Soc. Perkin I, 996–999.
Yamaguchi, I., Yokota, T., Murofushi, N., Takahashi, N. and Ogawa, Y. (1975c) Agr. Biol. Chem., 39, 2399–2403.
Yamaguchi, I., Yokota, T., Murofushi, N. and Takahashi, N. (1975d) Agr. Biol. Chem., 39, 2405–2410.
Yamane, H., Yamaguchi, I., Yokota, T., Murofushi, N., Takahashi, N. and Katsumi, M. (1973) Phytochemistry, 12, 255–261.
Yamane, H., Yamaguchi, I., Murofushi, N. and Takahashi, N. (1974) Agr. Biol. Chem., 38, 649–655.
Yamane, H., Murofushi, N. and Takahashi, N. (1975) Phytochemistry, 14, 1195–1200.
Yokota, T., Murofushi, N. and Takahashi, N. (1970) Tetrahedron Lett., 1489–1491.
Yokota, T., Murofushi, N., Takahashi, N. and Katsumi, M. (1971a) Phytochemistry, 10, 2943–2949.
Yokota, T., Murofushi, N., Takahashi, N. and Tamura, S. (1971b) Agr. Biol. Chem., 35, 573–582.
Yokota, T., Murofushi, N., Takahashi, N. and Tamura, S. (1971c) Agr. Biol. Chem., 35, 583–595.
Yokota, T., Yamazaki, S., Takahashi, N. and Iitaka, Y. (1974) Tetrahedron Lett., 2957–2960.
Yokota, T., Hiraga, K., Yamane, H. and Takahashi, N. (1975) Phytochemistry, 14, 1569–1574.
Yomo, H. (1960a) Hakko Kyoaki Shi, 18, 600–602.
Yomo, H. (1960b) Hakko Kyoaki Shi, 18, 603–606.
Yomo, H. and Iinuma, H. (1966) Planta, 71, 113–118.
Zeevaart, J.A.D. (1966) Plant Physiol., 41, 856–862.

Zeevaart, J.A.D. (1968) in: Biochemistry and Physiology of Plant Growth Substances, F. Wightman and G. Setterfield (eds.), Runge Press, Ottawa, pp. 1357–1370.
Zeevaart, J.A.D. (1969a) Planta, 84, 339–347.
Zeevaart, J.A.D. (1969b) Planta, 86, 124–133.
Zeevaart, J.A.D. (1969c) Neth. J. Agric. Sci., 17, 215–220.
Zeevaart, J.A.D. (1971) Plant Physiol., 47, 821–827.
Zeevaart, J.A.D. (1973) Planta, 114, 285–288.
Zweig, G. and DeVay, J.E. (1959) Mycologia, 51, 877–886.
Zweig, G., Yamaguchi, S. and Mason, G.W. (1961) Advanc. Chem. Ser., 28, 122–134.

Phytohormones and Related Compounds – A Comprehensive Treatise, Volume I
Editors: Letham, Goodwin and Higgins

4

Cytokinins

D.S. LETHAM

4.1. Introduction

The work leading to the discovery of cytokinins, and their recognition as plant hormones, began in the laboratories of Folke Skoog at the University of Wisconsin (Madison) about 1946. The discovery of cytokinins was a direct consequence of tissue culture studies by Skoog and Tsui (1948) and Jablonski and Skoog (1954). These workers found that when tobacco stem segments, consisting of cortical, vascular and pith tissues, are cultured on a medium containing auxin, the pith cells proliferate. In contrast, on the same auxin-containing medium, the cells of pith explants freed from all vascular elements were found to enlarge markedly but not to undergo cell division. Cell division did occur, however, when the pith explants had attached vascular strands or were placed in contact with vascular tissue (Jablonski and Skoog 1954). Extracts of the vascular tissue, coconut milk or malt extract were found by these workers to induce cell division in the tobacco pith. A particularly rich source of cell division promoting activity was later found to be autoclaved DNA from which a highly active compound was isolated; it was termed kinetin and was identified as 6-furfurylaminopurine (Miller et al. 1956). Kinetin induced cell division in tobacco pith at concentrations as low as 1 μg/l. The mechanism of its formation by degradation of DNA has been discussed by Skoog and Leonard (1968) and Scopes et al. (1976). Although kinetin did not appear to occur in living tissue, compounds with similar activity were clearly widely distributed in plants.

Initially, kinetin was thought to promote cytokinesis specifically and hence the term kinin was proposed for all compounds with similar activity (Miller et al. 1956). Subsequently this term has been replaced by cytokinin (Letham 1963c; Skoog et al. 1965) which has become generally accepted. It is now known that cytokinins, auxins and gibberellins each evoke a broad spectrum of growth and physiological responses; to some extent these spectra overlap. A cytokinin is best defined as a compound, which, in the presence of optimal auxin, induces cell division in tobacco pith or similar tissue cultures (e.g. soybean callus, carrot secondary phloem); in its other activities a cytokinin also resembles kinetin, the first known cytokinin.

The structure of kinetin was determined in 1955; 9 years elapsed before the structure of the first naturally occurring cytokinin (zeatin) was established. However a number of naturally occurring cytokinins have now been characterized and many synthetic analogues prepared.

Since the identification of kinetin, cytokinins have been of great significance to developmental botanists and the role of these compounds in plant development is discussed in Volume II. Today, however, cytokinins are also of interest to the molecular biologist and the animal biochemist. Cytokinins are known to occur in tRNA adjacent to the anticodon where they appear to influence translational processes. Some cytokinins also evoke responses in mammalian cells. For example they inhibit platelet aggregation (Kikugawa et al. 1973) and growth of tumour cells both in vitro (Fleysher et al. 1968; Fleysher 1972; Rathbone and Hall 1972b) and in vivo (Suk et al. 1970; Grindley et al. 1973). N^6-(3-methylbut-2-enyl)adenosine is an immunosuppressive agent (Hacker and Feldbush 1969) and also induces remissions in leukemia patients (Mittelman et al. 1975). Cytokinins can regulate the growth of phytohemagglutinin-transformed human lymphocytes (Gallo et al. 1969, 1972), possibly by involvement in cyclic AMP metabolism. Some cytokinin ribosides are effective competitive inhibitors of beef heart cyclic AMP phosphodiesterase (Hecht et al. 1974) and glucose 6-phosphate dehydrogenase from rat liver (Tritsch 1973). The basis of the latter inhibition is probably competition with NADP (the coenzyme) for its site on the enzyme. Zeatin and certain other cytokinins, but not adenine or adenosine, cause marked cell elongation in cultured fibroblasts; with zeatin the effect is discernible after 30 min and reaches a maximum after about 6 hr (Johnson et al. 1974).

While auxins and gibberellins appear to be of regulatory significance to plants only, the regulatory role of cytokinins now appears to extend to animal cells as well.

4.2. Identity and occurrence

Free cytokinins, extractable by solvents such as ethanol, have been detected in diverse higher plant genera, and also appear to be produced by bacteria, fungi and algae and secreted by certain insects. Cytokinins also occur as bases in some tRNA species of bacteria and of plant and animal tissues. In this section, the chemical structure of naturally occurring cytokinins is discussed, and their occurrence in microorganisms, higher plant tissues and tRNA is reviewed.

4.2.1. Identity of naturally occurring cytokinins

Naturally occurring cytokinins of high activity which have been isolated in a state of purity and unequivocally identified are listed in Table 4.1 (free cytokinins) and Table 4.2 (tRNA cytokinins), while structural formulae are presented in Figure 4.1. All compounds mentioned are adenine derivatives. Further free cytokinins of very weak activity are 6-(1-carboxy-2-hydroxy-propylamino)-9-ribofuranosylpurine, 6-(1,2-dicarboxyethylamino)-9-β-D-ribofuranosylpurine, and 6-(2,3,4-trihydroxy-3-methylbutylamino)purine (Letham 1973). Discadenine, a spore germination inhibitor isolated from the slime mould *Dictyostelium discoideum,* has been identified as 3-(3-amino-3-carboxypropyl)-6-(3-methylbut-2-enylamino)purine and found to possess cytokinin activity (Abe et al. 1976).

The first natural cytokinin to be isolated in a state of purity was the purine derivative termed zeatin, about 1.0 mg being obtained in crystalline form from 60 kg of sweet corn *(Zea mays)* kernels (Letham 1963b, 1964, 1966a). The earlier report that *N,N'*-diphenylurea, a compound with extremely weak cytokinin activity, occurs in coconut milk (Shantz and Steward 1955) has not been confirmed and its presence in other plants has not been demonstrated. Shantz and Steward (1955) consider it could have been derived

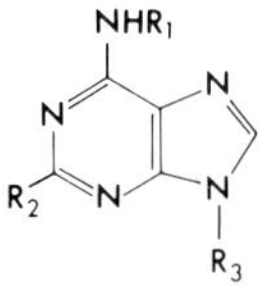

	R_1	R_2	R_3
I	$-CH_2(H)C=C(CH_3)CH_2OH$	H	H
II	$-CH_2(H)C=C(CH_3)CH_2OH$	H	ribofuranosyl
III	$-CH_2(H)C=C(CH_3)CH_2OH$	H	ribofuranosyl 5′-phosphate
IV	$-CH_2-CH=C(CH_3)-CH_3$	H	H
V	$-CH_2-CH=C(CH_3)-CH_3$	H	ribofuranosyl
VI	$-CH_2-CH_2-CH(CH_3)-CH_2OH$	H	H
VII	$-CH_2(H)C=C(CH_3)CH_2OH$	OH	H
VIII	$-CH_2-CH_2-C(OH)(CH_3)-CH_2OH$	H	H
IX	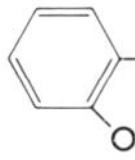 $-CH_2-$ (OH)	H	ribofuranosyl
X	$-CH_2(H)C=C(CH_2OH)CH_3$	H	ribofuranosyl
XI	$-CH_2-CH=C(CH_3)-CH_3$	CH_3S	ribofuranosyl
XII	$-CH_2-CH=C(CH_3)-CH_2OH$	CH_3S	ribofuranosyl

Figure 4.1. Structural formulae for naturally occurring cytokinins.

Table 4.1.

Free naturally occurring cytokinins of high activity and known structure, and the sources from which isolation in a state of purity was achieved. The Roman numerals correspond to those in Fig. 4.1

(I)	6-(4-Hydroxy-3-methylbut-*trans*-2-enylamino)purine, zeatin – Z *Zea mays* kernels (immature seed) (Letham 1966a); *Rhizopogon roseolus* culture filtrates (Miller 1967a); cotton ovules (Shindy and Smith 1975); unidentified ectendotrophic species of mycorrhizal fungus (Crafts and Miller 1974); coconut milk (Van Staden and Drewes 1975); sycamore sap (Purse et al. 1976); tomato xylem sap (Van Staden and Menary 1976). Note: *cis*-Zeatin [*(cis)*Z] and 2-methylthio-*cis*-zeatin, i.e. *(cis)*[2MeS]Z, have been isolated from *Corynebacterium fascians* culture filtrates (Scarbrough et al. 1973; Armstrong et al. 1976).
(II)	6-(4-Hydroxy-3-methylbut-*trans*-2-enylamino)-9-β-D-ribofuranosylpurine, zeatin riboside – [9R]Z *Zea mays* kernels (Letham 1968, 1973); *Rhizopogon roseolus* (Miller 1967a); cotton ovules (Shindy and Smith 1975); unidentified ectendotrophic species of mycorrhizal fungus (Crafts and Miller 1974); chicory root (Bui-Dang-Ha and Nitsch 1970); coconut milk (Letham 1968, 1974a); sycamore sap (Horgan et al. 1973); *Vinca rosea* gall tumour tissue (Miller 1974, 1975a); needles of *Picea sitchensis* (Lorenzi et al. 1975); tomato xylem sap (Van Staden and Menary 1976).
(III)	Zeatin riboside 5′-monophosphate – [9R-5′P]Z *Zea mays* kernels (Letham 1968, 1973).
(IV)	6-(3-Methylbut-2-enylamino)purine – iP *Corynebacterium fascians* (Helgeson and Leonard 1966; Rathbone and Hall 1972); sea water (Pedersen 1973); cotton ovules (Shindy and Smith 1975).
(V)	6-(3-Methylbut-2-enylamino)9-β-D-ribofuranosylpurine – [9R]iP Tobacco tissue, autonomous strain (Dyson and Hall 1972); cotton ovules (Shindy and Smith 1975).
(VI)	6-(4-Hydroxy-3-methylbutylamino)purine, dihydrozeatin – (diH)Z Immature lupin seeds (Koshimizu et al. 1967); cotton ovules (Shindy and Smith 1975); sycamore sap (Purse et al. 1976). The configuration of the asymmetric carbon has been established as *S* (Fujii and Ogawa 1972).
(VII)	2-Hydroxy-6-(4-hydroxy-3-methylbut-*trans*-2-enylamino)purine *Zea mays* kernels (Letham 1973).
(VIII)	6-(3,4,-Dihydroxy-3-methylbutylamino)purine *Zea mays* kernels (Letham 1973).
(IX)	6-(*O*-Hydroxybenzylamino)-9-β-D-ribofuranosylpurine *Populus robusta* leaves (Horgan et al. 1975).

Table 4.2.

Cytokinin ribosides of known structure which occur in tRNA hydrolysates and the source of the tRNA. Roman numerals correspond to those in Fig. 4.1

Compound		Source
II	6-(4-Hydroxy-3-methylbut-*trans*-2-enylamino)-9-β-D-ribofuranosylpurine, zeatin riboside – [9R]Z	tRNA from pea shoots* (Vreman et al. 1972), *Agrobacterium tumefaciens* (Chapman et al. 1976)
V	6-(3-Methylbut-2-enylamino)-9-β-D-ribofuranosylpurine – [9R]iP	tRNA from plants, animal tissues, bacteria and yeast (see references in Gauss et al. 1971 and Hall 1973)
X	6-(4-Hydroxy-3-methylbut-*cis*-2-enylamino)-9-β-D-ribofuranosylpurine, *cis*-zeatin riboside – *(cis)*[9R]Z	tRNA of sweet corn (Hall et al. 1967), wheat germ and tobacco callus tissue (Playtis and Leonard 1971; Dyson and Hall 1972), pea shoots* (Vreman et al. 1972), pea roots (Babcock and Morris 1970)
XI	6-(3-Methylbut-2-enylamino)-2-methylthio-9-β-D-ribofuranosylpurine – [2MeS-9R]iP	tRNA of wheat germ (Burrows et al. 1970), pea shoots (Vreman et al. 1972), bacteria (Gauss et al. 1971; Hall 1973)
XII	6-(4-Hydroxy-3-methylbut-2-enylamino)-2-methylthio-9-β-D-ribofuranosylpurine – [2MeS-9R]Z and *(cis)*[2MeS-9R]Z	tRNA of wheat germ (Burrows et al. 1970), tobacco callus (Burrows et al. 1971), pea shoots** (Vreman et al. 1974) and *Pseudomonas aeruginosa* (Thimmappaya and Cherayil 1974)

* Relative to *cis*-zeatin riboside, the amount of zeatin riboside is small.
** XII from pea tRNA was shown to be a mixture of *cis* and *trans* isomers.

from solvents and reagents used in the extraction of coconut milk. Zeatin was identified as 6-(4-hydroxy-3-methylbut-*trans*-2-enylamino)purine (Letham et al. 1964, 1967) and was shown to be identical (Letham and Miller 1965) to a cytokinin partially purified from *Zea mays* by Miller (1961). 9-β-D-Ribofuranosylzeatin and its 5′-monophosphate were also isolated crystalline from *Zea mays* kernels (Letham 1966b, c, 1968, 1973). Zeatin and [^{3}H] zeatin have been synthesized by unambiguous methods (Letham et al. 1969; Letham and Young 1971a).

Sources from which free zeatin and zeatin riboside were isolated in a state of purity are given in Table 4.1, but chromatographic studies suggest they are widely distributed in the plant kingdom. Thus cytokinins with the chromatographic and ultraviolet spectral characteristics of zeatin have been purified from sources such as plum fruitlets (Letham 1963a, 1964, 1966a), immature sunflower fruits (Miller and Witham 1964), sunflower leaves and root exudate (Klambt 1968), pumpkin seeds (Gupta and Maheshwari 1970), apple fruitlets (Letham and Williams 1969), *Cannabis* fruit (Rybicka and Engelbrecht 1974) and *Suillus punctipes,* a mycorrhizal fungus (Crafts and Miller 1974). By the same criteria, 6-(3-methylbut-2-enylamino)purine was tentatively identified in *Agrobacterium tumefaciens* culture filtrates (Upper et al. 1970). Zeatin and zeatin riboside have been found to cochromatograph with the cytokinin activity of numerous plant extracts. Both cytokinins have been isolated from coconut milk (Table 4.1).

A generally accepted system of abbreviations for the principal naturally-occurring cytokinins is very desirable. Previously proposed abbreviations have tended to result in confusion. Thus 6-(3-methylbut-2-enylamino)purine has been termed IPA (Whitaker and Kende 1974); this abbreviation and the closely related one, 2iPA, have also been applied to the riboside of this purine (Robins et al. 1967; Skoog and Armstrong 1970). A more definitive system of abbreviations is proposed in Tables 4.1 and 4.2 for the principal naturally occurring cytokinins. The cytokinin bases zeatin and 6-(3-methylbut-2-enyl-amino)purine [i.e. N^6-(Δ^2-isopentenyl)adenine] are assigned the abbreviations Z and iP respectively. Substitution in the purine ring of these basic compounds is indicated within brackets, i.e. [ ], while substitution or modification in the side chain is presented within parentheses, i.e. ( ). Thus the ribosides of Z and iP are [9R] Z and [9R] iP, R being adopted as a symbol for ribose. 6-(4-Hydroxy-3-methylbutylamino)-purine is termed (diH)Z; 6-(3-methylbut-2-enylamino)-2-methylthio-9-β-D-ribofuranosylpurine would be termed [2MeS-9R] iP. Using this nomenclature, the *cis* isomer of zeatin, which occurs as a base in tRNA, and the *cis* isomer of zeatin riboside would be designated *(cis)*Z and *(cis)* [9R] Z respectively. Zeatin riboside 5′-monophosphate is denoted by [9R-5′P] Z. In the remainder of this chapter, the abbreviations of Tables 4.1 and 4.2, which are based on this system, will be used.

With one exception, all naturally occurring cytokinins can be regarded as derivatives of iP; the exception is 6-(*o*-hydroxylbenzylamino)-9-β-D-ribofuranosylpurine, a compound isolated from mature leaves of poplar (Horgan et al. 1975). In this tissue it appears to occur together with Z and [9R] Z (Hewett and Wareing 1973c). Some cytokinins which are present in tRNA (Table 4.2) do not occur as free cytokinins in plants (Table 4.1). Thus *(cis)*Z occurs in tRNA, but unlike Z, it has not been found to occur in the free form except in *Corynebacterium fascians* where it may arise by turnover of tRNA (cf. section

4.6.1.2). Cytokinin bases in tRNA may contain a 2-methylthio group; again these compounds, [2MeS] iP and [2MeS] Z, have not been detected in the free form in plants. The evolutionary significance of the occurrence of Z and *(cis)*Z has been discussed (Kaminek 1974).

4.2.2. Occurrence of cytokinins in tRNA

Cytokinins are present in hydrolysates of tRNA from microorganisms, animals and plants, but these compounds do not occur in purified ribosomal RNA (Skoog et al. 1966; Letham and Ralph 1967). When tRNA is fractionated, cytokinin activity is associated with specific tRNA species which constitute a small proportion of the total tRNA. All tRNA species known to contain a cytokinin correspond to codons with the initial letter U; however, cytokinin activity is not found in all tRNA species of this type (Skoog and Armstrong 1970; Skoog 1973). In the case of *E. coli,* cytokinin-containing tRNA species correspond to all six amino acids (phenylalanine, leucine, serine, tyrosine, cysteine, and tryptophan) for which there are codons starting with uridine. The same situation appears to occur in *Lactobacillus acidophilus.* However in yeast, tRNA species for leucine, serine, tyrosine and cysteine have been found to contain cytokinins, but phenylalanine tRNA contains base Y, a cyclic derivative of guanine (see review by Skoog 1973). In all tRNA species of known sequence, the cytokinin is adjacent to the 3′ end of the anticodon. Base Y, *N*-(purin-6-ylcarbamoyl)threonine and other unusual modified nucleosides without cytokinin activity, are also located at this position (see review by Letham and Wettenhall 1977).

[9R] iP probably occurs in the tRNA of all species, but its derivatives exhibit some specificity in their occurrence. [2MeS-9R] iP has been found only in the tRNA of plants and bacteria. The hydroxylated derivative, *(cis)*[9R] Z, appears to occur only in plant tRNA in which [9R] Z is absent or is relatively a very minor nucleoside. Enzymic hydrolysates of plant tRNA typically contain *(cis)*[9R] Z, *(cis)*[2MeS-9R] Z and [9R] iP with lesser amounts of [2MeS-9R] iP. However, in the tRNA of *Agrobacterium tumefaciens,* [9R] Z is present in significant amounts while *(cis)*[9R] Z was not detected (Chapman et al. 1976). This is the only tRNA known in which [9R] Z makes a considerable contribution to the cytokinin activity. [9R] iP always occurs in the sequence -A-[9R]iP-A-suggesting that a triple A sequence is one requirement for specification of the site of [9R] iP formation.

Cytokinin moieties in tRNA are definitely of functional significance. Modification of these bases can markedly affect the ability of tRNA to function in protein synthesis. Chemical modification of the iP base of yeast serine tRNA (Fittler and Hall 1966) and the [2MeS] iP in *E. coli* phenylalanine tRNA (Faulkner and Uziel 1971) does not affect amino acid-acceptor activity. However the modification to the serine tRNA markedly reduces binding to the mRNA-ribosome complex, while the modified phenylalanine tRNA does not function in polypeptide synthesis. The latter modification is particularly interesting because it is reversible and reversion restores ability to function in protein synthesis. From *E. coli,* Gefter and Russell (1969) purified three species of suppressor tyrosine tRNA which differed only in the base adjacent to the 3′ end of the anticodon. At this position, species A contained an unmodified adenine, B possessed iP and C had

[2MeS] iP. These differences did not affect aminoacylation but did influence trinucleotide-dependent binding to ribosomes, the relative affinities being C>B>A. The three species also differed markedly in their ability to promote in vitro protein synthesis; A did not support synthesis, C was very effective, and B was intermediate in functional ability.

Cytokinins and other hypermodified bases adjacent to the anticodons probably prevent misreading of the codons by bases near anticodons and stabilize the single-stranded helical conformation of the anticodon loop (Fuller and Hodgson 1967; Leonard et al. 1969b; Parthasarathy et al. 1974).

It is not known whether the growth-promoting activity of cytokinins is related to their presence in tRNA. This possible relationship might be reflected in an increased cytokinin content of tRNA derived from rapidly growing cells. This has been observed in *E. coli* (Bartz et al. 1970b), but not in chick embryos (Maenpaa et al. 1972) or in pea root segments (Short and Torrey 1972a). For further discussion of the possible connection between the presence of cytokinins in tRNA and cytokinin mechanism of action, the reader is referred to Chapter 11 and to the recent review by Letham and Wettenhall (1977).

4.2.3. Occurrence and level of free cytokinins in higher plants

In this section, the occurrence of cytokinins in organs and exudates of higher plants is discussed briefly. In many studies, an attempt has been made to relate free cytokinin levels to growth, development and senescence in plants. Although such investigations are mentioned briefly in this section, they are usually discussed in greater detail in Volume II. In higher plants, root tips, xylem sap, developing fruits, tumour tissue and germinating seeds are particularly rich sources of cytokinins.

4.2.3.1. Roots, tubers and root nodules. Extracts of apices of sunflower roots are rich in cytokinins but extracts of the older root tissues contain no detectable cytokinin (Weiss and Vaadia 1965). Similarly the terminal 0- to 1-mm segment of pea roots contains about 40 times more free cytokinin than the segment 1–5 mm from the apex (Short and Torrey 1972a). These workers failed to detect cytokinin activity in segments further from the apex. Chromatographic studies suggested that two of the pea root tip cytokinins are Z and [9R] Z. Hence in roots, the highest level of free cytokinins is in the meristematic tip region. In the terminal mm of *Zea mays* roots, about 20% of the cytokinin activity is in the root cap (Feldman 1975).

Changes in cytokinin levels in radish roots during maturation have been assessed. Increased levels of two cytokinins (possibly [9R-5′P] Z and Z) were correlated with initiation of cambial activity. These cytokinins are distributed equally between the xylem and the outer regions of the root but a third cytokinin is localized in the xylem (Radin and Loomis 1971). The cytokinin level in rice roots declines markedly at flowering (Oritani and Yoshida 1971). In lateral roots of sugar maple, the cytokinin level is maximal just before bud burst (Dumbroff and Brown 1976).

Near the end of potato tuber dormancy, the cytokinin level rises in the tuber tissue. The outer regions of the tubers and particularly regions around the "eyes" are sites of cytokinin accumulation (Engelbrecht and Bielinska-Czarnecka 1972). Grating and cutting potato tuber tissue causes an increase in butanol-soluble cytokinins and these are probably associated with the stimulation of cell division caused by wounding (Conrad and Kohn 1975). Budding of sweet potato *(Ipomoea batatas)* tubers is accompanied by a marked rise in the cytokinin level in the tubers (Oritani and Yoshida 1971). For further discussion of the occurrence of cytokinins in roots and tubers, the reader is referred to Chapter 2, Volume II.

Root nodules contain unusually high cytokinin levels. Thus the cytokinin content of *Vicia faba* root nodules induced by *Rhizobium leguminosarum* is over 10 times that in roots (Henson and Wheeler 1976). Chromatographic evidence suggests [9R]Z is the principal cytokinin in these nodules. The cytokinin activity in extracts of *Pisum sativum* nodules appears to be due mainly to Z and [9R]Z (Syono and Torrey 1976). The extent to which the nodule cytokinins are derived from the invading bacteria (cf. section 4.2.5) is not known.

4.2.3.2. Exudates. Since the early work of Kulaeva (1962), Kende (1964, 1965), Loeffler and Van Overbeek (1964) and Nitsch and Nitsch (1965) which established the presence of cytokinins in xylem sap, these hormones have been detected in bleeding sap of diverse genera (see Chapter 2, Volume II, section 2.1.2). After decapitation of sunflower plants, the cytokinin content of the sap was found to remain constant over 4 days (Kende 1965; Kende and Sitton 1967); however, this is not in accord with the observations of Carr and Reid (1968) who observed a steady decline in cytokinin level. Since the findings of Kende and Sitton are based on bioassay of chromatographed extracts, they are probably more reliable than the latter observations. [9R]Z appears to be the principal cytokinin in the xylem sap of a number of woody plant species (Hewett and Wareing 1973a; Horgan et al. 1973; Jones 1973; Hewett and Wareing 1974; Alvim et al. 1976). During early vegetative growth, [9R]Z is probably the major cytokinin in xylem sap of tomato, but at anthesis [9R]Z and Z appear to be present in equal amounts (Davey and Van Staden 1976). Although leaves and buds may contain a diversity of cytokinins (see below), the cytokinin activity of xylem sap is usually due to only one or two compounds, frequently [9R]Z.

The cytokinin level in xylem sap increases markedly prior to, or at, bud burst (Luckwill and Whyte 1968; Hewett and Wareing 1973a; Hewett and Wareing 1974) and following exposure to conditions (short days) causing floral induction (Beever and Woolhouse 1973, 1974). Cytokinin level in sap is also affected by root stock (Skene and Antcliff 1972) and declines markedly prior to senescence (Sitton et al. 1972). Changes in cytokinin levels in xylem sap in relation to shoot development and physiology are discussed in detail in Chapter 4, Volume II (see also section 4.9). The origin of xylem sap cytokinins is also discussed in Chapter 4, Volume II.

The occurrence of cytokinins in phloem sap has now been established. Honeydew produced by aphids feeding on both flowering and vegetative *Xanthium* plants contains a cytokinin which cochromatographs with [9R]Z (Phillips and Cleland 1972). Aphids feed primarily on phloem sap and the honeydew produced is regarded as qualitatively similar to phloem sap. No cytokinin activity was detected in honeydew obtained from aphids feeding on a chemically defined diet. Honeydew from aphids feeding on flowering plants contains a much higher level of cytokinin than honeydew from vegetative plants

(Phillips and Cleland 1972). In the above study, the unlikely possibility that aphids synthesize cytokinin from precursors in the phloem sap was not eliminated. However, cytokinins have subsequently been detected in phloem sap obtained directly from plants. Thus cytokinin and also auxin and gibberellin were detected in phloem exudate of *Ricinus communis* plants (Hall and Baker 1972). A cytokinin, probably [9R-5′P] Z, was detected in the phloem sap of the inflorescence stalk of *Yucca flaccida* (Vonk 1974). During autumn, the major cytokinin in the phloem sap of *Salix babylonica* is probably a glucoside of zeatin (Van Staden 1976a).

As roots form on detached bean hypocotyls and petioles, cytokinins are exuded into the culture solution (Wheeler 1971). Petioles without roots but with a callus also exude cytokinins. Cytokinins are exuded into agar by excised apical buds of *Dahlia variabilis* (Kannangara and Booth 1974); four cytokinins were detected in the buds but only two appeared in the agar diffusate. It is unlikely that diffusion of these cytokinins into agar results simply from the emptying of the contents of the xylem present at bud excision.

4.2.3.3. Buds, leaves and petals. Detailed studies of cytokinin levels during development of buds and leaves are confined to species of *Populus* and *Acer*. Appreciable levels of cytokinin activity were detected in resting maple *(Acer platanoides)* buds by Engelbrecht (1971). Although cytokinins could not be detected in dormant *Populus* × *robusta* buds in midwinter (December), the cytokinin level increased markedly in late February and early March and reached a maximum about the time of natural bud break (Hewett and Wareing 1973a). As the buds expanded, a transient decrease in cytokinin level occurred. Excised shoots of poplar were also induced to bud burst at 20 °C under long days and the cytokinin level in the buds was assessed (Hewett and Wareing 1973a). Maximum cytokinin levels were detected about the time of bud burst. The cytokinin level of sycamore *(Acer pseudoplatanus)* buds is at a maximum just prior to bud burst (Hewett and Wareing 1974). The cytokinin content of sugar maple *(Acer saccharum)* buds is very low during winter but increases markedly when bud water content begins to increase prior to bud break (Taylor and Dumbroff 1975).

In *Populus* × *robusta* leaves, total cytokinin activity and diversity are at a maximum in young expanding leaves. As the leaves mature and senesce, the number and level of the cytokinins decrease (Hewett and Wareing 1973b). At least seven cytokinins are present in leaves of *Populus* × *robusta.* Two are probably Z and [9R] Z, while a third, possibly a cytokinin glucoside, is the principal cytokinin in fully expanded lower leaves and in yellow senescing leaves (Hewett and Wareing 1973b, c). Removal of the apex results in a marked increase in the level of this cytokinin in expanded leaves (Hewett and Wareing 1973b). A fourth cytokinin in *Populus* × *robusta* leaves has been identified as 6-(*o*-hydroxybenzylamino)-9-β-D-ribofuranosylpurine (Horgan et al. 1975). The identity of the three remaining cytokinins is not known.

Changes in cytokinin levels in leaves of *Populus tremula* and *Acer platanoides* have been reported by Engelbrecht (1971a). Both species have two cytokinin fractions; one was termed "fraction Z" and contained two cytokinins with the chromatographic properties of Z and [9R] Z, while the other, designated "fraction N", appeared to be a zeatin nucleotide. In buds of both species, fraction Z is dominant. However, the level of this fraction declines during leaf development, and in mature and senescent leaves, fraction N accounts for nearly all the cytokinin activity. Similarly in *Xanthium strumarium* plants, cytokinins present in young actively growing leaves and in buds are mainly in the form of bases and/or ribosides, while in mature leaves a high proportion of the cytokinin activity is probably due to nucleotides (Van Staden and Wareing 1972a).

As *Picea sitchensis* buds begin to swell, their content of zeatin riboside increases markedly and this riboside is the principal cytokinin in young rapidly growing needles (Lorenzi et al. 1975). The cytokinin level in rice leaves declines markedly as the leaves mature and senesce (Oritani and Yoshida 1971; 1973). Senescence of soybean leaves is also accompanied by a large fall in cytokinin level (Oritani and Yoshida 1973).

While the level of butanol-extractable cytokinins in proximal portions of *Streptocarpus molweniensis* leaves changes little during summer and autumn, the level in distal portions changes markedly (Van Staden 1973c). The increase in early summer may be associated with formation of the abscission layer which bisects the lamina, while the subsequent marked fall appears to be connected with induction of senescence of the distal portions of the leaves (Van Staden 1973c). The cytokinin level in the blade of bean leaf cuttings increases greatly soon after the petioles develop roots (Engelbrecht 1972). The longitudinal distribution of hormones in wheat leaves was studied by Wheeler (1973). Most of the cytokinin activity was in the apical fifth of the mature leaves.

In senescing poplar leaves, a compound which has been tentatively identified as a glucoside is the principal cytokinin (see above). Similarly, in senescing leaves of *Salix babylonica* and *Ginkgo biloba,* most of the cytokinin activity appears to be due to glucosides of zeatin (Van Staden 1976a, c). A glucoside of zeatin or a related compound also appears to be the dominant cytokinin in mature leaves of *Xanthium strumarium,* but it was not detected in buds or young leaves (Henson and Wareing 1976). These workers also compared the level of cytokinin activity in leaves of this species. On a fresh weight basis, the apical bud and first leaf possessed the highest level of cytokinin activity, the lowest level being present in the oldest leaves.

Changes in cytokinin level in rose petals have been assessed. The level increased as the flowers commenced to open and then decreased to a low level (Mayak et al. 1972). The cytokinin content of petals of a short-lived variety was less than that of a long-lived variety (Mayak and Halevy 1970). These and other observations implicate cytokinins in the regulation of petal senescence.

In relation to studies reported in this section, the following generalizations and comments are noteworthy. First, leaves contain a diversity of cytokinins, some of which may be glucosides. These cytokinins remain to be identified and it should be noted that crude preparations of β-glucosidase were used to hydrolyze them. This is the only evidence for a glucosidic moiety, the presence of which has therefore not been established unequivocally. These cytokinins may be identical to the *O*-glucosyl metabolites of zeatin which have been conclusively identified and synthesized chemically (Letham et al. 1977; see also section 4.7.1). Second, as leaves and petals mature and senesce, the cytokinin level declines markedly.

4.2.3.4. Fruits and seeds. Developing fruits have proved to be a rich source of cytokinins. Extracts with high activity have been prepared from many immature fruits (for review see Letham 1967b, and Chapter 3, Volume II).

The cytokinin activities of apple and plum fruitlet extracts have been studied in relation to the cell division periods of these fruits (Letham 1963c). With apple extracts, activity was greatest during the period of intense cell division following fertilization and declined just before the cessation of division. In the case of plum fruitlets, highest activity occurred at the onset of active cell division (Letham 1963c). Extracts of developing seeds of apple fruitlets were much more active than extracts of the fruitlet flesh (Letham and Williams 1969).

In tomato fruits, the level of butanol-extractable cytokinins was found to be greatest during the first two weeks following anthesis, that is, just before and during the early stages of rapid cell division (Abdel-Rahman et al. 1975). Later it declined very rapidly and then remained very low. The level of nucleotide-like cytokinins in tomato fruits was also high during the first two weeks of development. It then declined rapidly but subsequently increased (Abdel-Rahman et al. 1975). The cytokinin content of tomato fruits has also been studied by El-Beltagy et al. (1976) who observed a peak in extractable cytokinin when the fruit were about 20 g in weight. Parthenocarpic tomato fruits showed retarded development and a very low cytokinin content. With seeded fruits, reduction of foliage increased the cytokinin level very markedly (Varga and Bruinsma 1974).

Cytokinin levels in avocado seed have been studied during fruit development (Blumenfeld and Gazit 1970b). The cytokinin content of the endosperm was very high throughout the period this tissue was present and exceeded that of the seed coat and embryo. The level in these tissues was greatest two to three weeks after fruit set and then decreased markedly. Cytokinin activity was detected in methanol extracts of avocado fruit mesocarp only after acid hydrolysis (Gazit and Blumenfeld 1970). These authors suggested the extracts contained a bound cytokinin which was liberated by acid. The level of this activity was high in very immature fruit but declined markedly as the fruit developed.

The peak of cytokinin activity in cotton fruits occurs from the fourth to the ninth day after anthesis; however 15 days after anthesis, cytokinin activity is not detectable (Sandstedt 1971). The cytokinin level in pumpkin *(Cucurbita pepo)* seeds reaches its maximum about 11 days after pollination and then decreases rapidly as the seeds develop further (Gupta and Maheshwari 1970). The cytokinin content of developing pea seeds exhibits a pronounced maximum at 24 days after anthesis which coincides with maxima in growth rates of the whole seed and the embryo (Burrows and Carr 1970). In contrast with the above observations, the maximum cytokinin activity in ovules of *Ginkgo biloba* occurs prior to fertilization when the gametophytic cells are dividing actively (Banerjee 1968). The activity in the gametophyte tissue greatly exceeds that in the nucellus-integument tissues.

Several studies of cytokinin activity levels in developing seeds of monocotyledons have been reported. Cytokinin activity in sweet corn *(Zea mays)* kernels was found to be maximal about 11 days after pollination (Miller 1967b). Levels of activity were much higher in rice ears at the heading and milk stages than at the fully ripe stage (Oritani and Yoshida 1971). The cytokinin activity in barley caryopses decreased very rapidly after pollination (Michael and Seiler-Kelbitsch 1972). The activity level of wheat grains was maximal immediately after anthesis (Wheeler 1972).

A marked increase in the level of cytokinin activity in seed during germination, or in response to conditions causing germination, has been reported frequently (Barzilai and Mayer 1964; Van Staden and Wareing 1972b; Van Staden et al. 1972; Van Staden and Brown 1973; Brown and Van Staden 1973; Van Staden et al. 1973). These effects are referred to again in section 4.9.

4.2.3.5. Cytokinin-autotrophic tissue cultures and tumour tissue. Plant tissues which grow in vitro in the absence of exogenously supplied cytokinins appear to synthesize adequate amounts of endogenous cytokinins. A strain of soybean cotyledon callus cells capable of growth on a medium lacking cytokinin was shown to produce three cytokinins with the chromatographic characteristics of Z, [9R]Z and [9R-5′P]Z (Miura and Miller 1969a). This independent strain was detected only after 6 years of subculturing. Hence the independent cells probably were not present when the original cultures were prepared, but developed subsequently. Soybean root cells do not require exogenous cytokinin for growth in vitro and release cytokinins into the culture medium (Sargent and King 1974).

Several studies have been made using cytokinin-autotrophic tobacco tissue and evidence obtained for the production of cytokinins by such tissue (Dyson and Hall 1972; Tandeau de Marsac and Jouanneau 1972; Einset and Skoog 1973). The last mentioned investigation is of particular significance. A cytokinin-autotrophic strain of tobacco callus tissue was found to contain cytokinins with the chromatographic properties of Z, [9R]Z and iP. When the autotrophic tissue was supplied with [^{14}C]adenine, it produced labelled compounds with the chromatographic characteristics of Z and iP. However these products were not produced by the cytokinin-dependent tissue (Einset and Skoog 1973).

Cytokinin activity was detected in extracts of pea-root callus cultured in the absence of exogenous cytokinins (Short and Torrey 1972b). In the early period of culture, a peak in activity was associated with a high frequency in mitoses. However as growth proceeded, both cytokinin level and mitotic index decreased markedly. Avocado cotyledonous callus (Blumenfeld and Gazit 1971) and sycamore cells (Mackenzie and Street 1972) can also be cultured in the absence of exogenous cytokinin and both yielded cytokinin-containing extracts. A rapid rise in cytokinin activity of sycamore cells was found during the lag period which precedes active cell division (Mackenzie and Street 1972).

The occurrence and identity of cell division factors in plant tumour tissue has attracted considerable attention. Crown-gall tumour cells, caused by the infection of plants by *Agrobacterium tumefaciens,* can be obtained free from bacteria and can be cultured on synthetic media (White and Braun 1942; Braun and White 1943). Unlike normal cells, fully transformed tumour cells grow actively without exogenous auxin and cytokinin. The ability of extracts of crown-gall tumour cells to induce cell division (Braun and Naf 1954; Steward et al. 1955) suggested that tumour tissues produce adequate amounts of plant hormones including growth-promoting compounds other than auxins. Extracts of crown-gall tissue from *Parthenocissus tricuspidata* exhibited good activity in the soybean callus cytokinin bioassay and chromatographic evidence and other data suggested the major cytokinin was [9R]Z (Tegley et al. 1971). In contrast, Wood et al. (1969) considered the cytokinin activity in extracts of crown-gall tumour cells of *Vinca rosea* was not due to adenine derivatives. The active compounds were termed cytokinesins and one was assigned a 3,7-dialkyl-2-alkylthiopurin-6-one structure (Wood 1970). Glucose was considered to be present as a sugar moiety.

Unfortunately Wood and coworkers have failed to establish adequately the purity of their products and hence the structural features assigned by these workers may not apply to the active compounds. In view of this, it is not surprising that a meticulous reexamination of the cytokinins in *Vinca rosea* tumour tissue by Miller (1974, 1975a) yielded results at variance with those of Wood. Miller isolated in crystalline form a compound which is the major cytokinin and identified it unequivocally as [9R] Z. "Cytokinesin I" probably contains small amounts of [9R] Z (Miller 1975b).

False broomrape is a widely distributed root disorder of tobacco. Symptoms are white, succulent, tumorous outgrowths of the root. These outgrowths may develop into leafy shoots. Although investigators have suggested viruses and bacteria as possible causes, it now seems very probable that the disorder is a physiological one caused by hormonal imbalance involving abnormally high cytokinin-auxin ratios (Hamilton et al. 1972). These workers report exceptionally high cytokinin contents in false broomrape tissue.

4.2.4. Occurrence in algae

The occurrence of cytokinins in marine algae has been the subject of several studies. These hormones have been detected in extracts of the unicellular marine algae *Gymnodinium splendens* and *Phoeodactylum tricornutum* (Bentley-Mowat and Reid 1968), and *Cricosphaera elongata* and *C. carterae* (Hussain and Boney 1971). In the *Cricosphaera* spp., the level of cytokinins in cells in the motile phase exceeds that in the non-motile phase. Cytokinins have also been detected in the stipe and holdfast regions of the brown alga *Laminaria digitata* (Hussain and Boney 1969). Jennings (1969) detected cytokinins in the marine brown alga *Ecklonia radiata* and the red alga *Hypnea musciformis.* The cytokinin activity of commercial seaweed extract has also been studied (Brain et al. 1973).

Sea water surrounding the littoral algae in the *Fucus-Ascophyllum* zone contains cytokinins, the level varying greatly depending on the season (Pedersen and Fridborg 1972). One cytokinin in sea water has been identified as iP (Pedersen 1973). This cytokinin, kinetin or sea water from the *Fucus-Ascophyllum* zone promotes the growth of certain marine algae in axenic culture (Pedersen 1973). Part of the highly stimulating effect of sea water on algal growth appears to depend on its content of cytokinins.

Cytokinin activity has also been detected in extracts from certain freshwater algae including *Volvox carteri* (Van Staden and Breen 1973). The activity was not detectable in filtrates from vegetatively reproducing *Volvox* cultures, but was present in the filtrate from sexual male cultures. However the cytokinins detected did not appear to be involved in the induction of sexuality and Van Staden and Breen (1973) suggest they regulate the growth of the spheroids.

4.2.5. Production by bacteria and fungi; parasitic and symbiotic relationships with plants

Several studies of free cytokinins in bacteria have been reported. *Corynebacterium fascians* is a plant pathogen which causes a "fasciation" disease in certain dicotyledonous plants, e.g. the garden pea and chrysanthemum. Infected plants show swollen internodes and release of lateral buds from apical dominance. These symptoms are also produced by application of cytokinins to the stems of pea plants and this led to the suggestion that the disease is caused by cytokinins synthesized by the infecting bacteria (Thimann and Sachs 1966). These workers found cytokinin activity in cultures of *C. fascians.* Klambt et al. (1966) and Helgeson and Leonard (1966) identified iP in such cultures, but a more critical study by Rathbone and Hall (1972a) indicated that much of this cytokinin was derived by hydrolysis of tRNA during extraction. The results of these latter workers indicated that iP is a minor free cytokinin in *C. fascians* cultures and that much of the cytokinin activity is due to Z-like compounds. Subsequently, Scarbrough et al. (1973) identified *(cis)*Z as a free cytokinin in cultures of *C. fascians* which also contain *(cis)* [2MeS] Z (Armstrong et al. 1976). *Agrobacterium tumefaciens,* the causal organism of crown gall, also appears to produce iP (Upper et al. 1970). It is relevant to note that the cytokinin content of geranium stems infected with *C. fascians* is high relative to normal stems (Balazs and Sziraki 1974).

Root nodule formation, which is necessary for symbiotic nitrogen reduction in legumes, begins with infection of the root cortex by bacteria of the genus *Rhizobium.* These induce polyploid mitoses in the cortical cells and the nodule develops through continued mitotic activity of these polyploid cells. It appears that a cell division factor is necessary for nodule formation and is produced by the *Rhizobium* (Phillips and Torrey 1970). These workers showed that a cytokinin is produced by *Rhizobium japonicum,* the species which nodulates soybean roots. This cytokinin induces polyploid mitoses in cultured pea root segments. The cytokinin level in *R. japonicum* culture media increases markedly during the logarithmic phase of growth and then decreases (Phillips and Torrey 1972). These workers consider the amounts of cytokinin produced by *R. japonicum* would be sufficient to initiate the cortical cell divisions necessary for root nodule development. The cytokinin released by *R. japonicum* exhibits chromatographic properties similar to Z (Phillips and Torrey 1972). Other bacteria which have been reported to produce cytokinins include *Azotobacter pasphali* (Barea and Brown 1974), *Azotobacter chroococcum* (Coppola et al. 1971) and *Arthrobacter* spp. (Blondeau 1970).

The cytokinin activity of *Psychotria punctata* leaf discs with bacterial nodules appears to be markedly greater than that of discs lacking nodules (Edwards and LaMotte 1975). Some species in the genus *Ardisia* (e.g. *A. crenata* and *A. crispa*) have bacterial symbionts which live intercellularly in nodules on leaf margins, on shoot meristems and in the gynoeceum. These host plants have lost the ability to develop normally in the absence of the bacteria. Seedlings which are free of bacteria do not produce more than four leaves and eventually die. The symbiotic bacteria, which include *Chromobacterium lividum,* have been shown to produce cytokinin (Pereira et al. 1972). The bacterial symbiosis in *Ardisia* species is probably based on the ability of the bacteria to provide the host plant with the cytokinins it is unable to produce.

In mycorrhizae, an obvious growth modification is the enlargement of root cortical cells and this effect can be produced by treatment with cytokinin (Arora et al. 1959). These facts caused Miller (1967a) to examine the possibility that mycorrhizal fungi produce and release cytokinins. Both Z and [9R] Z were isolated in crystalline form from *Rhizopogon roseolus* cultures (Miller 1967a, 1968) and these cytokinins are also produced by *Suillus punctipes* and by an unidentified ectendotrophic species (Crafts and Miller 1974). However these workers found that none of 22 species of *Hebeloma* produce detectable amounts of cytokinins; cytokinin production by certain species of *Rhizopogon* was also not detectable. *N*-(Purin-6-ylcarbamoyl)threonine lacks cytokinin activity, but it is of interest that this purine, which occurs in tRNA, is also secreted into the culture medium by *Rhizopogon roseolus* (Lalone and Hall 1973). Cytokinins are extractable from soils which support plants growing in symbiosis with mycorrhizal fungi and *Rhizobium* species (Van Staden and Dimalla 1976).

Cytokinin production by mycorrhizal fungi could aid both the plant host and the fungus in a number of ways. These have been discussed by Miller (1971) and by Crafts and Miller (1974). One very probable advantage to the fungus would be the mobilization of nutrients to the region of the association. The promotion of the synthesis of phenolic compounds by cytokinins (see Miller 1969) might favour the plant host by protecting it against further fungal infection.

Abnormal growth of leaves or stems is a noticeable feature of plants infected by fungi of the genus *Taphrina. Taphrina deformans* causes the common leaf curl of peach, the infected leaves becoming thickened and distorted. Infection of lateral buds of cherry by *Taphrina cerasi* causes formation of a dense cluster of shoots termed a "witches' broom". Such abnormal growth of the host plant has been related to excessive synthesis of auxin by the infecting fungus or by the host tissue in response to infection. The presence of IAA in liquid cultures of *Taphrina deformans* and *T. cerasi* has been established (Crady and Wolf 1959; Hirata 1957; Somner 1961b; Perley and Stowe 1966). However, some observations suggest cytokinins may also be involved in induction of the abnormal growth associated with infection by *Taphrina* spp. Cytokinin activity in peach leaves infected with *T. deformans* is markedly greater than that of uninfected leaves (Taris and Avenard 1969; Sziraki et al. 1975). Somner (1961b) obtained the first evidence for the presence of cytokinins in cultures of *T. deformans.* Subsequent work using the specific soybean callus assay has established this beyond doubt (Johnston and Trione 1974). Chromatographic studies by these workers suggested that *T. cerasi* and *T. deformans* both produce Z, iP and their ribosides. The presence of Z in culture filtrates of *T. cerasi* has been confirmed by gas-liquid chromatography (Barthe and Bulard 1974).

Cytokinin production by powdery mildew *(Erysiphe graminis)* is probably responsible for the formation of "green islands" and for inhibition of root elongation and lateral root development in infected barley plants (Vizarova 1974; Vizarova and Minarcic 1974). These workers observed increased cytokinin levels in both barley leaves and roots following infection with mildew although the fungus does not invade the root system.

Infection of leaves with rust fungi (e.g. *Uromyces phaseoli* and *U. fabae*) markedly affects leaf growth and physiology in a variety of ways. Nutrients are mobilized into the area of infection which becomes a metabolic sink, while leaf senescence is retarded around the infection site with formation of a "green island". Growth of infected leaves is promoted, while development of young uninfected leaves is retarded (Pozsar and Kiraly 1966; Hammady et al. 1968; Shaw 1961, 1963). It is evident that rust infections and exogenously supplied cytokinin have very similar effects on leaf physiology and development. The level of cytokinin activity in rust-infected leaves is considerably greater than that in non-infected leaves (Kiraly et al. 1966, 1967; Dekhuijzen and Staples 1968). Cytokinins are extractable from uredospores and from rust mycelia (Dekhuijzen and Staples 1968). The above results suggest strongly that certain of the effects of rust infections are a consequence of cytokinin production by the rust fungi.

The abnormal tissue induced by *Plasmodiophora brassicae,* the cabbage clubroot pathogen (an obligate parasite), contains a much higher level of ether-extractable cytokinin activity than healthy roots (Katsura et al. 1966). Explants of tissue infected with *P. brassicae,* unlike those from uninfected roots, do not require exogenously supplied cytokinin for growth in vitro (Ingram 1969; Reddy and Williams 1970). Extracts of infected tissue cultures grown on cytokinin-free media contain significant cytokinin activity (Reddy and Williams 1970). Turnip clubroots contain a cytokinin which exhibits the chromatographic properties of Z (Dekhuijzen and Overeem 1971). The above observations suggest that cytokinins are produced by *P. brassicae* and that these hormones may play an important role in clubroot development.

A number of other species of phytopathogenic fungi also produce cytokinins. Noteworthy examples are *Monilia* spp. which have been shown to release cytokinins into their culture media; cytokinin production is particularly marked during formation of the fructification organs (Vizarova 1975).

4.2.6. Production by parasitic caterpillars

Cytokinin production by plant parasites is not limited to fungi and bacteria. Small caterpillars which infest leaves of deciduous trees also appear to produce cytokinins. In autumn, senescing leaves often have green areas surrounding small mining caterpillars. These "green islands" are rich in cytokinins although the rest of the leaf, and even green leaves on the same tree, may contain extremely low levels of cytokinin (Engelbrecht 1968; 1971b). Experiments with the caterpillars *Stigmella argyropeza* and *Stigmella argentipedella* indicate that the labial glands produce cytokinins which are probably responsible for formation of the green islands (Engelbrecht et al. 1969; Engelbrecht 1971b). The major cytokinin fractions in infected birch leaves exhibit the chromatographic properties of Z and [9R-5′P] Z (Engelbrecht 1971b). An endogenous cytokinin which co-chromatographed with zeatin was also obtained from larvae which infest leaves of the Japanese chestnut (Ohkawa 1974).

In the above studies, the caterpillars involved have mouthparts and there is the unlikely possibility that the cytokinin obtained from them was derived from their food. However the larvae which cause galls in *Erythrina latissima* leaves have no visible mouth parts and a blind intestine and are enclosed in a larval cavity. From these wasp larvae, cytokinins were extracted by Van Staden (1975); chromatography on Sephadex LH-20 indicated the presence of Z, [9R] Z and another cytokinin. Cytokinin derived from the larvae is probably involved in gall formation.

4.3. Isolation and identification

For detailed information regarding commonly used procedures, the reader is referred to the references in Table 4.1. In the present section only general trends and comments are presented briefly.

For extraction of cytokinins from plant tissues, the solvents usually employed are aqueous methanol or ethanol (final alcohol concentration about 75%). However extraction with water after attempted enzyme inactivation (Dekhuijzen and Gevers 1975) and extraction with ethyl acetate in the presence of suspended potassium phosphate (Miller 1975a) have recently been proposed. The latter technique appears to be a particularly convenient procedure. It should be emphasized that extraction of plant tissues with cold organic solvents does not inactivate phosphatase and other enzymes rapidly; even dropping tissue into alcohol at 65 °C may allow some enzyme action which can be considerable in the case of phosphatase. If the investigator is concerned with nucleotide cytokinins, phosphatase must be inactivated before extraction. This can be achieved by the procedure of Bieleski (1964, 1968) using methanol-chloroform-formic acid-water (12:5:1:2, by vol.) at –25 °C, Examples of the use of this type of procedure for cytokinin extraction are found in the following references: Letham (1966a), Parker and Letham (1973) and Gordon et al. (1974a).

At present there is no reliable information regarding the effectiveness of the above methods for extracting endogenous cytokinins from plant tissue. Considerable loss could occur by adsorption onto cell debris.

For purification of extracted cytokinins, solvent partition methods are useful and partition coefficients for naturally occurring cytokinins have been reported (Letham 1974b; see also Hemberg 1974). These should serve as a guide to the use of partition methods. However it should be remembered that the partition coefficients reported were determined with pure cytokinins in buffer solutions; the diverse compounds in crude plant extracts could modify the partition behaviour of cytokinins. Although [9R] Z is not extracted readily by ethyl acetate from dilute aqueous buffers (see partition data given by Letham (1974b) and Hemberg (1974)), a high concentration of potassium phosphate in the aqueous solution markedly enhances its extractability (Miller 1975a). Miller (1975a) used this procedure to isolate [9R] Z from tumour tissue.

Both paper and thin-layer chromatography (TLC) have been extensively used to purify naturally occurring cytokinins. All known naturally occurring cytokinins and their metabolites can be separated by TLC methods although use of both cellulose and silica gel layers and at least two solvents may be necessary to resolve some complex mixtures (Letham et al. 1977). It is noteworthy that even such closely related compounds as *O*-glucosylzeatin riboside and *O*-glucosyldihydrozeatin riboside can be separated by TLC (Letham et al. 1977). Usually silica gel yields spots which are more compact than those obtained with cellulose layers. Mixed layers of polyvinylpyrrolidone and calcium sulphate have also been used to separate cytokinins (Thomas et al. 1975a). When only small amounts of cytokinin (e.g. 30 μg) are available, chromatography on exhaustively washed paper is a very desirable final purification step prior to mass spectrometry (Letham 1973; Parker and Letham 1973, 1974).

Column chromatography has been of great value in isolating cytokinins. Polystyrene sulphonic acid resins (e.g. Dowex 50, Zeo-Karb 225) have been extensively used for initial purifications with crude extracts. Although such cation-exchange resins are usually used in the acid form which is strongly acidic, this is neither necessary nor desirable. Use of the resins in the ammonium ion form equilibrated to pH 3.0–3.4 is to be preferred. Cytokinin bases and ribosides are retained by the resins in this form without danger of hydrolysis and can be subsequently eluted with dilute aqueous ammonia (Letham 1973, 1974a). For small scale studies, cellulose phosphate (NH_4^+ form, equilibrated to pH 3.0–3.4) is preferable to polystyrene sulphonic acid resins (Parker and Letham 1973). Low recovery of cytokinins from the sulphonic acid resins, but not from cellulose phosphate columns, has been reported (Dekhuijzen and Gevers 1975; Vreman and Corse 1975).

Although cation-exchangers have proved very useful for preparing a cytokinin-rich fraction, they do not fractionate cytokinins. For column fractionation of cytokinins, the most widely used packing is Sephadex LH-20 (Armstrong et al. 1969; Hewett and Wareing 1973c). Other materials for column chromatography which may be of value are polyvinylpyrrolidone (Thomas et al. 1975b; Miller 1975a), neutral polystyrene resin (Armstrong and Skoog 1975) and silica gel (Hahn 1975). High-pressure liquid chromatography of cytokinins is without doubt a technique which will be of great value. Excellent rapid separations with mixtures of pure cytokinins have been reported (Pool and Powell 1974; Carnes et al. 1974; Challice 1975; Carnes et al. 1975; Thomas et al. 1975c). If coupled with field desorption mass spectrometry, as proposed by Cole et al. (1974), high-pressure liquid chromatography would be an elegant technique for cytokinin identification.

In all cases where structural formulae of naturally occurring cytokinins have been elucidated, mass spectrometry has been a key technique. During the work which led to elucidation of the structure of zeatin, the mass spectra of zeatin and other 6-aminopurines were determined and rationalized in terms of structural features (Shannon and Letham 1966). The results suggested that mass spectrometry would be a valuable technique in structural studies of cytokinins. Ions at m/e 148, 136, 135, 120, 119 and 108 are characteristic of 6-(substituted amino)purines. An excellent compilation of mass spectra of nucleosides, including cytokinin ribosides, has been prepared (Hecht et al. 1970c). The advances in instrumentation in mass spectrometry over the years 1963–1967 have greatly enhanced the value of the technique to cytokinin workers who may have very limited amounts of compound. In 1963, 0.3 mg of crystalline zeatin was required for one spectrum in an Atlas CH4 instrument; today 1 μg would be adequate in an AEI MS902.

The use of gas chromatography coupled to mass spectrometry (GC-MS) has proved to be of some value for cytokinin identification in fractions of crude plant extracts. Probably the greatest success is that reported by Shindy and Smith (1975) who identified (diH)Z, Z, iP, [9R]iP and [9R]Z in crude fractions prepared from cotton ovule extracts. Computer controlled GC-MS was employed in this study. In another successful application of the technique, [9R]Z was identified in a partially purified extract of bleeding sap from sycamore (Horgan et al. 1973). Retention times for TMSi derivatives of cytokinins during GLC have been recorded by several workers, but notably by Upper et al. (1970). The mass spectra of TMSi derivatives of cytokinin ribosides and glucosides have been recorded and rationalized in terms of structure (MacLeod et al. 1976). Because of the high temperatures involved, GLC of cytokinins should be employed with caution, especially when purifying new compounds. It is relevant to note that 6-benzylamino-3-β-D-glucopyranosylpurine undergoes rearrangement to the 9-glucoside during gas chromatography (Letham et al. 1975).

4.4. Bioassays

A diversity of cytokinin bioassays have been proposed. The sensitivities of 16 bioassays together with assay times and key references are listed in Table 4.3. These bioasssays fall into 6 groups: (a) to (d) are based on the ability of cytokinins to promote expansion of excised leaf or cotyledon tissue; (e) and (f) depend on promotion of growth of duckweed species; (g) and (h) utilize stem or coleoptile sections; (i) to (1) depend on the ability of cytokinins to induce cell division in tissue cultures; (m) and (n) are based on retardation of leaf senescence; the two remaining assays, (o) and (p), depend on pigment formation (chlorophyll and betacyanin respectively).

The tissue culture assays (i to l) possess the desirable features of high sensitivity, sterile conditions and specificity. The soybean callus assay (k) is probably the most popular tissue culture assay. An attractive aspect of the assay is the wide concentration range over which a linear relationship exists between response and concentration. For (k) this is 4–10,000 μg/l in the case of kinetin; in contrast the corresponding ranges for bioassays (i) and (l) are approximately 1–20 μg/l and 1–100 μg/l respectively.

An unsatisfactory feature of the tissue culture bioassays is the long assay time (21–35 days). Hence much effort has been devoted to the development of rapid bioassays. Eleven such assays (time required 0.8–4 days) are listed in Table 4.3. Unfortunately the rapid assays often lack specificity and are usually non-sterile. Thus gibberellins evoke responses, usually very weak, in assays (a), (b), (c), (d), (h) and (n), while high concentrations of inorganic salts are known to be active in (c), (m) and (n). Assays (a), (b), (g), (h), (m), (n) and (o) are non-sterile.

The *Amaranthus* betacyanin bioassay (p) is probably the rapid assay of choice. It can be made sterile and exhibits high specificity and sensitivity. However the radish cotyledon assay (c) has much to commend it. It is simpler to perform than the *Amaranthus* assay, can easily be made sterile and requires only a few millilitres of culture solution. Rozej and Chrominski (1976) consider it is preferable to (o) and (p). Furthermore the radish seedlings from which the cotyledons are derived develop in only 1–2 days. In contrast, the seedlings for the leaf senescence assays (m and n) require at least 6 days to develop. The assay (m) shows considerable specificity, inorganic salts at high concentrations being the only substances known to interfere. The oat-leaf assay (n) has one particularly useful feature. The volume of solution required (5 μl per leaf; 25 μl for sufficient replication) is much less than that required in all other bioassays; only 10^{-6} μmole of kinetin is required for detection. Unfortunately the bioassays (m) and (n) and other leaf senescence assays cannot be readily made sterile.

Bioassay (o), based on the ability of cytokinins to promote chlorophyll formation in cucumber cotyledons, requires a short assay time and exhibits excellent sensitivity and

Table 4.3.

Cytokinin bioassays

Bioassay	Key references	Normal assay time (days)	Lowest detectable conc. of kinetin ($\mu g/l$)
(a) Radish leaf disc expansion	Kuraishi (1959)	0.8	2–10
(b) Etiolated bean leaf disc expansion	Miller (1956, 1963)	2	200
(c) Radish cotyledon expansion	Letham (1968, 1971); Radin and Loomis (1971); Gordon and Letham (1975)	3	10
(d) *Xanthium* cotyledon expansion	Esashi and Leopold (1969)	4	20
(e) *Lemna minor* (growth in darkness)	Hillman (1957)	2	60
(f) *Spirodela*	Letham (1967a); Letham and Young (1971); McCombs and Ralph (1972)	14	10
(g) Etiolated pea stem section	Sommer (1961a)	1	10
(h) Wheat coleoptile	Rothwell and Wright (1967)	1	20–100
(i) Tobacco stem pith callus*	Murashige and Skoog (1962); Linsmaier and Skoog (1965); Skoog et al. (1967)	35	< 1
(j) Tobacco stem pith	Bottomley et al. (1963)	21	40
(k) Soybean callus	Miller (1963)	21	1
(l) Carrot phloem	Letham (1967a)	21	0.5
(m) Barley leaf senescence	Kende (1964, 1965); Carr and Burrows (1966)	2	3
(n) Oat leaf senescence	Thimann and Sachs (1966); Varga and Bruinsma (1973)	4	3–10
(o) Cucumber cotyledon chlorophyll	Fletcher and McCullagh (1971)	0.8	1
(p) *Amaranthus* betacyanin	Kohler and Conrad (1968); Bigot (1968); Biddington and Thomas (1973); Conrad (1974)	1–3	2–20

* Henceforth termed the "tobacco callus" assay.

specificity. It is surprising it has been used so little. Rapid cytokinin bioassays have a useful function provided their limitations as well as their advantages are appreciated.

The importance of avoiding autoclaving of labile compounds during preparation of media for cytokinin bioassay cannot be over-emphasized. Even the activities of some apparently stable compounds, e.g. *N,N'*-diphenylurea and 6-benzylamino-9-benzylpurine, change as a result of autoclaving (Skoog et al. 1967; Schmitz and Skoog 1970). Studies such as those of Shaw et al. (1971) which assess the cytokinin activity of purines by testing at only one concentration following autoclaving are of little value. To avoid autoclaving, solutions of the test compounds can be filter-sterilized; however this is cumbersome when dealing with a number of compounds over a range of concentrations. A very successful and convenient alternative is the use of dimethylsulphoxide solutions of the compounds for bioassay. Dimethylsulphoxide acts as a sterilizing agent and is an excellent solvent for purines. A concentration of 0.2% in culture media does not affect growth in the tobacco callus assay (Schmitz and Skoog 1970).

Several cytokinin bioassays not listed in Table 4.3 are worthy of mention. These include leaf senescence bioassays using wheat (Rothwell and Wright 1967), radish (Kefford et al. 1968), *Xanthium* (Osborne and McCalla 1961) and Chinese cabbage leaves (Letham 1967a). Other bioassays are based on the ability of cytokinins to promote lettuce seed germination (Miller 1958; Skinner et al. 1957), bud formation in the moss *Funaria hygrometrica* (Hahn and Bopp 1968; Szweykowska et al. 1969; Whitaker and Kende 1974), transpiration of excised oat leaves (Luke and Freeman 1967, 1968) and mobilization of amino acids in excised oat leaves (Dekhuijzen and Staples 1968). A recently proposed rapid assay of interest is based on the ability of cytokinins to promote uptake of either L-tryptophan or IAA by germinated spores of the water mould *Achlya* (LeJohn 1975).

Assessment of cytokinin levels in crude plant extracts by bioassay is obviously only an approximation. It is relevant to note, for example, that ABA at low concentrations can actually enhance cytokinin activity in the soybean callus assay; at high concentrations, ABA is of course inhibitory (Blumenfeld and Gazit 1970a). Glutathione is also known to inhibit growth in the soybean assay (Bergmann and Rennenberg 1975). Clearly, purely chemical methods for cytokinin estimation in plant extracts are desirable. A method for determining cytokinin ribosides involving periodate oxidation and [^{3}H]borohydride reduction has been proposed (Chen et al. 1974a). Single ion current monitoring has been used to determine 6-(*o*-hydroxybenzylamino)purine riboside in poplar leaves (Thompson et al. 1975). Further development of such methods is obviously worthwhile.

4.5. Structure–activity relationships

The identification of kinetin as 6-furfurylaminopurine in 1955 resulted in the synthesis of numerous related compounds which were tested for cytokinin activity. Later the elucidation of the structure of the naturally occurring cytokinin zeatin (Z) and the isolation of its riboside and ribotide stimulated interest in the cytokinin activity of 6-hydroxyalkenylaminopurines and their ribosides. Hence a great variety of purine derivatives have now been tested for cytokinin activity and numerous highly active 6-(substituted amino)-

purines characterized. Non-purine cytokinins are also known, notably certain urea derivatives. In this section, structure–activity relationships in these diverse groups of cytokinins are discussed. It should be noted that relative activities may depend on the bioassay used. Thus in the tobacco callus, the carrot phloem, the soybean callus and the radish cotyledon assays, the activity of Z markedly exceeds that of kinetin; however, in certain leaf senescence assays the converse applies (Letham 1967a). Similar structure–activity relationships appear to apply in the tobacco callus and radish cotyledon assays (Letham 1971). Unless stated otherwise, the discussion below usually concerns these two assays.

4.5.1. Substitution of adenine at various positions

In all monosubstituted adenines known to possess cytokinin activity, with the exception of 8-bromoadenine, the substituent is at position N_6. In early studies, 3-(3-methylbut-2-enyl)adenine (triacanthine) and certain 1-substituted adenines were reported to be active as cytokinins (Beauchesne and Goutarel 1963; Hamzi and Skoog 1964). However, in more critical experiments in which carefully purified compounds were used and in which heating was minimized during media preparation, the 1- and 3-substituted adenines were essentially inactive (Roqozinska et al. 1964; Leonard et al. 1966; Letham 1971). Heating resulted in activity due to formation of the N_6-isomers (Leonard et al. 1966). The 2-, 7-, 8- and 9-alkyl and benzyl adenines tested have all proved inactive (Skoog et al. 1967; Letham 1971). 7-Substituted adenines were not activated by heating; however 9-substituted adenines, like 1- and 3-substituted adenines, were converted to active compounds when heated (Skoog et al. 1967). These findings can be explained in terms of a scheme involving purine ring opening and reclosure (Skoog and Leonard 1968). According to this, 1-, 3- and 9-substituted adenines could yield N_6-substituted adenines, while 7-substituted adenines could not rearrange to these N_6 compounds. The mechanism of the rearrangement of 3-benzyladenine to N_6-benzyladenine has been elucidated by labelling studies (Leonard and Henderson 1975).

A very recent observation of interest is the weak cytokinin activity of 8-bromoadenine which equals that of 8-bromoadenosine and exceeds that of 8-bromoadenosine 3′:5′-cyclic monophosphate (Schmitz et al. 1975). The weak activity of the last mentioned compound and of 8-bromo-AMP had been reported previously (Wood and Braun 1973; Hagen and Marcus 1975; Gordon and Letham 1975). The activity of 8-bromoadenosine 3′:5′-cyclic monophosphate has been attributed to the possibility that it is a stable active form of adenosine 3′:5′-cyclic monophosphate; the activity of 8-bromo-AMP has been attributed to conversion to the cyclic nucleotide (Wood and Braun 1973). However the activity of these compounds may be due to the 8-bromoadenine moiety per se. Adenosine 3′:5′-cyclic monophosphate is inactive in the tobacco callus, soybean callus, radish cotyledon and leaf disc expansion assays (Dekhuijzen and Overeem 1972; Letham et al. 1972; Schmitz et al. 1975). For further discussion of the possible regulatory role of this compound in plant development see Chapter 7.

The effect of substituted adenines on cell division in moss protonemata depends on the site of substitution. While N_6-substituted adenines promote division, 1-, 3- and 9-(3-methylbut-2-enyl)adenines inhibit cell division (Szweykowska et al. 1971; Szweykowska and Korcz 1972).

Introduction of a methyl or benzyl group at the 1- or 3-position of the purine ring of 6-benzylaminopurine very markedly reduces or nearly eliminates cytokinin activity (Skoog et al. 1967). iP is inactivated by cyclization of the side chain to the 1-position (Leonard et al. 1966). An unsubstituted 1-position appears to be a critical requirement for cytokinin activity. A second substituent at the N_6- or 7-position of an N_6-monosubstituted adenine markedly lowers activity (Skoog et al. 1967; Letham 1971).

The effect on cytokinin activity of a substituent at position 2 has also been investigated. A 2-methyl, 2-methylthio- or 2-amino group lowers the activity of Z or iP only moderately while a 2-hydroxy group markedly reduces activity. However a 2-chloro substituent has only a small effect and the activity of 2-chlorozeatin is possibly slightly greater than that of Z (Hecht et al. 1970a; Dammann et al. 1974). When the N_6-substituent is saturated, the reduction in activity caused by 2-methylthiolation is considerable (Schmitz et al. 1972a). In promoting lettuce seed germination, the activities of 6-benzylaminopurine and its 2-fluoro and 2-amino derivatives are similar (Leonard et al. 1961).

A number of 8-substituted derivatives of iP have been tested for cytokinin activity but only one compound, 8-methyl-iP, is more active than iP (Dammann et al. 1974). Usually 8- and 2,8-substituted derivatives of iP possess lower cytokinin activity than unsubstituted iP. However substitution at the 8-position reduces activity less than substitution at position 2 (Dammann et al. 1974). In leaf senescence (chlorophyll retention) assays, the 8-methyl derivatives of 6-benzylaminopurine and kinetin exhibit greater activity than the unsubstituted compounds (Kulaeva et al. 1968). The senescence-retardation activity of 6-benzylamino-8-methylpurine exceeds that of the 8-ethyl, 9-methyl and 2-methyl derivatives of 6-benzylaminopurine (Zwahlen 1962). Substitution of the 8-position of Z and iP with a hydroxyl group causes a smaller decrease of cytokinin activity than hydroxylation at both the 2- and 8-positions (Chen et al. 1975).

The cytokinin activities of N_6,9-disubstituted adenines have been assessed by a number of workers. However their conclusions with regard to 9-alkyl derivatives tend to differ. Thus Elliott et al. (1972) reported that 6-benzylamino-9-butylpurine was more active than 6-benzylaminopurine (soybean callus and *Amaranthus* assays) while Hashizume et al. (1974) considered it was less active (tobacco callus assay). The closely related cytokinin, 6-benzylamino-9-propylpurine, has been reported by Fox et al. (1973) to be much less active than 6-benzylaminopurine (soybean and tobacco callus assays), but Hashizume et al. (1974) considered that its activity equalled that of 6-benzylaminopurine (tobacco callus assay). The cytokinin activities of the 9-methyl derivatives of Z (Shaw et al. 1968) and 6-benzylaminopurine (Kende and Tavares 1968) have been reported to be very similar to, or greater than, those of the unsubstituted compounds. The findings of Kende and Tavares were not confirmed by Fox et al. (1971). The lack of agreement indicated above may arise from two factors; firstly, the compounds for assay were sterilized by autoclaving in some investigations and by filtration in others; secondly, plant tissues derived from different species were employed in the assays reported and these tissues may vary in their ability to cleave the substituent at position 9. In relation to the above possibilities, it is relevant to note that the activity of N_6,9-dibenzyladenine is enhanced by autoclaving (Skoog et al. 1967), and that the methyl group of 6-benzylamino-9-methylpurine appears to be rapidly cleaved by plant tissue although the identity of the products has not been established (Fox et al. 1971). In considering the activities of cytokinins with 9-alkyl substituents, it appears appropriate to consider only those studies in which the cytokinins were not exposed to heating. If this is done, the effect of the 9-alkyl substituent is always to lower activity. In such studies, cytokinins have been tested with the following alkyl or related groups at position 9: methyl (Fox et al. 1971), propyl (Fox et al. 1973b), 3-hydroxypropyl (Fox and Chen 1968), cyclohexyl (Young and Letham 1969, Fox et al. 1973) and benzyl (Skoog et al. 1967).

Although 9-alkyl substituents depress cytokinin activity of adenine derivatives, a related substituent, the tetrahydropyran-2-yl group, does appear to enhance cytokinin activity when present at position 9. With stem segments (Nitsch and Nitsch 1967), leaf cuttings (Heide 1965) and developing grapes (Weaver et al. 1966), 6-benzylamino-9-(tetrahydropyran-2-yl)purine evokes a greater growth response than 6-benzylaminopurine. The tetrahydropyranyl derivative is almost as active as 6-benzylaminopurine in the tobacco callus assay (Skoog et al. 1967). 9-(Tetrahydropyran-2-yl)-iP is more active than iP in promoting the expansion of excised radish cotyledons and in retarding the senescence of tobacco leaf discs (Young and Letham 1969). The above tetrahydropyranyl derivatives are similar structurally to acetals and hence the tetrahydropyranyl substituent is readily cleaved by acid. The high activity of these purine derivatives may be associated with the lability of the tetrahydropyranyl group. 9-Methoxymethyl derivatives of purine cytokinins are also acetal-like compounds, but their activities are less than those of the parent cytokinins (Young and Letham 1969; Fox et al. 1973).

Considerable information has now been accumulated regarding the activities of cytokinin glycosides relative to those of the corresponding free bases. In the tobacco callus assay, purine cytokinin ribosides (i.e. 9-β-D-ribofuranosides) including [9R]Z and [9R]iP are appreciably less active than the corresponding free bases (Schmitz et al. 1972a). In the radish cotyledon bioassay, the activity of Z exceeds that of [9R]Z and [9R-5′P]Z (Letham 1973), while in the carrot phloem assay, Z is also more active than [9R-5′P]Z (Letham 1966c). In promoting bud formation in protonemata of *Funaria hygrometrica*, cytokinin bases including Z, are markedly more active than the corresponding ribosides (Whitaker and Kende 1974; Spiess 1975). In the soybean callus assay, the following activity sequence was established at low concentrations: 6-benzylaminopurine > 6-benzylaminopurine riboside > riboside 5′-monophosphate (Van Staden 1973b). The cytokinin activity of an N_6-substituted adenosine 5′-monophosphate exceeds that of the corresponding 3′:5′-cyclic monophosphate (Schmitz et al. 1975). While 6-benzylaminopurine and its riboside promote lettuce seed germination in darkness, the riboside 5′-monophosphate is inactive (Van Staden 1973b). Modification of the ribosyl moiety of [9R]iP by periodate oxidation followed by borohydride reduction (this gives a triol-ether group at position 9) causes a moderate reduction in cytokinin activity (Chen et al. 1974b).

Cytokinin activity of cytokinin glucosides has also been reported. The 7- and 9-glucopyranosides of 6-benzylaminopurine exhibit very weak activity relative to that of the free purine base in several systems (Guern and Hugon 1966; Wilson et al. 1974; Parker et al. 1975; Fox and McChesney 1974). However the 3-glucopyranoside of 6-benzylaminopurine is much more active than the 7- and 9-glucosides. In the tobacco and soybean callus assays its activity approaches that of 6-benzylaminopurine (Fox and McChesney 1974), while in the radish cotyledon assay it is at least as active as 6-benzylaminopurine (Parker et al. 1975; Letham et al. 1975). In an *Amaranthus* betacyanin assay, the 3-glucopyranoside is slightly less active than the parent cytokinin (Letham, unpublished results).

4.5.2. Structure of the substituent at position 6 in purine cytokinins

The nature of the atom attached directly to the 6-position of the purine ring markedly affects cytokinin activity. Thus replacement of the exocyclic nitrogen in N_6-substituted adenines by sulphur markedly reduces activity in the tobacco callus assay (Strong 1958; Skoog et al. 1967). In retarding senescence of barley leaf segments, 6-benzylthiopurine is possibly slightly more active than 6-benzylaminopurine at very low concentrations; however at higher concentrations 6-benzylaminopurine is markedly more effective (Kulaeva et al. 1965). In promoting expansion of radish leaf discs, 6-benzylthiopurine and 6-furfurylthiopurine are markedly less active than kinetin (Kuraishi 1959). Certain 6-alkyl- and 6-phenylalkylthiopurines promote lettuce seed germination, but are less effective than the corresponding 6-(substituted amino)purines (Skinner et al. 1956). This difference in activity is particularly marked when the chain length is not optimal (e.g., propyl).

A comparison has been made recently between the activities of compounds with four different linkages (NH, S, O, CH_2) between position 6 of the purine ring and 3-methylbut-2-enyl or benzyl groups (Henderson et al. 1975). In the former series of compounds, cytokinin activity varies with the linkage group thus: $NH > CH_2 > S >> O$; in the latter series the order is as follows: NH slightly $< CH_2 = O >> S$. Hence an appropriately substituted amino group or methylene group at position 6 of the purine ring appears to be required for high cytokinin activity.

A great diversity of N_6-substituted adenines have now been tested for cytokinin activity and the influence of substituent structure on activity is outlined below. For high activity, the substituent size is critical. Thus the activity of 6-*n*-alkylaminopurines increases with increase in chain length to an optimum of five carbons; further extension of the chain reduces activity and 6-*n*-decylaminopurine is almost inactive (Skoog et al. 1967). In the radish leaf disc expansion assay (Kuraishi 1959) and radish cotyledon assay (Letham 1971), similar trends are observed. Similarly in the homologous series of 6-*n*-alkylthiopurines, optimum activity is exhibited by the C_5 homologue (Skinner et al. 1956). Among 6-(ω-phenylalkylamino)purines, C_1 and C_2 alkyl groups are optimal (Kuraishi 1959). Although 6-methylaminopurine is inactive in the normal tobacco callus and soybean callus assays, it exhibits some activity in the presence of adenine (Miller 1962). Miller suggests the methyl compound has two types of action in soybean callus. Firstly, it is a very weakly active cytokinin, and secondly, it inhibits adenine-requiring processes. At concentrations needed for activity as a cytokinin, the inhibitory effect is pronounced and the growth response is not expressed. If, however, adenine is present, the inhibition is alleviated and the cytokinin activity of 6-methylaminopurine becomes evident.

At very high concentrations, adenine is weakly active in both the tobacco callus and soybean callus assays (Skoog et al. 1967; Miller 1968). The latter worker has carefully studied factors which affect the response to adenine in the soybean callus assay. Adenine exhibits a very narrow effective concentration range and the response to it is markedly influenced by the vitamin content of the culture medium and the nature of the auxin present. The observations of Miller (1968) suggest that adenine acts directly as a very weak cytokinin. Although adenine may function as a very weak cytokinin, it also appears to exert other regulatory activity. The ability of adenine to promote bud formation on internode segments of tobacco was established many years ago by Skoog and Tsui (1948). Using internode segments of *Plumbago indica* and tobacco pith tissue, it has been established that adenine and cytokinins such as zeatin and kinetin act synergistically to induce bud formation. Adenine could not be replaced by other bases and may be an important regulator of bud formation (Nitsch et al. 1967; Nitsch and Nitsch 1967).

The presence of unsaturation in the N_6-substituent of 6-(substituted amino)purines markedly enhances activity. The high activity of iP, which is much greater than that of kinetin, is due to the presence of the double bond and not to chain branching. Saturation of this bond yields 6-isopentylaminopurine which has essentially the same activity as 6-*n*-pentylaminopurine and kinetin (Skoog et al. 1967; Leonard et al. 1968, 1969a). The location of the double bond also influences activity because 6-(3-methylbut-3-enylamino)purine is less active than iP (Leonard et al. 1968). The influence of unsaturation is also apparent in the following relative activities: 6-geranylaminopurine > 6-decylaminopurine; 6-phenylaminopurine >> 6-cyclohexylaminopurine; Z > (diH)Z (Skoog et al. 1967; Matsubara et al. 1968; Letham 1972; Schmitz et al. 1972a). Schmitz et al. (1972a) observed that the extent to which saturation lowers activity depends on the presence or absence of substituents at positions other than N_6. This reduction in activity is marked for all free bases, very marked for 2-methylthiolated cytokinin ribosides, but only slight for non-methylthiolated ribosides.

The presence of hydroxyl groups on aliphatic N_6-substituents markedly influences activity. A hydroxyl group at the 4-position of either the isopentyl or 3-methylbut-2-enyl side-chain enhances activity, but in the 2-, 3- or 2- and 3-positions of the former, the hydroxyl group decreases activity (Leonard et al. 1969a). Z is markedly more active than iP in the tobacco callus (Munsche et al. 1968; Schmitz et al. 1972a), soybean callus (Miller 1968), carrot phloem (Letham 1967a), radish cotyledon (Letham 1972), *Amaranthus* betacyanin (Munsche et al. 1968; Bigot 1968; Conrad 1971) and *Spirodela* assays (Letham 1967a). The allylic hydroxyl group of Z is therefore a structural feature associated with high cytokinin activity in diverse bioassays. Z is also markedly more active than kinetin in the above assays and in the etiolated bean leaf disc assay (Letham 1967a). However in leaf senescence bioassays, the converse applies, both kinetin and 6-benzylaminopurine being more active than Z (Letham 1967a; Fawcett and Wright 1968; Varga and Bruinsma 1973). In inducing bud formation on stem segments, the activity of Z greatly exceeds that of other cytokinins (Nitsch and Nitsch 1967).

Z is the most active known naturally occurring cytokinin. The high activity of Z stimulated interest in the activity of closely related compounds. A number of N_6-(ω-hydroxyalkyl)- and N_6-(ω-hydroxyalkenyl)adenines have been tested for cytokinin activity (Fawcett and Wright 1968; Schmitz et al. 1972c; Letham 1972). All these compounds are less active than Z. *(cis)*Z and *(cis)*[9R]Z are much less active than Z and [9R]Z respectively (see also Vreman et al. 1974). Shift of the 3-methyl group of Z to position 2 in the side-chain also reduces activity markedly. Hence small changes in the structure of the zeatin side-chain diminish activity greatly.

The only modification of the Z side-chain which enhances activity is *O*-acylation (i.e. esterification of the -OH group). Thus *O*-formyl-, *O*-acetyl-, *O*-propionyl- and *O*-(indol-3-ylacetyl)-zeatin are slightly more active than zeatin in the tobacco callus assay (Schmitz et al. 1971, 1972b), while the activity of the *O*-acetyl compound exceeds that of zeatin in the radish cotyledon assay (Letham 1972). A chloro substituent at position 2 of the purine ring slightly enhances the activity of the *O*-acyl compounds and the resulting derivatives of Z are the most active known cytokinins in the tobacco callus assay (Schmitz et al. 1971). These last workers suggested that the esters probably provide a continuing supply of free Z within plant tissue. Although the size of the acyl group does not appear to affect activity, the orientation of the ester group markedly influences activity. Reversal of the orientation, as in methyl 2-methyl-4-(purin-6-ylamino)-*trans*-crotonate, reduces activity about a thousand-fold (Schmitz et al. 1971). Unlike *O*-acylzeatins, *O*-alkylzeatins are much less active than zeatin and activity decreases with increase in size of the alkyl group (Letham 1972).

In addition to those mentioned above, a number of other groups of N_6-substituted adenines have been synthesized and tested for cytokinin activity. These groups are discussed very briefly below. Optically active cytokinins (asymmetric carbon α to the exocyclic nitrogen) were compared in the tobacco callus assay (Yamada et al. 1972; Matsubara et al. 1974). The levorotatory compounds are more active than the dextrorotatory isomers. However, in the barley leaf senescence assay, this rule does not always apply (Koshimizu et al. 1968). A series of alkoxyalkylaminopurines and phenoxyalkylaminopurines were tested by Rothwell and Wright (1967). Of the alkoxyalkyl compounds tested, 6-(β-ethoxyethylamino)purine and 6-(β-propoxyethylamino)purine exhibited the highest activity; in the wheat coleoptile assay, both are more active than kinetin. In contrast the activity of the phenoxyalkyl-compounds is markedly less than that of kinetin in three assays. A study of the cytokinin activity of *N*-(purin-6-yl)amino acids indicated that a carboxylic acid group in the N_6-substituent diminishes activity markedly (Letham and Young 1971b). The activities of the purinylamino acids are less than those of the corresponding methyl esters.

Although Skoog et al. (1967) reported that 6-acylaminopurines (e.g. 6-benzoylaminopurine) are inactive, several workers have established that such compounds possess considerable cytokinin activity. Thus adenine amides including 6-benzoylaminopurine and 6-butyrylaminopurine are clearly active in the soybean callus, tobacco callus, leaf senescence and radish cotyledon assays (Letham et al. 1972; Dekhuijzen and Overeem 1972; Martin et al. 1973). A derivative of adenosine 3′:5′-cyclic monophosphate, 6-*N*,2′-*O*-dibutyryladenosine 3′:5′-cyclic monophosphate, is also active as a cytokinin due to the presence of the 6-acylamino group (Letham et al. 1972; Dekhuijzen and Overeem 1972).

Cytokinin activity has been determined for a group of ureidopurines related to the tRNA component, *N*-(purin-6-ylcarbamoyl)threonine (XIII). Although this naturally occurring compound is inactive as a cytokinin, the following analogues exhibit good activity in the soybean assay: 6-*n*-butylureidopurine (XIV), 6-*t*-butylureidopurine, 6-*o*-fluorophenylureidopurine and 6-*o*-chlorophenylureidopurine (XV), the two last-mentioned compounds being the most active with activity much greater than that of 6-phenylureido-

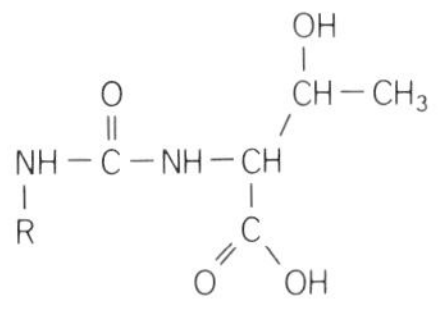

XIII

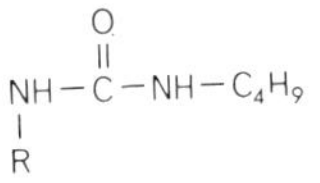

XIV

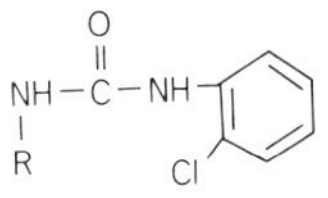

XV

R = purin-6-yl

purine (Dyson et al. 1972b). In the tobacco callus assay, 6-phenylureidopurine and 6-*o*-methylphenylureidopurine exhibit about 10% of the activity of 6-benzylaminopurine (McDonald et al. 1971). Molecular models indicate the overall shape and size of the *o*-chlorophenyl group are about the same as those of the threonyl group. Furthermore, the carbonyl in the threonine residue and the chlorine of the chlorophenyl group occupy the same position relative to the carbonyl in the ureido bridge of the molecules (Dyson et al. 1972). These workers suggest that XIII has a growth regulatory role in vivo, but when exogenously supplied it does not reach the active site.

Several ring substituents at the N_6 position confer marked cytokinin activity on adenine. 6-Benzylaminopurine is slightly more active than kinetin which approximately equals 6-(2-thenylamino)purine in activity. However, 6-(2-pyridyl-methylamino)purine and 6-cyclohexylaminopurine are both much less active (Skoog et al. 1967). The length of a methylene bridge linking the exocyclic nitrogen of adenine to a phenyl ring greatly affects activity. In the tobacco callus assay, the optimal length appears to be one methylene (Skoog et al. 1967); in the radish leaf disc expansion assay, compounds with one and two methylenes both exhibit optimal activity (Kuraishi 1959).

The effect of substitution in the phenyl ring of 6-benzylaminopurine has been assessed using the radish leaf disc expansion assay (Kuraishi 1959). *Ortho* substitution of methyl or chloro groups enhance activity, but all other substitutions yield derivatives less active than 6-benzylaminopurine. *Ortho* hydroxylation, of interest because of the natural occurrence of 6-(*o*-hydroxybenzylamino)purine riboside, reduces activity markedly in the soybean callus assay, but only slightly in a radish leaf disc senescence assay (Horgan et al. 1975). For the chlorobenzyl compounds, Kuraishi (1959) established the following activity sequence: *ortho*-Cl>H>*meta*-Cl>*para*-Cl. In the wheat senescence assay, the same sequence operates for the phenoxyethylaminopurines (Rothwell and Wright 1967).

A group of compounds were synthesized by Hecht et al. (1970b) to define the relationship between side-chain planarity and cytokinin activity. It was concluded that side-chain planarity is essential for high activity.

4.5.3. Non-purine cytokinins

A considerable number of compounds which lack a purine ring have been reported to possess cytokinin activity. The most active of these are phenylurea derivatives. The weak and variable activity found initially for samples of *N,N'*-diphenylurea (Bottomley et al. 1963) was due to inhibitory impurities which tended to mask the cytokinin activity (Bruce and Zwar 1966). Purification of *N,N'*-diphenylurea by recrystallization using several solvents in sequence yielded a product with reproducible activity at 10 mg/l in the tobacco pith assay.

Following the establishment of the cytokinin activity of *N,N'*-diphenylurea, a great number of related compounds were tested for cytokinin activity in the tobacco pith assay by Bruce and Zwar (1966). The minimum structural requirements for activity are satisfied by phenyl urea; these are an -NH-CO-NH- grouping and a planar ring. The activity of phenyl urea is increased markedly by substitution (effect of *meta>para>ortho*) of the phenyl group with electron-attracting groups (e.g. nitro, chloro). A second phenyl group in the *N'* position (i.e. on the opposite side of the ureido grouping) further enhances activity, but substitution of this additional phenyl ring generally reduces activity very markedly. Compounds with an -NH-CO-NH- grouping exhibit greater activity than the corresponding analogues with an -NH-CS-NH- linkage while other modifications to the "bridge" yield inactive compounds. The most active ureas tested by Bruce and Zwar (1966), e.g. *N*-3-chlorophenyl-*N'*-phenylurea and *N*-3-nitrophenyl-*N'*-phenylurea *showed* detectable activity at 4×10^{-7} *M* but were appreciably less active than kinetin. *N*-Phenyl-*N'*-4-pyridylurea also exhibits good activity (Bruce and Zwar 1966; Okamoto et al. 1974).

In addition to inducing cell division in tobacco stem pith, urea derivatives promote pea bud development (Kefford et al. 1966) and seed germination (Kefford et al. 1965). These compounds also retard leaf senescence, promote cucumber cotyledon expansion and inhibit root development (Kefford et al. 1973, and references therein). Sex expression in *Luffa acutangula* is altered by *N*-3-chlorophenyl-*N'*-phenylurea, the formation of female flowers being promoted (Bose and Nitsch 1970). This spectrum of activities clearly qualifies urea derivatives for classification as cytokinins. A good correlation was observed between ability to retard leaf senescence and to induce cell division (Kefford et al. 1973). This suggests that the cytokinin receptor site may be the same for senescence retardation and for induction of cell division. A number of thiourea derivatives have also been reported to retard leaf senescence (Karanov et al. 1968). It should be noted that urea cytokinins may not be active per se and may be metabolized to the active forms (see discussion in Burrows and Leworthy 1976).

Cytokinin activity, usually very weak, is also shown by the following compounds: aromatic amide derivatives (Okamoto et al. 1974), azaindene and azanaphthalene derivatives (Torigoe et al. 1972; Okamoto et al. 1974), 4- and 5-substituted benzimidazoles (Kuroishi and Yamaki 1967; Kohler and Conrad 1968; Torigoe et al. 1972), systemic benzimidazole fungicides such as methyl benzimidazol-2-yl-carbamate (Skene 1972a; Thomas 1974), 6-(substituted amino)-8-azapurines (Skoog et al. 1967; Letham 1971), 1-deaza and 3-deaza analogues of kinetin and iP (Rogozinska et al. 1973; Kroon et al. 1974), and 7-(substituted amino)pyrazolo-[4,3-*d*]pyrimidines such as 7-(3-methylbut-2-enylamino)-pyrazolo[4,3-*d*]pyrimidine, XVI (Hecht et al. 1971, 1975). The last mentioned compounds are of particular interest. The 3-position of these cytokinins is analogous to the 9-position in purines but consists of a carbon instead of a nitrogen atom, the imidazole ring being replaced by a pyrazole ring. These compounds are not likely to be ribosylated enzymically at the 3-position. Hence Hecht et al. (1971) suggest that 9-ribosylation is not a prerequisite for purine cytokinin action.

Other non-purine heterocyclic compounds which exhibit very weak cytokinin activity are 4-(substituted amino)-pyrazolo[3,4-*d*]pyrimidines (Letham 1971; Hecht et al. 1975), although in early studies they appeared to be inactive (Skinner and Shive 1957; Strong 1958). The related compound, 4-(3-methylbut-2-enylamino)pyrrolo[2,3-*d*]-pyrimidine, also exhibits feeble activity (Hecht et al. 1975). Linear benzo analogues of iP and 6-benzylaminopurine, in which a benzene ring is inserted between the pyrimidine and imidazole rings of the purine moiety, exhibit weak activity (Sprecker et al. 1976). These compounds with a heterotricyclic nucleus are fluorescent and may be useful in studies of cytokinin binding sites. Certain 7-acylaminoimidazo-[4,5-*b*]pyridines and 4-acylamino-imidazo[4,5-*c*]pyridines exhibit high cytokinin activity (Sugiyama et al. 1975); 7-benzoylaminoimidazo[4,5-*b*]pyridine is more active than kinetin in the tobacco callus bioassay.

The pyrimidine derivative, 4-benzylamino-6-methyl-pyrimidine, retards leaf senescence (Kulaeva et al. 1965) while 6-methyluracil also exhibits some cytokinin-like activity (Pozsar and Matolcsy 1968). However the activity reported for these pyrimidines is not sufficiently specific to permit their designation as cytokinins. Certain herbicidal phenyl-urethanes also resemble cytokinins in some properties (Mann et al. 1967).

Because benzothiazol-2-yloxyacetic acid promotes cell division in carrot phloem tissue (Steward and Shantz 1956) it has been classified erroneously as a cytokinin (Fox 1969). This compound is an auxin; it substitutes for IAA, not cytokinin, in promoting growth of carrot phloem tissue (Letham 1963c) and is inactive in the tobacco callus bioassay (Skoog et al. 1967).

4.5.4. *Cytokinin antagonists*

The availability of cytokinin antagonists would permit certain unsolved problems in cyto-kinin research to be apprcached in a new way. These antagonists would be of value in studies of mechanism of cytokinin action and of growth in systems which produce ade-quate endogenous cytokinins. The first known cytokinin antagonists were derivatives of benzylurea (Kefford et al. 1968). These workers used urea cytokinins, kinetin and 6-benzylaminopurine as inducers of chlorophyll retention in radish leaf tissue and observed that certain derivatives of benzylurea acted as antagonists of the chlorophyll retention caused by the cytokinins mentioned. The effective antagonists included *N*-benzyl-*N'*-3,4-dichlorophenylurea and *N*-benzyl-*N'*-phenylurea. Urea cytokinins and purine cytokinins exhibit the same regulatory effects in a variety of systems and this suggests that they act at the same site in the cell. The ability of benzylureas to antagonize both purine and urea cytokinins is in accord with this view.

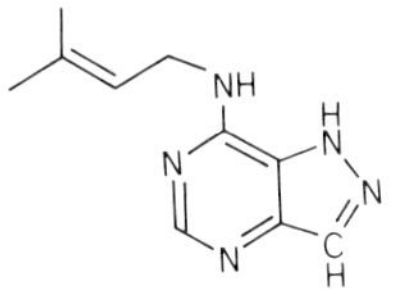

XVI

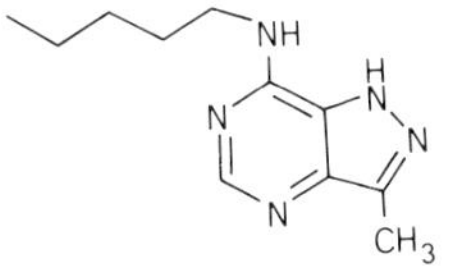

XVII

Purine cytokinin antagonists are also known. A series of 7-substituted 3-methylpyrazolo[4,3-*d*]pyrimidines were tested for cytokinin antagonist activity in the tobacco callus bioassay (Skoog et al. 1973). Several antagonists were found, the most effective being 3-methyl-7-*n*-pentylaminopyrazolo[4,3-*d*] pyrimidine (XVII). Structural features of the 7-substituent needed for high antagonist activity are similar to those of the 6-substituents of highly active purine cytokinins. However notable inconsistencies in this generalization are firstly, the appreciable increase in antagonist activity caused by replacing the isopentyl group with an *n*-pentyl substituent; secondly, introduction of a 2,3-double bond into the 7-isopentyl substituent converts the antagonist into a weak cytokinin. The pyrazolo[4,3-*d*]pyrimidine antagonists appeared to be competitive inhibitors of the cytokinins (Skoog et al. 1973). The inhibitory activity of the antagonists was alleviated, not only by iP and 6-benzylaminopurine, but also by *N,N'*-diphenylurea when used at high concentrations. This again suggests that the urea and purine cytokinins both act at the same site in plant cells.

In a study of growth kinetics of tobacco callus and suspension cultures supplied with XVII, Helgeson et al. (1974) failed to obtain inhibitor-dependent log phase growth rates. These workers suggest that compounds such as XVII do not antagonize cytokinins in a simple competitive manner. Hence the mechanism of action of pyrazolo[4,3-*d*]pyrimidines, a particularly interesting group of plant growth inhibitors, merits more critical study.

Certain pyrrolo[2,3-*d*] pyrimidines also appear to exhibit anticytokinin activity. The most active of these is 4-cyclopentylamino-2-methylthiopyrrolo[2,3-*d*] pyrimidine (Skoog et al. 1975). A related compound which exhibits anticytokinin activity at low concentrations is 4-furfurylamino-7-(β-D-ribofuranosyl)pyrrolo[2,3-*d*]pyrimidine (Iwamura et al. 1974); the 4-benzyl and *n*-pentenyl analogues of this riboside may exhibit stronger anticytokinin activity (Iwamura et al. 1975).

From the information presented in this section, it is evident that cytokinin activity is exhibited by a diversity of chemical structures. Hence we are challenged with the paradox of biological specificity being associated with molecules which apparently lack structural specificity. A similar paradox exists in auxin structure–activity relationships. Structure-activity studies have not provided any definite information concerning the mechanism of cytokinin action. However such studies suggest that purine cytokinin bases do not require activation by derivatization (e.g. glycosylation) at positions 7 or 9 in order to evoke a growth response (Hecht et al. 1975). The most significant consequence of structure-activity studies would appear to be the development of anticytokinins. These compounds will probably prove to be an extremely useful physiological tool.

4.6. Sites and pathways of biosynthesis

4.6.1. Free Cytokinins

4.6.1.1. Sites of biosynthesis. A number of observations indicate that roots, and in particular root apices, are sites of active cytokinin biosynthesis. Firstly, cytokinins are known to occur in xylem sap and their level in the sap is reduced markedly when the root system is subjected to stress or exposed to low temperatures (see section 4.9). Flooding is one form of stress which markedly reduces the cytokinin level in sap. During flooding of sunflower plants, Burrows and Carr (1969) observed a correlation between the metabolic activity of root apices and the cytokinin content of xylem sap. Secondly, the level of extractable cytokinins in root apices is very much greater than the level in more proximal regions of roots (Weiss and Vaadia 1965; Short and Torrey 1972a). Two cytokinins extracted from the root tips of sunflower plants are reported to be chromatographically identical to the two cytokinins of the xylem sap of this species (Kende and Sitton 1967). Thirdly, these workers found the level of the two cytokinins in sunflower root exudate did not decrease appreciably for four days after topping of the plants. Fourthly, root formation on petioles and treatment of the laminae with cytokinins have very similar effects on the physiology of detached leaves (see references given in review by Letham 1967b). Fifthly, the level of cytokinins in the petioles of detached bean leaves increases very markedly immediately after the petioles develop roots (Engelbrecht 1972). Although cytokinins are not detectable in the leaf blades at or before this time, their level in the blade is subsequently markedly greater than that in the petioles plus roots. Sixthly, in seedlings and cuttings, exogenous cytokinins will frequently substitute for roots in inducing growth and physiological responses (see Chapter 4, Volume II section 4.3.2). For example, excision of the roots of 56-hr-old *Avena* seedlings reduces markedly the growth of the coleoptiles; exogenous cytokinin completely restores the coleoptile growth rate in these derooted seedlings and stimulates auxin production by excised coleoptile tips (Jordan and Skoog 1971). These results suggest that coleoptile growth and auxin production in the coleoptile tip depend on a supply of root-produced cytokinins. Seventhly, rice root tips when cultured under sterile conditions secrete cytokinins into the culture medium (Yoshida and Oritani 1974). Hence roots, in addition to being organs of support and nutrient absorption, have the further role of supplying cytokinins and possibly other hormones to the shoot. It is likely that the cytokinins in xylem sap are synthesized in root apices (see also Chapter 4, Volume II, section 4.3.3.1).

Numerous observations indicate that roots are not the only site of cytokinin biosynthesis. Tomato, tobacco, pea and bean plants develop fruits even though the roots are removed at flowering and the formation of adventitious root primordia is prevented by periodic excision of the stem bases (Peterson and Fletcher 1973). Tobacco fruits reach maturity on stems without roots and produce viable seed. The appreciable degree of fruit and seed development under these conditions indicates developing fruits and seeds are not completely dependent on roots for their supply of cytokinins and other hormones (Peterson and Fletcher 1973). Other sites of biosynthesis probably exist and from the evidence outlined below one appears to be developing seed.

The cytokinin activities of developing apple (Letham and Williams 1969) and avocado seed (Blumenfeld and Gazit 1970; Gazit and Blumenfeld 1970) are markedly greater than those of the receptacles and mesocarp respectively. Extracts of developing apple seed are also much more active than extracts of all other parts of the developing apple fruit bud (Letham and Williams 1969). Avocado cotyledon tissue derived from seed grows in vitro without exogenous cytokinin and actually produces cytokinins (Blumenfeld and Gazit

1971). Parthenocarpic tomato fruits show retarded development and possess low cytokinin contents compared with seeded fruits (Varga and Bruinsma 1974). Excision of the developing seeds from apple fruitlet explants renders the growth of the explants in sterile culture more dependent on exogenous cytokinin (Letham and Bollard 1961). Further evidence that cytokinins are produced by seed is provided by experiments in which pea pods excised 10 days after anthesis were grown in sterile culture on media lacking cytokinin (Hahn et al. 1974). The pea seeds developed normally and reached full maturity; their cytokinin content rose dramatically during 18 days in culture (< 0.1–3.3 iP equivalents per seed).

Germinating seeds frequently yield extracts with appreciable cytokinin activity. Because exogenous cytokinin can replace the embryonic axis in promoting development of certain enzyme activities (e.g. amylase, protease) in cotyledons (Penner and Ashton 1967a, b; Gepstain and Ilan 1970; Locker and Ilan 1975), the axis appears to be a site of cytokinin biosynthesis in germinating seed. Unlike cytokinins, other hormones are unable to replace the stimulus of the embryonic axis. Regulation of degradation of storage carbohydrates and proteins in cotyledons by cytokinins produced in the axis appears to be a general phenomenon (Locker and Ilan 1975).

In addition to the root tip, the developing seed and the embryonic axis, another possible site of cytokinin biosynthesis is the developing bud. Cytokinin levels increase in buds of excised poplar twigs which lack roots about the time of bud burst (Hewett and Wareing 1973a). Hence cytokinins of developing poplar buds are not derived solely from the root and may be produced within the bud itself either by synthesis or by release from bound forms. From a study of the diffusible cytokinins in excised apical buds of *Dahlia variabilis* and of cytokinins in xylem exudates of this species, Kannangara and Booth (1974) suggested the apices are probable sites of cytokinin biosynthesis. However the experiments are inconclusive and can be interpreted in other ways. Experiments with the inhibitor hadacidin indicate that the cytokinin required for lateral pea bud development is synthesized locally in the bud itself (Kung-Woo Lee et al.. 1974). Hadacidin does not appear to be translocated, but it inhibits bud development and probably cytokinin biosynthesis when applied locally to lateral buds.

4.6.1.2. Pathways of biosynthesis. Although considerable information is available regarding biosynthesis of cytokinins in tRNA (see below) little is known of biosynthetic pathways leading to free cytokinins. tRNA is a potential source, but hydrolysis of tRNA to the mononucleotide level would presumably be necessary to release activity. Although such hydrolysis and release of cytokinin may be significant in dying, autolyzing cells of differentiating vascular tissues as suggested by Sheldrake (1973), it is unlikely that significant amounts of free cytokinins are released from tRNA in living plant cells. The tRNA of cytokinin-dependent tissue contains iP and its derivatives (Chen and Hall 1969; Burrows 1976). The inability of this tissue to grow on cytokinin-free media indicates physiologically significant amounts of cytokinins are not produced by tRNA degradation. Other evidence also suggests that free cytokinins in plants are biosynthesized by a pathway which does not involve degradation of tRNA. Firstly, Short and Torrey (1972) found that the activity of free cytokinins in pea root tips is 27 times the cytokinin activity in tRNA. Unless a very rapid turnover of tRNA occurs in meristematic tissue, these results indicate tRNA is not the principal source of free cytokinins (see Hall 1973; cf. Klemen and Klambt 1974). Secondly, lateral bud development in pea seedlings is dependent on cytokinin synthesis; studies with the inhibitor hadacidin indicate this synthesis takes place by a route similar to that for AMP and adenine (Kung-Woo Lee et al. 1974). Thirdly, if free cytokinins are derived simply by hydrolysis of tRNA, the following compounds present in tRNA would be expected to occur frequently as free cytokinins (see section 4.2.1): *(cis)*[9R]Z, [2MeS-9R]iP, and *(cis)*[2MeS-9R]Z; none have been identified as free naturally occurring cytokinins in plants. Fourthly, there is good evidence that the fungus *Rhizopogon roseolus* synthesizes *N*-(purin-6-ylcarbamoyl)-threonine, a compound closely related to cytokinins, by a means independent of its synthesis in tRNA (Laloue and Hall 1973). Fifthly, when callus cells derived from the sporogonium of the moss *Funaria hygrometrica* × *Physcomitrium piriforme* are supplied with labelled adenine, a labelled cytokinin with the chromatographic properties of iP is produced; however no labelled cytokinin can be isolated from the tRNA (Beutelmann 1973).

Although the free cytokinins of plants do not appear to be derived from tRNA, the free cytokinins of bacteria may arise from the turnover of tRNA. From the half-life of tRNA in *Agrobacterium tumefaciens,* Hahn et al. (1976) calculated that the amount of free cytokinin (iP) extracted from *A. tumefaciens* cultures could be entirely attributed to degradation of tRNA. The principal cytokinin base in tRNA of *A. tumefaciens* is iP (Chapman et al. 1976).

The enzyme reactions which yield free cytokinins independently of tRNA degradation are unknown. Transfer of an isopentenyl group from isopentenylpyrophosphate to adenine or adenosine, a reaction analogous to that which yields iP in tRNA (see section 4.6.2), could be envisaged as a likely basic step in cytokinin biosynthesis. However no evidence for its occurrence has been reported. Indeed the inhibition of cytokinin biosynthesis by hadacidin, and the inability of adenine to alleviate it, suggests a biosynthetic route involving IMP rather than AMP, adenosine or adenine (Kung-Woo Lee et al. 1974). Labelled adenine appears to be converted to free cytokinins in plant tissues, but the percentage yield is very low (Miura and Miller 1969b; Beutelmann 1973; Einset and Skoog 1973). The incorporation of labelled adenine into Z and [9R]Z in *Vinca rosea* tumour tissue is enhanced when the tissue is cultured on media containing ammonium chloride (Peterson and Miller 1976). Unfortunately these studies provide no information regarding the biosynthetic pathways which are independent of tRNA.

Although the route to free iP is not clear, iP and [9R]iP are precursors of [9R]Z. Labelled iP is converted to [9R]Z in very high yield by *Rhizopogon roseolus* (Miura and Miller 1969b), while [9R]iP is *trans* hydroxylated to yield [9R]Z in both *Rhizopogon roseolus* and sweet corn endosperm (Miura and Hall 1973).

4.6.2. Cytokinins in tRNA

In the initial studies concerning biosynthesis of tRNA cytokinins, labelled mevalonate was supplied to certain bacteria and to plant cells. When the mevalonate-requiring organisms, *Lactobacillus acidophilus* and *L. plantarum,* are grown in the presence of

$[2\text{-}^{14}C]$ mevalonate, the tRNA becomes labelled. All the radioactivity in the tRNA can be accounted for in the isopentenyl group of iP (Fittler et al. 1968a; Peterkofsky 1968). Similarly when $[2\text{-}^{14}C]$mevalonate was supplied to tobacco pith tissue in culture, the tRNA became labelled and all this radioactivity was in the iP of the tRNA (Chen and Hall 1969). However, when the tobacco pith was cultured under different conditions and labelled with $[2\text{-}^{14}C]$ mevalonate, a large proportion of the radioactivity in tRNA was due to *(cis)*Z (Murai et al. 1975). Hence, like other isoprenoid constituents of cells, the isopentenyl group of tRNA cytokinins is derived from mevalonate.

In vitro studies using cell-free systems have established that the isopentenyl side chain in iP is attached to adenine residues in tRNA at the macromolecular level. Treatment of tRNA with permanganate under mild conditions cleaves the isopentenyl side chains leaving adenine residues in place of iP. The resulting tRNA serves as a substrate for the attachment of the Δ^2-isopentenyl (i.e. 3-methylbut-2-enyl) group in vitro. Crude enzyme preparations from yeast, rat liver and tobacco pith were found to catalyze the attachment of this group which was derived from labelled mevalonate supplied to the in vitro system (Fittler et al. 1968b; Chen and Hall 1969). From yeast an enzyme has been purified that catalyzes in vitro the transfer of the Δ^2-isopentenyl group from Δ^2-isopentenyl pyrophosphate to receptor adenosine residues in homologous permanganate-treated tRNA (Kline et al. 1969). Δ^3-Isopentenyl pyrophosphate does not serve as a substrate. There is good evidence that the isopentenyl groups are attached only to those adenosine residues in the tRNA from which the Δ^2-isopentenyl side chain has been removed (Kline et al. 1969).

Similar transferase enzymes have been purified from *E. coli* (Bartz et al. 1970a; Rosenbaum and Gefter 1972). These enzymes were shown to transfer a Δ^2-isopentenyl group to *E. coli* tRNA species which naturally lack this substituent. The group was transferred to the adenosine adjacent to the 3′ end of the anticodon (Rosenbaum and Gefter 1972). Another homologous isopentenylating system has recently been prepared from *Lactobacillus acidophilus* (Holtz and Klambt 1975). These workers partially purified an enzyme which, in an in vitro system containing Δ^2-isopentenyl pyrophosphate, increased the content of Δ^2-isopentenyl groups in tRNA from 2.1 to 6.3 moles per cent; the nucleoside formed was [9R] iP.

The enzymic synthesis of *N*-(purin-6-ylcarbamoyl)threonine in tRNA has also been investigated. An enzyme has been purified from *E. coli* which can synthesize this base in tRNA when supplied with L-threonine, bicarbonate, ATP, Mg^{2+} and tRNA deficient in the base (Elkins and Keller 1974).

4.7. Metabolism

Studies of cytokinin metabolism are important for several reasons, but two are particularly significant. Firstly such studies may lead to recognition of the active form or forms of cytokinins. There is still no certain answer to the basic question: are cytokinins such as zeatin and 6-benzylaminopurine active per se or are metabolites the functional molecular species? Secondly, there is evidence for the existence of bound (storage?) forms of cyto-

Table 4.4.

A list of the principal studies of cytokinin metabolism in plants and the metabolites identified

Cytokinin, plant tissue and reference	Metabolites identified
6-BAP*, *Xanthium pennsylvanicum* leaf discs (McCalla et al. 1962).	N^6-benzyladenosine**, N^6-benzyladenylic acid, adenylic, guanylic and inosinic acids, adenine, guanine, urea and a ureide.
6-BAP, *Cicer arietinum* seedlings (Guern et al. 1968)	N^6-benzyladenosine**.
6-BAP, *Lemna minor* (Bezemer-Sybrandy and Veldstra 1971).	N^6-benzyladenosine 5′-phosphates.
6-BAP, *Solanum andigena* stolons and stems (Woolley and Wareing 1972a and b).	N^6-benzyladenosine.
6-BAP, soybean callus, potato tuber, tobacco tissue cultures (Dyson et al. 1972a; Fox et al. 1973a).	7-glucosyl-6-BAP**, N^6-benzyladenosine, N^6--benzyladenosine 5′-monophosphate.
6-BAP, radish cotyledons (Wilson et al. 1974; Letham et al. 1975).	7-glucosyl-6-BAP, 9-glucosyl-6-BAP, N_6--benzyladenosine, 3-glucosyl-6-BAP.
6-BAP and kinetin, *Acer pseudoplatanus* cell suspensions (Doree and Guern 1973; Guern et al. 1974).	N^6-benzyladenosine 5′monophosphate** and kinetin riboside 5′-monophosphate**, ureides, nucleotides of adenine.
Z***, bean axes (Sondheimer and Tzou 1971).	(diH)Z, (diH)[9R]Z, (diH)[9R–5′P]Z, [9R]Z, [9R–5′P]Z.
Z, ash embryos (Tzou et al. 1973).	[9R]Z**, [9R–5′P]Z, 5′-di- and tri-phosphates of [9R]Z.
Z, derooted radish seedlings (Parker et al. 1972; Parker and Letham 1973).	7-glucosylzeatin**, [9R–5′P]Z, [9R]Z, adenine, adenosine and AMP.
Z, radish roots (Gordon et al. 1974a).	7-glucosylzeatin, [9R–5′P]Z, [9R]Z.
Z, *Zea mays* seedlings (Parker and Letham 1974; Parker et al. 1973).	9-glucosylzeatin, [9R–5′P]Z, [9R]Z, adenine, adenosine, adenosine 5′-phosphate, and 7-glucosylzeatin (minor metabolite).
Z, de-rooted lupin seedlings (Parker et al. 1975; MacLeod et al. 1975).	*O*-β-D-glucopyranosylzeatin, β-[6-(4-hydroxy-3-methylbut-*trans*-2-enylamino)purin-9-yl]-alanine.
Z, *Populus spp.* leaves (Letham et al. 1977).	*O*-β-D-glucopyranosylzeatin, *O*-β-D-glucopyranosyldihydrozeatin, 9-β-D-ribofuranosyl-*O*-β--D-glucopyranosyldihydrozeatin, 9-β-D-ribofuranosyl-O-β-D-glucopyranosylzeatin.
iP, tobacco and *Acer* cell suspensions (Laloue et al. 1974).	5′-mono-, 5′-di- and 5′-tri-phosphates of [9R]iP; 7-glucosyl-iP in tobacco.

* 6-Benzylaminopurine.

** Indicates compound is the principal metabolite.

*** Zeatin; the system of abbreviations for naturally occurring cytokinins is that defined previously (see section 4.2.1).

kinins (see e.g. Gazit and Blumenfeld 1970; Borkowska and Borkowski 1975; Borkowska and Rudnicki 1975). The identity of these very weakly active or inactive compounds is not known; studies of cytokinin metabolites could lead to a knowledge of the structure of these potentially important substances which yield highly active cytokinins on hydrolysis.

During the last 6 years, cytokinin metabolism studies have attracted considerable attention. Numerous studies concerning metabolites of zeatin and 6-benzylaminopurine have now been reported and are summarized in Table 4.4. It is evident that these cytokinins are converted to a diversity of metabolites. This metabolism can be very rapid. Thus when labelled 6-benzylaminopurine was supplied to excised radish cotyledons for 1 hr, over 30% of the extracted radioactivity was in the form of metabolites; after the cotyledons had been transferred to buffer without cytokinin for several hours, 6-benzylaminopurine was no longer detectable (Wilson et al. 1974; Gordon et al. 1974b). The rapid metabolism of kinetin to unidentified polar metabolites in mung bean hypocotyl segments was enhanced by calcium ions (Lau and Yang 1975). Although 6-benzylaminopurine is rapidly metabolized by radish cotyledons, the corresponding amide, 6-benzoylaminopurine (a compound with good cytokinin activity, see section 4.5.2), exhibits considerable stability (Letham et al. 1972). Such a compound may be of value in investigations of cytokinin binding sites in intact tissue.

4.7.1. Glycoside metabolites

From Table 4.4 it is evident that 9-ribosides and riboside 5′-phosphates are very common metabolites of cytokinins. *E. coli* purine nucleoside phosphorylase catalyzes the formation of cytokinin 9-ribosides from the free bases and ribose 1-phosphate (Sivadjian et al. 1969). It is also possible to prepare enzymically nucleoside 5′-monophosphates from N_6-substituted adenosines and ATP, the kinase activity required being present in both yeasts and higher plants (Doree and Terrine 1973). These in vitro enzyme reactions suggest cytokinin nucleotide biosynthesis from the free bases may be a two-step process involving first a nucleoside phosphorylase to yield a cytokinin riboside and then a kinase to give the nucleotide. However in *Acer pseudoplatanus* cells at least, this does not appear to occur. Here the well established in vivo formation of cytokinin riboside 5′-monophosphates from the corresponding bases appears to involve the direct transfer of a ribose 5′-monophosphate group from 5-phosphoribosyl-pyrophosphate to the cytokinin free base catalyzed by a purine phosphoribosyl transferase (EC 2.4.2.7). Cell extracts of *A. pseudoplatanus* catalyze such a reaction from adenine, 6-methylaminopurine and 6-benzylaminopurine (Pethe-Sadorge et al. 1972; Doree and Guern 1973).

The formation of the 9-ribosides and 9-riboside 5′-phosphates of purine cytokinins is not surprising. However a number of cytokinin metabolites (Table 4.4) possess unusual structures and these include cytokinin glucosides. Glucoside metabolites of Z are the 7-glucoside (a compound originally termed raphanatin), the 9-glucoside, the *O*-glucoside, and three derivatives of the *O*-glucoside formed by reduction of the side chain and/or 9-ribosylation. Glucoside metabolites of 6-benzylaminopurine, which form in radish seedlings, are the 7-, 9- and 3-glucosides, the 3-glucoside being a minor metabolite (Wilson et al. 1974; Letham et al. 1975). However it is markedly more active than the 7- and 9-

glucosides (Letham et al. 1975). The 7-glucoside has also been detected as a metabolite of 6-benzylaminopurine in other plant tissues (Fox et al. 1973).

The 3- and 7-glucoside metabolites of cytokinins are unusual in two respects – the site of the glycosidic linkage and the sugar involved. The only previously known purine 7-glycosides were certain compounds related to vitamin B_{12} which contain a 7-ribosylpurine moiety. The 3-glucoside of 6-benzylaminopurine is the first compound with a glycosidic linkage at position 3 of a purine ring to be isolated from plant tissue. Only two compounds of this type have previously been purified from natural sources; these are 3-ribofuranosyluric acid and its 5′-phosphate, both obtained from beef blood. The identification of glucosides of Z and 6-benzylaminopurine was the first unequivocal evidence for the occurrence of purine glucosides in living tissues.

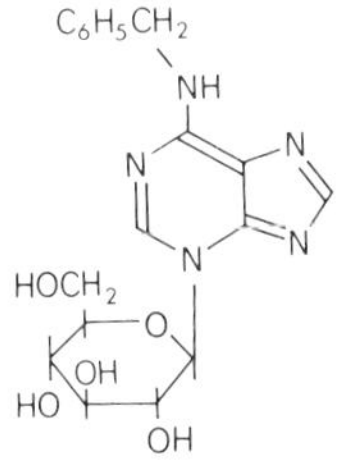

XVIII

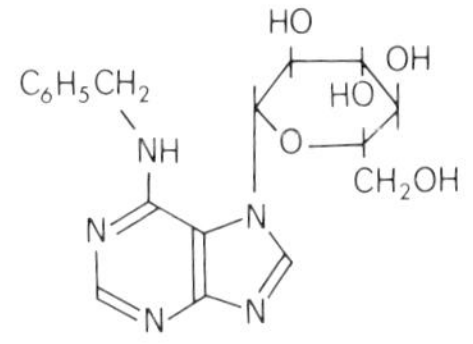

XIX

The structures of the sugar moieties of the cytokinin glucoside metabolites have been investigated. In the case of the 7- and 9-glucosides, the metabolites have been compared with glucopyranosides and glucofuranosides prepared by unambiguous synthesis (Cowley et al. 1975; Duke et al. 1975). The following structures have been established unequivocally for these *N*-glucoside metabolites: 6-benzylamino-3-β-D-glucopyranosylpurine (XVIII), 6-benzylamino-9-β-D-glucopyranosylpurine, 6-benzylamino-7-β-D-glucopyranosylpurine (XIX). 9-glucopyranosylzeatin (probably β) and 7-glucopyranosylzeatin (probably β). *O*-Glucoside metabolites of zeatin are also pyranosides with β configuration, e.g. XX (*O*-β-D-glucopyranosylzeatin) and XXII (Cowley et al. 1975; Duke et al. 1975; Letham et al. 1975; Parker et al. 1975; Letham et al. 1977). The synthesis of XX and the three related *O*-glucosides including XXII (Table 4.4) has been reported (Letham et al. 1977). A metabolite of zeatin formed by soybean tissue has been tentatively identified as *o*-β-D-glucosylzeatin (sugar ring structure not established) (Horgan 1975).

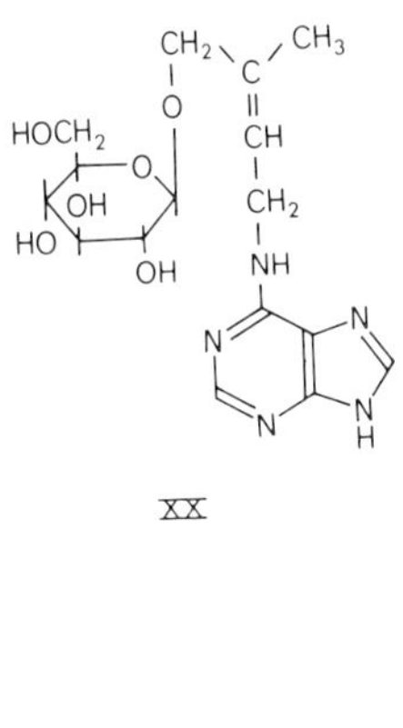

XX

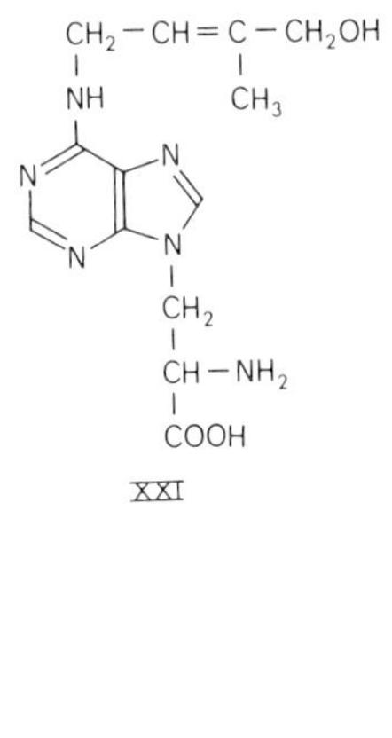

XXI

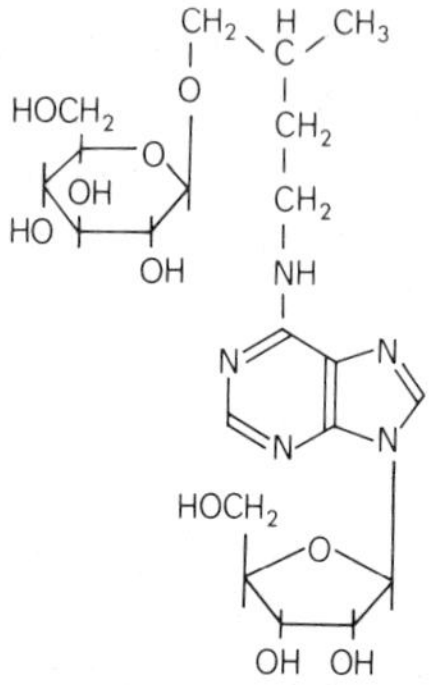

XXII

It should be noted that Deleuze et al. (1972) proposed a furanose structure for the 7-glucoside metabolite of 6-benzylaminopurine isolated from potato tuber tissue. However this assignment is based entirely on a comparison of fragment ion intensities in the mass spectrum of the metabolite (TMSi derivative) with those in the spectra of simple glucopyranosides and glucofuranosides. Hence it lacks an acceptable experimental basis. This potato metabolite has recently been isolated in this laboratory and shown to be the pyranoside XIX (Letham, unpublished data).

Cytokinin 7-glucosides appear to be extremely stable metabolites which persist in plant tissues for considerable periods after the supplied cytokinin is no longer detectable (Parker and Letham 1973; Fox et al. 1973; Gordon et al. 1974a). These glucosides may be storage forms of cytokinins, but their exact physiological significance, like that of other cytokinin metabolites, remains to be established. Cytokinin glucosides may well be the "bound" cytokinins, which are essentially inactive, that have been detected by workers such as Gazit and Blumenfeld (1970), Borkowska and Borkowski (1975) and Borkowska and Rudnicki (1975). A "bound" form of cytokinin detected in rice roots may be a zeatin glucoside but its exact structure has not been established (Yoshida and Oritani 1972). It should be noted that Van Staden (1976b) claims a glucoside of zeatin occurs in coconut milk. However the mass spectrum reported for the β-glucosidase hydrolysis product is very different from that of zeatin. For example, a peak at *m/e* 160 is very prominent in the mass spectrum of zeatin (Shannon and Letham 1966); it is absent in the spectrum reported by Van Staden (1976b).

Cytokinin 7- and 9-glucosides are much less active than the free cytokinins. However in the radish cotyledon bioassay, 6-benzylaminopurine and its 3-glucopyranoside exhibit similar activity (Letham et al. 1975).

The above discussion has concerned glucoside metabolites of the purine cytokinins. Recently it has been established that *N,N′*-diphenylurea is also converted to a glucoside in tobacco callus tissue (Burrows and Leworthy 1976).

4.7.2. Other metabolites formed in plants

Another unusual metabolite of Z is the purinyl amino acid, lupinic acid (XXI). This compound, β-[6-(4-hydroxy-3-methylbut-*trans*-2-enylamino)purin-9-yl]alanine, was identified as a metabolite of Z in *Lupinus angustifolius* seedlings (Parker et al. 1975; MacLeod et al. 1975). This is the first reported, naturally occurring purine derivative in which an amino acid is conjugated to a purine ring nitrogen atom.

Adenine and adenosine have frequently been identified as cytokinin metabolites (Table 4.4). Exogenous ABA has been reported to suppress the degradation of kinetin to adenine in *Rumex pulcher* leaves (Back et al. 1972). It is of interest that Mullins and Osborne (1970), in attempting to account for effects of ABA on plant development, had suggested that ABA had a cytokinin-sparing effect. It is also noteworthy that ABA suppressed reduction of the Z side chain in bean axis, the amounts of (diH)Z riboside and ribotide formed being appreciably reduced (Sondheimer and Tzou 1971).

The enzyme reactions involved in the conversion of cytokinins to adenine and adenosine have been studied by Hall and coworkers. An enzyme system capable of cleaving the isopentenyl group of [9R]iP to give adenosine was detected in tobacco pith tissue by Paces et al. (1971). The enzyme was subsequently purified and studied in greater detail using immature corn kernels as the source (Whitty and Hall 1974). iP and [9R]iP were the most reactive substrates, but Z and [9R]Z were also degraded. However the synthetic cytokinins, 6-benzylaminopurine and N_6-benzyladenosine, were not substrates for the enzyme. The enzyme reaction requires oxygen and hence the enzyme can be regarded as a cytokinin oxidase. In addition to adenine, a product of the oxidative cleavage of iP is the C_5 aldehyde, 3-methylbut-2-enal (Brownlee et al. 1975).

A purine nucleoside hydrolase which degrades adenosine and [9R]iP to adenine and iP respectively has been detected in corn kernels (Whitty and Hall 1974), while similar enzymes may occur in tobacco tissue (Paces et al. 1971) and barley leaves (Paces 1976).

4.7.3. Metabolism by mammalian enzymes and tissues

Adenosine deaminase (adenosine aminohydrolase) from animal sources degrades cytokinin ribosides to inosine (Doree and Guern 1967; Terrine et al. 1969, 1972; Hall et al. 1971). However the rate of conversion is much less than that for adenosine. The rate of degradation of cytokinin ribosides by adenosine aminohydrolases from chicken bone marrow and calf intestinal mucosa is influenced markedly by the structure of the N_6-substituent. While [9R]iP is slowly hydrolyzed, hydroxylated derivatives including [9R]Z are not degraded (Hall et al. 1971; Terrine et al. 1972). There is no convincing evidence at present to suggest that direct conversion of cytokinin ribosides to inosine occurs appreciably in higher plant tissues. However cytokinin degradation by deaminase type enzymes has been established in *Schizosaccharomyces pombe* (see discussion by Guern et al. 1974).

Cytokinin ribosides inhibit growth of animal tumour cells and induce remissions in leukemia patients. This has stimulated interest in their metabolism by animal tissues. In man, [9R]iP, unlike other modified nucleosides in tRNA, is not excreted but is metabolized. After intravenous administration of labelled [9R]iP to human subjects, 6-(3-methyl-3-hydroxybutylamino)purine, hypoxanthine and adenine were identified as metabolites. A large proportion of the radioactivity was excreted as non-UV absorbing compounds suggesting degradation of the purine moiety. However [9R]iP is not degraded to uric acid as are common purine ribosides (Chheda and Mittelman 1972). While [9R]iP is rapidly cleared from rabbit blood, N_6-(3-chlorobut-2-enyl)adenosine, which is not a substrate for adenosine aminohydrolase, is cleared only slowly. Adenosine aminohydrolase probably plays an important role in the metabolism of [9R]iP in blood (Hall and Mintsioulis 1973). Enzyme preparations from animal tissues convert [9R]iP and other cytokinin ribosides to the corresponding 5′-monophosphates. The enzyme involved appears to be adenosine kinase (Hacker 1970; Divekar and Hakala 1971; McLennan and Pater 1973).

4.7.4. Incorporation into RNA

The possible incorporation of cytokinin into tRNA is a controversial issue, but one of considerable interest, since such incorporation has been suggested as a mechanism of cytokinin action. A number of workers (Hall 1968; Richmond et al. 1970; Bezemer-Sybrandy and Veldstra 1971; Elliott and Murray 1972) have failed to detect incorporation of labelled cytokinins into tRNA, while Kende and Tavares (1968) found that the 9-methyl cytokinin, 6-benzylamino-9-methylpurine, was not incorporated. However Fox and Chen (1967) reported incorporation of 6-benzylaminopurine into tRNA where it appeared to occur as a nucleotide, while Burrows et al. (1971) identified N_6-benzyladenosine in the hydrolyzate of tRNA isolated from tobacco callus grown on 6-benzylaminopurine. These reports of incorporation have been criticized on the basis of purity of the tRNA preparations (Kende 1971 and references therein; see also Hall 1973) and the demonstrated difficulty of freeing tRNA from non-covalently bound 6-benzylaminopurine and its metabolites (Elliott and Murray 1972). The problem of cytokinin incorporation into tRNA has recently been reexamined critically by Walker et al. (1974) who supplied a mixture of [8-^{14}C]6-benzylaminopurine and [benzene-^{3}H]6-benzylaminopurine to tobacco tissue. This work established that there is incorporation of intact 6-benzylaminopurine, but the level is extremely low – one molecule per 10,000 tRNA molecules. It is possible this is due to incorporation into the -CCA "tail" of tRNA which undergoes turnover, or is the result of transcriptional errors. Such incorporation would probably be of no physiological significance.

In further work, Armstrong et al. (1976) established that labelled 6-benzylaminopurine is also incorporated intact into rRNA of tobacco callus. In fact the total labelled 6-benzylaminopurine recovered from the rRNA fraction greatly exceeded the total recovered from the tRNA fraction. In view of this, the apparent incorporation into tRNA should be reassessed to determine whether it could be accounted for by contamination of the tRNA with degraded tRNA.

4.8. Translocation

In the early studies of cytokinins, little evidence was obtained for the movement of exogenous cytokinins away from the site of application. Thus, when kinetin was applied to the surface of leaves (Mothes 1964) and stems (Sachs and Thimann 1964) its effects were extremely localized. Such experiments suggested that cytokinins were immobile. Evidence to the contrary was first derived from experiments involving the application of cytokinin to the cut ends of stems and petioles. Thus, Osborne and coworkers obtained evidence for the movement of 6-benzylaminopurine with predominantly basipetal polarity in petiole segments of *Phaseolus vulgaris* and observed that this movement was enhanced by the auxin IAA (Osborne and Black 1964; Black and Osborne 1965; Osborne and McCready 1965). Lagerstedt and Langston (1966) were able to confirm the results of Osborne and Black (1964), although they did find that the polar movement of kinetin was very dependent upon the age of the tissue and the concentration of IAA. Other instances of basipetal cytokinin movement have been reported. Seth et al. (1966) detected basipetal movement of kinetin applied to the stumps of decapitated *Phaseolus vulgaris* plants and demonstrated that the movement was stimulated by IAA. In the presence of IAA (10 μM), polar transport of kinetin occurred in segments of petioles, hypocotyls and roots of radish. Polarity was basipetal in petioles and hypocotyls, but acropetal in roots (Radin and Loomis 1974). Pieniazek (1964) observed bud break on apple seedlings below the point of application of 6-benzylaminopurine.

However basipetal movement of 6-benzylaminopurine and kinetin does not appear to be a general phenomenon. Fox and Weis (1965) obtained no evidence for polar transport in *Phaseolus vulgaris* petiole segments with or without IAA, and obtained similarly negative results using petioles of *Coleus blumei* (a result confirmed by Veen and Jacobs 1969), epicotyls of *Pisum sativum* and coleoptiles of *Avena sativa.* Lack of polarity was also indicated in experiments with petioles of *Xanthium pennsylvanicum* (Osborne and McCready 1965; Lagerstedt and Langston 1966), stem sections of *Lens* (Pilet et al. 1967; Pilet 1968), mesocotyl and coleoptile segments of *Zea mays* (El-Saidi 1972), and sections of *Cleome hassleriana* stamen filaments, gynophores and pedicels (Koevenig 1973). In a recent study of translocation of zeatin in *Coleus* petiole segments, Jacobs (1976) detected only apolar movement. The significance of the reported basipetal transport of cytokinin is uncertain. In some experiments where this was observed (e.g. Black and Osborne 1965), the cytokinin concentrations were about 400 times greater than optimal levels for callus growth, and were certainly unphysiological.

Using autoradiography, Lagerstedt and Langston (1967) confirmed that kinetin remained localized when applied to laminar areas of excised leaves, but they also demonstrated that when the kinetin was applied to the leaf surface directly over a main vein, it moved in the vascular system to the area distal to the applied spot, suggesting entry into the transpiration stream. In experiments involving intact or nearly intact plants, acropetal movement of labelled cytokinins has been observed, but this may be simply a consequence of entry into the transpiration stream. When [^{14}C] 6-benzylaminopurine was applied to the stems of intact seedlings in a superficial paste (Pilet 1968) or by injection (Guern et al. 1968), the radioactivity moved acropetally, and in the latter case a stimulation of growth was observed in the axillary buds above the injection sites. Lagerstedt and Langston (1967) found that [8-^{14}C] kinetin was readily absorbed by the roots of intact tobacco seedlings and that ^{14}C radioactivity moved throughout the plants. In addition, agar blocks that had been inserted into petiole notches produced positive results in cytokinin bioassays, but it was not established that kinetin itself was the active substance. Z supplied to the roots of radish seedlings was translocated as the riboside in the xylem sap (Gordon et al. 1974); however Z and 7-glucosylzeatin, but no [9R]Z, were detected in the cotyledons.

Cytokinin translocation in relation to lateral bud development has received considerable attention. The work of Guern and Sadorge (1967) indicated that the polarity of 6-benzylaminopurine movement in young plants of *Cicer arietinum* depended on the relative positions of the site of cytokinin application and a growing axillary shoot. Thus, in decapitated seedlings with a developing axillary shoot, injected 6-benzylaminopurine moved preferentially into the shoot, whether the site of injection on the main stem was above or below the axillary shoot. Morris and Winfield (1972) obtained a similar result with [^{14}C]kinetin applied to decapitated dwarf pea seedlings. Whether kinetin was applied to the root or upper stem, appreciable radioactivity was transported to the axillary buds. Treatment of the apical stump with IAA inhibited this transport, and instead label moved as it did in intact plants, accumulating in the stem and apical region. This suggests that the effect of an active apex on cytokinin transport can be mimicked by exogenous auxin. Although transport of label from intact root to shoot was demonstrated in this experiment, the chemical nature of the translocated activity was not determined.

Another recent study concerning relationships between cytokinins and apical dominance is that of Woolley and Wareing (1972a) who used decapitated plants of *Solanum andigena*. The top lateral buds of decapitated plants tend to develop into stolons, rather than leafy shoots, if an IAA-GA_3 mixture is applied to the apical stump. When [^{14}C]6-benzylaminopurine was supplied to the stems of such plants, the cytokinin and its riboside did accumulate to some extent in the tips of induced stolons, but showed an increased accumulation there as a result of removal of the IAA-GA_3 mixture from the apex. This increased cytokinin content may be associated with the resulting transformation of the stolon into a leafy shoot.

In further experiments using decapitated, two-node cuttings of *Solanum andigena* without roots, Woolley and Wareing (1972b) observed that application of IAA to the upper cut surface suppressed the accumulation, in the lateral buds, of 6-benzylaminopurine supplied at the base of the cutting. In contrast, IAA did not suppress, and may actually have enhanced, the accumulation of adenine in these buds. IAA did not appear to reduce the accumulation of 6-benzylaminopurine in lateral buds by diversion to the point of IAA application, but IAA did promote the formation of an unknown metabolite of the cytokinin in stem tissue (Woolley and Wareing 1972b). Hence auxin appears to regulate both the translocation and the metabolism of cytokinin.

Evidence for radial and tangential transport of cytokinins in stems has also been reported. Radial translocation of [^{14}C]kinetin and [^{14}C]GA_3 in *Salix viminalis* from the transpiration stream to the sieve-tube sap has been established by Bowen and Wareing (1969). Tangential movement of kinetin also occurs in this species. While IAA moves preferentially into the lower half of a horizontal stem, kinetin moves preferentially into the upper half when these substances are applied to a bark abrasion on the side of the stem (Lepp and Peel 1971).

The occurrence of endogenous cytokinins in xylem and phloem sap has been discussed previously (see section 4.2.3.2).

4.9. Effect of environmental and chemical stimuli on cytokinin levels

4.9.1. The effect of light

Light quality and photoperiod markedly alter cytokinin levels in plant tissues (Table 4.5). Of particular interest are the rapid changes caused by exposure of seeds and leaves to red light for 5–10 min. The rapidity of these effects raises the possibility that cytokinin may be released from bound and inactive forms in response to red light. However, very rapid biosynthesis cannot be excluded at present. Comparable studies with etiolated wheat leaves indicate that red light causes both synthesis of gibberellins and release of bound gibberellins (Loveys and Wareing 1971). It has been claimed (Van Staden and Wareing 1972b) that the increase in cytokinin levels in *Rumex* seed caused by red light can be reversed if the red light treatment is followed immediately by far-red irradiation for 20 min. However, the controls required to reach this conclusion were not incorporated into the experiment. Hence the claim of Van Staden and Wareing (1972b) that cytokinin levels in *Rumex* seed are under the control of phytochrome lacks an adequate experimental basis. Red – far-red reversibility does not appear to have been demonstrated satisfactorily for any of the reported rapid effects on cytokinin levels.

Red-light treatment of poplar leaves results in a pronounced and specific increase in the level of the cytokinin 6-(*O*-hydroxybenzyl)aminopurine riboside (Hewett and Wareing 1973d). The level of this cytokinin in poplar leaves rises to a maximum about 5 min after daybreak (Thompson et al. 1975). The total cytokinin content of attached poplar leaves was determined over a 24-hr period by Hewett and Wareing (1973d). Again a pronounced peak was detected at daybreak.

4.9.2. The effect of temperature

Change in temperature can markedly alter cytokinin levels and several studies are concerned with the effect of stratification on the level of cytokinin activity in seeds. Sugar maple seeds kept at 20 °C for various periods do not germinate when transferred to suitable germination conditions and no cytokinin activity could be detected in them (Van Staden et al. 1972). However, seed subjected to chilling was able to germinate and cytokinin activity was extractable from the stratified seed. The highest level of cytokinin activity was found in seed chilled for 20 days (Van Staden et al. 1972). Surprisingly no cytokinin was detected in the germinated seed. Stratification of dormant seed of *Protea compacta* (embryo dormant) and of *Leucadendron daphnoides* (coat-imposed dormancy) resulted in an increased level of butanol-extractable cytokinins (Brown and Van Staden 1973). In the former seed, these cytokinins appear to be either synthesized or released from a bound form; in the latter they are probably derived from cytokinins (?nucleotides) which are not extractable by butanol (Brown and Van Staden 1973). In contrast, no marked changes were detected in cytokinin level during stratification of sycamore seeds; only a gradual decline in level was observed (Webb et al. 1973).

Changes in root temperature affect cytokinin levels in xylem sap. A cytokinin of low R_f was present in sap of *Vitis vinifera* when the root temperature was maintained at 20 °C; it was absent in sap when the root temperature was held at 30 °C (Skene and Kerridge 1967). Cytokinin activity in sap of *Vitis vinifera* canes increased markedly when

Table 4.5.

The effect of light quality, photoperiod and red light on cytokinin levels

Photostimulus	Plant species	Period of exposure to stimulus	Effect on cytokinin level	Reference
Change in light quality (supplementation of fluorescent with incandescent)	Poplar seedlings (*Populus* × *robusta)*	3 weeks	Increased in apices, mature leaves and sap	Hewett and Wareing (1973d)
Change in light quality (red light at end of 16-hr light period)	Poplar seedlings (*Populus* × *robusta)*	4 hr	Increased 20-fold in leaves	Hewett and Wareing (1973d)
Change in light quality (light from tungsten-filament lamps at end of 16-hr photoperiod from fluorescent tubes)	*Xanthium strumarium* seedlings	throughout seedling development (9 weeks)	Marked increase in axillary buds	Tucker and Mansfield (1972, 1973)
Photoperiod (change to short days)	Potato plants *(Solanum tuberosum)*	4 days	Marked increase in entire plants	Langille and Forsline (1974); Forsline and Langille (1975)
Photoperiod (change to short days)	*Perilla frutescens* plants	35 days	Marked increase in xylem sap	Beever and Woolhouse (1973
Photoperiod (change to short days)	*Xanthium strumarium*	1 day	Marked decrease in leaves, buds and root exudate	Van Staden and Wareing (1972a); Henson and Wareing (1974)
Photoperiod (change to short days)	*Begonia* × *cheimantha*	8 days	Increase in leaves	Heide and Skoog (1967)
Photoperiod (change to long days)	*Solanum andigena*	throughout growth of plants	Increase in roots	Woolley and Wareing (1972c)
Red light after dark period	*Rumex obtusifolius* seed	10 min	Marked increase	Van Staden and Wareing (1972b)
Red light after dark period	*Lactuca sativa* seed	20 min	Marked increase in butanol extractable cytokinin	Van Staden (1973a)
Red light after dark period	*Populus* × *robusta* seedlings	10 min	Increase in leaves	Hewett and Wareing (1973d)
Red light after dark period	*Populus* × *robusta* detached leaves	5 min	Marked but transient increase	Hewett and Wareing (1973d)

the canes were subjected to cold-storage (Skene 1972b). A careful study of the effect of root temperature on hormones in *Zea mays* xylem sap has been reported by Atkin et al. (1973). The flux of both cytokinins and gibberellins from the root was greatest at a root temperature of 28 °C as was their concentration in the sap. As root temperature was lowered to 8 °C, shoot growth was reduced and cytokinin and gibberellin export also declined. However lowering temperature increased inhibitor export (Atkin et al. 1973). Supraoptimal root temperature reduced the cytokinin level in roots and leaves of apple seedlings (Gur et al. 1972).

The level of cytokinin activity in the apical regions of potato tubers increased markedly during storage at low temperature. Subsequent transfer of the tubers from the cold room to a moist chamber at room temperature further raised the level of cytokinin activity (Engelbrecht and Bielinska-Czarnecka 1972).

4.9.3. The effect of other factors including hormones

Certain chemicals including plant hormones are known to affect cytokinin levels. Germination of *Spergula arvensis* seed is promoted by treatment with ethylene and the process appears to be under phytochrome control (Van Staden et al. 1973). The breaking of dormancy in this seed by light and ethylene is accompanied by a marked increase in the level of cytokinin activity. This response occurs prior to germination (Van Staden et al. 1973). When dormant seeds of lettuce are allowed to imbibe GA_3 solution in darkness, germination is induced and the level of butanol-extractable cytokinins increases (Van Staden 1973a).

The growth retardant 2-chloroethyltrimethylammonium chloride (CCC) has the ability to elevate cytokinin levels in plant tissues. When applied to grape seedling roots, CCC induces swollen root tips and markedly increases the amount of cytokinin activity in xylem sap (Skene 1970). CCC appears to increase cytokinin biosynthesis in the root and the swelling of the root tips is possibly a consequence of the increased cytokinin level in this region (Skene 1970). When applied to *Chenopodium rubrum* plants during floral induction, CCC causes an increase in the level of cytokinins in apical buds. GA_3 or IAA, at concentrations which reverse the inhibitory effect of CCC on flowering, reduce the cytokinin activity in apical buds (Teltscherova 1970). CCC enhances markedly the cytokinin level in *Vicia faba* plants, especially at the time of shedding of flowers and immature pods (El-Antably 1976).

Dormancy of *Leucadendron daphnoides* seed is apparently due to the restricting effect of the seed coat on oxygen diffusion to the embryo. High oxygen tensions markedly promote germination and cause a large increase in butanol extractable cytokinins before germination is detectable (Van Staden and Brown 1973; Brown and Van Staden 1975).

Low nitrogen status reduces the level of cytokinin activity in roots and exudate of sunflower plants (Wagner and Michael 1971) and in roots of *Solanum andigena* plants (Woolley and Wareing 1972c). Application of potassium nitrate or ammonium sulphate to the soil increases markedly the cytokinin content of leaves and xylem sap of rice plants, nitrate being more effective than the ammonium fertilizer (Yoshida and Oritani 1974). The presence of compounds containing reduced nitrogen in the culture medium, markedly increases the cytokinin level in *Vinca rosea* tissue cultures (Peterson and Miller

1976). Phosphorus deficiency results in a large reduction in the cytokinin level in root exudate of tomato plants (Menary and Van Staden 1976).

Stress conditions of diverse types result in reduced levels of cytokinin activity in plant tissues and xylem sap. Thus when plant roots are subjected to water shortage (Itai and Vaadia 1965), flooding (Burrows and Carr 1969), excess salinity or osmoticum (Itai et al. 1968), low pH (Andreenko et al. 1964; Banko and Boe 1975) or heat treatment at 46 °C for 2 min (Itai et al. 1973; Itai and Benzioni 1974), cytokinin activity of the root exudate is reduced. When water stress of coffee trees is relieved, the cytokinin level in the sap increases very greatly (Browning 1973). Water stress applied to the shoots of tobacco plants caused by placing the plants in an air stream, reduces cytokinin activity in xylem sap and in leaves (Itai and Vaadia 1971). When detached leaves are wilted, their cytokinin content is also diminished, but this rises markedly when the leaves recover in a humid chamber (Itai and Vaadia 1971; see also Engelbrecht 1972). Cytokinin activity in leaves is reduced markedly when the leaves are subjected to heat shock (Engelbrecht 1972; Itai and Benzioni 1974; Benzioni and Itai 1975). Cytokinin metabolism is enhanced in leaves by both heat and water stress (Itai and Vaadia 1971; Itai and Benzioni 1974). The effect of stress conditions on cytokinin levels is discussed further in Chapter 11, Volume II.

In numerous studies reported in this section, the cytokinin level in xylem sap has changed in response to an environmental or chemical stimulus. Some of these changes have been discussed in relation to shoot development or physiology in Chapter 4, Volume II, where it is evident that root-produced cytokinins play a vital role in the growth and senescence of the aerial parts of the flowering plant.

References

Abdel-Rahman, M., Thomas, T.H., Doss, G.J. and Howell, L. (1975) Physiol. Plant., 34, 39–43.
Abe, H., Uchiyama, M., Tanaka, Y. and Saito, H. (1976) Tetrahedron Lett., pp. 3807–3810.
Alvim, R., Hewett, E.W. and Saunders, P.F. (1976) Plant Physiol., 57, 474–476.
Andreenko, S.S., Potapov, N.G. and Kosulina, L.G. (1964) Dokl. Akad. Nauk SSSR, 155, 35–37.
Armstrong, D.J. and Skoog, F. (1975) Plant Physiol., 55, 237–239.
Armstrong, D.J., Burrows, W.J., Evans, P.K. and Skoog, F. (1969) Biochem. biophys. Res. Commun., 37, 451–456.
Armstrong, D.J., Murai, N., Taller, B.K. and Skoog, F. (1976a) Plant Physiol., 57, 15–22.
Armstrong, D.J., Scarbrough, E., Skoog, F., Cole, D.L. and Leonard, N.J. (1976b) Plant Physiol., 58, 749–752.
Arora, N., Skoog, F. and Allen, O.N. (1959) Amer. J. Bot., 46, 610–613.
Atkin, R.K., Barton, G.E. and Robinson, D.K. (1973) J. exp. Bot., 24, 475–487.
Babcock, D.F. and Morris, R.O. (1970) Biochemistry, 9, 3701–3705.
Back, A., Bittner, S. and Richmond, A.E. (1972) J. exp. Bot., 23, 744–750.
Balazs, E. and Sziraki, I. (1974) Acta phytopath., 9, 287–292.
Banerjee, S.N. (1968) Bot. Mag. (Tokyo), 81, 67–73.
Banko, T.J. and Boe, A.A. (1975) J. Amer. Soc. Hort. Sci., 100, 168–172.
Barea, J.M. and Brown, M.E. (1974) J. appl. Bacteriol., 37, 583–593.
Barthe, P. and Bulard, C. (1974) Canad. J. Bot., 52, 1515–1518.
Bartz, J.K., Kline, L.K. and Soll, D. (1970a) Biochem. biophys. Res. Commun., 40, 1481–1487.
Bartz, J., Soll, D., Burrows, W.J. and Skoog, F. (1970b) Proc. nat. Acad. Sci. (Wash.), 67, 1448–1453.

Barzilai, E. and Mayer, A.M., (1964) Aust. J. biol. Sci., 17, 798–800.
Beauchesne, G. and Goutarel, R. (1963) Physiol. Plant., 16, 630–631.
Beever, J.E. and Woolhouse, H.W. (1973) Nature New Biology, 246, 31–32.
Beever, J.E. and Woolhouse, H.W. (1974) in: Mechanisms of Regulation of Plant Growth, R.L. Bieleski, A.R. Ferguson and M.M. Cresswell (eds.), Royal Society of New Zealand, Wellington, pp. 681–686.
Bentley-Mowat, J.A. and Reid, S.M. (1968) Ann. Bot. (Lond.), 32, 23–32.
Benzioni, A. and Itai, C. (1975) Physiol. Plant., 35, 80–84.
Bergmann, L. and Rennenberg, H. (1975) Naturwissenschaften, 62, 349–350.
Beutelmann, P. (1973) Planta, 112, 181–190.
Bezemer-Sybrandy, S.M. and Veldstra, H. (1971) Physiol. Plant., 25, 1–7.
Biddington, N.L. and Thomas, T.H. (1973) Planta, 111, 183–186.
Bieleski, R.L. (1964) Analyt. Biochem., 9, 431–442.
Bieleski, R.L. (1968) Plant Physiol., 43, 1297–1308.
Bigot, M.C. (1968) C.R. Acad. Sci. (Paris), 266, 349–352.
Black, M.K. and Osborne, D.J. (1965) Plant Physiol., 40, 676–680.
Blondeau, R. (1970) C.R. Acad. Sci. (Paris), 270, 3158.
Blumenfeld, A. and Gazit, S. (1970a) Plant Physiol., 45, 535–536.
Blumenfeld, A. and Gazit, S. (1970b) Plant Physiol., 46, 331–333.
Blumenfeld, A. and Gazit, S. (1971) Physiol. Plant., 25, 369–371.
Borkowska, B. and Rudnicki, R.M. (1975) Fruit Science Reports (Skierniewice, Poland), 11, No. 2, 1–16.
Borkowska, B. and Borkowski, J. (1975) Bull. Acad. Pol. Sci. (Ser. Sci. Biol), 23, 291–294.
Bose, T.K. and Nitsch, J.P. (1970) Physiol. Plant., 23, 1206–1211.
Bottger, M. (1974) Planta, 121, 253–261.
Bottomley, W., Kefford, N.P., Zwar, J.A. and Goldacre, P.L. (1963) Aust. J. biol. Sci., 16, 395–406.
Bowen, M.R. and Wareing, P.F. (1969) Planta, 89, 108–125.
Brain, K.R., Chalopin, M.C., Turner, T.D., Blunden, G. and Wildgoose, P.B. (1973) Plant Sci. Lett., 1, 241–245.
Braun, A.C. and White, P.R. (1943) Phytopathology, 33, 85–100.
Braun, A.C. and Naf, U. (1954) Proc. Soc. exp. Biol. (N.Y.) 86, 212–214.
Brautigam, E. and Engelbrecht, L. (1976) in: Proceedings of the Fourth Symposium on Plant Growth Regulators, Acta Universitatis Nicolai Copernici (Biologia XVIII), Torun, Poland, pp. 149–151.
Brown, N.A.C. and Van Staden, J. (1973) Physiol. Plant., 28, 388–392.
Brown, N.A.C. and Van Staden, J. (1975) Z. Pflanzenphysiol., 75, 31–37.
Browning, G. (1973) J. Hort. Sci., 48, 297–310.
Brownlee, B.G., Hall, R.H. and Whitty, C.D. (1975) Canad. J. Biochem., 53, 37–41.
Bruce, M.I. and Zwar, J.A. (1966) Proc. roy. Soc. B, 165, 245–265.
Brueske, C.H. and Bergeson, G.B. (1972) J. exp. Bot., 23, 14–22.
Bui-Dang-Ha, D. and Nitsch, J.P. (1970) Planta, 95, 119–126.
Burrows, W.J. (1976) Planta, 130, 313–316.
Burrows, W.J. and Carr, D.J. (1969) Physiol. Plant., 22, 1105–1112.
Burrows, W.J. and Carr, D.J. (1970) Physiol. Plant., 23, 1064–1070.
Burrows, W.J. and Leworthy, D.P. (1976) Biochem. biophys. Res. Commun., 70, 1109–1114.
Burrows, W.J., Armstrong, D.J., Kaminek, M., Skoog, F., Bock, R.M., Hecht, S.M., Dammann, L.G., Leonard, N.J. and Occolowitz, J. (1970) Biochemistry, 9, 1867–1872.
Burrows, W.J., Skoog, F. and Leonard, N.J. (1971) Biochemistry, 10, 2189–2194.
Carnes, M.G., Brenner, M.L. and Andersen, C.R. (1974) in: Plant Growth Substances 1973, Hirokawa Publishing Co., Tokyo, pp. 99–110.
Carnes, M.G., Brenner, M.L. and Andersen, C.R. (1975) J. Chromatogr., 108, 95–106.
Carr, D.J. and Burrows, W.J. (1966) Life Sci., 5, 2061–2077.
Carr, D.J. and Reid, D.M. (1968) in: Biochemistry and Physiology of Plant Growth Substances, F. Wightman and G. Setterfield (eds.), Runge Press, Ottawa, pp. 1169–1185.

Challice, J.S. (1975) Planta, 122, 203–207.
Chapman, R.W., Morris, R.O. and Zaerr, J.B. (1976) Nature (Lond.), 262, 153–154.
Chen, C.M. and Hall, R.H. (1969) Phytochemistry, 8, 1687–1695.
Chen, C.M., Hartnell, G.F., Smith, O.C. and Francis, W.S. (1974a) Analyt. Biochem., 60, 441–448.
Chen, C.M., Smith, O.C. and Hartnell, G.F. (1974b) Canad. J. Biochem., 52, 1154–1161.
Chen, C., Smith, O.C. and McChesney, J.D. (1975) Biochemistry, 14, 3088–3093.
Chheda, G.B. and Mittelman, A. (1972) Biochem. Pharmacol., 21, 27–37.
Cole, D.L., Leonard, N.J. and Cook, J.C. (1974) in: Recent Developments in Oligonucleotide Synthesis and Chemistry of Minor Bases of tRNA, Universytet Im. Adama Mickiewicza Press, pp. 153–174.
Conrad, K. (1971) Biochem. Physiol. Pflanz., 162, 327–333.
Conrad, K. (1974) Biochem. Physiol. Pflanz., 165, 531–535.
Conrad, K. and Kohn, B. (1975) Phytochemistry, 14, 325–328.
Coppola, S., Percuoco, G., Zoina, A. and Picci, G. (1971) Annali Microbiol., 21, 45.
Cowley, D.E., Jenkins, I.D., MacLeod, J.K., Summons, R.E., Letham, D.S., Wilson, M.M. and Parker, C.W. (1975) Tetrahedron Lett., pp. 1015–1018.
Crady, E.E. and Wolf, F.D. (1959) Physiol. Plant., 12, 526–533.
Crafts, C.B. and Miller, C.O. (1974) Plant Physiol., 54, 586–588.
Dammann, L.G., Leonard, N.J., Schmitz, R.Y. and Skoog, F. (1974) Phytochemistry, 13, 329–336.
Davey, J.E. and Van Staden, J. (1976) Planta, 130, 69–72.
Dekhuijzen, H.M. and Gevers, E.Ch.Th. (1975) Physiol. Plant., 35, 297–302.
Dekhuijzen, H.M. and Overeem, J.C. (1971) Physiol. Plant Pathol., 1, 151–161.
Dekhuijzen, H.M. and Overeem, J.C. (1972) Phytochemistry, 11, 1669–1672.
Dekhuijzen, H.M. and Staples, R.C. (1968) Contrib. Boyce Thompson Inst., 24, 39–52.
Deleuze, G.G., McChesney, J.D. and Fox, J.E. (1972) Biochem. biophys. Res. Commun., 48, 1426–1432.
Divekar, A.Y. and Hakala, M.T. (1971) Molec. Pharm., 7, 663–673.
Doree, M. and Guern, J. (1967) C.R. Acad. Sci. (Paris), Ser. D, 265, 29–32.
Doree, M. and Guern, J. (1973) Biochim. biophys. Acta (Amst.), 304, 611–622.
Doree, M. and Terrine, C. (1973) Phytochemistry, 12, 1017–1023.
Duke, C.C., Liepa, A.J., MacLeod, J.K., Letham, D.S. and Parker, C.W. (1975) J. Chem. Soc. Chem. Commun., pp. 964–965.
Dumbroff, E.B. and Brown, D.C.W. (1976) Canad. J. Bot. 54, 191–197.
Dyson, W.H. and Hall, R.H. (1972) Plant Physiol., 50, 616–621.
Dyson, W.H., Fox, J.E. and McChesney, J.D. (1972a) Plant Physiol., 49, 506–513.
Dyson, W.H., Hall, R.H., Hong, C.I., Dutta, S.P. and Chheda, G.B. (1972b) Canad. J. Biochem., 50, 237–243.
Edwards, W.J. and LaMotte, C.E. (1975) Plant Physiol., 56, 425–428.
Einset, J.W. and Skoog, F. (1973) Proc. nat. Acad. Sci. (Wash.), 70, 658–660.
El-Antably, H.M.M. (1976) Z. Pflanzenphysiol., 80, 29–35.
El-Beltagy, A.S., Patrick, J.P., Hewett, E.W. and Hall, M.A. (1976) J. Hort. Sci., 51, 15–30.
Elkins, B.N. and Keller, E.B. (1974) Biochemistry, 13, 4622–4628.
Elliott, D.C. and Murray, A.W. (1972) Biochem. J., 130, 1157–1160.
Elliott, D.C., Murray, A.W., Saccone, G.T. and Atkinson, M.R. (1972) in: Plant Growth Substances 1970, D.J. Carr (ed.), Springer, Berlin, pp. 459–466.
El-Saidi, M.T. (1972) Ann. Bot. (Lond.), 36, 99–107.
Engelbrecht, L. (1968) Flora Abt. A, 159, 208–214.
Engelbrecht, L. (1971a) Biochem. Physiol. Pflanz., 162, 547–558.
Engelbrecht, L. (1971b) Biochem. Physiol. Pflanz., 162, 9–27.
Engelbrecht, L. (1972) Biochem. Physiol. Pflanz., 163, 335–343.
Engelbrecht, L. and Bielinska-Czarnecka, M. (1972) Biochem. Physiol. Pflanz., 163, 499–504.
Engelbrecht, L., Orban, U. and Heese, W. (1969) Nature (Lond.), 223, 319–321.

Esashi, Y. and Leopold, A.C. (1969) Plant Physiol., 44, 618–620.
Faulkner, R.D. and Uziel, M. (1971) Biochim. biophys. Acta (Amst.), 238, 464–474.
Fawcett, C.H. and Wright, S.T.C. (1968) Phytochemistry, 7, 1719–1725.
Feldman, L.J. (1975) in: The Development and Function of Roots, J.G. Torrey and D.T. Clarkson (eds.), Academic Press, London, pp. 55–72.
Fittler, F. and Hall, R.H. (1966) Biochem. biophys. Res. Commun., 25, 441–446.
Fittler, F., Kline, L.K. and Hall, R.H. (1968a) BIochemistry, 7, 940–944.
Fittler, F., Kline, L.K. and Hall, R.H. (1968b) Biochem. biophys. Res. Commun., 31, 571–576.
Fletcher, R.A. and McCullagh, D. (1971) Planta, 101, 88–90.
Fleysher, M.H. (1972) J. med. Chem., 15, 187–191.
Fleysher, M.H., Hakala, M.T., Bloch, A. and Hall, R.H. (1968) J. med. Chem., 11, 717–720.
Forsline, P.L. and Langille, A.R. (1975) Physiol. Plant., 34, 75–77.
Fox, J.E. (1969) in: The Physiology of Plant Growth and Development, M.B. Wilkins (ed.), McGraw-Hill, New York, pp. 85–123.
Fox, J.E. and Chen, C. (1967) J. biol. Chem., 242, 4490–4494.
Fox, J.E. and Chen, C.M. (1968) in: Biochemistry and Physiology of Plant Growth Substances, F. Wightman and G. Setterfield (eds.), Runge Press, Ottawa, pp. 777–789.
Fox, J.E. and McChesney, J.D. (1974) in: Plant Growth Substances 1973, Hirokawa, Tokyo, pp. 468–479.
Fox, J.E. and Weis, J.S (1965) Nature (Lond.), 206, 678–679.
Fox, J.E., Sood, C.K., Buckwalter, B. and McChesney, J.D. (1971) Plant Physiol., 47, 275–281.
Fox, J.E., Cornette, J., Deleuze, G., Dyson, W., Giersak, C., Niu, P., Zapata, J. and McChesney, J. (1973a) Plant Physiol., 52, 627–632.
Fox, J.E., Sood, C.K. and McChesney, J.D. (1973b) Phytochemistry, 12, 1531–1533.
Fujii, T. and Ogawa, N. (1972) Tetrahedron Lett., pp. 3075–3078.
Fuller, W. and Hodgson, A. (1967) Nature (Lond.), 215, 817–821.
Gallo, R.C., Hecht, S.M., Whang-peng, J. and O'Hopp, S. (1972) Biochim. biophys. Acta (Amst.), 281, 488–500.
Gauss, D.H., Haar, F., Maelicke, A. and Cramer, F. (1971) Ann. Rev. Biochem., 40, 1045–1078.
Gazit, S. and Blumenfeld, A. (1970) Plant Physiol., 46, 334–336.
Gefter, M.L. and Russell, R.L. (1969) J. molec. Biol., 39, 145–157.
Gepstain, S. and Ilan, I. (1970) Plant Cell Physiol., 11, 819–822.
Gordon, M.E. and Letham, D.S. (1975) Aust. J. Plant Physiol., 2, 129–154.
Gordon, M.E., Letham, D.S. and Parker, C.W. (1974a) Ann. Bot. (Lond.), 38, 809–825.
Gordon, M.E., Wilson, M.M., Parker, C.W. and Letham, D.S. (1974b) in: Mechanisms of Regulation of Plant Growth, R.L. Bieleski, A.R. Ferguson and M.M. Cresswell (eds.), Royal Society of New Zealand, Wellington, pp. 773–780.
Grindley, G.B., Divekar, A.Y. and Hakala, M.T. (1973) Cancer Res., 33, 2459–2463.
Guern, J. and Hugon, E. (1966) C.R. Acad. Sci. (Paris), 262, 354–357.
Guern, J., Doree, M. and Sadorge, P. (1968) in: Biochemistry and Physiology of Plant Growth Substances, F. Wightman and G. Setterfield (eds.), Runge Press, Ottawa, pp. 1155–1167.
Guern, J., Doree, M., Terrine, C., Pethe-Sadorge, P. and Leguay, J.J. (1974) in: Plant Growth Substances 1973, Hirokawa, Tokyo, pp. 356–363.
Guern, N. and Sadorge, P. (1967) C.R. Acad. Sci. (Paris), 264, 2106–2109.
Gupta, G.R.P. and Maheshwari, S.C. (1970) Plant Physiol., 45, 14–18.
Gur, A., Bravado, B. and Mizrahi, Y. (1972) Physiol. Plant., 27, 130–138.
Hacker, B. (1970) Biochim. biophys. Acta (Amst.), 224, 635–638.
Hacker, B. and Feldbush, T.L. (1969) Biochem. Pharmacol., 18, 847–853.
Hagen, G.L. and Marcus, A. (1975) Plant Physiol., 55, 90–93.
Hahn, H. (1975) Physiol. Plant., 34, 204–207.
Hahn, H. and Bopp, M. (1968) Planta, 83, 115–118.
Hahn, H., de Zacks, R. and Kende, H. (1974) Naturwissenschaften, 61, 170.

Hahn, H., Heitmann, I. and Blumbach, M. (1976) Z. Pflanzenphysiol., 79, 143–153.
Hall, R.H. (1968) in: Biochemistry and Physiology of Plant Growth Substances, F. Wightman and G. Setterfield (eds.), Runge Press, Ottawa, pp. 47–56.
Hall, R.H. (1973) Ann. Rev. Plant Physiol., 24, 415–444.
Hall, R.H. and Mintsioulis, G. (1973) J. Biochem. (Tokyo), 73, 739–748.
Hall, R.H., Csonka, L., David, H. and McLennan, B. (1967) Science, 156, 69–71.
Hall, R.H., Alam, S.N., MacLennan, B.D., Terrine, C. and Guern, J. (1971) Canad. J. Biochem., 49, 623–630.
Hall, S.M. and Baker, D.A. (1972) Planta, 106, 131–140.
Hamilton, J.L., Lowe, R.H. and Skoog, F. (1972) Plant Physiol., 50, 303–304.
Hammady, M.E., Pozsar, B.I. and Kiraly, Z. (1968) Acta phytopathol., 3, 157–164.
Hamzi, H.Q. and Skoog. F. (1964) Proc. nat. Acad. Sci. (Wash.), 51, 76–83.
Hashizume, T., Iizuka, M., Matsubara, S., Yoshida, K. and Kobayashi, T. (1974) in: Plant Growth Substances 1973 Hirokawa, Tokyo, pp. 462–467.
Hecht, S.M., Leonard, N.J., Schmitz, R.Y. and Skoog, F. (1970a) Phytochemistry, 9, 1173–1180.
Hecht, S.M., Leonard, N.J., Schmitz, R.Y. and Skoog, F. (1970b) Phytochemistry, 9, 1907–1913.
Hecht, S.M., Gupta, A.S. and Leonard, N.J. (1970c) Analyt. Biochem., 38, 230–251.
Hecht, S.M., Bock, R.M., Schmitz, R.Y., Skoog, F., Leonard, N.J. and Occolowitz, J.L. (1971) Biochemistry, 10, 4224–4228.
Hecht, S.M., Faulkner, R.D. and Hawrelak, S.D. (1974) Proc. nat. Acad. Sci. (Wash.), 71, 4670–4674.
Hecht, S.M., Frye, R.B., Werner, D., Hawrelak, S.D., Skoog, F. and Schmitz, R.Y. (1975) J. biol. Chem., 250, 7343–7351.
Heide, O.M. (1965) Physiol. Plant., 18, 891–920.
Heide, O.M. and Skoog, F. (1967) Physiol. Plant., 20, 771–780.
Helgeson, J.P. and Leonard, N.J. (1966) Proc. nat. Acad. Sci. (Wash.), 56, 60–63.
Helgeson, J.P., Haberlack, G.T. and Hecht, S.M. (1974) in: Plant Growth Substances 1973, Hirokawa, Tokyo, pp. 485–493.
Hemberg, T. (1974) Physiol. Plant., 32, 191–192.
Henderson, T.R., Frihart, C.R., Leonard, N.J., Schmitz, R.Y. and Skoog, F. (1975) Phytochemistry, 14, 1687–1690.
Henson, I.E. and Wareing, P.F. (1974) Physiol. Plant., 32, 185–187.
Henson, I.E. and Wareing, P.F. (1976) J. exp. Bot., 27, 1268–1278.
Henson, I.E. and Wheeler, C.T. (1976) New Phytol., 76, 433–439.
Hewett, E.W. and Wareing, P.F. (1973a) Physiol. Plant., 28, 393–399.
Hewett, E.W. and Wareing, P.F. (1973b) Physiol. Plant., 29, 386–389.
Hewett, E.W. and Wareing, P.F. (1973c) Planta, 112, 225–233.
Hewett, E.W. and Wareing, P.F. (1973d) Planta, 114, 119–129.
Hewett, E.W. and Wareing, P.F. (1974) in: Mechanisms of Regulation of Plant Growth, R.L. Bieleski, A.R. Ferguson and M.M. Cresswell (eds.), Royal Society of New Zealand, Wellington, pp. 693–701.
Hillman, W.S. (1957) Science, 126, 165–166.
Hirata, S. (1957) Ann. Phytopathol. Soc. Jap., 22, 153–158.
Holtz, J. and Klambt, D. (1975) Hoppe-Seyler's Z. physiol. Chem., 356, 1459–1461.
Horgan, R. (1975) Biochem. Biophys. Res. Commun., 65, 358–363.
Horgan, R., Hewett, E.W., Purse, J.G., Horgan, J.M. and Wareing, P.F. (1973) Plant Sci. Lett., 1, 321–324.
Horgan, R., Hewett, E.W., Horgan, J.M., Purse, J. and Wareing, P.F. (1975) Phytochemistry, 14, 1005–1008.
Hussain, A. and Boney, A.D. (1969) Nature (Lond.), 223, 504–505.
Hussain, A. and Boney, A.D. (1971) Botanica Mar., 14, 17–21.
Ingram, D.S. (1969) J. gen. Microbiol., 56, 55–67.
Itai, C. and Benzioni, A. (1974) in: Mechanisms of Regulation of Plant Growth, R.L. Bieleski, A.R. Fergusion and M.M. Cresswell (eds.), Royal Society of New Zealand, Wellington, pp. 477–482.

Itai, C. and Vaadia, Y. (1965) Physiol. Plant., 18, 941–944.
Itai, C. and Vaadia, Y. (1971) Plant Physiol., 47, 87–90.
Itai, C., Richmond, A. and Vaadia, Y. (1968) Israel J. Bot., 17, 187–193.
Itai, C., Benzioni, A. and Ordin, L. (1973) Physiol. Plant., 29, 355–360.
Iwamura, H., Ito, T., Kumazawa, Z. and Ogawa, Y. (1974) Biochem. biophys. Res. Commun., 57, 412–416.
Iwamura, H., Ito, T., Kumazawa, Z. and Ogawa, Y. (1975) Phytochemistry, 14, 2317–2321.
Jablonski, J.R. and Skoog, F. (1954) Physiol. Plant., 7, 16–24.
Jacobs, W.P. (1976) Amer. J. Bot., 63, 571–577.
Jennings, R.C. (1969) Aust. J. biol. Sci., 22, 621–627.
Johnson, G.S., D'Armiento, M. and Carchman, R.A. (1974) Exp. Cell Res., 85, 47–56.
Johnston, J.C. and Trione, E.J. (1974) Canad. J. Bot., 52, 1583–1589.
Jones, O.P. (1973) J. Hort. Sci., 48, 181–188.
Jordan, W.R. and Skoog, F. (1971) Plant Physiol., 48, 97–99.
Kaminek, M. (1974) J. Theoret. Biol., 48, 489–492.
Kannangara, T. and Booth, A. (1974) J. exp. Bot., 25, 459–467.
Karanov, E., Ivanova, I. and Vasilev, G. (1968) Fiziol. Rast., 15, 430–435.
Katsura, K., Egawa, H., Toki, T. and Ishii, S. (1966) Ann. Phytopathol. Soc. Jap., 32, 123–129.
Kefford, N.P., Zwar, J.A. and Bruce, M.I. (1965) Planta, 67, 103–106.
Kefford, N.P., Bruce, M.I. and Zwar, J.A. (1966) Planta, 68, 292–296.
Kefford, N.P., Zwar, J.A. and Bruce, M.I. (1968) in: Biochemistry and Physiology of Plant Growth Substances, F. Wightman and G. Setterfield (eds.), Runge Press, Ottawa, pp. 61–69.
Kefford, N.P., Bruce, M.I. and Zwar, J.A. (1973) Phytochemistry, 12, 995–1003.
Kende, H. (1964) Science, 145, 1066–1067.
Kende, H. (1965) Proc. nat. Acad. Sci. (Wash.), 53, 1302–1307.
Kende, H. (1971) Int. Rev. Cytology, 31, 301–338.
Kende, H. and Sitton, D. (1967) Ann. N.Y. Acad. Sci., 144, 235–243.
Kende, H. and Tavares, J.E. (1968) Plant Physiol., 43, 1244–1248.
Kikugawa, K., Iizuka, K. and Ichino, M. (1973) J. med. Chem., 16, 358–364.
Kiraly, Z., Pozsar, B.I. and El-Hammady, M. (1966) Acta phytopathol., 1, 29–38.
Kiraly, Z., El-Hammady, M. and Pozsar, B.I. (1967) Phytopathology, 57, 93–94.
Klambt, D. (1968) Planta, 82, 170–178.
Klambt, D., Thies, G. and Skoog, F. (1966) Proc. nat. Acad. Sci. (Wash.), 56, 52–59.
Klemen, F. and Klambt, D. (1974) Physiol. Plant., 31, 186–188.
Kline, L.K., Fittler, F. and Hall, R.H. (1969) Biochemistry, 8, 4361–4371.
Koevenig, J.L. (1973) Canad. J. Bot., 51, 2079–2083.
Kohler, K.H. and Conrad, K. (1968) Flora Abt. A, 159, 293–298.
Koshimizu, K., Kusaki, T., Mitsui, T. and Matsubara, S. (1967) Tetrahedron Lett., pp. 1317–1320.
Koshimizu, K., Kobayashi, A., Fujita, T. and Mitsui, T. (1968) Phytochemistry, 7, 1989–1994.
Kroon, C., Salemink, C.A. and Rogozinska, J.H. (1974) in: Plant Growth Substances 1973, Hirokawa, Tokyo, pp. 480–484.
Kulaeva, O.N. (1962) Fiziol. Rast., 9, 229–239.
Kulaeva, O.N., Chernyshev, E.A., Kayutenko, L.A., Dolgaya, M.E., Vorobeva, I.P., Popova, E.A. and Klyachko, N.L. (1965) Fiziol. Rast., 12, 902–908.
Kulaeva, O.N., Cherkasov, V.M. and Tretyakova, G.S. (1968) Dokl. Akad. Nauk SSSR, 178, 1204–1207.
Kung-Woo Lee, P., Kessler, B. and Thimann, K.V. (1974) Physiol. Plant., 31, 11–14.
Kuraishi, S. (1959) Scientific Papers of the College of General Education (University of Tokyo), 9, 67–104.
Kuraishi, S. and Yamaki, T. (1967) Physiol. Plant., 20, 208–212.

Lagerstedt, H.B. and Langston, R.G. (1966) Physiol. Plant., 19, 734–740.
Lagerstedt, H.B. and Langston, R.G. (1967) Plant Physiol., 42, 611–622.
Laloue, M. and Hall, R.H. (1973) Plant Physiol., 51, 559–562.
Laloue, M., Terrine, C. and Gawer, M. (1974) FEBS Lett., 46, 45–50.
Langille, A.R. and Forsline, P.L. (1974) Plant Sci. Lett., 2, 189–191.
Lau, O. and Yang, S.F. (1975) Plant Physiol., 55, 738–740.
Le John, H.B. (1975) Canad. J. Biochem., 53, 768–778.
Leonard, E.O., Skinner, C.G. and Skive, W. (1961) Arch. Biochem., 92, 33–37.
Leonard, N.J. and Henderson, T.R. (1975) J. Amer. chem. Soc., 97, 4990–4999.
Leonard, N.J., Achmatowicz, S., Loeppky, R.N., Carraway, K.L., Grimm, W.A.H., Szweykowska, A.M., Hamzi, H.Q. and Skoog, F. (1966) Proc. nat. Acad. Sci. (Wash.), 56, 709–716.
Leonard, N.J., Hecht, S.M., Skoog, F. and Schmitz, R.Y. (1968) Proc. nat. Acad. Sci. (Wash.), 59, 15–21.
Leonard, N.J., Hecht, S.M., Skoog, F. and Schmitz, R.Y. (1969a) Proc. nat. Acad. Sci. (Wash.), 63, 175–182.
Leonard, N.J., Iwamura, H. and Eisinger, J. (1969b) Proc. nat. Acad. Sci. (Wash.), 64, 352.
Lepp, N.W. and Peel, A.J. (1971) Planta, 99, 275–282.
Letham, D.S. (1963a) Life Sci., 2, 152–157.
Letham, D.S. (1963b) Life Sci., 2, 569–573.
Letham, D.S. (1963c) N.Z. J. Bot., 1, 336–350.
Letham, D.S. (1964) Colloq. Int. C.N.R.S. (Paris), No. 123, 109–117.
Letham, D.S. (1966a) Phytochemistry, 5, 269–286.
Letham, D.S. (1966b) Life Sci., 5, 551–554.
Letham, D.S. (1966c) Life Sci., 5, 1999–2004.
Letham, D.S. (1967a) Planta, 74, 228–242.
Letham, D.S. (1967b) Ann. Rev. Plant. Physiol., 18, 349–364.
Letham, D.S. (1968) in: Biochemistry and Physiology of Plant Substances, F. Wightman and G. Setterfield (eds.), Runge Press, Ottawa, pp. 19–31.
Letham, D.S. (1971) Physiol. Plant., 25, 391–396.
Letham, D.S. (1972) Phytochemistry, 11, 1023–1025.
Letham, D.S. (1973) Phytochemistry, 12, 2445–2455.
Letham, D.S. (1974a) Physiol. Plant., 32, 66–70.
Letham, D.S. (1974b) Planta, 118, 361–364.
Letham, D.S. and Bollard, E.G. (1961) Nature (Lond.), 191, 1119–1120.
Letham, D.S. and Miller, C.O. (1965) Plant Cell Physiol., 6, 355–359.
Letham, D.S. and Ralph, R.K. (1967) Life Sci., 6, 387–394.
Letham, D.S. and Wettenhall, R.E.H. (1977) in: The Ribonucleic Acids, P.R. Steward and D.S. Letham (eds.), Springer, New York, pp. 129–193.
Letham, D.S. and Williams, M.W. (1969) Physiol. Plant., 22, 925–936.
Letham, D.S. and Young, H. (1971a) Phytochemistry, 10, 2077–2081.
Letham, D.S. and Young, H. (1971b) Phytochemistry 10, 23–28.
Letham, D.S., Shannon, J.S. and McDonald, I.R.C. (1964) Proc. Chem. Soc., p. 230.
Letham, D.S., Shannon, J.S. and McDonald, I.R.C. (1967) Tetrahedron Lett., 23, 479–486.
Letham, D.S., Mitchell, R.E., Cebalo, T. and Stanton, D.W. (1969) Aust. J. Chem., 22, 205–219.
Letham, D.S., Parker, C.W. and Gordon, M.E. (1972) Physiol. Plant., 27, 285–290.
Letham, D.S., Wilson, M.M., Parker, C.W., Jenkins, I.D., MacLeod, J.K. and Summons, R.E. (1975) Biochim. biophys. Acta (Amst.), 399, 61–70.
Letham, D.S., Parker, C.W., Duke, C.C., Summons, R.E. and MacLeod, J.K. (1977) Ann. Bot. (Lond.), 41, 261–263.
Linsmaier, E.M. and Skoog, F. (1965) Physiol. Plant., 18, 100–127.
Locker, A. and Ilan, I. (1975) Plant Cell Physiol., 16, 449–454.
Loeffler, J.E. and Van Overbeek, J. (1964) Colloq. Int. C.N.R.S. (Paris), No. 123, 77–82.

Lorenzi, R., Horgan, R. and Wareing, P.F. (1975) Biochem. Physiol. Pflanz., 168, 333–339.
Loveys, B. and Wareing, P.F. (1971) Planta, 98, 109–116.
Luckwill, L.C. and Whyte, P. (1968) Soc. Chem. Ind., Monograph No. 31, 87–101.
Luke, H.H. and Freeman, T.E. (1967) Nature (Lond.), 215, 874–875.
Luke, H.H. and Freeman, T.E. (1968) Nature (Lond.), 217, 873–874.
Mackenzie, I.A. and Street, H.E. (1972) New Phytol., 71, 621–631.
MacLeod, J.K., Summons, R.E., Parker, C.W. and Letham, D.S. (1975) J.C.S. Chem. Commun., pp. 809–810.
MacLeod, J.K., Summons, R.E. and Letham, D.S. (1976) J. org. Chem., 41, 3959–3967.
Maenpaa, P.H., Bernfield, M.R. and Atkin, R.K. (1972) Biochem. Pharm., 21, 2277–2280.
Mann, J.D., Yung, K., Storey, W.B., Pu, M. and Conley, J. (1967) Plant Cell Physiol., 8, 613–622.
Martin, J.H., Fox, J.E. and McChesney, J.D. (1973) Phytochemistry, 12, 749–752.
Matsubara, S., Koshimizu, K. and Fujii, T. (1974) in: Plant Growth Substances 1973, Hirokawa, Tokyo, pp. 456–461.
Matsubara, S., Koshimizu, K. and Nakahira, R. (1968) Sci. Rep. Kyoto Pref. Univ., No. 19, Ser. A, 19–24.
Mayak, S. and Halevy, A.H. (1970) Plant. Physiol., 46, 497–499.
Mayak, S., Halevy, A.H. and Katz, M. (1972) Physiol. Plant., 27, 1–4.
McCalla, D.R., Morre, D.J. and Osborne, D.J. (1962) Biochim. biophys. Acta (Amst.), 55, 522–528.
McCombs, P.J.A. and Ralph, R.K. (1972) Planta, 107, 97–109.
McDonald, J.J., Leonard, N.J., Schmitz, R.Y. and Skoog, F. (1971) Phytochemistry, 10, 1429–1439.
McLennan, B.D. and Pater, A. (1973) Canad. J. Biochem., 51, 1341–1346.
Menary, R.C. and Van Staden, J. (1976) Aust. J. Plant Physiol., 3, 201–205.
Michael, G. and Seiler-Kelbitsch, H. (1972) Crop Sci., 12, 162–165.
Miller, C.O. (1956) Plant Physiol., 31, 318–319.
Miller, C.O. (1958) Plant Physiol., 33, 115–117.
Miller, C.O. (1961) Proc. nat. Acad. Sci. (Wash.), 47, 170–174.
Miller, C.O. (1962) Nature (Lond.), 194, 787–788.
Miller, C.O. (1963) in: Modern Methods of Plant Analysis, Vol. 6, H.F. Linskens and M.V. Tracey (eds.), Springer-Verlag, Berlin, pp. 194–202.
Miller, C.O. (1967a) Science, 157, 1055–1057.
Miller, C.O. (1967b) Ann. N.Y. Acad. Sci., 144, 251–257.
Miller, C.O. (1968) in: Biochemistry and Physiology of Plant Growth Substances, F. Wightman and G. Setterfield (eds.), Runge Press, Ottawa, pp. 33–45.
Miller, C.O. (1969) Planta, 87, 26–35.
Miller, C.O. (1971) in: Mycorrhizae, E. Hacskaylo (ed.), United States Government Printing Office, Washington, D.C., pp. 168–174.
Miller, C.O. (1974) Proc. nat. Acad. Sci. (Wash.), 71, 334–338.
Miller, C.O. (1975a) Plant Physiol., 55, 448–449.
Miller, C.O. (1975b) Proc. nat. Acad. Sci. (Wash.), 72, 1883–1886.
Miller, C.O. and Witham, F.H. (1964) Colloq. Int. C.N.R.S. (Paris), No. 123, 1–6 (erratum).
Miller, C.O., Skoog, F., Okumura, F.S., von Saltza, M.H. and Strong, F.M. (1956) J. Amer. chem. Soc., 78, 1375–1380.
Mittelman, A., Evans, J.T. and Chheda, G.B. (1975) Ann. N.Y. Acad. Sci., 255, 225–234.
Miura, G. and Hall, R.H. (1973) Plant Physiol., 51, 563–569.
Miura, G.A. and Miller, C.O. (1969a) Plant Physiol., 44, 1035–1039.
Miura, G.A. and Miller, C.O. (1969b) Plant Physiol., 44, 372–376.
Morris, D.A. and Winfield, P.J. (1972) J. exp. Bot., 23, 346–355.
Mothes, K. (1964) Colloq. Int. C.N.R.S. (Paris), No. 123, 131–140.
Mullins, M.E. and Osborne, D.J. (1970) Aust. J. biol. Sci., 23, 479–485.
Munsche, D., Engelbrecht, L., Gohler, K.D. and Conrad, K. (1968) Flora Abt. A, 159, 268–273.
Murai, N., Armstrong, D.J. and Skoog, F. (1975) Plant Physiol., 55, 853–858.

Murashige, T. and Skoog, F. (1962) Physiol. Plant., 15, 473–497.
Nitsch, C. and Nitsch, J.P. (1967) Planta, 72, 355–370.
Nitsch, J.P. and Nitsch, C. (1965) Bull. Soc. Bot. Fr., 112, 11–18.
Nitsch, J.P., Nitsch, C., Rossini, L.M.E. and Bui Dang Ha, D. (1967) Phytomorphology, 17, 446–453.
Ohkawa, M. (1974) HortScience, 9, 458.
Okamoto, T., Shudo, K. and Isogai, Y. (1974) in: Plant Growth Substances 1973, Hirokawa, Tokyo, pp. 447–455.
Oritani, T. and Yoshida, R. (1971) Proc. Crop Sci. Soc. (Japan), 40, 325–331.
Oritani, T. and Yoshida, R. (1973) Proc. Crop Sci. Soc. (Japan), 42, 280–287.
Osborne, D.J. and Black, M.K. (1964) Nature (Lond.), 201, 97.
Osborne, D.J. and McCalla, D.R. (1961) Plant Physiol., 36, 219–221.
Osborne, D.J. and McCready, C.C. (1965) Nature (Lond.), 206, 679–680.
Paces, V. (1976) Biochem. biophys. Res. Commun., 72, 830–839.
Paces, V., Werstuik, E. and Hall, R.H. (1971) Plant Physiol., 48, 775–778.
Parker, C.W. and Letham, D.S. (1973) Planta, 114, 199–218.
Parker, C.W. and Letham, D.S. (1974) Planta, 115, 337–344.
Parker, C.W., Letham, D.S., Cowley, D.E. and MacLeod, J.K. (1972) Biochem. biophys. Res. Commun., 49, 460–466.
Parker, C.W., Wilson, M.M., Letham, D.S., Cowley, D.E. and MacLeod, J.K. (1973) Biochem. biophys. Res. Commun., 55, 1370–1376.
Parker, C.W., Letham, D.S., Wilson, M.M., Jenkins, I.D., MacLeod, J.K. and Summons, R.E. (1975) Ann. Bot. (Lond.), 39, 375–376.
Parthasarathy, R., Ohrt, J.M. and Chheda, G.B. (1974) Biochem. biophys. Res. Commun., 60, 211–218.
Pedersen, M. (1973) Physiol. Plant., 28, 101–105.
Pedersen, M. and Fridborg, G. (1972) Experientia (Basel), 28, 111–112.
Penner, D. and Ashton, F.M. (1967a) Biochim. biophys. Acta (Amst.), 148, 481–485.
Penner, D. and Ashton, F.M. (1967b) Plant Physiol., 42, 791–796.
Pereira, A.S.R., Houwen, P.J.W., Deurenberg-Vos, H.W.J. and Pey, E.B.F. (1972) Z. Pflanzenphysiol., 68, 170–177.
Perley, J.E. and Stowe, B.B. (1966) Plant Physiol., 41, 234–237.
Peterkofsky, A. (1968) Biochemistry, 7, 472–482.
Peterson, C.S. and Fletcher, R.A. (1973) Canad. J. Bot., 51, 1899–1905.
Peterson, J.B. and Miller, C.O. (1976) Plant Physiol., 57, 393–399.
Pethe-Sadorge, P., Signor, Y. and Guern, J. (1972) C.R. Acad. Sci. (Paris), Ser. D, 275, 2493–2496.
Phillips, D.A. and Cleland, C.F. (1972) Planta, 102, 173–178.
Phillips, D.A. and Torrey, J.G. (1970) Physiol. Plant., 23, 1057–1063.
Phillips, D.A. and Torrey, J.G. (1972) Plant Physiol., 49, 11–15.
Pieniazek, J. (1964) Acta agrobot. (Wars.), 16, 157–169.
Pilet, P.E. (1968) in: Biochemistry and Physiology of Plant Growth Substances, F. Wightman and G. Setterfield (eds.), Runge Press, Ottawa, pp. 993–1004.
Pilet, P.E., Guern, J. and Hugon, E. (1967) Physiol. Veg., 5, 261–270.
Playtis, A.J. and Leonard, N.J. (1971) Biochem. biophys. Res. Commun., 45, 1–5.
Pool, R.M. and Powell, L.E. (1974) in: Plant Growth Substances 1973, Hirokawa, Tokyo, pp. 93–98.
Pozsar, B.I. and Kiraly, Z. (1966) Phytopathol. Zeit., 56, 297–309.
Pozsar, B.I. and Matolcsy, G. (1968) Life Sci., 7, Part II, 699–704.
Purse, J.G., Horgan, R., Horgan, J.M. and Wareing, P.F. (1976) Planta, 132, 1–8.
Radin, J.W. and Loomis, R.S. (1971) Physiol. Plant., 25, 240–244.
Radin, J.W. and Loomis, R.S. (1974) Plant Physiol., 53, 348–351.
Rathbone, M.P. and Hall, R.H. (1972a) Planta, 108, 93–102.
Rathbone, P. and Hall, R.H. (1972b) Cancer Res., 32, 1647–1650.
Reddy, M.N. and Williams, P.H. (1970) Phytopathology, 60, 1463–1465.

Richmond, A., Back, A. and Sachs, B. (1970) Planta, 90, 57–65.
Robins, M.J., Hall, R.H. and Thedford, R. (1967) Biochemistry, 6, 1837–1848.
Rogozinska, J.H., Helgeson, J.P. and Skoog, F. (1964) Physiol. Plant., 17, 165–176.
Rogozinska, J.H., Kroon, C. and Salemink, C.A. (1973) Phytochemistry, 12, 2087–2092.
Rosenbaum, N. and Gefter, M.L. (1972) J. biol. Chem., 247, 5675–5680.
Rothwell, K. and Wright, S.T.C. (1967) Proc. roy. Soc. B, 167, 202–223.
Rozej, B. and Chrominski, A. (1976) in: Proceedings of the Fourth Symposium on Plant Growth Regulators, Acta Universitatis Nicolai Copernici (Biologia XVIII), Torun, Poland, pp. 167–171.
Rybicka, H. and Engelbrecht, L. (1974) Phytochemistry, 13, 282–283.
Sachs, T. and Thimann, K.V. (1964) Nature (Lond.), 201, 939–940.
Sandstedt, R. (1971) Physiol. Plant., 24, 408–410.
Sargent, P.A. and King, J. (1974) Canad. J. Bot., 52, 2459–2463.
Scarbrough, E., Armstrong, D.J., Skoog, F., Frihart, C.R. and Leonard, N.J. (1973) Proc. nat. Acad. Sci. (Wash.), 70, 3825–3829.
Schmitz, R.Y. and Skoog, F. (1970) Plant Physiol., 45, 537–538.
Schmitz, R.Y., Skoog, F., Hecht, S.M. and Leonard, N.J. (1971) Phytochemistry, 10, 275–280.
Schmitz, R.Y., Skoog, F., Hecht, S.M. and Bock, R.M. (1972a) Phytochemistry, 11, 1603–1610.
Schmitz, R.Y., Skoog, F., Hecht, S.M. and Leonard, N.J. (1972b) Plant Physiol., 50, 114–116.
Schmitz, R.Y., Skoog, F., Playtis, A.J. and Leonard, N.J. (1972c) Plant Physiol., 50, 702–705.
Schmitz, R.Y., Skoog, F., Vincze, A., Walker, G.C., Kirkegaard, L.H. and Leonard, N.J. (1975) Phytochemistry, 14, 1479–1484.
Scopes, D.I.C., Zarnack, U., Leonard, N.J., Schmitz, R.H. and Skoog, F. (1976) Phytochemistry, 15, 1523–1526.
Seth, A.K., Davies, C.R. and Wareing, P.F. (1966) Science, 151, 587–588.
Shannon, J.S. and Letham, D.S. (!966) N.Z. J. Sci., 9, 833–842.
Shantz, E.M. and Steward, F.C. (1955) J. Amer. chem. Soc., 77, 6351–6353.
Shaw, M. (1961) Canad. J. Bot., 39, 1393–1407.
Shaw, M. (1963) Ann. Rev. Phytopathol., 1, 259–294.
Shaw, G., Smallwood, B.M. and Steward, F.C. (1968) Experientia (Basel), 24, 1089–1090.
Shaw, G., Smallwood, B.M. and Steward, F.C. (1971) Phytochemistry, 10, 2329–2336.
Sheldrake, A.R. (1973) Biol. Rev., 48, 509–559.
Shindy, W.W. and Smith, O.E. (1975) Plant Physiol., 55, 550–554.
Short, K.C. and Torrey, J.G. (1972a) Plant Physiol., 49, 155–160.
Short, K.C. and Torrey, J.G. (1972b) J. exp. Bot., 23, 1099–1105.
Sitton, D., Itai, C. and Kende, H. (1967) Planta, 73, 296–300.
Sivadjian, A., Sadorge, P., Gawer, M., Terrine, C. and Guerne, J. (1969) Physiol. Veg., 7, 31–42.
Skene, K.G.M. (1970) J. exp. Bot., 21, 418–431.
Skene, K.G.M. (1972a) J. Hort. Sci., 47, 179–182.
Skene, K.G.M. (1972b) Planta, 104, 89–92.
Skene, K.G.M. and Antcliff, A.J. (1972) J. exp. Bot., 23, 283–293.
Skene, K.G.M. and Kerridge, G.H. (1967) Plant Physiol., 42, 1131–1139.
Skinner, C.G. and Shive, W. (1957) Plant Physiol., 32, 500–501.
Skinner, C.G., Claybrook, J.R., Talbert, F.D. and Shive, W. (1956) Arch. Biochem., 65, 567–569.
Skinner, C.G., Claybrook, J.R., Talbert, F. and Shive, W. (1957) Plant Physiol., 32, 117–120.
Skoog, F. (1973) Biochem. Soc. Symp., 38, 195–215.
Skoog, F. and Armstrong, D.J. (1970) Ann. Rev. Plant Physiol., 21, 359–384.
Skoog, F. and Leonard, N.J. (1968) in: Biochemistry and Physiology of Plant Growth Substances, F. Wightman and G. Setterfield (eds.), Runge Press, Ottawa, pp. 1–18.
Skoog, F. and Tsui, C. (1948) Amer. J. Bot., 35, 782–787.
Skoog, F., Strong, F.M. and Miller, C.O. (1965) Science, 148, 532–533.
Skoog, F., Armstrong, D.J., Cherayil, J.D., Hampel, A.E. and Bock, R.M. (1966) Science, 154, 1354–1356.

Skoog, F., Hamzi, H.Q., Szweykowska, A.M., Leonard, N.J., Carraway, K.L., Fujii, T., Helgeson, J.P. and Loeppky, R.N. (1967) Phytochemistry, 6, 1169–1192.
Skoog, F., Schmitz, R.J., Bock, R.M. and Hecht, S.M. (1973) Phytochemistry, 12, 25–37.
Skoog, F., Schmitz, R.Y., Hecht, S.M. and Frye, R.B. (1975) Proc. nat. Acad. Sci. (Wash.), 72, 3508–3512.
Somner, N.F. (1961a) Physiol. Plant., 14, 741–749.
Somner, N.F. (1961b) Physiol. Plant., 14, 460–469.
Sondheimer, E. and Tzou, D. (1971) Plant Physiol., 47, 516–520.
Spiess, L.D. (1975) Plant Physiol., 55, 583–585.
Sprecker, M.A., Morrice, A.G., Gruber, B.A. and Leonard, N.J. (1976) Phytochemistry, 15, 609–613.
Steward, F.C. and Shantz, E.M. (1956) in: The Chemistry and Mode of Action of Plant Growth Substances, R.L. Wain and F. Wightman (eds.), Academic Press, London, pp. 165–186.
Steward, F.C., Caplin, S.M. and Shantz, E.M. (1955) Ann. Bot. (Lond.), 19, 29–47.
Strong, F.M. (1958) Topics in Microbial Chemistry, John Wiley, New York, pp. 98–157.
Sugiyama, T., Kitamura, E., Kubokawa, S., Kobayashi, S., Hashizume, T. and Matsubara, S. (1975) Phytochemistry, 14, 2539–2543.
Suk, D., Simpson, C.L. and Mihich, E. (1970) Cancer Res., 30, 1429–1436.
Syono, K. and Torrey, J.G. (1976) Plant Physiol., 57, 602–606.
Sziraki, I., Balazs, E. and Kiraly, Z. (1975) Physiol. Plant Pathol., 5, 45–50.
Szweykowska, A. and Korcz, I. (1972) Planta, 108, 279–282.
Szweykowska, A., Schneider, J. and Prusinska, U. (1969) Acta Soc. Bot. Pol., 38, 139–142.
Szweykowska, A., Dornowska, E., Cybulska, A. and Wasiek, G. (1971) Biochem. Physiol. Pflanz., 162, 514–525.
Tandeau de Marsac, N. and Jouanneau, J.P. (1972) Physiol. Veg., 10, 369–380.
Taris, B. and Avenard, J.C. (1969) C.R. Acad. Sci. (Paris), Ser. D, 268, 1069–1072.
Taylor, J.S. and Dumbroff, E.B. (1975) Canad. J. Bot., 53, 321–331.
Tegley, J.R., Witham, F.H. and Krasnuk, M. (1971) Plant Physiol., 47, 581–585.
Teltscherova, L. (1970) Biol. Plant., 12, 134–138.
Terrine, C., Sadorge, P., Gawer, M. and Guern, J. (1969) Physiol. Veg., 7, 425–435.
Terrine, C., Doree, M., Guern, J. and Hall, R.H. (1972) in: Plant Growth Substances 1970, D.J. Carr (ed.), Springer, Berlin, pp. 467–475.
Thimann, K.V. and Sachs, T. (1966) Amer. J. Bot., 53, 731–739.
Thimmappaya, B. and Cherayil, J.D. (1974) Biochem. biophys. Res. Commun., 60, 665–672.
Thomas, T.H. (1974) Ann. appl. Biol., 76, 237–241.
Thomas, T.H. (1976) Physiol. Plant., 38, 35–38.
Thomas, T.H., Carroll, J.E., Isenberg, F.M.R., Pendergrass, A. and Howell, L. (1975a) J. Chromatogr., 103, 211–215.
Thomas, T.H., Carroll, J.E., Isenberg, F.M.R., Pendergrass, A. and Howell, L. (1975b) Physiol. Plant., 33, 83–86.
Thomas, T.H., Carroll, J.E., Isenberg, F.M.R., Pendergrass, A. and Howell, L. (1975c) Plant Physiol., 56, 410–414.
Thompson, A.G., Horgan, R. and Heald, J.K. (1975) Planta, 124, 207–210.
Torigoe, Y., Akiyama, M., Hirobe, M., Okamoto, T. and Isogai, Y. (1972) Phytochemistry, 11, 1623–1630.
Tritsch, G.L. (1973) Cancer Res., 33, 310–312.
Tucker, D.J. and Mansfield, T.A. (1972) Planta, 102, 140–151.
Tucker, D.J. and Mansfield, T.A. (1973) J. exp. Bot., 24, 731–740.
Tzou, D., Galson, E.C. and Sondheimer, E. (1973) Plant Physiol., 51, 894–897.
Upper, C.D., Helgeson, J.P., Kemp, J.D. and Schmidt, C.J. (1970) Plant Physiol., 45, 543–547.
Van Staden, J. (1973a) Physiol. Plant., 28, 222–227.
Van Staden, J. (1973b) Physiol. Plant., 29, 137–139.
Van Staden, J. (1973c) J. exp. Bot., 24, 667–673.

Van Staden, J. (1975) Plant Sci. Lett., 5, 227–230.
Van Staden, J. (1976a) Physiol. Plant., 36, 225–228.
Van Staden, J. (1976b) Physiol. Plant., 36, 123–126.
Van Staden, J. (1976c) Physiol. Plant., 38, 1–5.
Van Staden, J. and Breen, C.M. (1973) Plant Sci. Lett., 1, 325–330.
Van Staden, J. and Brown, N.A.C. (1973) Physiol. Plant., 29, 108–111.
Van Staden, J. and Dimalla, G.G. (1976) Planta, 130, 85–87.
Van Staden, J. and Drewes, S.E. (1975) Physiol. Plant., 34, 106–109.
Van Staden, J. and Menary, R.C. (1976) Z. Pflanzenphysiol., 78, 262–265.
Van Staden, J. and Wareing, P.F. (1972a) Physiol. Plant., 27, 331–337.
Van Staden, J. and Wareing, P.F. (1972b) Planta, 104, 126–133.
Van Staden, J., Webb, D.P. and Wareing, P.F. (1972) Planta, 104, 110–114.
Van Staden, J., Olatoye, S.T. and Hall, M.A. (1973) J. exp. Bot., 24, 662–666.
Varga, A. and Bruinsma, J. (1973) Planta, 111, 91–93.
Varga, A. and Bruinsma, J. (1974) J. Hort. Sci., 49, 135–142.
Veen, H. and Jacobs, W.P. (1969) Plant Physiol., 44, 1277–1284.
Vizarova, G. (1974) Phytopathol. Z., 79, 310–314.
Vizarova, G. (1975) Biol. Plant., 17, 380–382.
Vizarova, G. and Minarcic, P. (1974) Phytopathol. Z., 81, 49–55.
Vonk, C.R. (1974) Acta bot. neerl., 23, 541–548.
Vreman, H.J. and Corse, J. (1975) Physiol. Plant., 35, 333–336.
Vreman, H.J., Skoog, F., Frihart, C.R. and Leonard, N.J. (1972) Plant Physiol., 49, 848–851.
Vreman, H.J., Schmitz, R.Y., Skoog, F., Playtis, A.J., Frihart, C.R. and Leonard, N.J. (1974) Phytochemistry, 13, 31–37.
Wagner, H. and Michael, G. (1971) Biochem. Physiol. Pflanz., 162, 147–158.
Walker, G.C., Leonard, N.J., Armstrong, D.J., Murai, N. and Skoog, F. (1974) Plant Physiol., 54, 737–743.
Weaver, R.J., Van Overbeek, J. and Pool, R.M. (1966) Hilgardia, 37, 181–201.
Webb, D.P., Van Staden, J. and Wareing, P.F. (1973) J. exp. Bot., 24, 741–750.
Weiss, C. and Vaadia, Y. (1965) Life Sci., 4, 1323–1326.
Wheeler, A.W. (1971) Planta, 98, 128–135.
Wheeler, A.W. (1972) Ann. appl. Biol., 72, 327–334.
Wheeler, A.W. (1973) Planta, 112, 129–135.
Whitaker, B.D. and Kende, H. (1974) Planta, 121, 93–96.
White, P.R. and Braun, A.C. (1942) Cancer Res., 2, 597–617.
Whitty, C.D. and Hall, R.H. (1974) Canad. J. Biochem., 52, 789–799.
Wilson, M.M., Gordon, M.E., Letham, D.S. and Parker, C.W. (1974) J. exp. Bot., 25, 725–732.
Wood, H.N. (1970) Proc. nat. Acad. Sci. (Wash.), 67, 1283–1287.
Wood, H.N. and Braun, A.C. (1973) Proc. nat. Acad. Sci. (Wash.), 70, 447–450.
Wood, H.N., Braun, A.C., Brandes, H. and Kende, H. (1969). Proc. nat. Acad. Sci. (Wash.), 62, 349–356.
Woolley, D.J. and Wareing, P.F. (1972a) Planta, 105, 33–42.
Woolley, D.J. and Wareing, P.F. (1972b) New Phytol., 71, 781–793.
Woolley, D.J. and Wareing, P.F. (1972c) New Phytol., 71, 1015–1025.
Yamada, Y., Sekiya, J. and Koshimizu, K. (1972) Phytochemistry, 11, 1019–1021.
Yoshida, R. and Oritani, T. (1972) Plant Cell Physiol., 13, 337–343.
Yoshida, R. and Oritani, T. (1974) Proc. Crop Sci. Soc. (Japan), 43, 47–51.
Young, H. and Letham, D.S. (1969) Phytochemistry, 8, 1199–1203.
Zwahlen, K.D. (1962) Chem. Abst., 57, 13776–13777.

Phytohormones and Related Compounds – A Comprehensive Treatise, Volume I
Editors: Letham, Goodwin and Higgins

5

Ethylene

DAPHNE J. OSBORNE

5.1. Introduction

That certain air pollutants, smokes and fumes could modify the growth of plants was realised long before such evidence appeared in scientific literature. Puerto Rican pineapple growers and Phillipine mango growers at one time traditionally lit bonfires near their trees, knowing that the smoke helped to initiate and synchronise the flowering of their crops and ripening of their fruits. The damage caused to plants by leaks of "illuminating gas", the early town gas made from combusting coal, initiated much of the early work on air pollutants. In Germany, Girardin (1864) described how trees near faulty gas mains would shed their leaves prematurely; Molisch (1884) noted that abnormal geotropic responses and horizontal growth of seedlings could be related to the presence in the air of smokes or traces of illuminating gas. Flower growers observed that the traces of the gas escaping in their glass houses led to early fading and closure of blossoms (Knight and Crocker 1913), and in California, the citrus growers discovered that fumes from kerosene stoves could be used to ripen green lemon fruits (Sievers and True 1912). Although we now know that all these responses could be attributable to trace levels of ethylene in the air, it is doubtful whether any of the early observers could have suspected that this volatile hydrocarbon would later be shown to be a ubiquitous and highly active naturally occurring plant product with all the properties of a plant hormone.

5.2. Identification of the biologically active air pollutant

The first indication that ethylene might be the active air pollutant came from experiments of Dimitry Neljubow (1901) working in St. Petersburg. He found that pea plants grown in the gas-illuminated laboratory showed unusual curvature responses, stem swelling and a loss of response to gravity leading to a horizontal growth habit; plants grown in outside air, or air first passed over hot copper oxide, grew normally. When he exposed the normal pea plants to a trace of illuminating gas, they also developed symptoms similar to those of plants grown in the laboratory. He then tested the various components of coal gas for their effects on the growth of peas, and found that the concentrations of ethylene present in illuminating gas were sufficient to cause curvature responses essentially similar to those observed for the illuminating gas alone. Ethylene caused no permanent change in the growth habit for when ethylene was removed, the new growth again continued normally. Neljubow also noted that acetylene would similarly affect the growth of peas, but only at much higher (× 100–1000) concentrations.

Later, Crocker and Knight (1908) and Knight and Crocker (1913) identified ethylene as the active constituent of both illuminating gas and tobacco smoke, and Harvey (1913) determined that biological responses produced from burning paper could be attributed to ethylene. Denny (1924a, b) subsequently confirmed the suspicion that ethylene was the active substance in kerosene fumes causing the ripening of lemons. Concentrations of ethylene are of the order of 10 ppm in coal gas, as high as 4 ppm in smoke of burning stubble (McElroy 1960), and up to 400 ppm in the fumes of car exhausts (Abeles 1973). In contrast, mean levels in urban areas do not normally exceed 10–50 ppb and in rural areas may be less than 5 ppb.

5.3. Identification of ethylene as a natural plant product

Although fruit storers and shippers were aware that ripe and rotting fruit could in some way hasten the ripening of other fruit close to them, one of the first reports that the active agent might be a plant volatile was made by Cousins (1910). He wrote that ripe oranges release a product to the air that would hasten the ripening of bananas enclosed with them. Interest was renewed in this question when, in 1932, Elmer reported that potatoes stored in containers with ripe apples or pears showed an inhibition of shoot growth which he too attributed to a volatile product of the fruit. A year later Botjes (1933) demonstrated that a volatile from apples causing epinasty in tomato seedlings could be removed from the air by ethylene absorbants. Similar results were obtained by Huelin (1933) and Kidd and West (1933) who described how a volatile substance produced by ripe apples would stimulate the so-called "climacteric" of unripe fruit. It was not, however, until the work of Gane (1934) that indisputable evidence for ethylene production by plants was finally obtained. For four weeks he collected the air from Pearmain apples into bromine at −65 °C and then treated the brominated product (containing ethylene dibromide) with aniline, converting some to a compound identified as *N,N'*-diphenylethylenediamine, thereby showing that ethylene had been present in the original apple volatiles. Hence, ethylene became known as the "ripening" hormone. Subsequently ethylene has been identified as a natural product not only of fruits but of all plants parts including leaves, stems, flowers and roots (see Table I of Burg 1962).

5.4. Properties of ethylene

5.4.1. Physical properties

$CH_2{=}CH_2$ Mol. wt. 28.05 (I)

Ethylene (I) is a flammable, ethereal-smelling, colourless gas boiling at −103 °C. It absorbs light both in the ultraviolet and the infrared wavelengths. It is lighter than air (vapour density 0.978, air = 1.00). At normal pressures (760 mm Hg) the solubility in water at 0 °C is approximately 315 ppm and at 25 °C will approach 140 ppm. At 25 °C the diffusion equilibrium at 1 ppm between air and water is approximately 10,000 : 1. Ethylene functions as a rapid and non-toxic mammalian anaesthetic (Eger and Larsen 1964) at 85% in oxygen, but was reported to be lethal to mice at 95% in air.

5.4.2. Chemical properties

By virtue of the double bond, ethylene and other olefins readily react with halogens to form substituted hydrocarbons and other saturated addition products. Coordination compounds are formed with a variety of metals, and the reaction with mercuric ion is the basis of one of the most used methods of trapping ethylene in solution (Young et al. 1952). Solutions of mercuric perchlorate (0.25 *M* in 2 *M* perchloric acid), prepared from red mercuric oxide and perchloric acid, react rapidly with ethylene (and other olefins) in the air over a wide temperature range (0–25 °C) to form a coordination complex. Manometric determinations show that ethylene can be recovered quantitatively by the addition of Cl^- (4 *M* LiCl or HCl). Solutions of other mercuric salts can also be used to absorb ethylene (e.g. mercuric acetate) and although the efficiency in recovery is not necessarily so high, such solutions have certain advantages when a radioactive ethylene complex is to be counted in scintillation fluid (Yang et al. 1967).

Ethylene readily undergoes oxidation, and the reaction with $KMnO_4$, in which the ethylene is oxidized to ethylene glycol with the formation of MnO_2, is the chemical basis of a method for absorbing ethylene in horticulture. Alumina or silica gel impregnated with $KMnO_4$ is now commonly used for this purpose in packed flowers and fruits.

Another oxidation product of ethylene is the cyclic ether, ethylene oxide, which is toxic to plant tissues, and under certain conditions reduces the effects of applied ethylene. Although ethylene oxide delays closure of carnation flowers exposed to ethylene (Nichols 1968), there is no evidence that ethylene oxide is a specific inhibitor of ethylene in plants.

Ethylene will also react with ozone and oxides of nitrogen to form a number of volatile products common to smoke and other forms of air pollution. These compounds too cause toxic and non-specific injuries to plant tissues.

5.5. Assays of ethylene

5.5.1. Bioassays

Although many volatile compounds will induce growth effects in plants, the "triple response" of the etiolated pea plant developed by Neljubow (1901) is still a reliable test for ethylene if properly assessed. Plants enclosed in concentrations of 0.01 ppm show reductions in extension growth, while at 1 ppm horizontal growth followed by stem swelling in the elongating parts are characteristic responses after several days. Coupled with positive tests for the acceleration of abscission in debladed cotton cotyledons or bean petioles (Jackson and Osborne 1970) or the induction of leaf and stem epinasty in the tomato (Crocker 1948), bioassays can provide conclusive evidence for the presence of ethylene in the ambient air, but not of its concentration.

5.5.2. Physical assays

Although a variety of chemical and physical methods have been tested and used over the years with sensitivities of 50 ppm or more (see Abeles 1973), gas solid chromatography on columns of alumina or silica is by far the most sensitive and satisfactory method in use

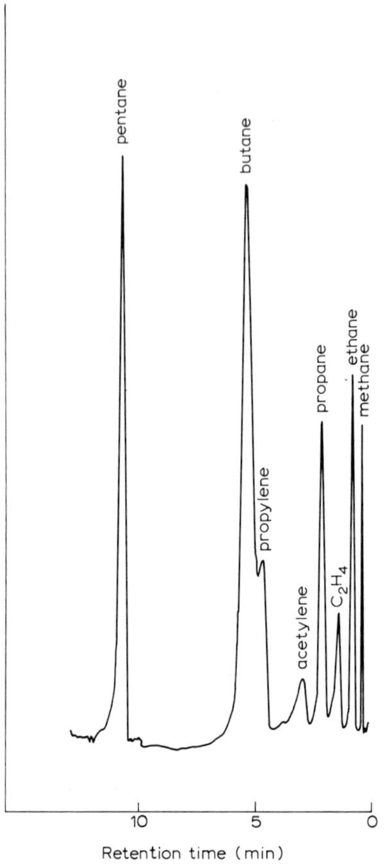

Figure 5.1. Typical scan from a Pye (Series 104) gas chromatograph fitted with a hydrogen-air flame ionization detector showing the separation of ethylene from other low-molecular-weight hydrocarbons on an activated silica gel column. (For details see Jackson and Osborne 1970.)

today. With properly adjusted instruments, with flame ionization detectors, concentrations as low as 1–10 ppb in 1 ml of air can be measured with accuracy. Depending on the column, the retention time is usually 1–5 min and the ethylene peak is clearly distinguishable from that of other low molecular weight hydrocarbons (Figure 5.1). Many determinations can be performed in a short time (Goeschl et al. 1967) making the method most suitable for biological investigations of plant material. Where the levels accumulated in ambient air spaces are too low to measure by direct injection, suitable collection and concentration procedures are first employed.

5.5.3. Methods of collection of ethylene from plant material

In most experimental situations, the plant material to be assayed is enclosed within a small chamber fitted with an injection port, and the ethylene accumulates. The septum is usually of rubber or silicone, and since most organic substances will produce ethylene during degradation, great care must be taken to ensure that the port does not contribute to ethylene within the vessel. After incubation of weighed amounts of the tissues for requisite periods of time, samples of the gas are withdrawn with a hypodermic syringe and injected directly into the gas chromatograph. From the known volume of the vials and the peak height of ethylene in comparison with that of prepared standards, the rate of production by the tissue can be calculated in nl/g fresh wt./hr. Under conditions where larger quantities of ethylene are required for linked chromatography and mass spectrometry (Turner et al. 1975), or when very small amounts of tissue are used (De Proft and De Greef 1975), trapping and concentration can be effected in small vessels or on columns of Porapak S held in liquid nitrogen. The gases are then volatilized by heating and swept directly into the gas chromatograph. Alternatively, the gases can be collected into mercuric perchlorate solutions and ethylene recovered again quantitatively for assay by the addition of LiCl. All these methods measure the levels of ethylene that are liberated from the tissues into the external air. Where large air cavities exist as in fruits, estimates of the levels of ethylene within the tissue can be obtained by direct sampling of the air space while the material is held submerged in water (Burg and Burg 1962). These internal levels must, in part, be determined by the rate of diffusion of the gas to the external air – i.e. number of stomata, permeability of cuticle, etc.

For small air spaces and where values for the levels dissolved in the cytoplasm are required, vacuum procedures must be employed. Again the tissue is submerged and a suitable trapping device attached for collecting the gas released when the system is subjected to reduced pressure (Beyer and Morgan 1970). Such procedures must cause some perturbation of the cytoplasmic organelles, and the extent to which a wound ethylene component (see section 5.8.2) is involved must be questioned (Blanpied 1971).

5.6. Levels of production

Although all parts of plants synthesize ethylene, by far the highest levels of production reported are associated either with senescent or ripening tissues or with immature dividing or expanding tissues ($>$ 1.0 nl/g/hr). Production by mature tissues is usually low ($<$ 0.1 nl/g/hr). An important feature of ethylene physiology however, is that tissues of *all* ages possess the capacity to produce high levels of ethylene either in response to wounding, stress, or the appropriate hormonal stimulus (see section 5.8). Ethylene production by germinating seed is not discussed in this Chapter but is covered in Chapter 8, Volume II.

5.6.1. Production by fruits

Many determinations have been made of ethylene production during development, maturation and ripening of fruits. The values in Table 5.1 refer to those of Looney et al. (1974) for peach. They show the generally high levels corresponding to the period of rapid cell division and enlargement (Stage I), followed by the low levels of the extended slow growing period (Stage II) and the very high levels again of the ripening and final expansion (Stage III). It is the second peak of ethylene production that is intimately associated with the initiation of the ripening process.

Table 5.1.

Ethylene production and fruit expansion during growth and development of the peach, *Prunus persica* (Halehaven) (After Looney et al. 1974)

Weeks from anthesis	Growth Stage	Increase in fruit diameter (mm)	Mean increase in fruit diameter per week	Ethylene production nl/g/h 0–24 hr	Mean ethylene production for growth stage
3				0.20	
4				0.16	
5	I	18	6.0	0.12	0.13
6				0.04	
7				0.02	
8				0.02	
9	II	12	2.4	0.03	0.02
10				0.03	
11				0.03	
12				0.01	
13				1.12	
14	III	14	7.0	1.55	5.36
15				8.05	

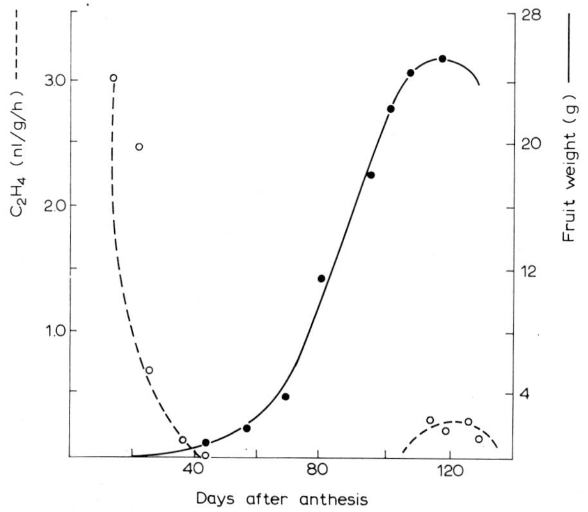

Figure 5.2. Relationship between rate of ethylene production and stage of fruit development in lychee (Adapted from Akamine and Goo 1973).

In certain other fruits (mango, lychee and avocado), the highest ethylene productions are associated with the early stages of fruit expansion and rates as high as 50–100 nl/g/hr have been recorded. After a period of low production during maturation, only a small rise in the rate occurs at fruit ripening (Figure 5.2) (Akamine and Goo 1973). The absence of appreciable increases in synthesis of ethylene at ripening (oranges may produce no more than 0.06 nl/g/hr) does not imply that ethylene is not intimately concerned with the initiation of the ripening process, for it is clear that the sensitivity of some fruits to the gas increases greatly at this stage (Burg and Burg 1965). A decrease in the level of other endogenous hormones, e.g. auxin, may be responsible for this increased sensitivity (Vendrell 1969).

In fruits with internal cavities such as the apple or melon, very high endogenous levels of ethylene may develop during the climacteric as shown in Table 5.2. Ethylene production by fruit in relation to fruit ripening is discussed further in Chapter 10, Volume II.

Table 5.2.

Internal levels of ethylene found in fruits at ripening (Burg and Burg, 1962)

Fruit	Maximum concentration (μl/l)
Apple	2,500
Canteloup melon	1,400
Passion fruit	530
Pear	80
Avocado	74
Tomato	30
Peach	21
Banana	2.1

5.6.2. Production by leaves

The production by leaves shows an early pattern similar to that of fruits, being high during the expansion period, and low when expansion is complete. As in fruits, senescing leaves appear to fall into two classes, those in which ethylene production increases to high levels (in most deciduous and evergreen species tested), and others (mainly herbaceous species) in which senescence is accompanied by little if any increase above the rate for mature leaves (Table 5.3).

Table 5.3.

Levels of ethylene production in examples of herbaceous and deciduous leaves*

	Ethylene production (nl/g/hr)						
	Herbaceous				Deciduous		Evergreen
Maturity of leaf	*Xanthium pennsylvanicum* Blade	*Phaseolus vulgaris* Blade	Pulvinus	*Sinapis alba* Cotyledon**	*Prunus serrulata* Blade	Petiole	*Euonymus japonica*
Young expanding	0.15–0.26	–	0.40	–	–	–	1st year (expanding) 0.15–0.30
Mature expanded	0.19–0.22	0.10	0.04	0.10	0.1	0.39	2nd year (mature) 0.01–0.02
Yellowing (senescent)	0.12–0.19	0.12	2.20	0.40	6.7	2.87	3rd year (abscinding) 0.7–1.4
Yellow	0.18	0.18	1.95	0.20	2.2	0.70	–

* Hallaway and Osborne (unpublished data).
** Cheah (1975).

5.6.3. Production by flowers

During opening of the bud, ethylene production in male flowers of the cucurbit *Echallum elaterium* increases only slightly, but as the petals senesce and fade, levels increase many-fold (Table 5.4) (Jackson 1971). Similar measurements have been made during the opening and fading of flowers of *Ipomoea* (Kende and Baumgartner 1974); they found a small peak in ethylene production during opening and expansion of the petals, and a second, much higher, peak associated with senescence. In this respect the pattern of ethylene production during development is similar to that of many fruits and leaves. An increase in ethylene production from less than 1 nl/g/hr to 35 nl/g/hr has been noted in the Vanda orchid (Burg and Dijkman 1967) and the rise has been attributed to the auxin transferred at pollination. From the experiments with male *Ecballium* flowers however, it is clear that the ethylene rise at fading resembles that at senescence of leaves, and the event can be hastened by, but is not dependent upon, pollination.

Table 5.4.

Production of ethylene (nl/g/hr) by male flowers of *Ecballium elaterium* during opening and senescence (Jackson 1971)

Half open	Fully open	Fading slightly	Fully faded (abscinding)	Abscinded
0.04	0.07	0.50	1.75	1.35

5.6.4. Production by seedling shoots and roots

Production by shoots of green or etiolated seedlings of *Pisum sativum* has been carefully monitored by Burg and Burg (1968). Highest production is found in the bud (0.30 nl/g/hr) and the expanding hook region (0.43 nl/g/hr) of etiolated plants, and in the expanding bud and youngest leaf in green plants (Table 5.5). In etiolated peas, Goeschl et al. (1967) calculated that 90% of the ethylene produced was evolved by the top 5 mm of the shoot, and that the rate reached as high as 6 nl/g/hr. Similar patterns, in which the highest production occurs in the most rapidly growing parts, have been observed for other plant shoots.

For roots, rates of 1.6 nl/g/hr have been reported for isolated radicles of 3-day-old Alaska pea seedlings (Fuchs and Lieberman 1968), but much lower constant rates of 0.12 nl/g/hr over 20 hr have been monitored by K. Heyes (personal communication) for intact roots of 2-day-old Meteor peas.

Table 5.5.

Ethylene production by segments from different regions of etiolated and green pea shoots

Ethylene (nl/g/hr)

Green peas*	14-day-old	Etiolated peas**	9-day-old
Apical bud	0.7	Apical bud	0.30
Apical leaf	1.9	Apical hook	0.43
Apical tendril	4.1	Sub-apical segment	0.09
		1st node below apical bud	0.34
1st internode below apical bud	0.2		
2nd internode below apical bud	0.2	2nd internode below apical bud	0.07
		2nd node below apical bud	0.22
3rd internode below apical bud	0.15	3rd internode below apical bud	0.04

* Burg (1968).
** Burg and Burg (1968).

5.6.5. Production by lower plants

Ethylene has been identified as a natural product from gymnosperms *(Ginkgo biloba)*, pteridophytes *(Pteridium)*, bryophytes *(Sphagnum squarrosum*, Polytrichum, Funaria* and *Marcantia)*, green algae *(Chara corallina, Chlorella protothecoides*)* and blue-green algae *(Anacystis nidulans*)*. Levels of production are of the same order as for mature tissues of higher plants.

* Sterile cultures from Natural Environment Research Council Culture Centre of Algae and Protozoa, Cambridge, U.K.

5.6.6. Production by fungi and bacteria

Using media containing corn steep liquor and glucose, Ilag and Curtis (1968) showed that many fungi will produce ethylene, and it seems that their ability to do so depends in part upon the nature of the substrate on which they grow. Lynch (1972) found *Mucor heimalis,* isolated from soil, produced ethylene in culture when supplied with glucose and methionine, while mycelium of *A. bisporus* produces significant amounts when grown on compost or on sterile compost agar, but not on glucose agar (T. Ward, unpublished). Although no specific role has been demonstrated for ethylene in the control of growth and development of a fungus, in cased laboratory cultures of *A. bisporus* (Turner et al. 1975), an ethylene production rising to 0.25 nl/g fresh weight compost/hr is clearly associated with a specific phase of expansion of each sporocarp. Further study, (T. Ward, unpublished) has established that the ethylene is a product of fungal metabolism and not of any associated bacteria introduced at the time of casing. Indeed, reports of ethylene production solely attributable to bacteria are few, but are increasing as more species are investigated. Freebairn and Buddenhagen (1964) found *Pseudomonas solanacearum* infections of bananas produced ethylene on a number of organic substrates, though not on a glucose medium, and Ilag and Curtis (1968) reported production from a *Streptomyces.* During isolation from waterlogged soils of microorganisms which might be responsible for the ethylene-type damage to crops in these conditions, Primrose and Dilworth (1976) and Primrose (1977a) found a wide range of bacteria, including strains of the common enteric bacterium, *E. coli,* which produced considerable quantities of the gas when methionine was supplied. As with fungi, therefore, ethylene production by bacteria seems closely linked to specific substrates.

5.6.7. Production by yeasts and mammalian tissues

Two soil yeasts, *Candida vartiovaaria* and *Trichosporon cutaneum,* have been shown to synthesize ethylene when supplied with a glucose – methionine medium (Lynch 1974). To date there is no compelling evidence to indicate that healthy mammalian tissues produce the gas.

5.7. Biosynthesis of ethylene

Despite the simplicity and sensitivity of ethylene assays and the uncomplicated chemistry of the end product, the intermediates in the biosynthetic pathway have proved difficult to unravel. This has been particularly true since most carbon sources lead to an increase in ethylene production. Where radioactive precursors (sugars, organic acids or peptides) have been supplied, it is clear from the specific activity of the recovered ethylene that most of those studied have been far back in the synthetic chain of events. Further, the enzymes concerned with the biosynthesis of ethylene have to date proved too labile to isolate.

5.7.1. Non-enzymic ethylene production

The formation of ethylene during the degradation or peroxidation of many kinds of organic molecules has led to extended studies of non-enzymic systems and the consideration that some of these model systems might be operative in vivo.

Lieberman and Mapson (1964) first showed that in air linolenic acid (II) or methionine (III) would liberate ethylene in the presence of Cu^{2+} and ascorbic acid (as a reducing agent), or in the presence of Cu^{+} alone; the concentrations of Cu^{+} (10^{-4}–10^{-5} *M*) are close to those found normally in plant tissues.

5.7.1.1. Linolenic acid:
$CH_3(CH_2 \cdot CH{=}CH)_3(CH_2)_7COOH$ (II).

For the formation of ethylene by non-enzymic reactions, linolenic acid (II) has first to be activated by light, visible or UV (253 nm), or by lipoxidases a form a presumed epoxide. It is possible that a similar peroxidative reaction could occur in apple fruits for a rise in lipoxidase activity was shown by Meigh et al. (1967) to precede the rise in ethylene production. However, [^{14}C]ethylene was not detected when [U-^{14}C]linolenic acid (Baur and Yang 1969a) or [1 or 2-^{14}C]propanal (CH_3CH_2CHO), a breakdown product of linolenic acid (Baur and Yang 1969b), was supplied to apple tissue, although both these substances are normal products in apples and both substances yield some [^{14}C]ethylene in the non-enzymic system. However, it should be noted that Galliard et al. (1968a) did find that the carbon of ethylene was derived from linolenic acid in apple *peel* extracts, although the incorporation was only one tenth of that with methionine (Mapson 1970).

Whereas linolenic acid yields ethane and ethylene in non-enzymic systems, methionine produces only ethylene. The addition of methionine to apple discs stimulates their production of ethylene in vivo, but no such stimulation is observed with linolenic acid. Some stimulation was observed in the extracts of apple *peel* supplied with linolenic acid by Galliard et al. (1968b) so this tissue may be unusual. Considering all the evidence, it seems unlikely that linolenic acid is a normal precursor in vivo, particularly as many compounds with $CH_3CH_2CH{=}CH$ terminal groupings do form ethylene in vitro, and the extent of this conversion to ethylene is extremely low, 0.02% (Abeles 1966).

5.7.1.2. Methionine:

$$\underset{\substack{\downarrow \\ NH_2}}{CH_3S \cdot CH_2 \cdot CH}_2CH_2COOH \text{ (III).}$$

In the production of ethylene from methionine (III) in the non-enzymic system, O_2 could be replaced by H_2O_2; the inclusion of catalase or peroxidase then abolished production. This suggested to Lieberman et al. (1965, 1966) that a copper enzyme (possibly a peroxidase) might be involved in the biosynthesis pathway in vivo. Both in the non-enzymic system (see Scheme 5.1) and with apple discs in vivo, [^{14}C]methionine yielded significant amounts of radioactive ethylene only when labelled at carbon 3 or 4. The efficiency of this conversion could be as high as 60%.

Products of the reaction were isolated from the non-enzymic system, separated chromatographically, and tested for their ability to enhance ethylene production. None was active, but evidence of thiol formation during the reaction led to the testing of methional ($CH_3S \cdot CH_2 \cdot CH_2 \cdot CHO$) which proved in vitro to be a better precursor than methionine. Methional is, however, inactive as an ethylene precursor when added to plant tissues in vivo. It is not clear whether it is inactive as a substrate per se, or whether it fails

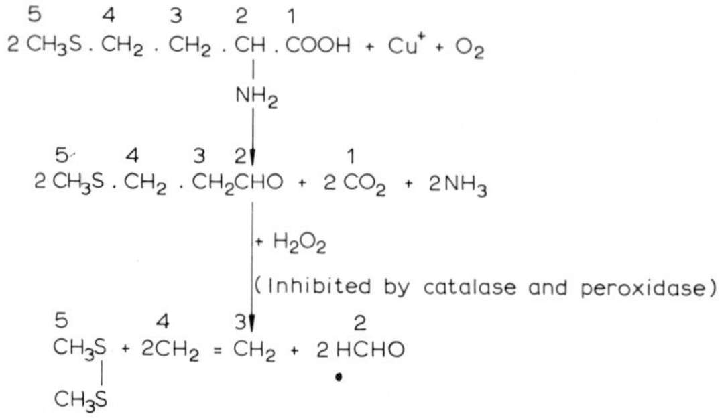

Scheme 5.1. Non-enzymic conversion of methionine to ethylene indicating the position from which the carbon atoms are derived.

to penetrate the tissue or reach the sites of synthesis. Two other intermediates, α-hydroxy-γ-methylthiobutyric acid and α-keto-γ-methylthiobutyric acid (KMBA) show precursor activity equal to that of methionine in both the non-enzymic system and with apple discs in vivo (Mapson and Wardale 1967). The apparent in vivo activity of KMBA is now attributed to an enzymic in vitro conversion in the buffer (see section 5.7.2). In a further in vitro but enzymic system consisting of precursor + peroxidase + cofactors Ku et al. (1967, 1969) found that both methional and KMBA led to ethylene formation, but methionine itself was quite inactive as a substrate. It is open to speculation therefore, how far in vitro studies are relevant to the synthesis of ethylene in vivo.

5.7.2. Biosynthetic pathway in higher plants

Whereas intact tissue and slices of tissue synthesize ethylene under aerobic conditions, activity is generally lost once the tissue is homogenised. There are however a number of cell-free systems that produce ethylene. Abeles and Rubinstein (1964) reported that a homogenate from etiolated pea seedlings evolved ethylene, but that production also occurred when protein was removed from solution provided flavine mononucleotide (FMN) was included. Later it was demonstrated by Yang et al. (1967) that this non-enzymic conversion of methionine to ethylene in the presence of FMN took place only in the light.

Mapson and his colleagues developed an in vivo system in which florets of cauliflower heads showed a considerably enhanced ethylene production on the addition of methionine (Mapson and Wardale 1967). Lieberman et al. (1966) tested 18 other amino acids in this system, but only methionine gave large (up to 100%) increases. However, when the florets were homogenised and the preparation separated into soluble and mitochondrial-microsomal particulate fractions, they found that the activity with added methionine, or with α-hydroxy-γ-methylthiobutyric acid, was confined to the particulate fractions. With methional, however, both the post-microsomal supernatant fraction and the particulate, fraction produced ethylene. Further analyses of the two fractions of the cauliflower floret system (Mapson and Wardale 1968) revealed that a glucose-oxidase (generating peroxide), a peroxidase, and two cofactors an ester of p-coumaric acid (Mapson and Mead 1968) and methanesulphinic acid (Mapson et al. 1969), were present in the extracts, and all of these were necessary for the conversion of α-hydroxy-γ-methylthiobutyric acid or

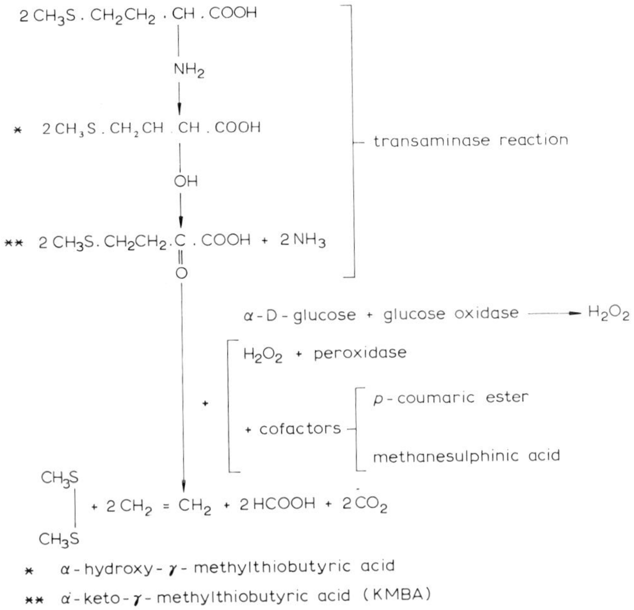

Scheme 5.2. Proposed scheme for biosynthesis of ethylene from methionine in cauliflower florets (Mapson 1969, 1970).

KMBA to ethylene. They proposed the pathway that is set out in Scheme 5.2. When small pieces of tissue are incubated in buffer solutions, as with the cauliflower floret material of Mapson et al. (1969), ethylene evolution on addition of KMBA has now been attributed to the development of an enzymic, but in vitro ethylene production system in the bathing medium. During the several hour incubations, peroxidases and cofactors for ethylene production leak from the tissue into the external solution in sufficient concentration to provide a highly active in vitro medium for converting KMBA to ethylene. Indeed, the conversion of KMBA in the presence of peroxidase and the cofactors in the complete absence of floret tissue can be readily demonstrated (Ku et al. 1969). When KMBA is added to florets in the absence of a bathing buffer, no ethylene production is obtained (Lieberman and Kunishi 1971).

Although many intermediates for the biosynthesis of ethylene have been studied, the summary given in Table 5.6 shows that only one of these, L-methionine, is known with certainty to be a precursor in vivo. It is also doubtful if peroxidase is involved in the intracellular conversion, for additions of catalase enhance, rather than reduce, ethylene production from "dry florets", and endogenous peroxidase activity does not parallel ethylene production (Kang et al. 1971).

The discovery that the fungal metabolite and phytotoxin, rhyzobitoxine, produced by certain strains of the root nodule bacterium *(Rhizobium japonicum)*, would inhibit the biosynthesis of ethylene from methionine (Owens et al. 1971) has provided a new tool for the study of the biosynthetic pathway. Inhibition of up to 75% was found in soybean (at 2 μM) and in apple disc tissue (at 12 μM). Inhibition was only partly reversed by the

Table 5.6.

Intermediates in ethylene production, in vitro and in vivo

Substance	Active in non-enzymic systems	Active when supplied to apple disc tissue or cauliflower florets in buffer solutions	Active in intra-cellular con-version (non-buffer systems) in vivo
Linolenic acid	+	+(?)	–
Propanal	+	–	–
Methionine (L-)	+	+	+*
(D-)	+	+	–
α-hydroxy-γ-methylthiobutyric acid	+	+	–
α-keto-γ-methylthiobutyricacid (KMBA)	+	+	–
Methional	+	–	–

* Also in banana tissue and pea stem segments (Burg and Clagett 1967), avocado (Baur et al. 1971) and *Phaseolus* leaves (Abeles 1972).

Table 5.7.

Inhibition of conversion of L[U-^{14}C]methionine to ethylene by L-canaline and 2,4-dinitrophenol (DNP) in plugs of apple tissue (Murr and Yang 1975)

Treatment	Ethylene produced	
	nCi	Per cent inhibition
Methionine (0.2 μCi)	4.1	0
Methionine (0.2 μCi) + canaline (7 μM)	1.9	54
Methionine (0.2 μCi) + canaline (70 μM)	0.5	88
Methionine (0.5 μCi)	10.3	0
Methionine (0.5 μCi) + canaline (7 μM)	5.7	45
Methionine (0.5 μCi) + canaline (70 μM)	2.7	74
Methionine (1.0 μCi)	13.8	0
Methionine (1.0 μCi) + canaline (7 μM)	14.4	–4
Methionine (1.0 μCi) + canaline (70 μM)	7.6	45
Methionine (1.0 μCi)	15 .0	0
Methionine (1.0 μCi) + DNP (10 μM)	13.0	12
Methionine (1.0 μCi) + DNP (60 μM)	7.4	50
Methionine (1.0 μCi) + DNP (100 μM)	3.9	78

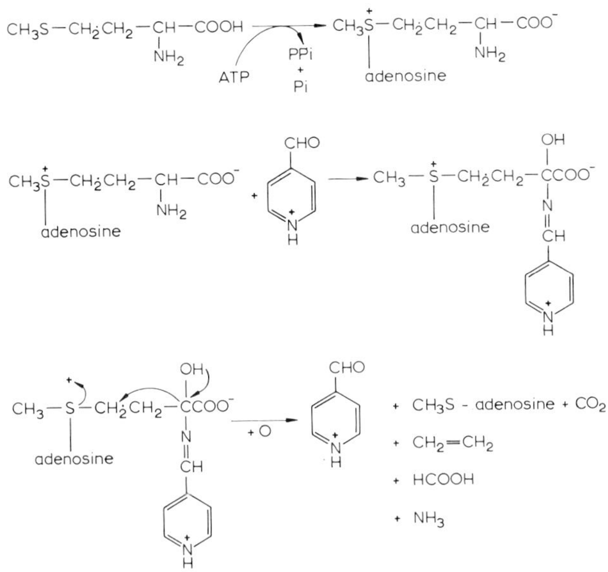

Scheme 5.3. A proposed mechanism of ethylene formation from methionine (Murr and Yang 1975).

addition of methionine or KMBA, so that the block must occur at some step in the pathway between methionine (and KMBA if this is an intermediate) and ethylene itself.

In spinach, rhizobitoxine is an effective inhibitor of the pyridoxal phosphate dependent β-cystathionase cleavage of β-cystathionine (Giovanelli et al. 1971). The possibility exists, therefore, that methionine might be cleaved in a similar reaction involving pyridoxal phosphate as a cofactor. Since L-canaline (2-amino-4-aminooxybutyric acid), another inhibitor of pyridoxal phosphate dependent reactions, and 2,4-dinitrophenol (DNP, a respiratory uncoupler) are both highly effective in inhibiting conversion of L-methionine to ethylene in apple tissue (Table 5.7), it may be that pyridoxal phosphate and ATP are both involved in the conversion of methionine to ethylene (Murr and Yang 1975). A possible scheme involving an initial ATP-dependent formation of *S*-adenosylmethionine (Burg 1973) and a final γ-elimination step fits many of the known facts for ethylene biosynthesis as set out in Scheme 5.3.

All attempts to isolate the enzymes concerned with methionine conversion to ethylene have so far proved unsuccessful. This may be because the enzymes concerned are readily inactivated by homogenisation, or because inhibitors of the system are present in cells. Sakai and Imaseki (1973) have isolated a protein of molecular weight 112,000, which inhibits auxin-induced ethylene biosynthesis from etiolated mung beans, and Sakai (1975) has demonstrated the presence of several high-molecular-weight, trypsin-inactivated, inhibitor proteins in dry seeds of other legumes, including bean, pea and soybean.

5.7.3. Biosynthetic pathway in micro-organisms

Biale (1940) and Miller et al. (1940) were the first to report the production of ethylene by a fungus, *Penicillium digitatum*. Since then numerous studies have been made of precursors in liquid cultures, and many sugars and Krebs cycle intermediates are known precursors for ethylene biosynthesis. It seems, however, that part of the production can occur outside the mycelium in the external medium. The conversion may be enzymic, through leakage of enzymes and substrates into the culture (Lynch 1974) analogous to that of the KMBA conversion in cauliflower floret bathing buffer (Lieberman and Kunishi 1971). Non-enzymic conversions in the medium may also occur, reminiscent of the non-enzymic conversion with FMN in extracts of the pea.

Although *Mucor heimalis* produces ethylene from methionine (Lynch 1972), other fungi, e.g. *Penicillium digitatum,* reputedly do not do so (Ketring et al. 1968). In their short-term experiments with radioactive substrates, Chou and Yang (1973) found glutamic acid and α-ketoglutaric acid to be effective substrates for ethylene production, radioactive ethylene being produced when carbons 3 or 4 were labelled. Succinate, the immediate precursor of α-ketoglutarate, and glutamine are both poor precursors compared with glutamic acid. Chou and Yang (1973) have proposed that, in contrast with higher plants, glutamic acid is the likely immediate precursor for ethylene in *Penicillium digitatum.* Owens et al. (1971) showed that whereas ethylene production in plant tissue is inhibited by rhyzobitoxine, production by *P. digitatum* is unaffected. The apparent difference between plants and fungi in the in vivo precursor to ethylene may have important implications in plant pathology (Williamson 1950).

In bacteria (*E. coli,* Primrose 1976), methionine, the sulphur-containing metabolites cysteine and homocysteine, or the methionine derivative KMBA can all function as the ethylene precursor. The possibility that KMBA is the intermediate in in vivo ethylene production from methionine via a peroxidative reaction accords with the high levels produced by mutants with low catalase activity. Like apple tissue (Table 5.7) production from methionine is inhibited by L-canaline (70 μM), suggesting a pyridoxal phosphate mediated reaction. Canaline does not, however, inhibit conversion of KMBA to ethylene by *E. coli* (Primrose 1977b). In cell free systems, flavin-producing strains of *E. coli* will also convert KMBA to ethylene in the light. Both enzymic and photochemical conversion are therefore possible in these bacteria.

5.7.4. Subcellular localisation of ethylene biosynthesis

The subcellular site of ethylene production has proved as elusive as the enzymes that control the biosynthetic reactions. The possibility that mitochondria (in 35,000 × *g* pellet) were the site received considerable attention (Ram Chandra and Spencer 1962). The fact that the reaction is temperature insensitive (Spencer and Meheriuk 1963), that it does not occur with intact or purified mitochondria (Ku and Pratt 1968; Stinson and Spencer 1970), and does occur in similar particulate fractions from animal tissues that do not synthesise ethylene in vivo (Gibson 1963) makes this localisation very doubtful.

The ethylene production from KMBA or methional by particulate preparations of spinach chloroplasts has been studied by Elstner and Konze (1974). In light, the production of H_2O_2 and the superoxide radicle ion O_2^- is assumed to lead to the formation of the OH^- radicle, and the subsequent oxidation of precursors to ethylene. The potential involvement of chloroplasts in such a reaction is made more likely by the inhibition of ethylene production by catalase or superoxide dismutase. Also, these preparations will evolve ethylene in the dark (Elstner and Konz 1975) in the presence of NADPH as an electron donor with ferredoxin as cofactor. However, since methionine is not converted to ethylene in this system in either light or dark,this pathway can at most account for only part of the in vivo reaction.

5.7.5. Metabolism of ethylene

For many years ethylene was considered to be an end product of a biosynthetic pathway, although experiments with samples of radioactive ethylene (probably containing impurities), and with non-sterile plant material, indicated that there might be a small fixation and conversion to soluble products such as organic acids and amino acids. Using deuterated ethylene, both Beyer (1972) and Abeles et al. (1972) could detect no exchange of deuterium from C_2D_4 with other plant constituents, nor any *cis-trans* conversion of CH_2D_2. This suggested that no incorporation of ethylene had occurred in their experimental tissue, and that any metabolism of the molecule was unlikely. More recently, a fixation process, albeit still on a very small scale, now appears to have been established (Beyer 1975). Supplying chromatographically pure $[^{14}C]C_2H_4$ of high specific activity (22 mCi/mmole) to etiolated pea seedlings grown under aseptic conditions, he determined that $^{14}CO_2$ and two as yet unidentified metabolites were formed. The biological importance of this fixation has yet to be determined, but the possibility of an involvement in the so-called autocatalytic effect of ethylene in initiating the production of high levels of ethylene in ripening fruits may be relevant (Mapson and Robinson 1966).

5.8. Factors regulating ethylene biosynthesis

5.8.1. Hormonal control of ethylene biosynthesis

The first indication that the synthesis of ethylene might be modified by another hormone came from the observation of Zimmerman and Wilcoxon (1935) that plants treated with herbicidal concentrations of the synthetic auxin 2,4-dichlorophenoxyacetic acid (2,4-D) produced a volatile which induced epinastic responses in other plants enclosed with them. Nearly 20 years later, Morgan and Hall (1962, 1964) showed that when plants were sprayed with 2,4-D or the natural auxin indol-3-ylacetic acid (IAA), they would subsequently produce greatly enhanced levels of ethylene.

Considerable evidence now links the rates of ethylene production of growing or expanding tissues with their levels of endogenous auxin. For example, in the green (Burg 1968) or etiolated pea shoot (Burg 1973; Burg and Burg 1968), the highest levels of production (Table 5.5) correspond with the regions of highest diffusible or extractable IAA (Scott and Briggs 1960). Other evidence of an association comes from the levels of ethylene production in developing leaves, which are relatively high during the period of expansion when auxin levels are also high (Wetmore and Jacobs 1953) and in fruits which produce high levels of ethylene (Looney et al. 1974) in the first stages of enlargement (Figure 5.2) when they contain a relatively high content of auxin (Nitsch 1970). An interesting and intimate correlation has been observed in floral buds of cucurbits. Higher endogenous auxin levels are found in the apices of female plants than male plants (Galun et al. 1965) and the ethylene production of such female buds is three times higher than that of the males (Rudich et al. 1972).

Applications of auxin to isolated plant parts invariably enhance the evolution of ethylene. As the tissue matures, however, the total levels produced with and without added auxin are lower both in leaves (Jackson and Osborne 1972) and in fruits (Lieberman and Kunishi 1975). Working with excised segments from seedling tissues of etiolated pea shoots supplied with [^{14}C]IAA, Kang et al. (1971) showed that the IAA-induced ethylene production and growth of the segments parallelled the level of free [^{14}C]IAA in the tissue; the level of free [^{14}C]IAA in the tissue was inversely related to the extent of conjugation to [^{14}C]IAA-aspartate. Ethylene production following spray treatments of etiolated pea shoots with solutions of IAA in vivo is also related to the levels of auxin applied (Figure 5.3). An important feature of the auxin regulated ethylene biosynthesis is that although the rate is enhanced within 60 min or less, the decay of the enhanced ethylene production may not occur until many hours after the endogenous auxin level has returned to normal.

The mechanism by which the auxin stimulates the production of ethylene is of considerable interest. In mung bean hypocotyl segments supplied with [^{14}C]methionine, the specific activity of the ethylene produced was similar in control and auxin-stimulated treatments; Sakai and Imaseki (1972) therefore concluded that the effect of auxin was to stimulate the enzyme system converting methionine to ethylene, and not to enhance the formation of methionine. The presence of L-canaline and DNP inhibits both normal and

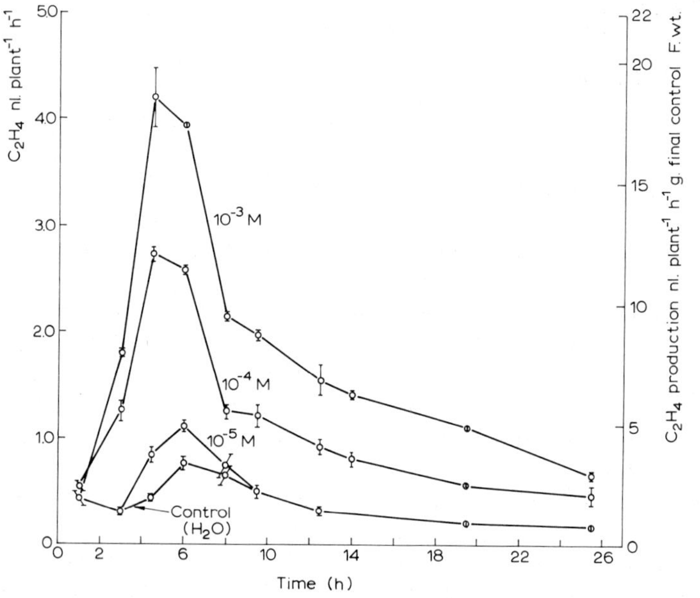

Figure 5.3. Time course of ethylene production by etiolated shoots of pea seedlings following a single spray with water or different concentrations of IAA. (Atack and Osborne, unpublished).

auxin-enhanced ethylene productions in etiolated peas (Owens et al. 1971; Murr and Yang 1975), also indicating that the IAA-induced pathway is probably no different from that which is normally functional. Incubation of segments of the subhook region with inhibitors of RNA synthesis (25 μg/ml actinomycin D, 25 μg/ml cordycepin, or 3 μg/ml α-amanitin) or of protein synthesis (100 μM cycloheximide) all inhibit IAA-induced ethylene production to a greater or lesser extent, whereas inhibitors of DNA synthesis over a range of concentrations do not. Although data from inhibitor experiments of this kind are always difficult to evaluate, in total the results are suggestive that new protein and RNA synthesis is necessary for the stimulation and maintenance of the pathway (Lieberman and Kunishi 1975). However, observations that cycloheximide (10 μM) did not inhibit auxin-induced ethylene production of pea root tips when low concentrations of IAA were supplied (1 μM) led Steen and Chadwick (1973) to propose that auxin had two effects on ethylene biosynthesis; one is an immediate and cycloheximide-insensitive stimulation occurring at low auxin (and therefore independent of new protein synthesis), and a second, which occurs after a lag period of 1–2 hr, is sensitive to protein synthesis inhibitors, and is dependent upon high auxin concentrations (10–100 μM).

Inhibitors of protein synthesis can prevent the rise in ethylene production at fruit ripening (Frenkel et al. 1968; Lieberman and Kunishi 1975), so the active synthesis of new enzyme protein may well be involved in ripening. In these instances, ethylene itself initiates the rise in ethylene production as well as increasing the incorporation of amino acids into protein (Hulme et al. 1971).

The literature abounds with reports that other hormones, gibberellins, cytokinins and abscisic acid, will regulate ethylene biosynthesis. It seems that gibberellin (Atack and Osborne, unpublished) and cytokinin will both retard the rate of metabolism of applied IAA. There is clear evidence that some increase in ethylene production occurs when cytokinins alone are added to plant tissues, but when cytokinin and auxin are supplied together (Figure 5.4) to mung bean hypocotyl segments (Lau and Yang 1973) production of ethylene is greatly increased, and can be directly related to a cytokinin suppression of the conjugation of auxin to give IAA-aspartate and the maintenance of high levels of free IAA. A clear enhancement of auxin-induced ethylene production in the presence of kinetin can be observed also in intact *Xanthium* leaves (Table 5.8). From such results it is reasonable to infer that the rate of metabolism of endogenous IAA would also be reduced, and ethylene production increased, by an enhanced level of endogenous cytokinin. Other factors may be involved, however, since kinetin was not found to enhance auxin-induced ethylene production in excised leaf discs of *Xanthium* (Table 5.8). According to Lau and Yang (1975) the effects of cytokinin may be mediated by Ca^{2+}, which not only enhanced uptake of the cytokinin, but also enhanced its rate of metabolism. They suggest that Ca^{2+} may play a role in ethylene biosynthesis and that release of Ca^{2+} is facilitated by the cytokinin. However, a role of Ca^{2+} in the control of IAA metabolism might equally be postulated.

Abscisic acid does not normally increase ethylene production except where tissue is approaching senescence. In bean explants (Jackson and Osborne 1972) the rise in ethylene which follows upon the abscisic acid treatment of abscission zones can be directly related to an enhanced rate of senescence and a precocious increase in senescent-phase ethylene production.

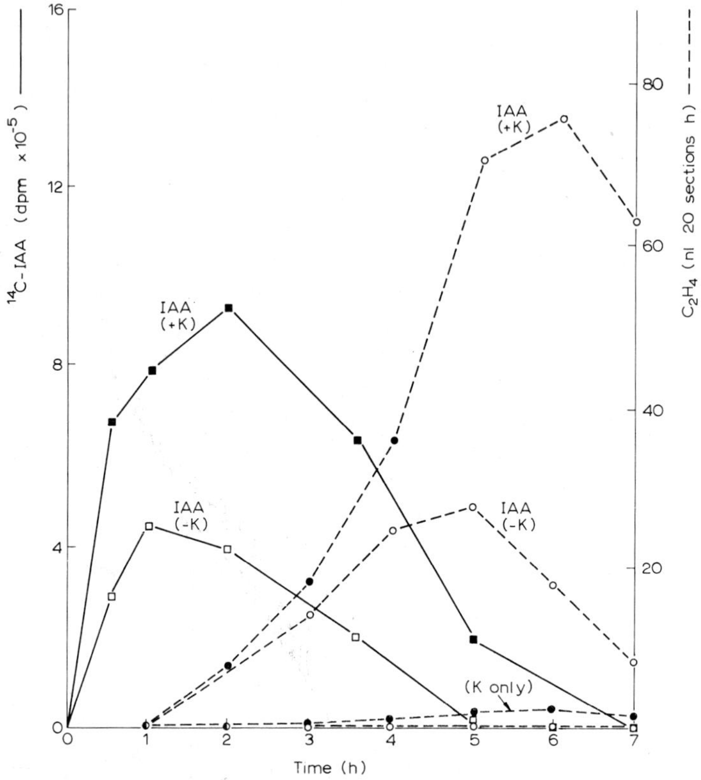

Figure 5.4. Relationship between the level of extractable [^{14}C]IAA and the ethylene production of segments of etiolated mung bean hypocotyl supplied with [^{14}C]IAA with or without the addition of kinetin (K). Bottom line is for the treatment (⊖k) (⊖IAA). (After Lau and Yang 1973).

Table 5.8.

Effect of kinetin and IAA on ethylene production in mature leaves and leaf discs of *Xanthium pennsylvanicum*

Production by whole leaves and leaf discs was measured 7–15 hr and 9–15 hr, respectively, after treatment. Differences between replicate samples were < 5%

	Ethylene production (nl/g/hr)			
	H_2O	Kinetin (20 mg/l)	IAA (176 mg/l)	Kinetin + IAA
Whole leaves	0.22	0.29	2.29	5.55
Leaf discs	0.89	1.87	4.44	6.12

(Hallaway and Osborne, unpublished data.)

5.8.2. Wound ethylene production

Although it is clear that the production of ethylene by young and growing tissue is readily stimulated by auxin, stimulations are not observed with post-climacteric apple fruit (Lieberman and Kunishi 1975) or in yellow and senescent leaves of *Prunus* or *Euonymus* (Hallaway and Osborne, unpublished). Similarly, although wound-enhanced ethylene production of several-fold can occur within 30–60 min in young tissue, e.g., after the handling of etiolated pea seedlings (Irvine and Osborne 1973) or following contact stimulation in pea tendrils (Jaffe 1970), the wound ethylene production from senescent fully yellow leaves is negligible (Hallaway and Osborne, unpublished). Age of the tissue is therefore important in determining whether the different types of ethylene biosynthesis can occur.

Apparently, any kind of touch, mechanical pressure or wounding, can provide the intracellular environment that leads to a rapid increase in ethylene synthesis, and this wound ethylene production can play an important role in regulating growth and development in plants. For example, the infestation of fruits of the fig, *Ficus sycomorus,* by a specific wasp causes sufficient damage and wound ethylene production to initiate enlargement and ripening of the fruit (Zeroni et al. 1972). If a growing shoot encounters mechanical pressure, for example when a pea shoot makes contact with the stopper of the flask in which it is growing, or when the shoot grows through a column of glass beads (Goeschl et al. 1966), ethylene production can be increased 10-fold. The shoot subsequently develops the short and swollen growth habit with tightly closed hook first recorded by Neljubow (1901) for peas exposed to illuminating gas. The developmental advantage of the resulting increased area-to-load ratio in the apex of seedlings emerging through the soil has been pointed out by Goeschl et al. (1966).

From studies of a possible cofactor of ethylene biosynthesis in *Phaseolus* leaves, Osborne et al. (1972) proposed that such a factor might be compartmentalised in a membrane-bound organelle, perhaps the chloroplast or the vacuole. Perturbation of the tissue could cause disorientation of this membrane, rendering it more permeable, and releasing some of the enclosed cofactor to the cytoplasm, thereby enhancing ethylene biosynthesis. If the normal membrane is subsequently reformed (either by new synthesis or reassembly) further release of cofactor would be precluded. This type of control system would account for the temporary nature of wound ethylene production. An auxin-enhanced permeability of the same organelle membrane could account for part of the auxin-enhanced ethylene production. In senescing tissue, however, where the rate of new membrane synthesis is generally low, and membranes become progressively impaired, a total leakage of cofactor could lead to a large and terminal senescence-rise in ethylene production which would not be enhanced further either by wounding or by auxin.

The evidence from studies with [U-^{14}C]methionine in leaves of *Phaseolus* indicates that the precursors for normal and wound ethylene synthesis are probably similar, both being derived, at least in part, from methionine (Abeles and Abeles 1972).

The sensitivity of ethylene responses to an environmental change is illustrated by the enhanced production of ethylene which occurs when plant organs are placed horizontal. This has been recorded after 6 hr in sunflower leaves (Palmer 1973), after 4 hr in tomato petioles (Leather et al. 1972) and within 90 min in the nodes of the inflorescence stalk of the grass *Echinochloa colonum* (Osborne and Wright 1977). The increased ethylene production in *Echinochloa* occurs on the lower side of the tissue (Osborne 1974) and may relate to an enhanced level of auxin on that side; a higher proportion of ^{14}C does accumulate on the lower side during basipetal transport of [^{14}C]IAA in horizontal nodes which suggests this may occur. However, it is not established that the ^{14}C is still present as IAA, and the increase is not observed until some hours after curvature has been initiated. The displacement of starch statoliths in the geoperceptive statocyte cells of root and shoot when they are placed horizontal could cause sufficient perturbation and intracellular "wounding" to alter the permeability in a membrane containing a cofactor for ethylene production, with a resulting increase in ethylene production. Again, however, the time interval between the displacement of statoliths (2.5 min) and the increased ethylene production (90 min) seems too long for an endogenous wound response. Furthermore, since inhibitions (Warner and Leopold 1971) and accelerations (Musgrave and Walters 1974) of elongation growth by ethylene are measurable within 6–10 min of exposing a tissue to the gas, and gravity induced curvatures in horizontally orientated nodes occur by 30 min, it is unlikely that the ethylene produced at 90 min plays a regulatory role in the growth response. New evidence (Wright and Osborne 1977) suggests that the asymmetric production of ethylene by nodes of *Echinochloa* in response to gravity is instead directly related to an asymmetric distribution of endogenous auxin in the tissue, which occurs within 30 min of horizontal orientation. Experiments with longitudinally halved nodes (Wright et al. 1977) have established that the rise in auxin on the lower side results from a net local increase and not to a redistribution from adjacent tissues. The increased ethylene production by the lower side is therefore a good example of auxin-controlled ethylene synthesis in vivo.

5.8.3. Light control of ethylene production

Extension of shoots of etiolated plants is rapidly arrested and expansion growth of the plumular bud enhanced when they are exposed to white light or to light in the red region of the spectrum. In an early report, Burg and Burg (1966) observed that light-grown pea plants in the presence of IAA produced ethylene at a slower rate than corresponding etiolated plants. Goeschl et al. (1967) found a correlation between ethylene production and plant growth in response to light. Working with a broad-band red light source, they showed that a single exposure of 60 sec decreased ethylene evolution from the upper part of the epicotyl and enhanced expansion of the plumule within 4 hr of irradiation. The effect on ethylene production and plumule expansion was maximal after 9 hr, and the rates of both responses returned to their original values by 48 hr. They also showed that the red light response was reversible by a following exposure to a far red-enriched source, indicating that phytochrome might be involved in the control of ethylene production. Kang and Ray (1969) and Kang and Burg (1972) confirmed that plumules or segments of etiolated pea plants (with or without added auxin) produced less ethylene after exposure to red light than those kept in total darkness (Table 5.9). Reversibility of the red light response by subsequent exposure to far red illumination has been clearly demonstrated by Imaseki et al. (1971) in segments of etiolated rice epicotyls (Figure 5.5). Their results imply the involvement of phytochrome in ethylene production. The results

Table 5.9.

Ethylene production in response to IAA by subapical sections of etiolated pea shoots during 4 hr of darkness 8 hr after 5 min illumination with red light (1250 ergs/cm² /sec) (Kang and Burg 1972)

IAA conc. *(M)*	Ethylene production (nl/g/hr)	
	Dark	After red light
0	0.30	0.15
10^{-7}	0.32	0.19
2×10^{-7}	0.47	0.30
10^{-6}	0.82	0.56
10^{-5}	2.60	1.04

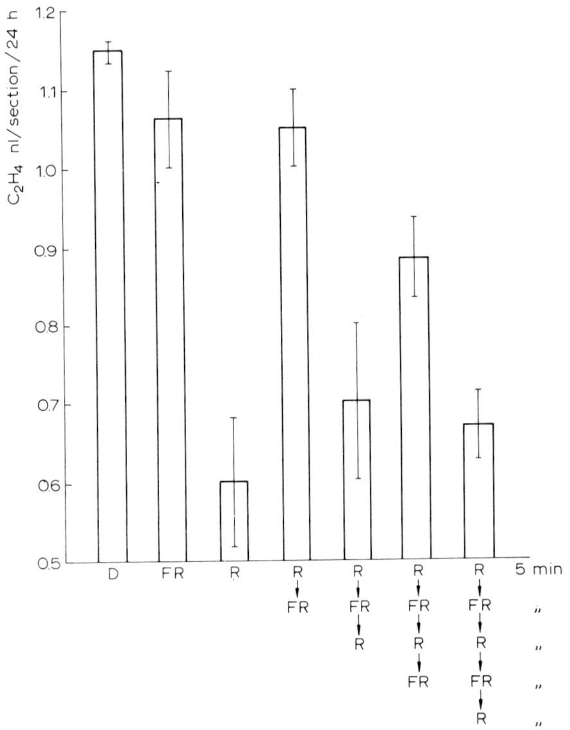

Figure 5.5. Inhibition of ethylene production in segments of rice coleoptile by exposure to 5 min red light (800 ergs/cm² /sec) and the reversal by 5 min far red light (18 kilo ergs/cm² /sec). (After Imaseki et al. 1971.)

are, however, for long periods of incubation (24 hr) and short-term experiments could be more convincing. Preliminary reports by De Proft and De Greef (1975) suggest that rapid changes controlled by light can occur in ethylene production; they noted a 40% reduction approximately 1.5 hr after a 5 min illumination of *Phaseolus* seedlings with red light.

From the work of Craker et al. (1973) it would appear that light sources other than red and far-red may also regulate ethylene production. With sections of etiolated soybean seedlings they found enhanced ethylene production over 24–25 hr illumination by blue regions of the spectrum corresponding to those of certain high irradiation growth responses (Figure 5.6). They too suggest that a high energy receptor (perhaps phytochrome) may be involved in the regulation of ethylene production. Curiously, however, unlike other reports, their results do not indicate a reduction in ethylene production below the dark values at the red wave lengths 600–660 nm.

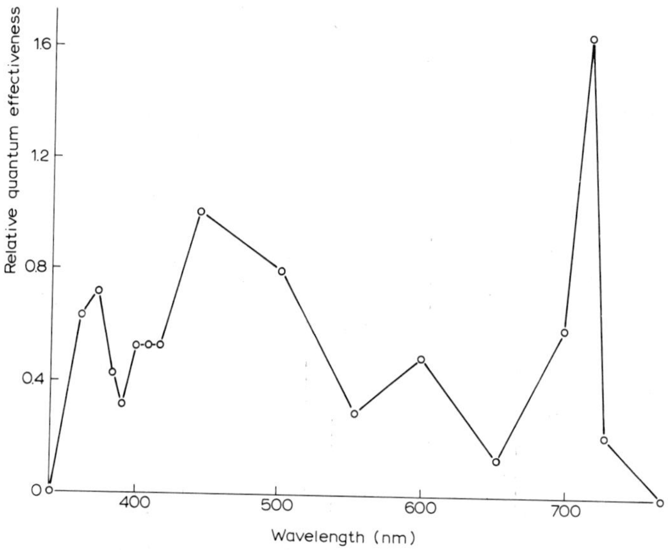

Figure 5.6. Action spectrum for ethylene production in segments of etiolated soybean hypocotyl. Measurements were made after 24 hr irradiation of the tissue. (After Craker et al. 1973.)

An involvement of phytochrome in modulating the properties of membranes has frequently been proposed (Hendricks and Borthwick 1967; Satter et al. 1970; Marmé and Schafer 1972), and a regulation of the membrane permeability of an organelle enclosing the proposed cofactor for ethylene evolution (Osborne et al. 1972) by different spectral regions of light could afford a sensitive mechanism for a rapid control of ethylene biosynthesis.

5.8.4. Effect of temperature on ethylene production

5.8.4.1. High temperature. The enzymes concerned in ethylene biosynthesis appear to be relatively labile compared with those of the respiratory pathway. Burg and Thimann (1959) with apples and Hansen (1942) with pears have shown a decrease in ethylene production above 30 °C falling to practically zero by 40 °C. The failure of plum fruits to ripen at high temperature has been related to an impaired ethylene formation, since normal ripening can be induced by added ethylene (Uota 1955). A possible further indication of the lability of enzymes concerned with ethylene biosynthesis comes from work with seeds of low viability. On germination of rape *(Brassica napus)*, aged seeds produced little ethylene compared with those of a high viability, freshly harvested stock (Takayanagi and Harrington 1971). The slow germination of the low viability seeds can, however, be considerably enhanced if they are supplied with ethylene, indicating that whereas other enzyme systems in the seeds are functional, those of the ethylene biosynthesis pathway may be impaired.

5.8.4.2. Low temperature. Although a Q_{10} of 2.8 is reported for ethylene production in apples between 10 ° and 25 °C (Burg and Thimann 1959), it is clear that exposure to temperatures below 10 °C can cause large increases in subsequent ethylene formation when the fruit is returned again to higher temperatures (Hansen 1942). This effect of cold has been cited as the cause of premature ripening of pears following cold (8 °C) nights (Wang et al. 1971). An effect of a cold treatment (5 °C for up to 6 hr) on the ethylene production of attached leaves of *Phaseolus* has been clearly shown by Wright (1974). Within 1 hr of returning the plants to 25 °C, ethylene production of the leaves rose to a level 20 times higher than that of controls maintained at 25 °C. (Figure 5.7). The enhanced ethylene production was accompanied by an increase in leakage of electrolytes from the tissues (Wright 1974), suggesting a temperature induced phase-change in membrane lipids during chilling, and an increase in membrane permeability. Such changes could also cause enhanced leakage of a cofactor of ethylene biosynthesis from within a cytoplasmic organelle (Osborne et al.1972).

5.8.5. Effects of water stress on ethylene production

There seems little doubt that plant water deficits and incipient wilting in whole plants can result in a doubling or trebling of ethylene production in cotton leaves, the rates declining again when the water potential returns to normal (McMichael et al. 1972) (Figure 5.8). Many-fold increases in ethylene production have also been observed during periods of water stress in excised citrus leaves (Ben-Yehoshua and Aloni 1974): provided the stress did not exceed 10 hr, and the relative water content of the leaf did not fall below 50–60% of controls, ethylene production returned to normal within 24 hr of supplying water. The increased ethylene production associated with plant water stress can be correlated closely both with the subsequent premature abscission of leaves (Jordan et al. 1972) and young fruits (Lipe and Morgan 1973) in cotton, and with a reduced growth rate and leaf and flower abscission in *Vicia faba* (El-Beltagy and Hall 1974). The water potential of avocado fruits can be important in determining the time of the increase in

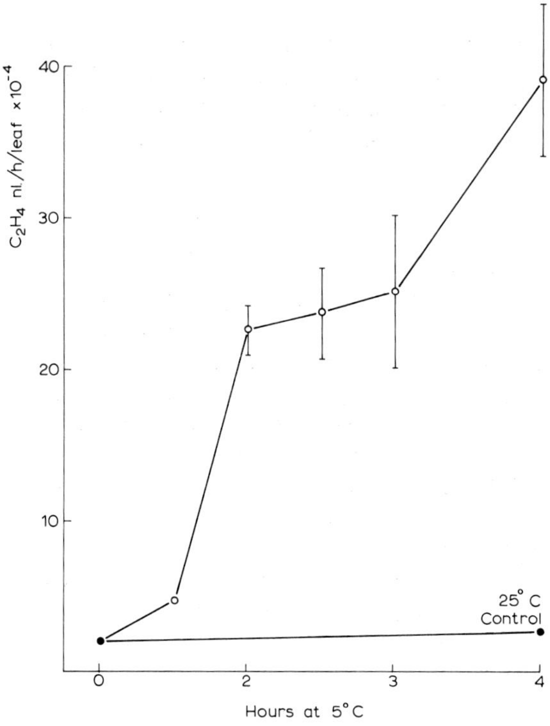

Figure 5.7. Ethylene production in leaves of *Phaseolus* during 1 hr at 25 °C following upon the exposure of plants in darkness to 5 °C (45% RH) for different lengths of time. (After Wright 1974.)

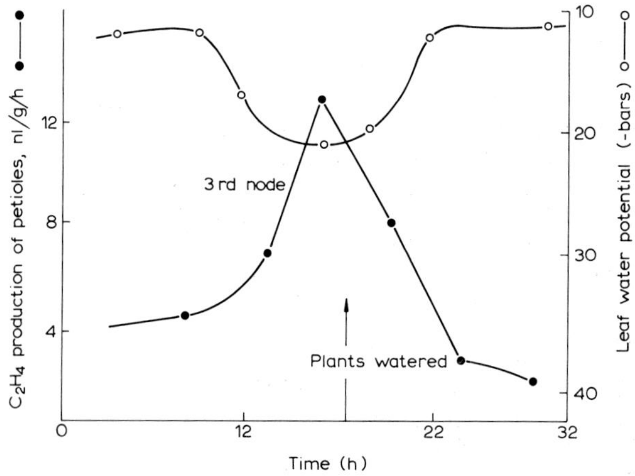

Figure 5.8. Production of ethylene by cotton petioles at different leaf water potentials. (After Mc-Michael et al. 1972.)

ethylene production (and hence the time of ripening) as shown by Adato and Gazit (1974). There could therefore be some practical advantage in keeping harvested fruit at a relatively high humidity to retard ripening during storage. The effects of water stress on ethylene production are discussed further in Chapter 11, sections 11.2.3 and 11.4.3, Volume II.

5.9. Ethylene as a hormone

Of all the hormones recorded for biological systems, ethylene is probably the most remarkable. The fact that it is a volatile sets it apart from the more usual hormonal substances of animals and plants. In some respects it can be considered a plant pheromone since the ethylene produced by one plant can affect the growth of another enclosed within the same container. Unlike a pheromone, ethylene, in concert with other hormones, also determines the growth and development of the individual that produces it. Although so much has been known for so long of the morphological effects that ethylene can induce, so little is yet certain of either the site of action in the cell or the pathway of biosynthesis. Now that it has been established that there is within plants a positional differentiation of target cells with special responses to ethylene (Wright and Osborne 1974; Osborne and Sargent 1976a, b; Osborne 1977), the elucidation of the physiology and biochemistry of ethylene as a hormone becomes even more provoking.

References

Abeles, F.B. (1966) Nature (Lond.), 210, 23–25.
Abeles, F.B. (1972) Ann. Rev. Plant Physiol., 23, 259–292.
Abeles, F.B. (1973) Ethylene in Plant Biology, Academic Press, New York.
Abeles, A.L. and Abeles F.B. (1972) Plant Physiol., 50, 496–498.
Abeles, F.B. and Rubinstein, B. (1964) Plant Physiol, 39, 963–969.
Abeles, F.B., Ruth, J.M., Forrence, L.E. and Leather, G.R. (1972) Plant Physiol., 49, 669–671.
Adato, I. and Gazit, S. (1974) Plant Physiol., 53, 45–46.
Akamine, E.K. and Goo, T. (1973) J. Amer. Soc. Hort. Sci., 98, 381–383.
Baur, A. and Yang, S.F. (1969a) Plant Physiol., 44, 189–192.
Baur, A. and Yang, S.F. (1969b) Plant Physiol., 44, 1347–1349.
Baur, A.H., Yang, S.F., Pratt, H.K. and Biale, J.B. (1971) Plant Physiol., 47, 696–699.
Ben-Yehoshua, S. and Aloni, B. (1974) Plant Physiol., 53, 863–865.
Beyer, E.M. (1972) Plant Physiol., 49, 672–675.
Beyer, E.M. (1975) Nature (Lond.), 255, 144–147.
Beyer, E.M. and Morgan, P.W. (1970) Plant Physiol., 46, 352–354.
Biale, J.B. (1940) Science, 91, 458–459.
Blanpied, G.D. (1971) HortScience, 6, 132–134.
Botjes, J.O. (1933) Tijdschr. Plantenziekten, 39, 207–211.
Burg, S.P. (1962) Ann. Rev. Plant Physiol., 13, 265–302.
Burg, S.P. (1968) Plant Physiol., 43, 1503–1511.
Burg, S.P. (1973) Proc. nat. Acad. Sci. (Wash.), 70, 591–597.
Burg, S.P. and Burg, E.A. (1962) Plant Physiol., 37, 179–189.
Burg, S.P. and Burg, E.A. (1965) Bot. Gaz. (Chic.), 126, 200–204.
Burg, S.P. and Burg, E.A. (1966) Proc. nat. Acad. Sci. (Wash.), 55, 262–269.

Burg, S.P. and Burg, E.A. (1968) Plant Physiol., 43, 1069–1074.
Burg, S.P. and Dijkman, M.J. (1967) Plant Physiol., 42, 1648–1650.
Burg, S.P. and Thimann, K.V. (1959) Proc. nat. Acad. Sci. (Wash.), 45, 335–344.
Cheah, K.R. (1975) Ph.D. Thesis, Cambridge, England.
Chou, T.W. and Yang, S.F. (1973) Arch. Biochem., 157, 73–82.
Cousins, H.H. (1910) Jamaica Dept. Agr. Ann. Rept., 7.
Craker, L.E., Abeles, F.B. and Shropshire, W. (1973) Plant Physiol., 51, 1082–1083.
Crocker, W. (1948) Growth of Plants, Reinhold, New York.
Crocker, W.C. and Knight, L.I. (1908) Bot. Gaz. (Chic.), 46, 259–276.
Denny, F.E. (1924a) J. Agr. Res., 27, 757–771.
Denny, F.E. (1924b) Bot. Gaz. (Chic.), 77, 322–329.
De Proft, M. and De Greef, J.A. (1975) In: Abstract, 73A, 22nd Easter School in Agricultural Science, Sutton Bonington, University of Nottingham, H. Smith (ed.).
Eger, E.I. and Larsen, C.P. (1964) Brit. J. Anaseth., 36, 140–149.
El-Beltagy, A.S. and Hall, M.A. (1974) New Phytol., 73, 47–60.
Elmer, O.H. (1932) Science, 75, 193.
Elstner, E.F. and Konze, J.R. (1974) FEBS Lett., 45, 18–21.
Elstner, E.F. and Konze, J.R. (1975) Z. Naturforsch, 30c, 58–63.
Freebairn, H.T. and Buddenhagen, I.W. (1964) Nature (Lond.), 202, 313–314.
Frenkel, C., Klein, I. and Dilley, D.R. (1968) Plant Physiol., 43, 1146–1153.
Fuchs, Y. and Lieberman, M. (1968) Plant Physiol., 43, 2029–2036.
Galliard, T., Hulme, A.C., Rhodes, M.J.C. and Wooltorton, L.S.C. (1968a) FEBS Lett., 1, 283–286.
Galliard, T., Rhodes, H.J.C., Wooltorton, L.S.C. and Hulme, A.C. (1968b) Phytochemistry, 7, 1465–1470.
Galun, E., Izhar, S. and Atsmon, D. (1965) Plant Physiol., 40, 321–326.
Gane, R. (1934) Nature (Lond.), 134, 1008.
Gibson, M.S. (1963) Biochim. biophys. Acta, 78, 528–530.
Giovanelli, J., Owens, L.D. and Mudd, J.H. (1971) Biochim. biophys. Acta, 227, 671–678.
Girardin, J.P.L. (1864) Jahresb. Agr., 7, 199–200.
Goeschl, J.D., Rappaport, L. and Pratt, H.K. (1966) Plant Physiol., 41, 877–885.
Goeschl, J.D., Pratt, H.K. and Bonner, B.A. (1967) Plant Physiol., 42, 1077–1080.
Hansen, E. (1942). Bot. Gaz. (Chic.), 103, 543–558.
Harvey, E.M. (1913). Bot. Gaz. (Chic.), 56, 438–442.
Hendricks, S.B. and Borthwick, H.A. (1967) Proc. nat. Acad. Sci. (Wash.), 58, 2125–2130.
Huelin, F.E. (1933) Brit. Dept. Sci. Ind. Res., Food Invest. Board Rept. (1932) 51–2, H.M. Stationery Office.
Hulme, A.C., Rhodes, M.J. and Wooltorton, L.S. (1971) Phytochemistry, 10, 749–756.
Ilag, L. and Curtis, R.W. (1968) Science, 159, 1357–1358.
Imaseki, H., Pjon, C.J. and Furuya, M. (1971) Plant Physiol., 48, 241–244.
Irvine, R.F. and Osborne, D.J. (1973) Biochem. J., 136, 1133–1135.
Jackson, M.B. (1971) D. Phil. Thesis., Oxford, England.
Jackson, M.B. and Osborne, D.J. (1970) Nature (Lond.), 225, 1019–1022.
Jackson, M.B. and Osborne, D.J. (1972) J. exp. Bot., 23, 849–862.
Jaffe, M.J. (1970) Plant Physiol., 46, 631–633.
Jordan, W.R., Morgan, P.W. and Davenport, T.L. (1972) Plant Physiol., 50, 756–758.
Kang, B.G. and Burg, S.P. (1972) Plant Physiol., 50, 132–135.
Kang, B.G. and Ray, P.M. (1969) Planta, 87, 206–216.
Kang, B.G., Newcomb, W. and Burg, S.P. (1971) Plant Physiol., 47, 504–509.
Kende, H. and Baumgartner, B. (1974) Plant Physiol., 116, 279–289.
Ketring, D.L., Young, R.E. and Biale, J.B. (1968) Plant Cell Physiol., 9, 617–631.
Kidd, F. and West, C. (1933) Brit. Dept. Sci. Ind. Res., Food Invest. Board Rept. (1932) 55, H.M. Stationery Office.

Knight, L.I. and Crocker, W.C. (1913) Bot. Gaz. (Chic.), 55, 337–371.
Ku, H.S. and Pratt, H.K. (1968) Plant Physiol., 43, 999–1001.
Ku, H.S., Yang, S.F. and Pratt, H.K. (1967) Arch. Biochem., 118, 756–758.
Ku, H.S., Yang, S.F. and Pratt, H.K. (1969) Phytochemistry, 8, 567–573.
Lau, O.L. and Yang, S.F. (1973) Plant Physiol., 51, 1011–1014.
Lau, O.L. and Yang, S.F. (1975) Plant Physiol., 55, 738–740.
Leather, G.R., Forrence, L.E. and Abeles, F.B. (1972) Plant Physiol., 49, 183–186.
Lieberman, M. and Kunishi, A.T. (1971) Plant Physiol., 47, 576–580.
Lieberman, M. and Kunishi, A.T. (1975) Plant Physiol., 55, 1074–1078.
Lieberman, M. and Mapson, L.W. (1964) Nature (Lond.), 204, 343–345.
Lieberman, M., Kunishi, L., Mapson, L.W. and Wardale, D.A. (1965) Biochem. J., 97, 449–459.
Lieberman, M., Kunishi, L., Mapson, L.W. and Wardale, D.A. (1966) Plant Physiol., 41, 376–382.
Lipe, J.A. and Morgan, P.W. (1973) Plant Physiol., 51, 949–953.
Looney, N.E., McGlasson, W.B. and Coombe, B.G. (1974) Aust. J. Plant Physiol., 1, 77–86.
Lynch, J.M. (1972) Nature (Lond.), 240, 45–46.
Lynch, J.M. (1974) J. gen. Microbiol., 83, 407–411.
McElroy, J.J. (1960) Calif. Agr., 14, 3.
McMichael, B.L., Jordan, W.R. and Powell, R.D. (1972) Plant Physiol., 49, 658–660.
Mapson, L.W. (1969) Biol. Rev., 44, 155–187.
Mapson, L.W. (1970) Endeavour, 106, 29–33.
Mapson, L.W. and Mead, A. (1968) Biochem. J., 108, 875–881.
Mapson, L.W. and Robinson, J.E. (1966) J. Food Technol., 1, 215–225.
Mapson, L.W. and Wardale, D.A. (1967) Biochem. J., 102, 574–585.
Mapson, L.W. and Wardale, D.A. (1968) Biochem. J., 107, 433–442.
Mapson, L.W., Self R. and Wardale, D.A. (1969) Biochem. J., 111, 413–418.
Marmé, D. and Schafer, E. (1972) Z. Pflanzenphysiol., 67, 192.
Meigh, D.F., Jones, J.D. and Hulme, A.L. (1967) Phytochemistry, 6, 1507–1515.
Molisch, H. (1884) Sitzungsber. Kais Acad. Wiss (Wein), 90, 111–196.
Morgan, P.W. and Hall, W.C. (1962) Physiol. Plant., 15, 420–427.
Morgan, P.W. and Hall, W.C. (1964) Nature (Lond.), 204, 99.
Murr, D.P. and Yang, S.F. (1975) Plant Physiol., 55, 79–82.
Musgrave, A. and Walters, J. (1974) Planta, 121, 51–56.
Neljubow, D.N. (1901) Beit. Bot. Zentralbl., 10, 128–138.
Nichols, R. (1968) J. Hort.. Sci., 43, 335–349.
Nitsch, J.P. (1970) in: The Biochemistry of Fruits and their Products. Vol. 1, A.C. Hulme (ed.), Academic Press, London, pp. 427–472.
Osborne, D.J. (1974) in: Mechanisms of Regulation of Plant Growth. R.L. Bieleski, A.R. Ferguson and M.M. Cresswell (eds.), The Royal Society of New Zealand, Wellington, pp. 645–654.
Osborne, D.J. (1977) Sci. Prog. (Oxford), 64, 53–65.
Osborne, D.J. and Sargent, J.A. (1976a) Planta, 130, 203–210.
Osborne, D.J. and Sargent, J.A. (1976b) Planta, 132, 197–204.
Osborne, D.J. and Wright, M. (1977) Proc. roy. Soc. B, 199, 551–564.
Osborne, D.J., Jackson, M.B. and Milborrow, B.V. (1972) Nature New Biology, 240, 98–101.
Owens, L.D., Lieberman, M. and Kunishi, A. (1971) Plant Physiol., 48, 1–4.
Palmer, J.H. (1973) Physiol. Plant., 28, 188–193.
Primrose, S.B. (1976) J. gen. Microbiol., 95, 159–165.
Primrose, S.B. (1977a) J. gen. Microbiol., 96.
Primrose, S.B. (1977b) J. gen. Microbiol., 96.
Primrose, S.B. and Dilworth, M.J. (1976) J. gen. Microbiol., 93, 177–181.
Ram Chandra, G. and Spencer, M. (1962) Nature (Lond.), 194, 361–364.
Rudich, J., Halevy, A.H. and Kedar, N. (1972) Plant Physiol., 49, 998–999.
Sakai, S. (1975) Plant Cell Physiol., 16, 529–532.
Sakai, S. and Imaseki, H. (1972) Planta, 105, 165–173.

Sakai, S. and Imaseki, H. (1973) Planta, 113, 115–128.
Satter, R.L., Marinoff, P. and Galston, A.W. (1970) Amer. J. Bot., 57, 916–926.
Scott, T.K. and Briggs, W.R. (1960) Amer. J. Bot., 47, 492–499.
Sievers, A.F. and True, R.H. (1912) U.S. Dept. Agr. Bur. Plant Ind. Bull., 232.
Spencer, M. and Meheruik, M. (1963) Nature (Lond.), 199, 1077–1078.
Steen, D.A. and Chadwick, A.V. (1973) Plant Physiol., 52, 171–173.
Stinson, R.A. and Spencer, M. (1970) Canad. J. Biochem., 48, 541–546.
Takayanagi, K. and Harrington, J.F. (1971) Plant Physiol., 47,521–524.
Turner, E.M., Wright, M., Ward, T., Osborne, D.J. and Self, R. (1975) J. gen. Microbiol.., 90, 167–176.
Uota, M. (1955) Proc. nat. Acad. Sci. (Wash.), 65, 231–243.
Vendrell, M. (1969) Aust. J. Biol. Sci., 22, 601–610.
Wang, C.Y., Millenthin, W.M. and Hansen, E. (1971) J. Amer. Soc. Hort. Sci., 96, 122.
Warner, H.L. and Leopold, A.C. (1971) Biochem. biophys. Res. Commun., 44, 989–994.
Wetmore, R.H. and Jacobs, W.P. (1953) Amer. J. Bot., 40, 272–276.
Williamson, C.E. (1950) Phytopathology, 40, 205–209.
Wright, M. (1974) Planta, 120, 63–69.
Wright, M. and Osborne, D.J. (1974) Planta, 120, 163–170.
Wright, M. and Osborne, D.J. (1977) Biochem. Physiol. Pflanz., 197, 474–492.
Wright, M., Mousdale, D. and Osborne, D.J. (1977) Biochem. Physiol., Pflanz., (in press).
Yang, S.F., Ku, H.S. and Pratt, H.K. (1967) J. biol. Chem., 242, 5274–5280.
Young, R.E., Pratt, H.K. and Biale, J.B. (1952) Analyt.. Biochem., 24, 551–555.
Zeroni, M., Ben-Yehoshua, S. and Galil, J. (1972) Plant Physiol., 50, 378–381.
Zimmerman, P.W. and Wilcoxon, F. (1935) Contrib. Boyce Thompson Inst., 7, 209–229.

Phytohormones and Related Compounds – A Comprehensive Treatise, Volume I
Editors: Letham, Goodwin and Higgins

6

Abscisic acid

B.V. MILBORROW

6.1. Introduction

The presence of growth-inhibitory compounds in plants has been known since the 1930's but the systematic investigation of these materials began when Bennet-Clark et al. (1952) combined the technique of paper chromatography with growth bioassays for the analysis of extracts of plants. They found that the extracts consistently contained growth-inhibitory activity which chromatographed at about R_f 0.6 in their system and they named this fraction "inhibitor β". These preliminary experiments were repeated on a variety of species by many groups of workers using different chromatographic systems. However, the R_f values of the inhibitory fractions from different plants were not compared in any one system, neither was one plant's inhibitor tested in a range of chromatographic systems: a "Tower of Babel" situation developed. It was impossible to know whether any one inhibitor was identical with those isolated from other plants.

A highly active inhibitor of coleoptile elongation was isolated in crystalline form from cotton bolls (Ohkuma et al. 1963), sycamore leaves (Cornforth et al. 1965b) and yellow lupin fruits (Rothwell and Wain 1964) and shown to be the same compound. A structure was proposed by Ohkuma et al. (1965) and this was confirmed by synthesis by Cornforth et al. (1965a). It is now known as abscisic acid (ABA) (I).

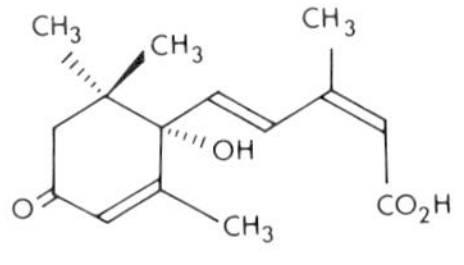

I. (+)-*(S)*-abscisic acid
(revised absolute configuration)

Since the synthetic material became widely available, most of the physiological investigations have been carried out with this racemic compound (± ABA).

Subsequently (+)-ABA has been isolated in crystalline form from peas *(Pisum sativum)* (Isogai et al. 1970), yellow lupin *(Lupinus luteus)* (Koshimizu et al. 1966) and avocado *(Persea gratissima)* (Milborrow, unpublished) and identified by combinations of thin-layer and gas-liquid chromatography, mass spectrometry and optical methods, from a large

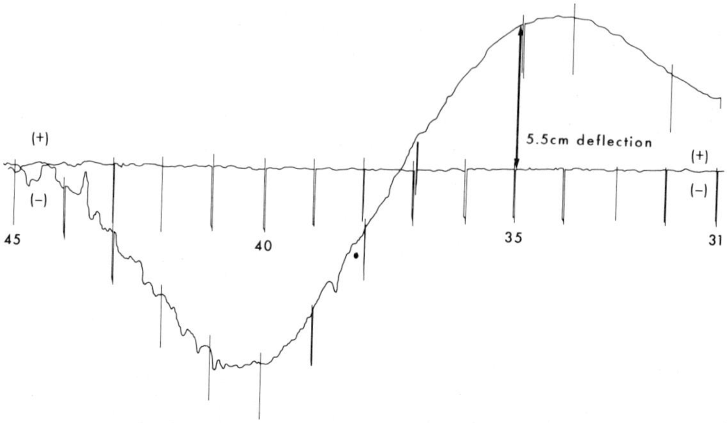

Figure 6.1. Optical rotatory dispersion (ORD) curve of a sample of (+)-abscisic acid isolated from an avocado fruit. At the instrument setting used, 5.5 μg/ml in acidic methanol (10-mm path length) gives a deflection of 5.5 mm from the base line at 35 wave number. The instrument trace has not had the Verdet correction applied.

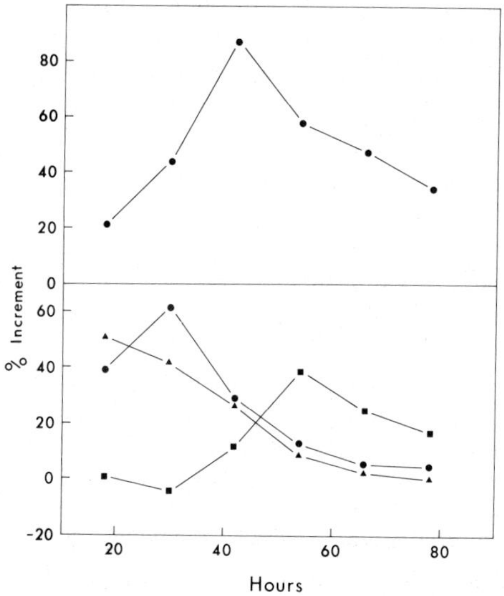

Figure 6.2. The changing sensitivity of wheat coleoptiles to gibberellic acid ▲, kinetin ●, and indole-acetic acid ■ (lower graphs). The data are replotted from Wright (1961) together with a measure of growth rate (upper graph). Growth rate is expressed as per cent increase over 6-hr periods. In the lower graph, per cent increase over controls is plotted against coleoptile age.

variety of vascular plants. Apart from precursors, metabolites and conjugates (+)-abscisic acid appears to be the only member of its class of hormone. All isolates of naturally occurring ABA have shown the same positive, optical rotatory dispersion curve (Figure 6.1) and, within the limits of experimental error, have the same specific rotation. Natural ABA, therefore, consists entirely of the (+)-enantiomer.

Although ABA is a potent inhibitor of growth when applied to growing, intact plants, or pieces of plant, this does not necessarily mean that it acts as an inhibitor of the normal, endogenous growth. However, it has now been found to have a variety of effects on plant metabolism (Milborrow 1974a) and the burden of the evidence suggests that it does inhibit growth in vivo.

Growth processes of many tissues often respond to two or more of the growth-promoting hormones and ABA is almost invariably found to antagonise the action of the growth-promoting hormones to which the tissue responds. The work of Wright (1961, 1968) has shown that developing wheat coleoptiles pass through a sequential pattern of sensitivity to gibberellin, cytokinin and auxin as the cells mature (Figure 6.2). At all growth stages ABA antagonises the growth-promoting activity of the stimulatory hormones.

Apart from its inhibitory effects on growth, ABA plays a part in the regulation of the abscission of leaves and young fruit, the dormancy of buds and seeds, and the ripening of fruit.. One of its most striking effects is its ability to cause the closure of stomata and this function is associated with a rapid 40-fold increase in the ABA content of the leaves which occurs when they wilt or suffer stress from waterlogging or chilling. Thus ABA participates in the hormonal regulation of many aspects of plant metabolism.

6.2. Methods of estimation of abscisic acid

6.2.1. Bioassays

Bioassay methods for abscisic acid most frequently exploit its growth-inhibitory properties; indeed the first 3 isolations of crystalline material were guided, at least in part, by measurement of its inhibitory effects on coleoptile growth. More recently other physiological effects have been exploited in bioassay procedures; these include its ability to cause stomatal closure and its inhibition of seed germination and amylase synthesis (Aspinall et al. 1967).

Often the "assay" systems have been used in investigations of the effects of environmental factors and of biologically active molecules on the growth and physiology of the plant material under study. While such studies are useful to define the interactions of several growth regulators applied simultaneously, they should not be carried out with a tissue which has been specifically chosen and treated for its large and selective response to one kind of active molecule.

The typical response of a growing plant part to an application of ABA is a sharp decrease in growth at a characteristic concentration. When the response is plotted graphically, the region of maximum change is followed by a plateau as the maximum effective concentration is approached asymptotically (Figures 6.3, 6.4).

If the logarithm of the concentration is used as the abscissa then the shape of the

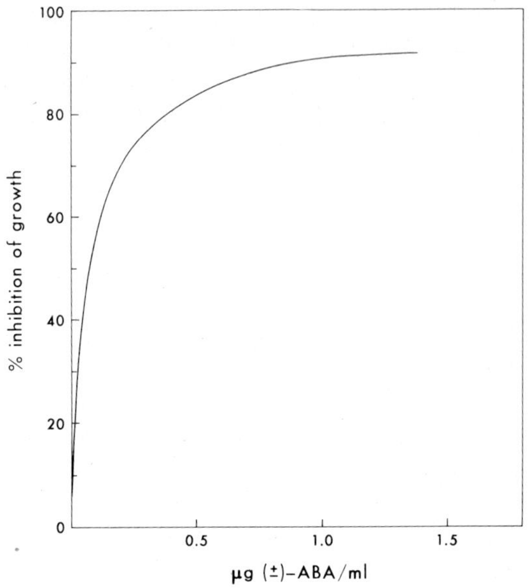

Figure 6.3. Theoretical curve for the inhibition of coleoptile growth. The curve is derived from the linear probit/log dosage relationship shown in Figure 6.7.

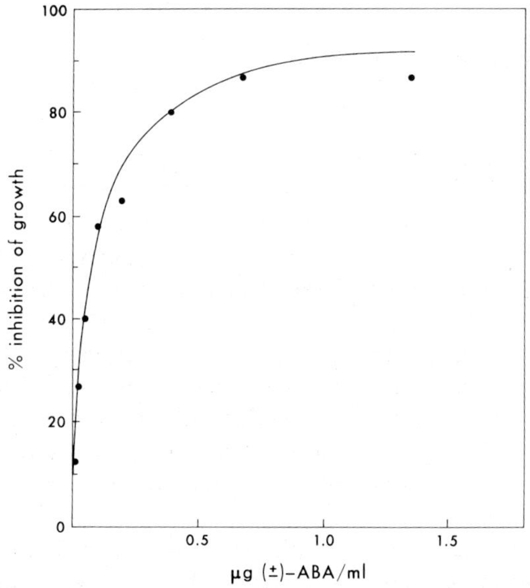

Figure 6.4. Data from an experiment in which the inhibition of coleoptile elongation of dissected wheat embryos by a sample of synthetic, racemic abscisic acid was measured. The percentage inhibition is plotted against the concentration of ABA. The theoretical curve from Fig. 6.3 is superimposed.

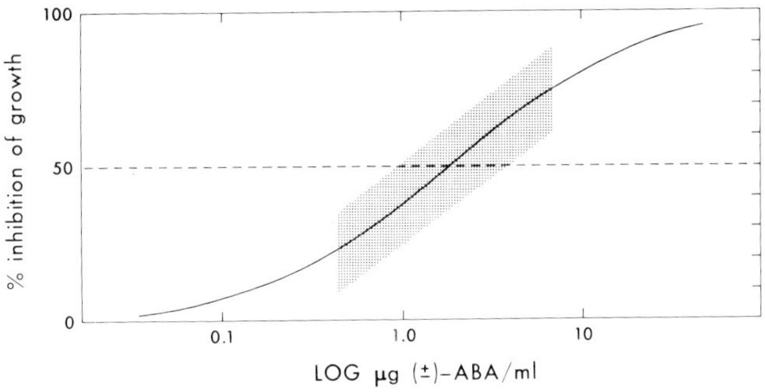

Figure 6.5. Theoretical curve for the inhibition of coleoptile growth. The same data as shown in Figs. 6.3 and 6.7 are plotted against log concentration of ABA. The value for 50% inhibition of growth (the point at which measurements are most accurate) is shown as a broken line. The shaded area indicates the region where the dosage/response relationship is almost linear.

dosage – response curve is sigmoid (Figures 6.5 and 6.6). A further refinement is to plot probit per cent inhibition against log concentration: this transforms the sigmoid dosage – response curve into a straight line (Figures 6.7 and 6.8) which can be used most conveniently for comparisons.

The most frequently used method of assay in plant physiology is to test one concentration of an extract and compare its effect with a reference curve constructed from measurements of a few concentrations of a suitable standard. It can readily be appreciated that this gives a reasonably accurate result if the concentration of the active ingredient of the extract happens to fall on the part of the curve where the response is nearly directly related to log dosage (shaded in Figure 6.5) but not if the single concentration used is outside this region. Results from such a bioassay are only relatively quantitative and unless the traditional methods of microbiology and biochemistry are used (where dilution series of an unknown and a standard are compared) then physical methods of measurement, which give a truly quantitative estimate, are preferable.

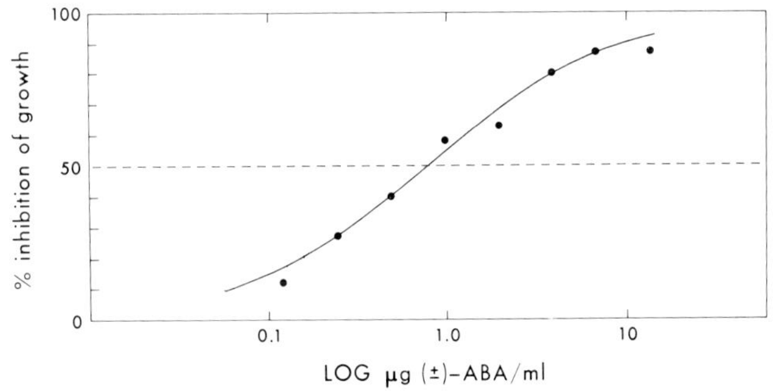

Figure. 6.6. The same data as in Figs. 6.4 and 6.8 plotted against log concentration. The value for 50% inhibition is shown as a broken line.

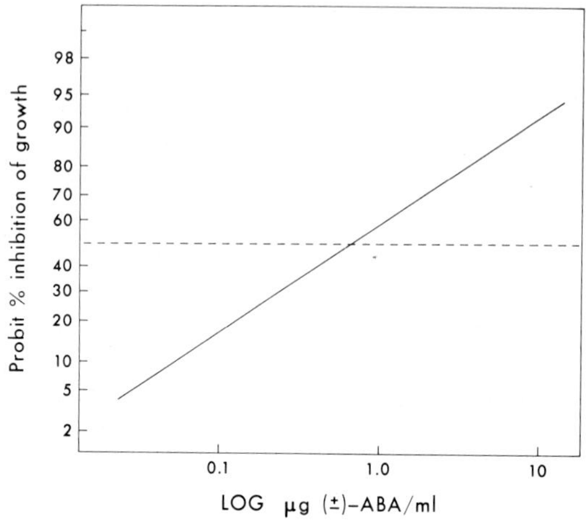

Figure 6.7. Theoretical curve for the inhibition of coleoptile growth. A linear relationship is obtained when probit percentage inhibitions of growth in many bioassays are plotted against log concentration. The values from this curve are plotted against differently transformed coordinates in Figs. 6.3 and 6.5.

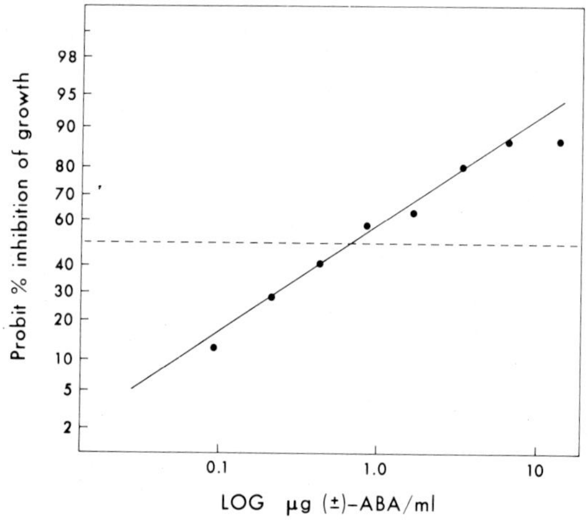

Figure 6.8. Data from the same experiment as shown in Figs. 6.4 and 6.6 in which the inhibition of coleoptile elongation of dissected wheat embryos by a sample of synthetic (±)-ABA was measured. Probit percentage inhibition of growth is plotted against log concentration. This transformation of the data gives a linear relationship and consequently the concentration at which 50% inhibition of growth occurs can be interpolated with greater accuracy than in Figs. 6.4 and 6.6.

The decision as to which bioassay material to choose is subject to a number of constraints. Among the important factors that affect the choice of a bioassay system can be listed: cost, convenience, availability of plant material and apparatus, speed of response, specificity, difficulty of manipulation, sensitivity, variability and amount of substance available for testing or comparison. The choice will be influenced by local conditions and the purpose for which the information is intended.

Most of these constraints apply to the assays of the other growth regulators but the problems of sensitivity and specificity associated with abscisic acid are particularly acute. The most obvious and most frequently used effect is its inhibition of growth but this is an extremely unspecific response and shared by almost all compounds at sufficiently high concentrations; thus many components of an extract will, if sufficiently abundant, mimic or enhance the effect of ABA.

6.2.1.1. Inhibition of growth. The sensitivity of assays for growth-promoting hormones is often enhanced by a pretreatment that lowers the amounts present in the test objects. Here again, abscisic acid poses a special dilemma because such pretreatment lowers the growth rate so that there is no growth to inhibit and stimulation of growth by the addition of growth promoters antagonises the effect to be measured with a consequent loss of precision. The presence of growth-promoting hormones in an extract will, of course, distort the result if they are not removed or destroyed.

A number of different physiological effects of ABA on various plant organs have been exploited as assays and these have particular advantages and disadvantages. The original isolation of abscisin II (ABA) from cotton fruit was guided during the earlier stages by an abscission accelerating assay using stem explants of cotton seedlings, whereas the isolation of "sycamore dormin" (ABA) relied on the growth inhibitory effect that the dormancy inducing hormone was assumed to have. As with abscission, the role of ABA in causing dormancy is not as decisive as was once throught (Saunders et al. 1975) and the induction of dormant buds is too slow and imprecise an effect to be used quantitatively (El Antably et al. 1967).

The inhibition of cereal coleoptile growth has been the most frequently used test. The techniques are familiar from its widespread usage in auxin work but the growth of the untreated control is so slight that IAA has to be added to produce an accurately measurable response. Coleoptile growth can be stimulated by auxins, and sometimes gibberellins, but it is quite susceptible to the inhibitory effects of a wide range of compounds.

6.2.1.2. Inhibition of seed germination. Seed germination and seedling growth is inhibited by ABA and has been extensively used for the comparison of the effectiveness of a number of synthetic analogues of ABA when plenty of material was available (Oritani and Yamashita 1970a, b; Tamura and Nagao 1969). Quite a lot of compound is required and the test takes 2–5 days. Dissected wheat embryos have been employed because less material is required in comparison with whole seeds and the isolated embryos germinate much faster and so a result can be obtained during a much shorter time than with the intact seed. The special advantage of the use of seeds is that an antagonist can be applied to metabolically quiescent tissues, or even soaked into the cells in organic solvents (Milborrow 1963), so that the capacity of a compound to maintain a dormant state is measured with the minimum of complications caused by penetration phenomena and metabolism.

The response of the embryos to ABA is antagonised slightly by gibberellins and cytokinins but IAA has a negligible effect. In this test both cell division and cell enlargement are inhibited whereas in the coleoptile test only the latter process occurs. The cotton explant assay depends on cellular differentiation and the other widely used growth assay, which observes the effect of ABA on the growth of duckweed *(Lemna)*, is principally a measure of cell division (Van Overbeek et al. 1967; Van Staden and Borman 1970).

6.2.1.3. Inhibition of Lemna multiplication. During the last few years researchers have become increasingly aware of the potential interference from microorganisms in growth regulator experiments and the *Lemna* test is the one that can most conveniently be carried out in aseptic conditions. Its disadvantages are: susceptibility to interaction with cytokinins, long duration (6–7 days), large volumes of solution and the necessity for the plantlets to grow in the light which can isomerise or inactivate ABA and its analogues.

6.2.1.4. Inhibition of α-amylase synthesis. The aleurone layer of barley synthesises α-amylase in response to gibberellins (Nicholls and Paleg 1963) and this production is inhibited by ABA (Varner and Johri 1968). The reducing sugar produced can be measured colorimetrically (Sivori et al. 1971) and very little compound is required; the tissues are non-growing initially so metabolic degradation of the compounds is probably slight. The quantitative measurement of the reducing sugar can be time consuming.

6.2.1.5. Stomatal closure. The discovery of the capacity of abscisic acid to close stomata (Mittelheuser and Van Steveninck 1969; Jones and Mansfield 1970) has been suggested as an assay by Tucker and Mansfield (1971). They used epidermal strips from the lower

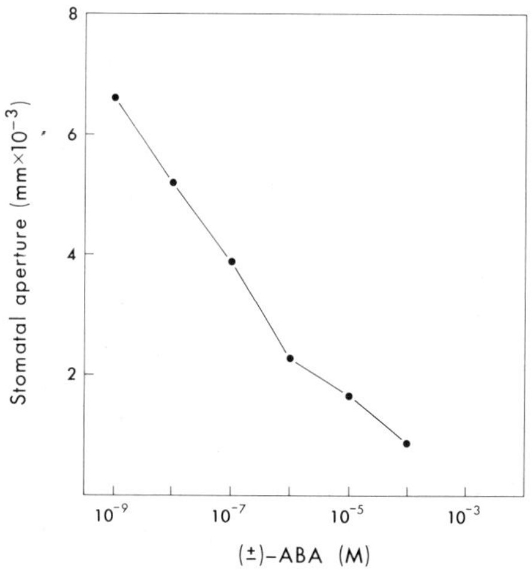

Figure 6.9. Measurements of stomatal apertures of *Commelina communis* leaf epidermis as a bioassay for abscisic acid. Redrawn from the data of Tucker and Mansfield (1971).

surface of *Commelina communis* leaves and the great advantage of the method is that the stomata close within minutes in an appropriate solution of ABA (Figure 6.9). Other growth regulators have little effect but light and CO_2 interact and many compounds can be expected to cause closure by their toxic action at high concentrations. Bacterial attack and metabolism are, of course, unlikely to pose serious problems in such short term tests.

6.2.2. Physical methods of measurement

Similar constraints to those influencing the choice of bioassay apply to the choice of physical methods: cost, availability of apparatus and experimental material. Physical methods are often less sensitive than bioassays but can be considerably more specific. Those commonly used and their characteristics are discussed below.

6.2.2.1. Ultraviolet absorption. The detection level is relatively poor, and the sample has to be extremely pure, e.g. at least two chromatographic separations are necessary. The specificity depends on the position of the absorption maximum of the free acid, but because this wavelength and the shape of the absorption curves are pH dependent, this characteristic can be exploited to give a high degree of specificity. However, a number of chemically similar structures have very similar spectra.

6.2.2.2. Optical rotary dispersion and circular dichroism. The ORD and CD spectra of (+)-ABA are highly specific but the usefulness of the methods to identify and assay the hormone in extracts is limited by the UV absorption of the solution and the presence of other optically active materials. (Cornforth et al. 1966). Plant extracts require a considerable degree of purification, e.g. at least one and usually two chromatographic separations, before they can be subjected to ORD and CD analysis. The loss of material which occurs during these manipulations is a disadvantage and one that becomes progressively more acute at lower concentrations.

6.2.2.3. Racemate dilution method. This method is analogous to the inverse of isotope dilution and can be used to determine the amount of (+)-ABA present in a plant without the complication of unknown losses during extraction (Milborrow 1967). However, it suffers from the combined disadvantages of the UV and the ORD procedures. A carefully measured amount of synthetic, and therefore racemic, abscisic acid is added to an extract (a pilot run is helpful to give an approximate measure of the amount of (+)-ABA extractable from a given weight of plant material) so that the ratio of natural, optically-active compound to racemate is not less than 1–10 (otherwise the UV absorption of the racemic ABA would be so great that the optical activity of the natural material could not be measured accurately). The large amount of added (±)-ABA facilitates extraction and when the extract has been purified to such an extent that ABA is the only UV-absorbing material detectable (i.e. when the UV absorption curve is identical with that of the pure compound), the total amount can be calculated from the known specific absorption ϵ = 21 500, λ_{max} 260 nm in acidic methanol). The amount of optically active (+)-ABA derived from the plant material and present in the solution can then be measured by spectropolarimetry ([α]289 nm = + 24,000°). The difference between total ABA and optically active (+)-ABA present gives the amount of racemic ABA extracted. The percentage loss of racemic ABA during the extraction procedure can then be calculated. The same percentage loss of the (+)-ABA must also have occurred and so the original (+)-ABA content of the plant can be calculated.

The method depends on the ABA being obtained free of UV-absorbing and optically active impurities but crystallisation must not be allowed to occur because the excess free (+)-ABA and the racemate crystallise separately. (±)-ABA (m.p. 190–191 °C) is a racemic compound and behaves differently from the crystalline (+)-ABA (m.p. 160–162 °C). The racemate must not be added to living plant material because metabolism cannot be allowed to occur as this also might lead to preferential loss of (+) or (–) material and so invalidate the result.

If labelled (±)-ABA is available, then the losses during extraction can be monitored by adding a known quantity of the radioactive racemate to an extract. The radioactivity in a sample of the final purified extract can then be determined and the percentage recovery used to correct the value for the (+)-ABA measured by spectropolarimetry. The racemic ABA added acts as a "scavenger" and aids extraction – the percentage recoveries of ABA are higher when larger amounts are present in an extract.

6.2.2.4. Gas – liquid chromatography. Abscisic acid has to be esterified to make it sufficiently volatile to pass through a GLC column. Methylation with diazomethane is the method most frequently used but the mono trimethylsilyl ester has also been employed. The tertiary hydroxyl group cannot be acetylated but there is a report (Davis et al. 1968) that under forcing conditions ABA can be doubly trimethylsilylated. However, the reaction is not completely understood and some rearrangement may occur (Most et al. 1970).

Some degree of preliminary fractionation of the acid fraction is usually necessary before an extract is prepared for GLC and although flame ionisation detectors are frequently used, the signal of abscisic acid esters is almost swamped by the presence of other materials. Me-ABA gives a very strong response in an electron capture detector (Seeley and Powell 1970) and its signal is readily identified as few contaminants in the extracts affect the detector to the same degree. Injections of 100 pg give a detectable response. However most workers use flame ionisation detectors.

Saunders and co-workers have developed an ingenious method for measuring the losses which occur during the preliminary steps of purifications by adding a known amount of 2-*trans*-abscisic acid (II) to a sample of the extraction medium (Lenton et al. 1971). Abscisic acid and its 2-*trans* isomer have quite dissimilar retention times in most GLC systems (Table 6.1) and hence the percentage recovery of 2-*trans*-ABA from the extract can be determined. The assumption is then made, and tests have verified that it holds, within the limits of sensitivity in most experiments, that the percentage recovery of ABA is the same as that for 2-*trans*-ABA and so the true concentration of ABA within the plant can be calculated. The most dangerous sources of error are two: (i) the percentage recovery may be closely related to the amount of material present so although, for example, 30% of a large amount of added 2-*trans*-ABA may be recovered in a pilot experiment, the trace amounts of ABA in an extract may suffer a disproportionately higher percentage loss and (ii) Abscisic acid and 2-*trans*-abscisic acid are interconverted in solution

Table 6.1.

Retention times in GLC systems for methyl-ABA and methyl-2-*trans*-ABA

	Relative Retention Times			
Temp.	Me-ABA	Me-2-*trans*-ABA	Column	Source of data
200 °C	12.0 min	16.5 min	2% SE 30 on Gas-Chrom Q60–80 mesh	Seeley and Powell (1970)
210 °C	13.0 min	17.5 min	2% Epon 1001 on AW-DMCS Chromosorb W	Lenton et al. (1968)
160 °C	10.5 min	14.7 min	2% SE 33 on Gas-Chrom Q	Gaskin and MacMillan (1968)
198 °C	5.0 min	8.3 min	1% XE-60 on Gas-Chrom Q100–120 mesh	Zeevaart and Milborrow (1976)
202 °C	4.9 min	6.4 min	3% SE-30 on Gas-Chrom Q100/120 mesh	Zeevaart and Milborrow (1976)
185 °C	4.4 min	5.6 min	1% OV-17 on Gas-Chrom Q100/120 mesh	Zeevaart and Milborrow (1976)

II. 2-*trans*-abscisic acid

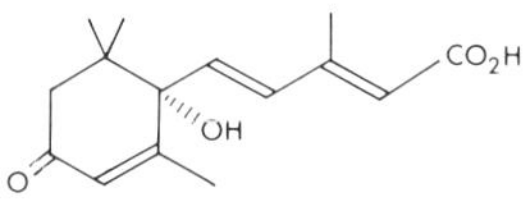

by light to a 1:1 equilibrium mixture, consequently great care has to be exercised to keep the standard 2-*trans*-ABA solution and all the extracts in semi-darkness throughout.

6.2.2.5. Mass spectral measurements. The cracking pattern of methyl abscisate has been documented (Ohkuma et al. 1965; Gaskin and MacMillan 1968; Most et al. 1970) and analysed in detail (Gray et al. 1974). Combined GLC-mass spectrometry gives the most certain identification of ABA in a plant extract but preparative GLC has been strangely neglected and little work has been done with this method; this is even more surprising when one considers the ease with which even the methyl ester of abscisic acid can be hydrolysed (Cornforth et al. 1965a).

6.2.2.6. Derivatisation. Extractions of ABA designed to lead to identification or measurement start with the killing and extraction of a sample of plant material followed by the partitioning of the extract and isolation of the ether-soluble acid fraction. This fraction is purified by one or more chromatographic separations which are carried out in systems which exploit sequentially the different physical characteristics (partition coefficient, adsorption, pK_a, etc.) of the component compounds.

However, an alternative and more reliable method for the final steps of an extraction is to form various derivatives, whereby the ABA derivative comes to have properties which are quite different from those of the impurities and so can be separated clearly. The formation of the methyl ester is frequently used for the purification of abscisic acid and the methyl abscisate, or the free acid, can be reduced with borohydride to give an equal mixture of the isomeric diols (III and IV) which are more polar than the parent compound and can be separated from it and from each other by chromatography. The disadvantage of these procedures for bioassay tests is that the ester is inactive (but it can readily be hydrolysed by strong alkali) and the 1′,4′-diols autooxidise back to ABA. An alternative method is to treat the extract of the chromatograms that contain ABA with acetic anhydride in pyridine so that any acetylatable impurities become less polar and separate readily on further chromatography; ABA cannot be acetylated and so it re-chromatographs at the same R_f.

The formation of derivatives is a useful technique to aid separation but some features of the molecules can be exploited to provide simultaneously a means of identification. The light-catalysed isomerisation of ABA to an equilibrium mixture with its 2-*trans* isomer is an example of this. A sample of ABA can be photoisomerised and either the isomeric acids can be separated by TLC or the methyl esters by GLC.

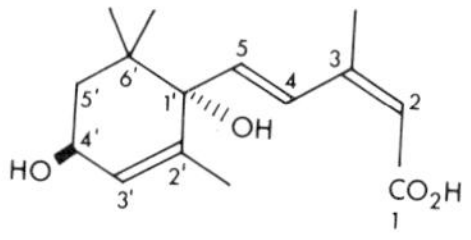

III

1′,4′-*trans*-diol of (+)-abscisic acid (abscisic acid numbering)

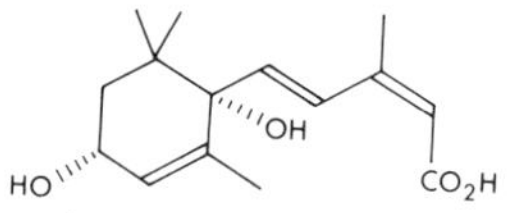

IV

1′,4′-*cis*-dial of (+)-abscisic acid

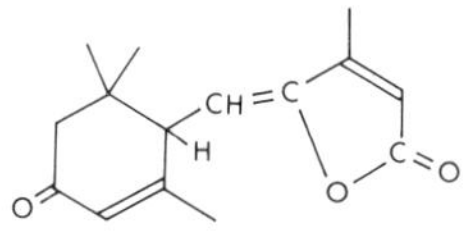

V

Lactone of ABA

6.2.2.7. Colour test. When abscisic acid is treated first with a strong acid, it rearranges to a lactone (V), and then on further treatment with a strong base, a magenta coloured compound is formed transiently (Mallaby and Ryback 1972). Only ABA and to a slight extent its 2-*trans*-isomer have so far been found to give this colour reaction. Unfortunately other acids present in plant extracts interfere strongly and although less than 0.1 μg of pure ABA gives a clearly recognisable colour, the test cannot be used with anything like this order of sensitivity for impure samples. The colour fades rapidly and so far no procedure has been found which stabilises it sufficiently to allow it to be used for quantitative measurement.

The various tests for abscisic acid make use of specific features of the molecule and the more of these characteristics shown by a presumed sample the more certain one can be that it does contain ABA.

6.3. Occurrence

ABA has now been satisfactorily identified in such a large number of flowering plants, both mono- and dicotyledons, that it can be safely assumed to be present in all species (Table 6.2). It has also been identified in several gymnosperms (Table 6.2), a fern, a horsetail and a moss. There then appears to be a taxonomic division in the Bryophyta because, so far, attempts to identify ABA in liverworts have failed. Its place as a growth inhibitor seems to be taken by lunularic acid (Valio et al. 1969; Pryce 1972). Attempts to isolate ABA from algae have also failed and lunularic acid (VI) has been shown to be present (Pryce 1971, 1972), so it appears that this dihydrostilbene acid may have some of the functions in algae that ABA has in plants. Lunularic acid is discussed further in Chapter 7.

Table 6.2.

A selection of determinations of (+)-abscisic acid from plants

This list is by no means exhaustive and incorporates a few values from some of the papers cited. Other data are incorporated in tables elsewhere in the text

Species		μg ABA/ kg fresh wt.	Source	Reference
Dicotyledons				
Cocklebur	*Xanthium strumarium*	306	Apical buds	Tucker and Mansfield (1972)
		19 174	Axillary buds	
		374	Mature leaves	
Peach	*Prunus persica*	60	July xylem sap	Davidson and Young (1974)
		10	November	
Spinach	*Spinacea oleracea*	80 to 110	Normal leaves	Zeevaart (1974)
		1 011	Wilted leaves	
Tomato	*Lycopersicon esculentum*	106.5	Upper shoot	Tal and Gavish (1973)
		116.7	Grown under saline conditions	
Beech	*Fagus sylvaticus*	60	July buds	Wright (1975)
		30	August buds	
		100	October buds	
		310	November buds	
		130	December buds	
		60	February buds	
		20	April buds	
Coffee	*Coffea arabica*	25 to 75	Flower buds	Browning (1973)
Silky Oak	*Grevillea robusta*	53	Nectar	Milborrow (1969a)
Avocado	*Persea gratissima*	500	Seed	Milborrow and Robinson (1973)
		5 950	Fruit	
		500	Stem	
		20 to 540	Leaf (turgid)	
		550 to 2670	Leaf (wilted)	
		27	Root	
Monocotyledons				
Rye grass	*Lolium perenne*	12 to 4766	Leaves (long days)	Hanks (1974)
		8 to 3976	Leaves (short days)	
Gladiolus	*Gladiolus grandiflorus*	110 to 120	Corms (dormant)	Ginzberg (1973)
		23 to 28	Corms (non-dormant)	
Gymnosperms				
Yew	*Taxus baccata*		Fruit	Le Page-Degivry et al. (1969)
			Seed	
Silver fir	*Abies alba*		Aphid exudate	Milborrow (1969a)
Pine	*Pinus radiata*		Stems	Jenkins and Shepherd (1972)
Sitka spruce	*Picea sitchensis*	20 to 30 ($\mu g/l/m^2$)	Leaf washings	Good (1974)
Pteridophytes				
Fern	*Anemia phyllitidis*		2-*trans*-ABA in spores and chloronemata	Cheng and Schraudolf (1974)
Bracken	*Pteridium aquilinum*		Rhizomes	Milborrow (1969a)
Horsetail	*Equisetum arvense*		Shoots	cited by Milborrow (1974a)
Bryophytes				
Moss				cited by Pryce (1972)

VI. Lunularic acid

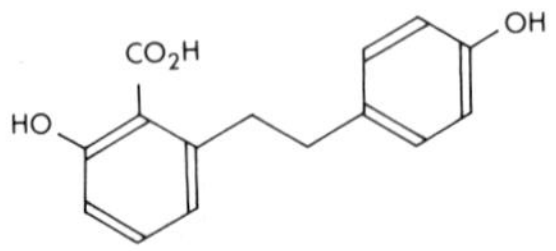

A number of different parts of higher plants have been extracted and found to contain (+)-ABA; these include leaves, fruit, seed, buds, roots and stems. Some parts have been dissected in greater detail; for example sycamore stems have been separated into xylem and bark and both found to contain ABA. Phloem and xylem exudate and nectar have all been found to contain ABA. Avocado seeds have been divided into cotyledons and embryos, wheat seeds into embryo and endosperm, and ABA was present in all of these. Lime, yew, avocado, grape and ash have been found to contain ABA in the fruit coat or pulp and in the seeds.

Concentrations vary from 3 μg/kg fresh weight in a water plant to 10,000 μg/kg in avocado fruit pulp and 20,000 μg/kg in dormant cocklebur *(Xanthium)* buds (Table 6.2). Most tissues contain between 20 and 100 μg/kg but changes in concentration in an organ occur during the course of development. A few such patterns have been reported (Fig. 6.10) (Davis and Addicott 1972). Changes in ABA levels in relation to stem and leaf growth (Chapter 2), fruit development (Chapter 3), flower formation (Chapter 6), senescence and abscission (Chapter 7), seed dormancy and germination (Chapter 8), bud dormancy (Chapter 9) and fruit ripening (Chapter 10) are discussed in Volume II.

One of the criteria in the classical definition of a hormone is that a compound should be synthesised in one part of an organism and transported to another part where it exerts its effect. Abscisic acid poses something of a dilemma in this regard. In several experiments the ^{14}C-labelled compound applied to a plant, has been shown to be redistributed

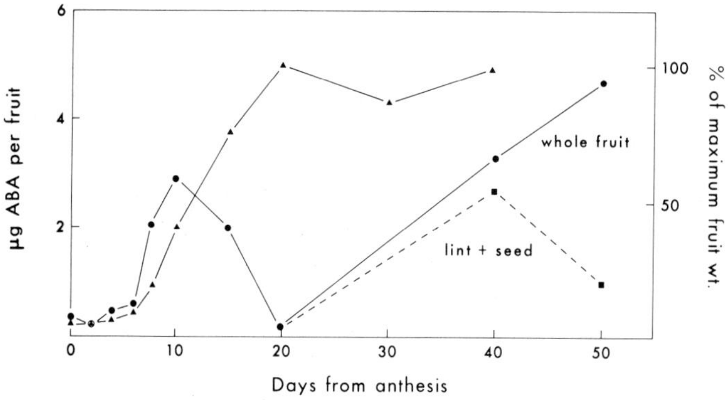

Figure 6.10. The abscisic acid content of developing cotton fruit. The ABA was measured by flame ionisation detection after GLC of the trimethylsilyl derivative of the acid and expressed as μg/fruit, ●. The percentage of maximum fruit weight during development is also shown, ▲. The square symbols show the ABA content of the seeds and lint when these can be separated. Data redrawn from Davis and Addicott (1972).

quite rapidly (Hocking et al. 1972). Furthermore, it has been isolated from phloem and xylem sap (Bowen and Hoad 1968) and shown to be moved from cell to cell (Milborrow 1968; Pilet 1975a; Dörffling and Böttger 1968); it can be expected, therefore, to be moved by more than one transport mechanism. Although such movement could account for the presence of ABA in all the different parts of plants in which it has been sought, it appears that it can also be biosynthesised throughout the plant (Milborrow and Robinson 1973). The difficulty experienced in obtaining significant incorporation of label from mevalonic acid into ABA requires the use of quite large amounts of plant material and consequently this constraint has precluded the precise localisation of biosynthesis to other than broadly defined organs such as leaves, stems, roots, etc.

Biosynthesis of ABA from labelled mevalonic acid has been demonstrated to occur in plastids (Milborrow 1974b) isolated from ripening avocado fruit and this tissue gives the highest percentage incorporation of mevalonate into ABA that has been found. This is possibly because the permeability of the cellular membranes increases as the climateric ripening processes begin and possibly because less sterol biosynthesis occurs at this stage than during fruit growth and so a smaller proportion of the labelled precursor is incorporated into sterols. An attempt has recently been made to measure the concentration of the newly biosynthesised ABA in the various organelles isolated from an avocado fruit that had been supplied with [^{14}C]mevalonic acid. The [^{14}C]ABA was found in similar concentrations in the mitochondrial, microsomal, chloroplastic and supernatant fractions indicating either that it redistributes rapidly or that in this senescent tissue the organelles are more permeable than normal.

The biosynthesis of ABA in non-green parts of a plant can be accounted for by the presence of leucoplasts in such tissues. The colourless, central flesh of an avocado incorporates label from mevalonate into ABA as rapidly as the green, outer flesh. Plastids from the two regions also synthesise ABA equally effectively.

That ABA is synthesised only in plastids is difficult to establish. The failure of supernatant fractions to incorporate [^{14}C]mevalonic acid into ABA could be due to the release of phosphatases and other degradative enzymes. However, rapid incorporation of labelled mevalonate into steroids (which occurs principally in the cytosole) has been found in tomato shoots which have yielded insignificant amounts in ABA. Furthermore, AMO 1618, a potent inhibitor of steroid biosynthesis, inhibited the incorporation of mevalonate into steroids by avocado fruit almost 20-fold whereas it stimulated the incorporation into ABA and carotene 3-fold (Milborrow 1976). These observations suggest that by far the most important, if not the sole, site of ABA biosynthesis is in the plastids.

Movement of ABA from plastids would be established by detecting it in materials which lack plastids. It has been determined (Table 6.2) in xylem and phloem sap and in the exudate from aphids whose mouthparts were embedded in the phloem. There is always a suspicion that cutting or probing the tissues could damage the organelles, but the presence of ABA in the nectar collected from silky oak flowers (Milborrow 1969a) establishes that the ABA must have passed out of the plastids.

Small quantities of 2-*trans*-ABA (Davis et al. 1968; Gaskin and MacMillan 1968; Seeley and Powell 1970) have been isolated from plants and in the case of wild rose leaves (±)-[2-^{14}C]ABA was added to the extraction medium and the amount of natural (+)-2-

trans-ABA was determined by optical rotary dispersion measurements (Milborrow 1970). The (+)-2-*trans* isomer was present at a concentration equivalent to 4.4% of the (+)-ABA but only 0.9% of the ^{14}C was in the 2-*trans* isomer. This amount of ^{14}C defines the maximum amount of the ABA that could have isomerised to 2-*trans*-ABA during the purification, or been in the form of the *trans* isomer. Consequently the majority of the (+)-2-*trans*-ABA isolated must have been present in the leaves. It probably arises by a photolytically catalysed isomerisation (Mousseron-Canet et al. 1966).

Apart from its degradation products, abscisic acid occurs in plants in a bound form, namely (+)-abscisyl-β-D-glucopyranoside. The content of this compound rises when plants recover from wilting and there is a complementary fall in the content of free abscisic acid (Figure 6.11) (Hiron and Wright 1973). In most plant tissues the amount of ABA that is released by base-catalysed hydrolysis (usually tentatively identified as glucose ester) is between 10% and 30% of the content of free acid, but in the peel of oranges matured on the tree, the ABA released by hydrolysis exceeds the content of free acid ten-fold (Goldschmidt et al. 1973). It is not known at present whether the ABA in the form of the glucose ester can be hydrolysed back to the free acid within the plant and so play a further part in regulating metabolism or whether ester formation sequesters it permanently.

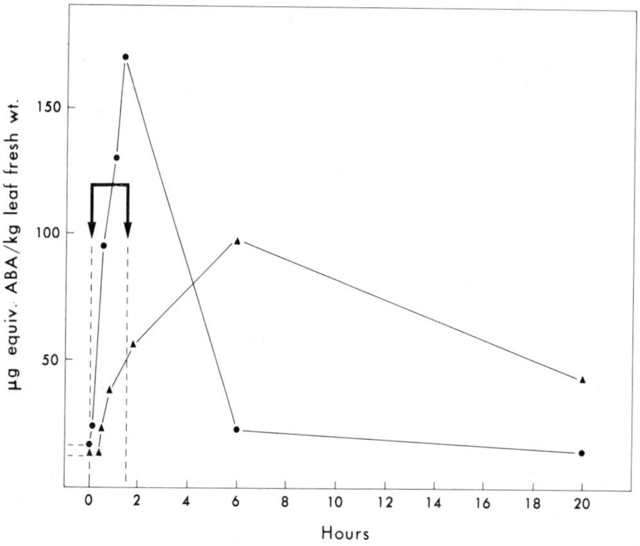

Figure 6.11. The abscisic acid content of dwarf bean *(Phaseolus vulgaris)* during a wilt-recovery cycle. The seedlings were stressed for 90 min. by exposure to a draught of warm air (between arrows). They were then allowed to recover. Free ABA (●) and the ABA released by alkaline hydrolysis (▲) were estimated by bioassay. Data redrawn from Hiron and Wright (1973).

6.4. Structure–activity relationships

6.4.1. Analysis of assumptions

The analysis of the structure–activity relationships of any group of compounds is based on a number of assumptions. The most important of these are: that the activity of all putative analogues is attributable to the same mode of action; that differences in rates of penetration to the receptor sites do not affect the manifestation of activity, and that metabolism neither activates inactive compounds that are applied nor inactivates active ones. Clearly, the validity of the structure–activity relationships deduced depends on how closely the behaviour of the compounds matches these assumptions. Penetration into the cells can be certain to have occurred only with active compounds and the other assumptions can be verified only by a considerable experimental analysis.

The studies of [2-^{14}C]ABA show that it can reach the site of action on, or in, the cells quite rapidly (Varner and Johri 1968; Cummins and Sondheimer 1973) and the close structural similarity of most of the synthetic analogues to the parent molecule suggest that their rates of movement to the active site are similar. Indeed, most of them are less polar than ABA and some are non-ionic and so they can be expected to enter cells even more readily than ABA.

The metabolism of ABA is now better known than that of any other plant hormone but the metabolism of very few closely related analogues has been studied so the activity, or otherwise, of most compounds described here is, of necessity, tentative.

The most commonly assayed biological property of ABA is its inhibitory effect on plant growth. This test has the disadvantages that it requires one or two days to give a clear result and it is extremely unspecific. Often it is impossible to decide whether a relatively inactive analogue is active by virtue of its having a weak action of the same kind as ABA or whether its inhibitory effects are caused by the acidity or toxicity of the high concentrations necessary to cause a measurable effect. Finally, it is necessary to have a suitable range of analogues so that the effects of any molecular change are not compounded with changes produced by other, simultaneous, alterations.

6.4.2. Optical isomerism

The unnatural (–)-enantiomer of ABA was first resolved from the synthetic racemate by crystallisation and distillation of the mother liquors after removal of some of the (+)-enantiomer as a crystalline brucine salt (Cornforth et al. 1967). When the separate enantiomers were assayed on dissected wheat embryos (Figure 6.12) they were found to be equally potent (Milborrow 1968). This was a most unexpected result because unnatural mirror-image forms are almost invariably unacceptable to active sites. For instance, inversion of just one of the eight optical centres of GA_1 (that at C-3, which carries a hydroxyl group) renders the molecule inactive. The result of the wheat embryo assay has been confirmed and amplified by Sondheimer et al. (1971).who have also resolved (±)-[2-^{14}C] ABA. They assayed the growth-inhibitory activity of the two enantiomers in a bean embryo and a barley embryo assay and found that, although the (+) was up to three times more active than the (–), the latter was still extremely potent. In a barley half-seed α-amylase, and a cotton explant abscission assay they reported that the (+) and (–) were

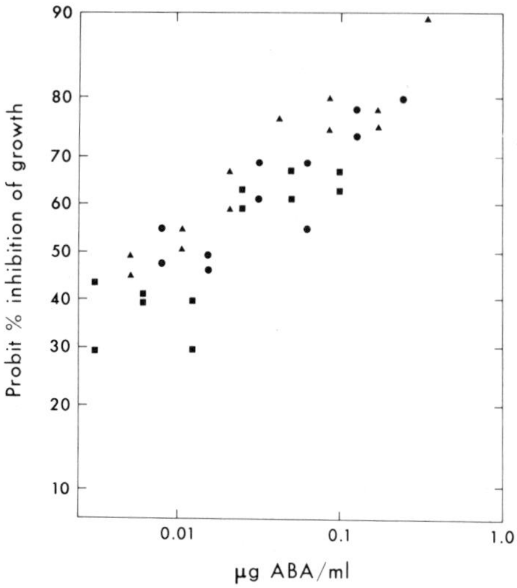

Figure 6.12. The equal growth-inhibitory activity of (+)-ABA (■),(-)-ABA(●), and (±)-ABA (▲) on the coleoptile growth of excised wheat embryos. The 10 excised embryos per petri dish were incubated at 20 °C for 48 hr and mean coleoptile length is plotted as probit percentage control length (2.5 mm).

equally effective. However later work by Cummins and Sondheimer (1973) and Kriedemann et al. (1972) has shown that the (—) enantiomer has virtually no ability to cause the closure of barley leaf stomata. The comparison of the intrinsic activity of the two antipodes is less exact in assays which use actively growing tissues because (+)- and (—)-ABA are metabolised differently and Sondheimer et al. (1971) found some indication that even the rates of penetration of the two were different. The (+)-[2-^{14}C] ABA appeared to enter into the bean axes more readily than (—)-[2-^{14}C]ABA but whether this was caused by selective uptake of the former or the maintenance of a steeper concentration gradient for entry of the (+) by conversion into M-1 (subsequently identified as phaseic acid) could not be established.

The equal, or at least similar, biological activity of the two mirror-image forms of ABA focusses attention on the stereochemical structure of the compound. Although the specific rotation of abscisic acid is one of the highest that has been recorded, the molecule is surprisingly symmetrical, as can be seen from a three-dimensional model or from an unconventional projection of the structural formula (Figure 6.13).

The only differences across the plane of symmetry A-B and the features that give the

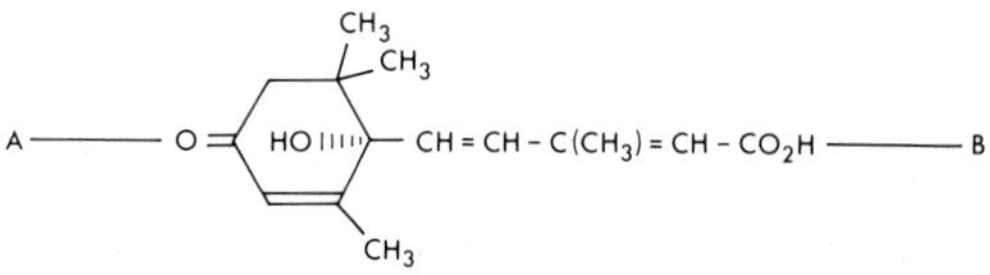

Figure 6.13. Projection of (+)-abscisic acid (revised absolute configuration).

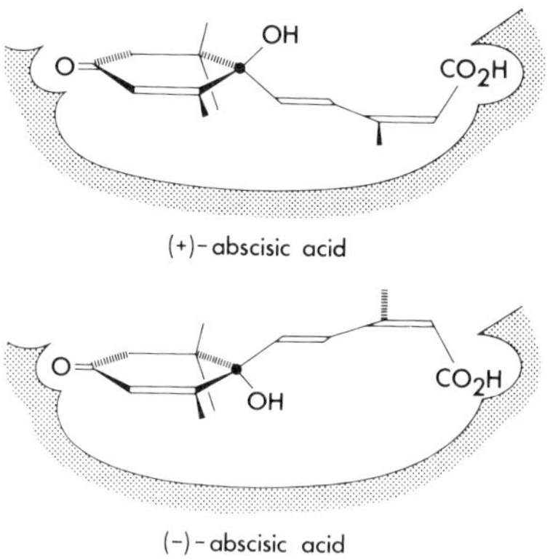

Figure 6.14. Arrangement of the natural (+)- and the unnatural (–)-enantiomer of abscisic acid in a hypothetical active site. As depicted here the 1′-hydroxyl group is not required for the molecule to exert its effect.

molecule its optical activity, are that C-6′ carries two methyl groups whereas C-2′ carries one methyl group and there is a double-bond at C-2′.

Two alternative hypotheses have been proposed to account for the activity of the (–)-enantiomer: (i) the rings and side chains of the (+) and (–) lie on the active site in the same orientation but the 1′-hydroxyl group has no role, as shown in Figure 6.14; (ii) the opposite face of the cyclohexenyl ring of the (–) can lie on the active site with its 1′-hydroxyl group in the same position as that occupied by the 1′-hydroxyl group of the (+) but with the geminal 6′-methyl groups and 2′-methyl groups transposed (Figure 6.15).

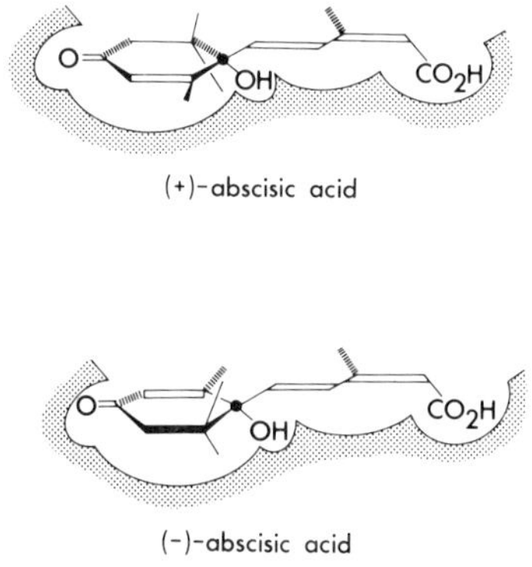

Figure 6.15. Arrangement of the natural (+)- and the unnatural (–)-enantiomer of abscisic acid in a hypothetical active site which requires the presence of a 1′-hydroxyl site. For the 1′-hydroxyl groups of opposite stereochemistry to interact with the same point it is necessary for opposite faces of the cyclohexenyl ring to lie against the active site. This can occur if the 2′-methyl and the 6′-geminal-dimethyl groups are transposed.

Hypothesis (i) appeared tenable because 1′-desoxy abscisic acid methyl ester in a wheat embryo test had about half the inhibitory activity of methyl abscisate (Milborrow 1968) although in a rice seedling test the methyl ester was almost inactive. The former result suggested that the 1′-hydroxyl group was unnecessary for activity but recently R. Mallaby has synthesised ^{14}C-labelled 1′-desoxy ABA (VII) and found that it is rapidly oxidised in air to ABA (Mallaby 1974). Furthermore, its apparent growth inhibitory activity in a wheat embryo assay was related to the amount of [^{14}C]ABA formed that was present in the growing tissues. Thus it appears that desoxy-ABA is inactive and so hypothesis (ii) is the more attractive.

There has been considerable debate during recent years whether the rapid action of plant hormones, which appears to be primarily an effect on cell membranes, can account for the effects on protein synthesis and other responses which occur more slowly. The regulatory molecules may affect these qualitatively different cellular functions in two quite different ways and so might quite reasonably be expected to satisfy different molecular requirements for activity at the two sites.

In the case of abscisic acid, it now appears quite likely that different modes of action do have differing structural requirements. The growth processes of germination affected by ABA are inhibited equally, or almost so, by the (+) and by the (–) enantiomers; on the other hand Cummins and Sondheimer (1973) have found that the (–) enantiomer has little effect on the adjustment of the stomatal apertures of barley leaves – a rapid reaction that is operated by an ion pump in the guard cells and is regulated by (+)-ABA (Raschke 1975).

6.4.3. Activity of analogues

6.4.3.1. Requirement of the carboxyl group for activity. The carboxyl group is necessary for activity because the methyl ester is essentially inactive in oat mesocotyl growth (Milborrow 1967), and stomatal closure (Kriedemann et al. 1972) assays. When esters of otherwise active acids are tested for inhibitory activity in long-time tests they are often more potent than the parent acids. It is possible that this occurs because hydrolysis produces a continuous supply of the active moiety, generated over some days, and this slow generation of the active ingredient produces a greater effect than the same quantity of free acid delivered instantaneously. The 1-alcohol (VIII) has been variously reported to have 1% of the activity of the acid (Oritani and Yamashita 1970b) in a rice seedling test but greater activity (Taylor and Burden 1972) in a cress seed test; these may represent results of tests in which differing rates of metabolic oxidation occurred.

The isomeric 1′,4′-diols (III, IV) and abscisic aldehyde (IX) are readily auto-oxidised in air to give ABA (Milborrow and Garmston 1973) so it is impossible to discriminate between the activity of the compounds applied and the ABA formed from them. The epidioxide(X) has been found to be converted into ABA by a non-biological process (Sondheimer et al. 1969).

6.4.3.2. The dienoic side chain. Light catalyses the isomerisation of the 2-*cis* double bond of ABA to yield a 1:1 equilibrium mixture with the 2-*trans* isomer, and the 2-*trans*-

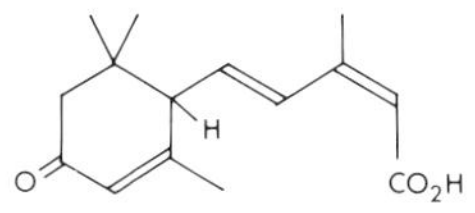

VII

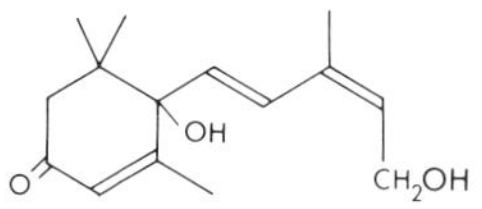

VIII

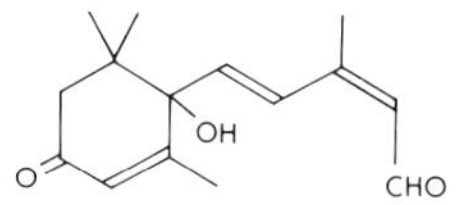

IX

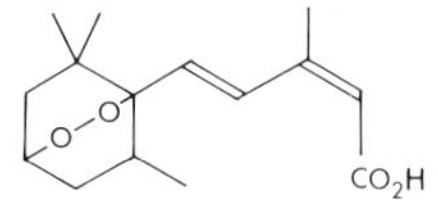

X

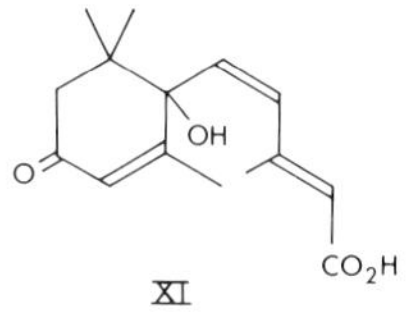

XI

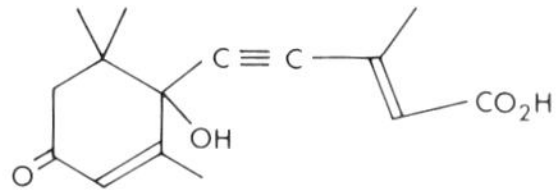

XII

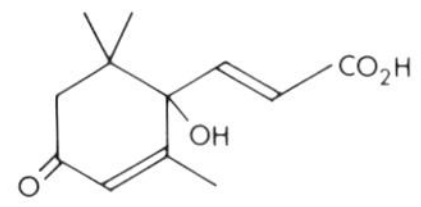

XIII

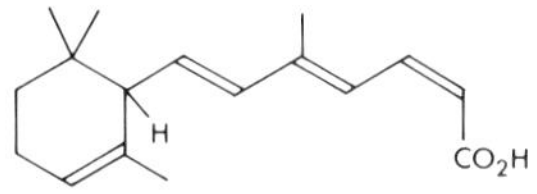

XIV

ABA (*t*-ABA) is virtually inactive inexperiments carried out in darkness. This suggests that the 2-*trans*-ABA cannot be isomerised enzymatically to ABA and (±)-2-*trans*-[2-^{14}C] ABA fed to tomato seedlings in darkness was not isomerised to ABA (Milborrow 1970).-

The 4-double bond has to be *trans* otherwise the molecule is almost inactive (XI, activity 2% of ABA) (Ohkuma 1965) and a 4–5 acetylenic bond (XII) lowers activity to a similar degree (3%) (Ohkuma 1966). Molecules lacking one side chain *cis* vinyl group (XIII) (Mousseron-Canet et al. 1970), or with a longer side chain are inactive, e.g. XIV (Oritani and Yamashita 1970b) and XV (Oritani and Yamashita 1972).

The requirement of the side chain methyl group at C-3 has not been satisfactorily established. Abscisic aldehyde lacking the methyl group at C-3 (XVI) had 10% of the activity of ABA in the rice seedling assay in which abscisic alcohol had only 1% activity (Oritani and Yamashita 1970a, b), whereas in the cress seed germination test abscisic aldehyde and abscisic alcohol were more active than ABA; unfortunately XVI was not included in this latter test. Substitution of the C-3 methyl by an ethyl group has not been carried out in ABA but it has in a closely related compound. The methyl ester of the acid XXI is more potent than ABA in some tests (Tamura and Nagao 1969) and has been shown to be converted into ABA; however when the side chain carries an ethyl group at C-3 (XVII), the activity is reduced to 10% that of ABA. Again, interference with a structural feature present in ABA reduces growth inhibitory activity.

Several compounds (XVIII, Ohkuma 1966; V, Mallaby and Ryback 1972) with lactone and other rings on the side chain are inactive. It appears that removal or elaboration of the methyl group at C-3 reduces biological activity and, apart from compounds that can either be readily oxidised or hydrolysed to a carboxylic acid, any other alteration to the side chain renders the compound inactive.

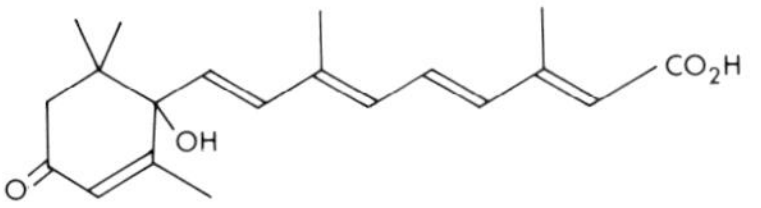

XV

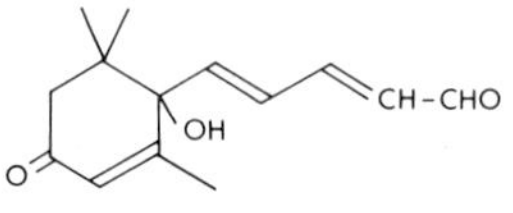

XVI

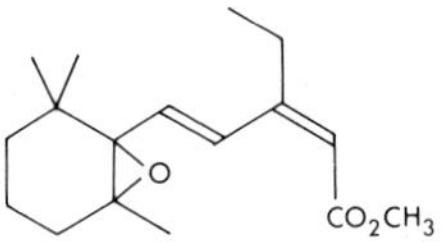

XVII

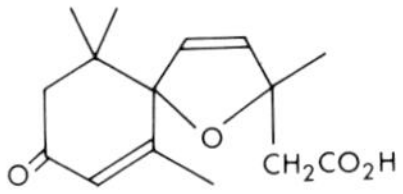

XVIII

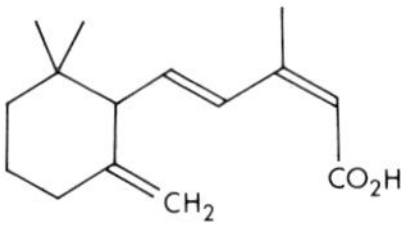

XIX

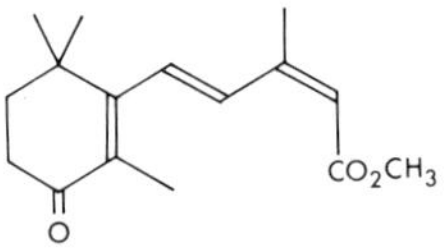

XX

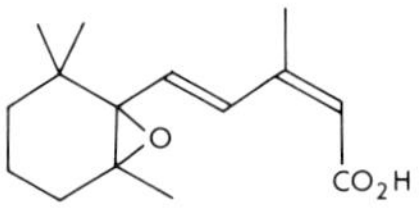

XXI

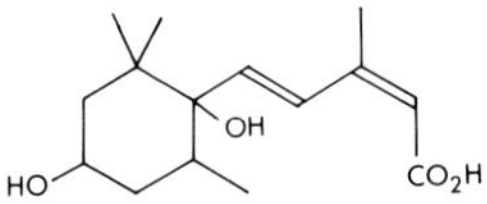

XXII

6.4.3.3. *Structural requirements of the ring system.* All alternations to the ring system have retained the three methyl groups except for XIX, a des-keto derivative in which the methyl at C-2′ is replaced by an exocyclic methylene group (Mousseron-Canet et al. 1970). The ethyl ester showed 42% of the activity of ABA in a rice seedling assay but the loss of activity occasioned by the presence of the exocyclic double bond may have been aggravated by the absence of the 4′-ketone group (Tamura and Nagao 1969). In contrast to this, the free acid showed 0.08% of the activity of ABA in a wheat coleoptile test so metabolic activation (isomerisation to a α-double bond and a methyl at C-2′) may be responsible for the discrepancy between the results of the two experiments.

Compounds containing α- or β-unsaturated rings without a ketone group show activity; possibly β rings are oxidised by the pathway suggested for xanthoxin (Scheme 6.1). On

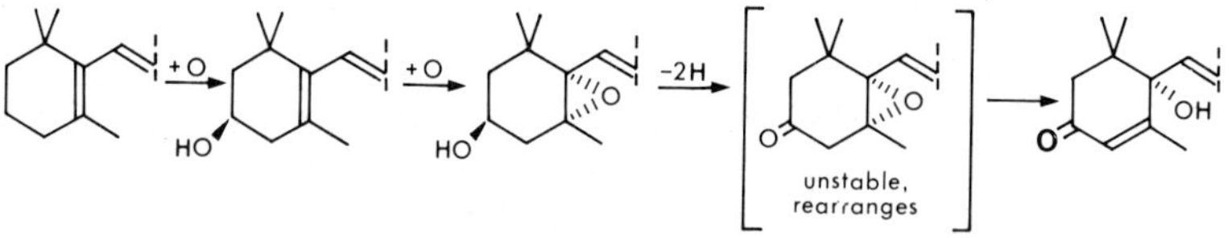

Scheme 6.1

the other hand, oxidation of the equivalent α-rings, such as occurs to give the carotenoid lutein, could give rise to ABA by another route, for each step of which there is a close biochemical analogy (Scheme 6.2).

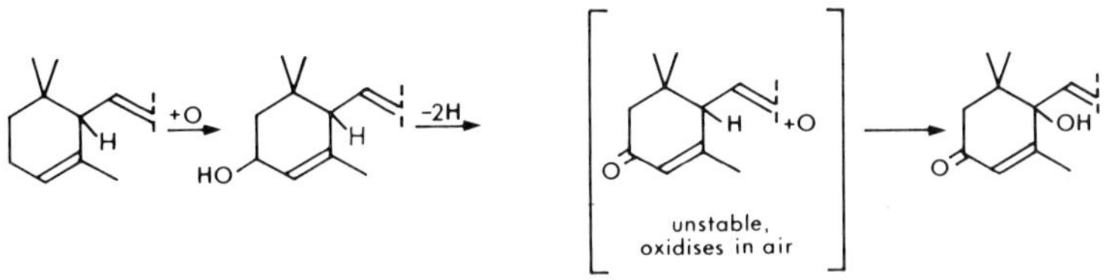

Scheme 6.2

The 3′-ketone XX which would be unlikely to allow metabolism to ABA is inactive (Oritani and Yamashita 1970a). 4′-Methoxy compounds are also inactive so it appears that a 4′-ketone group is required for activity.

^{14}C-labelled compound XXI (Milborrow and Noddle 1970) and [^{14}C]-xanthoxin (Taylor and Burden 1973) have been shown to be converted into ABA by plants, consequently the hydroxylation of the 4′ position and its oxidation to ketone can be accomplished quite readily. Other groups at the 4′ position which would not allow rapid metabolism to a ketone exhibit considerably less activity. Attempts to isolate labelled XXI from avocado tissue synthesising ABA from labelled mevalonate failed and even when a "cold trap" of XXI was added to a replicate sample, the XXI recovered was not radioactive. Thus, although it appears not to be a natural intermediate, it is possible that XXI is metabolised to ABA by the same enzymes that form ABA endogenously, possibly operating on the substrates in a different order.

Reduction of ABA with borohydride gives, in addition to the isomeric 1′,4′-diols, a dihydrodiol lacking the ring double bond (XXII). It is inactive, as are other saturated ring compounds that cannot be easily rearranged to give a 2′-double bond.

6.4.4. The relationship between biodegradation and structure

It is instructive at this point to consider how plants inactivate abscisic acid. If, as appears probable, the glucose ester (Milborrow 1970) is formed as a way of "locking up" excess of the active hormone, then in this compound it is the carboxyl group that is shielded. The other degradation route to phaseic and dihydro phaseic acids (see sections 6.6.3 and 6.6.4) first saturates the ring double bond and then reduces the 4′-ketone group. In all three instances a feature of the molecule that is apparently essential for activity is attacked.

The foregoing indicates that the structural requirements for activity of the same type as abscisic acid are very narrowly and precisely limited; hardly any change from the parent molecule can be tolerated. This contrasts with the gibberellins (38 known in plants) and cytokinins (12 known in plants) and provides an interesting commentary on the occurrence of only one active molecule in the abscisic acid class.

6.5. Biosynthesis

6.5.1. Precursors of Abscisic Acid

The arrangement of the 15 carbon atoms that comprise the skeleton of ABA suggests that the molecule is made up of three five-carbon isoprene units (Scheme 6.3) and that it obeys the isoprene rule. Each of these units is derived from the six-carbon mevalonic acid (XXIII; the natural (+) enantiomer only is incorporated) by a series of enzyme reactions that have been intensively studied and whose detailed stereochemistry is perhaps as well understood as that of any other metabolic pathway (Popják and Cornforth 1966). The reactions proceed with the junction of the isoprenoid precursors leading to the formation of C_{10}, C_{15}, C_{20}, C_{25}, C_{30} and C_{40} compounds made by the fusion of two, three, four, six and eight isoprene residues respectively. Although the presence of 15 carbon atoms in the molecule of ABA suggests that it consists of three isoprene residues, there is no certainty that it was derived by the condensation of only three residues. The carbon skeleton and the positions of the oxygen containing substituents of ABA are very similar to the carbon skeletons of terminal parts of some C_{40} carotenoids and which have analogous oxygen substituents. Whether or not some, most, or all of the ABA present in leaves is derived by cleavage of a carotenoid or by a direct synthesis involving intermediates with a maximum of fifteen carbon atoms is by no means settled yet, but mevalonic acid would be an early precursor for both routes.

6.5.2. Rate of formation of abscisic acid

The incorporation of labelled mevalonate into ABA was demonstrated by Noddle and Robinson (1969) who found that ripening fruits gave the highest percentage incorporation of any tissue studied. Of the fruits that have been used (banana, tomato, orange and avocado), the last gives the highest percentage incorporation of added mevalonate into ABA. It is possible that most of the ABA present in the fully ripe fruit is synthesised during the early phases of the climateric rise in respiration rate and this occurs at a time when other isoprenoid compounds are being made in decreasing amounts. It should not be overlooked that the highest percentage incorporation of mevalonate into ABA, and some of the highest concentrations of ABA that have been found, occur in the fruit with the most rapid and sharply defined climateric, namely the avocado. There is now some evidence (Coombe and Hale 1973) that ABA is involved in the regulation of the ripening processes of grapes and so the high concentrations and rapid synthesis of ABA by fruit can be related to a regulatory function (for further discussion see Chapter 10, Volume II).

The rapid increase in ABA content which occurs when plants wilt, and the slow decrease after they regain turgor (see Chapter 11, Volume II) can be interpreted in terms of metabolic instructions (Milborrow 1974a). These instructions must ultimately be expressed in terms of rates of synthesis, degradation and transport but the net effects can be defined as: (a) an instruction to start the synthesis of the "extra" ABA, (b) to stop when the required concentration is present, and (c) an instruction to commence the degradation when turgor is restored. So far an identification of only one of these instructions has been suggested – the stop message. The basis for this is an experiment in which a quantity of racemic [^{14}C]abscisic acid was fed to cut tomato shoots so that the overall internal concentration was about 600 μg/kg., i.e. greater than the maximum concentration induced by water loss. The shoots were then wilted. After 3 hr the free ABA extracted was subjected to a combination of UV absorption and optical rotatory dispersion spectrometry and the ^{14}C content was measured. Calculations showed that less endogenous, unlabelled (+)-ABA was present than in the extracts of unwilted control plants. Wilting normally causes the synthesis of the "extra" (+)-ABA until the concentration reaches approximately 400 μg/kg and the favoured interpretation of the result of this experiment is that the rapid biosynthesis is "switched off" by the presence of large amounts of added ABA. In other words, ABA itself operates a feedback loop to inhibit its own formation when the required concentration has been reached (Milborrow, unpublished data).

Harrison and Walton (1975) have repeated the experiments with beans and suggested that the results can be interpreted in terms of an equally rapid rate of metabolism (i.e. a start message for rapid degradation of ABA to balance synthesis once a high ABA content has been reached). They base their conclusions on the metabolism of [^{14}C]ABA to phaseic and dihydrophaseic acids of the same specific activity in wilted and turgid leaves. Their results show clearly that turnover of ABA continues during the period of wilt; however, they did not measure conjugated ABA and, as Figure 6.11 (Hiron and Wright 1973) shows, this form of ABA is a major sink (if not a source) in beans.

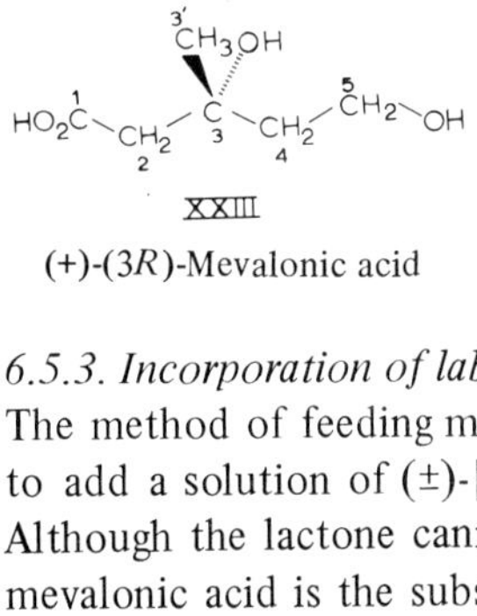

XXIII

(+)-(3*R*)-Mevalonic acid

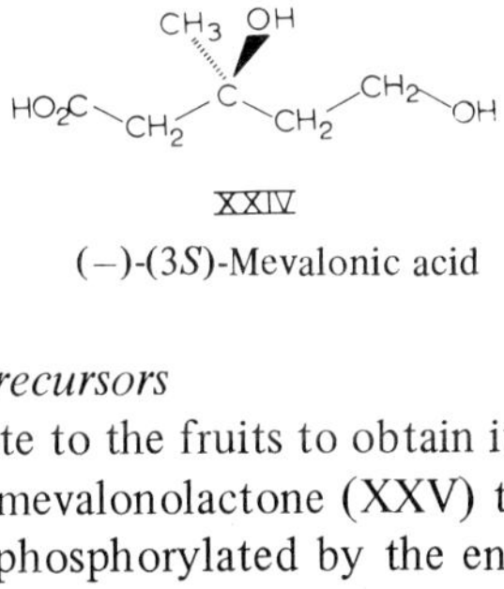

XXIV

(–)-(3*S*)-Mevalonic acid

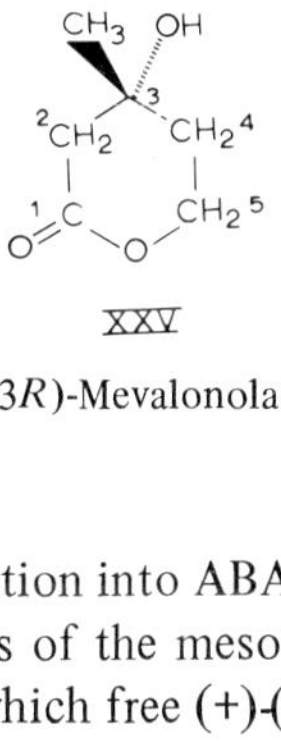

XXV

(+)-(3*R*)-Mevalonolactone

6.5.3. Incorporation of labelled precursors

The method of feeding mevalonate to the fruits to obtain its incorporation into ABA was to add a solution of (±)-[2-^{14}C] mevalonolactone (XXV) to thin slices of the mesocarp. Although the lactone cannot be phosphorylated by the enzyme for which free (+)-(3*R*)-mevalonic acid is the substrate (mevalonate phosphokinase), a 30-fold greater incorporation is sometimes obtained with mevalonolactone in comparison with the free acid.

The most probable explanation of this observation is that the lactone molecule penetrates into the cells more readily than the ionised free acid, and it is then hydrolysed to give a higher concentration within the cells than when the cells are bathed in the free acid. The unnatural (–)-enantiomer (XXIV) is not metabolised.

6.5.4. Isolation of radiochemically pure ABA

After a 12–24 hr incubation, the fruits are homogenised in methanol or acetone, the ether-soluble acid fraction is separated after the extracting solvent has been evaporated, and the acids are subjected to silica gel thin-layer chromatography in toluene – ethyl acetate – acetic acid (50:30:4, by vol.). After elution and methylation with diazomethane, the methyl ABA is chromatographed in hexane – ethyl acetate (3:1, v/v). If the TLC plates are not overloaded, and are subjected to multiple development, then the methyl abscisate obtained by this procedure is radiochemically pure; confirmation that any ^{14}C present in the extract has indeed been incorporated into ABA is afforded by thoroughly reducing the methyl abscisate with ice-cold borohydride. This produces an equal mixture of two isomeric diol esters which can be well separated from impurities, any ABA remaining, and from each other, by rechromatography in hexane – ethyl acetate (2:1, v/v). Autoradiography of such plates shows that label is present in the diols only.

Avocado fruits are so rich in ABA that when the extract is chromatographed on fluorescent silica gel plates the ABA zone can be seen as a zone of quenched fluorescence. Autoradiograms of such plates can show the distribution of label and this can be compared with the distribution of the quenched fluorescence. Such comparison gives one of the most sensitive methods for detecting the presence of labelled impurities and the presence of label in the ABA.

6.5.5. The derivation of the hydrogen atoms of ABA from mevalonic acid

6.5.5.1. The hydrogen atoms of C-5 of mevalonic acid. No complete chemical degradation method is known for the carbon atoms of ABA although a large part of the ring has been isolated as a substituted glutaric acid (Ryback 1972). The hydrogen atoms, on the other hand, can be isolated with a considerable degree of selectivity (Milborrow 1969b, 1972b). The origin, in mevalonic acid, of the carbon-borne hydrogen atoms in ABA has been established by feeding mixtures of stereospecifically-tritiated and ^{14}C-labelled mevalonic acid to avocado fruits and analysing the $^{14}C{:}^{3}H$ ratios in the ABA biosynthesised. Of the three mevalonate residues that go to make the carbon skeleton of ABA, only one carries hydrogen on what had been C-5 of mevalonate and this carbon (C-5 of ABA) carries only one hydrogen. Thus one (or possibly neither) of the two C-5

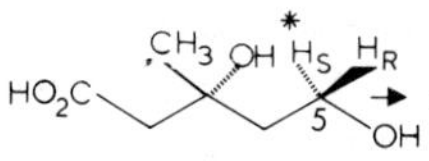

Mevalonic acid showing the stereochemistry of the 5-(*S*) hydrogen atom

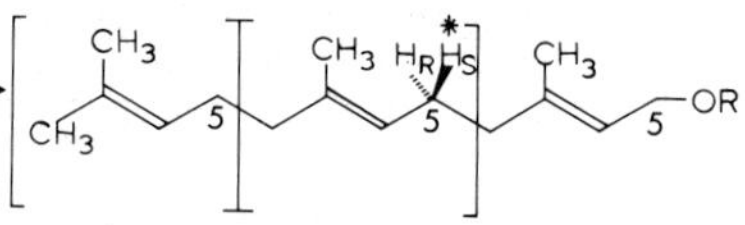

Isoprene unit

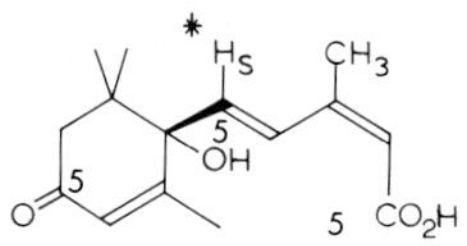

Abscisic acid showing the positions of the carbon atoms derived from C-5 of mevalonic acid. Only C-5 of ABA carries a hydrogen atom derived from those of C-5 of mevalonic acid.

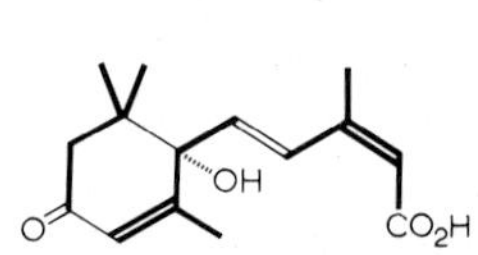

The three isoprene units that comprise the carbon skeleton of abscisic acid shown in heavy outline.

Scheme 6.3

hydrogens of mevalonate could be retained during biosynthesis to become the C-5 hydrogen of ABA and, by analogy with carotenoids, this was expected to be the *pro*-5-*(S)* hydrogen atom. Mevalonate carrying tritium in this position, synthesised by Cornforth and Ross (1970), was mixed in a defined ratio with [2-^{14}C] mevalonate and incorporated into ABA by an avocado fruit. The normalised ^{14}C:^{3}H ratio of 3.0:3.0 in the mevalonic acid fell to approximately 3:1 in the purified ABA (Milborrow 1972b), as expected, showing that the *pro*-5-*(S)* hydrogen of mevalonate is retained as the C-5 hydrogen of ABA and the *pro*-5-*(R)* was presumably lost during the formation of the 4-double bond (Scheme 6.3).

Although by analogy with carotenoids the hydrogen from C-5 of mevalonate was expected to be retained in ABA in this position, a more conclusive proof of its location

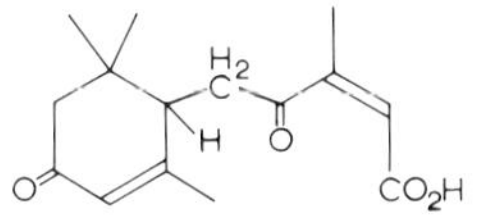

XXVI

is required than the occurrence of the expected ^{14}C:^{3}H ratio. The hydrogen from C-5 of ABA is not readily removed and the only reactions that could be carried out to define the position of the C-5 hydrogen were the formation of the lactone (Mallaby and Ryback 1972) which did not affect the ^{14}C:^{3}H ratio, and then the enolisation of the compound (XXVI) formed transiently during the rearrangement of the lactone by base. As expected, this treatment removed some of the tritium from C-5 (Table 6.3).

Table 6.3.

Incorporation of (+)-[2-^{14}C,5-*S*-^{3}H$_1$] mevalonic acid into ABA by avocado fruit (data from Milborrow 1972b)

	^{14}C:^{3}H ratios Expected	^{14}C:^{3}H ratios Observed
Mevalonic acid	3.0:3.0	–
Methyl abscisate	3.0:1.0	3.0:0.88
Lactone (V)	3.0:1.0	3.0:0.87
Exchanged lactone	3.0:0	3.0:0.16

6.5.5.2. The hydrogen atoms of C-4 of mevalonic acid. Two tritium atoms from C-4 of mevalonate would be expected to be retained, one at C-2 and the other at C-5′ of ABA. No 4-*(S)* tritium was incorporated and the ^{14}C:^{3}H ratios showed that tritium from the *pro*-4-*(R)* position was retained at both positions. Furthermore the exchange of six carbon-bound hydrogen atoms, which include the C-5′ hydrogen atoms, removed half the tritium, as expected. Popják and Cornforth (1966) have shown that, during the condensation of isoprene units, the formation of the double bond, between what had been C-3 and C-4 of mevalonate, occurs with the retention of the 4-*(R)* tritium and loss of the *pro*-4-*(S)* when the 3–4 double bond is formed in a *trans* configuration. When the bond is formed in the *cis* configuration the 4-*(S)* tritium is retained and the 4-*(R)* tritium is lost, as in rubber hydrocarbon (Archer et al. 1966). If the same stereochemistry applies in ABA biosynthesis, then the 2-double bond of ABA (between what had been C-3 and C-4 of mevalonate), which carries the 4-*(R)* tritium from mevalonate, is formed in a *trans* configuration (Scheme 6.4).

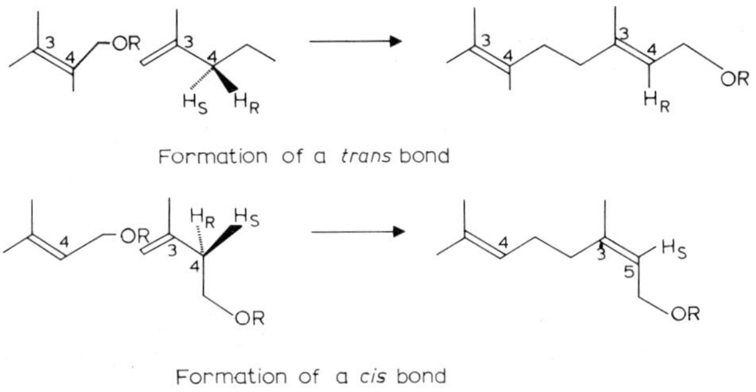

Scheme 6.4. Retention of the *pro-(R)* and *pro-(S)* hydrogen atoms from C-4 of mevalonate during the formation of *trans* and *cis* double bonds respectively between what had been the C-3 and C-4 atoms of mevalonic acid.

If the tritium from the 4-*(S)* position of mevalonate had been retained at C-2 of ABA then this would have shown that the 2-double bond was formed in the *cis* configuration and thereby differentiated between ABA biosynthesis and carotenoid biosynthesis. As the results (Table 6.4) indicate that the bond was formed *trans* it must have been isomerised at a later stage of biosynthesis (Robinson and Ryback 1969). This could occur during the conversion of a carotenoid into ABA or during a direct synthesis. The 2-*trans* isomer of ABA has virtually no growth-inhibitory activity and although it is formed photolytically from ABA, and is present in excised leaves, no attempts to demonstrate the enzymic isomerisation of [2-^{14}C]-2-*trans*-ABA to ABA have been successful. Furthermore the 2-*trans*-ABA isolated from avocado fruit biosynthesising ABA from labelled mevalonolactone was unlabelled. All this evidence suggests that 2-*trans*-ABA is not a precursor of ABA. The ^{14}C:^{3}H ratios of the ABA show that a second 4-*(R)* hydrogen from mevalonate is retained in ABA and its removal by base-catalysed exchange is consistent with it being at C-5′. The stereochemistry at this position has not yet been determined.

Table 6.4.

The incorporation of tritium from [2-^{14}C,4-*R*-$^{3}H_1$-, but not from [2-^{14}C, 4-*S*-$^{3}H_1$] mevalonic acid, into two positions in abscisic acid (data from Robinson and Ryback 1969)

	Expected ^{14}C:^{3}H ratio for the retention of 4-(*R*) tritium at C-2 and C-5′of ABA (normalised)	Observed ^{14}C:^{3}H ratios (normalised)
Mevalonic acid (4*R*)	3.0:3.0	–
Mevalonic acid (4*S*)	3.0:3.0	–
Me-ABA from (4*R*)	3.0:2.0	3.0:1.93
Me-ABA from (4*S*)	3.0:0.0	3.0:0.05
Me-ABA (4*R*) after exchange of the 5′ hydrogen atoms	3.0:1.0	3.0:0.92
Me-ABA (4*S*) after exchange of the 5′ hydrogen atoms	3.0:0.0	3.0:0.02

6.5.5.3. The hydrogen atoms of C-2 of mevalonic acid. The expected ^{14}C:^{3}H ratios in ABA synthesised from 2-^{14}C and 4-(*R*)-$^{3}H_1$,4-(*S*)-$^{3}H_1$ and 5-(*S*)-$^{3}H_1$ mevalonates would be obtained only if the three mevalonate molecules that go to make the carbon skeleton of ABA are derived equally from the same pool and if there had been no degradation of the mevalonate to acetate. This being so it was possible to investigate the fate in ABA of hydrogen from C-2 of mevalonate. One or other would be expected to occur at C-4 and C-3′ and both in one of the methyl groups.

The usual kind of ^{14}C:^{3}H ratio experiment cannot be carried out with C-2 tritiated mevalonate because of two complications. Firstly, in plants the activity of isopentenyl pyrophosphate (IPP)/dimethylallyl pyrophosphate (DMAPP) isomerase is considerably greater than that of the condensing enzyme (Goad 1970) and the isomerase is readily reversible. Its action "scrambles" the stereospecificity of the tritium and other isotopes of hydrogen on what had been C-2 of mevalonate. Secondly, the number of possible labelling patterns requires the isolation and measurement of tritium at each position.

The "scrambling" occurs when IPP is isomerised to DMAPP (Scheme 6.5). A hydrogen atom from the medium is added to the tritium-labelled methylene group to convert it into a methyl group. The methyl group spins so that the three hydrogen atoms become stereochemically equivalent and on reversal of the enzyme action any one of the three can be removed. If the same hydrogen atom is removed as was added then the original configuration is retained (Retention, Scheme 6.5). If the original unlabelled hydrogen atom is removed then the tritium comes to occupy the opposite position (Inversion, Scheme 6.5). If the tritium is removed from the molecule then the relative ^{14}C content of the sample increases but no stereochemical difference can be detected (Loss, Scheme 6.5). Thus, on average, all the IPP that has been cycled through DMAPP will be

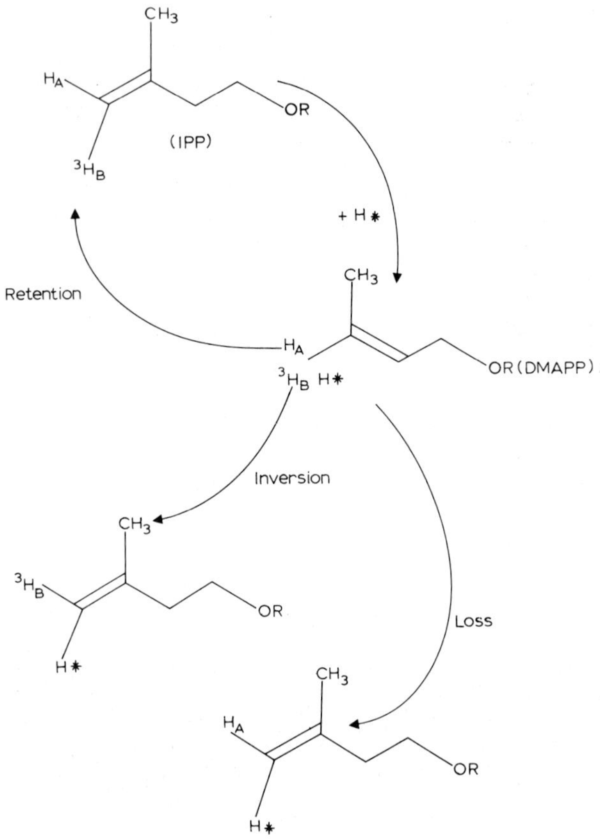

Scheme 6.5. Racemisation of stereochemically tritiated isopentenyl pyrophosphate (IPP) by the action of IPP/DMAPP isomerase.

racemic with regard to the tritium at C-2. Any isotope effect will tend to reduce the loss of tritium but will not affect the racemisation.

One of the positions in ABA occupied by hydrogen from C-2 of mevalonate is susceptible to base-catalysed exchange; this is C-3′ of ABA. Hence the preferential retention of the *pro*-2-*(R)* or *pro*-2-*(S)* tritium at this position can be determined by the comparison of ^{14}C:^{3}H ratios of the original samples of ABA derived from [(2*R*)-2-$^{3}H_1$] or [(2*S*)-2-$^{3}H_1$] mevalonates with those of the samples of ABA after base-catalysed exchange. The tritium transferred to the water can also be measured.

The preferential retention of tritium from the *pro*-2-*(R)* or *pro*-2-*(S)* position of mevalonate at C-4 of ABA can be determined from the change in ^{14}C:^{3}H ratios which occur when the ABA is rearranged by strong acid to the lactone (Mallaby and Ryback 1972). The remaining tritium resides in one of the 6′-methyl groups and is provided by both the *pro*-2-*(R)* and *pro*-2-*(S)* hydrogen atoms of mevalonate.

Using these methods it was found that relatively more tritium from the [(2*R*)-2-

Table 6.5.

Incorporation of (+)-[2-^{14}C,2-R-^{3}H$_1$]- and (+)-[2-^{14}C,2-S-^{3}H$_1$]-mevalonic acid into ABA by avocado fruit (data from Milborrow 1972b)

	Expected ratios for the retention of the 2-*(R)* hydrogen of mevalonate at C-4 and C-3′	Observed ^{14}C:^{3}H ratios
Mevalonic acid (2*R*)	(3.0:3.0)	–
Methyl abscisate (sample)	3.0:3.0	3.0:2.28
MeABA after exchange of 3′ hydrogen atoms	3.0:2.0	3.0:1.55
Lactone (V) (hydrogen at C-4 removed)	3.0:1.0	3.0:1.04
Aqueous exchange medium	3.0:1.0	3.0:0.584
Mevalonic acid (2*S*)	(3.0:3.0)	–
Methyl abscisate (sample)	3.0:1.0	3.0:1.63
Methyl abscisate (sample)	3.0:1.0	3.0:1.44
Lactone (V) (hydrogen at C-4 removed)	3.0:1.0	3.0:1.05
Aqueous exchange medium	3.0:0	3.0:0.23
Repeat of experiment with iodoacetamide (2.0 m*M*)		
Methyl abscisate (2*R*)	3.0:3.0	3.0:2.55
Methyl abscisate (2*S*)	3.0:1.0	3.0:1.12
Me ABA after exchange (2*R*)	3.0:2.0	3.0:1.99
Me ABA after exchange (2*S*)	3.0:1.0	3.0:0.92
Lactone (V) (2*R*)	3.0:1.0	3.0:1.19
Lactone (V) (2*S*)	3.0:1.0	3.0:0.90

^{3}H$_1$]mevalonate was incorporated into the C-4 and C-3′ positions of ABA than from the [(2*S*)-2-^{3}H$_1$]mevalonate (Table 6.5). However, the ratios were obscured by considerable racemisation and so an attempt was made to lessen the activity of the isomerase by adding iodoacetamide, a well-known inhibitor of the enzyme, so that it became rate-limiting but was not completely inhibited. A 5 m*M* solution of this compound reduced the racemisation without inhibiting the overall synthesis of ABA and the ratios were correspondingly more precise (Table 6.5).

Apart from the intrinsic interest of the above results, they allow two further deductions to be made. First, in all stereochemical choices the analogous hydrogen atoms of mevalonate are retained by the enzymes that synthesise ABA and carotenoids. Second, the stereochemistry of the four hydrogens at C-4 and C-5 in the isoprenoid precursor of ABA (C-2 and C-5 of two mevalonic acid molecules respectively) is known and so the formation of the 4-double bond can be deduced to occur by the elimination of two atoms in a *trans* configuration (Scheme 6.6).

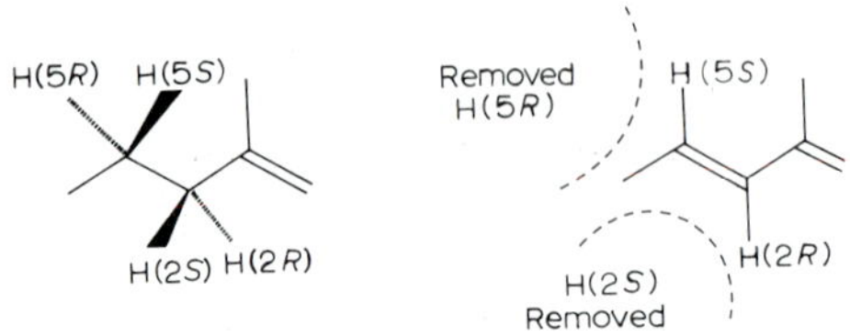

Scheme 6.6. The *trans* elimination of protons on C-4 and C-5 of ABA

6.5.5.4. The 3′-methyl group of mevalonic acid. Three of the four methyl groups of ABA can be expected to be derived from the methyl group of mevalonic acid. If mevalonate, with ^{14}C- and ^{3}H-labelled methyl groups, is incorporated by avocado fruit into ABA which is isolated under mildly acidic conditions, it has the same ^{14}C:^{3}H ratio (3.0:3.0) as the precursor. This then falls to 3:2 on exchange in base showing that, as expected, the methyl group at C-2′ is derived from the methyl group of mevalonate but also that there are two other such methyl groups in ABA. Kuhn – Roth oxidation of the exchanged abscisic acid forms acetic acid from the methyl groups and the carbon atoms to which they are attached. The ^{14}C:^{3}H ratio of this acetate confirmed that the remaining label occurred in other methyl groups. Thus one methyl group of each isoprene residue of ABA is derived, as expected, from the 3′-methyl group of mevalonic acid (Table 6.6).

Table 6.6.

The ^{14}C:^{3}H ratios of abscisic acid biosynthesised from (±)-[3′-^{14}C,3′-^{3}H$_1$]-mevalonic acid by avocado fruit (data from Milborrow 1975c)

	Expected ^{14}C:^{3}H ratio	Observed ^{14}C:^{3}H ratio
Mevalonic acid	3.0:3.0	–
Me ABA Expt. 1	3.0:3.0	3.0:2.908
Expt. 2		3.0:3.165
Exchanged Me ABA Expt. 1	3.0:2.0	3.0:1.998
Expt. 2		3.0:1.995
Acetic acid formed from methyl groups	3.0:1.80	3.0:2.107

6.5.6. The direction of cyclisation in abscisic acid

The stereochemical determination of the biosynthetic origin of the 6′-geminal methyl groups of ABA (i.e. which one is derived from C-2, and which from C-3′, of mevalonic acid) enables the direction of cyclisation in the precursor to be deduced. The investigation requires a method to discriminate between the two 6′-methyl groups and this is done by feeding ABA to tomato shoots which oxidise it, via the unstable Metabolite C, to phaseic acid (see section 6.6.3). Only the (+)-enantiomer of ABA is converted into phaseic acid and only one isomer of phaseic is formed; the enzyme, therefore, stereospecifically oxidises one methyl group only. The absolute configuration of phaseic acid has recently been determined by an NMR analysis of the two epimeric dihydrophaseic acids formed from it by reduction with borohydride (Milborrow 1975b). The 6′-methyl group of (+)-ABA *(pro-S)* on the same face of the cyclohexenyl ring as the tertiary hydroxyl group is the one incorporated into the oxymethylene bridge.

The investigation further requires a method of determining whether the *pro-(R)*-6′-geminal methyl group of ABA (i.e. the one remaining as a methyl group in phaseic acid) is derived from C-2 or C-3′ of mevalonate. This is done by subjecting the phaseic acid (isolated from tomato plants fed ABA biosynthesised from [2-^{14}C,2-$^{3}H_2$]mevalonate by avocado fruit) to a modified Kuhn – Roth procedure so that the methyl groups, and the carbons to which they are attached, are converted into acetic acid. The only position in phaseic acid that can be derived from C-2 of mevalonic acid and that can contribute to the acetic acid is the remaining 6′-methyl group *(pro-R)* of abscisic acid present in phaseic acid. If, therefore this methyl group is derived from C-2 of mevalonic acid, then phaseic acid biosynthesised, via ABA, from [2-^{14}C]mevalonic acid should give ^{14}C-labelled acetic acid. If the acetic acid were unlabelled then the *pro-(R)*-6′-methyl group of ABA must have been derived from the methyl group at C-3 of mevalonic acid.

When the acetic acid was distilled, converted into *p*-bromophenacyl acetate and re-crystallised it was found to carry ^{14}C and ^{3}H in almost the correct proportion (Table 6.7), so proving that the *pro-(S)*-6′-methyl derived from the 3-methyl of mevalonic acid was hydroxylated, and the remaining *pro-(R)*-6′-methyl of ABA was derived from the C-2 of mevalonic acid and gave rise to the labelled acetate.

Table 6.7.

The occurrence of ^{14}C and ^{3}H in acetate derived from the 6′-methyl group of phaseic acid (ABA numbering) that had been biosynthesised, via ABA, from [2-^{14}C,2-$^{3}H_2$] mevalonate (data from Milborrow 1975a)

	^{14}C dpm	^{3}H dpm	Normalised ^{14}C:^{3}H ratio
Methyl phaseate before oxidation by a modified Kuhn – Roth procedure	3786	7401	3.0:3.0
Dpm. in *p*-bromophenacyl acetate after two recrystallisations	294.8	943.2	1.0:1.637
Expected ratio in *p*-bromophenacyl acetate			1.0:2.0

This stereochemical arrangement must arise from the juxtaposition of the double bonds of the precursor during cyclisation as shown in Scheme 6.7.

An identical arrangement of the analogous geminal methyl groups of β-carotene has been deduced by Bu'Lock et al. (1970) from an analysis of trisporic acid – a fungal hormone (see Chapter 7) produced by oxidation of β-carotene (Scheme 6.8).

The foregoing describes the overall stereochemistry of biosynthesis between mevalonic

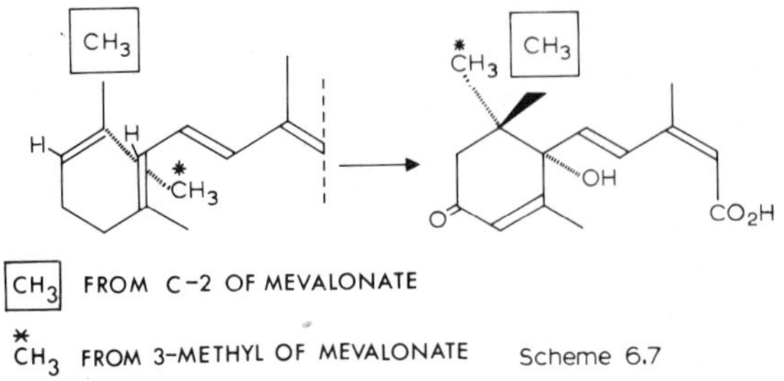

Scheme 6.7

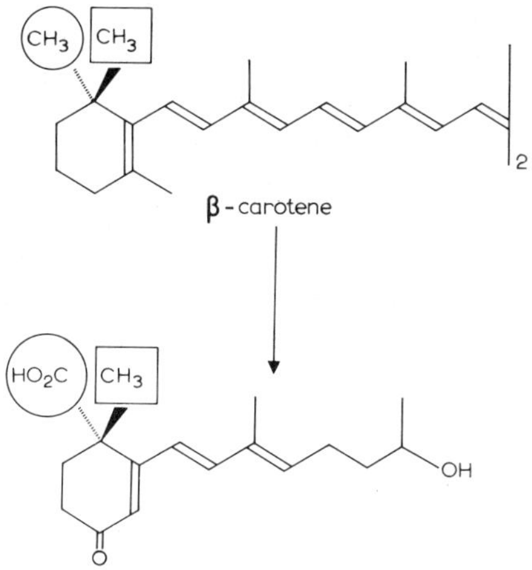

Scheme 6.8. The methyl enclosed by a circle of derived from C-3′ of mevalonate, whereas that enclosed by a square is from C-2 of mevalonate.

acid and ABA but the identities of the intermediates are unknown; indeed even the pathway has not been established. There are two possibilities – the carotenoid pathway, and the "direct synthesis" pathway with C_{15} and smaller precursors (see section on xanthoxin, 6.9).

6.5.7. Localisation of ABA biosynthesis

It was relevant to consider some aspects of the location of ABA biosynthesis in the section on occurrence of ABA (6.3). Further information is now presented.

Biosynthesis of abscisic acid has been demonstrated in several different parts of plants by feeding labelled mevalonate. Apart from fruit tissues, incorporation has been demonstrated in seeds (cotyledons and embryos of avocado, wheat embryo and endosperm), roots (sunflower), and leaves and stems of avocado (Milborrow and Robinson 1973). Thus, it appears that, like other plant hormones, the sites of synthesis are widespread. Of particular interest are the regulatory mechanisms in the different organs. As described later, leaves of avocado increase their content of ABA 40-fold on wilting but fruit tissues are unaffected. Stems and roots show a slight response to wilting, while in the seed of the ripening fruit the rate of synthesis of ABA is barely detectable. The physiological state of the tissue is probably important because Davis and Addicott (1972) have reported marked increases in ABA content during two stages of development of cotton bolls, and Loveys et al. (1975) were able to induce increased incorporation of [^{14}C]mevalonate by callus tissue culture from grape (*Vitis vinifera* cv. Cabernet Sauvignon) when grown on agar containing a high concentration of mannitol.

6.5.8. Formation of ABA in isolated chloroplasts

The incorporation of labelled mevalonic acid into abscisic acid by leaves is very slight, whereas the incorporation into steroids and other related compounds is considerable. This relationship is reminiscent of the small proportion of mevalonate incorporated into carotenoid by intact leaves. These features, taken together with the similarities of the carbon skeletons, stereochemistry of biosynthesis and possible biogenetic relationships of carotenoids and ABA, suggested that both could be formed in the same cellular compartment. Carotenoid biosynthesis has been found to occur in chloroplasts (Goodwin 1958) so an attempt was made to find ABA biosynthesis from labelled mevalonate using plastids from a tissue which makes ABA under conditions of turgor and wilt, namely avocado mesocarp. Intact membranes of plastids present a strong barrier to the entry of mevalonate (Goodwin 1958) and so samples of chloroplasts, isolated intact, were lysed at the start of the incubation. Samples of ABA analysed as radiochemically pure methyl ester, or diol methyl esters, contained label derived from mevalonic acid (Table 6.8).

Table 6.8.

Incorporation of [2-$^{3}H_2$] mevalonic acid by intact and lysed chloroplasts isolated from ripening avocado fruit (data from Milborrow 1974b)

	ABA isolated as methyl ester (μg)	Dpm in Me-ABA	Dpm in *cis*-diol	Dpm in *trans*-diol
Intact chloroplasts (2 hr)	20.8	440	187	175
Lysed chloroplasts (2 hr)	22.2	1000	380	450
Intact chloroplasts (17 hr)	23.6	546	238	201
Lysed chloroplasts (17 hr)	24.2	2860	1376	1197

The incorporation of mevalonate into ABA by chloroplasts identifies a site of synthesis but the failure of the extra-chloroplastic soluble fractions from the same fruits to carry out the synthesis does not exclude the possibility that some ABA can be synthesised outside the chloroplasts in vivo. On the other hand the extremely small amount of labelled ABA formed in tomato shoots during short term experiments, compared with the amount of label incorporated into steroids, which are mostly synthesised in the extra-chloroplastic cytoplasm, suggests that there is little or no synthesis of ABA in the cytosole.

The plastids used were obtained from the green and the colourless parts of the avocado mesocarp, but both parts and plastids from both parts, were equally effective at incorporating mevalonate into ABA. Thus there is no requirement for the plastids to be green in order to synthesise ABA.

The discovery of the plastids as the site of ABA biosynthesis may provide an explanation for the presence of chloroplasts in all functioning stomatal guard cells. ABA is a potent regulator of the hydroactive closure of stomata and it can be speculated that a function of the guard cell chloroplasts is to provide a supply of ABA.

6.6. Metabolism

6.6.1. *(+)-L-Abscisyl-β-D-glucopyranoside*

The first experiments carried out with (±)-[2-^{14}C]ABA measured transport of the labelled compounds through petiole segments of bean and sycamore (Milborrow 1967). Analysis of the tissue after 12 hr incubation showed that three labelled metabolites were present: A, small quantities of an ether-soluble neutral compound; B, a polar, water-soluble conjugate, and C, an ether-soluble acidic compound more polar than ABA.

B was identified as (+)-L-abscisyl-β-D-glucopyranoside (XXVII) – a compound characterised previously from yellow lupin by Koshimizu et al. (1966). Glucose ester formation appears to be a rather unspecific detoxification process because not only is the (+) enantiomer of racemic ABA converted into the base-labile conjugate, but the unnatural (–) and also both enantiomers of the isomeric 2-*trans*-ABA are also accepted by the esterification mechanism. Analyses of many shoot, leaf and fruit tissues have shown that between 10% and 50% of the total (+)-ABA in a tissue is present as a base-hydrolysable conjugate, but in ripening oranges up to 10 times more (+)-ABA occurs in a conjugated form than as the free acid (Goldschmidt et al. 1973).

At present only two characterisations of a conjugate of abscisic acid have been carried out and both yielded L-abscisyl-β-D-glucopyranoside. There is a danger that all base-labile ABA will be imperfectly identified as glucose esters and small amounts of other adducts will be overlooked.

6.6.2. *Methyl abscisate*

When the first sample of glucose ester (isolated from tomato shoots fed with (±)-ABA) was acetylated and dissolved in methanol prior to mass spectrometry, it underwent methanolysis to give tetraacetyl glucose and methyl abscisate. It has now been established that L-abscisyl-β-D-glucopyranoside is attacked by methanol in neutral or basic conditions (Milborrow and Mallaby, in press) but not in acidic conditions as originally reported, to give glucose and methyl abscisate. Metabolite A was identified as methyl abscisate and, so far, it has been found only as an artefact of extraction when the extracts have been prepared with methanol. No methyl abscisate has been found in extracts prepared with acetone as the solvent. The facility with which this reaction occurs means that all previous determinations of abscisyl-β-D-glucopyranoside contents made after methanol extraction are of uncertain value and many reports of neutral inhibitors in extracts prepared using methanol as a solvent may be explained by the presence of adventitious methyl abscisate derived from this source.

6.6.3. Metabolite C and phaseic acid

The first isolation of Metabolite C (XXVIII) gave a crystalline solid with a strong ORD spectrum similar to that of (+)-ABA (Milborrow 1969b). On methylation it rearranged to a compound which was identified as the methyl ester of phaseic acid (isolated previously from beans (MacMillan and Pryce 1968)) and which showed a quite different and weak ORD spectrum. The original structure proposed for phaseic acid (XXIX) could not be reconciled with the rearrangement of a metabolite of ABA but a structure proposed for the unstable Metabolite C (XXVIII) not only accounted for the rearrangement but indicated a new structure for phaseic acid (XXX) (Milborrow 1969b). There was a crucial difference between the two alternative structures for phaseic acid (XXIX and XXX) that could be tested by NMR analysis. In structure XXIX the methyl group attached to C-2′ of ABA becomes part of an epoxide ring, and both of the geminal methyl groups at C-6′ remain. On the other hand, one of the 6′-geminal methyl groups of ABA would be expected to disappear during the formation of XXX and the methyl group at C-2′ would remain intact if the new structure were correct (Scheme 6.9). It is fortunate that the

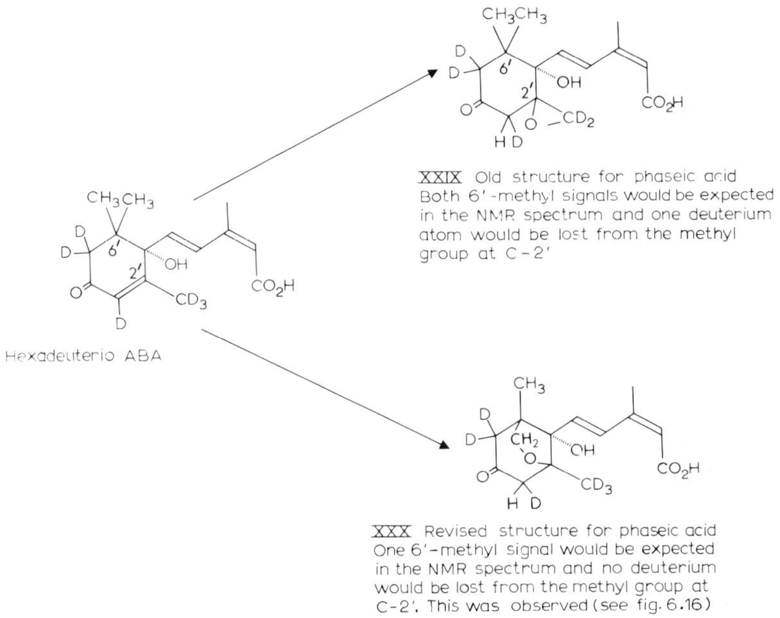

Scheme 6.9. Metabolism of hexadeuterio-ABA to phaseic acid confirms structure XXX.

hydrogen atoms of the C-2′ methyl group of ABA exchange with the medium in alkaline aqueous solutions and so can be replaced by deuterium if N sodium deuterioxide is used. When such deuteriated ABA was fed to tomato plants it was oxidised to phaseic acid and mass spectrometry showed that no deuterium had been lost from the methyl group at C-2′ and the NMR signals of one of the geminal C-6′ methyl groups had disappeared (Figure 6.16). Thus one of the C-6′ methyl groups had been oxidised as expected, thereby confirming that structure XXX was the correct one (Milborrow 1971).

is esterified and the conjugate then undergoes further metabolism. This is thought not to occur because the highly unstable Metabolite C was isolated as a free acid and the specific activities decrease along the sequence: ABA; phaseic acid; dihydrophaseic acid epimers, both of free acids and conjugates. Furthermore, the specific activity of each free acid is higher in comparison with the specific activity of the respective acid moiety in the conjugate. These results show that metabolism proceeds according to Scheme 6.10, not by the metabolism of bound materials.

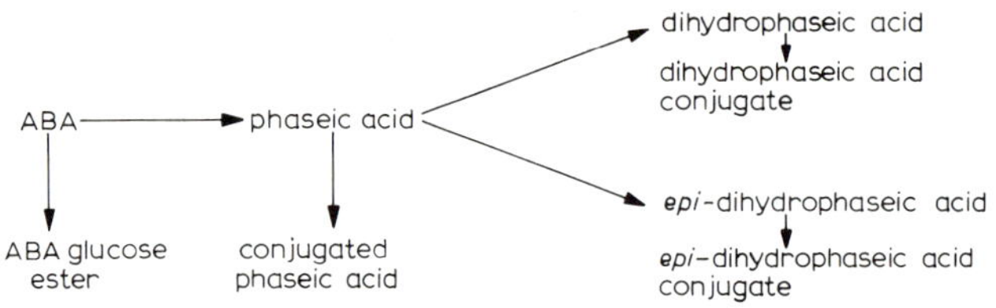

Scheme 6.10. Pathway of [^{14}C]ABA metabolism deduced from the specific activities of the compounds.

Pathway of [^{14}C] ABA metabolism deduced from the specific activities of the compounds.

6.6.4. The epimeric dihydrophaseic acids

The presence of the two epimeric compounds (XXXI and XXXII) in one plant is an unusual feature and other possible explanations for their occurrence were examined. It was thought that the 4′-hydroxyl group (ABA numbering, see III) could be epimerised chemically during the isolation procedure but both dihydrophaseic acids are stable and are not interconverted by strong bases or acids, or under the conditions employed to extract them from plants. There was no significant conversion of the alternate epimer when XXXI and XXXII were fed to bean shoots and the ^{14}C-labelled acids must have penetrated into the cells in these experiments because a proportion of both XXXI and XXXII was converted into water-soluble base-labile conjugates (Zeevaart and Milborrow 1976). Although there appears to be negligible metabolic interconversion of XXXI and XXXII fed to plants, alcohol dehydrogenase enzymes, such as would be expected to form XXXI and XXXII from phaseic acid, are often quite readily reversible and it is possible that the reversal of the enzyme mechanisms responsible for the production of the dihydrophaseic epimers could lead to a gradual, indirect interconversion via oxidation to phaseic acid (Scheme 6.11).

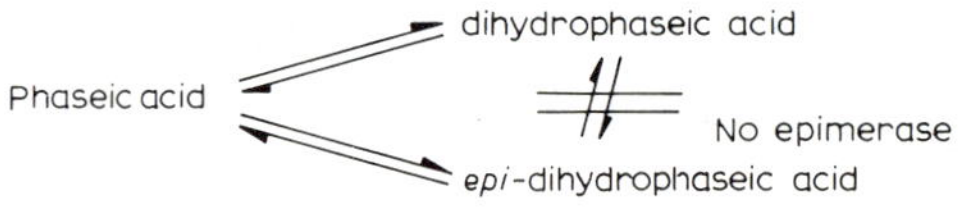

Scheme 6.11 Possible interconversion of the epimeric dihydrophaseic acids via phaseic acid.

The physiological role played by XXXI and XXXII and their further metabolism is unknown at present.

When (±)-[2-^{14}C,4′-^{18}O] ABA was fed to bean seedlings it was converted into phaseic acid and into the epimeric dihydrophaseic acids; both of these carried ^{18}O in their 4′-hydroxyl groups where it is stable. The unmetabolised ABA had lost a small amount (30%) of ^{18}O and the phaseic acid nearly all (81%), indicating that the 4′-ketone oxygen in this compound can exchange even more rapidly with the medium than can that of ABA (Gray et al. 1974; Zeevaart and Milborrow 1976). The equal, and high, ^{18}O contents of XXXI (43%) and XXXII (41%) demonstrated that they were formed rapidly from phaseic acid and that neither lost ^{18}O by a mechanism which inverted the stereochemistry at C-4′. The 4′-ketone group of phaseic acid must, therefore, be reduced to give the two epimeric acids XXXI and XXXII. The former is usually the major product but XXXII can occur at from 2% to 50% of the concentration of XXXI.

It is possible that XXXII is formed by the same enzyme that reduces XXX to XXXI (and hence whose active site would have to accommodate the non-chiral 4′-ketone group with the cyclohexyl ring in both possible orientations). On the other hand, two enzymes could operate with opposite stereochemistry on the same substrate to produce the two epimers.

The possibility that the small amount of *epi*-dihydrophaseic acid arises from bacterial reduction of phaseic acid while the major proportion of dihydrophaseic acid is formed by plant enzymes was investigated by growing bean plants from seed under sterile conditions. Labelled ABA was metabolised to both epimers in the usual proportions, showing that *epi*-dihydrophaseic acid is a plant product (Zeevaart and Milborrow 1976).

Most of the experiments on XXXI and XXXII were carried out on plants which had been fed with (±)-[2-^{14}C]ABA, but phaseic, dihydrophaseic and *epi*-dihydrophaseic acids occur naturally as free acids and as base-hydrolysable conjugates. This was demonstrated by GLC analysis of the acid fraction and the acid fraction released by alkaline hydrolysis from extracts of untreated bean seeds and seedlings (Zeevaart and Milborrow 1976).

6.7. Transport of ABA

6.7.1. Methods of measurement

Experiments which yield information on the movement of ABA in plants are of three kinds: (i) the time taken for a tissue to respond to an application of ABA can give a measure of the time taken for it to reach the site at which it exerts its effect, (ii) the determination of the presence of ABA in transport pathways, (iii) direct measurement of the rate of movement of labelled compound in whole plants or pieces of plant.

6.7.2. Penetration of ABA into cells

In a number of tests the response of a tissue to ABA has been detectable within 10 min. The data of Mittelheuser and Van Steveninck (1971) and Cummins et al. (1971) show that stomatal closure begins within less than 10 min after the application of ABA via the transpiration stream so the ABA must have passed through several cells during this period. Rapid responses to ABA are discussed in detail in Chapter 12, Volume II. Some rapid effects of ABA could be accounted for by an effect on the cell membrane (see Chapter 12, Volume II). Hence a more certain indication that a compound has penetrated into plant cells is afforded by its presence in the transport system out of the organ to which it was applied, or by the formation of metabolites. These latter methods have not yet been used to measure rate of penetration into cells over short time intervals.

6.7.3. The presence of ABA within the transport system

Abscisic acid has been detected in almost all tissues of vascular plants in which it has been sought; it is also biosynthesised (section 6.5) in several different organs but in spite of this ubiquity of formation it seems to be transported quite extensively. Hoad (1967) has isolated ABA from phloem exudate of willow *(Salix viminalis)* and castor bean *(Ricinus communis)* and the honeydews excreted by aphids feeding in the phloem of these and a number of other species also contain ABA (Hoad 1973a). Hoad determined the ABA as methyl abscisate by an electron capture detector after GLC and identified it in samples of phloem exudate as small as 0.01 ml. With apparatus of such sensitivity he was able to observe an increase in ABA content immediately after the onset of wilting in the leaves and by 6 hr the ABA concentration rose to about 20 times that prior to moisture stress (Hoad 1973b). This suggests that ABA moves in the plant quite rapidly and it may be transported to buds and other growing parts from leaves which are wilting under drought conditions.

ABA has also been isolated from xylem sap of willow (Lenton et al. 1968) so this route also may serve to distribute the compound in the plant.

6.7.4. Movement of ABA through excised plant parts

6.7.4.1. Direct measurement of transport through petiole explants. The movement of [^{14}C]IAA through the parenchymatous cells of 10-mm segments of bean *(Phaseolus vulgaris)* and *Coleus* petioles has been studied by McCready and Jacobs (1963) and their coworkers. They find that [^{14}C]IAA moves basipetally some 100 times faster than it moves acropetally; there is a powerful metabolically controlled polarity which influences the velocity of transport in the two directions. The same explant system was used by Milborrow (1968) to find that (±)-[2-^{14}C]ABA was transported but with a much reduced polarity: basipetal transport was some 2–3 times faster than acropetal transport. Dörffling and Böttger (1968) carried out similar experiments with *Coleus* but bioassayed the ABA rather than estimated it by radioactivity; they confirmed that polarity was slight, compared with IAA, and was exhibited by explants from young leaves but not by those from old ones.

Dörffling and Böttger calculated the rate of transport of ABA through their explants at between 24 and 35 mm/hr using the same extrapolation procedure that McCready and Jacobs has used for IAA in bean petioles. A comparison of the results indicates that ABA is transported 4–6 times faster than IAA in the basipetal direction and considerably faster than this in the acropetal direction. A difference of this magnitude in the basipetal direction is unlikely to be accounted for by the higher temperature (27 °C) used for the *Coleus* experiments in comparison with those using beans (25 °C) and is also unlikely to be a species difference.

A possible explanation is that whereas McCready and Jacobs assayed [^{14}C]IAA that had passed *through* the explants and had been trapped in agar "receiver" blocks, Dörffling and Böttger measured the total amount of ABA that had passed out of the explants. An application of ABA to one end of an explant could lead to a "shunting" of the endogenous ABA into a receiver block before any of the applied material had passed through the explant and so might give a spuriously high value for the rate of movement.

Ingersoll and Smith (1970), however, have obtained similarly high rates, 20–30 mm/hr, for the rate of transport of (±)-[^{14}C] ABA through cotton explants, their experiments being carried out at 30 °C. Later work by these authors (1971) gave a more refined estimate of 22.4 mm/hr at 30 °C. They administered a pulse of (±)-[^{14}C] ABA followed by unlabelled material and found that 70% of the [^{14}C] ABA passed through the explants within 6 hr; this far exceeds the proportion of IAA transported. The continuous destruction of [^{14}C] IAA by decarboxylation and other degradative reactions as it passed through the petiole sections would give a low apparent velocity, even in "pulse and chase" experiments. The higher apparent velocity of ABA transport in the basipetal direction in comparison with IAA requires further detailed investigation. The reasons for the wider discrepancy between the rates of movement in the acropetal direction is even more obscure.

In contrast to these earlier results are those reported by Veen (1975) also using *Coleus* petiolar sections. (±)-[1-^{14}C] ABA was not transported in a polar manner and the data also show that it was not transported as rapidly as the synthetic auxin NAA (9.7 mm/hr); in fact the pattern of movement suggested that it moved by passive diffusion. Thus several features of the transport of ABA through petiole segments (polarity, velocity relative to IAA, even the occurrence of metabolically controlled transport) have given conflicting results and the explanation of the differences cannot be resolved at present.

The transport of ABA and other compounds through petiolar explants has been considered to be a metabolically regulated cell-to-cell transfer, rather than movement in the phloem, because the 3- to 10-mm sections used are believed to be too short for sieve tubes to function. The velocities (20–30 mm/hr) calculated for these compounds in explants are lower than those calculated for sucrose and other solutes in intact phloem (100–1500 mm/hr); however some residual phloem transport cannot be discounted.

6.7.4.2. Transport of ABA in root segments. Wilkins et al. (1972), in an elegant series of experiments, demonstrated that the geotropic response of root tips of maize *(Zea mays)* was mediated, in part, by a growth inhibitor produced in the root tip. This material is transported basipetally (i.e. out of the root tip and back towards the shoot) and is redistributed laterally under the influence of gravity. Although it has not been conclusively identified, indirect evidence, for example an identical pattern of effects produced by applied ABA (Kundu and Audus 1974; Wilkins and Wain 1974; Pilet 1975a), indicates that the root growth inhibitor is abscisic acid.

Wilkins et al. have shown that the endogenous growth inhibitor is transported basipetally from its point of origin in the maize root cap and Pilet (1975a) has shown that the same pattern of translocation and growth responses occur when (±)-ABA is applied to the decapitated tip segments. Basal application of ABA had no effect on growth so it appears that in the terminal 15-mm root segment transport of ABA is principally basipetal. Hartung and Behl (1975a) used similar root segments from runner bean *(Phaseolus coccineus)* primary roots and applied agar blocks containing [^{14}C] ABA to apical or basal ends. They observed clear acropetal polarity and by separating the central stele from the cortex they were able to demonstrate that transport in the cortex was slow and non-polar while transport in the central stelar cylinder was polar and considerably faster than in the cortex. After a 9-hr incubation, the ratio (corrected for tissue volume) of label in the stele to that in the cortex was 12:1, and the radial transport was strongly inhibited by low temperature and so was probably metabolically controlled. Illumination of the root segments during the transport period stimulates acropetal and reduces basipetal transport (Hartung and Behl 1975b) – a further indication of the physiological complexity of the transport phenomenon. ABA transport in roots in relation to the geotropic response is also discussed in Chapter 5, Volume II.

6.7.5. Movement of ABA in intact plants

Experiments carried out by Hocking et al. (1972), using applications of (±)-[^{14}C] ABA to the surface of primary leaves of intact bean plants, are in close agreement with the conclusions drawn from the foregoing work. They found that radioactivity moved out of the leaf and was distributed fairly evenly throughout the plant within 24 hr. One curious feature was that the root nodules became some 8 times more heavily labelled per unit fresh weight than any other part of the plant. Redistribution of labelled material occurred both upwards and downwards from the leaf to which the [^{14}C] ABA was applied but steam-girdling prevented any movement of label downwards past the killed zone. Girdling of the stem above the primary leaf which was exporting label reduced the amount of radioactivity reaching the upper parts of the plants but did not stop it entirely. A substantial proportion of the applied ABA was metabolised. Similar experiments have been carried out by Shindy et al. (1973) who applied (±)-[^{14}C] ABA to the cotyledons of cotton *(Gossypium hirsutum)* seedlings. Basipetal transport predominated and although the export of ABA from the treated leaf appears to be slower than in bean seedlings, the difference can probably be accounted for by more rapid flow of assimilates out of the mature primary leaf of the bean. A chromatographic analysis of the cotton seedling extracts confirmed the presence of ABA. The pattern of redistribution of [^{14}C] ABA in these two plants is in accord with the results of Hoad which suggest that ABA moves extensively in the phloem and to a small extent in the xylem sap.

Translocation of ABA in intact plants is also discussed in Chapter 11, Volume II.

6.7.6. Form transported

The form in which ABA is transported within the plant has not been definied unambiguously. Chromatographic analyses of aphid exudate and of agar receiver blocks into which labelled material has been passed have shown the presence of ABA, but the possibility has not been excluded that ABA is moved through the phloem mainly as a glucose ester and that it is hydrolysed to free ABA and glucose during its passage through the aphids or through damaged tissue at the cut surface of explants. Hydrolysis of of ABA glucose ester has been observed to occur rapidly in leaf homogenates (Milborrow 1970).

Although ABA was the only labelled compound present in receiver blocks in explant experiments, phaseic acid and abscisyl-β-D-glucopyranoside were isolated from the petiole segments themselves. Hocking et al. (1972) found that labelled metabolities were present in other parts of intact bean seedlings when [^{14}C] ABA had been applied to the primary leaves, but it has not yet been established whether these metabolites were transported or formed in situ from the transported [^{14}C] ABA.

6.8. Effects of the environmental conditions on ABA levels

A change in the environmental conditions under which a given plant is growing will cause an adjustment of many constituents; some of these changes will be primary and an adjustment of a regulatory compound, such as ABA, can be expected to bring about a sequence of further responses. Which of these responses can be attributed directly to alterations in ABA or xanthoxin metabolism, and which are due to other indirect effects, cannot be decided with certainty at present. As a first step, the changes in ABA content brought about by environmental factors can be defined.

6.8.1. Effects of stress

It has been known for many years that a period of water deficit inhibits the growth of plants, even after water becomes readily available again. In addition, the stomata often remain closed for some days after turgor has been regained. Wright and Hiron (1969) have provided an explanation for these phenomena; under conditions of water deficit there is a rapid increase in the content of a growth-inhibitory compound that they were able to identify as ABA. The increased content of ABA may be up to 40 times greater than the 10–20 μg/kg present in normal, turgid leaves and its formation is almost complete within an hour of the commencement of the stress.

Waterlogging, cold treatment and certain pathological conditions also stimulate ABA production but it is not known whether it is the stress conditions themselves, or the slight water loss that they cause, which triggers the response. The effect of an increased ABA content in wilted leaves spreads to the rest of the plant because the ABA content of phloem exudate from turgid and wilted castor bean plants has been measured by Hoad (1973b) who found that the ABA was considerably higher in the latter. The mean ABA content of sap from the plants before drought was 105 μg/l and this rose to 1020 μg/l during drought; 24 hr after the water supply had been renewed the content was 292 μg/l.

The capacity of different parts of a plant to synthesise ABA in response to wilting has been investigated by Milborrow and Robinson (1973) who found that while the ABA contents of avocado leaves increased 40-fold, the contents of stems doubled. Roots also showed a slight response but the ripening mesocarp, although synthesising ABA rapidly, was unaffected by water loss. It appears, therefore, that the biosynthetic mechanism is regulated differently in the various tissues and that the leaves probably play a major role in the ABA economy of the whole plant although ABA can be synthesised in many tissues.

Hiron and Wright (1973) measured the diffusive resistance of wilted dwarf bean *(Phaseolus vulgaris)* plants and were able to correlate the values, a measure of stomatal aperture, with the ABA content not only during the wilting phase but also during the recovery phase. This, together with the known effects of ABA on growth and stomatal aperture, suggests that these processes are regulated by the large amounts of ABA formed as a response to water deficit. Other experiments of Hiron and Wright amplify this. When a continuous draught of warm air caused bean seedlings to wilt, their ABA content rose and diffusive leaf resistance increased concurrently. The plants then regained their turgor. After the restoration of the normal water content, the content of free ABA fell (Figure 6.11) while the content of alkali-hydrolysable ABA conjugate rose, and successive wilt-recovery cycles boosted the content still further. Thus the formation of the conjugate represents a normal inactivation mechanism but whether or not the process is reversible has not yet been established. At least a proportion of the wilt-induced increase in ABA is attributable to de novo synthesis because a much greater proportion of labelled mevalonate is incorporated into ABA in wilted tomato shoots than in turgid ones (Milborrow and Noddle 1970). Further discussion of the effect of water stress on ABA levels is found in Chapter 11, Volume II.

6.8.2. Other factors affecting ABA content

Light regime and quality have an effect on ABA content but it is less certain whether the changes produced are a direct response to a primary change in ABA content. There is also an interaction between the effect of light on growth and the ability of plants to synthesise ABA in response to a water deficit. Simpson and Saunders (1972) measured the ABA contents of tall and dwarf pea *(Pisum sativum)* plants grown in darkness; they contained 6.6 and 10.7 μg/kg respectively and these values rose to 34.0 and 69.2 μg/kg on wilting. In contrast to this, the plants grown with a daily photoperiod contained 61.1 and 51.1 μg/kg respectively and after wilting these values rose to 359 and 434 μg/kg. The slight differences between each pair indicates that ABA content is not closely related with dwarf habit, but the differences between the light- and dark-grown plants show that the ABA synthesising system is altered dramatically by the growth conditions.

Tucker and Mansfield (1972) grew Cocklebur *(Xanthium strumarium)* plants with and without supplementary red/far-red illumination from tungsten lamps. Under the red/far-red treatment, the plants failed to branch and their axillary buds contained the highest concentration of ABA so far recorded. Plants grown without the supplementary red/far-red illumination branched profusely and contained less ABA. There were similar changes in cytokinin contents. Thus quality of light is an important factor in controlling ABA metabolism. Zeevaart (1974) determined ABA and xanthoxin contents of spinach *(Spinacea oleracea)* growing under long or short days and found negligible changes in the latter and smaller changes in the content of the former, either as conjugate or free acid. There was evidence of a very slight diurnal fluctuation but this could be an indirect effect of water balance. The author concludes that there is no correlation between xanthoxin and ABA contents of spinach with growth and flowering responses.

The effect of photoperiod on ABA levels is also discussed in Chapter 6 (section 6.5.2) and Chapter 9 (section 9.3), Volume II.

6.9. Xanthoxin

6.9.1. The occurrence and identification of xanthoxin

It was noticed by Simpson and Wain (1961) that plants grown in the light contained more inhibitor than dark-grown ones and the content of inhibitory material increased after illumination; furthermore the action spectrum of phototropism resembled the absorption spectrum of carotenoids. The possibility that ABA was derived from a carotenoid was suggested by the chemical similarity between ABA and the terminal rings of some carotenoids (Ohkuma et al. 1965).

Taylor and Smith (1967) followed up these observations by photolysing a mixture of carotenoids and finding that a neutral inhibitor was formed. They found that the xanthophyll violaxanthin (XXXIII) was the source of this inhibitor and they isolated and characterised the active component (Taylor and Burden 1970). It was named xanthoxin and is defined as a mixture of the 2-*cis* and 2-*trans* isomers of a 15 carbon aldehyde (XXXVI, XXXVII). Two other degradation products (XXXIV and XXXV) were also characterised.

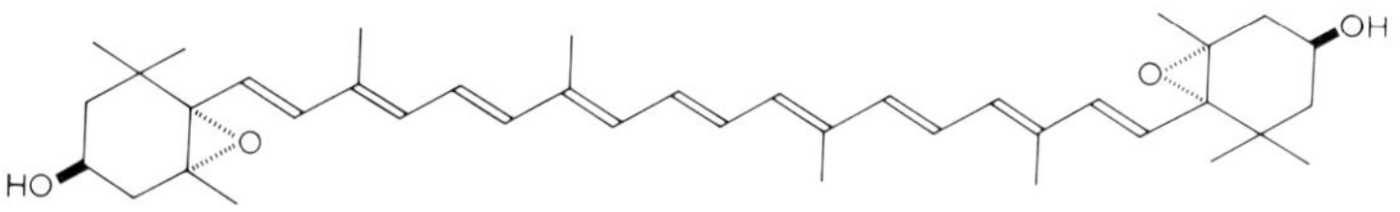

XXXIII.
Violaxanthin

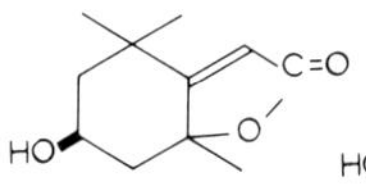

XXXIV.
Loliolide

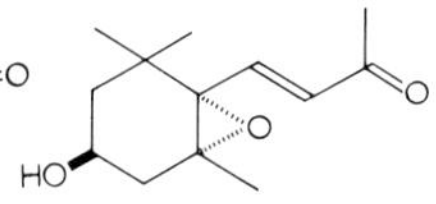

XXXV.
"Butenone"

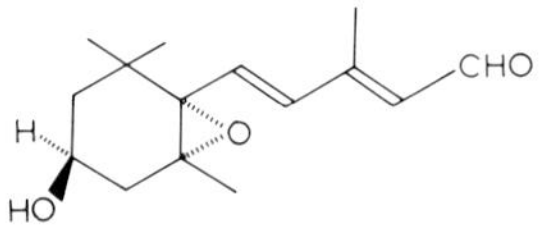

XXXVI.
2-*trans*-xanthoxin

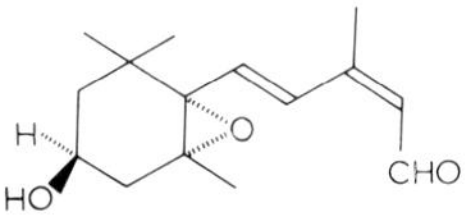

XXXVII.
2-*cis*-xanthoxin

Only the 2-*cis* isomer of xanthoxin is strongly inhibitory to growth, more so than ABA in some tests, and 2-^{14}C-labelled XXXVII has been shown to be rapidly metabolised to (+)-ABA by plant tissues (Taylor and Burden 1973). Thus the inhibitory activity of xanthoxin can probably be attributed to its conversion into ABA. Xanthoxin has been isolated from the shoots and leaves of a number of higher plants, including two fern species (Table 6.9), but it was not detected in seeds of maple, rose hips and pea roots (Firn et al. 1972). It is particularly interesting that it could not be detected in a liverwort *(Marchantia)* and a brown alga *(Fucus)* because attempts to identify abscisic acid in plants of these classes have also failed.

Three questions are raised by these experiments: (i) Is ABA derived from carotenoids? (ii) Is xanthoxin derived from carotenoids in vivo by a photolytic process? (iii) Is xanthoxin a precursor of ABA? These questions are discussed below.

Table 6.9.

Contents of xanthoxin isomers in plant extracts estimated by the GLC of *O*-acetyl derivatives (data from Firn 1974)

Plant species	Tissue extracted	2-*cis* Xanthoxin µg/kg	2-*trans* Xanthoxin µg/kg
Bean *(Phaseolus vulgaris)* cv. Canadian Wonder	1st leaves	48	174
Pea *(Pisum sabivum)* cv. Alaska	Shoots	35	49
Buttercup (*Ranunculus* sp.)	Shoots	13	21
Bracken *(Pteridium aquilinum)*	Pinnules	5	48
Fern *(Nephrolepsis exaltata)*	Pinnules	48	71

The experiments with labelled mevalonic acid do not differentiate between a pathway for the biosynthesis of ABA involving carotenoids (40 carbon atoms) and one with 15-carbon and smaller precursors, but an experiment carried out by the late D.R. Robinson indicates that ABA is formed in avocado fruit by a pathway which does not involve a carotenoid.

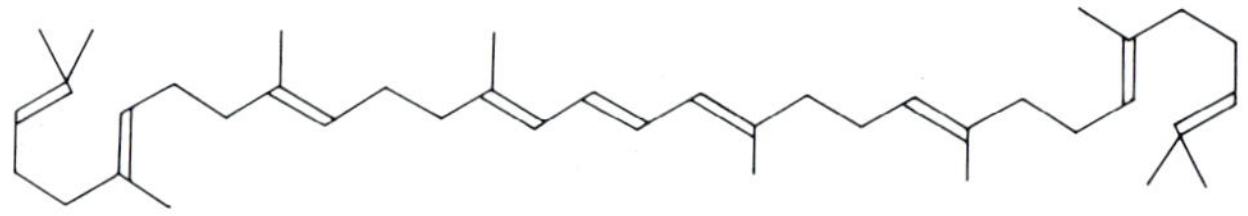

XXXVIII

Phytoene (the "all *trans*" isomer)

The uncyclised precursor of carotenoids, phytoene (XXXVIII), was labelled with ^{14}C and it was fed to an avocado fruit with [2-^{3}H]-mevalonolactone. The [^{3}H]mevalonate was incorporated into ABA, but although the phytoene was incorporated into carotene, no ^{14}C was present in the ABA (Scheme 6.12). As slices of mesocarp tissue were used, the phytoene may not have penetrated into the cellular compartment in which the ABA was formed or, alternatively, other isomers of phytoene may be involved. The experiment does not rule out the possibility that a portion of the ABA in an illuminated leaf is formed by photolysis of violaxanthin.

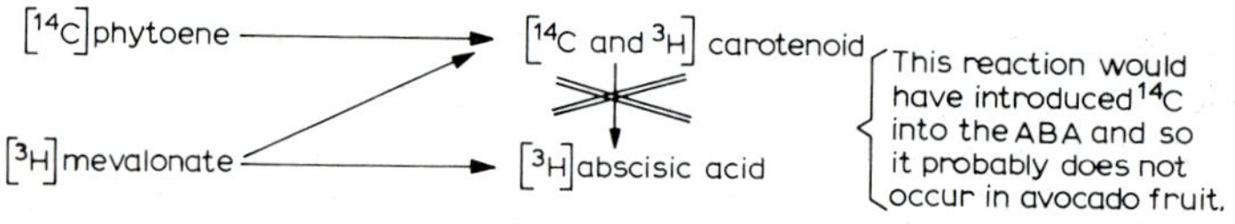

Scheme 6.12.

The presence of XXXIV and XXXV and the presence of both the 2-*cis* and 2-*trans* isomers of xanthoxin in leaves in similar ratios to those found in the photolysis experiments indicates that at least some violaxanthin may be degraded by light in vivo.

The synthesis of ABA by wheat leaves wilted in darkness and the incorporation of [^{14}C] mevalonate into ABA by avocado fruit incubated in darkness shows that light is not an obligatory requirement for the formation of the ABA. Also Firn and Friend (1972) have shown that violaxanthin can be oxidised to give xanthoxin (XXXVII) and a trace of XXXV by a soybean lipoxygenase system; this result provides a model of a possible enzymic reaction for the formation of xanthoxin from violaxanthin without the direct involvement of a photolytic cleavage.

This in no way divorces the biosynthesis of xanthoxin from the indirect effects of illumination. Indeed Firn (1974) has demonstrated that xanthoxin synthesis in etiolated pea seedlings is regulated by phytochrome which in turn shows that xanthoxin synthesis is under enzymic control in this system, rather than being brought about entirely by photolysis. He was also able to show that, while xanthoxin contents were much higher in seedlings illuminated with red light, the kinetics of red light inhibition of growth permitted a rejection of the hypothesis that xanthoxin controlled the rate of internode elongation in dwarf peas (Table 6.10).

Table 6.10.

The relative growth rates of dwarf pea internodes after exposure to red light and xanthoxin content (data from Firn 1974)

Time after 10-min exposure to red light (hr)	Relative growth rate	Relative xanthoxin content
0	100	100
3.5	46	100
9.5	73	95
24	113	175

While the direct experimental evidence available at present suggests that photolytic conversion of violaxanthin into ABA is not the major pathway of ABA biosynthesis, the possibility that ABA can be formed by plants, via carotenoid intermediates, by enzymic processes cannot be discounted.

Xanthoxin could be an intermediate for either mechanism of formation of ABA and it must be emphasised that xanthoxin could well be an intermediate in a separate biosynthetic pathway involving 15-carbon, and smaller precursors. However, the presence of 2-*trans*-xanthoxin in larger amounts than the 2-*cis*-isomer in most plants is in accord with the suggestion that xanthoxin is derived from a carotenoid source.

6.9.2. Experiments with (+)-[^{14}C] xanthoxin

The synthesis of (+)-2-*cis*-[2-^{14}C] xanthoxin has been carried out by Taylor and Burden (1973) who added the necessary two-carbon moiety to the optically active "butenone" (XXXV) formed during the oxidation of natural violaxanthin. When fed to cut shoots of tomato and bean, some 11% and 7% respectively of the xanthoxin were recovered as abscisic acid after 8 hr. Furthermore, the metabolite of ABA, phaseic acid, also became labelled. Thus it has been established that exogenous 2-*cis*-xanthoxin can be metabolised to ABA, but the physiological significance of this remains unclear. In this connection Burden et al. (1971) have shown that the ABA content of dark-grown dwarf pea seedlings was very similar to that of similar seedlings treated with red light, whereas both the the 2-*cis*- and the 2-*trans*-xanthoxin contents of the illuminated seedlings were some seven times higher than those of the dark-grown ones.

The metabolism of (+)-2-*trans*-[2-^{14}C]xanthoxin (Taylor and Burden 1973) fed to tomato shoots was qualitatively different from that of the 2-*cis*-isomer in that most of the 3.3% of the [^{14}C]2-*trans*-ABA was present in a bound form. This compares with 4.0% bound and 10.8% free acid for [^{14}C]ABA. Furthermore 4.4% of the [^{14}C]2-*cis*-xanthoxin was recovered as phaseic acid whereas 4% of the [^{14}C]2-*trans*-xanthoxin was present as bound 2-*trans*-xanthoxin acid (XXXIX).

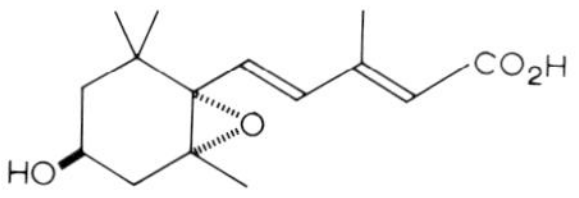

XXXIX

6.9.3. *Xanthoxin content of plants*

The greater abundance of the apparently biologically inactive 2-*trans*-xanthoxin in comparison with the 2-*cis*-isomer is one of the most intriguing problems in this area of in-

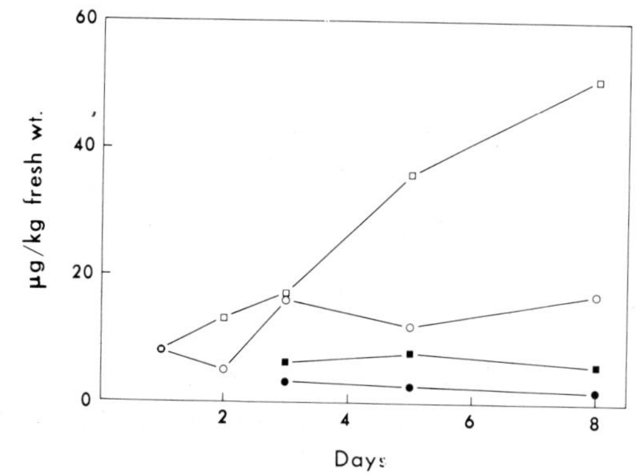

Figure 6.17. Content of xanthoxin isomers in pea seedlings grown in complete darkness or red light. ○, 2-*cis*-xanthoxin in red light grown plants. ●, 2-*cis*-xanthoxin in dark grown plants. □, 2-*trans*-xanthoxin in red light grown plants. ■, 2-*trans*-xanthoxin in dark grown plants. The scale of the abscissa refers to the time from the commencement of red illumination. Data from Anstiss et al. (1975).

vestigation. In pea seedlings there appears to be a greater build up of 2-*trans*-xanthoxin after illumination than of the 2-*cis* isomer (Figure 6.17). This could be accounted for by the conversion of the 2-*cis* isomer into ABA but in the experiments of Burden et al. (1971) the ABA content of light- and dark-grown seedlings was similar while the 2-*cis*-xanthoxin content (measured by bioassay) was higher in the former. The function of xanthoxin in the hormone economy of plants remains enigmatic.

6.10. References

Anstis, P.J.P., Friend, J. and Gardner, D.C.J. (1975) Phytochemistry, 14, 31–35.

Archer, B.L., Barnard, D., Cockbain, E.G., Cornforth, J.W., Cornforth, R.H. and Popjak, G. (1966) Proc. roy. Soc., B. 163, 519–523.

Aspinall, D., Paleg, L.G. and Addicott, F.T. (1967) Aust. J. biol.Sci., 20, 869–882.

Bennet-Clark, T.A., Tambiah, M.S. and Kefford, N.P. (1952) Nature (Lond.), 169, 542–543.

Bowen, M.R. and Hoad, G.V. (1968) Planta, 81, 64–70.

Browning, G. (1973) J. Hort. Sci., 48, 297–310.

Bu'Lock, J.D., Austin, D.J., Snatzke, G. and Hruban, L. (1970) J. Chem. Soc. chem. Commun., 255–256.

Burden, R.S., Firn, R.D., Hiron, R.W.P., Taylor, H.F. and Wright, S.T.C. (1971) Nature (Lond.), 234, 95–96.

Cheng, C.Y. and Schraudolf, H. (1974) Z. Pflanzenphysiol., 71, 366–369.

Coombe, B.C. and Hale, C.R. (1973) Plant Physiol., 51, 6629–6634.

Cornforth, J.W. and Ross, P. (1970) J. Chem. Soc. chem. Commun., 1395–1396.

Cornforth, J.W., Milborrow, B.V. and Ryback, G. (1965a) Nature (Lond.), 206, 715.

Cornforth, J.W., Milborrow, B.V., Ryback, G. and Wareing, P.F. (1965b) Nature (Lond.), 205, 1269–1270.

Cornforth, J.W., Milborrow, B.V. and Ryback, G. (1966) Nature (Lond.), 210, 627–628.

Cornforth, J.W., Draber, W., Milborrow, B.V. and Ryback, C. (1967) J. Chem. Soc. chem. Commun., 114–116.

Cummins, W.R. and Sondheimer, C. (1973) Planta, 111, 365–369.

Cummins, W.R., Kende, H. and Raschke, K. (1971) Planta, 99, 347–351.

Davis, L.A. and Addicott, F.T. (1972) Plant Physiol., 49, 644–648.

Davis, L.A., Heinz, D.E. and Addicott, F.T. (1968) Plant Physiol., 43, 1389–1394.

Davison, R.M. and Young, H. (1973) Planta, 109, 95–98.

Dörffling, K. and Böttger, M. (1968) Planta, 80, 299–308.

El Antably, H.M.M., Wareing, P.F. and Hillman, J. (1967) Planta, 73, 74–90.

Firn, R.D. (1974) in: Plant Growth Substances 1973, Hirokawa, Tokyo, pp. 1225–1230.

Firn, R.D. and Friend, J. (1972) Planta, 103, 263–266.

Firn, R.D., Burden, R.S. and Taylor, H.F. (1972) Planta, 102, 115–126.

Gaskin, P. and MacMillan, J. (1968) Phytochemistry, 7, 1699–1701.

Ginzberg, C. (1973) J. exp. Bot., 24, 558–566.

Goad, L.J. (1970) Biochem. Soc. Symp., 29, 45–77.

Goldschmidt, E.E., Goren, R., Even-Chen, Z. and Bittner, S. (1973) Plant Physiol., 51, 879–882.

Good, J.E.G. (1974) Planta, 116, 45–54.

Goodwin, T.W. (1958) Biochem. J., 70, 612–617.

Gray, R.T., Mallaby, R., Ryback, G. and Williams, V P. (1974). J. Chem. Soc. Perkin, II, 919–924.

Hanks, G.R. (1974) Z. Pflanzenphysiol., 71, 63–66.

Harrison, M.A. and Walton, D.C. (1975) Plant Physiol., 56, 250–254.

Hartung, W. and Behl, R. (1975a) Planta, 122, 53–59.

Hartung, W. and Behl, R. (1975b) Planta, 122, 61–65.

Hiron, R.W.P. and Wright, S.T.C. (1973) J. exp. Bot., 24, 769–781.
Hoad, G.V. (1967) Life Sci., 6, 1113–1118.
Hoad, G.V. (1973a) Proc. Res. Inst. Pom. Skierniewice, 3, 17–30.
Hoad, G.V. (1973b) Planta, 113, 367–372.
Hocking, T.J., Hillman, J.R. and Wilkins, M.B. (1972) Nature (Lond.), 235, 124–125.
Ingersoll, R.B. and Smith, O.E. (1970) Plant Physiol., 45, 576–578.
Ingersoll, R.B. and Smith, O.E. (1971) Plant Cell Physiol., 12, 301–309.
Isogai, Y., Komoda, Y. and Okamoto, T. (1970) Chem. Pharm. Bull., 18, 1872–1879.
Jenkins, P.A. and Shepherd, K.R. (1972) New Phytol., 71, 501–511.
Jones, R.J. and Mansfield, T.A. (1970) J. exp. Bot., 21, 714–719.
Koshimizu, K., Fukui, H., Mitsui, T. and Ogawa, Y. (1966) Agr. Biol. Chem., 30, 941–943.
Kriedemann, P.E., Loveys, B.R., Fuller, G.L. and Leopold, A.C. (1972) Plant Physiol., 49, 842–847.
Kundu, K.K. and Audus, L.J. (1974) J. exp. Bot., 25, 479–489.
Lenton, J.R., Bowen, M.R. and Saunders, P.F. (1968) Nature (Lond.), 220, 86–87.
Lenton, J.R., Perry, V.M. and Saunders, P.F. (1971) Planta, 96, 271–280.
Le Page-Degivry, M.T., Bulard, C. and Milborrow, B.V. (1969) C.R. Acad. Sci. (Paris), Ser. D, 271, 482–484.
Loveys, B.R., Brien, C.J. and Kriedemann, P.E. (1975) Physiol. Plant., 33, 166–170.
Mallaby, R. (1974) in: Studies on the Chemical and Biological Synthesis of Abscisic Acid, Ph.D. Thesis, University of London.
Mallaby, R. and Ryback, G. (1972) J. Chem. Soc. Perkin, H, 919–921.
MacMillan, J. and Pryce, R.J. (1968). J. Chem. Soc. chem. Commun., 124–126.
McCready, C.C. and Jacobs, W.P. (1973), New Phytol., 62, 19–34.
Milborrow, B.V. (1963) Nature (Lond.), 199, 716–717.
Milborrow, B.V. (1967) Planta, 76, 93–113.
Milborrow, B.V. (1968) in: Biochemistry and Physiology of Plant Growth Substances, F. Wightman and G. Setterfield (eds.), Runge Press, Ottawa, pp. 1531–1545.
Milborrow, B.V. (1969a) Sci. Prog. (Oxford), 57, 533–558.
Milborrow, B.V. (1969b) J. Chem. Soc. chem. Commun., 966–967.
Milborrow, B.V. (1970) J. exp. Bot., 21, 17–29.
Milborrow, B.V. (1971) in: Aspects of Terpenoid Chemistry and Biochemistry, T.W. Goodwin (ed.), Academic Press, London, pp. 137–151.
Milborrow, B.V. (1972a) in: Plant Growth Substances 1970, D.J. Carr (ed.), Springer, Heidelberg, pp. 281–290.
Milborrow, B.V. (1972b) Biochem. J., 128, 1135–1146.
Milborrow, B.V. (1974a) Ann. Rev. Plant Physiol., 25, 259–307.
Milborrow, B.V. (1974b) Phytochemistry, 13, 131–136.
Milborrow, B.V. (1974c) in: Plant Growth Substances 1973, Hirokawa, Tokyo, pp. 384–395.
Milborrow, B.V. (1975a) Phytochemistry, 14, 123–128.
Milborrow, B.V. (1975b) Phytochemistry, 14, 1045–1053.
Milborrow, B.V. (1975c) Phytochemistry, 14, 2403–2405.
Milborrow, B.V. (1976) in: Perspectives in Experimental Biology, Vol. 2, N. Sunderland (ed.), Pergamon Press, New York, pp. 111–124.
Milborrow, B.V. and Garmston, M. (1973) Phytochemistry, 12, 1597–1608.
Milborrow, B.V. and Noddle, R.C. (1970) Biochem. J., 119, 727–734.
Milborrow, B.V. and Robinson, D.R. (1973) J. exp. Bot., 24, 537–548.
Mittelheuser, C.J. and Van Steveninck, R.F.M. (1969) Nature (Lond.), 221, 281–282.
Mittelheuser, C.J. and Van Steveninck, R.F.M. (1971) Planta, 97, 83–86.
Most, B.H., Gaskin, P. and MacMillan, J. (1970) Planta, 92, 41–49.
Mousseron-Canet, M., Mani, J.C., Dalle, J.P. and Olivé, J.L. (1966) Bull. Soc. Chim. Fr., 3874–3878.
Mousseron-Canet, M., Mani, J.C., Durand, B., Nitsch, J., Dornand, J. and Bonnafous, J.C. (1970) C.R. Acad. Sci. (Paris) Ser. D, 270, 1936–1939.

Nicholls, P.B. and Paleg, L.G. (1963) Nature (Lond.), 199, 823–824.
Noddle, R.C. and Robinson, D.R. (1969) Biochem. J., 112, 547–548.
Ohkuma, K. (1965) Agr. Biol. Chem., 29, 962–964.
Ohkuma, K. (1966) Agr. Biol. Chem., 30, 434–437.
Ohkuma, K., Lyon, J.L., Addicott, F.T. and Smith, O.E. (1963) Science, 142, 1592–1593.
Ohkuma, K., Addicott, F.T., Smith, O.E. and Thiessen, W.E. (1965) Tetrahedron Lett., 2529–2534.
Oritani, T. and Yamashita, K. (1970a) Agr. Biol. Chem., 34, 108–114.
Oritani, T. and Yamashita, K. (1970b) Agr. Biol. Chem., 34, 1184–1190.
Oritani, T. and Yamashita, K. (1972) Agr. Biol. Chem., 36, 362–369.
Pilet, P.E. (1975a) Planta, 122, 299–302.
Pilet, P.E. (1975b) Plant Sci. Lett., 5, 137–140.
Popják, G. and Cornforth, J.W. (1966) Biochem. J., 101, 553–568.
Pryce, R.J. (1971) Planta, 97, 354–357.
Pryce, R.J. (1972) Phytochemistry, 11, 1759–1761.
Raschke, K. (1975) Ann. Rev. Plant Physiol., 26, 309–340.
Robinson, D.R. and Ryback, G. (1969) Biochem. J., 113, 895–897.
Rothwell, K. and Wain, R.L. (1964) in: Regulateurs Naturels de la Croissance Végétale, C.N.R.S., Paris, pp. 363–385.
Ryback, G. (1972) J. Chem. Soc. chem. Commun., 1190–1191.
Saunders, P.F., Harrison, M.H. and Alvin, R. (1975) in: Plant Growth Substances 1973, Hirokawa, Tokyo, pp. 871–881.
Seeley, S.D. and Powell, L.E. (1970) Analyt. Biochem., 35, 530–533.
Shindy, W.W., Asmundson, C.M., Smith, O.E. and Kumamoto, J. (1973) Plant Physiol., 52, 443–447.
Simpson, G.M. and Saunders, P.F. (1972) Planta, 102, 272–276.
Simpson, G.M. and Wain, R.L. (1961) J. exp. Bot., 12, 207–216.
Sivori, E.M., Sonvico, V. and Fernandez, N.O. (1971) Plant Cell Physiol., 12, 993–996.
Sondheimer, E., Michniewicz, B.M. and Powell, L.E. (1969) Plant Physiol., 44, 205–209.
Sondheimer, E., Galson, E.C., Chang, Y.P. and Walton, D.C. (1971) Science, 174, 829–831.
Tal, M. and Gavish, U. (1973) Aust. J. Agr. Res., 24, 353–361.
Tamura, S. and Nagao, M. (1969) Agr. Biol. Chem., 33, 1357–1360.
Taylor, H.F. and Burden, R.S. (!970) Nature (Lond.), 277, 302–304.
Taylor, H.F. and Burden, R.S. (1972) Proc. roy. Soc. B, 180, 317–346.
Taylor, H.F. and Burden, R.S. (1973) J. exp. Bot., 24, 873–880.
Taylor, H.F. and Smith, T.A. (1967) Nature (Lond.), 215, 1513–1514.
Tinelli, E.T., Sondheimer, E. and Walton, D.C. (1973) Tetrahedron Lett., 139–140.
Tucker, D.J. and Mansfield, T.A. (1971) Planta, 98, 157–163.
Tucker, D.J. and Mansfield, T.A. (1972) Planta, 102, 140–151.
Valio, I.F.M., Burden, R.S. and Schwabe, W.W. (1969) Nature (Lond.), 223, 1176–1178.
Van Overbeek, J., Loeffler, J.E. and Mason, M.I.R. (1967) Science, 156, 1497–1499.
Van Staden, J. and Borman, C.H. (1970) J. S. Afr. Bot., 36, 9–12.
Varner, J.E. and Johri, M.M. (1968) in: Biochemistry and Physiology of Plant Growth Substances, F. Wightman and G. Setterfield (eds.), Runge Press, Ottawa, pp. 793–814.
Veen, H. (1975) Acta Bot. Neer., 24, 55–65.
Walton, D.C. and Sondheimer, E. (1972) Plant Physiol., 49, 285–289.
Wilkins, M.B., Gibbons, G.S.B. and Shaw, S. (1972) in: Plant Growth Substances 1970, D.J. Carr (ed.), Springer, Heidelberg, pp. 717–724.
Wilkins, H. and Wain, R.L. (1974) Planta, 121, 1–8.
Wright, S.T.C. (1961) Nature (Lond.), 190, 699–700.
Wright, S.T.C, (1968) in: Biochemistry and Physiology of Plant Growth Substances, F. Wightman and G. Setterfield (eds.), Runge Press, Ottawa, pp. 521–542.
Wright, S.T.C. (1975) J. exp. Bot., 26, 161–174.
Wright, S.T.C. and Hiron, R.W.P. (1969) Nature (Lond.), 224, 719–720.
Zeevaart, J.A.D. (1974) Plant Physiol., 53, 644–648.
Zeevaart, J.A.D. and Milborrow, B.V. (1976) Phytochemistry, 15, 493–500.

Phytohormones and Related Compounds – A Comprehensive Treatise, Volume I
Editors: Letham, Goodwin and Higgins

7

Naturally-occurring Plant Growth Regulators other than the Principal Hormones of Higher Plants

D.S. LETHAM

7.1. Introduction

Today auxins, gibberellins, cytokinins, abscisic acid and ethylene are regarded as the principal hormones of higher plants. However numerous other naturally occurring organic compounds are known to promote or inhibit plant growth markedly when supplied exogenously at low concentrations. In this chapter, the identity and biochemistry of compounds of this type are discussed together with the responses they evoke from plants. Nearly all the compounds dealt with may be assigned to one of the following groups:

(1) Compounds produced by lower plants (fungi, algae, liverworts and ferns) to regulate their development. With the exception of lunularic acid, all the compounds discussed regulate sexual reproduction.

(2) Compounds produced by microorganisms which promote higher plant growth.

(3) Compounds which occur in higher plants and promote or inhibit plant growth when supplied exogenously, usually to excised tissues.

Many of the compounds in the third category probably do not normally regulate growth in intact plants. This observation applies particularly to growth inhibitors such as lycoricidinol, lycoricidine, cucurbic acid, *cis*-4-cyclohexene-1,2-dicarboximide, asparagusic acid, *N,N*-dimethyltryptophan and the terpenoid dilactones discussed in sections 7.9.7 and 7.9.8. The only observed responses to these inhibitors are in plant species quite unrelated to the species in which they are known to occur. It is not possible to relate these compounds to any particular aspect of development in the plants where they occur. It is also uncertain if phenolic acids, flavonoids and coumarins, compounds which are widely distributed in higher plants, exert a regulatory role in vivo. Marked responses may be evoked by these substances when supplied exogenously, but it does not follow that the same responses are induced by the endogenous compounds. However, strigol, batatasins and G inhibitors of *Eucalyptus* are compounds produced by higher plants and they definitely appear to regulate plant development in vivo. Strigol is an extremely potent germination stimulant for seeds of certain plant parasites. The first attempts to isolate this compound were reported in 1949; determination of the structure of strigol in 1972 and its synthesis in 1974 were great achievements of the organic chemist.

Two types of growth regulators not discussed herein are, firstly, compounds with regulatory activity which have been detected in plants but are of unknown structure and secondly, microbial metabolites with only phytotoxic activity. However phytotoxins of known structure (e.g. fusicoccin), which under some circumstances promote growth, are reviewed. In the present chapter, it is not possible to discuss growth regulatory compounds usually regarded as vitamins. However it should be emphasized that the vitamins thiamine and *myo*-inositol are essential for the active growth of many tissues in vitro. Thus, in the case of carrot phloem tissue, the only organic compounds required for rapid in vitro growth are sucrose, a cytokinin, an auxin, *myo*-inositol and thiamine (Letham 1967). Thiamine was the only vitamin found essential for growth of tobacco pith tissue (Linsmaier and Skoog 1965). Because there is evidence that thiamine is synthesized in leaves of higher plants and that it moves to the roots where it is required for development, some workers regard it as a hormone (Kruszewski and Jacobs 1974). These workers also found that it moves with strong basipetal polarity in petiole sections of tomato.

During the last 10 years considerable progress has been made in elucidation of the structure and function of compounds which regulate sexual reproduction in fungi and algae. This work is without doubt the highlight of this chapter. Studies of hormonal control of sexual reproduction in fungi and algae have not received the recognition which they merit. This elegant work based on sound physiology and sophisticated modern chemistry has yet to be discussed in the Annual Review of Plant Physiology; the work attracted almost negligible attention at the international conferences on plant growth substances (1963, 1967, 1970 and 1973); in a recent review entitled "fifty years of plant hormone research" (Thimann 1974) fungal and algal hormones are not even mentioned. If this chapter helps to rectify this oversight it will have been worth writing.

7.2. Sexual hormones of fungi and algae

Sexual hormones in several genera of fungi have now been isolated and identified; they are discussed in this section together with identified sexual hormones of algae. Numerous other fungal hormones of unknown structure (e.g. the conjugation hormones of the basidiomycete *Tremella mesenterica* (Reid 1974)) appear to occur, but systematic discussion of these partially characterized compounds is beyond the scope of this chapter.

7.2.1. Hormones involved in mating in yeast

Haploid cells of the yeast *Saccharomyces cerevisiae* can be one of two alternative mating types termed *a* and α. Cells of both types can reproduce vegetatively by mitotic division and grow indefinitely in the haploid condition. When in close contact, however, cells of opposite mating type conjugate to yield a diploid zygote which, by mitotic division, gives rise to a clone of stable diploid cells. Cells of type α excrete a diffusible substance which inhibits division in cells of the opposite mating type *a* and causes these cells to elongate (Duntze et al. 1970). This substance, termed α factor, appears to be involved in the mating process. Factor α specifically inhibits the synthesis of DNA in *a*-type cells by interfering with initiation of DNA replication (Bucking-Throm et al. 1973).

The α factor has been purified from culture filtrates of α cells (Duntze et al. 1973). It is a low-molecular-weight peptide consisting of 13 amino acids all of which occur normally in proteins (Duntze et al. 1973; Stotzler and Duntze 1973). The factor readily forms a stable copper complex and was isolated in this form. The biosynthesis of α factor has been studied (Scherer et al. 1974). It is formed mainly during the stationary phase of the growth curve and appears to be produced by ribosomal translation of mRNA and not by an extraribosomal mechanism of peptide synthesis.

Other factors which may regulate the mating processes in *S. cerevisiae* have been detected (Yanagishima 1969, 1971; Shimoda and Yanagishima 1973; Yanagishima et al. 1974). These workers have shown that *a* and α cells also produce mating type-specific factors which are extractable by methylene chloride and which cause enlargement of opposite mating type cells specifically. A factor produced by α-type cells which causes expansion of *a* type cells has been identified as *n*-octanoic acid and termed α-hormone (Sakurai et al. 1974a, b). α Type cells also secrete a compound which enhances sexual agglutinability of *a* type cells. The factor was partially purified and appears to be a peptide (Sakurai et al. 1974a).

7.2.2. Trisporic acids

The majority of fungi in the order Mucorales are heterothallic, zygospore formation occurring only between individuals of opposite mating types, termed (+) and (–), of the same species. When a (+) and a (–) strain grow into close proximity, sexual hyphae (zygophores) are formed in both mycelia without physical contact occurring between the two sexual partners. The aerial zygophores grow towards each other (zygotropism), contact in mated pairs and then swell at their tips to form progametangia. Cross-wall formation then occurs in these to yield gametangia which finally fuse to give rise to zygospores. By experiments in which the (+) and (–) mating types were separated by a collodion membrane, Burgeff (1924) showed that diffusible substances are responsible for the formation of the zygophores in *Mucor mucedo*. This was the first demonstration of chemical control of development of sex organs in fungi.

Sexual reproduction in Mucorales is often accompanied by a very marked increase in carotenoid content and this effect is particularly striking in *Blakeslea trispora* (Van den Ende 1968, and references therein). From a mixed culture of (+) and (–) strains of this species, three closely related terpenoid carboxylic acids, termed trisporic acids A, B and C, were isolated by Caglioti et al. (1964). These acids markedly stimulated carotenoid synthesis in a (–) strain of the fungus, but had little effect on the (+) strain (Van denEnde 1968).

Purification of zygophore-inducing substances of *Mucor mucedo* commenced about 1956. Considerable progress was made by Plempel (1963a, b; 1965) and in 1969 an active compound purified from mixed cultures of the (+) and (–) strains (Gooday 1968a) was identified as trisporic acid C (Austin et al. 1969). From culture fluid derived from mixed cultures of the (+)- and (–)-strains of *Blakeslea trispora,* Van den Ende (1967, 1968) isolated two compounds which were active in the induction of zygophores in both the (+) and (–) strains of *Mucor mucedo.* The compounds were identified as trisporic acids B and C (Van den Ende 1968). A bioassay based on zygophore counts with (–)-*Mucor mucedo* detected 0.01 μg of trisporic acid C (Gooday 1968b; Bu'Lock et al. 1972). Trisporic acid-induced zygophore production is accompanied by suppression of formation of the asexual sporangiophores.

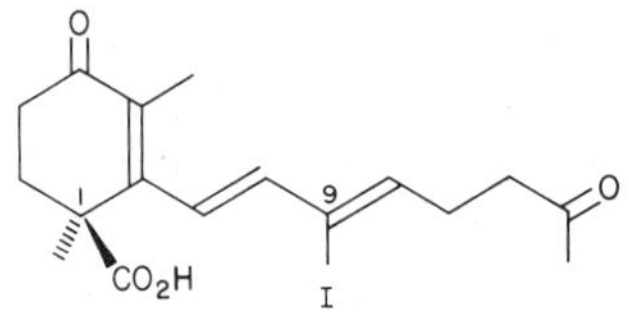

I

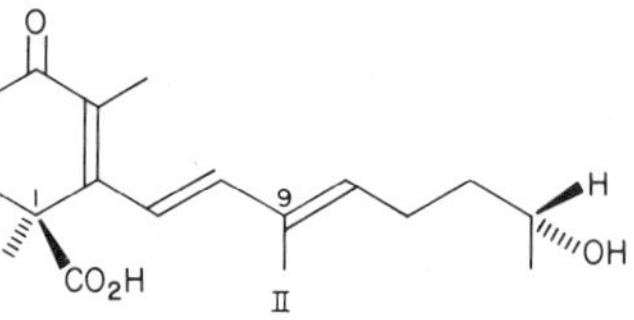

II

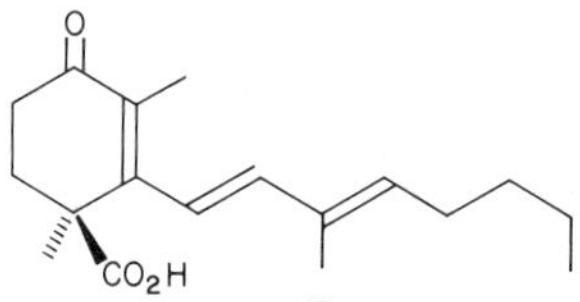

III

The structures of trisporic acid B (I) and trisporic acid C (II) have been determined (Gaglioti et al. 1966, Cainelli et al. 1967, Bu'Lock et al. 1970), while a probable structure (III) has been presented for trisporic acid A (Bu'Lock et al. 1972), which appears to be a minor hormone in *Blakeslea trispora.* Natural trisporic acids B and C both comprise two geometrical isomers, 9-*cis* and 9-*trans;* these four compounds all have a 7-*trans* structure. The 9-*cis/trans* pairs are separable as the methyl esters, from which the free acids have been regenerated for bioassays (Bu'Lock et al. 1972). The 9-*trans* isomer of trisporic acid B methyl ester has been elegantly synthesized (Edwards et al. 1971).

The 9-*cis* and 9-*trans* isomers of trisporic acids B and C were tested for hormone activity using both (+) and (–) *Mucor mucedo.* Both mating types responded to all of these trisporic acids, but the response from the (–) type exceeded that from the (+) type with each acid. Trisporic acid A was considerably less active than acids B and C (Bu'Lock et al. 1972).

The metabolism of methyl esters of trisporic acids of *Blakeslea trispora* has been investigated (Bu'Lock et al. 1972). The (+) and (–) strains metabolized these esters differently; in (–) mycelium, ester hydrolysis and -COOH reduction predominated, but in (+) mycelium, oxidative degradation was the principal modification.

The biosynthesis of trisporic acids in mixed (+) and (–) cultures of *Blakeslea trispora* has been studied (Bu'Lock 1973; Bu'Lock et al. 1974a). They appear to be formed by cleavage of β-carotene; retinal, the C_{18} ketone IV, and trisporols (the primary alcohols corresponding to trisporic acids) are intermediates in the biosynthesis.

Biosynthesis of trisporic acids in heterothallic species normally requires the (+) and (–) mycelia to be in diffusion contact via the substrate. The identity of the substances

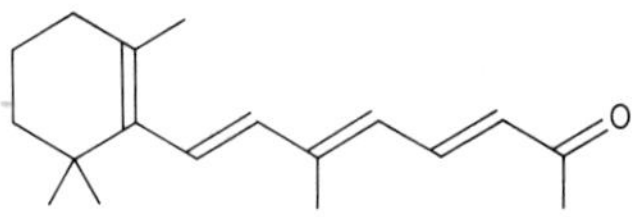

IV

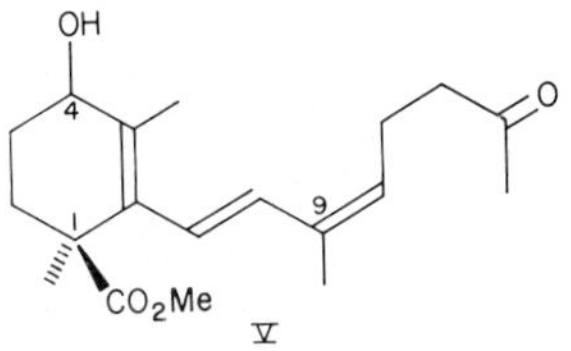

V

involved in the interaction between mating types is of great interest. Separate (+) and (–) *Blakeslea trispora* cultures were found to produce small amounts of trisporic acids when stimulated by certain factors derived from the opposite mating type (Sutter et al. 1973). These workers suggested the factors were precursors of the trisporic acids. Subsequent experimental work is in accord with this view. Isolated mycelia of (+) and (–) mating types are considered to produce specific "prohormones", termed P^+ and P^- respectively; only mycelium of opposite mating type is able to convert the prohormones into trisporic acids. Bu'Lock et al. (1974b) have chemically characterized the prohormones of *Blakeslea trispora.* P^+ possesses structure V with predominantly *cis* stereochemistry at C-9. P^- is a mixture of closely related compounds of which the trisporols (VI, VII) are the most

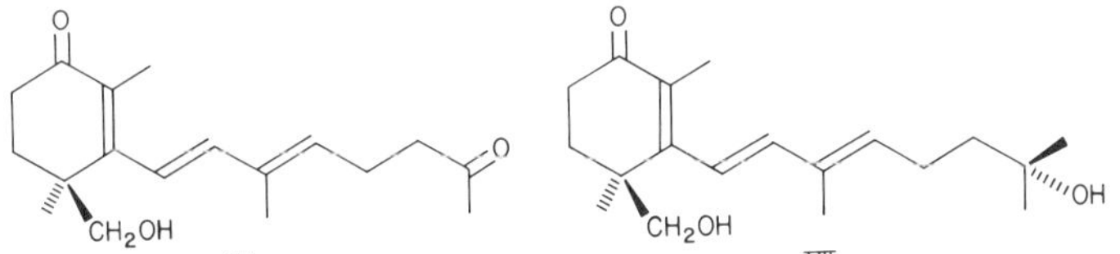

VI VII

active. The prohormones, like the trisporic acids formed in mixed cultures, are synthesized via the C_{18} ketone IV. Hence, in summary, synthesis of trisporic acids in mated cultures involves precursor production and secretion into the substrate by both (+) and (–) strains, and then sex-specific conversion to the acids by the mating partners. Production of precursors appears to be stimulated by trisporic acids suggesting that the hormone synthesis is self-amplifying (Werkman and Van den Ende 1973, 1974).

Since trisporic acids are made by both (+) and (–) mating types, Sutter et al. (1973) have suggested they are internal regulators and not compounds regulating sexual reproduction after translocation from one type of cell to another. However trisporic acids are released into the culture medium. Hence by diffusion through the medium they can induce zygophores and coordinate the activities of hyphae at some distance from those involved in the primary sexual interaction. However it is noteworthy that zygophore formation in mated *Mucor mucedo* cultures may also be induced by hormones other than trisporic acids. These unidentified compounds are volatile (Mesland et al. 1974).

Little is known of the mechanism of zygotropism. Two compatible zygophores grow steadily towards each other from distances up to 2 mm apart, and finally meet and adhere. Volatile hormones are almost certainly responsible for zygotropism which would require two complementary substances. Involvement of readily oxidised volatiles has been indicated (Plempel 1962) but they have yet to be identified. However it is likely the "zygotropic hormones" are in fact the P^+ and P^- hormones as suggested by Bu'Lock et al. (1974b).

7.2.3. Sirenin

The male and female gametangia of the water mould *Allomyces* are borne on the same thallus and discharge their motile gametes into the ambient liquid. The orange-coloured male gametes are attracted to the larger and colourless female gametes. Fusion of a male with a female gamete yields a motile, biflagellate zygote. Female gametes secrete a substance which attracts the male gametes, production of the compound beginning before the female gametes emerge from the gametangia. This attractant termed sirenin was purified by Machlis et al. (1966). The work was dependent on an assay in which solutions to be tested were placed on one side of a membrane which contacted a suspension of male gametes on the other. The male gametes attached to the membrane in proportion to the concentration of sirenin (Machlis 1958, 1973a). Sirenin was extracted with methylene dichloride from water in which large numbers of female gametes had been cultured. To obtain a suspension composed almost entirely of female gametes, the normal hermaphroditic habit of *Allomyces* needed to the circumvented. This was achieved by selecting a hybrid (derived from crosses between *Allomyces sarbuscula* and *A. javanicus*) in which over 95% of the gametangia produced were female (Machlis et al. 1966).

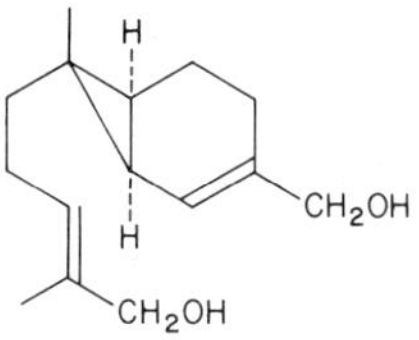

VIII

The chemotactic hormone sirenin is a sesquiterpene of structure VIII in which there are two primary allylic alcohol moieties (Nutting et al. 1968). DL-Sirenin was synthesized (Bhalerao et al. 1970) and later D- and L-sirenin were prepared (Plattner and Rapoport 1971). L-Sirenin was active at 0.3 n*M* but D-sirenin and several analogues of sirenin were inactive (Machlis 1973b). The side chain allylic alcohol group is essential for activity and the stereochemistry of the attachment of the six carbon moiety to the cyclopropane ring is also critical for function. When sirenin is taken up by male gametes, it is converted into a chemotactically inactive compound (Carlile and Machlis 1965; Machlis 1973b).

7.2.4. Antheridiol

In most species of *Achlya* (water moulds), oogonia and antheridia are produced on the same thallus. However there are at least two species *(Achlya ambisexualis, Achlya bisexualis)* which are heterothallic and when mycelia of opposite sex grow in close proximity, antheridial branches will develop on one ("male") and grow toward the oogonial initials of the other ("female"). A hormone (termed hormone A) produced by the vegetative hyphae of the female initiates the sexual reaction by inducing antheridial branches on the male. The sexually activated male then secretes hormone B which induces formation of oogonial initials in the female (Raper 1940, 1942 and 1957). Other hormones (C and D) were also detected by Raper but are not discussed herein as they have not been characterized chemically.

Purification of hormone A was first attempted by Raper and Haagen-Smit (1942) who obtained an amorphous substances active at a concentration of 1 ng/l. From 2,000 l of *Achlya bisexualis* culture liquid, McMorris and Barksdale (1967) obtained 10 mg of crystalline hormone A which they termed antheridiol. Structural studies indicated that antheridiol was a steroid with the formula IX (Arsenault et al. 1968) and this was con-

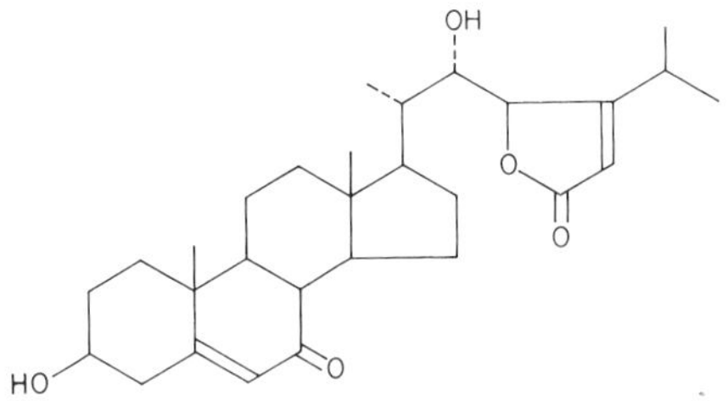

IX

firmed by chemical synthesis (Edwards et al. 1969). This synthesis yielded two of the four possible isomers of IX which arise from the presence of asymmetric carbons at positions 22 and 23. One of the synthetic compounds had hormonal activity equal to that of antheridiol. Further structural studies indicated that antheridiol possesses (22*S*, 23*R*)-stereochemistry (Green et al. 1971).

Antheridiol was active at a concentration of 6 ng/l (10^{-11} *M*) when assayed with *Achlya ambisexualis.* It elicits a sequence of responses from the male plant; these are formation of antheridial branches, a chemotropic response from the branches which appear to grow towards a source of antheridiol, the development of the antheridia, and secretion of hormone B. The concentration of antheridiol determines the number of branches initiated; with a high concentration as many as 40 branches may be initiated on a single hypha about 3 mm in length. The time which elapses between addition of antheridiol and the appearance of branches is surprisingly short–about 40 min when an optimal concentration is used (Barksdale 1969). Stereoisomers of antheridiol tested for ability to induce branching in *Achlya* were much less active than antheridiol (Barksdale et al. 1974).

Two compounds (termed oogoniol-1 and oogoniol-2) with hormone B activity have been isolated from culture liquids of *Achlya heterosexualis.* Like antheridiol, both are 7-keto steroids; however they do not possess a lactone ring in the side chain as does antheridiol (McMorris et al. 1975).

7.2.5. Chemotactic factors of algae

Cellular chemotaxis also occurs in algae. Female gametes of the marine brown alga *Ectocarpus siliculosus* secrete an attractant for the male gametes. This was purified (Muller 1968) and identified as 1-(but-1-enyl)cyclohepta-2,5-diene, X (Muller et al. 1971). This highly volatile compound was removed from the cultures by a stream of purified air, condensed at –80 °C and purified by gas chromatography. It has been named ectocarpene and synthesized (Jaenicke et al. 1973).

Spermatozoids of the brown seaweed *Fucus serratus* are attracted by the eggs. The attractant has been identified as the conjugated hydrocarbon 1,3,5-octatriene and termed fucoserratene (Muller and Jaenicke 1973). A comparison of the hormone with all four

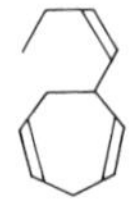

X

stereoisomers of 1,3,5-octatriene, established that fucoserratene is 1,*trans*-3,*cis*-5-octatriene (Jaenicke and Seferiadis 1975).

A third chemotactic hormone has been isolated from *Cutleria multifida* eggs, termed multifidene and identified as 4-vinyl-5-(but-*cis*-1-enyl)cyclopentene (Jaenicke et al. 1974). It appears to be responsible for the attraction of large numbers of male gametes to the female gametes.

7.3. Antheridogens

Sexual reproduction of ferns occurs on a small heart-shaped gametophyte (a prothallus) produced by germination of the asexual spore. Prothalli produce male and female organs termed antheridia and archegonia respectively. A substance produced by the prothalli of *Pteridium aquilinum* (bracken fern) was found to induce formation of antheridia of this fern (Dopp 1950). The antheridia arise from single cells of the prothalli by a series of unequal cell divisions. The natural substance extracted by Dopp which could initiate this cell division was later termed antheridogen A* and was purified (Pringle et al. 1960). The effect of environmental factors on the level of the antheridogen in *Pteridium aquilinum* culture media has been carefully assessed by Voeller (1964a). Addition of coconut milk to the culture medium markedly enhanced this level; auxin, gibberellin and cytokinin could not substitute for coconut milk (Voeller 1964a).

Purified antheridogen A induced antheridia on the prothalli of the fern *Onoclea sensibilis* at a concentration of less than 0.1 μg/l of culture medium. The compound was not species-specific and nearly all tested members of the Aspidiaceae, Pteridaceae and Blechnaceae responded to antheridogen A by forming normal antheridia under conditions in which none were formed in control cultures. However, some families, e.g. Schizaeaceae and Osmundaceae, appeared to be unresponsive (Voeller 1964a). Unfortunately the structure of antheridogen A was not elucidated, but it appeared to be a complex carboxylic acid (Pringle 1961). It has been proposed that antheridogen A should now be referred to as A_{Pt} (Naf 1968).

The prothalli of *Anemia phyllitidis* and *Lygodium japonicum,* two species of the family Schizaeaceae, also produce antheridogens which have been termed A_{An} and A_{Ly}; A_{Pt}, A_{An} and A_{Lv} are different chemical entities (Naf 1968). A_{An} induced antheridia on *A. phyllitidis* and *L. japonicum* but was inactive towards tested species of the families Polypodiaceae, Osmundaceae and Cyatheaceae (Naf 1968). Because A_{An} was stable and could be detected by two sensitive bioassays, it was selected by Endo et al. (1972) for isolation and characterization. The active compound, purified from the culture media of

* The term antheridiogen has also been proposed for compounds which induce antheridium formation (Endo et al. 1972).

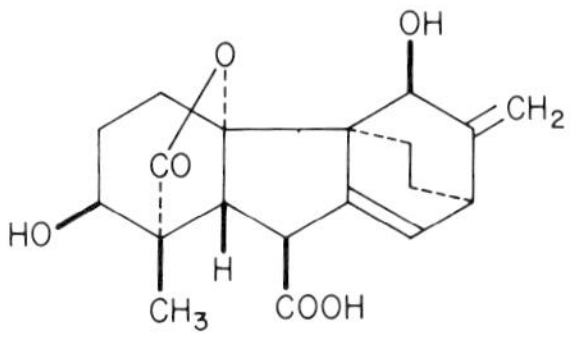

XI

Anemia phyllitidis prothalli, induced antheridia at 10 μg/l (Endo et al. 1972). A_{An} has been assigned structure XI (Nakanishi et al. 1971). An antheridogen from *Anemia hirsuta* appears to possess an identical structure (Zanno et al. 1972). A second antheridogen from *A. phyllitidis* has been detected by Schraudolf (1972) but its structure is not known. An unidentified antheridogen derived from *Ceratopteris thalictroides* differed from A_{Pt} and A_{An} in biological activity and chromatographic properties (Schedlbauer 1974, 1976).

A_{An} is probably the first member of a new class of diterpenes. In structure A_{An} is closely related to the gibberellins. Hence it is not surprising that some gibberellins exhibit antheridogen activity and certain antheridogens are active in some gibberellin bioassays. Thus, gibberellic acid induced antheridia in the six tested species of the family Schizaeaceae including *L. japonicum* and *A. phyllitidis* (Schraudolf 1966a; Naf 1968), but was inactive with species of seven other fern families (Voeller 1964a, b). GA_3 and six other gibberellins showed no effect upon antheridium induction in *Onoclea sensibilis,* the fern species which responds markedly to A_{Pt} (Voeller 1964a). In *A. phyllitidis* gibberellins other than GA_3 are also active in antheridogen induction; the following order of activity has been established: $GA_7 > GA_4 > GA_9 > GA_3 > GA_1 > GA_5 > GA_8$ (Schraudolf 1966b). Helminthosporol and related compounds also show weak activity (Schraudolf 1967).

Although the antheridogen A_{An} was inactive in the dwarf pea bioassay for gibberellins, it was effective in the gibberellin assay based on starch degradation in barley half-seeds (Naf 1968). In further studies with pure samples of A_{An}, the antheridogen was found to promote dwarf corn growth and barley amylase production, but was less active than GA_3 (Sharp et al. 1975). In the dwarf corn bioassay, the effects of A_{An} and GA_3 were approximately additive. In the presence of GA_3, A_{An} promoted lettuce hypocotyl extension (Sharp et al. 1975).

7.4. Lunularic acid

Growth of the liverwort *Lunularia cruciata* is very sensitive to photoperiod and under long days it rapidly becomes dormant. The day-length response is controlled by phyto-

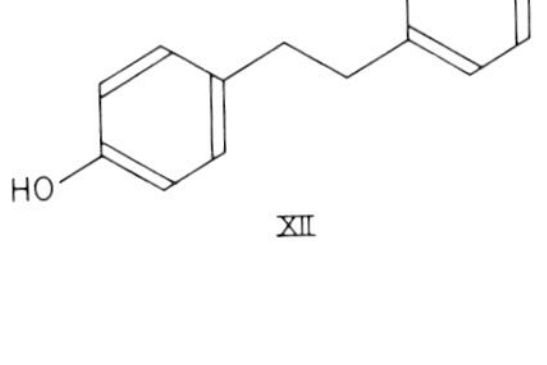

XII

chrome and induction of dormancy appears to be due to an increased level of an inhibitor (Valio et al. 1969, and references therein).

The inhibitor termed lunularic acid was purified and shown to be identical with the dihydrostilbene derivative dihydrohydrangeic acid, XII (Valio et al. 1969; Valio and Schwabe 1970). Lunularic acid inhibited growth of *Lunularia* gemmae at 0.1 mg/l of culture solution. When daylength conditions were changed, the level of the acid in *Lunularia* thalli altered rapidly. Although the inhibitor was present in thalli under short days, the content increased very rapidly when the thalli were placed under continuous light; subsequent transfer back to short days resulted in a rapid fall in the inhibitor level (Valio and Schwabe 1970).

All liverworts and all fresh-water and marine algae examined by TLC and GLC contained lunularic acid; however it was not detectable in mosses or in Pteridophytes (Pryce 1972a). The acid was detected in only one angiosperm species, *Hydrangea macrophylla* (the garden hydrangea). It is of interest that ABA has not been detected in liverworts and

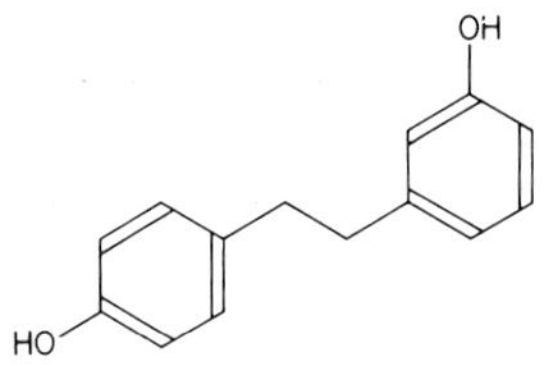

XIII

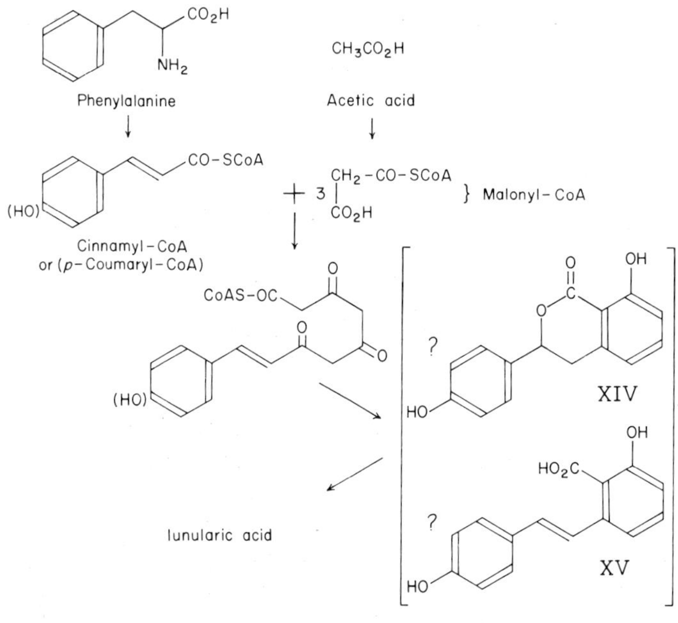

Scheme 7.1. The biosynthetic pathway for lunularic acid.

algae; in these lower plants, lunularic acid appears to be the biological equivalent of ABA.

The metabolism and biosynthesis of lunularic acid have been investigated. Like other plant stilbene derivatives, the acid can be biosynthesized by the phenylpropanoid-polymalonate pathway (Scheme 7.1); however it is uncertain whether hydrangenol (XIV) or the isomeric hydrangeic acid (XV) or both compounds, are intermediates (Pryce 1971). In the liverwort *Lunularia cruciata,* one of the primary metabolites of lunularic acid is the simple decarboxylation product, lunularin (XIII). This is a normal constituent of the liverwort (Pryce 1972b). Metabolism of lunularic acid proceeded at very similar rates and yielded the same products in both light and darkness and it appears that the observed variation in amount of extractable lunularic acid with change in photoperiod (Valio and Schwabe 1970) is probably due to regulation of biosynthesis (Pryce 1972b). Lunularic acid decarboxylase from the liverwort *Conocephalum conicum* has been extracted and some of its properties determined (Pryce and Linton 1974).

7.5. Fungal metabolites which promote growth of higher plants

A number of fungal metabolites have the ability to affect the growth of higher plants, and compounds which promote growth are discussed herein. Numerous substances which exhibit only inhibitory or phytotoxic activity are not reviewed.

7.5.1. Helminthosporol and related compounds

A metabolite produced by the pathogenic fungus *Helminthosporium sativum* was shown by Tamura et al. (1963) to promote the growth of rice seedlings. The active compound, termed helminthosporol, was isolated by these workers and shown to be the sesquiterpene XVI.

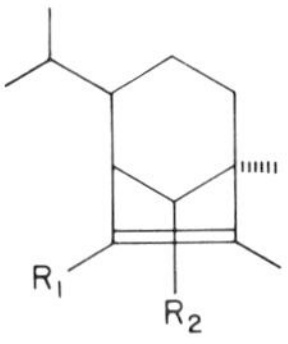

XVI – XVIII

XVI R_1 = CHO R_2 = CH_2OH
XVII R_1 = COOH R_2 = CH_2OH
XVIII R_1 = CHO R_2 = CHO

The growth regulatory activity of helminthosporol (H-ol) and related compounds has been investigated extensively. These compounds resemble gibberellins in activity but are usually less effective than GA_3. H-ol promotes elongation of the leaf sheaths of rice and dwarf maize seedlings but not of oat or wheat seedlings (Hashimoto et al. 1967). H-ol also promotes hypocotyl and primary root elongation in intact seedlings, expansion of radish leaf discs and elongation of oat coleoptile and mesocotyl segments (Hashimoto et al. 1967; Hashimoto and Tamura 1967a; Katsumi et al. 1967). In addition, H-ol breaks dormancy of artichoke tubers (Okuda 1966), induces α-amylase activity in barley half-seeds (Briggs 1966; Okuda et al. 1967) and promotes seed germination (Hashimoto and Tamura 1967b) and flowering (Suge and Rappaport 1968).

The activities of a number of compounds related to H-ol have been assessed. Helminthosporic acid (XVII, H-acid) exhibits high activity which in some assays exceeds that of H-ol (Hashimoto and Tamura 1967a; Hashimoto et al. 1967). In promotion of hypocotyl elongation of cucumber seedlings, relative activities were dihydro-H-acid>H-acid>H-ol (Katsumi et al. 1967). H-acid acted synergistically with ethylene to promote the growth of rice seedlings and increased the lengths of both coleoptiles and first-leaf sheaths. H-ol was less effective in this system (Suge•1974). From a study of the activities of compounds related to H-ol and H-acid, Coombe et al. (1974) concluded the hydroxymethyl group was not necessary for activity.

It has been suggested that the biological activity of H-acid is a consequence of the similarity of its structure to the C and D rings of GA_3 (Briggs 1966; see also Coombe et al. 1974). However Kato et al. (1968) found pseudogibberellin A_1 inhibited gibberellin-induced growth of rice seedlings, but not H-ol induced growth; helminthosporic dicarboxylic acid inhibited H-ol induced growth but not GA_3-induced growth. Hence it was concluded the site of H-ol action was different from that of GA_3. The activity of H-ol is discussed further in Chapter 3, section 3.6.3.

It is relevant to note that the closely related compound XVIII, a dialdehyde, was isolated from culture filtrates of the fungus *Helminthosporium sativum* (Strobel 1974, and references therein). This compound termed helminthosporal is a phytotoxin which causes necrosis of roots of cereals and inhibits barley seed germination and growth of seedlings. The phytotoxicity of helminthosporal appears to be due to inhibition of mitochondrial electron flow and oxidative phosphorylation (Taniguchi and White 1967).

7.5.2. Cotylenins

From culture filtrates of an unidentified fungus ("strain 501–7w", now believed to be a *Cladosporium* sp.), two compounds were isolated initially which promoted the growth of Chinese cabbage, radish and cucumber cotyledons. These substances were termed cotylenins A and B (Sassa 1970; Sassa et al. 1970; Sassa 1971). In promoting enlargement of excised radish cotyledons, cotylenins A and B were as active as kinetin. However, unlike cytokinins, A and B were inactive in a leaf senescence assay (Sassa 1970). Further compounds with similar activity, cotylenins C, D and E, were later purified (Sassa and Togashi 1973; Sassa 1974; Sassa et al. 1975b).

The cotylenins, which are unusual glycosides, appear to possess a common aglycone moiety, termed cotylenol, with the structure XIX (Sassa 1972; Sassa and Togashi 1973; Sassa 1974). Cotylenol, which has also been isolated from the fungal culture filtrates, has a fused ring system identical to that of fusicoccin (see section 7.5.7). The stereochemistry of cotylenol which is a novel diterpene, has recently been elucidated (Sassa et al. 1975a). The cotylenins possess unusual sugar moieties attached to the oxygen at position 9 of the ring system. In cotylenin E, this is a 6-*O*-methylglucopyranosyl group (Sassa and Togashi 1973; Sassa et al. 1975b). However in cotylenins A and C, the sugar is 6-*O*-methylglucopyranose with a C_5 moiety containing an epoxide grouping (Sassa 1974; Sassa et al. 1975b).

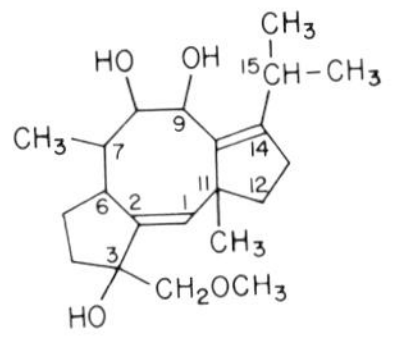

XIX

7.5.3. Radiclonic acid

From mycelial extracts of an unidentified fungus, Sassa et al. (1973a) purified a compound, termed radiclonic acid, which promoted root growth of Chinese cabbage and rice seedlings. When Chinese cabbage seedlings were cultured in an optimal concentration (25 mg/l) of the acid, the length of their roots was double that of the control seedlings (Sassa et al. 1973a).

Radiclonic acid is a hydroxydienoic acid of structure XX (Sassa et al. 1973b).

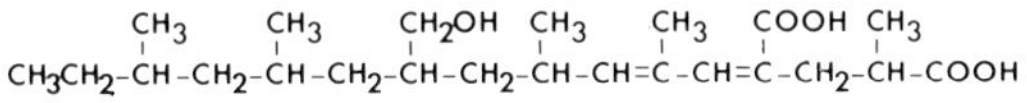

XX

7.5.4. Sclerin and related compounds

Extraction of the mycelia of *Sclerotinia libertiana* yielded a compound termed sclerin which promoted lipase activity in germinating castor beans and α-amylase activity in germinating rice seedlings (Satomura and Sato 1965). Sclerin also promoted the growth of rice and other seedlings, both shoot elongation and root growth being promoted. Sclerin was shown to be the aromatic anhydride XXI (Tokoroyama et al. 1968), and was found to promote oxidative phosphorylation and amino acid incorporation into protein in both plant and animal tissues (Satomura 1974). Two related compounds, termed sclerotinin A and B, were isolated from culture filtrates of *Sclerotinia sclerotiorum* and assigned structures XXII and XXIII respectively. Both compounds, which are dihydroisocoumarin derivatives, promoted the growth of rice seedlings (Sassa et al. 1968a, b).

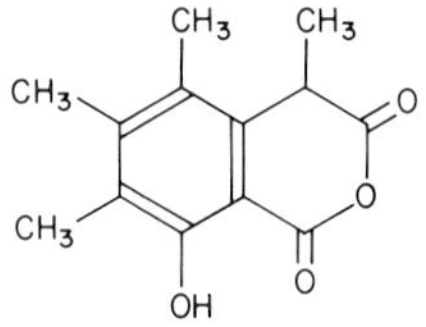

XXI

Two additional metabolites which promoted growth of rice seedlings were isolated from *Sclerotinia sclerotiorum* termed sclerone and isosclerone, and assigned structures XXIV and XXV respectively (Suzuki et al. 1968; Morita and Aoki 1974).

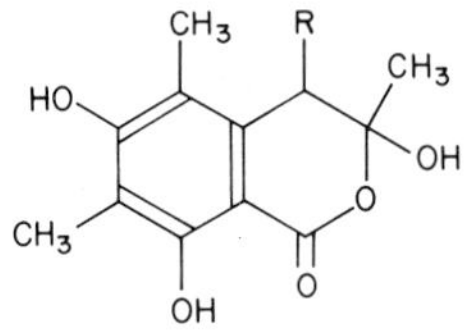

XXII , XXIII

XXII R = CH_3
XXIII R = H

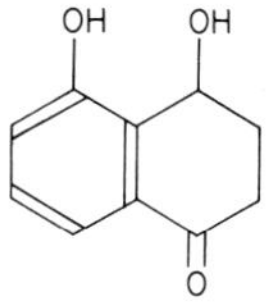

XXIV

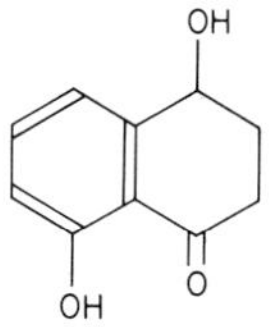

XXV

7.5.5. Pestalotin

From the culture filtrate of the fungus *Pestalotia cryptomeriaecola,* Kimura et al. (1971) isolated a gibberellin synergist termed pestalotin. This compound was assigned the structure 6-(1-hydroxypent-1-yl)-4-methoxy-5,6-dihydro-α-pyrone (XXVI). A number of synthetic analogues of pestalotin, in which the hydroxypentyl group was replaced by other structures, did not exhibit any synergistic activity with GA_3 when applied to rice seedlings (Kimura and Tamura 1974). Pestalotin when applied alone did not affect the growth of rice seedlings, but when applied in combination with GA_3, pestalotin promoted elongation of the second leaf sheath. GA_3-induced release of reducing sugars in rice endosperms was enhanced by pestalotin (Tamura et al. 1974).

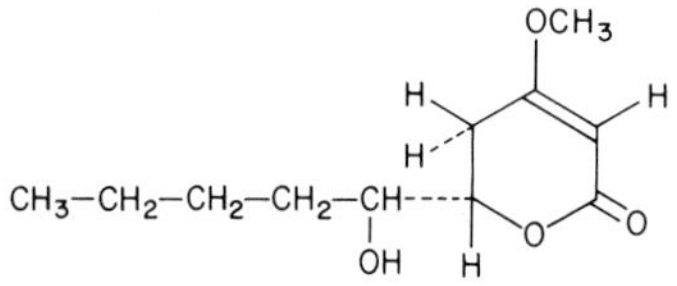

XXVI

7.5.6. β-Phenyllactic acid

From the culture medium of *Exobasidium symploci-japonicae* (a plant pathogenic fungus), a root growth-promoting compound was isolated by Tamura and Chang (1965), who identified it as L-β-phenyllactic acid. The acid promoted root elongation of rice and lettuce seedlings and elongation of the hypocotyls of lettuce seedlings. A number of synthetic hydroxy acids also promote root growth (Mikami et al. 1970).

7.5.7. Fusicoccin

Fusicoccum amygdali causes wilting and dehydration of shoots on infected almond and peach trees. Since leaves are affected some distance from the site of fungal infection, toxins produced by the pathogen were considered to be responsible for the lesions (Graniti 1962; Strobel 1974). From cultures of the fungus, a toxin termed fusicoccin was purified, which, when introduced into the xylem of almond shoots, evoked effects similar to those produced by infection with *F. amygdali* (Ballio et al. 1964). Fusicoccin, a diterpene glucoside, has been assigned structure XXVII (Ballio et al. 1968; Barrow et al. 1968, 1971). It possesses a triple ring system with a central eight-membered ring similar to that of the ophiobolins which are also phytotoxic fungal metabolites (Nozoe et al. 1965). Fusicoccin is a phytotoxin which lacks species-specificity; it induced wilting in all 75 plant species tested (Chain et al. 1971). The positions of labelled atoms in fusicoccin biosynthesized from [1-^{13}C]- and [2-^{13}C]acetate have been determined by Barrow et al. (1975) using [^{13}C]NMR spectroscopy; the results indicate that fusicoccin is formed by direct cyclization of a precursor such as geranylgeraniol pyrophosphate.

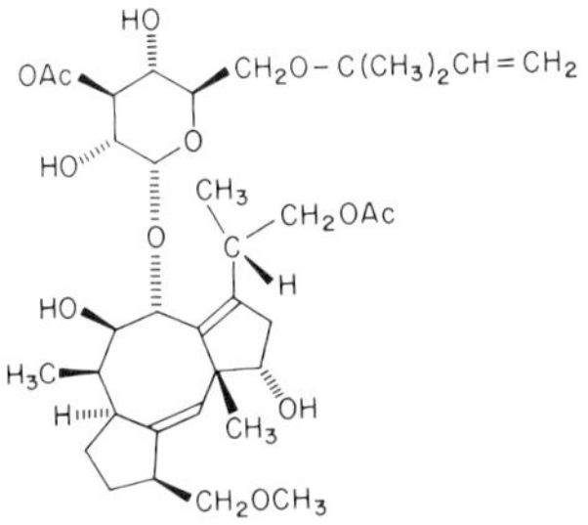

XXVII

Fusicoccin is a very effective promoter of cell enlargement in a variety of plant tissues. At an optimal concentration (about 10 μM), the cell enlargement induced in etiolated pea internode segments by fusicoccin is considerably greater than that evoked by optimal concentrations of IAA and other auxins (Lado et al. 1973). The response from fusicoccin is pronounced after contact with the tissue for only 2 hr. Unlike IAA, fusicoccin does not induce ethylene formation at high concentrations (Lado et al. 1973). Fusicoccin at concentrations less than 1 μM together with auxin promoted elongation of *Avena* coleoptile segments in an additive manner (Yamagata and Masuda 1975). The rate of cell elongation induced by fusicoccin in *Avena* coleoptiles exceeded that evoked by optimal auxin (Cleland 1976). The toxin also promoted elongation of roots of intact lettuce seedlings (Yamagata and Masuda 1975), and stimulated elongation of root segments especially in the presence of potassium ions (Lado et al. 1976).

The expansion of discs of leaves of a number of species is also promoted by fusicoccin, a response not evoked by IAA (Lado et al. 1972). Cell enlargement induced in excised squash and radish cotyledons by fusicoccin is markedly greater than that caused by cytokinins (Marre et al. 1974a). The response to fusicoccin was readily detectable after incubation for 2 hr with radish cotyledons.

The stimulation of cell enlargement by fusicoccin in leaf tissue, excised cotyledons, pea internode segments, *Avena* coleoptiles and root segments is accompanied by a rapid decrease in the pH of the medium (Marre et al. 1973a, b, 1974a; Cleland 1976; Lado et al. 1976).

In the case of radish cotyledons, the decrease in pH was detectable 5 min after addition of the toxin; with root segments the effect was detectable after 1 min. A hyperpolarization of the transmembrane potential difference and an increase in K^+ uptake appears to be associated with the proton release induced by fusicoccin (Marre et al. 1974b, c; Cocucci et al. 1976). Evidence for electrogenic coupling between fusicoccin-stimulated H^+ efflux and cation uptake has been obtained (Pitman et al. 1975). It has been suggested that the promotion of cell enlargement by fusicoccin is a consequence of a decrease in the pH of the cell wall which causes cell wall loosening (Marre et al. 1973a, 1974a; Yamagata and Masuda 1975).

Fusicoccin causes opening of stomata in both light and darkness, the opening being accompanied by an increase in the potassium content of guard cells (Turner 1973). However, leaves treated with fusicoccin whose stomata were fully open, and leaves treated in addition with ABA, whose stomata were closed, both wilted, whereas the corresponding controls without fusicoccin did not wilt (Chain et al. 1971). This indicates that stomatal opening is a secondary effect of fusicoccin and is not the cause of wilting. Chain et al. (1971) suggest the toxin decreases resistance of the plasmalemma to passage of water causing a greater flux of water from treated cells.

Fusicoccin induces germination of dormant wheat seeds at 1 μM and is more active than GA_3 and 6-benzylaminopurine in causing light-requiring lettuce seeds to germinate (Lado et al. 1974). The toxin also relieves almost completely the inhibitory effect of ABA on germination of radish and lettuce seed. Both fusicoccin and GA_3 promote germination of dormant *Haplopappus gracilis* seeds; however fusicoccin acts more rapidly and is more effective than GA_3 (Galli et al. 1975). The promoting effect of fusicoccin on embryo growth of decoated seed is accompanied by proton release into the medium. It is likely that the mode of action of the toxin in stimulating seed germination is basically the same as that involved in promotion of cell expansion in excised cotyledons and stem and coleoptile sections (Lado et al. 1975).

7.5.8. Malformin

Malformin, a cyclic peptide produced by the fungus *Aspergillus niger,* induces severe malformations on the stems and petioles of a number of plant species and causes root curvature (see references given by John and Curtis (1974)). Malformin is cyclo-D-cysteinyl-D-cysteinyl-L-valyl-D-leucyl-L-isoleucyl (Bodanszky and Stahl 1974).

Although malformin has plant growth inhibitory activity, it is also known to promote plant elongation. The cyclic peptide stimulated elongation of etiolated *Phaseolus vulgaris* cuttings in the light, the optimum concentration being 1 μM (John and Curtis 1974). In darkness, however, malformin inhibitied elongation of the etiolated cuttings. Malformin promoted elongation in spite of the fact that it enhanced ethylene production (John and Curtis 1974).

In further studies using *Phaseolus aureus,* green cuttings exhibited a greater growth response to malformin than etiolated cuttings; malformin promoted elongation in light only (Curtis 1976). The peptide also stimulated the growth of intact *P. aureus* seedlings in light. Most of the growth stimulation occurred in the upper 1 cm of the stem (Curtis 1976). When GA_3 and malformin were supplied together, the growth induced was additive.

The activity of malformin was inhibited by hydroxyproline and the inhibition was alleviated by proline (Buckhout and Curtis 1976). When [^{14}C]malformin was supplied to *P. vulgaris*, about 70% of the radioactivity was associated with the cell wall fraction (Ciarlante and Curtis 1974).

7.6. Phenolic carboxylic acids and cinnamic acid

After it was demonstrated that 2,4-dichlorophenol and other simple monophenols promoted the enzymic oxidation of IAA in vitro (Goldacre et al. 1953), the possibility that naturally occurring phenolic compounds might regulate IAA oxidase in vivo attracted great attention. In this section the ability of phenolic acids to act as regulators of IAA oxidase is discussed together with other properties and activities of these compounds. Flavonoids, another group of phenolic compounds which may regulate IAA oxidase activity, are discussed in section 7.7.

7.6.1. Occurrence and biochemistry of phenolic acids

Phenolic carboxylic acids are of very widespread occurrence in plants. Since these well-known compounds are discussed in text-books of plant biochemistry, their chemistry will not be dealt with herein.. The biosynthesis of phenolic acids has been adequately reviewed recently (Stafford 1974) and the following biosynthetic sequence is now well established: phenylalanine → *trans*-cinnamate → *p*-coumarate → caffeate → ferulate → 5-hydroxyferulate → sinapate. Phenylalanine ammonia lyase (PAL) catalyzes the deamination of phenylalanine to *trans*-cinnamate. It is noteworthy that exogenous ABA increases PAL activity in excised bean axes (Walton and Sondheimer 1968). The level of the enzyme activity in plant tissue also appears to be controlled by phenolic acids (see below), while the enzyme is inhibited in vitro by various flavonoids (Attridge et al. 1971).

In addition to occurring as free acids, phenolic acids are also present in plants as glucose esters and amides. In wheat and lupin seedlings and seeds of *Lilium* spp. and barley, some phenolic acids are largely in the form of glucose esters or other derivatives (Asen and Emsweller 1962; El-Basyouni and Towers 1964; Mashtakov et al. 1971; Van Sumere et al. 1972), while in other tissues these acids are entirely in "bound forms" (Niemann 1971). Hydroxycinnamic acids supplied to plant tissues were converted to the corresponding glucose esters and *O*-glucosides (Harborne and Corner 1961; Runeckles and Woolrich 1963; Van Sumere et al. 1972). Hydroxybenzoic acids were also converted to these derivatives. 2,3-, 2,4- and 2,5-Dihydroxybenzoic acids yielded the 3-, 4-, and 5-*O*-glucosides respectively with free *o*-hydroxyl groups (Cooper-Driver et al. 1972).

Changes in the contents of phenolic acids and other phenols have been assessed during wheat seedling growth (El-Basyouni and Towers 1964), tobacco flower and capsule development (Sheen 1973), *Citrullus colocynthis* and apple fruit development (Das et al.

1967; Macheix 1974), and seed germination (Van Sumere et al. 1972). Fertilization induced a very rapid increase in chlorogenic acid in the tobacco ovary, while the ratio of growth-promoting phenolic acids to growth inhibiting acids declined markedly during *Citrullus colocynthis* fruit development. During the active stages of peach bud development, the chlorogenic acid content increased noticeably (Kriventsov and Anandaradzha 1973).

Zones of paper chromatograms of plant extracts have often been found to inhibit growth of *Avena* coleoptile sections or to be inhibitory in other bioassays. In some cases, the activity of these zones appears to be at least partly due to their content of phenolic acids such as salicylic and *p*-coumaric acids (Koves 1957; Varga 1957, 1958; Vieitez et al. 1967; Gesto et al. 1967 and references therein). The endogenous inhibitors of wheat rust uredospores have been identified as the *cis* and *trans* isomers of methyl ferulate, while the inhibitors of bean and other rust uredospores are *cis*- and *trans*- methyl 3,4-dimethoxycinnamate (Sussman and Douthit 1973).

7.6.2. The effect of phenolic acids on IAA oxidase and IAA-induced growth

Naturally occurring phenolic acids can markedly affect IAA-induced growth of sections of coleoptiles, internodes and hypocotyls. These compounds can also act synergistically with IAA, or antagonize IAA, in the *Avena* curvature test. Generally, phenolic acids with two or three adjacent hydroxyl groups (e.g. caffeic, chlorogenic and gallic acids) enhance the action of IAA but not 2,4-D, while phenolic acids with a single hydroxyl group (e.g. *p*-coumaric acid) antagonize IAA [see numerous references cited by Tomaszewski and Thimann (1966) and by Zenk and Muller (1963)]. There are of course certain exceptions to the above statement. For example, although 4-hydroxybenzoic acid inhibits IAA-induced growth of *Avena* first-internode sections (Nitsch and Nitsch 1962), it promotes growth of coleoptile sections at low concentrations both in the absence and presence of IAA with which it acts synergistically (Vieitez et al. 1967). Caffeic acid has also been reported to promote coleoptile section growth in the absence of exogenous IAA, but this has been attributed to interaction with endogenous IAA (Thimann et al. 1962).

A close parallel was observed by Tomaszewski and Thimann (1966) between the effects of a phenolic acid on IAA decarboxylation apparently within plant tissue and its effects on tissue growth under identical conditions; the parallels held for *Avena* coleoptiles and etiolated and green pea stem sections. Compounds which inhibited growth enhanced this oxidative decarboxylation, while acids which promoted growth suppressed decarboxylation of IAA. Correlations between effects of phenolic acids on growth and effects on IAA oxidase in vitro had been established previously (see e.g. Zenk and Muller 1963). Because of the above observations, the view is now widely held that phenolic acids enhance or decrease the growth-promoting activity of exogenous IAA by regulating the IAA level in tissue through an effect on the rate of IAA oxidation within the tissue. However the validity of this conclusion in the case of experiments with pea stem and root segments has been questioned by Andreae and Collet (1968). Their work, which tends to be overlooked, indicates that the effect of phenolic compounds on activity of IAA supplied to tissue segments can be attributed to a direct effect on the IAA concentration in the culture solutions rather than in the tissues. Regulation of IAA oxidation by exoenzymes appears to be involved according to Andreae and Collet (1968).

Although it is clear from the above that exogenous phenolic acids can markedly affect oxidative degradation of IAA supplied to plant tissues in vitro, it is uncertain whether endogenous phenolic compounds regulate auxin levels in plant tissues. However certain observations suggest that this does occur. Thus tissue explants oxidize less IAA if they have previously been induced to synthesize more polyphenolic compounds, particularly chlorogenic acid (Runkova et al. 1972). Green pea epicotyls which contain endogenous polyphenols degrade less IAA than etiolated epicotyls which synthesize monophenols only (Tomaszewski 1964). Chlorogenic acid increases the endogenous IAA level in roots of *Lens culinaris* (Pilet 1964).

The molecular size of naturally occurring phenolic IAA oxidase inhibitors ranges from simple phenolic acids (e.g. chlorogenic, ferulic, caffeic) to high molecular weight polyphenol-protein complexes (mol. wt. up to 200,000, see Schneider and Wightman 1974). Hydroxylated coumarins and flavonoids are other naturally occurring compounds which can regulate IAA oxidation (see sections 7.7 and 7.8).

Regulation of IAA oxidation may explain the ability of phenolic acids to markedly affect bud formation in tobacco tissue cultured in the presence of IAA and kinetin (Lee and Skoog 1965a). 4-Hydroxybenzoic acid was the most active of all the tested compounds, inducing marked bud formation and inhibiting callus growth at 2 *μM* and higher concentrations. Salicylic acid also induced bud formation; it was less effective than 4-hydroxybenzoic acid but more active than the 3-hydroxy isomer. Protocatechuic acid (3,4-dihydroxybenzoic acid) did not induce buds and suppressed bud formation on media containing 4-hydroxybenzoic acid. Caffeic, ferulic and sinapic acids also failed to induce buds and promoted callus growth. Ferulic acid inhibited bud formation when added to a medium which caused their development. Bud formation in tobacco tissue occurs when the cytokinin to auxin ratio is high, and the response to phenolic acids can be accounted for in terms of an effect on IAA oxidation (Lee and Skoog 1965b).

The response of tobacco cells in suspension culture to IAA was increased by exogenous ferulic acid which prevented decomposition of IAA in the medium and increased the accumulation of indolylacetyl aspartate in the cells (Rekoslavskaya et al. 1974).

7.6.3. Other responses to phenolic acids

A number of growth responses to phenolic acids do not appear to be a consequence of regulation of the IAA oxidase system. These responses and effects on other enzyme systems are now discussed.

Certain phenolic acids have the ability to affect seed germination. Caffeic acid at low concentrations promoted lettuce and barley seed germination but was without any appreciable effect at the higher concentrations tested. Other acids, notably ferulic and *o*-coumaric, weakly inhibited germination (Van Sumere et al. 1972). Evidence which suggests that endogenous free ferulic acid may be a germination regulator has been discussed by these authors.

Salicylic acid retarded senescence of radish leaf discs at 1 *μM* and acted synergistically with kinetin (Karanov 1972). This acid also induced flowering in *Lemna gibba* cultured under short days, the optimum concentration being 6 *μM* (Cleland and Ajami 1974).

However this may be a consequence of the ability of the acid to act as a chelating agent (Oota 1975). Salicylic acid as the iron chelate also causes elongation of the pedicels of this species (Dekoch et al. 1974).

A number of pure hydrolysable tannins (galloyl derivatives of glucose) are very effective inhibitors of GA_3-induced growth of dwarf pea seedlings (Corcoran et al. 1972). The inhibition was completely reversed by increased amounts of GA_3 and the tannins did not affect endogenous growth. Hence these compounds appear to qualify as gibberellin antagonists. Terchebin was the most effective tannin tested, but 1,6-di-*O*-galloyl glucose and 1,2,3,4,6-penta-*O*-galloyl glucose also exhibited marked inhibitory activity. Hypocotyl growth induced by GA_3 in cucumber seedlings was also inhibited by tannins; however growth induced by IAA was enhanced by the tannins. Unidentified inhibitors in certain plant extracts, notably carob extracts (see Corcoran 1970), exhibit activity similar to that of tannins (Corcoran et al. 1972). The ability of tannins, including condensed tannins, to inhibit the activity (dwarf pea assay) of gibberellins other than GA_3 has been studied recently (Green and Corcoran 1975). The tannins were particularly inhibitory against GA_4 and GA_{14}. The role of tannins in plants is unclear, but it has been suggested that they may play a regulatory role by inhibiting gibberellin-induced growth (Green and Corcoran 1975). Tannins have also been reported to both inhibit and stimulate root growth and germination (see numerous references listed by Corcoran et al. 1972).

Certain phenolic acids, particularly *p*-coumaric acid and 4-hydroxybenzoic acid, can promote auxin-induced rooting of cuttings although these compounds enhance IAA oxidation (Roy et al. 1972, and references therein). Differences in the rooting ability of cuttings may be due at least partially to differences in the content of phenolic compounds (Roy et al. 1972). Ferulic acid promotes rooting of *Phaseolus vulgaris* and willow cuttings and appears to act independently of auxin. Ferulic acid did not affect the level of endogenous auxin in the basal region of the bean cuttings before root emergence (Michniewicz et al. 1976). The ability of phenolic acids to promote rooting of cuttings is discussed further in Chapter 2, section 2.7.7, of Volume II.

Phenolic acids can markedly inhibit potassium and phosphate ion uptake by excised roots (Glass 1973, 1974a, b). These acids depolarize cell membrane potentials and the inhibition of ion uptake appears to be caused by change in membrane permeability (Glass and Dunlop 1974; Glass 1975). These workers suggest that inhibition of ion uptake by phenolic acids is of importance in allelopathy, the phenomenon in which one plant species inhibits development of others by release of specific compounds, including phenolics, into the environment.

The effect of polyphenolic acids on enzyme activity is not limited to IAA oxidase. Caffeic and chlorogenic acids also inhibit peroxidase-catalyzed oxidation of glutathione (Stonier and Yang 1973). This inhibition could be of regulatory significance. The activity of PAL, a key enzyme in aromatic biosynthesis, appears to be controlled by end products of the shikimic acid pathway, including cinnamic, *p*-coumaric and ferulic acids (see references in Engelsma 1974; Johnson et al. 1975). Some recent observations indicate that polyphenolic acids mediate in the photocontrol of PAL activity. Irradiation with UV light causes a partial conversion of *trans*-hydroxycinnamic acids to the *cis*-isomers in the hypocotyls of gherkin seedlings grown in darkness. The *trans*-acids inhibit development of the

PAL activity, and their conversion to the much less inhibitory *cis*-acids provides a likely explanation for the increase of PAL activity in hypocotyls irradiated with UV light (Engelsma 1974). The increase in PAL activity induced by blue light in gherkin hypocotyls is also probably initiated by *trans-cis* isomerisation of hydroxycinnamic acids.

Phenolic acids are known to release respiratory control, and inhibit respiration when incubated with isolated mitochondria (Demos et al. 1975). Ferulic acid stimulates transpiration in bean plants and increases stomatal aperture (Michniewicz and Rozej 1974).

7.6.4. trans-Cinnamic acid

trans-Cinnamic acid, a biosynthetic precursor of phenolic acids, competitively inhibits the growth-promoting effect of auxins in the pea stem section assay (Van Overbeek et al. 1951) and is usually regarded as a general antiauxin. Other effects of auxins which are inhibited by *trans*-cinnamic acid include promotion of elongation of oat first-internode sections (Nitsch and Nitsch 1962), induction of hyponasty of immature leaf blades (Lippincott and Lippincott 1971) and promotion of RNA synthesis (Masuda and Tanimoto 1967; Masuda 1968). However the inhibition by the acid of pollen germination could not be overcome by IAA (Goss 1968). In this and in certain other cases, the growth regulatory activity of *trans*-cinnamic acid does not appear to depend on antiauxin action. The acid has also been reported to inhibit growth of rice seedlings (Chandramohan et al. 1973), induce sporulation of *Piricularia* spp. (Narayanarao et al. 1972), inhibit germination of *Colletotrichum falcatum* spores (Srinivasan and Narasimhan 1971), promote root formation on bean cuttings in the presence of IAA (Basu 1971) and stimulate seed germination (Palesiko et al. 1966).

7.7. Flavonoids

Since the chemical structure (Harborne 1965, 1972) and biosynthesis (Grisebach 1968; Smith 1973; Stafford 1974) of flavonoids have been discussed recently, these subjects are not dealt with herein. However, factors affecting the level of flavonoids in plant tissues, their metabolism and possible regulatory role are briefly discussed (see also Chapter 9, section 9.3).

7.7.1. Flavonoid content and metabolism

Environmental factors and hormonal stimuli markedly influence the flavonoid content of plant tissue. Ethylene enhances light-induced anthocyanin biosynthesis in *Sorghum vulgare* seedlings (Craker 1975) while application of GA_3 to light-grown pea seedlings reduces the quercetin content in mature leaves. The latter change was detectable 24 hr after treatment and before any detectable growth response (Moore and Pecket 1972). Deoxyisoflavone synthesis in soybean tissue cultures is promoted by cytokinin, the effect being detectable after a lag period of only 6 hr (Miller 1972). Kinetin stimulates anthocyanin formation in dark-grown seedlings of *Brassica oleracea* (red cabbage) and when applied for only 15 min the response is similar to that evoked by 5 min of red light (Pecket and Bassim 1974). A most interesting observation is that the effect of kinetin is nullified by exposure to far-red light. Kinetin has also been reported to promote anthocyanin formation in petals of *Impatiens balsamina* (Klein and Hagen 1961). Abscisic acid and the ethylene-releasing compound Ethrel act synergistically with sucrose to increase markedly the level of both total phenolics and anthocyanins in leaf discs of grapevine (Pirie and Mullins 1976). These workers suggest that ABA and sugar supply promote anthocyanin accumulation in grape berries in vivo. Photoperiod (Manai and Paulet 1973) and nutrient deficiency (Krause and Renzik 1972) also markedly affect flavonoid content. The effect of red light on the flavonoid content of pea seedlings is discussed in detail later in this chapter in relation to IAA oxidase activity.

The level of an inhibitor (naringenin) extracted from peach flower buds was correlated with emergence of the buds from rest (Hendershott and Walker 1959a, b). However the significance of this correlation has been questioned by Dennis and Edgerton (1961). They showed that naringenin is confined to the bud scales and is not present in the flower parts until about sepal opening. As the buds expand, the ratio of bud scales to primordia declines, resulting in an apparent decrease in inhibitor level. The results of Corgan (1965) are in accord with this view and indicated that cessation of peach bud rest is not accompanied by any immediate change in naringenin level. However a high content of inhibitors including naringenin appears to be associated with bud failure in peach (Malcolm 1972).

Interesting changes in naringenin content during tomato fruit ripening have been reported (Wardale 1973). While there is very little change in the phenolic compounds present in the flesh of the fruit during ripening, naringenin increases markedly in the skin at the onset of the climacteric. The increase in naringenin is accompanied by an increase in the production of ethylene in the skin. In an in vitro system, naringenin stimulates production of ethylene from 4-methylmercapto-2-oxobutyric acid (Wardale 1973) and this flavonoid may be involved in regulation of ethylene biosynthesis in the ripening tomato fruit. Changes in the content of flavonoids during pea leaf development have also been investigated. As the leaves mature, the quercetin content decreases while that of kaempferol increases (Moore and Pecket 1972). In young green pea plants, the highest concentration of quercetin and kaempferol glucosides is in the fully expanded leaves, while the content in internode tissue is relatively low (Bottomley et al. 1966).

The subcellular distribution of flavonoids has been investigated. In etiolated tissues of *Avena sativa,* certain flavonoids are contained primarily in the etioplasts, whereas in green leaves, the flavonoids are found mainly in the vacuole and cell wall and not in the chloroplasts (Weissenbock 1973).

The metabolites formed from flavonols by cell suspension cultures of *Cicer arietinum* and *Glycine max* have been investigated (Hosel et al. 1972). Kaempferol and quercetin are catabolized via 2,3-dihydroxyflavanones to 4-hydroxybenzoic acid and 3,4-dihydroxybenzoic acid respectively. These metabolites are derived from the B ring of the flavonoid. Flavonols are attacked by peroxidases and their degradation in plants is probably partially mediated by these enzymes (Schreiber 1975, and references therein). 3-Hydroxyflavone, the simplest known substrate, yields on enzymic cleavage with horseradish peroxidase salicylic, phenylglyoxylic and probably benzoic acids (Schreiber 1975).

In plant tissues, flavonoids occur mainly in the form of glycosides, usually *O*-glycosides but *C*-glycosides are also known. Thus [^{14}C]naringenin supplied to *Spirodela polyrhiza* was converted into apigenin-7-*O*-glucoside and apigenin-8-*C*-glucoside (vitexin) and into luteolin-7-*O*-glucoside and luteolin-8-*C*-glucoside (Wallace and Grisebach 1973).

7.7.2. Flavonoids, IAA oxidase and photomorphogenisis

Like phenolic acids, flavonoids can markedly affect the activity of IAA oxidase. Stenlid (1963) examined the effect of flavonoids (flavones, isoflavones, flavanones and anthocyanidins) on pea root IAA oxidase in vitro. Activity depends on the B-ring hydroxylation pattern; 3′,4′-dihydroxyflavonoids inhibit the enzyme, but 4′-hydroxyflavonoids stimulate the activity. Thus apigenin and naringenin enhance, while quercetin and rutin inhibit, the activity of the enzyme. A hydroxyl group at position 7 of the A ring markedly increases the stimulatory activity. It is noteworthy that the dihydrochalcone glucoside, phloridzin, is a particularly effective cofactor, while the related compound sieboldin with *o*-diphenolic substitution of the B ring is a strong inhibitor of pea root IAA oxidase (Stenlid 1968). IAA oxidases from other plant tissues respond similarly to flavonoids. Thus lupin leaf IAA oxidase is stimulated by apigenin and inhibited by rutin and quercetin (Kornelyuk and Volynets 1971).

Some observations suggest that flavonoids function in vivo as regulators of IAA oxidase. Hillman and Galston (1957) showed that the IAA oxidase activity of pea buds is greatly suppressed by exposure of the seedlings to red light and that this effect is reversed by far-red light. The diminished IAA oxidase activity is due to enhanced production of a dialyzable inhibitor. Because phytochrome-mediated promotion of synthesis of an IAA-oxidase inhibitor could be causally related to growth control by light, the identification of the inhibitor was attempted. Quercetin derivatives, quercetin-3-triglucoside (XXVIII) and quercetin-3-(*p*-coumaroyltriglucoside) (XXIX) were considered responsible for the

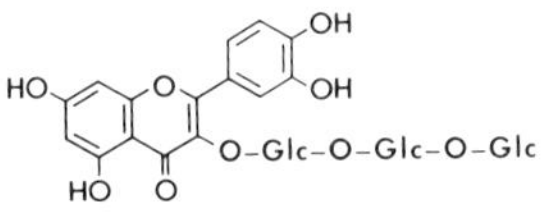

XXVIII

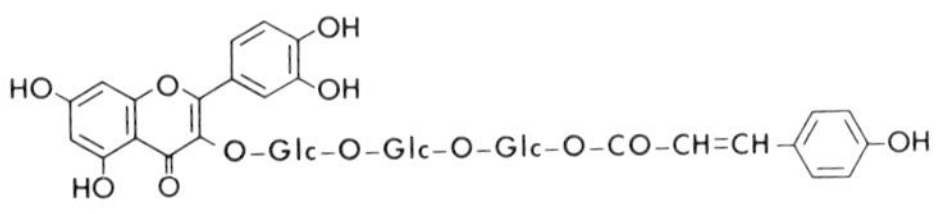

XXIX

inhibitory activity by Furuya et al. (1962), while the two analogous kaempferol conjugates in etiolated peas were promoters of IAA oxidase at low and physiological concentrations. At higher concentrations, however, these kaempferol conjugates inhibited the oxidase (Furuya et al. 1962). The kaempferol conjugate XXX was also purified from red

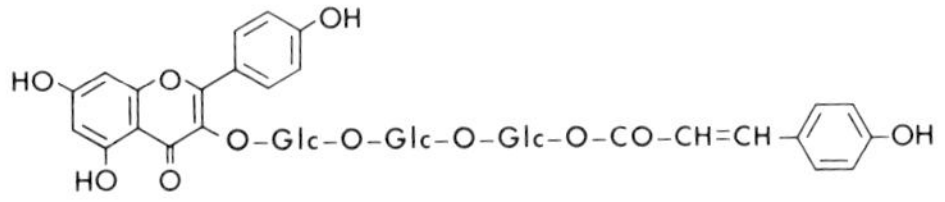

XXX

light-treated pea buds by Mumford et al. (1961) who regarded it as an IAA-oxidase inhibitor produced in response to red light. This work appears to contradict the findings of Furuya et al. (1962).

However further investigation of the response of pea seedlings to red light by Bottomley et al. (1966) supported the work of Furuya et al. and showed that red light markedly enhances the level of XXIX, but not of the kaempferol analogue, in the terminal buds, the response being suppressed by far-red light. In contrast, in the internode tissue, red light induces an increase in kaempferol-3-triglucoside and its *p*-coumaroylester, the effect being reversed by far-red (Russell and Galston 1967). GA_3 prevents red-light inhibition of growth of the internode sections, and also inhibits both IAA inactivation and kaempferol-3-triglucoside synthesis (Russell and Galston 1969). Production of an inhibitor of IAA oxidase (quercetin derivative) in pea buds and a cofactor (kaempferol derivative) in internode tissue in response to red light, may, as suggested by Galston (1969), be involved in the promotion of leaf growth and inhibition of stem growth induced by red light. However definite evidence in support of this proposition is lacking. Thus, comparative kinetics of growth and kaempferol responses to red light (Russell and Galston 1967) fail to provide evidence of a regulatory role for this flavonoid in growth inhibition of pea internodes. Significant correlations between flavonoid concentrations or ratios in pea buds and growth rate were not found by Smith and Harper (1970).

Red light induces hook opening of seedlings and this is thought to result from a red light-induced inhibition of ethylene production by hook tissue (see references in Kang and Ray 1969). Like IAA oxidase, the peroxidase systems which catalyze ethylene formation are affected markedly by phenolic compounds. Kang and Ray (1969) suggest that red light-induced formation of phenolic compounds such as flavonoids suppresses ethylene production and thereby causes hook opening.

7.7.3. Flavonoids and ATP production

At the biochemical level, another effect of certain flavonoids is inhibition of ATP production in plant mitochondria (Stenlid 1970, and references therein). Of the common flavonoids, quercetin and kaempferol are particularly effective as inhibitors; flavones are more inhibitory than flavanones and glycosides are less effective than the corresponding aglycones. In contrast, substituted cinnamic acids (e.g. caffeic and *p*-coumaric acids) are inactive. Kaempferol appears to act specifically on the phosphorylation mechanism and not on electron transport (Koeppe and Miller 1974).

This flavonoid is also a potent inhibitor of both cyclic and noncyclic photophosphorylation in isolated pea chloroplasts (Arntzen et al. 1974). The inhibition of photophosphorylation in isolated chloroplasts by phloridzin (Winget et al. 1969) and by phloretin (Uribe 1970) has also been reported. When incubated with lupin chloroplasts, quercetin greatly decreases the Hill reaction rate (Prokhorchik et al. 1972). At very low concentrations (10^{-6}–10^{-5} M) it promotes, and at higher concentrations it decreases, the cyclic phosphorylation rate (Prokhorchik et al. 1972).

7.7.4. Flavonoids and plant growth

Like the phenolic acids, flavonoids can markedly affect IAA-induced elongation of oat first internode sections. Certain flavonoids including quercetin, fisetin and the leucoanthocyanin, goratensidin, markedly enhance this response; however, apigenin, genistin and naringenin are inhibitory (Nitsch and Nitsch 1962). These growth responses can be attributed to an effect on the IAA oxidase system. Elongation of wheat coleoptile sections is markedly suppressed by certain flavonoids including naringenin and phloridzen (Hancock et al. 1961; Correa 1969). Flavonoids including naringenin and naringenin-7-glucoside appear to account for the inhibitory activity, detected by coleoptile bioassays, in certain zones of chromatograms of plant extracts (Hancock et al. 1961; Dennis and Edgerton 1961; Corgan 1967; Turetskaya et al. 1968). An inhibitor detected in this way in willow extract was identified as chalconaringenin-2′-glucoside (XXXI) (Turetskaya et al. 1968).

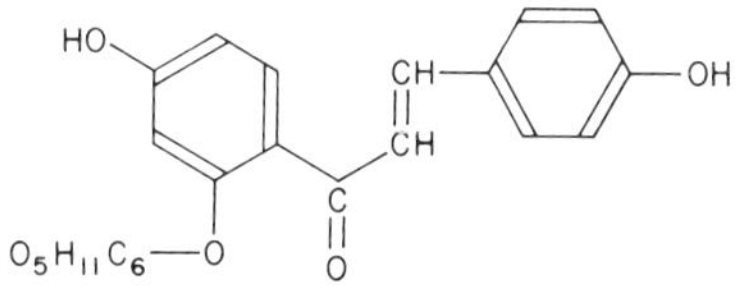

XXXI

Certain anthocyanidins at concentrations of 10^{-6} to 10^{-4} *M* promote root growth of wheat seedlings and markedly alleviate the inhibition of root growth caused by auxin (Stenlid 1962). Phloretin and naringenin also stimulate root growth, especially in the presence of auxins (Stenlid 1968). A correlation was found between ability to uncouple oxidative phosphorylation and root growth-promoting activity; however there was no parallel between effects on IAA oxidase and root growth (Stenlid 1968). It is relevant to note that Bachelard and Stowe (1962) observed a correlation between ability of cuttings to root and anthocyanin formation in the leaves. Although exogenously supplied naringenin failed to delay opening of peach buds (Dennis and Edgerton 1961), the bud dormancy-breaking activity of applied gibberellin was suppressed by concomitant application of naringenin (Phillips 1962).

Leucoanthocyanins have been reported to promote cell division in carrot phloem tissue cultures (Steward and Shantz 1956, 1959). However this rather weak response may be due simply to chelation of iron or to an effect on the IAA oxidase system. Hence it is of doubtful significance. However, a more striking growth response to an exogenous flavonoid in an in vitro system has recently been reported by Jones (1976). Phloridzin, a dihydrochalcone occurring in *Malus* spp., very markedly promotes the growth of cultured apple shoot tips. Jones suggests endogenous phloridzin may be involved in regulation of growth of apple trees.

Leaf senescence is retarded by certain flavonoids. Quercetin strongly delays the senescence of leaf discs of radish, while rutin acts synergistically with kinetin to retard senescence of these discs (Karanov 1972).

7.7.5. *Phenolic inhibitors and ABA*

In this section and also in section 7.6, a number of phenolic compounds which inhibit growth have been discussed. It is relevant to consider briefly their physiological relationship to ABA. Because PAL activity is enhanced by exogenous ABA, it has been suggested that ABA may regulate the biosynthesis of cinnamic acid and phenolic acid inhibitors (Kefeli and Kadyrov 1971). A number of aspects of plant growth are similarly affected by both ABA and phenolic inhibitors. Thus, both inhibit seed germination and growth of coleoptile and stem sections; both promote rooting of cuttings. However ABA is usually effective at much lower concentrations. Phenolic compounds regulate IAA oxidase activity and hence possibly influence IAA levels in tissue; ABA in contrast may affect gibberellin levels and metabolism (Chapter 3; Railton and Wareing 1973), but does not regulate IAA oxidase activity in vitro (Kornelyuk and Volynets 1971). ABA may act at the translational level to regulate enzyme synthesis (but see Chapter 11); phenolic compounds appear to influence enzyme activity directly. Certain phenolic compounds are very effective inhibitors of ATP production; ABA has not been shown to regulate at this level.

Although exogenously supplied phenolic compounds can affect growth and influence the activity of enzymes in vitro, it has yet to be established that endogenous phenolic compounds exert a regulatory role.

7.8. Coumarin and simple coumarin derivatives

A number of naturally occurring unsaturated lactones exhibit growth regulatory activity at low concentrations. The most notable of these is coumarin. However, a number of simple derivatives of coumarin also occur in plants and the structural formulae of the principal compounds of this type are presented in Figure 7.1. In this section, the biochemistry and activity of simple coumarins are discussed while other unsaturated lactones are considered in section 7.9.

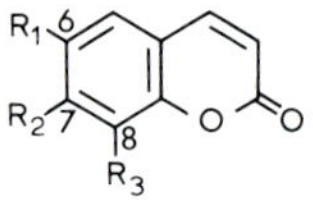

Fig. 7.1. Structural formulae of common coumarins including those mentioned in the text

	R_1	R_2	R_3
Coumarin	H	H	H
Umbelliferone	H	OH	H
Skimmin	H	OGlc	H
Herniarin	H	OCH_3	H
Daphnetin	H	OH	OH
Daphnin	H	OGlc	OH
Esculetin	OH	OH	H
Cichoriin	OH	OGlc	H
Esculin	OGlc*	OH	H
Scopoletin	OCH_3	OH	H
Scopolin	OCH_3	OGlc	H

* Glc denotes a glucosyl moiety.

Coumarins are frequently considered to be simply inhibitors of plant growth (see e.g. Thimann 1972). However, it is now evident that coumarins can also promote numerous phases of plant development at low and physiological concentrations. In this respect coumarins differ from the inhibitor ABA which rarely stimulates developmental processes. Although exogenous coumarins can both promote and inhibit growth, it is still not absolutely certain that endogenous coumarins act in a regulatory role.

7.8.1. Occurrence and coumarin content of plant tissues

Coumarins are widely distributed in plants and frequently occur as glucosides; however, the occurrence of a particular coumarin is often restricted. Coumarin itself is prominent in the Leguminoseae and Gramineae and is considered to be responsible for the smell of newly mown hay. Scopoletin is a widely distributed coumarin and is often the dominant coumarin in plants of the Solanaceae. The occurrence of coumarins in higher plants has been tabulated by Neish (1965) and will not be discussed further here. However it is noteworthy that coumarin and certain of its derivatives including umbelliferone occur in fungi (Van Sumere et al. 1957; Austin and Clarke 1966).

In tobacco plants, very marked increases (in some cases 20-fold) in the levels of scopoletin and/or scopolin accompany viral infection, bacterial infection, mineral and nitrogen deficiency, cold treatment, and stress induced by UV and X-ray irradiation (Wender 1970, and references therein). In tobacco tissue cultures, kinetin promotes formation of the glucoside scopolin, while high IAA levels favour production of the aglycone scopoletin (Sargent and Skoog 1960; Skoog and Montaldi 1961). Scopolin increased markedly in tobacco leaves, stems and roots after the plants had been sprayed with 2,4-D, the increase in leaves being 30-fold (Wender 1970). Maleic hydrazide also caused a similar effect. The level of coumarin in barley embryos decreases during the first few hours of germination (Van Sumere et al. 1972). Relative to the germinating seeds, the non-germinating seeds of *Melilotus alba* have a high level of free coumarin (Stoker 1964). During germination of *Coumarouna odorata* seeds which contain a high level of endogenous coumarin, coumarin was reported to remain constant per unit dry weight in the cotyledons and to increase in the embryos (Valio 1973). However, the specificity of the method for determining coumarin would seem questionable.

7.8.2. Biosynthesis and metabolism

Coumarins are formed from L-phenylalanine which is derived from the shikimic acid pathway. The steps in the conversion of phenylalanine to coumarin in white sweet clover *(Melilotus alba)* are indicated in Scheme 7.2 (Kosuge and Conn 1959, 1961; Stoker and Bellis 1962). In this sequence, the β-glucoside of *o*-coumaric acid (2-β-D-glucosyloxy-*trans*-cinnamic acid) isomerizes to its *cis* isomer, the β-glucoside of coumarinic acid. The *trans-cis* isomerization is induced by light and appears to be non-enzymic (Edwards and Stoker 1967). Sweet clover contains a β-glucosidase which, unlike emulsin, attacks only slowly the β-glucoside of *o*-coumaric acid, but readily hydrolyzes the β-glucoside of coumarinic acid (Kosuge and Conn 1961). The free coumarinic acid spontaneously lactonizes to coumarin. "Bound coumarin" has been shown to be the β-glucoside of coumarinic acid (Kosuge 1961; Brown 1963). In some plant species, coumarin is largely

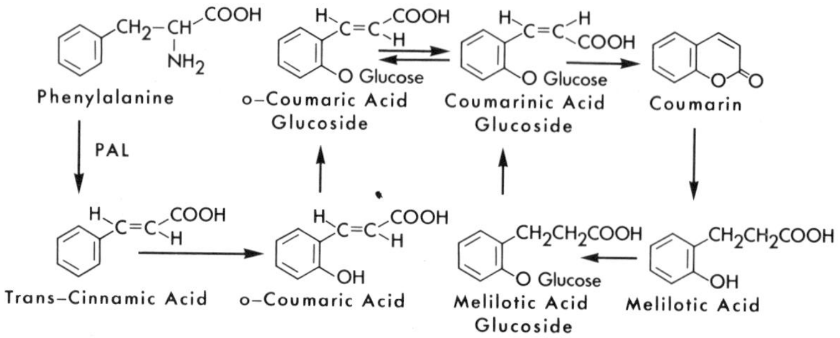

Scheme 7.2. The biosynthetic pathway for coumarin in *Melilotus alba.*

in this "bound" form and much of the coumarin isolated from some plants is probably an artefact derived from coumarinic acid glucoside. The biosynthesis of coumarin has been investigated in other plant species and the results obtained are consistent with the occurrence of the pathway of Scheme 7.2 (Brown 1962, 1963). When radioactive coumarin was supplied to excised sweet clover shoots, it was converted via 3,4-dihydrocoumarin to melilotic acid (Scheme 7.2) and to the glucoside of this acid (Kosuge and Conn 1959, 1962). Melilotic acid glucoside appears to be further metabolized slowly to the glucosides of *o*-coumarinic acid and *o*-coumaric acid (Stoker and Bellis 1962). The principal metabolite of coumarin formed by barley embryos was identified as *o*-coumaric acid glucoside (Van Sumere and Kint 1964).

The biosynthesis of hydroxylated coumarins has also been investigated. The proposed pathway for umbelliferone biosynthesis is as follows (Austin and Meyers 1965): *trans*-cinnamic acid → 4-hydroxy-*trans*-cinnamic acid (*p*-coumaric acid) → 4-glucosyloxy-*trans*-cinnamic acid → 4-glucosyloxy-2-hydroxy-*trans*-cinnamic acid → 2,4-diglucosyloxy- *trans*-cinnamic acid → 2,4-diglucosyloxy-*cis*-cinnamic acid → 7-glucosyloxycoumarin (skimmin) → 7-hydroxycoumarin (umbelliferone). Herniarin is formed by the following partial sequence (Brown 1963): *p*-coumaric acid → 4-methoxy-*trans*-cinnamic acid → 2-glucosyloxy-4-methoxy-*trans*-cinnamic acid → *cis*-isomer → 7-methoxycoumarin (herniarin). The biosynthesis of scopoletin in tobacco tissue cultures appears to proceed principally according to the pathway presented in Scheme 7.3. As in the synthesis of herniarin,

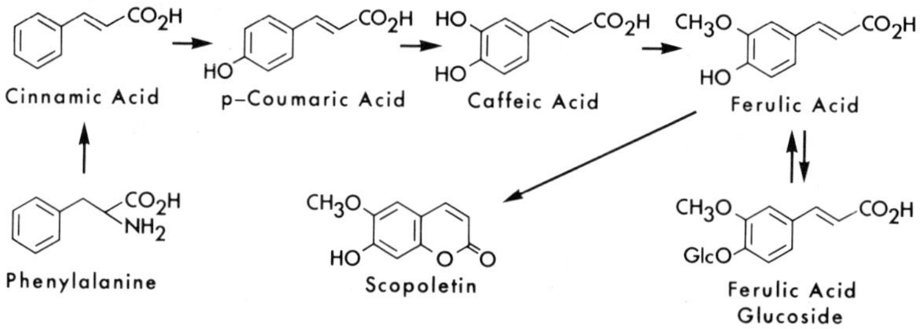

Scheme 7.3. The biosynthetic pathway for scopoletin in tobacco tissue cultures.

methylation precedes cyclization (Fritig et al. 1970). Conversion of ferulic acid to scopoletin no doubt involves *ortho* hydroxylation and *trans-cis* isomerization. In the above syntheses, the coumarin is derived by cyclization of the corresponding cinnamic acid derivative. However, the major route to esculetin and daphnetin appears to be via *p*-coumaric acid; caffeic acid does not appear to be an intermediate (Sato and Hasegawa 1972a) as had been suggested by Neish (1965). This suggests that in the cases of esculetin and daphnetin, hydroxylation of the aromatic ring may occur after closure of the lactone ring.

An interesting aspect of coumarin glucoside biosynthesis is the role of trans-glucosylases. An enzyme in *Daphne odora* is capable of hydrolyzing daphnin (daphnetin-7-glucoside) to daphnetin and of transferring the glucosyl residue from a second molecule of daphnin to the liberated aglycone to yield daphnetin-8-glucoside (Sato and Hasegawa 1971). The enzymic transfer of glucose from cichoriin (esculetin-7-glucoside) to esculetin to produce esculin (esculetin-6-glucoside) has also been reported (Sato and Hasegawa 1972b).

Hydroxylated coumarins such as scopoletin are subject to peroxidase-catalyzed oxidation. In the presence of free H_2O_2 and peroxidase, stoichiometric oxidation occurs; however, in the absence of H_2O_2, a reductant, IAA and molecular oxygen can initiate the oxidation. Different oxidation products are formed in the two systems (Miller et al. 1975).

7.8.3. The effects of coumarins on plant growth

Coumarin is usually regarded as a growth inhibitor. It inhibits the auxin-induced curvature of slit pea stems and the elongation of wheat *(Triticum aestivum)* and *Avena* coleoptile segments (Thimann and Bonner 1949; Goren and Tomer 1971). Coumarin also inhibits growth of roots of intact wheat seedlings decreasing the elasticity of cell walls (Burstrom 1954). With excised maize roots, coumarin inhibits longitudinal growth of the elongating region but promotes transverse cell expansion (Svensson 1972). Coumarin and scopoletin inhibit the growth of *Phleum pratense* roots and scopoletin prevents trichoblast differentiation. The inhibitory effect of scopoletin on cell division in these roots is greater than that of coumarin (Avers and Goodwin 1956).

Coumarin inhibited the growth of rice, mung bean, lettuce and clover seedlings. Inhibition was accompanied by very pronounced tissue swelling at the base of the stem and at the root tip; this was due to cell enlargement (Hara et al. 1973). Elongation of plumular hook sections of mung bean was inhibited by coumarin which also induced very marked thickening of the sections. Coumarin can inhibit red light-induced opening of the hook of *Phaseolus* hypocotyls. The inhibitory concentration of coumarin causes ethylene release and the amounts of ethylene so produced can duplicate the effect of coumarin (Morgan and Powell 1970).

Coumarin markedly inhibits seed germination and is regarded by Mayer and Shain (1974) as a "hormonal regulator of germination". The inhibition of lettuce seed germination caused by coumarin resembles the natural dormancy of some lettuce varieties in that it is relieved by red light (Berrie et al. 1968). However, the validity of some experiments leading to this conclusion has been questioned (White et al. 1972).

Although coumarins are usually regarded as growth inhibitors, these compounds can also promote growth. Thus, coumarin inhibits growth of *Avena* coleoptiles (see above), but at very low concentrations elongation is promoted (Thimann and Bonner 1949). Coumarin retarded senescence of leaf tissue (Knypl 1967) and promoted expansion of leaf discs of *Chenopodium album* at 10 μM (Miller and Meyer 1951). Coumarin caused enlargement of epidermal cells of *Phleum pratènse* roots (Avers and Goodwin 1956), and stimulated elongation of excised segments of sunflower hypocotyls, pea epicotyls and bean hypocotyls (Neumann 1959, 1960). Pretreatment with coumarin for a short period markedly promoted subsequent growth of sunflower hypocotyl and pea epicotyl segments in an optimal concentration of IAA; a synergistic interaction between IAA and coumarin was indicated (Neumann 1960). Although the GA_3-induced elongation of bean epicotyl sections was suppressed by coumarin, the kinetin inhibition of elongation was alleviated markedly by coumarin (Hogetsu et al. 1974). These workers suggest these responses are probably linked with the ability of coumarin to inhibit cellulose synthesis (see below).

Certain of the above studies with intact seedlings and excised tissues indicate coumarin can promote both longitudinal cell expansion (elongation) and transverse cell expansion. The promotion of expansion of isolated plant cells by coumarin is particularly striking. In a suspension culture of soybean root cells, coumarin caused spherical cells to enlarge symmetrically, but only a part of each rod-shaped cell expanded in response to coumarin while the remainder of the cell retained its original form. The shape of rod-shaped cells was thus altered markedly by coumarin (Hara et al. 1973). Marked expansion of isolated leaf mesophyll cells with change in cell shape was also caused by coumarin (Harada et al. 1972). In plumular hook sections of mung bean, coumarin markedly inhibited cellulose synthesis but appeared to have little effect on pectin and hemicellulose formation (Hara et al. 1973). These workers suggest that coumarin-induced cell expansion is a consequence of inhibition of cellulose microfibril formation.

Daphnetin, but not coumarin, promoted growth of wheat roots at low concentrations (Burstrom 1954). Root formation on bean cuttings was promoted markedly by coumarin (Basu 1972) and by umbelliferone (Vazquez 1973). Coumarin is also a very effective promoter of tuber initiation on excised shoots of *Solanum tuberosum* cultured in vitro (Stallknecht 1972). Although a coumarin concentration of 10^{-3} M inhibited growth of artichoke tuber tissue, low concentrations (10^{-8}–10^{-5} M) promoted tissue growth (Duplessy-Graillott 1962; Bagni and Fracassini 1971).

Seed germination and seedling development of some species was promoted by low concentrations of coumarin (Knypl 1964, and references therein), 0.3 μM being optimal for barley seed germination and seedling growth. Scopoletin and 4-hydroxycoumarin induced germination of *Striga asiatica* (witchweed) seed but numerous other coumarins tested were inactive (Worsham et al. 1962). Coumarin also promoted growth of *Coumarouna* embryos (Valio 1973). Germination of uredospores of *Puccinia graminis* (wheat rust) was markedly promoted by coumarin, umbelliferone and daphnetin (Van Sumere et al. 1957).

Since certain coumarins possess phenolic hydroxyl groups, they can influence the oxidation of IAA; methyl umbelliferone promotes oxidation while scopoletin can inhibit it in vitro (Andreae and Collet 1968). Scopoletin modifies horseradish peroxidase to yield a stable form of the enzyme which is less active in reactions with molecular oxygen and IAA (Sirois and Miller 1972). Control of IAA oxidation by coumarins has been suggested as a regulatory mechanism in plant growth and there is limited evidence to support this view (Miller et al. 1975). It is noteworthy that the inhibition of cucumber radicle growth by coumarin is accompanied by an increase in the IAA oxidase and peroxidase activity of tissue extracts (Goren and Tomer 1971).

7.8.4. Structure–activity relationships

The ability of coumarins to inhibit growth of *Avena* roots has been assessed. Coumarin, 8-methylcoumarin, daphnetin and 4-methyldaphnetin all exhibited similar activity; all other tested compounds were much less inhibitory. These included 3-, 4-, 5-, 6- and 7-methyl coumarin, and umbelliferone, esculetin, scopoletin and herniarin (Goodwin and Taves 1950).

Numerous coumarin derivatives have been tested for ability to inhibit seed germination (Mayer and Evenari 1952; Berrie et al. 1968; Van Sumere et al. 1972). Mono-substitution by methoxyl at positions 4 and 8 did not result in appreciable reduction in activity. However, hydroxylation and other substitutions usually markedly reduced activity. In inhibiting both barley and lettuce seed germination, coumarin was markedly more effective than the naturally occurring derivatives umbelliferone, esculetin, herniarin and scopoletin (Van Sumere et al. 1972). Only one compound appears to be more effective than coumarin; tetrachlorocoumarin has been reported to be more active than coumarin in inhibiting wheat seed germination. (Mayer and Evenari 1952). The coumarin inhibition of lettuce seed germination can be overcome by exogenous gibberellin and it has been suggested that coumarins are in fact anti-gibberellins (Berrie et al. 1968). Their action is possibly a consequence of the structural similarity to the moiety of gibberellin with the lactone bridge. 3,4-Dihydrocoumarin has been found to interact with gibberellin in a manner which indicates competition for an active site (Berrie et al. 1968).

7.9. Other unsaturated lactones and chelidonic acid

In addition to simple coumarins, several other unsaturated lactones merit discussion. These are considered in this section together with the related compound chelidonic acid, a γ-pyrone. Chrysartemin A and B and chlorochrymorin are unsaturated lactones which are not dealt with in this section. Since these compounds were isolated as factors which promoted rooting of cuttings, it is appropriate to include them in section 7.10.1.

7.9.1. Protoanemonin

This compound (XXXII), the lactone of γ-hydroxyvinylacrylic acid, occurs in *Ranunculus* spp. and is liberated enzymically from a glucoside (ranunculin XXXIII) present in these species (Hill and Van Heyningen 1951). Protoanemonin gradually polymerizes to a dimer, anemonin. The occurrence of ranunculin and protoanemonin in *Clematis flammula* has been reported (Font and Pascual 1966) and thus these compounds are not confined to the family Ranunculaceae.

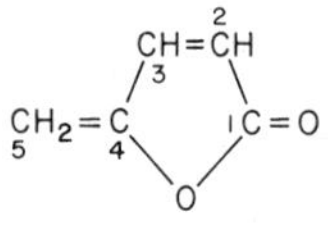

XXXII

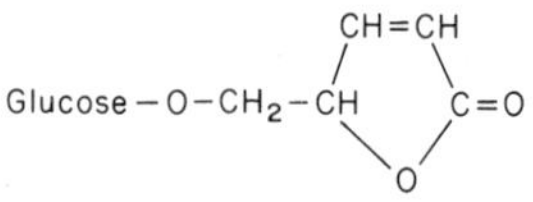

XXXIII

Protoanemonin inhibits auxin curvature of slit pea stems and elongation of *Avena* coleoptile segments, but at low concentrations elongation is promoted (Thimann and Bonner 1949). In the *Avena* response, protoanemonin is markedly more active than coumarin both as a promoter and an inhibitor of growth. The growth of roots and mitosis in roots is also inhibited by protoanemonin (Erickson and Rosen 1949).

The biosynthesis of protoanemonin has been investigated only recently (Suga and Hirata 1973). It is formed by condensation of glycerate with malonate with loss of carbon dioxide and subsequently lactonization and dehydration. C-1 and C-2 of XXXII are derived from malonate, while C-3, C-4 and C-5 represent the glycerate moiety.

7.9.2. Parasorbic acid

For over a century it has been known that the berries of the rowan (mountain ash, *Sorbus aucuparia*) contain a lactone termed parasorbic acid (Hofmann 1859). The structure was established as XXXIV (Kuhn and Jerchel 1943; Haynes and Jones 1946) with a (5*S*)-configuration (Kuhn and Kum 1962; Lukes et al. 1962). Structurally-related compounds known to occur naturally are XXXV and XXXVI (Crombie 1955; Hlubucek and Robertson 1967). Parasorbic acid inhibits seed germination and cell enlargement in *Avena* coleoptiles and etiolated pea stems (Kuhn et al. 1943; Larsen 1947; Thimann and Bonner 1949).

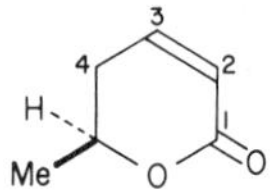

XXXIV

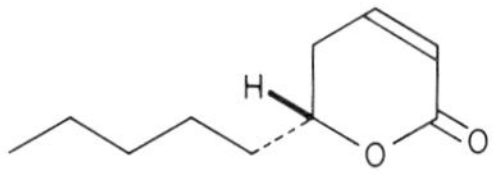

XXXV

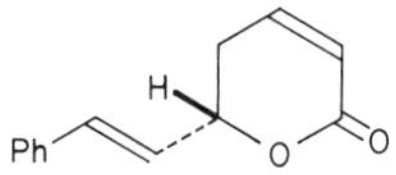

XXXVI

The biosynthesis of parasorbic acid has been studied in ripening rowanberries attached to the tree (Crombie and Firth 1968). The acid is produced by an acetate-malonate pathway and any pathway involving direct formation from glucose can be of only minor importance.

7.9.3. Psoralen

For maximal percentage germination, *Psoralea subacaulis* seeds require both scarification and leaching. The leachate was inhibitory to embryo growth of *P. subacaulis* and to seed germination and root growth of unrelated species (Baskin et al. 1967). The inhibitor was located in the seed coat but not in the embryo. From the leachate, a furanocoumarin was isolated and identified as psoralen (XXXVII). This compound, previously isolated from other species (see references in Baskin et al. 1967), appeared to account for the inhibitory activity. Lettuce seed germination and seedling growth were markedly inhibited by psoralen and the related compounds angelicin (XXXVIII) and xanthotoxin (XXXIX) which also occur in plants (Rodighiero 1954).

The biosynthesis of furanocoumarins has been discussed in detail by Floss (1972). These compounds are derived from umbelliferone by isoprenylation adjacent to the hydroxyl group, closure of the furan ring and elimination of an isopropyl group.

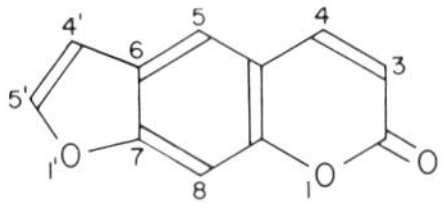

XXXVII

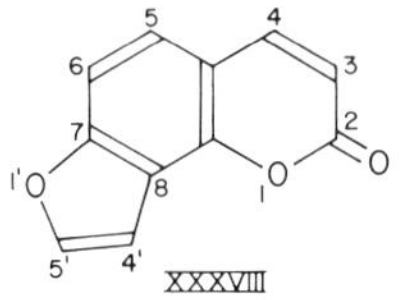

XXXVIII

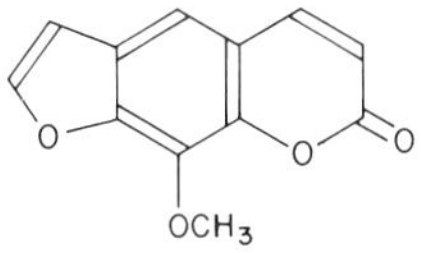

XXXIX

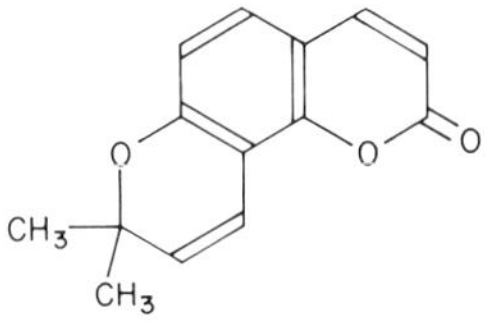

XL

7.9.4. *Seselin*

The pyranocoumarin, seselin (XL), has been isolated from citrus roots and shown to occur in other species of Rutaceae (Tomer et al. 1969). Seselin is probably the IAA-oxidase cofactor detected previously in citrus roots (Goldschmidt et al. 1968; Goren and Tomer 1971). Seselin inhibited radicle growth of cucumber and other seedlings, the inhibition being accompanied by an increase in the peroxidase and IAA oxidase activity of the tissue. Coumarin was much less effective than seselin in suppressing GA_3-induced release of reducing sugars from barley endosperm (Goren and Tomer 1971).

7.9.5. *Heliangine*

An inhibitor which was transported from the leaf to the stem of sunflower plants *(Helianthus annuus)* was detected by Shibaoka (1961). The movement of this compound, which inhibited growth of sunflower epicotyls, was promoted by illumination of the leaf. A substance with the same properties was purified from leaves of Jerusalem artichoke *(Helianthus tuberosus)*, termed heliangine, and assigned structure XLI (Shibaoka 1961;

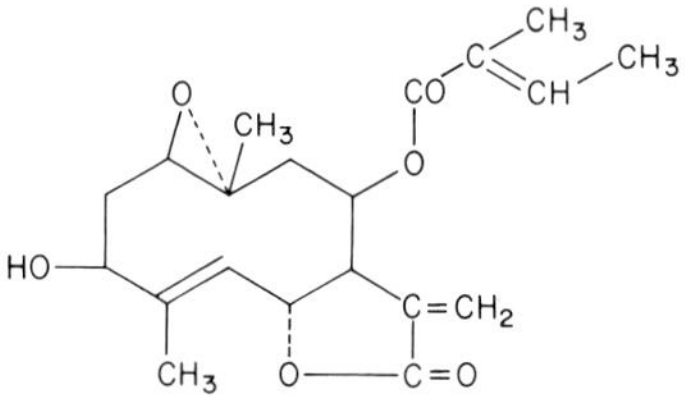

XLI

Morimoto et al. 1966). This structure was confirmed by X-ray crystallography of dihydroheliangine monochloroacetate (Nishikawa et al. 1966). Heliangine was the first germacrane-type sesquiterpene found to have plant growth regulatory activity; it is the tiglic acid ester of a C_{15} ring compound with an $\alpha\beta$-unsaturated lactone moiety.

Heliangine inhibited growth of *Avena* coleoptiles and the GA_3-induced elongation of stem segments of *Phaseolus angularis.* However, the IAA-induced elongation of these stem segments was not inhibited (Yamaki et al. 1966). An unexpected property of the inhibitor was its ability to promote root formation on mung bean stem cuttings and sunflower callus tissue (Yamaki et al. 1966; Shibaoka et al. 1967b). A number of pyrimidines, but not purines, inhibited the promotion of rooting on bean cuttings (Shibaoka et al. 1967a).

7.9.6. Chelidonic acid

This compound, 4-pyrone-2,6-dicarboxylic acid (XLII), was first isolated from *Chelidonium majus* about 1839 and is now known to occur in a wide variety of plants (Ramstad 1953) including grasses (Bough and Gander 1972). Chelidonic acid inhibited growth of pea stem segments at 10 μM, but in the presence of IAA, inhibition was observed only at considerably higher concentrations and at 10 μM a marked promotion of growth occurred (Leopold et al. 1952). In the presence of auxin, chelidonic acid also promotes growth of other tissues including *Avena* coleoptiles.

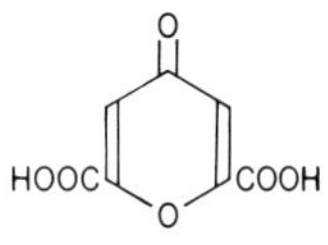

XLII

Phenylalanine, tyrosine, shikimate, pyruvate, aspartate and acetate were poor precursors of chelidonic acid, whereas glucose, ribose and carbon dioxide were good precursors (Bohm 1966; Malcolm and Gear 1971). The acid appears to be produced by condensation of a three-carbon unit and a four-carbon unit both derived from photosynthetic metabolism (Malcolm and Gear 1971).

7.9.7. Sesquiterpenoid dilactones

The sesquiterpenoid dilactones, elephantin (XLIII), elephantopin (XLIV) and vernolepin (XLV) have been reported to be strong inhibitors of extension growth of wheat coleoptile sections (Sequeira et al. 1968). However, vernolepin, purified from *Vernonia hymenolepis* (Kupchan et al. 1968), was the most effective inhibitor and at 18 μM suppressed growth by 25% in the coleoptile assay. The respiration rate of inhibited tissues was identical to that of the control coleoptile sections. After inhibited coleoptile sections had been washed with water and then transferred to IAA solution, the tissue exhibited a growth response to auxin (Sequeira et al. 1968). This indicates vernolepin is not simply a toxic compound.

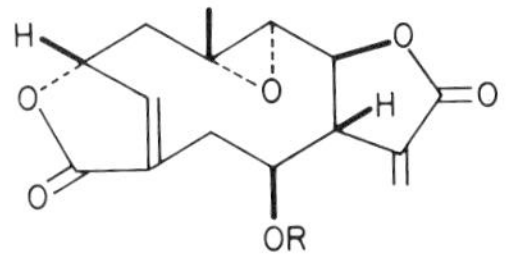

XLIII , XLIV

XLIII R = COCH=C(CH$_3$)–CH$_3$

XLIV R = COC(CH$_3$)=CH$_2$

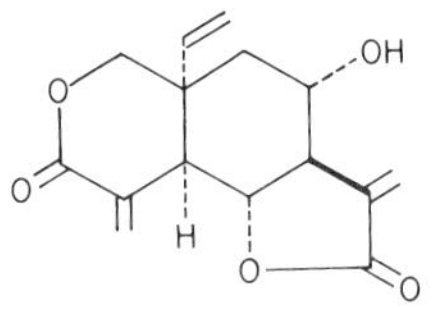

XLV

Elephantin and elephantopin are sesquiterpenes of the germacrane type (Kupchan et al. 1966) and are hence structurally similar to heliangine.

7.9.8. Norditerpenoid dilactones

Partially purified extract of bark of *Podocarpus* spp. was found by Galbraith et al. (1970) to inhibit markedly cell expansion in pea stem segments. From the extract, two very effective inhibitors were isolated and termed podolactones A and B which were assigned structures XLVI and XLVII respectively. Podolactone A also inhibited cell division in Jerusalem artichoke tissues (Galbraith et al. 1970).

The ability of a number of related norditerpene dilactones to inhibit growth of pea stem segments was also assessed (Sasse et al. 1972). Inumakilactones A (XLVIII) and B (XLIX), compounds purified by Ito et al. (1971), were particularly effective inhibitors and when tested with hook segments were more active than ABA and podolactones A and B.

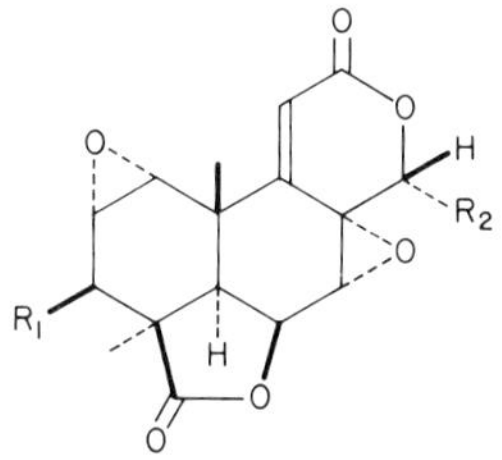

XLVI – XLIX

XLVI R$_1$ = H, XLVII R$_1$ = OH: R$_2$ = –C(OH)(CH$_2$OH)–CH$_3$

XLVIII R$_1$ = OH, R$_2$ = –CH(OH)–CH$_3$

XLIX R$_1$ = OH, R$_2$ = –CH=CH$_2$

Nagilactones A, B, C, D, E and F are also norditerpenoid dilactones with strong growth inhibitory activity. These compounds were purified from the seed or root bark of *Podocarpus nagi* (Hayashi et al. 1968; Hayashi et al. 1972) and are structurally very similar to the podolactones. In the *Avena* coleoptile section growth assay, all the nagilactones were inhibitory but nagilactone F (L) was the most active and caused a 50% inhibition at

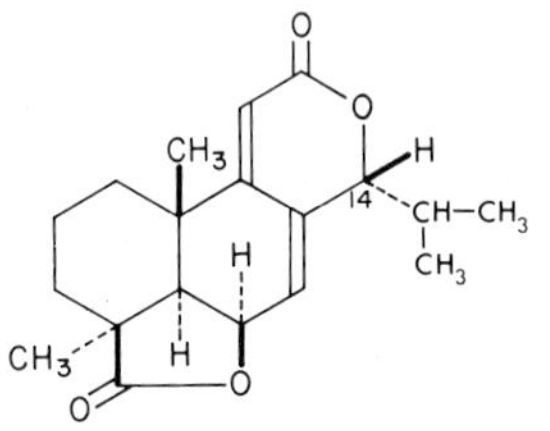

L

10^{-7} *M*. While nagilactones E and F exhibited only inhibitory activity, lactones A, B, C and D, which possess an α-pyrone ring, promoted coleoptile section growth at low concentrations (10^{-8}–10^{-7} *M*). Competitive experiments indicated the lactones acted as antiauxins when functioning as inhibitors (Hayashi and Sakan 1974).

7.9.9. Xanthinin

From leaves of cocklebur *(Xanthium pennsylvanicum)*, Bonde and Khudairi (1954) extracted an inhibitor which suppressed the growth of *Avena* coleoptile sections and intact seedlings. A crystalline inhibitor was purified (Geissman et al. 1954) and assigned structure LI (Deuvel and Geissman 1957; cf. Minato and Horibe 1965). Hence xanthinin is a sesquiterpenoid α-methylene-γ-lactone with a single 7-membered carbocyclic ring. It resembles the rooting factors chrysartemins A and B in structure (see section 7.10.1).

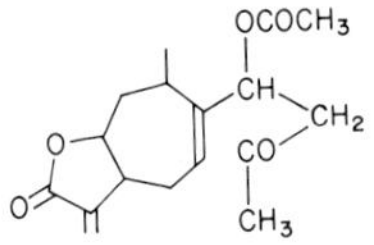

LI

A related compound termed xanthatin also occurs in young *Xanthium* leaves, and appears to be formed from xanthinin by enzymic deacetylation (Khan 1975). Xanthatin has been reported to possess a CH_3·CO·CH=CH- side chain but is otherwise identical to xanthinin. Both xanthinin and xanthatin inhibit development of axillary buds when applied to the adjacent leaves (Khan 1975).

7.10. Regulators of root formation on cuttings

Although IAA plays a key role in the initiation of roots, other naturally occurring compounds can markedly promote or inhibit root development on cuttings. The effect of simple phenolic acids has been discussed previously (see section 7.6). It is now known that certain cyclic terpenes of unusual structure may also regulate initiation of roots on cuttings; one such compound (heliangine) has already been noted. Other compounds are discussed briefly below; in addition, dihydroasparagusic acid (see section 7.13.6) can promote adventitious root formation on mung beans (Kuhnle et al. 1975).

7.10.1. Chrysartemins A and B and chlorochrymorin

Using the mung bean rooting assay, Hess (1964) found that the content of root-promoting substances in the juvenile form of *Hedera helix,* which roots readily as cuttings, was greater than that in the mature form, which is difficult to root. Similarly, easy-to-root varieties of *Chrysanthemum* were richer in these factors than difficult-to-root varieties. The promoters detected appeared to act as cofactors of IAA. A number of active compounds were detected by Hess (1964) and those from *Chrysanthemum* were partially purified. The principal active fraction ("cofactor 4") appeared to be a mixture of four oxygenated terpenoids.

Further work by Osawa et al. (1971) identified two root promoting compounds which were regarded as components of "cofactor 4". These were extracted from *Chrysanthemum morifolium* and were termed chrysartemin A (LII) and chrysartemin B (LIII). In structure these two compounds resemble xanthinin. Both are α-methylene-γ-lactones, a structural feature previously noted in heliangine which also promotes rooting of cuttings. However, chrysartemin A was considerably more active than compound B or heliangine (Osawa et al. 1971). Unlike heliangine, however, A and B were only weak inhibitors of IAA-induced *Avena* coleoptile growth.

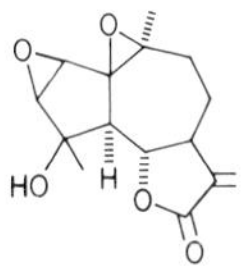

LII

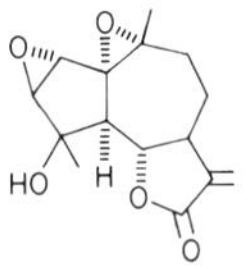

LIII

From juvenile plants of *Chrysanthemum morifolium,* a third promoter of rooting of mung bean cuttings was purified by Osawa et al. (1973). It was identified as the chlorine-containing sesquiterpene lactone LIV and termed chlorochrymorin.

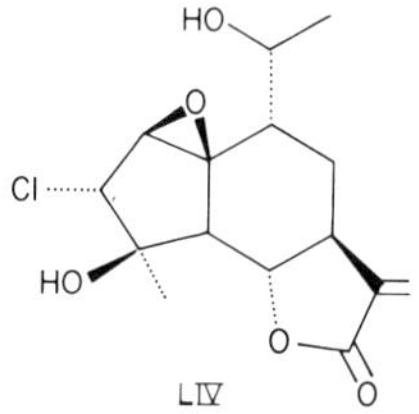

LIV

7.10.2. Portulal

From leaves of *Portulaca grandiflora,* a compound was isolated which inhibited IAA-induced elongation of *Avena* coleoptile segments and also suppressed adventitious root formation of *Raphanus* hypocotyl cuttings (Mitsuhashi and Shibaoka 1965). The active compound was assigned structure LV and termed portulal (Yamazaki et al. 1969). Portulal, an aldehyde, is a novel type of bicyclic diterpene as it contains a perhydroazulene nucleus.

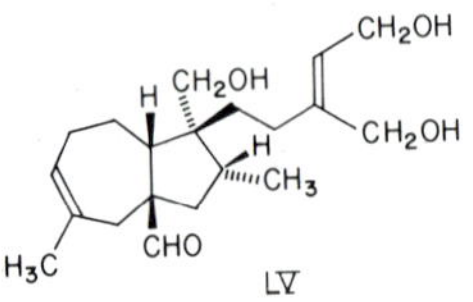

LV

7.10.3. Rooting inhibitors of Eucalyptus

Rooting inhibitors detected in *Eucalyptus grandis* may regulate root formation on stem cuttings of this species (Paton et al. 1970). Three closely related inhibitors termed G1, G2 and G3 ("G" derived from *grandis*) were purified from the adult tissue and assigned structures LVI, LVII and LVIII respectively (Crow et al. 1971; Sterns 1971; Nicholls et al. 1972). G1 and G2 are diastereomers. All three inhibitors contain an unusual peroxide linkage and are structurally related to compounds such as leptospermone (LIX) isolated from the steam-volatile oils of *Eucalyptus* spp. At a concentration of 1 m*M*, G1, G2 and G3 completely inhibited root formation on cuttings of mung bean, *E. grandis* and *E. deglupta.*

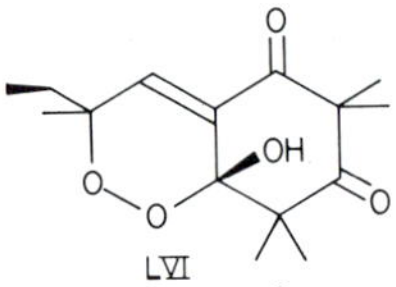

LVI

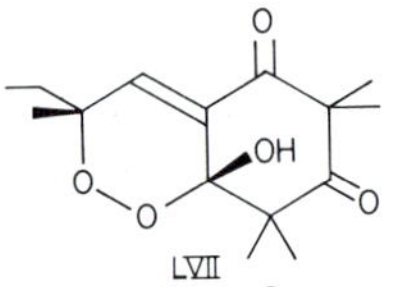

LVII

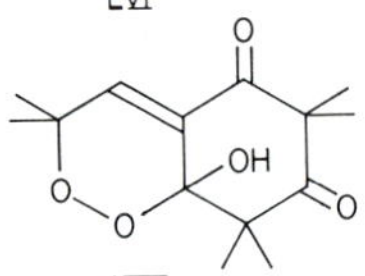

LVIII

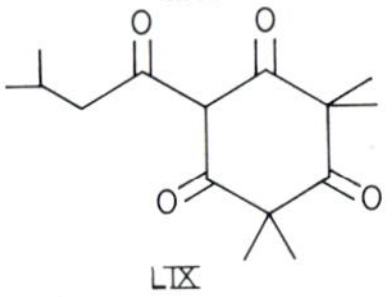

LIX

These rooting inhibitors are found in extracts of adult leaves and bark, but do not appear to occur in juvenile tissue which roots readily. Furthermore, the inhibitors are not transported through juvenile tissue to any appreciable extent (Paton and Willing 1974). These workers suggest that one function of juvenile tissue is to protect the root system of the seedling from G inhibitors present in the adult tissue. Juvenile leaves which lack G inhibitors tend to wilt while the adult leaves with these inhibitors seldom do so. Paton and Willing (1974) suggest G inhibitors have a role in the resistance of adult leaves to water deficiency. G inhibitors are discussed further in Chapter 2, section 2.7.7, Volume II.

7.11. Fatty acids, their esters and related compounds

A diversity of compounds of this type influence plant growth when supplied exogenously and are discussed below. In addition compounds extracted from rape pollen which promote bean seedling growth and which are regarded as lipids, have been termed brassins (Mitchell et al. 1970; Mitchell and Gregory 1972; Mandava et al. 1973). However there is no convincing evidence that the observed activity was due to a new type of plant hormone as claimed by Mitchell and coworkers. Milborrow and Pryce (1973) who have carefully evaluated the findings of Mitchell et al. consider the promotion of bean seedling growth could be due to gibberellin. The active compounds do not appear to have been obtained in a state of purity and hence the conclusion of Mandava and Mitchell (1972) that brassins are glucose esters of fatty acids appears premature. Because of the uncertainty regarding the significance of brassins, these compounds are not discussed further.

7.11.1. Oleanimins

Even in the presence of auxin plus gibberellin, excised sections of pea stems elongate markedly less than the corresponding zones in the intact plant. Part of this "missing growth" was found to be restored by a variety of lipid substances supplied as stable aqueous emulsions (Stowe 1960). The most effective compounds were the fatty acid esters, methyl myristate, methyl palmitate, methyl oleate, methyl linoleate, methyl arachidonate, triolein and trilinolein. While these compounds were effective at 10–50 μM in the presence of IAA plus GA_3, the free fatty acids were inactive.

In further studies with pea stem sections, Stowe and Obreiter (1962) found that a number of natural oils (e.g. olive and linseed oil) were as effective as the fatty acid esters in promoting elongation. The fat soluble vitamins E and K, were also active and their activity at low concentrations (about 3 μM) exceeded that of the acid esters. In contrast, vitamins A and D_2 and β-carotene were inactive (Stowe and Obreiter 1962). Farnesol (LXI), insect juvenile hormone (LX), C_{14}, C_{15} and C_{16} alkyl nitriles, and C_{16} and C_{18}

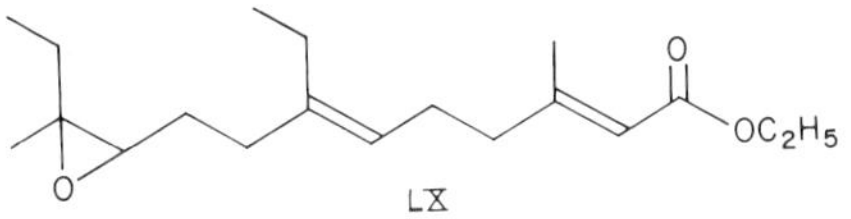

LX

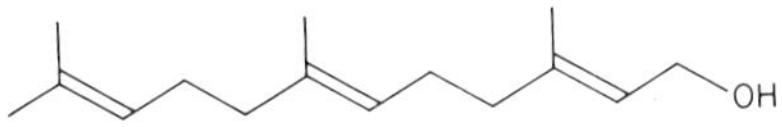

LXI

alkyl acetylenes were also effective promoters of elongation (Stowe and Hudson 1969). Alkyl chlorides, bromides and iodides, alkyl benzenes and dialkyl ethers and sulphides are also active provided the length of the molecules exceeds 2.0 nm; the optimum length is about 2.9 nm (Stowe and Dotts 1971). These workers have proposed the term "oleanimin" for the lipids which promote elongation of pea stem sections in the presence of auxin; structural features which appear necessary for activity are a molecular length of over 2.0 nm and a low polarity (Stowe and Dotts 1971; Stowe 1972). These workers suggest that oleanimins act by altering the properties of a regulatory membrane which is modified by the ability of these regulators to force lecithin molecules apart.

Oleanimins stimulate respiration of pea stem sections (Penny and Stowe 1965, 1966) and induce a period of ethylene formation in them after a lag period of about 1 hr (Iwata and Stowe 1973). However, the effect of oleanimins on pea stem growth does not appear to be mediated by ethylene. Growth, respiration and ethylene production are all promoted by oleanimins at the same low concentrations and with very similar kinetics. It has been suggested that all three responses are regulated by oleanimins at a single locus in a membrane, probably a respiration-modulating membrane (Iwata and Stowe 1973).

The ancient Egyptians used olive oil to promote the ripening of figs, a small drop applied to the eye of the fruit being very effective. The oil causes fig fruit enlargement and stimulates ethylene production and respiration (Hirai et al. 1966, 1967, 1968). Stowe (1974) suggests that olive oil contains the oldest known plant growth substance, presumably triolein. Oleanimins also promote ripening of fig fruits probably as a result of promotion of ethylene production (Stowe 1974).

7.11.2. Other esters and long-chain alcohols

From dwarf pea seedlings grown under red light, α-stearoylglycerol was isolated as an inhibitor which suppressed the stem elongation of these seedlings especially in the presence of GA_3. An application of 1 μg to the apical bud produced complete inhibition of the response of GA_3. However, the glyceride did not inhibit elongation of *Avena* coleoptile segments (Komoto et al. 1972b).

Methyl caprate (methyl ester of C_{10} acid) strongly inhibited lateral bud development of decapitated tobacco plants (Tso 1964). The methyl esters of C_8, C_{12} and C_{14} fatty acids were also strongly inhibitory. Esters of C_8–C_{12} fatty acids can be used to selectively inhibit the terminal meristem without damaging axillary buds or foliage of a variety of plants including *Chrysanthemum* (Cathey et al. 1966; Steffens and Cathey 1969). Aliphatic alcohols (C_9 to C_{11}) are also very effective inhibitors of axillary and terminal bud growth of tobacco (Steffens et al. 1967) and selectively inhibit the terminal bud of some species (Steffens and Cathey 1969). Long chain esters and alcohols appear to penetrate readily into the meristematic tissues of buds but not into other regions of the plant (Steffens et al. 1967).

A crude preparation of long-chain aliphatic alcohols prepared from tobacco leaves and apical tissues was found to promote elongation of *Avena* first-internode sections (Vlitos and Crosby 1959). These workers also reported that C_{18}–C_{22} aliphatic alcohols exhibited similar activity. However Kefford (1962) detected only very small and variable responses to 1-docosanol with *Avena* first-internode sections and with excised tobacco stem pith. In *Avena* coleoptile assays, the alcohol was inactive.

7.11.3. Fatty acids

Compounds extracted from pea seedlings, which inhibited their response to GA_3, were identified as palmitic, oleic and fumaric acids and methyl hydrogen succinate (Komoto et al. 1972a). However they were all much less active than ABA and α-stearoylglycerol (cf. Komoto et al. 1972b).

Capric acid, a growth inhibiting substance purified from dormant iris bulbs, lauric and myristic acids inhibited growth of *Avena* coleoptile segments (Ando and Tsukamoto 1974). Germination of mustard seeds was inhibited by certain saturated fatty acids, notably nonanoic acid (Le Poidevin 1965). Linoleic acid inhibited unrolling of etiolated barley leaf segments (Menhenett and Carr 1973).

Fatty acids (C_6 –C_{10}) are widely distributed in plants, but seed and dormant buds contain very high levels relative to actively growing tissues (Berrie et al. 1975). These acids inhibited oat seed germination, nonanoic acid (C_9) being the most active. This acid also inhibited lettuce seed germination; it is noteworthy the inhibition could be alleviated by red light or by exogenous cytokinin or gibberellin (Berrie et al. 1975). The acids possess chromatographic properties similar to inhibitor-β and may contribute to this complex.

A number of fatty acids, particularly pentanoic (C_5) and nonanoic, strongly inhibit gibberellin-induced α-amylase production in embryo-free half seeds of barley (Buller et al. 1976). Inhibition by nonanoic acid was complete at 1 mM, but non-existent at 10 μM; in contrast ABA inhibits completely at 0.1 mM but not at all at 0.01 μM. Unlike ABA, fatty acids exhibit a threshold effect which could be important in "all or nothing" phenomena such as germination (Buller et al. 1976).

7.12. Other growth-promoting compounds derived from higher plants

7.12.1. Phaseolic acid

A keto-carboxylic acid which stimulated elongation of dwarf pea seedlings was isolated from bean seeds by Redemann et al. (1968). It was termed phaseolic acid and two alternative structures (LXII, LXIII) were proposed for the compound. Phaseolic acid also promoted elongation of dwarf maize mutants and pea stem segments; it induced α-

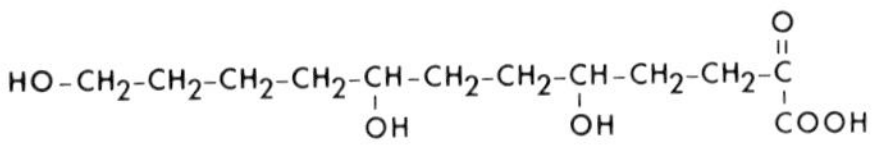

LXII

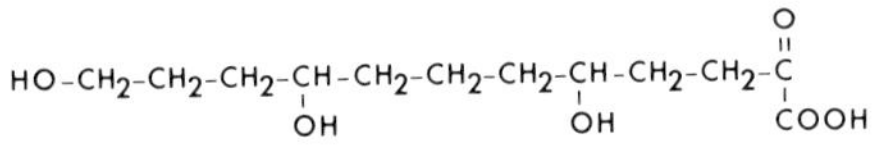

LXIII

amylase synthesis in the barley endosperm gibberellin assay (Redemann et al. 1968). Hence phaseolic acid evokes certain responses characteristic of gibberellins but the effective concentrations are about 1,000 times those of GA_3. However phaseolic acid exhibited activity approximately equal to that of kinetin in the barley leaf senescence assay.

7.12.2. Strigol and strigyl acetate

Seeds of several species of *Striga,* like those of certain other angiosperm parasites, will germinate readily only when they are near a host root. The stimulation is due to compounds exuded by the root and absorbed by the seed (Brown et al. 1949, and references therein). Purification of germination stimulants of this type proved to be an extremely difficult task. Unsuccessful attempts were made to isolate and identify the *Striga hermonthica* germination stimulant produced by *Sorghum vulgare* (Brown et al. 1949; Brown et al. 1952a), the *Orobanche minor* germination stimulant secreted by linseed roots (Brown et al. 1951; Brown et al. 1952b), and the witchweed *(Striga lutea)* germination stimulant produced by *Zea mays* (Worsham et al. 1964).. A complex of stumulants appeared to be produced by *Sorghum vulgare* (Brown et al. 1952a) and by maize roots (Sutherland 1960).

A crystalline, highly active compound, which induced germination of *Striga lutea* seed, was finally purified from cotton root exudates by Cook et al. (1966). The crystalline compound, termed strigol, caused a 50% germination of *S. lutea* seeds at 10^{-11} *M* in water, and a second non-crystalline stimulant (stimulant B) was at least as active.

Strigol has been shown to possess the structure LXIV (Cook et al. 1972; Coggon et al. 1973) and this has been confirmed by syntheses (Heather et al. 1974; MacAlpine et al.

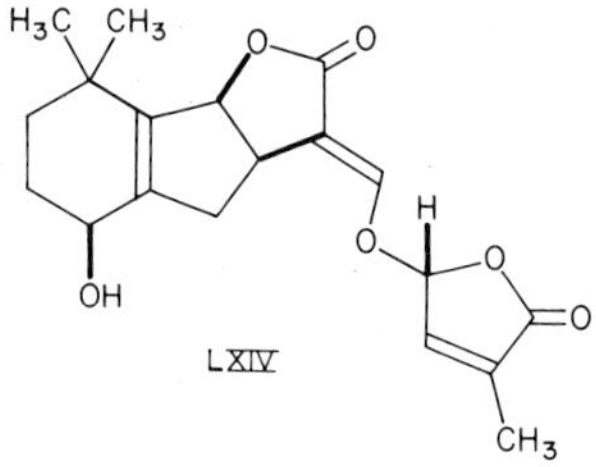

LXIV

1974). Stimulant B is strigyl acetate. The extreme potency of strigol and the probable occurrence of chemically similar germination stimulants in other plants (Cook et al. 1972) combine to make strigol a growth regulator of great interest. This unsaturated lactone may represent a new class of phytohormones. The only other compounds known to cause germination of *Striga* seed are certain cytokinins and coumarins (Worsham et al. 1962), thiourea and allylthiourea (Brown and Edwards 1945) and D-xylulose (D-xyloketose) (Brown et al. 1949).

7.12.3. Polyamines

Aliphatic diamines and polyamines are of widespread occurrence in plants. Spermine, spermidine, putrescine and cadaverine were found to promote the growth in vitro of explants of dormant artichoke *(Helianthus tuberosus)* tubers (Bertossi et al. 1965; Bagni 1966). An optimal concentration of spermine or spermidine (0.1 m*M*) induced a growth response similar to that caused by IAA (2 μM). Polyamines have been reported to promote protein synthesis in vitro by ribosomes from dormant artichoke tubers (Cocucci and Bagni 1968; see also discussion of Bagni and Fracassini 1974). It has been suggested that the effect of IAA on growth of artichoke tissue results at least partly from IAA-induced polyamine synthesis (Cocucci and Bagni 1968).

Polyamines are present in only trace amounts in dormant artichoke tubers but the level increases about 10–20-fold during active growth which can be induced by IAA. The polyamine content of cultures of plant tumour tissue is reported to be much greater than that of cultures of normal tissue (Bagni and Fracassini 1974). Accumulation of agmatine and putrescine in higher plants grown under conditions of potassium deficiency is a well-established phenomenon (Smith 1971). For a discussion of the biosynthesis and metabolism of polyamines in plants, the reader is referred to the excellent reviews of Smith (1971, 1975).

The growth promoting activity of polyamines in both plant and animal tissues may be associated with their ability to regulate nucleic acid metabolism and protein synthesis. A number of very relevant observations have been made with animal and bacterial systems. A considerable proportion of intracellular polyamines is associated with ribosomes and these amines promote association of ribosomal subunits, and the binding of aminoacyl-tRNA to ribosomes. There is a close correlation between the intracellular concentration of spermidine and the rate of RNA synthesis in both rat liver and *E. coli* (Stevens 1970). Spermidine stabilizes ribosomes (Hardy and Turnock 1971) while spermine promotes binding of ribosomes to endoplasmic reticulum membranes in vitro (Khowaja and Raina 1970). Polyamines promote aminoacyl-tRNA formation (Igarashi and Takeda 1970) and stimulate tRNA methylation in vitro (Young and Srinivasan 1971).

The ability of polyamines to promote oxidation of NADPH by spinach flavoprotein-iron-sulphur protein complexes is a noteworthy biochemical response in plant systems (see Smith 1975). Another is the increase in phenylalanine ammonia lyase activity induced by polyamines (especially spermine) in pea pods (see Smith 1975).

7.12.4. Amino acids which promote growth of plant tumours

Several unusual basic amino acid derivatives have been isolated from crown-gall tumours and occur in markedly lower amounts in normal plant tissues (Lippincott et al. 1972; Johnson et al. 1974). These include octopine [N_2-(1-carboxyethyl)-L-arginine], lysopine [N_2-(1-carboxyethyl)-L-lysine] and nopaline [N_2-(1,3-dicarboxypropyl)-L-arginine]. These compounds and also carnosine promoted crown-gall tumour growth on bean leaves (Lippincott et al. 1972; Lippincott and Lippincott 1975). A crown-gall tumour growth factor extracted from bean leaves inoculated with *Agrobacterium tumefaciens* was identified as γ-aminobutyric acid (Peters et al. 1974).

Octopine and lysopine promoted bud formation in the moss *Pylaisiella selwynii.* Cytokinins also induced buds but these failed to develop into normal gametophores (Spiess et al. 1972).

7.12.5. Dihydroconiferyl alcohol and 4-hydroxybenzyl alcohol

Removal of cotyledons from lettuce seedlings was observed to decrease markedly the response of the hypocotyl cells to gibberellin. This suggests that a factor produced by the cotyledons interacts with gibberellin to promote hypocotyl elongation (Kamisaka 1973). The active compound was purified and identified as dihydroconiferyl alcohol (LXV)

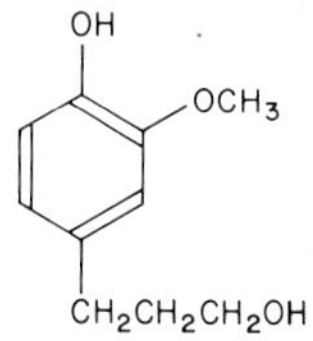

LXV

(Shibata et al. 1974) which is probably an endogenous growth regulator in lettuce seedlings. The activity of structurally-related compounds has been reported by Shibata et al. (1975). If the phenolic hydroxyl is blocked by a methoxy or glucosyl group, activity is abolished. Common phenolic acids (e.g. chlorogenic acid) are inactive.

Gibberellin-induced elongation of cucumber hypocotyls requires a factor supplied by the cotyledons (Katsumi et al. 1965). However dihydroconiferyl alcohol does not appear to be this cucumber factor (Sakurai et al. 1974).

Dihydroconiferyl alcohol also acts synergistically with IAA to promote elongation of cucumber hypocotyl segments (Sakurai et al. 1974). The effect of the alcohol is particularly pronounced when the concentration of IAA is suboptimal. In this cucumber hypocotyl system, dihydroconiferyl alcohol appears to act as an inhibitor of IAA oxidase (Sakurai et al. 1975).

The related compound, 4-hydroxybenzyl alcohol, has been isolated from muskmelon *(Cucurbita moschata)* seedling extracts and found to be a very effective cofactor for IAA oxidase (Mumford et al. 1963). The alcohol inhibited IAA-induced growth of oat coleoptiles by 18% at 8 μM.

7.12.6. Sterols

Plant sterols are usually regarded as secondary metabolites with little physiological significance. However, some interesting effects of these compounds on plant growth are beginning to accumulate. Thus, sitosterol stimulated the growth of dwarf pea plants (Kopcewicz 1969) and promoted initiation of flower buds in *Chrysanthemum* (Biswas et al. 1967). Cholesterol did not evoke these responses. However, sitosterol has been reported to inhibit the response of dwarf pea plants to GA_3 (Komoto et al. 1972b). Estrogenic compounds affect vegetative growth, sex expression and flowering (see references in Grunwald 1975). The content of estrogen-like substances in some plants appears to be correlated with reproductive development (see Chapter 6, Volume II, section 6.7). Exogenous auxin and cytokinin markedly affect sterol levels in plant tissues (Brain and Lockwood 1976; Lockwood and Brain 1976).

A compound which enhanced IAA-induced elongation of *Avena* coleoptile segments was extracted from peanut plants by Kimura et al. (1975). It was identified as stigmasteryl-β-D-glucoside by these workers. Free sterols were inactive in the growth assay.

The biochemistry and physiology of phytosterols has been reviewed recently (Grunwald 1975). The biosynthesis and metabolism of glycosides of sterols has recently attracted considerable attention (Wojciechowski et al. 1976).

7.13. Other growth-inhibiting compounds derived from higher plants

A considerable number of growth-inhibiting compounds, which do not fall into any of the previous categories, have been extracted from higher plants and are discussed below. The batatasins very probably exert a regulatory role in vivo. However there is no evidence that the other compounds discussed below regulate growth in the plant tissues from which they were extracted.

Compounds not discussed below but which merit mention are α-tomatine which inhibits markedly IAA-induced elongation of wheat coleoptile segments (Roddick 1975) and diterpenoid alkaloids which inhibit pea cambial growth (Waller and Burstrom 1969).

7.13.1. Batatasins

When vigorous growth of the stem of yam *(Dioscorea batatas)* ceases, the plant forms bulbils by thickening of axillary buds. The mature bulbils require stratification for at least three months before their dormancy is broken. ABA has been isolated from yam bulbils, but ABA is not an effective inducer of dormancy in bulbils after stratification (Hashimoto and Tamura 1969). However, three new growth inhibitors, termed batatasins I, II and III, were isolated from the dormant bulbils and found to induce dormancy effectively in stratified yam bulbils (Hashimoto et al. 1972). Bulbils treated with batatasins regained the ability to sprout after additional stratification.

While ABA is distributed uniformly throughout the bulbil tissues, the three batatasins are localized in the outer layers (Hasegawa and Hashimoto 1973). The level of batatasin I was found to increase with development of dormancy. When the bulbils were stratified, the batatasin content, and particularly the content of batatasin I, decreased; however, the level of ABA remained constant (Hasegawa and Hashimoto 1973, 1975; Hashimoto and Hasegawa 1974). After transfer to sprouting beds at 23 °C, the batatasin content of stratified bulbils decreased very rapidly but that of non-stratified bulbils showed little change (Hasegawa and Hashimoto 1975). The sprouting of yam bulbils was suppressed by GA_3. GA_3 treatment increased the content of batatasins but not that of ABA (Hasegawa and Hashimoto 1974). These workers suggest that the inhibition of sprouting caused by GA_3 is a consequence of increased batatasin levels in the bulbils. Consideration of the above observations indicates that batatasins are involved in the regulation of dormancy in yam bulbils. Batatasin I also inhibits growth of *Avena* coleoptile sections but is much less active than ABA (Hashimoto and Hasegawa 1974).

Batatasin I, a phenanthrene derivative, has been assigned structure LXVI (6-hydroxy-2,4,7-trimethoxyphenanthrene) by Letcher (1973). Batatasin III (LXVII) is the bibenzyl derivative, 3,3′-dihydroxy-5-methoxy-bibenzyl (Hashimoto et al. 1974). Substituted phenanthrenes and bibenzyls have previously been purified from plant sources, but this is the first evidence that such compounds may be important as growth regulators.

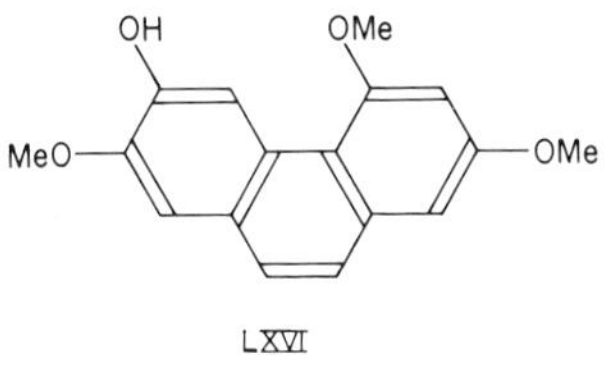

LXVI

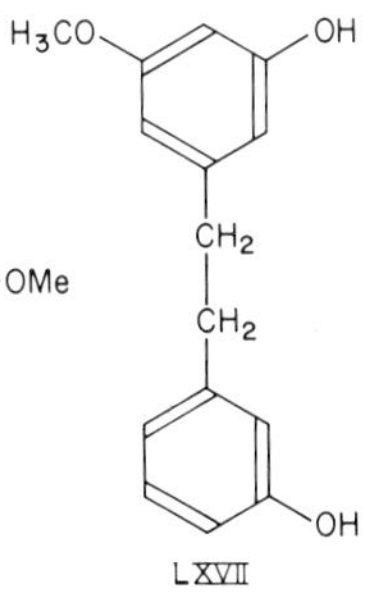

LXVII

7.13.2. Lycoricidinol and lycoricidine

Extracts of bulbs of *Lycoris radiata* strongly inhibit the growth of *Avena* coleoptile sections. This activity is due to two inhibitors which were purified and termed lycoricidinol (LXVIII) and lycoricidine (LXIX) (Okamoto et al. 1968, 1972). Narciclasine which occurs in daffodil bulbs (Piozzi et al. 1968) is identical with lycoricidinal. Lycoricidinol

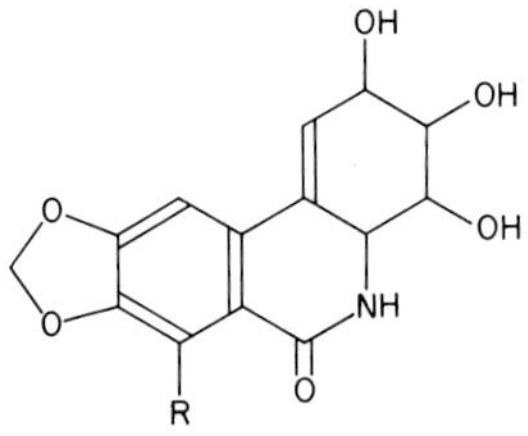

LXVIII, LXIX

LXVIII R = OH
LXIX R = H

and lycoricidine both inhibited growth of *Avena* coleoptile sections and rice seedlings at 10 mg/l; the former compound inhibited cell division very markedly in tobacco pith tissue cultures (Okamoto et al. 1968, 1972). Lycoricidinol at 10^{-5} *M* markedly inhibited growth of pea-stem hook segments and was more active than ABA in this system (Sasse et al. 1972). Basipetal transport of IAA in *Citrullus edulis* hypocotyl sections was inhibited by lycoricidinol (Kaldeway and Ginkel 1974).

7.13.3. Cucurbic acid and related compounds

From seed of pumpkin *(Cucurbita pepo)*, compounds which inhibit elongation of the second-leaf sheath of rice seedlings were purified (Koshimizu et al. 1974). These inhibitors were assigned the structures LXX [cucurbic acid, 2-(*cis*-pent-2-enyl)-3-hydroxycyclopentylacetic acid], LXXI and LXXII, the two last-mentioned compounds being cucurbic acid

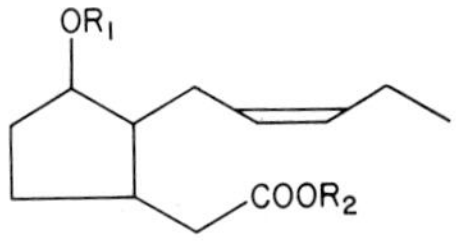

LXX – LXXII

LXX	R_1 = H	R_2 = H
LXXI	R_1 = Glc	R_2 = H
LXXII	R_1 = Glc	R_2 = CH_3

glucoside and methyl cucurbate glucoside. The methyl ester (LXXII) was markedly more active than the other two inhibitors while methylation of the carboxyl group of cucurbic acid enhanced its activity. The glucosides LXXI and LXXII possessed activity comparable to that of the aglycones. Although the above compounds were effective inhibitors of rice-leaf sheath elongation, they were essentially inactive in the *Avena* coleoptile straight-growth assay (Koshimizu et al. 1974).

7.13.4. 1-Acetoxy-2,4-dihydroxy-n-heptadeca-16-ene

In a study of the development of the avocado fruit, Blumenfeld and Gazit (1969) detected a new growth inhibitor which differed chromatographically from the "β-inhibitor" zone usually attributed to ABA. Chromatographic studies suggested the new inhibitor or a similar compound was also present in extracts of other tissues, e.g. mango seed (Blumenfeld and Gazit 1969). The inhibitor from avocado was purified and identified as 1-acetoxy-2,4-dihydroxy-*n*-heptadeca-16-ene, LXXIII (Bittner et al. 1971).

$$H_2C{=}HC(CH_2)_{11}\underset{\displaystyle OH}{CH}{-}CH_2{-}\underset{\displaystyle OH}{CH}CH_2OCOCH_3$$

LXXIII

This compound inhibited markedly the growth of wheat coleoptiles and soybean callus tissue (Bittner et al. 1971). The inhibited tissue resumed active growth when transferred to inhibitor-free medium. The deacetylated compound was inactive. The level of the inhibitor in avocado fruit during development was studied by Gazit and Blumenfeld (1972). The level rose in both the embryo and mesocarp as the fruit developed; increase in inhibitor activity and rate of growth were negatively correlated. In contrast, the level of ABA in the mesocarp remained constant during fruit development.

7.13.5. cis-4-Cyclohexene-1,2-dicarboximide

Ether extracts of sugar beet *(Beta vulgaris)* seed contain a compound which inhibits markedly the germination of lettuce seed. This was purified and identified as *cis*-4-cyclohexene-1,2-dicarboximide (LXXIV) which at 0.2 m*M* caused 48% inhibition of germination (Mitchell and Tolbert 1968).

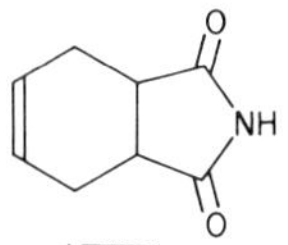

LXXIV

GA$_3$ did not alleviate the inhibition. However when the inhibitor was removed from the seed by washing, germination occurred. The inhibitor had little effect on the growth of *Avena* coleoptile sections, but markedly reduced α-amylase synthesis in whole seeds of barley (Mitchell and Tolbert 1968).

Some analogues of *cis*-4-cyclohexene-1,2-dicarboximide were tested for ability to inhibit lettuce seed germination. Aliphatic *N*-substitution enhanced activity, *N-n*-butyl- and *N-n*-amyl-4-cyclohexene-1,2-dicarboximide being particularly active (Mitchell and Tolbert 1968).

7.13.6. Asparagusic acid and derivatives

From extracts of etiolated asparagus *(Asparagus officinalis)*, five compounds which inhibited markedly the growth of lettuce and other seedlings were purified. These substances were asparagusic acid (LXXV), dihydroasparagusic acid (LXXVI), *S*-acetyldihydro-

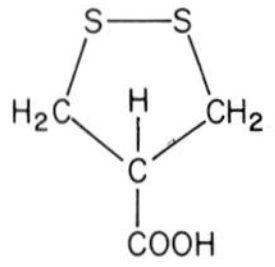

LXXV

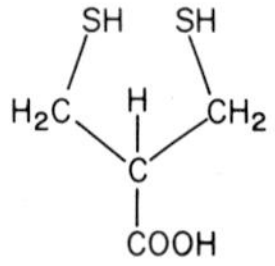

LXXVI

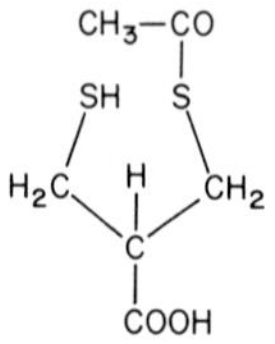

LXXVII

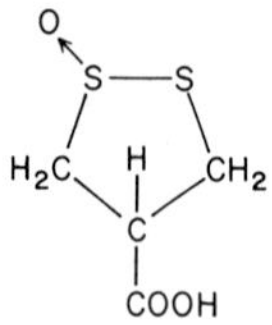

LXXVIII

asparagusic acid (LXXVII), asparagusic acid-*anti-S*-oxide and *-syn-S*-oxide (LXXVIII) (Yonagawa et al. 1972, 1973). The inhibitory activity of asparagusic acid was very similar to that of ABA (Kitahara et al. 1972).

7.13.7. N,N-Dimethyl-L-tryptophan

In extracts of the seed of jequirity beans *(Abrus precatorius)* a compound was detected which inhibited growth of germinating lettuce and other seeds. This compound also inhibited auxin- and kinetin-stimulated ethylene production by lettuce seedlings. It was identified as *N,N*-dimethyl-L-tryptophan (Mandava et al. 1974). A number of related indole compounds present in jequirity beans were inactive. The compound also inhibited IAA-induced growth and ethylene production in wheat coleoptile sections (Anderson et al. 1975).

7.13.8. 4,8,13-Duvatriene-1,3-diol

Removal of the small upper leaves of tobacco plants induces growth of the resting axillary buds and it was suggested that the young leaves may produce an inhibitor which suppresses development of lateral shoots. Extracts of rapidly growing tobacco leaves were found to inhibit growth of wheat coleoptiles but the activity was not due to ABA (Cutler and Cole 1974). The principal active compound purified by Cutler and Cole (1974) was identified as 4,8,13-duvatriene-1,3-diol (LXXIX) (Springer et al. 1975). This is the first report of a cembrene compound with plant growth regulatory activity.

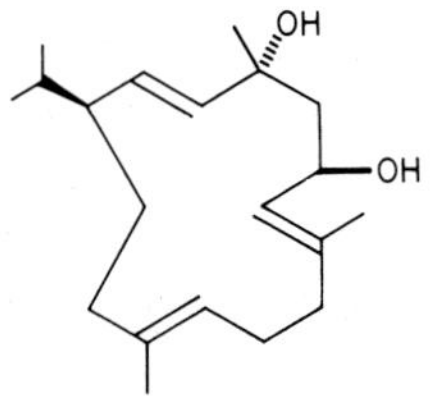

LXXIX

7.13.9. Monoepoxylignanolide

The dispersal units of *Algilops ovata* contain a compound which inhibits germination of lettuce seed. The active compound was assigned structure LXXX and inhibits germination in light but is inactive in darkness (Lavie et al. 1974).

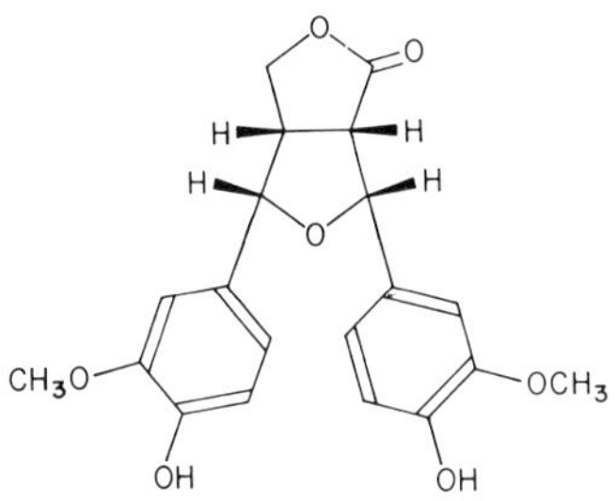

LXXX

7.14. Acetylcholine

By a combination of chromatography and bioassays, acetylcholine (ACh) has been detected in a variety of plants (Jaffe 1970, and references therein). In mung bean seedlings *(Phaseolus aureus),* the highest concentrations were found in tissues undergoing active growth, namely, buds and secondary roots. Red light caused a very rapid increase in the efflux of ACh from secondary root tips and an increase in the endogenous level, the response being negated by far-red light (Jaffe 1970). Exogenous ACh, supplied for 4 min in the dark, was able to substitute for red light in suppressing the formation of secondary roots, in increasing H^+ efflux and in causing the root tips to adhere to a negatively-charged glass surface. Yunghams and Jaffe (1972) also reported that red light and ACh increased oxygen uptake but decreased ATP concentrations in secondary root tips of etiolated mung bean, the effects of red light being reversed by far-red. Because of these observations, Jaffe suggested that ACh was a "local hormone" which regulated phytochrome-mediated responses. Jaffe and Thoma (1973) also obtained evidence that ACh may regulate the red light-induced increase in uptake of acetate by mung bean roots, while Hartmann (1971) considered the ACh level in moss callus cultures was under phytochrome control.

Although ACh may play a regulatory role in phytochrome-controlled processes of mung-bean roots, ACh certainly does not appear to be a general mediator in such processes (Satter et al. 1972; Kasemir and Mohr 1972; White and Pike 1974). Furthermore the attachment of mung bean root tips to glass caused by ACh is probably simply a consequence of its behaviour as a cation and not as a hormone-like substance (Tanada 1972).

Sporulation is induced at the growing edge of cultures of the fungus *Trichoderma viride* by blue light. Although ACh alone did not induce sporulation of this fungus in the dark, in the presence of the acetylcholinesterase inhibitor eserine, sporulation was induced by ACh which was thus able to mimic photoinduction (Gressel et al. 1971). The ranges of effective ACh and eserine concentrations were surprisingly narrow.

ACh has been reported to promote growth of wheat seedlings when supplied to the roots (Dekhuijzen 1973). The response was very dependent on the pH of the medium. Although ACh has been reported to promote elongation of *Avena* coleoptile segments (Evans 1972), a more critical study showed the response was not due to ACh (Evans 1974). However, ACh does appear to promote the germination of light-requiring seeds of various weed species (Holm and Miller 1972). ACh or eserine suppressed floral induction of the duckweed, *Lemna gibba.* In the presence of these compounds and when exposed to continuous light, *L. gibba* responded rhythmically to chilling just as it does under alternating long light and short dark periods (Oota and Hoshino 1974). Recently ACh has been reported to inhibit ethylene production in etiolated bean tissues and to partially prevent IAA-induced opening of bean hypocotyl hooks (Parups 1976).

The work of Jaffa (1970) suggested that mung bean roots contain a cholinesterase which regulates the ACh content. Such an esterase has been partially purified and characterized (Riov and Jaffe 1973a, b; Fluck and Jaffe 1974). The mung bean root cholinesterase was strongly inhibited by AMO 1618 and was associated with the cell wall.

7.15. Adenosine 3′:5′-cyclic monophosphate

Many animal hormones function by elevating the intracellular level of adenosine 3′:5′-cyclic monophosphate (cAMP) which then acts as a second messenger to evoke a variety of physiological effects within the cell. cAMP modulates the activity of a number of specific protein kinases in animal cells. In turn these kinases regulate the activity of various cellular processes by phosphorylation of key proteins (Jost and Rickenberg 1971). The important role of cAMP in transcription of catabolite-repressible operons in bacteria has been established (Pastan and Perlman 1972). cAMP acts as a chemotactic factor causing cell aggregation in the slime mould *Dictyostelium discoideum* (Konijn et al. 1969); the nucleotide also appears to regulate morphogenesis of this microorganism (Gerisch et al. 1972). cAMP probably regulates developmental processes in fungi. Thus, it appears to play an important role in the induction of fruiting bodies in *Coprinus macrorhizus* (Ono and Ishikawa 1973). cAMP enhances differentiation of chloronema filaments of the moss *Funaria hygrometrica* (Handa and Johri 1976).

The important regulatory role of cAMP in animal tissues and microorganisms provoked conjecture that it occurred in higher plants as a regulator of developmental processes and a mediator of phytohormone action. Observations relevant to this hypothesis are outlined below.

7.15.1. The possible occurrence of cAMP in higher plants; translocation

Numerous attempts to demonstrate the presence of cAMP in higher plants have been reported. These were usually based on one of the following methods: (1) prelabelling the intracellular ATP pools with radioactive adenine and subsequent detection of a labelled metabolite with the chromatographic characteristics of cAMP; (2) radioimmunoassay methods; (3) firefly luminescence assays; (4) assays involving cAMP-binding proteins; (5) assays using cAMP-activated protein kinases. The numerous claims to have detected cAMP in higher plants using these methods have been critically assessed recently by Lin (1974). It was justifiably concluded that there was no conclusive evidence for the presence of cAMP in higher plants.

In a careful investigation involving extensive chromatographic purification and finally a specific protein kinase assay, Amrhein (1974) failed to detect cAMP in a variety of plant tissues. It is noteworthy that these included *Avena sativa* coleoptiles in which Brewin and Northcote (1973a) claimed to detect cAMP by applying a protein binding assay to a crude tissue extract. A number of other critical investigations have also failed to establish the presence of cAMP in plant tissues (Keates 1973b; Alvarez et al. 1974; Niles and Mount 1974a). The occurrence in plant extracts of compounds which interfere in the usual cAMP assays has been clearly established by recent studies (Bressan et al. 1976). In an investigation using purely chemical methods, Brown and Newton (1973) claimed to have demonstrated the presence of cAMP in bean seedlings. It is unfortunate the identification was not confirmed by a protein-kinase assay on the purified nucleotide. Although cAMP is known to occur in the green alga *Chlamydomonas reinhardi* (Amrhein and Filner 1973) where it may be involved in flagellar function (Rubin and Filner 1973), its occurrence in higher plants has not been established.

Because of the probable inadequacy of methods for detection and estimation of cAMP in plant tissues, it is pointless to discuss reports claiming that exogenous plant hormones and environmental factors affect cAMP levels. However it has been claimed that auxin causes a rapid increase in the cAMP level in *Avena* coleoptiles (Brewin and Northcote 1973a), that exposure to light markedly increases the cAMP content of *Avena* etioplasts (Wellburn et al. 1973) and that incubation of dormant artichoke tissue slices in aerated water results in a rapid decline in cAMP level (Giannattasio et al. 1974a). The recent work of Ownby et al. (1975) does not substantiate the results of Brewin and Northcote (1973a).

Although cAMP is not generally accepted as an endogenous regulator, it is of interest to note that exogenous labelled cAMP undergoes polar transport in coleoptile sections of oats *(Avena sativa)* and corn *(Zea mays)* with basipetal to acropetal ratios of 4.0 and 3.2 respectively (Gordon et al. 1973). The transport of cAMP is very rapid indeed.. In corn coleoptile segments, the linear velocity of basipetal transport is 183 mm/hr which is much greater than that of IAA and one of the highest yet recorded for growth regulatory compounds (Sherwin and Gordon 1974). Unlike IAA transport, the velocity of cAMP transport was increased by 2,4-dinitrophenol and by cyanide. The data suggest an extra-cellular mechanism of cAMP transport, or transport in the cell wall or on the outer surface of the membrane (Sherwin and Gordon 1974). If cAMP does occur in plants, the high velocity of its transport suggests it could have an intercellular function.

7.15.2. Responses evoked by exogenous cAMP in higher plant tissues

Exogenously supplied cAMP has been reported to stimulate cell enlargement, seed germination and flowering and to increase the activity of certain enzymes in plant tissues. cAMP has been reported to promote growth of one plant tissue culture. The cAMP derivative 8-bromo-cAMP, which is more resistant to enzymic degradation than is cAMP (Wood and Braun 1973), induced cell division and cell differentiation in other excised plant tissues. All these responses are summarized in Table 7.1. cAMP and seed germination are discussed further in Chapter 8, section 8.7.3, Volume II.

Responses to exogenous cAMP not listed in this table which are worthy of note are

Table 7.1.

Responses evoked by cAMP in higher plants

Responses evoked and species involved	Relationship to activity of other hormones	Key references
A. Enhancement of enzyme activity or synthesis		
α-Amylase in barley half seeds*	very low activity relative to GA_3	Galsky and Lippincott 1969
α-Amylase in barley aleurone	very low activity relative to GA_3	Barton et al. 1973
α-Amylase in rice endosperm	activity less than GA_3	Roy et al. 1973
Protease and acid phosphatase in barley half seeds*	very low activity relative to GA_3	Nickells et al. 1971 Gilbert and Galsky 1972
ATPase in barley half seeds*	very low activity relative to GA_3	Earle and Galsky 1971
Nitrate reductase in mung bean seedlings	–	Higgins et al. 1974
Isocitrate lyase in hazel seeds	mimics GA_3, but less active	Potempa and Galsky 1973
Tryptophan oxygenase activity in *Cicer arietinum* seedlings	mimics IAA, but less active	Azhar and Murti 1971
B. Cell enlargement		
Avena coleoptile segments	in presence of glucose mimics IAA and GA_3; low activity relative to IAA	Hartung 1972, 1973
Lettuce hypocotyls	activity less than GA_3	Kamisaka et al. 1972
Artichoke tuber tissue	acted synergistically with auxin; activity mimics GA_3	Kamisaka et al. 1973
C. Growth in tissue cultures		
Sunflower stem tissue cultures	replaces auxin and cytokinin	Truelsen et al. 1974
Tobacco pith tissue cultures**	mimics cytokinin, but less active	Wood and Braun 1973 Hager and Marcus 1975
D. Differentiation in excised tissues		
Lettuce pith**	mimics cytokinin	Basile et al. 1973

E. Seed germination		
Lettuce (cv. Grand Rapids), darkness	acted synergistically with GA_3 (no effect alone)	Kamisaka and Masuda 1971
Lettuce (cv. Spartan Lake), darkness	active in absence of GA_3; low activity relative to GA_3	Hall and Galsky 1973
Weeds of various spp., darkness	mimics GA_3	Holm and Miller 1972
Rice, darkness	mimics GA_3; activity approx. equal to that of GA_3	Roy et al. 1973
Phacelia tanacetifolia, light	mimics GA_3 but less active	Safran and Galsky 1974
F. Flowering		
Lemna gibba cultured on glucose medium***		Oota 1972
Lemna gibba cultured on NH_4NO_3 medium		Oota and Kondo 1974
Lemna gibba under inductive long days		Kessler and Steinberg 1973

* The half-seeds lacked an embryo.
** 8-Bromo-cAMP was used in these experiments; cAMP was inactive in tobacco pith and much less active than 8-bromo-cAMP in lettuce pith.
*** cAMP may alleviate catabolite repression.

discussed below and include retardation of petiole abscission (Salomon and Mascarenhas 1971) and promotion of stomatal opening (Rustagi and Sankhla 1974). In the former response cAMP mimicked IAA. Formation of polyribosomes in barley embryos was enhanced by both cAMP and GA_3, but cAMP acted more rapidly than GA_3 (Rao and Khan 1975). Chromatin of oat coleoptiles isolated in the presence of cAMP was more active in RNA synthesis than chromatin isolated in its absence (Salomon and Mascarenhas 1972). cAMP (0.1 μM) caused a very rapid enhancement of RNA synthesis in protoplasts isolated from maize seedlings (Tarantowicz-Marek and Kleczkowski 1975). The effect of cAMP was similar to that of GA_3 but the cyclic nucleotide acted more rapidly. Finally, cAMP reduced the number of crown-gall tumours which developed on bean leaves (Babula and Galsky 1975).

It is evident that cAMP often appears to mimic gibberellin and occasionally IAA or cytokinin (see Table 7.1). Because of this it has been suggested that cAMP functions in vivo as a "second messenger" as it does in animal systems. However at least three observations are relevant in this connexion. Firstly, the effect of exogenous cAMP is usually small relative to that of the phytohormone mimicked and is evident only at relatively high concentrations. With only two growth responses (promotion of sunflower tissue growth and germination of rice seed), cAMP exhibits activity comparable to that of a plant hormone. Secondly, ADP, a common nucleotide, can replace cAMP in some instances, e.g. production of α-amylase and protease in barley half seeds (ADP was actually more effective than cAMP) and promotion of lettuce (cv. Spartan Lake) seed germination. Thirdly, in barley endosperm, cAMP has been reported to induce gibberellin biosynthesis which results in synthesis of α-amylase (Kessler and Kaplan 1972). There is no convincing evidence at present to indicate that cAMP is a "second messenger", that is, a mediator in plant hormone action.

6-*N*,2′-*O*-Dibutyryladenosine 3′:5′-cyclic monophosphate (DBcAMP) is often more active than cAMP in evoking certain responses in animal systems. This difference in activity has also been observed in plant tissues. Thus DBcAMP promotes α-amylase production in aleurone layers (Pollard 1971) and cell expansion in artichoke tissue (Kamisaka and Masuda 1970; Kamisaka 1972) to a greater degree than cAMP. This greater activity in plant tissues could be a consequence of the resistance of DBcAMP to enzymic degradation, enhanced uptake by the tissue, or the cytokinin activity of DBcAMP. This derivative of cAMP has a butyrylaminopurine moiety and is thus an amide of adenine. Such compounds have strong cytokinin activity (Letham et al. 1972; Dekhuijzen and Overeen 1972). Thus, although cAMP and cGMP are inactive in the radish cotyledon cytokinin bioassay, DBcAMP and 6-butyrylaminopurine both exhibit appreciable activity (Letham et al. 1972; Gordon and Letham 1975). DBcAMP induces betacyanin synthesis in *Amaranthus* seedlings in darkness (Rast et al. 1973; Guidice de Nicola et al. 1975). These workers consider that this is a consequence of the cAMP moiety of DBcAMP. However it is more probable that the activity of DBcAMP in this system is due to its ability to function as a cytokinin. Cytokinins are known to induce betacyanin formation in *Amaranthus* species.

Some workers (see Table 7.1) observed cytokinin-like responses to 8-bromo-cAMP which they associated with the resistance of this compound to cAMP phosphodiesterase.

However, 8-bromoadenosine and 8-bromoadenine are more active than 8-bromo-cAMP in the tobacco callus cytokinin bioassay (Schmitz et al. 1975) and the three compounds possess similar activity in the *Amaranthus* betacyanin assay (Elliott and Murray 1975). The activity of 8-bromo-cAMP may be due to the cytokinin activity of the 8-bromoadenine moiety per se.

cAMP-dependent protein kinases have not been detected in higher plants. Protein kinase activity has been detected in nuclei plus chloroplast preparations from Chinese cabbage leaves (Ralph et al. 1972) and in soybean callus tissue (Elliott 1973; Elliott and Murray 1974), while protein kinases have been partially purified from pea shoots (Keates 1973a). Protein kinases are also associated with purified plant ribosomes (Ralph et al. 1972; Keates and Trewavas 1974). cAMP did not modify the activity of any of these protein kinases in vitro. Hence, unlike animal protein kinases, plant protein kinases do not appear to be stimulated by cAMP. The only enzyme from higher plants known to be affected by cAMP is a 5′(3′)-ribonucleotide phosphohydrolase purified from wheat seedling leaves (Polya and Ashton 1973; Polya 1974) and potato (Polya 1975). cAMP and cGMP are both powerful competitive inhibitors of the enzyme in vitro.

7.15.3. Enzymes of cAMP biosynthesis and metabolism in higher plants

Demonstration of the presence in higher plants of adenylate cyclase, the enzyme which catalyzes the formation of cAMP from ATP, and of a phosphodiesterase which hydrolyzes cAMP specifically would obviously be of great significance.

The occurrence of adenylate cyclase in higher plants has not been established. Wood et al. (1972) claimed to have detected the enzyme in the membrane and cell wall fraction of *Vinca rosea* tumour cells; Janistyn (1972) reported its presence in maize coleoptile homogenates while Giannattasio and Macchia (1973) claimed it occurs in artichoke tissue. However, the activity reported in these studies was so low it is of doubtful significance and the "cAMP" formed was not properly characterized. Indeed the labelled product obtained by Wood et al. (1972) could well be adenosine 5′-phosphorofluoridate. An enzyme present in barley seeds appears to catalyze the formation of this fluoridate from ATP, Mg^{2+} and fluoride, components of the incubation mixture of Wood et al. The nucleotide derivative is also formed nonenzymically when this mixture is heated to 100 °C for a few minutes (Alvarez et al. 1974). The chromatographic system which Wood et al. (1972) used to detect cAMP does not separate cAMP from adenosine 5′-phosphorofluoridate (Alvarez et al. 1974). In other studies, adenylate cyclase could not be detected in plant tissues (Lin and Varner 1972; Alvarez et al. 1974).

cAMP phosphodiesterase activity has been detected in a variety of higher plant tissues (see Lin 1974 for numerous references). The enzymes from pea seedlings (Lin and Varner 1972) and soybean callus tissue (Brewin and Northcote 1973b) have been considerably purified; their properties are briefly outlined below and compared with those of other diesterases. The pea enzyme cleaves both 3′:5′- and 2′:3′-cyclic monophosphate linkages of purine and pyrimidine cyclic nucleotides. Hence, it is unlike the corresponding enzymes from animal tissues and microorganisms which hydrolyze nucleoside 3′:5′-cyclic monophosphates specifically. The enzyme from peas also differs in terms of its hydrolysis products. While cAMP phosphodiesterases from animal tissues and microorganisms yield exclusively AMP (5′) from cAMP (3′:5′), the pea enzyme produces both AMP (5′) and adenosine 3′-monophosphate (ratio 1:7) from cAMP and only adenosine 3′-monophosphate from adenosine 2′:3′-cyclic monophosphate. The pH optimum of the pea enzyme was 5.4 (cf. animal and bacterial enzymes, pH optimum ca. 8) and its activity was not affected by methylxanthines, such as caffeine and theophylline, which are very effective inhibitors of mammalian cAMP phosphodiesterases.

The cAMP phosphodiesterase from soybean tissue (Brewin and Northcote 1973b) resembles the pea enzyme in many properties. DBcAMP was a very effective inhibitor of the soybean enzyme even at concentrations of 5 μM and a non-dialyzable inhibitor of the enzyme was found to occur in the soybean tissue from which it was partially purified. At least 35% of the soybean enzyme activity was associated with the plasmalemma.

cAMP phosphodiesterases have also been purified recently from carrot root tissue (Niles and Mount 1974b), potato tuber (Ashton and Polya 1975) and artichoke tuber tissue (Giannattasio et al. 1974b). These enzymes hydrolyze adenosine 2′:3′-phosphate in addition to cAMP and resemble the pea diesterase in their properties. Like the pea and soybean cyclic nucleotide phosphodiesterases, they are inhibited by inorganic phosphate (see also Shimoyama et al. 1972). It is noteworthy that phytohormones do not affect the activity of the soybean, artichoke or carrot diesterases.

The cyclic nucleotide phosphodiesterase of *Chlamydomonas reinhardi* has been investigated; surprisingly it closely resembles the enzymes from other microorganisms and animal tissues and is dissimilar to the cyclic nucleotide phosphodiesterases of higher plants (Fischer and Amrhein 1974).

With the exception of the cyclic nucleotide phosphodiesterases from *Vinca rosea* tumour cells (Wood et al. 1972) and *Amaranthus* (Elliott and Murray 1975), plant enzymes which hydrolyze cAMP are not appreciably inhibited in vitro by the methylxanthines, caffeine and theophylline. Furthermore such compounds do not prevent degradation of exogenously supplied cAMP (Keates 1973b). However, it is of interest that exogenous caffeine and theophylline frequently promote growth although high concentrations are sometimes required. Thus, theophylline induced cell division in tobacco pith tissue cultures (Wood et al. 1972), enhanced the stimulating effect of cAMP on cell expansion in artichoke tuber tissue (Kamisaka et al. 1973), promoted seed germination (Hall and Galsky 1973; Safran and Galsky 1974), stimulated radish cotyledon expansion (Gordon and Letham 1975), and at 10^{-8} M induced elongation of *Avena* coleoptile segments (Hartung 1973). Theophylline has also been reported to enhance acid phosphatase activity in barley endosperm (Gilbert and Galky 1972) and promote betacyanin formation in *Amaranthus* seedlings (Rast et al. 1973; Elliott and Murray 1975). Many of the above responses have been attributed to inhibition of cAMP phosphodiesterase, but this seems improbable because of the general inability of methylxanthines to inhibit appreciably plant cyclic nucleotide phosphodiesterases in vitro. The molecular basis of the responses is not known at present.

The cyclic nucleotide phosphodiesterases so far detected in plants lack substrate specificity. Their principal function may be to hydrolyze the nucleoside 2′:3′ cyclic monophosphates produced from RNA by plant ribonucleases (see Lin and Varner 1972). It is not known at present whether a phosphodiesterase which specifically cleaves nucleoside 3′:5′-cyclic monophosphates, including cAMP, also occurs in higher plants.

References

Alvarez, R., Moore, T.C. and Vandepeute, J. (1974) Plant Physiol., 53, 144–148.

Amrhein, N. (1974) Planta, 118, 241–258.

Amrhein, N. and Filner, P. (1973) Proc. nat. Acad. Sci. (Wash.), 70, 1099–1103.

Anderson, J.D., Mandava, N. and Garrett, S. (1975) Plant Cell Physiol., 16, 233–236.

Ando, T. and Tsukamoto, Y. (1974) Phytochemistry, 13, 1031–1032.

Andreae, W.A. and Collet, G. (1968) in: Biochemistry and Physiology of Plant Growth Substances F. Wightman and G. Setterfield (eds.), Runge Press, Ottawa, pp. 553–561.

Arntzen, C.J., Falkenthal, S.V. and Bobick, S. (1974) Plant Physiol., 53, 304–306.

Arsenault, G.P., Biemann, K., Barksdale, A.W. and McMorris, T.C. (1968) J. Amer. chem. Soc., 90, 5635–5636.

Asen, S. and Emsweller, S.L. (1962) Phytochemistry, 1, 169–174.

Ashton, A.R. and Polya, G.M. (1975) Biochem. J., 149, 329–339.

Attridge, T.H., Stewart, G.R. and Smith, H. (1971) FEBS Lett., 17, 84–86.

Austin, D.J. and Clarke, D.D. (1966) Nature (Lond.), 210, 1165–1166.

Austin, D.J. and Meyers, M.B. (1965) Phytochemistry, 4, 255–262.

Austin, D.J. Bu'Lock, J.D. and Gooday, G.W. (1969) Nature (Lond.), 223, 1178–1179.

Avers, C. and Goodwin, R.H. (1956) Amer. J. Bot., 43, 612–620.

Azhar, S. and Murti, C.R.K. (1971) Indian J. Biochem. Biophys., 8, 210–213.

Babula, M.J. and Galsky, A.G. (1975) Plant Cell Physiol., 16, 357–360.

Bachelard, E.P. and Stowe, B.B. (1962) Nature (Lond.), 194, 209–210.

Bagni, N. (1966) Experientia (Basel), 22, 732–733.

Bagni, N. and Fracassini, D.S. (1971) Experientia (Basel), 27, 1239–1241.

Bagni, N. and Fracassini, D.S. (1974) in: Plant Growth Substances 1973, Hirokawa, Tokyo, pp. 1205–1217.

Ballio, A., Chain, E.B., De Leo, P., Erlanger, B.F., Mauri, M. and Tonolo, A. (1964) Nature (Lond.), 203, 297.

Ballio, A., Brufani, M., Casinovi, C.G., Cerrini, S., Fedeli, W., Pellicciari, R., Santurbano, B. and Vaciago, A. (1968) Experientia (Basel), 24, 631–635.

Barksdale, A.W. (1969) Science, 166, 831–837.

Barksdale, A.W., McMorris, T.C., Seshadri, R., Arunachalam, T., Edwards, J.E., Sundeen, J. and Green, D.M. (1974) J. gen. Microbiol., 82, 295–299.

Barrow, K.D., Barton, D.H.R., Chain, E.B., Ohnsorge, U.F.W. and Thomas, R. (1968) J. Chem. Soc. chem. Commun., 1198–1200.

Barrow, K.D., Barton, D.H.R., Chain, E.B., Ohnsorge, U.F.W. and Thomas, R. (1971) J. Chem. Soc. C., 1265–1274.

Barrow, K.D., Jones, R.B., Pemberton, P.W. and Phillips, L. (1975). J. Chem. Soc. Perkin Trans., I pp 1405–1410.

Barton, K.A., Verbeek, R., Ellis, R. and Khan, A.A. (1973) Physiol. Plant., 29, 186–189.

Basile, D.V., Wood, H.N. and Braun, A.C. (1973) Proc. nat. Acad. Sci. (Wash.), 70, 3055–3059.

Baskin, J.M., Ludlow, C.J., Harris, T.M. and Wolf, F.T. (1967) Phytochemistry, 6, 1209–1213.

Basu, R.N. (1971) Curr. Sci., 40, 427–429.

Basu, R.N. (1972). J. exp. Bot., 23, 357–365.

Berrie, A.M.M., Don, R., Buller, D., Alam, M. and Parker, W. (1975) Plant Sci. Lett., 6, 163–173.

Berrie, A.M.M., Parker, W., Knights, B.A. and Hendrie, M.R. (1968) Phytochemistry, 7, 567–573.

Bertossi, F., Bagni, N., Moruzzi, G. and Caldarera, C.M. (1965) Experientia (Basel), 21, 80–81.

Bhalerao, U.T., Plattner, J.J. and Rapoport, H. (1970) J. Amer. chem. Soc., 92, 3429–3433.

Biswas, P.K., Paul, K.B. and Henderson, J.H.M. (1967) Nature (Lond.), 213, 917–918.

Bittner, S., Gazit, S. and Blumenfeld, A. (1971) Phytochemistry, 10, 1417–1421.

Blumenfeld, A. and Gazit, S. (1969) Israel J. Bot., 18, 217–219.

Bodanszky, M. and Stahl, G.L. (1974) Proc. nat. Acad. Sci. (Wash.), 71, 2791–2794.

Bohm, B.A. (1966) Arch. Biochem., 115, 181.
Bonde, E.K. and Khudairi, A.K. (1954) Physiol. Plant., 7, 66–71.
Bottomley, W., Smith, H. and Galston, A.W. (1966) Phytochemistry, 5, 117–123.
Bough, W.A. and Gander, J.E. (1972) Phytochemistry, 11, 209–213.
Bressan, R.A ., Ross, C.W. and Vandepeute, J. (1976) Plant Physiol., 57, 29–37.
Brewin, N.J. and Northcote, D.H. (1973a) J. exp. Bot., 24, 881–888.
Brewin, N.J. and Northcote, D.H. (1973b) Biochim. biophys. Acta (Amst.), 320, 104–122.
Brain, K.R. and Lockwood, G.B. (1976) Phytochemistry, 15, 1651–1654.
Briggs, D.E. (1966) Nature (Lond.), 210, 418–419.
Brown, S.A. (1962) Canad. J. Biochem. Physiol.., 40, 607–618.
Brown, S.A. (1963) Phytochemistry, 2, 137–144.
Brown, R. and Edwards, M. (1945) Nature (Lond.), 155, 455.
Brown, E.G. and Newton, R.P. (1973) Phytochemistry, 12, 2683–2685.
Brown, R., Johnson, A.W., Robinson, E. and Todd, A.R. (1949) Proc. roy. Soc. B, 136, 1–12.
Brown, R., Greenwood, A.D., Johnson, A.W., Long, A.G. and Tyler, G.J. (1951) Biochem. J., 48, 564–568.
Brown, R., Greenwood, A.D., Johnson, A.W., Lansdown, A.R., Long, A.G. and Sutherland, N. (1952b) Biochem. J., 52, 571–574.
Brown, R., Johnson, A.W., Robinson, E. and Tyler, G.J. (1952a) Biochem. J., 50, 596–600.
Buckhout, T.J. and Curtis, R.W. (1976) Nature (Lond.), 260, 435–436.
Bucking-Throm, E., Duntze, W., Hartwell, L.H. and Manney, T.R. (1973) Exp. Cell Res., 76, 99–110.
Buller, D.C., Parker, W. and Reid, J.S.G. (1976) Nature (Lond.), 260, 169–170.
Bu'Lock, J.D. (1973) Pure Appl. Chem., 34, 435–461.
Bu'Lock, J.D., Austin, D.J., Snatzke, G. and Hruban, L. (1970) J. Chem. Soc. chem. Commun., 225–226.
Bu'Lock, J.D., Drake, D. and Winstanley, D.J. (1972) Phytochemistry, 11, 2011–2018.
Bu'Lock, J.D., Jones, B.E., Taylor, D. Winskill, N. and Quarrie, S.A. (1974a) J. gen. Microbiol., 80, 301–306.
Bu'Lock, J.D., Jones, B.E. and Winskill, N. (1974b) J. Chem. Soc. chem. Commun., 708–709.
Burgeff, H. (1924) Bot. Abh., 4, 5–135.
Burstrom, H. (1954) Physiol. Plant., 7, 548–559.
Caglioti, L., Cainelli, G., Camerino, B., Mondelli, R., Prieto, A., Quilico, A., Salvatori, T. and Selva, A. (1964) Chim. Ind. (Milan), 46, 961.
Caglioti, L., Cainelli, G., Camerino, B., Mondelli, R., Prieto, A., Quilico, A., Salvatori, T. and Selva, A. (1966) Tetrahedron Lett., 22, Suppl. 7, 175–187.
Cainelli, G., Grasselli, P. and Selva, A. (1967) Chim. Ind. (Milan), 49, 628–629.
Carlile, M.J. and Machlis, L. (1965) Amer. J. Bot., 52, 478–483.
Cathey, H.M., Steffins, G.L., Stuart, N.W., Zimmerman, R.H. (1966) Science, 153, 1382–1383.
Chain, E.B., Mantle, P.G. and Milborrow, B.V. (1971) Physiol. Plant Pathol., 1, 495–514.
Chandramohan, D., Purushothaman, D. and Kothandaraman, R. (1973) Plant Soil, 39, 303–308.
Ciarlante, D. and Curtis, R.W. (1974) in: Plant Growth Substances 1973, Hirokawa, Tokyo, pp 396–403.
Cleland, R.E. (1976) Planta, 128, 201–206.
Cleland, C.F. and Ajami, A. (1974) Plant Physiol., 54, 904–906.
Cocucci, S. and Bagni, N. (1968) Life Sci., 7, 113–120.
Cocucci, M., Marre, E., Denti, A.B. and Scacchi, A. (1976) Plant Sci. Lett., 6, 143–156.
Coggon, P., Luhan, P.A. and McPhail, A.T. (1973). J. Chem. Soc. Perkin Trans., 2, pp 465–469.
Cook, C.E., Whichard, L.P., Turner, B., Wall, M.E. and Egley, G.H. (1966) Science, 154, 1189–1190.
Cook, C.E., Whichard, L.P., Wall, M.E., Egley, G.H., Coggon, P., Luhan, P.A. and McPhail, A.T. (1972). J. Amer. chem. Soc., 94, 6198–6199.
Coombe, B.G., Mander, L.N., Paleg, L.G. and Turner, J.V. (1974) Aust. J. Plant Physiol., 1, 473–481.
Cooper-Driver, G., Corner-Zamoditis, J.J. and Swain, T. (1972) Z. Naturforsch., B 27, 943–946.

Corcoran, M.R. (1970) Plant Physiol., 45, 531–534.
Corcoran, M.R., Geissman, T.A. and Phinney, B.O. (1972) Plant Physiol., 49, 323–330.
Corgan, J.N. (1965) Proc. Amer. Soc. Hort. Sci., 86, 129–132.
Corgan, J.N. (1967) Hort Science, 2, 105–106.
Correa, N.S. (1969) Phyton, 26, 125–134.
Craker, L.E. (1975) Phytochemistry, 14, 151–153.
Crombie, L. (1955) J. Chem. Soc., p. 2535.
Crombie, L. and Firth, P.A. (1968) J. Chem. Soc., (C) 2852–2855.
Crow, W.D., Nicholls, W. and Sterns, M. (1971) Tetrahedron Lett., 18, 1353–1356.
Curtis, R.W. (1976) Plant Physiol., 57, 365–368.
Cutler, H.G. and Cole, R.J. (1974) Plant Cell Physiol., 15, 19–28.
Das, V.S.R., Rao, J.V.S. and Rao, K.N. (1967) Z. Pflanzenphysiol., 56, 466–468.
Dekhuijzen, H.M. (1973) Planta, 111, 149–156.
Dekhuijzen, H.M. and Overeen, J.C. (1972) Phytochemistry, 11, 1669–1672.
Dekock, P.C., Grabowska, F.B. and Innes, A.M. (1974) Ann. Bot. (Lond.), 38, 903–908.
Demos, E.K., Woolwine, M., Wilson, R.H. and McMillan, C. (1975) Amer. J. Bot., 62, 97–102.
Dennis, F.G. and Edgerton, L.J. (1961) Proc. Amer. Soc. Hort. Sci., 77, 107–116.
Deuel, P.G. and Geissman, T.A. (1957) J. Amer. chem. Soc., 79, 3778–3783.
Dopp, W. (1950) Ber. Deut. Bot. Ges., 63, 139–147.
Duntze, W., MacKay, V. and Manney, T.R. (1970) Science, 168, 1472–1473.
Duntze, W., Stotzler, D., Bucking-Throm, E. and Kalbitzer, S. (1973) Europ. J. Biochem., 35, 357–365.
Duplessy-Graillot, G. (1962) C.R. Soc. Biol. (Paris), 156, 1064–1068.
Earle, K.M. and Galsky, A.G. (1971) Plant Cell Physiol., 12, 727–732.
Edwards, J.A., Mills, J.S., Sundeen, J. and Fried, J.H. (1969) J. Amer. chem. Soc., 91, 1248–1249.
Edwards, J.A. Schwarz, V., Fajkos, J., Maddox, M.L. and Fried, J.H. (1971) J. Chem. Soc. chem. Commun., pp 292–293.
Edwards, K.G. and Stoker, J.R. (1967) Phytochemistry, 6, 655–661.
El-Basyouni, S. and Towers, G.H.N. (1964) Canad. J. Biochem., 42, 203–210.
Elliott, D.C. (1973) Proc. Aust. Biochem. Soc., 6, 41.
Elliott, D.C. and Murray, A.W. (1974) in: Plant Growth Substances 1973, Hirokawa, Tokyo, pp 691–698.
Elliott, D.C. and Murray, A W. (1975) Biochem. J., 146, 333–337.
Endo, M., Nakanishi, K., Naf, U., McKeon, W. and Walker, R. (1972) Physiol. Plant., 26, 183–185.
Engelsma, G. (1974) Plant Physiol., 54, 702–705.
Erickson, R.O. and Rosen, G.W. (1949) Amer. J. Bot., 36, 317–322.
Evans, M.L. (1972) Plant Physiol., 50, 414–416.
Evans, M.L. (1974) Ann. Rev. Plant Physiol., 25, 195–223.
Fischer, U. and Amrhein, N. (1974) Biochim. biophys. Acta (Amst.), 341, 412–420.
Floss, H.G. (1972) Recent Adv. Phytochem., 4, 143–164.
Fluck, R.A. and Jaffe, M.J. (1974) Plant Physiol., 53, 752–758.
Font, J. and Pascual, J. (1966) Anales Real. Soc. Espan. Fis. Quim. (Madrid) Ser. B, 62, 705–707; Chem. Abst. 66, 8798.
Fritig, B., Hirth, L. and Ourisson, G. (1970) Phytochemistry, 9, 1963–1975.
Furuya, M., Galston, A.W. and Stowe, B.B. (1962) Nature (Lond.), 193, 456–457.
Galbraith, M.N., Horn, D.H.S., Sasse, J.M. and Adamson, D. (1970) J. Chem. Soc. chem. Commun., 170–171.
Galli, M.G., Sparvoli, E. and Caroi, M. (1975) Plant Sci. Lett., 5, 351–357.
Galsky, A.G. and Lippincott, J.A. (1969) Plant Cell Physiol., 10, 607–620.
Galston, A.W. (1969) in: Perspectives in Phytochemistry, J.B. Harborne and T. Swain (eds.), Academic Press, London, pp 193–204.
Gazit, S. and Blumenfeld, A. (1972) Physiol. Plant., 27, 77–82.

Geissman, T.A., Deuel, P., Bonde, E.K. and Addicott, F.A. (1954) J. Amer. chem. Soc., 76, 685–687.
Gerisch, G., Malchow, D., Riedel, V., Muller, E. and Every, M. (1972) Nature New Biology, 235, 90–92.
Gesto, M.D.V., Vazquez, A., Mendez, J., Vieitez, E. and Seoane, E. (1967) Phytochemistry, 6, 1687–1693.
Giannattasio, M. and Macchia, V. (1973) Plant Sci. Lett., 1, 259–264.
Giannattasio, M., Mandato, E. and Macchia, V. (1974a) Biochem. biophys. Res. Commun., 57, 365–371.
Giannattasio, M., Sica, G. and Macchia, V. (1974b) Phytochemistry, 13, 2729–2733.
Gilbert, M.L. and Galsky, A.G. (1972) Plant Cell Physiol., 13, 867–873.
Giudici de Nicola, M., Amico, V. and Piatteli, M. (1975) Phytochemistry, 14, 989–991.
Glass, A.D.M. (1973) Plant Physiol., 51, 1037–1041.
Glass, A.D.M. (1974a) in: Mechanisms of Regulation of Plant Growth, R.L. Bieleski, A.R. Ferguson and M.M. Cresswell (eds.), Royal Society of New Zealand, Wellington, pp. 159–164.
Glass, A.D.M. (1974b) J. exp. Bot., 25, 1104–1113.
Glass, A.D.M. (1975) Phytochemistry, 14, 2127–2130.
Glass, A.D.M. and Dunlop, J. (1974) Plant Physiol., 54, 855–858.
Goldacre, P.L., Galston, A.W. and Weintraub, R.L. (1953) Arch. Biochem., 43, 358–373.
Goldschmidt, E.E., Goren, R. and Monselise, S.P. (1968) Planta, 72, 213–222.
Gooday, G.W. (1968a) Phytochemistry, 7, 2103–2105.
Gooday, G.W. (1968b) New Phytol., 67, 815–821.
Goodwin, R.H. and Taves, C. (1950) Amer. J. Bot., 37, 224–231.
Gordon, S.A., Cameron, E. and Shen-Miller, J. (1973) Plant Physiol., 52, 105–110.
Gordon, M.E. and Letham, D.S. (1975) Aust. J. Plant Physiol., 2, 129–154.
Goren, R. and Tomer, E. (1971) Plant Physiol., 47, 312–316.
Goss, J.A. (1968) Amer. J. Bot., 55, 73–77.
Graniti, A. (1962) Phytopathol. Mediterr., 1, 182–185.
Green, D.M., Edwards, J.A., Barksdale, A.W. and McMorris, T.C. (1971) Tetrahedron, 27, 1199–1203.
Green, F.B. and Corcoran, M.R. (1975) Plant Physiol., 56, 801–806.
Gressel, J., Strausbauch, L. and Galun, E. (1971) Nature (Lond.), 232, 648–649.
Grisebach, H. (1968) Recent Adv. Phytochem., 1, 379–406.
Grunwald, C. (1975) Ann. Rev. Plant Physiol., 26, 209–236.
Hagen, G.L. and Marcus, A. (1975) Plant Physiol., 55, 90–93.
Hall, K.A. and Galsky, A.G. (1973) Plant Cell Physiol., 14, 565–571.
Hancock, C.R., Barlow, H.W.B. and Lacey, H.J. (1961) J. exp. Bot., 12, 401–408.
Handa, A.K. and Johri, M.M. (1976) Nature (Lond.), 259, 480–482.
Hara, M., Umetsu, N., Miyamoto, C. and Tamari, K. (1973) Plant Cell Physiol., 14, 11–28.
Harada, H., Ohyama, K. and Cheruel, J. (1972) Z. Pflanzenphysiol., 66, 307–324.
Harborne, J.B. (1965) in: Plant Biochemistry, J. Bonner and J.E. Varner (eds.), Academic Press, New York, pp 618–640.
Harborne, J.B. (1972) Recent Adv. Phytochem., 4, 107–141.
Harborne, J.B. and Corner, J.J. (1961) Biochem. J., 81, 242–250.
Hardy, S.J.S. and Turnock, G. (1971) Nature New Biology, 229, 17–19.
Hartmann, E. (1971) Planta, 101, 159–165.
Hartung, W. (1972) Z. Pflanzenphysiol., 67, 380–382.
Hartung, W. (1973) Z. Pflanzenphysiol., 68, 329–335.
Hasegawa, K. and Hashimoto, T. (1973) Plant Cell Physiol., 14, 369–377.
Hasegawa, K. and Hashimoto, T. (1974) Plant Cell Physiol., 15, 1–6.
Hasegawa, K. and Hashimoto, T. (1975) J. exp. Bot., 26, 757–764.
Hashimoto, T. and Hasegawa, K. (1974) In: Plant Growth Substances 1973, Hirokawa, Tokyo, pp. 150–156.
Hashimoto, T. and Tamura, S. (1967a) Plant Cell Physiol., 8, 35–45.
Hashimoto, T. and Tamura, S. (1967b) Plant Cell Physiol., 8, 197–200.

Hashimoto, T. and Tamura, S. (1969) Bot. Mag. (Tokyo), 82, 69–75.
Hashimoto, T., Sakurai, A. and Tamura, S. (1967) Plant Cell Physiol., 8, 23–24.
Hashimoto, T., Hasegawa, K. and Kawarada, A. (1972) Planta, 108, 369–374.
Hashimoto, T., Hasegawa, K., Yamaguchi, H., Saito, M. and Ishimoto, S. (1974) Phytochemistry, 13, 2849–2852.
Hashimoto, T. and Tamura, S. (1969) Bot. Mag. (Tokyo), 82, 69–75.
Hayashi, Y. and Sakan, T. (1974) in: Plant Growth Substances 1973, Hirokawa, Tokyo, pp. 525–532.
Hayashi, Y., Takahashi, S., Ona, H. and Sakan, T. (1968) Tetrahedron Lett., pp. 2071–2076.
Hayashi, Y., Yokoi, J., Watanabe, Y. and Sakan, T. (1972) Chem. Lett., 759–762.
Haynes, L.J. and Jones, E.R.H. (1946) J. Chem. Soc., 954–957.
Heather, J.B., Mittal, R.S.D. and Sih, C.J. (1974) J. Amer. chem. Soc., 96, 1976–1977.
Hendershott, C.H. and Walker, D.R. (1959a) Science, 130, 798–800.
Hendershott, C.H. and Walker, D.R. (1959b) Proc. Amer. Soc. Hort. Sci., 74, 121–129.
Hess, C.E. (1964) in: Regulateurs Naturels de la Croissance Vegetale. C.N.R.S., Paris, pp. 517–527.
Higgins, T.J.V., Goodwin, P.B. and Carr, D.J. (1974) Aust. J. Plant Physiol., 1, 1–8.
Hill, R. and Van Heyningen, R. (1951) Biochem. J., 49, 332–335.
Hillman, W.S. and Galston, A.W. (1957) Plant Physiol., 32, 129–135.
Hirai, J., Hirata, N. and Horiuchi, S. (1966) J. jap. Soc. Hort. Sci., 35, 354–360.
Hirai, J., Hirata, N. and Horiuchi, S. (1967) J. jap. Soc. Hort. Sci., 36, 36–44.
Hirai, J., Hirata, N. and Horiuchi, S. (1968) J. jap. Soc. Hort. Sci., 37, 20–29.
Hlubucek, J.R. and Robertson, A.V. (1967) Aust. J. Chem., 20, 2199–2206.
Hofmann, A.W. (1859) Justus Liebigs Ann. Chem., 110, 129–140.
Hogetsu, T., Shibaoka, H. and Shimokoriyama, M. (1974) Plant Cell Physiol., 15, 265–272.
Holm, R.E. and Miller, M.R. (1972) Weed Sci., 20, 209–212.
Hosel, W., Shaw, P.D. and Barz, W. (1972) Z. Naturforsch., 27, 946–954.
Igarashi, K. and Takeda, Y. (1970) Biochim. biophys. Acta (Amst.), 213, 240–243.
Ito, S., Sunagawa, M., Kodama, M. and Honma, H. (1971) J. Chem. Soc. chem. Commun., 91–93.
Iwata, T. and Stowe, B.B. (1973) Plant Physiol., 51, 691–701.
Jaenicke, L. and Seferiadis, K. (1975) Chem. Ber., 108, 225–232.
Jaenicke, L., Akintobi, T. and Marner, F.J. (1973) Justus Liebigs Ann. Chem., 1252–1262.
Jaenicke, L., Muller, D.G. and Moore, R.E. (1974) J. Amer. chem. Soc., 96, 3324–3325.
Jaffe, M.J. (1970) Plant Physiol., 46, 768–777.
Jaffe, M.J. and Thoma, L. (1973) Planta, 113, 283–291.
Janistyn, B. (1972) Z. Naturforsch., B 27, 872.
John, W.W. and Curtis, R.W. (1974) Experientia (Basel), 30, 1392–1393.
Johnson, C., Attridge, T. and Smith, H. (1975) Biochim. biophys. Acta (Amst.), 385, 11–19.
Johnson, R., Guderian, R.H., Eden, F., Chilton, M., Gordon, M.P. and Nester, E.W. (1974) Proc. nat. Acad. Sci. (Wash.), 71, 536–539.
Jones, O.P. (1976) Nature (Lond.), 262, 392–393.
Jost, J.P. and Rickenberg, H.V. (1971) Ann. Rev. Biochem., 40, 741–774.
Kaldewey, H. and Ginkel, U. (1974) in: Plant Growth Substances 1973, Hirokawa, Tokyo, pp. 1079–1089.
Kamisaka, S. (1972) in: Plant Growth Substances 1970, D.J. Carr (ed.), Springer, Berlin, pp. 654–660.
Kamisaka, S. (1973) Plant Cell Physiol., 14, 747–755.
Kamisaka, S. and Masuda, Y. (1970) Naturwissenschaften, 57, 546.
Kamisaka, S. and Masuda, Y. (1971) Plant Cell Physiol., 12, 1003–1005.
Kamisaka, S., Sakurai, N. and Masuda, Y. (1973) Plant Cell Physiol., 14, 183–193.
Kamisaka, S., Sano, H., Katsumi, M. and Masuda, Y. (1972) Plant Cell Physiol., 13, 167–173.
Kang, B.G. and Ray, P.M. (1969) Planta, 87, 217–226.
Karanov, E.N. (1972). in: Hormonal Regulation in Plant Growth and Development, H. Kaldewey and Y. Vardar (eds.), Verlag Chemie, Weinheim, pp. 467–472.
Kasemir, H. and Mohr, H. (1972) Plant Physiol., 49, 453–454.

Kato, J., Katsumi, M., Tamura, S. and Sakurai, A. (1968) in: Biochemistry and Physiology of Plant Growth Substances, F. Wightman and G. Setterfield (eds.), Runge Press, Ottawa, pp. 347–359.
Katsumi, M., Purves, W.K., Phinney, B.O. and Kato, J. (1965) Physiol. Plant., 18, 550–556.
Katsumi, M., Tamura, S. and Sakurai, A. (1967) Plant Cell Physiol., 8, 399–407.
Keates, R.A.B. (1973a) Biochem. biophys. Res. Commun., 54, 655–661.
Keates, R.A.B. (1973b) Nature (Lond.), 244, 355–357.
Keates, R.A.B. and Trewavas, A.J. (1974) Plant Physiol., 54, 95–99.
Kefeli, V.I. and Kadyrov, C.S. (1971) Ann. Rev. Plant Physiol., 22, 185–196.
Kefford, N.P. (1962) Aust. J. Biol. Sci., 15, 304–311.
Kessler, B. and Kaplan, B. (1972) Physiol. Plant., 27, 424–431.
Kessler, B. and Steinberg, N. (1973) Physiol. Plant., 28, 548–553.
Khan, A.A. (1975) Bot. Rev., 41, 391–420.
Khawaja, J.A. and Raina, A. (1970) Biochem. biophys. Res. Commun., 41, 512–518.
Kimura, Y. and Tamura, S. (1974) Agr. Biol. Chem., 38, 875–876.
Kimura, Y., Katagiri, K. and Tamura, S. (1971) Tetrahedron Lett., 3137–3140.
Kimura, Y., Tietz, A. and Tamura, S. (1975) Planta, 126, 289–292.
Kitahara, Y., Yanagawa, H. and Kato, T. (1972) Plant Cell Physiol., 13, 923–925.
Klein, A.O. and Hagen, C.W. (1961) Plant Physiol., 36, 1–9.
Knypl, J.S. (1964) Naturwissenschaften, 51, 117–118.
Knypl, J.S. (1967) Acta soc. bot. polon., 36, 589–603.
Koeppe, D.E. and Miller, R.J. (1974) Plant Physiol., 54, 374–378.
Komoto, N., Ikegami, S. and Tamura, S. (1972a) Agr. Biol. Chem., 36, 2547–2553.
Komoto, N., Noma, M. Ikegami, S. and Tamura, S. (1972b) Agr. Biol. Chem., 36, 2555–2561.
Konijn, T.M., Van de Meene, J.G.C., Chang, Y.Y., Barkley, D.S. and Bonner, J.T. (1969) J. Bacteriol., 99, 510–512.
Kopcewicz, J. (1969) Naturwissenschaften, 56, 287.
Kornelyuk, V.N. and Volynets, A.P. (1971) Dokl. Akad. Nauk. Beloruss. SSR, 15, 949–951; Chem. Abst., 76, 107.
Koshimizu, K., Fukui, H., Usuda, S. and Mitsui, T. (1974) in: Plant Growth Substances 1973, Hirokawa, Tokyo, pp. 86–92.
Kosuge, T. (1961) Arch. Biochem., 95, 211–218.
Kosuge, T. and Conn, E.E. (1959) J. biol. Chem., 234, 2133–2137.
Kosuge, T. and Conn, E.E. (1961) J. biol. Chem., 236, 1617–1621.
Kosuge, T. and Conn, E.E. (1962) J. biol. Chem., 237, 1653–1656.
Koves, E. (1957) Acta Biol. Szeged, 3, 179–187.
Krause, J. and Renzik, H. (1972) Z. Pflanzenphysiol., 68, 134–143.
Kriventsov, V.I. and Anandaradzha, K. (1973) Fiziol. Biokhim. Kult. Rast., 5, 75–80; Chem. Abst., 79, 2820.
Kruszewski, S.P. and Jacobs, W.P. (1974) Plant Physiol., 54, 310–311.
Kuhn, R. and Jerchel, D. (1943) Ber. Deut. Chem. Ges., 76, 413–419.
Kuhn, R. and Kum, K. (1962) Chem. Ber., 95, 2009–2011.
Kuhn, R. Jerchel, D., Moewus, F., Moller, E.F. and Lettre, H. (1943) Naturwissenschaften, 31, 468.
Kuhnle, J.A., Corse, J. and Chan, B.G. (1975) Biochem. Physiol. Pflanzen, 167, 553–556.
Kupchan, S.M., Aynehchi, Y., Cassady, J.M., McPhail, A.T., Sim, G.A., Schnoes, H.K. and Burlingame, A.L. (1966) J. Amer. chem. Soc., 88, 3674–3676.
Kupchan, S.M., Hemingway, R.J., Werner, D., Karim, A., McPhail, A.T. and Sim, G.A. (1968) J. Amer. chem. Soc., 90, 3596–3597.
Lado, P., Pennacchioni, A. and Caldogno, F.R. (1972) Physiol. Plant Pathol., 2, 75–85.
Lado, P., Caldogno, F.R., Pennacchioni, A. and Marre, E.. (1973) Planta, 110, 311–320.
Lado, P., Rasi-Caldogno, F. and Colombo, R. (1974) Physiol. Plant, 31, 149–152.
Lado, P., Rasi-Caldogno, F. and Colombo, R. (1975). Physiol. Plant., 34, 359–364.
Lado, P., De Michelis, M.I., Cerana, R. and Marre, E. (1976) Plant Sci. Lett., 6, 5–20.

Larsen, P. (1947) Amer. J. Bot., 34, 349–356.
Lavie, D., Levy, E.C., Cohen, A., Evenari, M. and Guttermann, Y. (1974) Nature (Lond.), 249, 388.
Lee, T.T. and Skoog, F. (1965a) Physiol. Plant., 18, 386–402.
Lee, T.T. and Skoog, F. (1965b) Physiol. Plant., 18, 577–585.
Leopold, A.C., Scott, F.I., Klein, W.H. and Ramstad, E. (1952) Physiol. Plant., 5, 85–90.
Le Poidevin, N. (1965) Phytochemistry, 4, 525–526.
Letcher, R.M. (1973) Phytochemistry, 12, 2789–2790.
Letham, D.S. (1967) Planta, 74, 228–242.
Letham, D.S., Parker, C.W. and Gordon, M.E. (1972) Physiol. Plant., 27, 285–290.
Lin, P.P. (1974) Adv. Cyclic Nucleotide Res., 4, 439–461.
Lin, P.P. and Varner, J.E. (1972) Biochim. biophys. Acta (Amst.) 276, 454–474.
Linsmaier, E.M. and Skoog, F. (1965) Physiol. Plant., 18, 100–127.
Lippincott, B.B. and Lippincott, J.A. (1971) Amer. J. Bot., 58, 817–826.
Lippincott, B.B. and Lippincott, J.A. (1975) Plant Physiol., 56, 213–215.
Lippincott, J.A., Lippincott, B.B. and Chang, C. (1972) Plant Physiol., 49, 131–137.
Lockwood, G.B. and Brain, K.R. (1976) Phytochemistry, 15, 1655–1660.
Lukes, R., Jary, J. and Nemec, J. (1962) Collect. Czech. Chem. Commun., 27, 735–741.
MacAlpine, G.A., Raphael, R.A., Shaw, A., Taylor, A.W. and Wild, H.J. (1974) J. Chem. Soc. chem. Commun., 834–835.
Macheix, J.J. (1974) Physiol. Veg., 12, 25–33.
Machlis, L. (1958) Physiol. Plant., 11, 181–192.
Machlis, L. (1973a) Plant Physiol., 52, 524–526.
Machlis, L. (1973b) Plant Physiol., 52, 527–530.
Machlis, L., Nutting, W.H., Williams, M.W. and Rapoport, H. (1966) Biochemistry, 5, 2147–2152.
Malcolm, H.D.R. (1972) in: Plant Growth Substances 1970, D.J. Carr (ed.), Springer, Berlin, pp. 767–774.
Malcolm, M.J. and Gear, J.R. (1971) Canad. J. Biochem., 49, 412–416.
Manai, H. and Paulet, P. (1973) C.R. Acad. Sci. (Paris) Ser. D., 276, 315–317.
Mandava, N. and Mitchell, J.W. (1972) Chem. Ind. (Lond), 930–931.
Mandava, N., Sidwell, B.A., Mitchell, J.W. and Worley, J.F. (1973) Ind. Eng. Chem. Prod. Res. Develop., 12, 138–139; Chem. Abst., 79, 225.
Mandava, N., Anderson, J.D. and Dutky, S.R. (1974) Phytochemistry, 13, 2853–2856.
Marre, E., Lado, P., Rasi-Caldogno, F. and Colombo, R. (1973a) Plant Sci. Lett., 1, 179–184.
Marre, E., Lado, P., Rasi-Caldogno, F. and Colombo, R. (1973b) Plant Sci. Lett., 1, 185–192.
Marre, E., Colombo, R., Lado, P. and Rasi-Caldogno, F. (1974a) Plant Sci. Lett., 2, 139–150.
Marre, E., Lado, P., Ferroni, A. and Denti, A.B. (1974b) Plant Sci. Lett., 2, 257–265.
Marre, E., Lado, P., Rasi-Caldogno, F., Colombo, R. and De Michelis, M.I. (1974c) Plant Sci. Lett., 3, 365–379.
Mashtakov, S.M., Volynets, A.P. and Laman, N.A. (1971) Fiziol. Rast., 18, 358–364.
Masuda, Y. (1968) in: Biochemistry and Physiology of Plant Growth Substances, F. Wightman and G. Setterfield (eds.), Runge Press, Ottawa, pp. 699–710.
Masuda, Y. and Tanimoto, E. (1967) Plant Cell Physiol., 8, 459–465.
Mayer, A.M. and Evenari, M. (1952) J. exp. Bot., 3, 246–252.
Mayer, A.M. and Shain, Y. (1974) Ann. Rev. Plant Physiol., 25, 167–193.
McMorris, T.C. and Barksdale, A.W. (1967) Nature (Lond.), 215, 320–321.
McMorris, T.C., Seshadri, R., Weihe, G.R., Arsenault, G.P. and Barksdale, A.W. (1975) J. Amer. chem. Soc., 97, 2544–2545.
Menhenett, R. and Carr, D.J. (1973) Aust. J. biol. Sci., 26, 527–537.
Mesland, D.A.M., Huisman, J.G. and Van den Ende, H. (1974) J. gen. Microbiol., 80, 111–117.
Michniewicz, M. and Rozej, B. (1974) Naturwissenschaften, 61, 33.
Michniewicz, M., Rozej, B. and Lamparska, K. (1976) in: Proceedings of the Fourth Symposium on Plant Growth Regulators, Acta Universitatis Nicolai Copernici (Biologia XVIII) Torun, Poland, pp. 95–101.

Mikami, Y., Takahara, H., Iimura, H., Suzuki, A. and Tamura, S. (1970) Agr. Biol. Chem., 34, 977–979.
Milborrow, B.V. and Pryce, R.J. (1973) Nature (Lond.), 243, 46.
Miller, C.O. (1972) Plant Physiol., 49, 310–313.
MIller, C.O. and Meyer, B.S. (1951) Plant Physiol., 26, 631–634.
Miller, R.W., Sirois, J.C. and Morita, H. (1975) Plant Physiol., 55, 35–41.
Minato, H. and Horibe, I. (1965) J. Chem. Soc., 7009–7017.
Mitchell, J.W. and Gregory, L.E. (1972) Nature New Biology, 239, 253–254.
Mitsuhashi. M. and Shibaoka, H. (1965) Plant Cell Physiol., 6, 87–98.
Mitchell, J.W., Mandava, N., Worley, J.F. and Plimmer, J.R. (1970) Nature (Lond.), 225, 1065–1066.
Mitchell, E.D. and Tolbert, N.E. (1968) Biochemistry, 7, 1019–1025.
Moore, K. and Pecket, R.C. (1972) Ann. Bot. (Lond.), 36, 109–113.
Morgan, P.W. and Powell, R.D. (1970) Plant Physiol., 45, 553–557.
Morimoto, H., Sanno, Y. and Oshio, H. (1966) Tetrahedron Lett., 22, 3173–3179.
Morita, T. and Aoki, H. (1974). Agr. Biol. Chem., 38, 1501–1505.
Muller, D.G. (1968) Planta 81, 160–168.
Muller, D.G. and Jaenicke, L. (1973) FEBS Lett., 30, 137–139.
Muller, D.G., Jaenicke, L., Donike, M. and Akintobi, T. (1971) Science, 171, 815–817.
Mumford, F.E., Smith, D.H. and Castle, J.E. (1961) Plant Physiol., 36, 752–756.
Mumford, F.E., Stark, H.M. and Smith, D.H. (1963) Phytochemistry, 2, 215–220.
Naf, U. (1968) Plant Cell Physiol., 9, 27–33.
Nakanishi, K., Endo, M., Naf, U. and Johnson, L.F. (1971) J. Amer. chem. Soc., 93, 5579–5581.
Narayanarao, A., Manibhushanrao, K. and Suryanarayanan, S. (1972) Proc. Indian Acad. Sci. Sect. B, 75, 140–144.
Neish, A.C. (1965) In: Plant Biochemistry, J. Bonner and J.E. Varner (eds.), Academic Press, New York, pp. 581–617.
Neumann, J. (1959) Science, 129, 1675–1676.
Neumann, J. (1960) Physiol. Plant., 13, 328–341.
Nicholls, W., Crow, W.D. and Paton, D.M. (1972) in: Plant Growth Substances 1970, D.J. Carr (ed.), Springer-Verlag, Berlin, pp. 324–329.
Nickells, M.W., Schaefer, G.M. and Galsky, A.G. (1971) Plant Cell Physiol., 12, 717–725.
Niemann, G.J. (1971) Acta Bot. Neer., 20, 397–400.
Niles, R.M. and Mount, M.S. (1974a) Plant Physiol., 54, 372–373.
Niles, R.M. and Mount, M.S. (1974b) Phytochemistry, 13, 2735–2740.
Nishikawa, M., Kamiya, K., Takabatake, A., Oshio, H., Tomiie, Y. and Nitta, I. (1966) Tetrahedron, 22, 3601–3606.
Nitsch, J.P. and Nitsch, C. (1962) Ann. Physiol. Veg., 4, 211–225.
Nozoe, S., Morisaki, M., Tsuda, K., Litaka, Y., Takahashi, N., Tamura, S., Ishibashi, K. and Shirasaka, M. (1965) J. Amer. chem. Soc., 87, 4968–4970.
Nutting, W.H., Rapoport, H. and Machlis, L. (1968) J. Amer. chem. Soc., 90, 6434–6438.
Okamoto, T., Torii, Y. and Isogai, Y. (1968) Chem. Pharm. Bull., 16, 1860–1864.
Okamoto, T., Torii, Y. and Isogai, Y. (1972) in: Plant Growth Substances 1970, D.J. Carr (ed.), Springer-Verlag, Berlin, pp. 311–317.
Okuda, M. (1966) Bot. Mag. (Tokyo), 79, 222–225.
Okuda, M., Kato, J. and Tamura, S. (1967) Planta, 72, 289–291.
Oota, Y. (1972) Plant Cell Physiol., 13, 195–199.
Oota, Y. (1975) Plant Cell Physiol., 16, 1131–1135.
Oota, Y. and Hoshino, T. (1974) Plant Cell Physiol., 15, 1063–1072.
Oota, Y. and Kondo, T. (1974) Plant Cell Physiol., 15, 403–411.
Osawa, T., Suzuki, A. and Tamura, S. (1971) Agr. Biol. Chem., 35, 1966–1972.
Osawa, T., Suzuki, A., Tamura, S., Ohashi, Y. and Sasada, Y. (1973) Tetrahedron Lett., 5135–5138.
Ownby, J.D., Ross, C.N. and Key, J.L. (1975) Plant Physiol., 55, 346–351.

Palesiko, N.B., Shubert, T.A. and Ovcharov, K.E. (1966) Fiziol. Rast., 13, 842–849.
Parups, E.V. (1976) Physiol. Plant., 36, 154–156.
Paton, D.M. and Willing, R.R. (1974) in: Plant Growth Substances 1973, Hirokawa Publishing Co., Tokyo, pp. 126–132.
Paton, D.M. Willing, R.R., Nicholls, W. and Pryor, L.D. (1970) Aust. J. Bot., 18, 175–183.
Pastan, I. and Perlman, R.L. (1972) Adv. Cyclic Nucleotide Res., 1, 11–16.
Pecket, R.C. and Bassim, T.A.H. (1974) Phytochemistry, 13, 1395–1399.
Penny, D. and Stowe, B.B. (1965) Plant Physiol., 40, 1140–1145.
Penny, D. and Stowe, B.B. (1966) Plant Physiol., 41, 360–365.
Peters, K.E., Lippincott, J.A. and Studier, M. (1974) Phytochemistry, 13, 2383–2386.
Phillips, I.D.J. (1962) J. exp. Bot., 13, 213–226.
Pilet, P.E. (1964) Phytochemistry, 3, 617–621.
Piozzi, F., Fuganti, C., Mondelli, R. and Geriotti, G. (1968) Tetrahedron, 24, 1119–1131.
Pirie, A. and Mullins, M.G. (1976) Plant Physiol., 58, 468–472.
Pitman, M.G., Schaefer, N. and Wildes, R.A. (1975) Planta, 126, 61–73.
Plattner, J.J. and Rapoport, H. (1971) J. Amer. chem. Soc., 93, 1758–1761.
Plempel, M. (1962) Planta, 58, 509–520.
Plempel, M. (1963a) Naturwissenschaften, 50, 226.
Plempel, M. (1963b) Planta, 59, 492–508.
Plempel, M. (1965) Planta, 65, 225–231.
Pollard, C.J. (1971) Biochim. biophys. Acta (Amst.), 252, 553–560.
Polya, G.M. (1974) Proc. nat. Acad. Sci. (Wash.), 71, 1299–1303.
Polya, G.M. (1975) Biochim. biophys. Acta (Amst.), 384, 443–457.
Polya, G.M. and Ashton, A.R. (1973) Plant Sci. Lett., 1, 349–357.
Potempa, L.A. and Galsky, A.G. (1973) Plant Cell Physiol., 14, 989–995.
Pringle, R.B. (1961) Science, 133, 284.
Pringle, R.B., Naf, U. and Braun, A.C. (1960) Nature (Lond.), 186, 1066–1067.
Prokhorchik, R.A., Volynets, A.P. and Stolovich, V.N. (1972) Dokl. Akad. Nauk. Beloruss. SSR, 16, 755–758; Chem. Abst., 77, 56.
Pryce, R.J. (1971) Phytochemistry, 10, 2679–2685.
Pryce, R.J. (1972a) Phytochemistry, 11, 1759–1761.
Pryce, R.J. (1972b) Phytochemistry, 11, 1355–1364.
Pryce, R.J. and Linton, L. (1974) Phytochemistry, 13, 2497–2501.
Railton, I.D. and Wareing, P.F. (1973) Planta, 112, 65–69.
Ralph, R.K., McCombs, P.J.A., Tener, G. and Wojcik, S.J. (1972) Biochem. J., 130, 901–911.
Ramstad, E. (1953) Helv. pharm. Acta, 28, 45–57.
Rao, V.S. and Khan, A.A. (1975) Biochem. biophys. Res. Commun., 62, 25–30.
Raper, J.A. (1940) Amer. J. Bot., 27, 162–173.
Raper, J.A. (1942) Amer. J. Bot., 29, 159–166.
Raper, J.A. (1957) Symp. Soc. exp. Biol., 11, 143–165.
Raper, J.A. and Haagen-Smit, A.J. (1942) J. biol. Chem., 143, 311–320.
Rast, D., Skrivanova, R. and Bachofen, R. (1973) Phytochemistry, 12, 2669–2672.
Redemann, C.T., Rappaport, L. and Thompson, R.H. (1968) in: Biochemistry and Physiology of Plant Growth Substances, F. Wightman and G. Setterfield (eds.), Runge Press, Ottawa, pp. 109–124.
Reid, I.D. (1974) Cand. J. Bot., 52, 521–524.
Rekoslavskaya, N.I., Gamburg, K.Z. and Gamanets, L.V. (1974) Fiziol. Rast., 21, 721–727.
Riov, J. and Jaffe, M.J. (1973a) Plant Physiol., 51, 520–528.
Riov, J. and Jaffe, M.J. (1973b) Plant Physiol., 52, 233–235.
Roddick, J.G. (1975) J. exp. Bot., 26, 749–756.
Rodighiero, G. (1954) Giorn. Biochim., 3, 138.
Roy, T., Ghose, B. and Sircar, S.M. (1973) J. exp. Bot., 24, 1064–1068.
Roy, B.N., Roychoudhury, N., Bose, T.K. and Basu, R.N. (1972) Phyton, 30, 147–151.

Tamura, S., Sakurai, A., Kainuma, K. and Takai, M. (1963) Agr. Biol. Chem., 27, 738–739.

Tamura, S., Kimura, Y., Inoue, T., Hirota, A., Katagiri, K. and Suzuki, A. (1974). in: Plant Growth Substances 1973, Hirokawa, Tokyo, pp. 193–200.

Tanada, T. (1972) Plant Physiol., 49, 860–861.

Taniguchi, E. and White, G.A. (1967) Biochem. biophys. Res. Commun., 28, 879–885.

Tarantowicz-Marck, E. and Kleczkowski, K. (1975) Plant Sci. Lett., 5, 417–423.

Thimann, K.V. (1972) in: Plant Physiology vol. VIB F.C. Steward (ed.), Academic Press, New York, pp. 3–332.

Thimann, K.V. (1974) Plant Physiol., 54, 450–453.

Thimann, K.V. and Bonner, W.D. (1949) Proc. nat. Acad. Sci. (Wash.), 35, 272–276.

Thimann, K.V., Tomaszewski, M. and Porter, W.L. (1962) Nature (Lond.), 193, 1203.

Tokoroyama, T., Kamikawa, T. and Kubota, T. (1968) Tetrahedron, 24, 2345–2355.

Tomaszewski, M. (1964) in: Regulateurs Naturels de la Croissance Végétale, C.N.R.S., Paris, pp. 335–351.

Tomaszewski, M. and Thimann, K.V. (1966) Plant Physiol., 41, 1443–1454.

Tomer, E., Goren, R. and Monselise, S.P. (1969) Phytochemistry, 8, 1315–1316.

Truelsen, T.A., De Langhe, E. and Verbeek-Wyndaele, R. (1974) Arch. Int. Physiol. Biochim., 82, 109–114.

Tso, T.C. (1964) Nature (Lond.), 202, 511–512.

Turetskaya, R., Kefeli, V., Kutacek, M., Vackova, K., Tschumakovski, N. and Krupnikova, T. (1968) Biol. Plant., 10, 205–221.

Turner, N.C. (1973) Amer. J. Bot., 60, 717–725.

Uno, I. and Ishikawa, T. (1973) J. Bacteriol., 113, 1249–1255.

Uribe, E.G. (1970) Biochemistry, 9, 2100–2106.

Valio, I.F.M. (1973) J. exp. Bot., 24, 442–449.

Valio, I.F.M., Burdon, R.S. and Schwabe, W.W. (1969) Nature (Lond.), 223, 1176–1178.

Valio, I.F.M. and Schwabe, W.W. (1970) J. exp. Bot., 21, 138–150.

Van den Ende, H. (1967) Nature (Lond.), 215, 211–212.

Van den Ende, H. (1968) J. Bacteriol., 96, 1298–1303.

Van Overbeek, J., Blondeau, R. and Horne, V. (1951) Amer. J. Bot., 38, 589–595.

Van Sumere, C.F. and Kint, J. (1964) Arch. int. Physiol. Biochim., 72, 706–708.

Van Sumere, C.F., Van Sumere-De Preter, C., Vining, L.C. and Ledingham, G.A. (1957) Canad. J. Microbiol., 3, 847–862.

Van Sumere, C.F., Cottenie, J., de Greef, J. and Kint, J. (1972) Recent Adv. Phytochemistry, 4, 165–221.

Varga, M. (1957) Acta Biol. Szeged, 3, 212–223.

Varga, M. (1958) Acta Biol. Szeged, 4, 41–49.

Vazquez, A. (1973) Plant Sci. Lett., 1, 433–438.

Vieitez, E., Seoane, E., Gesto, M.D.V., Vazquez, A., Mendez, J., Carnicer, A. and Areses, M.L. (1967) Phytochemistry, 6, 913–920.

Vlitos, A.J. and Crosby, D.G. (1959) Nature (Lond.), 184, 462–463.

Voeller, B.R. (1964a) in: Regulateurs Naturels de la Croissance Vegetale, C.N.R.S., Paris, pp. 665–684.

Voeller, B.R. (1964b) Science, 143, 373–375.

Wallace, J.W. and Grisebach, H. (1973) Biochim. biophys. Acta, 304, 837–841.

Waller, G.R. and Burstrom, H. (1969) Nature (Lond.), 222, 576–578.

Walton, D.C. and Sondheimer, E. (1968) Plant Physiol., 43, 467–469.

Wardale, D.A. (1973) Phytochemistry, 12, 1523–1530.

Weissenbock, G. (1973) Ber. Deut. Bot. Ges., 86, 351–364.

Wellburn, A.R., Ashby, J.P. and Wellburn, F.A.M. (1973). Biochim. biophys. Acta (Amst.), 320, 363–371.

Wender, S.H. (1970). Recent Adv. Phytochemistry 3, 1–29.

Werkman, B.A. and Van den Ende, H. (1973) Archiv. Mikrobiol., 90, 365–374.

Werkman, B.A. and Van den Ende, H. (1974) J. gen. Microbiol., 82, 273–278.
White, J.C., Hillman, J. and Phillips, I.D.J. (1972) J. exp. Bot., 23, 987–995.
White, J.M. and Pike, C.S. (1974) Plant Physiol., 53, 76–79.
Winget, G.D., Izawa, S. and Good, N.E. (1969) Biochemistry, 8, 2067–2074.
Wojciechowski, Z.A., Zimowski, J. and Zielenska, M. (1976) Phytochemistry, 15, 1681–1683.
Wood, H.N. and Braun, A.C. (1973) Proc. nat. Acad. Sci. (Wash.), 70, 447–450.
Wood, H.N., Lin, M.C. and Braun, A.C. (1972) Proc. nat. Acad. Sci. (Wash.), 69, 403–406.
Worsham, A.D., Klingman, G.C. and Moreland, D.E. (1962) Nature (Lond.), 195, 199–201.
Worsham, A.D., Moreland, D.E. and Klingman, G.C. (1964) J. exp. Bot., 15, 556–567.
Yamagata, Y. and Masuda, Y. (1975) Plant Cell Physiol., 16, 41–52.
Yamaki, T., Shibaoka, H., Syono, K., Morimoto, H. and Oshio, H. (1966) Bot. Mag. (Tokyo), 79, 339–341.
Yamazaki, S., Tamuva, S., Marumo, F. and Saito, Y. (1969) Tetrahedron Lett., 359–362.
Yanagawa, H., Kato, T. and Kitahara, Y. (1973) Tetrahedron Lett., 1073–1075.
Yanagawa, H., Kato, T., Kitahara, Y., Takahashi, N. and Kato, Y. (1972) Tetrahedron Lett., 2549–2552.
Yanagishima, N. (1969) Planta, 87, 110–118.
Yanagishima, N. (1971) Physiol. Plant., 24, 260–263.
Yanagishima, N., Shimoda, C., Tsuboi, M., Hagiya, M., Takao, N., Sakurai, A. and Tamura, S. (1974) in: Plant Growth Substances 1973, Hirokawa, Tokyo, pp. 173–184.
Young, D.V. and Srinivasan, P.R. (1971) Biochim. biophys. Acta (Amst.), 238, 447–463.
Yunghams, H. and Jaffe, M.J. (1972) Plant Physiol., 49, 1–7.
Zanno, P.R., Endo, M., Nakanishi, K., Naf, U. and Stein, C. (1972) Naturwissenschaften, 59, 512.
Zenk, M.H. and Muller, G. (1963) Nature (Lond.), 200, 761–763.

Phytohormones and Related Compounds – A Comprehensive Treatise, Volume I
Editors: Letham, Goodwin and Higgins

8

Phytohormones and Subcellular Structural Modification

T.J.V. HIGGINS and J.V. JACOBSEN

8.1. Introduction

There is considerable evidence that structural modification of cells and rearrangement of subcellular components accompanies the responses to hormones in plants (Fowke and Setterfield 1969; Jones 1969a, c and d; Valdovinos et al. 1971; Vigil and Ruddat 1973; Gawlik and Shen-Miller 1974; Srivastava et al. 1975). Certain structural configurations may mediate hormone action, but they may also be the result of a sequence of events set in motion by the hormone. Therefore, kinetic analyses of the events following hormone treatment are important in attempting to determine cause and effect relationships in hormone action. If the number of measurable hormone-dependent events are numerous but independent of each other then, theoretically, it should be a relatively simple matter to study any individual event and to pinpoint the primary site of action of the hormone for that event. The auxin-mediated increase in plant cell expansion may be such a case involving many independent responses. However, there are likely to be many individual biochemical steps which lead to each response and hence, determination of the mechanism of action of the hormone at the molecular level presents a formidable task. If this is true, as it is likely to be (see Chapter 12, Volume II), then it will be very difficult to determine whether or not there is a requirement for a subcellular structural arrangement for hormone action and to determine the primary site of action of the regulator and thus a simpler system must be sought. On the other hand, if the hormone-dependent events are both numerous and interdependent, then a study of the mechanism of action may become far too complex to analyze even with a systems approach and the hormone-plant system is best viewed in the short term as one in which the hormone is used as a probe with which to study the developmental biology of the plant. In the long term, it is hoped that such study will yield an integrated picture from which it may then be possible to devise experiments to study the mechanism of action of the hormone. In this and subsequent chapters, it will become apparent that most studies were initiated with mechanism of action in mind but that they ended up, at best, as descriptions of development. Structural modification of cellular components following hormone treatment of plant tissues is an example of this type of situation in that there have been many reports of structural changes, but few cases where these changes could be clearly linked to a primary action of the hormone or shown to be required for hormone action. Nevertheless, such changes are still of particular interest in a study of the mechanism of action of hormones, since they pose the question of the importance of subcellular structural differentiation in development.

It has been suggested that structural organization or reorganization of the cell is important in responses to hormones and subsequent development (Jones 1969d; Tata 1970; Chen et al. 1976). In his work on the induction of α-amylase in barley *(Hordeum vulgare)* aleurone cells by gibberellic acid (GA_3), Jones (1969a) applied a centrifugal force to barley aleurone layers which did not influence most cellular activities including respiration, RNA or total protein synthesis, however, it did impair the production of α-amylase in response to GA_3. He interpreted these findings as showing that the arrangement of the cellular components was important for hormone action. Tata (1970) suggested that formation of the endoplasmic reticulum may be important in the responses to such polypeptide hormones as thyrotropin. His data are based on kinetic analysis of membrane formation and protein synthesis. Ornithine decarboxylase (ODC) is induced in mammalian cells by a variety of stimuli, including hormones, and increased enzyme activity appears to be regulated in part by the subcellular structure (Chen et al. 1976). It was found that a range of drugs capable of disrupting the organization of microtubules and microfilaments also inhibited ODC induction within 30 min. This rapid response is interpreted to suggest that structural integrity is a primary requirement for certain hormonal responses. Thus, it is possible that hormones may bring about, or require, certain structural conditions in the cytoskeleton as a preliminary to their action.

In the sections to follow, many structural changes which accompany or follow hormone responses are mentioned. The electron microscope has been valuable in detecting many of these changes but it still has insufficient resolution for some of the potentially very important phenomena, e.g. changes in receptors (their number, configuration and location); changes in membranes with respect to structure, composition and permeability; changes in cell wall structure and formation; changes in ribosome structure and distribution and changes in chromatin structure, especially in relation to control of transcription. The limited information which is available on these topics and the more extensive literature on relatively long-term and ultrastructural changes will be discussed.

8.2. Membranes and receptor sites

8.2.1. Receptor sites for hormones

Receptors for plant hormones are likely to participate in the early events of responses to hormones and are therefore worthy of mention in any discussion of primary action of these compounds. However, even at the present time, when such a great deal of effort has been expended on searching for plant hormone receptors, the evidence for their existence is such that it is reasonable to ask whether such molecules exist at all, and, if so, in what form they are likely to occur. The great variety of processes which are affected by such a few compounds suggests that a number of receptors, mediators or recognition factors must exist in order to define priorities between the various processes competing for the one hormone. It is quite a different matter to provide plausible explanations which could describe how any or all of these receptors might operate.

Unlike the extensive literature on receptors for mammalian hormones (Cuatrecasas 1974), there is relatively little information available on receptors for plant hormones. So far, the term receptor as applied to plant hormones has not been defined and definition should not be attempted until further information on the nature of receptors is available. However, based upon the mammalian data, it is envisaged that plant hormone receptors are recognition or binding proteins with high affinity and low capacity for the hormone, and that they are located on or in responsive cells. The receptors are presumed to be activated by, or combine with, the hormone and then, either alone or in complex, to exercize metabolic control.

The mammalian model is probably misleading since it covers very specific responses, in highly specialized cell- or tissue-types, to hormones with a limited range of targets. Nevertheless, it has served to introduce techniques and build hypotheses in the plant studies. In mammalian cells, receptors for insulin, other polypeptide and non-polypeptide hormones are found on the outside of the plasmalemma (Cuatrecasas 1974) and in the cytoplasm (Jensen and DeSombre 1972). Hypotheses regarding polypeptide hormone interaction with receptors usually involve a conformational change in the receptor molecule (thought to be at least in part a protein) which may result in an inversion of the receptor-hormone complex in the membrane. This may or may not result in the entry of the hormone into the cell but it is thought to alter the shape and thus the activity of other proteins in the membrane. One such protein is adenyl cyclase, a change in whose activity results in the synthesis of cyclic AMP. Such a series of events, while largely speculative, could account for many of the experimental observations made so far. There is considerable evidence that some of these events do take place, but their interrelationships are probably far more complex than that outlined above.

Subtle structural changes at the molecular level are envisaged for steroid-receptor interactions with the chromatin and it is postulated that certain regions of the DNA are modified by the addition of non-histone proteins and/or the removal of histones to expose selected regions of the genome for transcription. The most extensively documented example of this involves the effect of estrogen on the chick oviduct (O'Malley and Means 1974). Evidence for nuclear acceptors has also been found (see Jensen and DeSombre 1972). The majority of these acceptors appear to depend upon cytoplasmic binding of hormone and will not accept hormone which has not formed a complex with a cytoplasmic receptor. However, there may also be some nuclear receptors with very high affinity for the hormone alone. The cytoplasmic receptor complex is 8S, the nuclear complex is 5S, but they show some immunological cross-reaction and may in fact be the same molecule in two forms (see also Chapter 10, section 10.3.1.5).

According to the strict criteria laid down by Cuatrecasas (1974), receptors (as defined for polypeptide hormone receptors) have not been shown to exist in plant cells. However, a number of attempts have been made to demonstrate plant hormone receptors and have included at least three different approaches. The first involves studies in which the receptive tissue or cell-type has been examined, usually by autoradiography of thin sections of the tissue, following treatment with the radioactive hormone for a limited period of time (Liao and Hamilton 1966; Veen 1966; Brandes and Kende 1968; Villiers 1968; Musgrave et al. 1969; Barendse 1974). In the second approach, direct interaction between the hormone and the putative macromolecular receptor has been studied. Such macromolecules have included enzymes (Sarkissian 1970; VanDerWoude et al. 1972; Kasamo and Yamaki 1976), RNA (McCalla et al. 1962; Bendana et al. 1965; Kobayashi 1974), DNA (Fellenberg 1969; Bamberger 1971; Nakamura and Takahashi 1974) and ribosomes or ribosomal proteins (Berridge et al. 1970; Fox and Erion 1975) (see Chapter 10 for further examples and discussion). The third and most recent approach has been to attempt to demonstrate the existence of receptors based upon current concepts of such molecules in mammalian cells. Highly specific recognition factors are being sought in defined subcellular fractions (usually membranous) (Hertel et al. 1972; Stoddart et al. 1974; Oostrom et al. 1975; Batt et al. 1976) and the general aim of these studies seems to have been the rigorous proof of specific binding of the hormone to the presumptive receptor. This has usually involved structural specificity of the hormone. Little attention has been paid to the nature of the receptors or to the physiological consequences of binding. An alternative form to the third approach has been to partially purify the presumed receptors or recognition factors and to test them for regulatory activity in complex biochemical assays such as transcription (Matthysse and Phillips 1969; Matthysse and Abrams 1970; Hardin et al. 1970 and 1972; Venis 1971; Mondal et al. 1972). Certain features of this approach have been unsatisfactory, especially the poor characterization of the recognition or binding event and its specificity with respect to molecules related to the hormone and, also, the great complexity of the physiological assays that have been used to monitor the "bound" or recognized hormone. A good case can be made for combining the two forms of the third approach and a cursory examination of the data suggests that some of the well-defined binding reactions may be related to the apparent hormone-complex-dependent enhancement of transcription that has been claimed by a number of workers.

However, this may be a naive interpretation of the way in which plant cells perceive and react to hormones. If the hormone is perceived as a signal at the outer surface of the cell by a highly specific recognition factor and if that signal is transmitted, for example, as a change in membrane permeability rather than as a physical transfer of the hormone-factor complex, then it probably would not be possible to monitor hormone binding with an in vitro physiological assay. In such a case, monitoring hormone recognition and its associated response would require an intact cell or possibly even a complete tissue. In any event, the structural and biochemical state of the cell seems to be highly critical during recognition of regulating compounds. Therefore, it seems worthwhile to examine the available data for evidence that strict integrity of the structure of the cell and tissue is required for hormone recognition and response. Such an outcome would imply that there is a sequence of interconnected events which is dependent upon the maintenance of the cellular continuum. Alternatively, if the recognition factor and hormone can be isolated and used to activate other processes of cell metabolism in an in vitro assay, it would imply that recognition and response are separate events which must involve re-arrangement of the intracellular structure to ensure specific translocation of regulators from recognition to activation sites. Combinations, further refinements and complexities of the above two models are likely to represent the real situation but, for the present, these simplified views serve as a basis for further analysis. Discussion of the possible role of some hormone-receptor complexes in transcription and translation assays is presented in Chapters 10 and 11. A thorough and critical review of hormone receptors in plants has been presented by Kende and Gardner (1976).

8.2.1.1. Gibberellins. Musgrave et al. (1969) have used elongating pea *(Pisum sativum)* internodes and both active and inactive gibberellins to show that excised, but otherwise intact, organs bind active GAs only in those tissues which responded to the hormone. This has been taken as preliminary evidence for the existence of some recognition factor(s) in the responsive tissue. The nature of these factors could not be determined since disruption of the tissue destroyed the "binding" capacity. This may reflect complexities of the receptor system which may require a structurally or biochemically functional integrity or it may merely reflect lability as shown by receptors for sex hormones (Jensen and DeSombre 1972). A major drawback in the pea studies was the non-saturable nature of the binding which is quite unlike the binding of the polypeptide sex hormones of mammalian systems (Jensen and DeSombre 1972). However, binding in pea had many features suggesting physiological significance such as decreased binding of the hormone in light (which decreases the rate of growth) and decreased binding of biologically inactive gibberellins. The barley aleurone layer responded to the addition of GA_1 and GA_5 by producing α-amylase. Accumulation of the hormone was unrelated to effectiveness in stimulating enzyme induction but closely correlated with its ability to be converted to polar metabolites (Musgrave et al. 1972). The biologically inactive methyl ester of GA_5 was accumulated and metabolized to the greatest extent. To further complicate the picture, an intermediate state between pea and barley has been discovered (Barendse 1974). *Pharbitis nil* seedlings did not accumulate the biologically inactive methyl ester of GA_3 but they did accumulate GA_3 in the light and dark. However, uptake was maximal in the light when biological effectiveness was much reduced. The enhanced uptake appeared to be correlated with increased metabolism of GA_3. Thus, although the early pea work showed promise as a system to study hormone perception at the cellular level, subsequent work has shown that so-called binding may in fact reflect the level of metabolizing enzymes as well as "receptors".

The first circumstantial evidence for the existence of a GA receptor was provided by Johri and Varner (1968) who showed that isolated nuclei of pea epicotyls responded to GA_3 by increasing RNA synthesis only when they were isolated from the tissue in the presence of hormone, i.e. the nuclei did not respond if GA_3 was added only after isolation. Whereas these data might be interpreted to reflect, among many other possibilities, an effect of GA_3 on the integrity of the nuclear envelope which might influence the loss of components such as cofactors from the nucleus necessary for nucleic acid synthesis, they may also reflect interaction of GA_3 with a cytoplasmic or nuclear "factor" or receptor capable of regulating some aspect of transcription. If the "factor" was of cytoplasmic origin then GA_3 may be involved in facilitating its transfer across the nuclear membrane to some site on the chromatin, polymerases or acceptor molecule. Preliminary data suggested that a buffer extract of pea epicotyls contained a proteinaceous fraction which bound GA_1 (Stoddart et al. 1974) with a poor degree of structural specificity. There was also little apparent correlation between responsive (elongating) tissue and level of hormone binding.

Further work should resolve whether these two sets of pea "factors" are related in any way, and provide some answers to the questions raised at the end of section 8.2.1 about the possibility of separating recognition and response to plant hormones into a series of partial reactions.

8.2.1.2. Cytokinins. Cytokinin uptake and accumulation was enhanced in moss *(Funaria hygrometrica)* caulonema cells which later developed gametophyte buds in response to cytokinin (Brandes and Kende 1968). It was suggested that this may represent receptor activity in responsive cells and the data are consistent with this view since selectivity of uptake of cytokinin over adenine was shown and removal of the hormone reversed the bud formation. This system deserves further analysis along the lines described in section 8.2.1.1 to determine whether there is binding to a macromolecule, a membrane or to a metabolizing enzyme. The latter appears unlikely since the hormone could be washed out and the response abolished. The nature of the response to "bound" hormone must be elucidated in biochemical terms before further interpretation is possible.

Preliminary evidence has been obtained to indicate that nuclei of tobacco *(Nicotiana tabacum)*, soybean *(Glycine max)* and pea contain a protein which interacted specifically with cytokinins (Matthysse and Abrams 1970). This protein bound to the chromatin in the presence of the hormone, but could be separated from it in the absence of hormone as judged by sucrose gradient centrifugation in the presence of salt. Precise data showing that the hormone and recognition factor did in fact bind together was not obtained. The presumed hormone-receptor complex interacted with the chromatin to increase transcription using *E. coli* polymerase suggesting de-repression of regions of the DNA, but it could also interact with purified DNA to produce the same result. This implies recognition of new initiation sites on the DNA. On the other hand, it may reflect enhanced non-specific binding of the polymerase to the DNA. The major weakness in this and most other studies on plant hormone receptors is that there is no convincing data on the physical interaction of the hormone with the receptor. It is very hard to assess the validity of the extensive biological investigations in the absence of knowledge about the specificity of the hormone-receptor interaction. Furthermore, McCombs and Ralph (1972) could not confirm this work. In the case of the mammalian female sex steroids, there is some evidence to suggest that there is tissue-specific binding of hormone receptor to the chromatin and tissue non-specific binding to DNA. The latter is otherwise apparently specific in that binding is a high affinity reaction and that there is a limited number of binding sites (O'Malley and Means 1974). However, the nature and significance of binding to DNA remains to be resolved and the majority of evidence points to the involvement of the chromatin proteins in recognition of the hormone receptor. Again it is tempting to speculate that the in vivo "binding" described for the moss *(Funaria hygrometrica)* (Brandes and Kende 1968) may be a reflection of the nuclear interaction described by Matthysse and Abrams (1970) but any conclusion awaits further work.

Cytokinin binding to ribosomes has been reported in Chinese cabbage *(Brassica pekinensis)* leaves (Berridge et al. 1970) and in wheat *(Triticum aestivum)* germ (Fox and Erion 1975). Fox and Erion (1975) determined affinity and number of binding sites while these workers and Berridge and colleagues showed that affinity for the ribosomes was a function of the biological activity of the cytokinin. However, Berridge et al. (1972) were unable to find any effect of kinetin on in vitro protein synthesis but it should be remembered that they used non-initiating systems. A preliminary report (Klämbt 1976) showed that, under specialized conditions of extensive dialysis of the cell-free components, cytokinins added in vitro stimulated (15%) amino acid incorporation. This data has not been confirmed. Thus, no convincing biological significance has so far been attached to the cytokinin-binding phenomenon.

8.2.1.3. Auxin. The search for auxin receptor proteins has been more extensive than for any of the other hormones. Matthysse and Phillips (1969) showed that nuclei isolated from tobacco and soybean tissue culture cells, in the presence of the auxin, 2,4-D, incorporated more [^{14}C] ATP into RNA than nuclei from untreated tissue or nuclei supplied with the hormone after isolation. The continued presence of the hormone was required and it was suggested that, in the nucleus, there was a factor which somehow interacted with auxin to enhance RNA synthesis. Nuclei from pea buds responded to the in vitro addition of hormone suggesting that the nuclei retained a factor which was auxin-responsive. The factor was isolated from a chromatin supernatant of all three tissues and shown tentatively to be a protein with a mol. wt. of 20,000 which aggregated readily into polymers. The protein alone had no effect on [^{14}C] ATP incorporation into RNA in vitro but stimulated its incorporation in the presence of auxin. It was suggested that there was a specific interaction between the "hormone-protein complex" and some proteins of the chromatin but there was no data to show that the hormone bound to the presumed factor, nor was there convincing data on the subcellular location of the factor. It is conceivable that auxin could interact with a transcription factor and regulate its location and activity within the cell as discussed below but these data do not justify such a claim.

Extensive studies on auxin action in soybean hypocotyls in Cherry's laboratory have resulted in the isolation from soybean cotyledons of a crude mixture of non-dialyzable, ammonium sulphate-precipitable materials which promoted incorporation of [^{3}H] UMP into RNA synthesized by chromatin-bound RNA polymerase on control chromatin but not on auxin-treated chromatin (Hardin et al. 1970). This was interpreted to mean that, in vivo, auxin was involved in transferring the factor(s) to the chromatin-polymerase system. Using an *E. coli* polymerase-saturated assay, it was shown that template availability might also be stimulated by the factor. Partial purification of the crude factor by gel filtration resolved two peaks of activity, one of which stimulated endogenous polymerase activity and the other of which stimulated the *E. coli* polymerase-saturated system. The latter is somewhat similar to the results of Matthysse and Phillips (1969), except that in the case of the soybean factor (Hardin et al. 1970) the continued presence of hormone was not required for the response.

Further partial characterization of the soybean endogenous polymerase-stimulating factor (Hardin et al. 1972) resulted in a crude fraction from partially purified subcellular components considered to be enriched for plasma membranes. It was shown to stimulate solubilized polymerase activity slightly. It was possible to release this factor by treatment of the membrane fraction with physiologically active but not inactive auxins and the authors suggested that the factor bound to or somehow modified RNA polymerase, thus modifying the quality of RNA synthesis (see Chapter 10). The nature of the factor is unknown as is its interaction with the polymerase or the DNA and confirmation of the data is required. If it can be released from the membranes by auxin it is interesting and suggests that it may be a recognition factor. It is surprising that it is active while still attached to the membrane and suggests that the act of release mediated by auxin does not alter its activity. How the factor is transported to the nucleus is of interest and has not been studied so far. The relationship between the factor from the membranes and the two activities previously reported by Hardin et al. (1970) is not clear, and especially interesting is the question of whether it still contained an activity similar to that of the factor isolated by Matthysse and Phillips (1969).

A transcription assay was also used by Venis (1971) to monitor the activity of crude protein preparations from corn *(Zea mays)* and pea, which bound to immobilized 2,4-D. A fraction which bound was partially characterized and shown to contain a limited number of polypeptide bands by electrophoresis. The fraction appeared to be soluble in vivo. It did not require auxin for activity in the assay, so that "binding" between the two components was not likely. The role of auxin in the behaviour of the factor was unclear. A subsequent study of auxin binding to enriched membrane fractions from corn did not concern itself with physiological assays of binding (Batt et al. 1976; Batt and Venis 1976). It has been stressed previously that thorough analysis of hormone binding, with respect to specificity, dissociation constant and capacity, should ideally be combined with an examination of the biological consequences of that binding. It is surprising that this has not been attempted in the case of the auxin-binding factor of corn, since physical interaction and physiological assay have been documented separately in the same laboratory.

A promising study on auxin receptors has been made by Mondal et al. (1972). They have isolated and partially purified a protein from the nucleoplasm of coconut *(Cocos nucifera)* endosperm nuclei which binds [^{14}C]IAA by equilibrium dialysis and which stimulates incorporation of β,γ-[^{32}P]ATP and [^{3}H]UMP into RNA synthesized by purified coconut nucleoplasmic polymerase, an initiation factor (B), and coconut DNA as template. New RNA products were synthesized as determined by gel electrophoresis and hybridization studies. The factor and IAA may both react with the DNA although this has not been convincingly demonstrated. There was no apparent effect on polymerase or initiation factor activity. This study was more informative than many others in that an attempt was made to demonstrate binding of the hormone to the protein factor and the complex was then tested in a suitable assay. The system might now be used to answer some of the following questions. Does the purified factor occur only in the nucleus or is it present elsewhere in the cell? Does the factor or the factor-IAA complex react with the DNA and, if so, what is the nature of the interaction? How is the factor transferred from the site of binding to the site of activity? Is the factor found in endosperm nuclei identical to binding factors in other parts of the plant? Is there heterogeneity within the factor with respect to affinity or any other aspect of binding?

Even though the foregoing investigations were on different tissues, there is marked inconsistency in the results. This may reflect inherent experimental difficulties, non-specific or inadequate assay procedures, artefacts generated by isolation or other procedures, or it may reflect heterogeneity of receptors (some cytoplasmic, some nuclear), only one of which was assayed in a particular system. It would be preferable to first find a suitable simple assay for receptors such as binding rather than the transcription of DNA, the complexity of control of which is imperfectly understood at present.

The preliminary findings of Hertel et al. (1972) on the possible existence of an auxin receptor in homogenates of corn coleoptiles points to such a simple system. They have demonstrated saturation of binding, a K_m of $10^{-5}-10^{-6}$ M and a tentative localization in a plasma membrane-rich fraction. Using inactive analogues and antagonists of auxin such as the transport inhibitor, 2,3,5-triiodobenzoic acid, they showed some specificity of binding. The binding site was heat-labile and sensitive to the ionic detergent, sodium dodecyl sulphate. These data have been confirmed and extended by other workers who showed that two binding reactions with high affinity exist and that the lower affinity site was the most specific (Batt et al. 1976; Batt and Venis 1976). The highly specific, lower affinity site was considered to be on the plasma membrane while the relatively non-specific, higher affinity site was tentatively assigned to the Golgi body membrane or endoplasmic reticulum. Developing pea buds were also shown to contain particulate capacity to bind IAA (Jablanovic and Nooden 1974). The binding showed some specificity, although kinetics of the reaction were not studied. There was a good correlation between a physiological response (release from apical dominance) and a reduced binding capacity for the hormone. The binding of [1-^{14}C] IAA and [2-^{14}C] IAA to isolated nuclei of light- and dark-grown pea apical buds has been studied by Tautvydas and Galston (1972). They found that the nuclei contained a limited number of binding sites which were saturated at low but physiological levels of hormone. Whether the bound compound was IAA is not known. This is important in view of the lability of IAA in the presence of cell-free isolates. These approaches to the study of hormone perception are very promising, and suggest that recognition of the hormone may soon be linked to a defined response.

Two enzymes, tentatively localized in a plasma membrane-rich fraction, have been shown to respond to IAA. VanDerWoude et al. (1972) showed that cellulose synthetase was activated to a slight extent in vitro by IAA. Similarly, a salt-activated ATPase was stimulated in vitro by IAA (Kasamo and Yamaki 1976). The relationship between these membrane-associated events and those described above (Hardin et al. 1972; Hertel et al. 1972; Batt et al. 1976; Batt and Venis 1976) remain to be elucidated.

The importance of determining definitively the existence of plant hormone receptors in cells cannot be overstressed, and from the above summary is is clear that there is some evidence for their existence. The best example of a modification of a presumptive recognition factor by a hormone is provided by the data of Hardin et al. (1972), since it involves a possible interaction at the plasmalemma, followed presumably by transport to the nucleus and alteration in the transcription of DNA. It thus represents a system worthy of further study, particularly with respect to the nature of the receptor, its abundance in receptive and non-receptive cells, the influence of hormone treatment on its distribution and the nature of the interaction with the hormone. A renewed effort should be made to define plant hormone receptors using a carefully chosen system which combines some of the desirable biochemical characteristics of coconut endosperm, the physiological responsiveness of pea epicotyl and the exciting prospects for chemical identification of the receptor from soybean hypocotyls. From a physiological point of view, such a system may answer some of the following questions:

(a) how do cells in certain physiological states recognize and utilize the hormone? Receptors may function in describing the disposition of the cell and also in scavenging the molecule;

(b) does the hormone per se contain sufficient information to set in motion the complex series of events usually associated with its action? Interaction or activation of a large molecule such as a protein or some other macromolecular complex might mediate the hormone action;

(c) is the hormone transported from the site of recognition and, if so, how is this transfer effected? Such a process might be mediated by a series of macromolecules with high affinity for each other and for the target;

(d) do the receptors of one hormone influence the behaviour of other receptors? When the receptors, e.g. for auxin, are activated by binding, are the receptors for other hormones affected? It is also worth considering the possibility that the number of auxin binding sites may change under such conditions;

(e) are there multiple receptors, primary, secondary etc. conferring different specifications on the hormone which might explain the multiple actions of plant hormones?;

(f) is there a master receptor which binds the hormone, thus activating a series of second messengers, each controlling separate physiological events?

Substantial questions thus remain unanswered but some general conclusions can be drawn from the available data. Studies of in vivo accumulation of hormones in receptive tissue have resulted in a complicated picture suggesting that such accumulation may sometimes reflect storage of inactive forms of the hormone. However, it should be remembered that the pea epicotyl (Musgrave et al. 1969) and the moss protonemata (Brandes and Kende 1968) systems may also yield valuable information on the structural features which prevail during hormone recognition. The altered subcellular structure of the barley aleurone layer which resulted in lack of physiological response to GA_3 (Jones 1969d) has not been studied further, but provides a good opportunity for further research on the role of structure in hormonal responses. The vast majority of the evidence collected so far suggests that structural integrity of the cell is not necessary for hormone recognition, and that binding factors can be isolated from cell homogenates. There is some preliminary evidence to show that hormone-modified factors or hormone-factor complexes possess new regulatory activities not previously present in the factor or hormone alone.

8.2.2. Plasma membranes and permeability; secretion phenomena

As mentioned in section 8.2.1, the plasma membrane has been viewed, particularly in mammalian cells, as a very likely location for receptor sites, especially for the polypeptide hormones. The interaction of hormones with these sites probably brings about a change in the form of some of the membrane constituents, and therefore a change in their function. The best-documented case to date is the relationship between insulin and adenyl cyclase (see Hollenberg and Cuatrecasas 1975). Although some of the physical characteristics of the insulin receptor are known, the details of the interaction with the hormone are not yet elucidated, nor is it known whether the receptor undergoes major modification as a result of binding the hormone. Receptors for small molecules such as steroids are considered to be of cytoplasmic origin, and the mechanism of uptake of the hormones is still somewhat controversial (Gorski and Gannon 1976). Steroid uptake is generally considered to be a diffusion process, although there is some evidence for facilitated transport. Thus, it is anticipated that relatively little modification of the plasma membrane is required for steroid uptake and binding. It is envisaged that the plant cell membrane presents little barrier to the uptake of phytohormones, and that recognition factors may also be found in the cytoplasm (Oostrom et al. 1975). However, in two well-documented cases, the binding site for auxin has been localized in the plasma membrane (Hertel et al. 1972; Batt and Venis 1976). If recognition factors occur exclusively on the outer cell surface or in the cytoplasm as well, then the perception of plant hormones may be quite different to the way in which mammalian cells respond to steroids, and may more closely resemble the mechanism for polypeptide hormone recognition. Thus, it is likely that structural features of the plasma membrane may be important for recognition, and that changes may occur subsequent to the recognition event. These could involve decreased or increased affinity for further hormone and regulation of the recognition of other hormones. Although the fine structure of plasma membranes is largely not understood, many features of the protein and lipid composition and the structural interrelationships which exist between the two are rapidly being elucidated (Singer 1974; Marchesi et al. 1976). Transport systems for ions, non-electrolytes and macromolecules will probably not be completely understood until the molecular organization of the membrane is described. Current concepts of membrane transport involve carrier systems or channels usually thought to consist of conformational changes in the arrangement of the membrane components. Hormonal control of sugar, amino acid and electrolyte transport has been reported for a number of mammalian cell types (Chambers et al. 1965, 1968 and 1970; Tews et al. 1970; Morgan and Whitfeld 1973; Livingston and Lockwood 1975), and the implication is that certain structural features of the plasma membrane were altered to facilitate the new transport rates. Indirect evidence for such changes has been presented by Chang et al. (1974), who showed that two membrane proteins were phosphorylated by an endogenous protein kinase of the fat cell membranes in response to low levels of cAMP. ATP inhibited phosphorylation and the insulin-induced increase in glucose transport, and it was concluded that insulin-dependent enhanced glucose transport was mediated through cAMP-dependent changes in phosphoproteins of the membrane. Although much of this evidence is circumstantial, the hypothesis which has emerged is that insulin reacts with a cell membrane receptor. This results is increased adenyl cyclase activity leading to increased levels of cAMP, which regulate a membrane-bound protein kinase. The protein kinase phosphorylates at least two membrane polypeptides considered from other evidence to be involved in glucose transport. This complex series of events takes place largely on the cell membrane.

There are many instances of changed permeability in plant cells in response to hormones. Some of these appear to occur early in the response, others much later. In all cases, however, the only manifestation of hormone action is the final measured activity of molecules either inside or outside the cell, and there is no evidence to date that would directly implicate the hormone in any structural or biochemical change in the plasma membrane. However, such changes have not been ruled out and in fact are quite likely. Electrochemical measurements have provided good evidence that the plasma membrane controls transport of electrolytes (Higinbotham 1974) and auxin has been shown to alter membrane potentials in plant cells (Etherton 1970). Thus, plant hormones may regulate electrogenic pumps considered to be localized in the plasma membrane. Some reports of permeability changes, considered to be evoked by plant hormones, will be discussed in the next sections and, although it is catalogue-like, it serves to illustrate the wide variety of systems studied.

8.2.2.1. Gibberellins and abscisic acid. The best-studied system for the hormonal control of secretion of macromolecules (enzymes and carbohydrates in this case) is that of the barley aleurone layer. Three to eight hours after GA_3 treatment, α-amylase, protease, phosphatase, β-glucanase, ribonuclease, pentosanase, peroxidase, esterase and glucosidase, phytase, ATPase, GTPase and soluble carbohydrates are released from the cells into the incubating medium (Paleg 1960; Yomo 1960 and 1961; McLeod and Millar 1962; Briggs 1963; Varner 1964; Chrispeels and Varner 1967a; Harmey and Murray 1968; Pollard and Singh 1968; Taiz and Jones 1970; Jacobsen and Knox 1972; Ashford and Jacobsen 1974a). Some of these, e.g. protease and α-amylase, involved GA_3-dependent de novo protein synthesis as well as postulated hormone-dependent release (Filner and Varner 1967; Jacobsen and Varner 1967). Others such as β-glucanase, phosphatase and ribonuclease are not dependent upon added GA_3 for synthesis (Chrispeels and Varner 1967a; Taiz and Jones 1970; Jones 1971; Bennett and Chrispeels 1972; Ashford and Jacobsen 1974a) but are thought to require the hormone for release. Jones (1971) has shown that release of β-1,3-glucanase was triggered by GA_3 application for only the first 4 hr of incubation. Release, (appearance in the incubating medium) required that antimycin A-sensitive respiration be unimpaired during these 4 hr and it also required continuous RNA and protein synthesis, although these may not be required for enzyme appearance inside the cell. There is also some tentative evidence to suggest that ethylene may be involved in α-amylase release but not in its synthesis (Jones 1968), although Jacobsen (1972) found that ethylene had a slight promotive effect on synthesis which might account for the effect on release. The release of protein and carbohydrate is controlled with respect to time (Pollard and Singh 1968; Jacobsen and Knox 1972) and direction of secretion (Taiz and Jones 1970; Jacobsen and Knox 1973), and therefore reflects specific, metabolically controlled events and not general breakdown of the whole cell and release of its contents.

The mechanism of release of such large molecules is not well defined. The enzymes α-amylase and acid phosphatase have been localized cytochemically on both sides of the plasmalemma (Jacobsen and Knox 1973; Ashford and Jacobsen 1974a) and outside the plasmodesmata (Ashford and Jacobsen 1974a). Enzyme transport might be effected by a form of reverse pinocytosis using vesicles which fuse with the membrane (Vigil and Ruddat 1973), but there is not general agreement that vesicles with such a function exist in aleurone cells (Van der Eb and Nieuwdorp 1967; Jones 1969c; Gibson and Paleg 1972 and 1975; Jones 1972; Jacobsen and Knox 1973; Firn 1975; Obata and Suzuki 1976). If secretion is mediated by transport and fusion of vesicles with the plasma membrane, then major structural changes within the cell membrane are to be envisaged. Transport may also be effected through a permease and carrier system, but again there is no evidence for this.

In a very elegant series of experiments, Varner and Mense (1972) have separated the process of release of α-amylase into two steps. The first, which they term secretion, involves the movement of α-amylase across the plasmalemma and is energy-dependent. It is not directly dependent upon GA_3, ABA, nor upon protein or nucleic acid synthesis. The second step, called release, is ion-dependent and has an inconclusive Q_{10} of 1.4. They suggest that release is diffusion-limited. The data of Ashford and Jacobsen (1974a) would indicate that release is dependent upon enzymatic digestion of the wall which should have a Q_{10} of 2. These data are probably not incompatible, since Varner and Mense (1972) made their measurements 19–24 hr after GA_3 treatment, at which time digestion of the wall was well progressed, and the process of release probably involves both the enzymatic digestion of the wall as well as the diffusion of the secreted molecule, which might account for the intermediate Q_{10} recorded by Varner and Mense. None of these studies throw any light on the actual secretion step itself and the membrane changes that might be involved. The barrier presented by the cell wall itself seems to be a major one in barley and wheat aleurone cells, and permeability to enzymes and other large molecules appears to be effected by enzymatic hydrolysis of the wall (Fulcher 1972; Ashford and Jacobsen 1974a). The kinetic data of Pollard and Singh (1968) may not fit this pattern, since they found that certain soluble carbohydrates were released within 3 hr of treatment followed by ATPase, GTPase and phytase. However, the cellular location of these molecules was not determined and their study may not be directly comparable to those mentioned above, since they used a different variety and different imbibition conditions.

Pentosan and the associated cell wall synthesis appear to be under the control of GA in barley aleurone (Johnson and Chrispeels 1973). In this case the activity of the membrane-bound arabinosyl transferase was inhibited by GA_3. The hormone did not influence uptake of precursors or the release of polymerized products. Therefore, if one interprets rather broadly from these three separate instances, i.e. the reduced activity or synthesis of a wall synthesizing enzyme by GA_3 (Johnson and Chrispeels 1973), increased synthesis of α-amylase and appearance in the medium (Varner and Mense 1972) and the demonstration, by cytochemical methods, that phosphatase was released into the medium from aleurone layers treated with GA_3 as a result of wall dissolution rather than effects on synthesis or on the plasma membranes (Ashford and Jacobsen 1974a), it is possible to speculate that GA_3 does not directly influence the transport of macromolecules across the plasma membrane. Reduced levels of wall precursors and increased levels of certain enzymes in the medium could reflect altered enzyme amounts or activity and cell wall hydrolysis.

Solubilization and release of inorganic ions from the aleurone of barley and wheat have been studied with respect to hormonal and substrate dependence, kinetics of release and requirements for RNA and protein synthesis and oxidative phosphorylation (Eastwood and Laidman 1971; Melcher and Varner 1971; Jones 1973). The release of these ions into the medium may be a reflection of increased phytase activity, however, GA_3 did not affect phytin hydrolysis in barley aleurone, and it is therefore unlikely that released phosphate came from phytin (Jacobsen and Zwar, unpublished). Gibberellin control of the membrane transport of ions in aleurone tissue has not been demonstrated. The role of the hormone may be indirect, in that wall hydrolysis may be required for release either because it presents a physical barrier or because it binds the solubilized ions after secretion. The role of hormones, if any, in controlling secretion across the plasma membrane might be resolved using techniques similar to those devised for enzymes by Varner and Mense (1972). Energy dependence and RNA and protein synthesis during ion release (Jones 1973) may also reflect the requirements for the induction of wall hydrolases, and not just energy-dependent transport processes.

Another approach to the study of hormonal control of membrane transport has been that taken by Paleg's group. Artificial membranes of known composition have been synthesized and their permeability to various ions and sugars in response to GA_3 tested (Wood et al. 1972; Wood and Paleg 1972 and 1974). They conclude that GA_3 could influence the permeability of these micelles to small molecules, although specificity for the gibberellin structure was very low. It is difficult to assess the validity of the results obtained with such membranes, especially when one considers the complexity of the natural membrane compared with those made artifically. However, this system may contribute to the characterization of the role of GA in membrane transport by allowing the study of transport in isolation from other cellular processes.

Abscisic acid (ABA), both as an antagonist of GA and in its own right, has been shown to affect the levels of accumulated or secreted ions, water, sugars and macromolecules (Milborrow 1974). Whether or not the hormone acts directly upon the transport systems operating in the plasma membrane has not been established. A few of the responses to ABA in various systems are noted below with some of their implications. It will become apparent that the hormone does influence, with considerable selectivity, the accumulation or secretion of certain molecules, but that the regulatory mechanisms are not understood.

ABA inhibited the appearance of α-amylase in the medium of GA_3-treated aleurone layers (Chrispeels and Varner 1967b; Jacobsen 1972). However, since Varner and Mense (1972) showed that ABA did not affect the secretion step, ABA may inhibit synthesis and release via an effect on the synthesis of α-amylase and some of the hydrolytic enzymes involved in wall digestion. On the other hand, ABA may influence the packaging or transport of the vesicles presumed to contain the enzyme (section 8.2.2.1).

ABA has been shown to influence the flux of inorganic ions in a number of tissues including expanding leaves (Mansfield and Jones 1971), elongating coleoptiles (Reed and Bonner 1974) and growing roots (Shaner et al. 1975). A selective inhibitory effect of the hormone on K^+ uptake was shown, while the uptake of Cl^-, proline and 3-*O*-methylglucose was unaffected (Reed and Bonner 1974). Further selectivity was demonstrated by Mansfield and Jones (1971), who showed that, within expanding leaf tissue, certain cells (the guard cells of stomata) were most responsive to the hormone, and Horton and Moran (1972) confirmed that ABA inhibited accumulation of K^+ in guard cells. Other evidence that ABA may act on cell membrane transport was provided by Shaner et al. (1975), who showed that there was partial depolarization of the transmembrane potential difference in maize roots. They concluded that ABA had no effect on passive diffusion potential and therefore must affect an electrogenic pump, however, they were unable to show any changes in the Mg^{2+}- or K^{+}-stimulated ATPase.

ABA has also been implicated in the control of water movement other than that brought about through stomatal closure. The hormone has been shown to affect both hydraulic conductivity and diffusional permeability of water in root and pith cells (Glinka and Reinhold 1971), while the action of ABA, on diffusional permeability only, appeared to require metabolic energy (Glinka and Reinhold 1972). ABA promoted root exudation through an enhanced hydraulic conductivity (Glinka 1973). The mechanism whereby the hormone influenced the cellular permeability to water remains to be determined.

The wide variety of systems in which ABA appears to affect accumulation and secretion of substances would suggest that the hormone affects transport mechanisms at the level of the plasma membrane. However, there is no evidence of a direct effect of ABA on the plasma membrane and in one instance, where secretion of an enzyme across the plasma membrane was studied (Varner and Mense 1972), it was found that ABA did not affect the membrane transport step.

8.2.2.2. Auxin and ethylene. There is extensive evidence that auxin influences the extracellular appearance of cell wall constituents. Whether this effect is on both synthesis and secretion is not always clear. The association between auxin-stimulated Golgi activity and wall deposition showed strong positive correlations (Fowke and Setterfield 1969; Gawlik and Shen-Miller 1974), but these do not directly implicate auxin in the transport of the cell wall constituents across the plasmalemma.

Auxin promoted the de novo synthesis of cellulase in pea epicotyls (Verma et al. 1975), and this was followed by cell wall dissolution, implying that the enzyme must cross the plasmalemma, if one accepts that cellulase is involved in wall hydrolysis. It is not known whether auxin affects the secretion event. Similarly, enzymes, including cellulase, involved in abscission of flowers, fruit and leaves cause cell wall dissolution (Horton and Osborne 1967; Abeles 1969), and are under the control of auxin and ethylene. There is tentative evidence to suggest that ethylene may control both synthesis and secretion of cellulase (Abeles and Leather 1971; Rasmussen 1973). If it does affect transport of macromolecules, ethylene would occupy a relatively unique position amongst the hormones. However, this claim is not proven and there is the added complexity that further analysis of the cellulase indicated that there were at least two isozymes with different isoelectric points, only one of which was secreted in large amounts (Reid et al. 1974). It was not determined whether the gas altered the membrane transport of the enzyme directly or indirectly through increased synthesis. In a similar study, auxin was shown to promote cellulose synthetase activity (Shore and Maclachlan 1973), and there was circumstantial, but no direct, evidence to suggest that it may influence secretion (Shore and Maclachlan 1975).

There are many reports of changes in the levels of small molecules such as ions, amino acids, water, sugars and nucleotides in response to hormone treatment. Again these studies may be divided into rapid and long-term responses. The latter may often reflect changes in response to new growth, and are thus difficult to interpret. It is essential that some selectivity such as differential influx or efflux of molecules be demonstrated. General changes in permeability may reflect leakiness.

In short-term experiments, Kang and Burg (1971) found that auxin influenced the exchange of tritiated water in pea stem segments. The response was small (10–15% above control values), but it represents a rapid response and thus merits further attention. Dowler et al. (1974) could not confirm these results using pea tissue, and concluded that auxin did not influence membrane permeability to water and that the cuticlar barrier may have confused Kang and Burg's interpretation.

One of the currently studied aspects of auxin physiology is the ability of the hormone to promote excretion of hydrogen ions (Cleland 1973; Marré et al. 1973; Rayle 1973). External application of hydrogen ions to auxin-responsive tissue resulted in growth promotion (Bonner 1934a), and isolated walls kept under tension also responded to acid solution (Rayle et al. 1970). Hager et al. (1971) proposed that auxin might activate ATPases in the plasma membrane, and that these might act as proton pumps. So far, the evidence for such a mechanism is scant, but in an unconfirmed report Kasamo and Yamaki (1974 and 1976) have shown a small stimulatory effect of auxin on isolated Mg^{2+}-activated ATPase from mung bean *(Phaseolus aureus)* hypocotyls, and the demonstration of an early auxin-induced increase in membrane negative potential which is sensitive to phosphorylation uncouplers provided circumstantial evidence that an electrogenic pump was involved in the proton extrusion (Marré et al. 1974). Cleland (1975) has provided further evidence which is consistent with auxin causing proton extrusion through an electrogenic pump mechanism rather than via a K^+–H^+ exchange electroneutral mechanism. Etherton (1970) has shown that auxin increases the membrane negative potential of *Avena* coleoptiles, while Tanada (1973) has shown indirectly that auxin decreases the membrane negative potential of barley root tips but increases the potential of mung bean root tips. There is thus some further evidence that auxin influences the charge distribution across membranes. However, the effect is not specific for auxin, since cytokinins have been shown to increase transmembrane potential of squash *(Cucurbita maxima)* cotyledons (Marré et al. 1974) and proton excretion in squash (Marré et al. 1974) and fenugreek *(Trigonella foenum-graecum)* cotyledons (Rijven 1976).

Further study of the membrane-associated ATPase of plant cells is clearly warranted in view of the auxin binding to plasma membrane-rich fractions (Hertel et al. 1972; Kasamo and Yamaki 1976; Batt et al. 1976) and the auxin-induced release of a factor from the plasmamembrane-rich fraction of soybean hypocotyl which is thought to promote RNA synthesis (Hardin et al. 1972). There may be a relationship between the above, apparently unrelated, phenomena.

Another approach to the involvement of auxin and ethylene with plant membranes is to study membrane components and artificial membranes (Mehard et al. 1970; Kennedy and Harvey 1972; Iwata and Stowe 1973). Stowe and colleagues have studied the enhancement, by various lipids, of auxin effects on respiration and growth in a number of systems (Stowe and Dotts 1971; Iwata and Stowe 1973). A minimum chain length of 2 nm was required for activity. As a probe of the mechanism of auxin action in membranes, the approach has been of limited use in interpreting physiological responses and has yielded only some speculative conclusions, which may, however, be useful if related closely to plasma membrane structure.

Changes in the permeability to ions in developing fruit in response to ethylene have been regarded as responses to ripening rather than to the hormone (Abeles 1972), and other systems for study have been sought. Senescing flower tissue, while promising, has not yielded clear-cut evidence that ethylene affects permeability as a primary response (Hanson and Kende 1975), however, these authors have made an important advance in attempting to differentiate between effects on the plasmalemma and the tonoplast. It was shown that ethylene promoted efflux of ions and small organic molecules from an unknown compartment, within a time period of 6–8 hr. No selectivity of effect was demonstrated, which may indicate a general breakdown of internal cellular membranes, although it was suggested that this may occur only in a few cells.

All 5 plant hormones have also been shown to interact with ions, especially calcium, in bringing about certain physiological responses (Leopold et al. 1974). Calcium has been shown to antagonize the effect of ethylene in leaf abscission (Poovaiah and Leopold 1973a), to enhance the effect of cytokinins in leaf senescence (Poovaiah and Leopold 1973b) and to be necessary for the polar transport of auxin (Dela Feunte and Leopold 1973). The implication here is that certain aspects of membrane integrity are essential to hormone action, and that protein stabilization or de-stabilization is affected by Ca^{2+}. There is now scope for an extensive investigation of ion – hormone interactions in many biochemical processes.

8.2.2.3. Cytokinins. In the cell wall of the oomycete, *Achlya,* calcium is bound to a glycopeptide (LeJohn et al. 1974). Cytokinins are effective in releasing this bound calcium and also in stimulating uptake of the ion. This elegant demonstration of ion storage and redistribution in the fungal cell has not been extended to higher plants, although similar controls may exist. By analogy with the data of Leopold et al. (1974), it is possible to postulate that the events following hormone activation or suppression of cellular processes involve ionic changes which may then lead to membrane modification, particularly in the case of calcium (Epstein 1961; Rains and Floyd 1970).

There is extensive evidence that the cytokinins influence permeability to certain ions in vegetative and tuberous tissues with reasonably short lag times (Sugiura 1963; Penot and Hourmant 1969; Rains 1969; Leonard and Hanson 1972; Penot and Hourmant 1974). Evidence has accumulated for selective effects such as enhanced influx of K^+ and Rb^+ with no effect on Na^+ and NH_4^+ (Ilan and Reinhold 1963). Similar selectivity was shown for enhanced uptake of polar non-electrolytes without any effect on non-polar species (Feng 1973). There are examples of inhibition of influx of phosphate (Sugiura 1963) and of sodium (Ilan 1971), which again may be taken to demonstrate selectivity of effects in hormone action depending on the tissue and process being studied.

Cytokinins have been shown to increase transmembrane potential of squash cotyledons (Marré et al. 1974) and to enhance H^+-ion extrusion in squash (Marré et al. 1974) and fenugreek cotyledons (Rijven 1976), thus implicating the hormone in membrane-associated events and possibly in electrogenic pump activity.

8.2.2.4. Summary and conclusions. In summary, although it has been possible to demonstrate binding of at least one plant hormone (auxin) to plasmalemma-rich fractions

(Hertel et al. 1972; Batt et al. 1976), it has been extremely difficult to link this binding to any subsequent response. An obvious first choice of response is altered structural characteristics of the plasma membrane and, although these changes would be expected to be very subtle, it is reasonable to expect that they could be monitored by biochemical or physiological assay. Since the basic structural features of the plasma membrane are not understood, it is almost impossible to design experiments to test whether that structure is altered. An indirect approach to the question is to examine certain functions associated with the membrane, in this case transport of a range of molecules, but again it is very difficult to assess the data, since the transport processes themselves are little understood. It has emerged that the massive transport of molecules from the barley aleurone layer in response to GA_3 probably does not reflect any direct effect of the hormone on the plasma membrane, if the results obtained for α-amylase secretion (Varner and Mense 1972) are of general applicability. On the other hand, potassium uptake by stomatal guard cells may be controlled by ABA acting upon the membrane, and this system warrants further study. The cytokinin- and auxin-mediated extrusion of protons also appears to offer excellent potential for further analysis. The available data would suggest that these hormones have a direct effect on some feature of membrane structure or activity. The ATPases which are considered to be associated with plasma membranes (Lai and Thompson 1972; Hodges et al. 1972) deserve careful scrutiny from the point of view of auxin action. Similarly, ionophores as isolated from mammalian mitochondrial membranes (Green et al. 1975) might be expected to occur in other membranes including plant organelles and possibly the plasma membrane. These small molecules could play a role in hormonal control of transport processes.

8.3. Hormones and long-distance transport

Long-distance transport is taken to represent long-distance movement of solutes, which probably involves traversing new barriers such as the phloem and xylem, and therefore loading and unloading through a few sets of plasma membranes. This area is confounded by disagreement on the precise form of solute movement both across the barrier from parenchyma into the transporting cell and the mechanism of movement in the phloem. There is a large body of literature relating hormones to the movement of photosynthates, especially sucrose (Booth et al. 1962; Seth and Wareing 1967; Bowen and Wareing 1971; Patrick and Wareing 1972 and for a comprehensive review see Patrick 1976).

The available evidence has been variously interpreted to show that hormones are definitively (Patrick 1976) or only peripherally (Wardlaw 1974) involved in some unresolved aspect of solute (sugar) transfer over long distances. There are three major steps in long-distance transport where hormones might affect a rate-limiting process, vein loading (source activity), the transport channel itself or vein unloading (sink activity). Many of the comments made on membrane transport (section 8.2.2.3) apply to the vein loading and unloading question.

Cytokinins have been implicated in directing the accumulation of metabolites in treated tissue (Mothes and Engelbrecht 1961; Muller and Leopold 1966; Schenk 1974). It is not clear whether the hormone affected the "source-sink" relationship or the translocation process. Cytokinins have been shown to suppress hydrolytic enzyme levels especially those involved in protein and RNA hydrolysis (Shibaoka and Thimann 1970; see also Chapters 10 and 11), and thus maintenance of tissue and cell integrity might be assumed to alter the "sink" activity. Similarly, experiments with auxin on the directed transport of metabolites and nutrients suggested that the hormone affected both "sink" activity and the translocation process during root formation in *Phaseolus* cuttings (Altman and Wareing 1975). In other experiments with decapitated pea and bean stems it was shown that auxin enhanced accumulation of metabolites at the point of application (Davies and Wareing 1965) without having any measurable effect on selected aspects of "sink" activity such as respiration and assimilate metabolism (Patrick and Wareing 1972). It has been suggested by elimination, therefore, that auxin directly affects the translocation process either through maintenance of the integrity of the phloem tissue (Patrick and Wareing 1972) or through active stimulation of some as yet unknown transport process (Mullins 1970). On the other hand, Wardlaw and Moncur (1976) interpreted their data on IAA movement in wheat stem as showing that the hormone could have no effect on the translocation step. Their data was based largely upon the finding that low temperature greatly decreased auxin movement while having little effect on rate of assimilate movement. While this rules out an effect of auxin movement on assimilate transport in the short term, it does not clarify whether auxin has any effect per se on the sieve tubes. The present understanding of the mechanism of assimilate transport in plants is such that we cannot readily relate hormone action with any particular aspect of translocation.

In conclusion, it can be said that transport processes and their regulation have told us little about hormone action, and the converse is also true. This is another example of the use of a process to study hormone action when that process is incompletely understood. However, the use of hormones as probes with which to study physiological processes is to be encouraged, since it is likely to lead to an understanding of both the physiological phenomena and the hormonal response.

8.4. Arrangement and redistribution of cellular components

The cellular property of "competence to respond" to a plant hormone is no doubt determined by a complex combination of structural and chemical attributes and interrelationships. It may be impossible to separate the various components of the system and still assay for "competence to respond". However, various pertubations have been attempted. We present some of the results and interpret them within the context of whether a particular arrangement of cellular components is necessary for hormone action or whether any existing arrangement allows hormone recognition. In the latter the hormone may mediate a rearrangement which is required for further response. This should entail an examination of the existing biochemical and structural status of responsive tissue and a comparison with non-responsive cells. However, a systematic survey of a wide range of cellular activities, processes and structure has rarely been attempted with the purpose of delineating what constitutes a competent cell. An exception is perhaps the study of the barley aleurone by Jones (Jones 1969a, b, c and d; Jones and Price 1970). Individual processes or structures have tended to be examined from the point of view of overall control.

8.4.1. Microtubules and microfilaments

Microtubules are considered to play a large role in the structural maintenance and development of eucaryotic cell shape and orientation (Hepler and Palevitz 1974; Snyder and McIntosh 1976). They may, in part, regulate the orientation of the deposition of cellulose in certain plant cell walls (Hepler and Palevitz 1974), possibly by guiding secretory vesicles in the cytoplasm (Chrispeels 1976) to sites of unloading. Microfilaments have been less studied in plant cells, but they have been implicated in generating cytoplasmic streaming (Nagai and Rebhun 1966; O'Brien and Thimann 1966) and contributing to the fibrillar material in phloem sieve elements (Wooding 1967), where they were thought to facilitate solute flow. Recent appraisal of the literature suggests that the data linking microfilaments and phloem fibrils is not very strong (Hepler and Pavevitz 1974). The bulk of evidence suggests that microtubules and microfilaments regulate cell orientation, shape and intracellular movement, and should therefore be studied in relation to hormone action.

The induction of ornithine decarboxylase in mammalian cells by hormones was dependent upon microtubule arrangement and activity (Chen et al. 1976). This was assessed by using colchicine and vinblastine, drugs which are known to disrupt the integrity of microtubules. Although Chrispeels (1972) showed that colchicine and cytochalasin B did not affect secretion of α-amylase from barley aleurone or extensin from carrot *(Daucus carota)* phloem parenchyma, he did not test the effect of these agents on the induction of α-amylase synthesis by GA_3. Atsmon (1972) found that colchicine did not affect the induction of the enzyme by GA_3 but that the elongation of γ-irradiated wheat coleoptile was strongly inhibited. Therefore, although the mammalian study suggests that cell structure and intracellular motility are important in recognition of certain hormonal stimuli, and Jones (1969d) presented evidence consistent with this idea in aleurone, the mechanism has not been explained.

Stimulation of cell elongation and cell expansion by hormones has naturally led to an extensive study of changes associated with the cell wall and its biogenesis (see section 8.5). Synthesis and deposition of cellulose are therefore usually primary targets for research. Changes in the arrangement of cellulose fibrils in response to plant hormones have been reported for IAA, ethylene, kinetin and gibberellic acid in a number of growing tissues (Probine 1964; Veen 1970; Apelbaum and Burg 1971; Ridge 1973; Shibaoka 1974). In all cases the observed hormonal effect seemed to be the result of induced growth, and there was no evidence that the existing fibrils were reoriented prior to cell elongation or expansion. The orientation of microtubules has also been shown to change in hormone-treated *Vigna angularis* (Shibaoka 1974), and it was postulated that such changes were related to the observed elongation or expansion growth induced by GA_3 and kinetin. The cytokinins promoted cell expansion and this was inhibited by colchicine, suggesting that microtubules were involved in the hormone-dependent cell expansion. In the same tissue, auxin promoted cell elongation but colchicine did not alter the response, and the microtubules were randomly oriented. The kinetics of microtubule rearrangement and cell growth were not closely studied and so do not enable one to assess the cause and effect relationship with the hormone. GA_3 promoted elongation of *Lactuca sativa* hypocotyls but colchicine inhibited the response (Sawhney and Srivastava 1974). The orientation of microtubules was not examined, and it is not possible to conclude that they played any role in the response.

In the auxin-promoted cell elongation of *Triticum vulgare* coleoptile segments, it was noted that a range of microtubule disorganizers, urea, vinblastine sulfate, colchicine and cupric ion, inhibited the growth response due to the hormone (Lawson and Weintraub 1975). In contrast to the above, microtubules appeared in cell suspension cultures of carrot only after removal of auxin and cessation of wall synthesis activity (Halperin and Jensen 1967). The majority of the available data shows a correlation between hormone-induced growth and microtubule orientation. However, there is scant information on the subject and it is difficult to suggest a significant role for microtubules in regulating hormonal responses. There is even less data available on the role of microfilaments. Auxin promoted protoplasmic streaming in *Avena* (Sweeney and Thimann 1937) and microfilaments may be involved, since cytochalasin B inhibited the response and the filaments became condensed on treatment with the drug (Cande et al. 1973).

It is clear that the role of microtubules and microfilaments has not been systematically studied in relation to the activity of plant hormones. In most cases the interpretation has relied heavily on data obtained using cytochalasin B and colchicine although their specificity of action on the two organelles is far from proven (Bos and Emmelat 1974; Hepler and Palevitz 1974).

8.4.2. The endoplasmic reticulum and polysomes

The endoplasmic reticulum (ER) is considered to provide the intracellular topography necessary for protein biosynthesis, modification and possibly transport to the plasmalemma or Golgi apparatus. The latter may provide further modification prior to export or transport to other organelles (Palade 1975). There is some controversy over the role of the ER in carbohydrate biosynthesis and export to the wall of plant cells (Chrispeels 1976). This may be resolved by the demonstration (Shore and Maclachlan 1975) that certain enzymes, especially those associated with cellulose biosynthesis, are synthesized on ER in response to auxin treatment, but that the majority of cellulose synthesis occurs at the plasma membrane. ER formation, maintenance and modification is thus a candidate for hormonal control.

In most ultrastructural studies of plant cells following hormone treatment, changes have been noted in the form and extent of the ER and its attached ribosomes (rough ER) at various times after the addition of hormone. The aleurone layer of barley has been extensively studied both during germination and during GA_3-stimulated hydrolase synthesis (Paleg and Hyde 1964; Van der Eb and Nieuwdorp 1967; Jones 1969c and d; Jones and Price 1970; Vigil and Ruddat 1973). Jones (1969c) described a stimulation of ER proliferation and formation of stacks of lamellae within two hours of treatment with GA_3. This was confirmed by Vigil and Ruddat (1973), who showed that similar changes occurred during germination. Earlier studies had also noted changes in membranes of GA_3-treated aleurone (Paleg and Hyde 1964) and in the rough ER of germinating barley seeds (Van der Eb and Nieuwdorp 1967). The proliferation of ER was associated with increased levels of ribosomes (not counted) and some of these were in polysomal configuration (Jones 1969c; Vigil and Ruddat 1973). Later the ER stacks became distended (Jones 1969d) and fragmented, and were observed to vesiculate along the plasma membrane region in a polar fashion (Vigil and Ruddat 1973). The ER stacks and vesicles were interpreted to represent the sites of GA_3-induced protein synthesis and secretion, respectively, since most of the hydrolytic enzymes formed were secreted and were thus likely to be made on rough ER. Whether the vesicles represent a mechanism for the secretion of induced enzymes (Vigil and Ruddat 1973) is questionable, since Jones (1972) could find no evidence for a particulate form of α-amylase during the synthetic and secretion stages following hormone induction. However, it is likely that his results were obscured either by large amounts of wall-bound enzyme, which was already secreted and would, therefore, be soluble, or possibly by extraction techniques, which disrupted the vesicles. Firn (1975) has presented convincing data that the intracellular enzyme is in particulate form. Gibson and Paleg (1972) obtained evidence that α-amylase and protease were in particulate form within wheat aleurone cells. The source of the membrane used for packaging and transport is likely to be the ER (Gibson and Paleg 1976).

Biochemical studies on the appearance of the rough ER using [^{14}C-methyl] choline as a membrane precursor indicated that GA_3 stimulated the synthesis of microsomal membranes selectively beginning about 4 hr after hormone treatment (Evins and Varner 1971). ABA inhibited the GA_3-induced stimulation of choline incorporation. Thus there was a good correlation between the ultrastructural data (Jones 1969c and d; Vigil and Ruddat 1973) and the biochemical observations. Further studies strengthened the biochemical data by showing that GA_3 stimulated the activity of two particulate enzymes involved in lecithin synthesis (phosphorylcholine cytidyl transferase and phosphorylcholine glyceride transferase) (Johnson and Kende 1971). GA_3 was also shown to have a rapid effect on the turnover of CTP in wheat aleurone, indicating a possible role in membrane precursor metabolism (Collins et al. 1972). Based upon correlative use of protein and RNA synthesis inhibitors, Johnson and Kende (1971) showed that lecithin and lecithin-synthesizing enzymes were induced by GA_3 prior to GA_3-induced hydrolase production. Confirmation of the effect of GA_3 on the rapid stimulation of the lecithin-synthesizing enzymes was provided by Ben-Tal and Varner (1974) who showed, however, that these enzymes appeared to be activated rather than de novo synthesized using cordycepin and amino acid analogs to inhibit RNA and protein synthesis respectively. The discrepancy between the two sets of data using two sets of inhibitors was not resolved.

Phospholipid synthesis in all cell fractions (not only the ER-rich fraction), was shown to be stimulated within 5 hr by GA_3 using labelled orthophosphate as precursor (Koehler and Varner 1973). This stimulation was sensitive to inhibition by ABA as well as protein and RNA synthesis inhibitors. There was no effect on acetate incorporation into lipids or membrane and no net increase in the phospholipid content of aleurone layers. This might be interpreted to mean that GA_3 stimulated phospholipid turnover rather than net synthesis, alternatively, the synthesis of a minor phospholipid species might be stimulated.

The synthesis and degradation of lipids of barley aleurone were studied by Firn and Kende (1974), who showed that GA_3 had no qualitatitve or quantitative effect during the first 12 hr of incubation. Similarly, they found no effect on the incorporation of [^{3}H] glycerol into various lipid fractions, and they concluded that the stimulation of ER synthesis seemed unlikely in view of their data. Furthermore, the level of ER in wheat aleurone has been shown to be independent of GA_3 during the early stages of hydrolase synthesis (Varty and Laidman 1976) or to be unchanged by the hormone (Gibson and

Paleg 1976). Thus, while the ultrastructural data for barley seemed very conclusive, that obtained biochemically has been more difficult to interpret. The best conclusions that can be drawn for barley are that GA_3 may influence turnover and redistribution of membrane components which already exist in some form, perhaps in spherosomes, in the aleurone cell, and that it does not promote significant de novo synthesis of membrane components, but that it probably promotes construction. In wheat, GA_3 does not seem to have a role in synthesis or construction of ER.

The ultrastructural data showed that ribosomes were attached to the newly formed ER (Jones 1969b; Vigil and Ruddat 1973), and Jones (1969b) noted that many of these were in polysome configuration. Evins (1971) confirmed these findings using biochemical techniques, and presented evidence to show that there were increased numbers of ribosomes and polysomes and increased rates of protein synthesis on these polysomes. In other experiments, GA_3 has been shown to increase total extractable ribosomes (Holmes and Speakman 1973). Contrary to these findings, however, it has been shown by analytical techniques that there is no increase in total RNA of aleurone layers during 24 hr of incubation with GA_3, and that there is no requirement for rRNA synthesis during enzyme induction (Varner et al. 1965; Zwar and Jacobsen 1972; Jacobsen and Zwar 1974a). The discrepancy between the above data is difficult to explain except on the basis of differential extraction of RNA from tissue which has been treated with GA_3. This and other alternatives are discussed in Chapter 10. The marked increase in rough ER of aleurone cells following GA_3 treatment may reflect redistribution of existing ribosomes to newly constructed or already existing ER. This would be consistent with the idea that GA_3 promotes the synthesis of such hydrolytic enzymes as protease and α-amylase on rough ER, which was compiled from existing ribosomes and intracellular membrane or membrane components. Steroid hormone action in mammalian cells is also believed to involve membrane rearrangement and synthesis prior to increased rates of protein synthesis (Tata 1970). In a diverse range of other tissue and cell types, it has been shown that the appearance of elaborated and organized ER often correlates with hormone-promoted events such as cell elongation, abscission and neoplastic growth.

Ultrastructural studies of a growing tissue which responded to GA_3 (lettuce hypocotyl) revealed that the hormone accelerated various cellular processes including the appearance of rough ER (Srivastava et al. 1975). Although it was implied from this data that certain physiological conditions (both structural and biochemical) must exist prior to hormone responsiveness, this was not particularly well-documented, since no novel features of the cells were demonstrated prior to or after hormone treatment. There is abundant ER in hybrid tumour tissue culture cells which produce auxin, however, in auxin-requiring cells to which the hormone was applied, there was no such marked response (Lipetz 1971). Thus, it appears as if factors other than auxin are required for ER formation in these cells. In callus tissue of *Cichorium intybus* roots, on the other hand, auxin induced extensive polysome and rough ER formation (Gwozdz et al. 1974). Ethylene rapidly induced rough ER formation in the abscission zone of tobacco flower pedicels (Valdovinos et al. 1971), which suggested a possible site of synthesis of enzymes involved in wall digestion. There is no information at present on the synthesis or turnover of this ER or on those enzymes which may be synthesized thereon. In the radial expansion of pea epicotyl induced by ethylene, Sargent and Osborne (1975) have noted changes in the size of ER profiles. Within 6–12 hr of hormone treatment, long sheets of rough ER were apparent, while the number of short ER profiles were greatly reduced. They also found greater numbers of smooth vesicles near the plasma membrane, and concluded that the two membranes were fusing at certain sites. Ethylene treatment of pea epicotyls resulted in reduced incorporation of [1-^{14}C]glycerol, into phospholipids within 3 hr (Irvine and Osborne 1973). To reconcile these two pieces of data one must conclude from the ultrastructural data that there is relatively little net loss or gain of membrane material, while the labelling data must be interpreted in terms of reduced turnover of total phospholipids or inhibition of synthesis of a minor fraction. Thus, there may be a superficial resemblance between the mechanisms of ER formation in GA_3-treated aleurone layers and ethylene-treated pea epicotyl in that the components of the membrane may not be de novo synthesized but assembled from existing phospholipids within the responding cells.

In the plant literature there is relatively little data on movement of polysomes from the free pool to membranes, although this might be expected to be an important control point regulating protein synthetic activity as has been suggested by Tata (1971). Changes in the distribution of free and membrane-bound polysomes have been sought in cytokinin-stimulated expansion of radish *(Raphanus raphanistrum)* cotyledons, but no effect could be demonstrated (Gordon 1973). However, in a long-term (7-day) study, McCombs and Ralph (1972) showed that cytokinins caused a decrease in the membrane-bound polysomes of *Spirodela oligorrhiza* with an increase in the free population. Polysomal profiles are known to be stabilized or enhanced under certain conditions of hormone treatment (Trewavas 1968; Davies and Larkins 1973; Travis et al. 1973; Short et al. 1974). In cytokinin-dependent soybean tissue culture cells, it appears as if cytokinin may promote polysome formation from existing RNA components in these cells (Short et al. 1974), thus effecting a reorganization of the protein synthetic machinery. The subcellular location of these polysomes was not described, although from the procedures used it is likely that only the free polysomes were studied. See Chapter 11 for further analysis of polysome data.

A systematic study of the redistribution of free monosomes into polysomes attached to ER has been made by Davies (1976). He has shown that auxin treatment of pea epicotyl sections results in decreased monosomes and a concomitant increase in membrane-bound polysomes within 3 hr of hormone treatment. There are no quantitative differences in ribosome levels between the treated and untreated tissues, suggesting that existing ribosomes are mobilized onto the membranes. The possibility that new ribosomes, which arise through an enhanced rate of synthesis of ribosomes, which then become ER-bound combined with an equal rate of degradation of existing ribosomes, could, however, give the same result.

There are numerous studies from which circumstantial evidence can be compiled to support the view that cytokinins, GA and ethylene may effect redistribution of ribosomes from the soluble to the membrane phase of plant cells. The gibberellin-dependent appearance of rough ER as shown by ultrastructural techniques (Jones 1969b) and discussed above in relation to ER synthesis serves as one example. There is some controversy over whether the ribosomes are derived from the cytoplasm (Jacobsen and Zwar 1974a) or from new RNA synthesis (Evins 1971) (see Chapter 10). The source of the ribosomes is less well-documented in the ethylene example (Sargent and Osborne 1975), but it is likely that the expansion of pea epicotyls which results in reassembled membrane (ER) may also involve mobilization of the ribosomes deteced by electron microscopy. Similar incomplete data is avilable on the cytokinin-promoted rough ER formation in senescing wheat leaves (Shaw and Manocha 1965). In this system kinetin slowed the onset of senescence, and in so doing increased the proportion of membrane-bound polysomes.

In summary, the ER has been shown to undergo extensive alteration with time in response to plant hormones, and this includes extension, stacking swelling, fragmentation, vesiculation, orientation close to certain organelles and the plasma membrane and becoming associated with polysomes and ribosomes. These phenomena are positively correlated with and usually precede such hormone-mediated events as enzyme secretion, abscission, growth or slowed senescence. The data are consistent with major effects upon protein biosynthesis, modification, packaging and transport.

8.4.3. Golgi apparatus

The Golgi apparatus is considered to be concerned with (a) the transport of lipids, (b) assembly, modification and transport of proteins and (c) the synthesis and transport of polysaccharides (Dauwalder et al. 1972). Some specific examples include biosynthesis of cell wall polysaccharide components in the Golgi from labelled galactose in corn root (Dauwalder and Whaley 1974) and, in pea epicotyls, labelled glucose was associated with a Golgi-rich fraction, and the radioactivity could be followed into vesicles (Ray et al. 1969). A glucan synthetase was also associated with the organelles. The Golgi was shown to be the site of glycosylation and possibly hydroxylation of the cell wall glycoprotein, extensin (Gardiner and Chrispeels 1975; Chrispeels 1976).

The participation of the Golgi apparatus in cell wall synthesis is well-documented (Northcote 1972), and there is a vast literature on the effects of plant hormones on cell walls, including their synthesis, modification and degradation (see section 8.5). The present discussion will be limited to direct observations on Golgi using ultrastructural and biochemical techniques. The Golgi bodies have been shown to increase in number in lettuce hypocotyl elongating in response to gibberellic acid (Srivastava et al. 1975), and in coleoptiles of *Avena* (Gawlik and Shen-Miller 1974) elongating in response to auxin treatment. Similarly, during auxin-induced cell expansion of Jerusalem artichoke *(Helianthus tuberosus)*, there was a marked increase in the number of Golgi bodies and associated vesicles, which correlated strongly with cell wall deposition (Fowke and Setterfield 1969). Another class of vesicles, the plasmalemmasomes, were not positively correlated with cell wall synthesis. ABA inhibited the growth of onion *(Allium cepa)* root sections, but did not influence the activity or number of Golgi bodies (Risueno et al. 1971). Wall deposition continued, and the final thickness suggested that normal wall synthesis proceeded in the absence of elongation.

In the non-growing barley aleurone layers, GA_3 had no visible effect on Golgi body activity until late in the response, when enzyme secretion was at a maximum and cell wall degradation was occurring (Jones 1969d; Vigil and Ruddat 1973). Jones (1969d) noted that vesicles proliferated from the Golgi 16–19 hr after hormone treatment, and he speculated that the Golgi might be involved in enzyme secretion; however, secretion begins long before 16 hr. Similarly, the Golgi were thought to be fusing with the plasma membrane in Vigil and Ruddat's study, thus implying a role for them in export of GA_3-induced intracellular components. There is no direct biochemical evidence to substantiate their conclusions at present.

Two features of the reported effect of plant hormones on Golgi activity deserve careful study. The first is the role played by the organelle in cell wall synthesis, and especially in transporting those enzymes such as cellulose synthetase to the plasma membrane as suggested by Shore and Maclachlan (1975). The second is the rapid response to auxin noted by Gawlik and Shen-Miller (1974). The known transport capacity of the organelle and the postulated requirements for intracellular factors including protein (Rayle et al. 1970; Cleland 1975) in inducing rapid cell elongation suggest that the role of Golgi in this response should be studied further.

8.4.4. Other cellular organelles

Ultrastructural work carried out on tissues treated with hormones for extended periods has usually detected numerous changes in the arrangement and the integrity of cell components. Most of these probably do not reflect primary effects of the phytohormone and are thus difficult to interpret. A few such studies with comprehensive data are mentioned below.

The aleurone grain of barley aleurone cells undergoes marked changes in shape and size during GA_3 treatment (Jones 1969c and d; Jones and Price 1970). The first detectable α-amylase produced by these cells was associated with the aleurone grain membrane (Jacobsen and Knox 1973), and it was speculated that some of the newly organized ER (Jones 1969c) might arise from extensions of the aleurone grain membrane (Paleg and Hyde 1964; Jacobsen and Knox 1973). These data are consistent with the idea that α-amylase is synthesized on rough ER. The aleurone grains swelled during incubation with GA_3; however, the inclusions within the grain were not affected. Later the grains became smaller, and it was presumed that they mobilized their contents for the synthesis of hydrolytic enzymes (Jones 1969c and d). Spherosomes seemed to be associated with the GA_3-induced swelling of aleurone grains, and they were seen to form a regular array around the grains (Jones 1969c). As the aleurone grains became vacuolate the spherosomes were reduced in number. About the same time, plastids and microbodies were seen to increase; the plastids divided and showed evidence of increased differentiation over controls (Jones 1969d). "Spiny vesicles" were also detected late during incubation. Many of the above changes were correlated with the phase of active secretion of hydrolytic enzymes (Jones 1969d). This comprehensive ultrastructural study by Jones has highlighted a number of changes occurring in a plant cell which is not growing, and which has a specialized function in providing substrates and degradative enzymes for mobilization of the starchy reserves of the barley endosperm.

The auxin-induced cell enlargement of Jerusalem artichoke appeared to be accompanied by a general stabilization of plasma membrane and tonoplast, and by a decrease in the number of multivesicular structures associated with the plasma membrane (Fowke and Setterfield 1969). On the other hand, herbicidal levels of auxin were shown to disrupt cellular membranes, particularly those of the chloroplast (Hallam 1970).

Ethylene has been shown to enhance the formation of ER in cells of the abscission layer of tobacco pedicels and to cause the membranes of microbodies to lose their integrity within 5 hr of exposure to the gas. However, other cellular membranes were unaffected at this stage, and it was only when cell wall degradation was advanced that disorganization of the cytoplasmic components became evident (Valdovinos et al. 1972). Thus, there appeared to be selective modification of various cellular membranes and their organelles by ethylene with time in these cells. Kinetin was shown to preserve the membranes of ER vesicles, chloroplasts, nuclei and Golgi following excision of wheat leaves, thus slowing down senescence and degeneration (Shaw and Manocha 1965).

It can be concluded that a variety of organelles undergo modification in the presence of hormones, and in most cases this probably represents secondary responses. That is, the changes reflect the end-result of a series of metabolic activities, only one of which was probably under direct control of the hormone.

8.4.5. Enzyme distribution

Redistribution or release of enzymes from one cellular compartment into another might be expected to be involved in control of growth and differentiation, and thus might be regulated by plant hormones. One such example is provided by the data of Shore and Maclachlan (1975), who have shown that IAA may alter the intracellular distribution of cellulose synthetase by moving it from Golgi bodies to smooth vesicles containing ER enzyme markers. There is some preliminary evidence to suggest that GA_3 may influence the intracellular location of glycolate oxidase based upon experiments carried out on barley leaves to which GA_3 had been added during homogenization (Lips and Roth-Bejerano 1974). GA_3 was shown to facilitate retention of the enzyme in microbodies during centrifugation through discontinuous sucrose gradients. The significance of these findings remains to be seen. It is likely that the hormone may have influenced the integrity of the microbody membrane, thus preventing loss of the enzyme from the organelle during extraction.

The ethylene control of abscission has been correlated with increased levels of extra-cellular cellulase, and it was suggested that the hormone controlled secretion of the enzymes (Abeles and Leather 1971; Rasmussen 1973) as well as synthesis (Horton and Osborne 1967). A further refinement of cellulase characterization showed that ethylene primarily promoted the synthesis of a basic isozyme (Reid et al. 1974). There is no convincing data implicating ethylene in the control of cellulase secretion in a direct fashion. Most of the control exerted by the hormone would appear to be on enzyme synthesis.

The redistribution of existing enzymes from one part of the cell to another would appear to be an efficient mechanism of enzyme regulation, but from the available data it is unlikely that it is one which is widely utilized.

8.5. Cell walls

Out understanding of the effect of hormones on cell wall metabolism is strictly constrained by lack of fundamental knowledge of the structure and biosynthesis of the wall itself. The basic constituents of the cell walls of dicotyledons, monocotyledons and gymnosperms studied to date show remarkable conservation within groups and considerable overlap between groups (Burke et al. 1974). The physics, chemistry and biochemistry of plant cell walls have been extensively studied, and models depicting possible arrangements of cell wall constituents have been constructed (Northcote 1972; Keegstra et al. 1973; Preston 1974). Plant hormones may be a useful probe, like pathogenesis (Albersheim and Anderson-Prouty 1975), with which to study the formation and dissolution of walls, although such an approach has had limited application so far. The questions usually asked are related to mechanism or mode of hormone action as mediated by cell wall changes, particularly in cell elongation of *Avena* coleoptiles and pea internodes (Cleland 1971), in cell expansion of decapitated auxin-treated pea epicotyls (Fan and Maclachlan 1966), wall dissolution of cereal aleurone (Paleg and Hyde 1964; Jones 1969d; Taiz and Jones 1970; Ashford and Jacobsen 1974a and b) and wall dissolution during abscission (see Abeles 1972).

The walls of plant cells have not been easy to characterize chemically except by complete hydrolysis of the polymers to their constituent subunits. Nevertheless, major fractions have been described (see Preston 1974) and are listed below:

(1) pectins (extracted by boiling in water for 12 hr or by hot ammonium oxalate);
(2) hemicellulose (extracted usually in 4 *N* KOH at room temperature);
(3) cellulose (the residue of the above and often extracted with 72% H_2SO_4);
(4) protein (rich in hydroxyproline).

However, the purity of the original cell wall preparation is often questionable. The fractions obtained are thus very ill-defined, and there is considerable overlap, especially between pectins and hemicellulose, depending on which variation in extraction is followed. The nature of the bonds between fractions is not clearly understood, and the relationship between fractions is therefore not clear. Preston (1974) believes that the majority of bonds are of three types (a) hydrogen bonds, (b) salt bridges and (c) Van der Waals forces, with a few if any covalent bonds, while Keegstra et al. (1973) postulated that covalent bonds were most common, and they put little emphasis on other linkages except for H^+ bonding between xyloglucan and cellulose. From the foregoing it is obvious that the fine structure of the wall is controversial, and theories range from the general notion of a cellulose superstructure containing a gel matrix of unspecified interrelationships, but which was thought to reflect a massive macromolecule (Albersheim 1965) to Lamport's notion (1970), that a hydroxyproline-rich protein formed the backbone of the wall and was involved in cross-linking some of the primary wall polysaccharides. A protein – glycan network, in general, is now supported by the studies of Albersheim and colleagues (Bauer et al. 1973; Keegstra et al. 1973; Talmadge et al. 1973). A potentially more specific dissection of the plant cell wall has been initiated by these workers who have used enzymes in combination with chemical treatment to digest wall components secreted by cell suspension cultures, and from this to deduce the nature of the interaction between the various subfractions. The emerging picture of the dicotyledonous cell wall shows that the cellulose microfibrils are coated with a single layer of xyloglucan by hydrogen bonding. The xyloglucan of one fibril may be linked to the xyloglucan of another fibril by covalent cross-linking through pectic polymers. A similar arrangement has been proposed for the walls of monocotyledonous plants, except that the xyloglucan backbone is replaced by an arabinoxylan backbone (Burke et al. 1974). It should be emphasized that the above are simple models of the cell wall based upon analyses of very few cell types, and it is likely that further analysis will complicate the picture. The models of the wall provide valuable material for hypotheses which can be tested experimentally, and within this framework the action of hormones is a possible probe. Most of the work with hormones and cell walls has been carried out without the benefit of the current models, not to mention the absence of direct knowledge of the chemical interrelationships and crystalline array of the walls. However, this lack of knowledge of the structure of the plant cell wall has not meant that this area is without an extensive literature and its attendant postulates, theories and speculations. Indeed, there has been remarkable activity in this area, mostly because of the interest in auxin-promoted cell elongation in coleoptiles, epicotyls and hypocotyls.

8.5.1. The effects of auxin on cell walls during elongation

Only a brief summary of some of the current areas of biochemical investigation into this most active research field will be attempted here, and the reader is referred to two reviews (Heyn 1940; Cleland 1971) which are of great assistance in obtaining a perspective of auxin involvement in cell elongation. See Chapter 12, Volume II for details of rapid effects of auxin on cell walls.

Early studies on cell elongation in response to auxin usually involved *Avena* coleoptile segments which were incubated in the test solution for 24 hr or longer. Growth was measured and related to whatever further additions (especially inhibitors of various physiological functions) had been made for the full 24-hr period. From studies of this nature it was possible to conclude that elongation involved respiration (Bonner 1936), water uptake (Kelly 1947), wall synthesis (Bonner 1934b; Ray and Baker 1962), RNA and protein synthesis (Key 1969). The question of the role of auxin in these events was continuously posed, and the temporal sequence of events was obviously important if the primary target of the hormone was to be determined. With the introduction of methods which could measure small changes in growth rate within minutes of such changes (Kohler 1956; Ray and Ruesink 1962; Penny 1969; Nissl and Zenk 1969; Evans and Ray 1969; Dela Fuente and Leopold 1970), it became possible to examine the role of auxin in detail (see Chapter 12, Volume II). Auxin was shown to enhance the rate of elongation of *Avena* coleoptiles within 10 min (Kohler 1956), and this lag period was deemed independent of RNA and protein synthesis as determined by the use of actinomycin D and cycloheximide (Evans and Ray 1969). Similarly, the kinetics of the response of the dicotyledonous plants, pea (Penny 1969; Barkley and Evans 1970), lupin *(Lupinus angustifolius)* (Penny 1969) and soybean (Vanderhoef and Stahl 1975) have been studied. A biphasic growth response had been demonstrated to occur following auxin application (Kohler 1956) and this has since been confirmed in many tissues. The rapid effect of auxin on cell elongation is considered to be mediated by wall loosening, new wall synthesis and increased turgor pressure (Cleland 1971). Cell wall modification by auxin has received intensive study by a great number of investigators, and the interaction between wall modification and cell turgor has also been studied (see Loescher and Nevins 1973). In general terms, the mode of cell wall elongation in response to auxin has been studied from the following three major points of view: physical changes in the wall, enzymatic or non-enzymatic wall loosening and turnover of wall components and synthesis of new wall material.

8.5.1.1. Physical changes in the wall. Physical changes in the wall are brought about by unknown mechanisms, and are measured usually by physical techniques which are adapted from studies of natural fibers, rubber and synthetic polymers. These studies have a major advantage in that very rapid changes may be measured. Heyn (1940) proposed that cell extensibility could be divided into two components, plasticity and elasticity. Changes in plasticity were found to be more closely related to changes in growth rate than were changes in elasticity. This was verified by Cleland (1967) using more sophisticated techniques. More recently, Cleland and colleagues and Masuda and colleagues have studied, in greater detail, the physical nature of stretching of the cell wall. They have searched for a description of growth based upon measurements of the physical properties of isolated fresh, frozen-thawed or methanol-killed tissue pieces. Cleland (1971) has reviewed the considerable progress made in this area, and has pointed out the limitations of the physical measurements and the uncertainty which surrounds the nature of the properties being measured. He has concluded that descriptions of extension which were based upon the behaviour of synthetic polymers were only partially capable of dealing with the observations on cell walls, and he found it necessary to include a property called "creep" (Cleland 1971), which was difficult to treat mathematically. Despite these stated limitations, some useful generalizations about the physical stretching of walls were made, and studies on the effects of auxin, CO_2 and low pH all suggest that the extension which is promoted by these substances can be correlated at least in part with the measured changes in the physical characteristics of the wall. Interpretation of the available data is hampered by the uncertainty of the wall parameters being measured. This can only be resolved by determining the structure of the cell wall. However, it was concluded that auxin might be acting through an H^+ ion extrusion mechanism (Rayle and Cleland 1970; Yamagata et al. 1974), and in support of this some extension could be obtained by low pH in the absence of auxin, CO_2 or cellular metabolism (Rayle et al. 1970). Limitations of the techniques have been discussed (Cleland 1967; Rayle et al. 1970; Masuda et al. 1974), and it has been concluded that, at best, these physical measurements provide a basis with which to study some of the very earliest effects of auxin on cell extension, and should not be used to study some of the secondary events in which very complex reactions occur. Amongst the latter might be classified hydrolysis of polysaccharides, biosynthetic reactions, protein metabolism and RNA synthesis. It has been concluded that studies on properties such as plasticity and viscoelastic extension have provided only partial explanations of cell elongation, yet they have allowed some characterization of early extension events and made modelling of elongation a possibility.

8.5.1.2. Enzymatic or non-enzymatic wall loosening? Two general approaches to the study of cell wall loosening, enzymatic breakage of constraining bonds with associated release of some reaction intermediates and non-enzymatic or acid breakage of constraining bonds with release of some reaction-intermediates, have formed the basis of the theories presented to account for rapid cell elongation caused by auxin. A currently popular model of rapid auxin action suggests that the hormone acts by some unknown mechanism to promote acidification of the cell wall and surrounding medium (Hager et al. 1971; Cleland 1973; Marré et al. 1973; Rayle 1973; Yamamoto et al. 1974). Whether the acidification is achieved by proton extrusion, organic acid extrusion or hydroxyl uptake has not been elucidated. That the hormone acts through acidification is based upon indirect experiments such as those carried out by Rayle and Cleland (Rayle and Cleland 1970; Rayle et al. 1970; Rayle and Cleland 1972), who showed that acid solutions could mimic some of the auxin-induced changes in walls such as extensibility of *Avena* measured by the Instrom technique, the time course of cell expansion as measured by growth rate studies, a similar Q_{10} between 15 and 25 °C and the absence of stored growth. They also showed that killed sections which had been frozen and thawed would respond to acid conditions (pH 3.0–3.6) by elongating if a constant weight was applied to replace turgor. This system was independent of auxin, and subsequent studies have provided extensive circumstantial evidence to suggest that there is considerable similarity between the effects of auxin and acidification on cell wall extension (Yamagata et al. 1974; Jacobs and Ray 1975 and 1976; Marré et al. 1975). Thus, an hypothesis has emerged which proposes that auxin causes cell elongation through a mechanism involving acidification of the wall environment. There are some data which bring this proposal into question, and they are fully discussed in Chapter 12, Volume II.

If one assumes that the cell loosening which occurs in response to auxin treatment is mediated or assisted by acidification, then two major questions present themselves. The first is: how does auxin induce the acidification process, and the second is: how do acid conditions cause wall loosening? The first question awaits further analysis of auxin interaction with the plasma membrane, cytoplasm and the regulation of membrane transport, all of which are probably involved in the "primary" action of acidification. In attempting to answer the second question, in vitro systems similar to that described by Rayle et al. (1970) were set up in which killed sections of *Avena* were treated with pronase to digest available protein or with detergent (SDS) to denature protein (Rayle and Cleland 1972). Sections treated thus still responded to acid treatment if turgor was replaced by a weight. These authors concluded that unspecified bonds in the cell wall were acid-labile and that they were ruptured by acid treatment, thus allowing wall extension. Using a somewhat similar excised system derived from pea epicotyl epidermis, Yamagata et al. (1974) concluded that the acid response could be mediated by enzymatic breakage of constraining bonds. They found that fresh or frozen-thawed tissues responded to acid, but that pronase treatment of the epidermis or immersing it in boiling methanol (considered to remove protein) resulted in a greatly decreased elongation response. As further evidence for an enzymatic role in elongation they found that nojirimycin (an inhibitor of β-glucosidases) inhibited the extension of frozen-thawed sections. Their conclusion, that auxin-induced acidification of epidermal walls could result in activiation of wall-bound enzymes capable of hydrolyzing polysaccharide cross-links, was consistent with other data obtained with exogenous polysaccharide-degrading enzymes (Masuda and Wada 1967; Wada et al. 1968), extraction of endogenous degradative enzymes (Tanimoto and Masuda 1968; Heyn 1970; Nevins 1970; Johnson et al. 1974) and the addition of nojirimycin to elongating *Avena* segments (Nevins 1975).

Serious limitations have, however, been found in the interpretations of most of the studies on enzyme involvement in wall extension. Exogenous application of enzyme mixtures, containing β-1,3-glucanase and cellulase, which was shown to rapidly stimulate elongation of *Avena* (Masuda and Wada 1967) or *Pisum* (Wada et al. 1968), could not be repeated by other workers (Cleland 1968; Ruesink 1969; Nevins 1970). In addition, auxin stimulation of the activities of β-1,3- and β-1,6-glucanases and hemicellulases (Tanimoto and Masuda 1968) or dextranase (Heyn 1970) are too slow to account for rapid changes in extensibility. However, endogenous growth (not stimulated by application of hormones) of *Phaseolus* hypocotyl (Nevins 1970) or *Pisum* epicotyl (Murray and Bandurski 1975) was correlated with high β-glucosidase activity, and it was shown (Johnson et al. 1974) that an exo-β-galactosidase activity of *Avena* cell walls was increased by auxin or low pH, and that other exo-glycosidases of the wall had acidic pH optima, thus making them possible candidates as regulators of wall loosening. But Evans (1974), who was able to confirm that auxin increased the activity of exo-β-galactosidase and exo-β-glucosidase slightly within 30 min, also showed that the activities of both exo-enzymes could be reduced by 43% and 75% respectively with sugar aldonolactones (inhibitors of glycosidases), without affecting elongation promoted by auxin or acid. On the other hand, Nevins (1975) concluded that exo-β-glucanases could regulate auxin-induced cell elongation based upon the use of nojirimycin, which inhibited the response of *Avena* to the hormone.

Based upon current concepts of the cell wall structure, in which hydrogen bonding of xyloglucan or arabinoxylan polymers to cellulose and covalent bonding of the pectic polymers to the xyloglucan or arabinoxylan play a dominant role in maintaining structure of the wall, there are two types of bond breakage which might allow wall loosening. The first is the hydrogen bonding of the xyloglucan to the cellulose but there is no evidence to show that auxin affects this relationship. Valent and Albersheim (1974) showed that acid pH would not alter the hydrogen bonding of xyloglucan to cellulose in an in vitro system using short polymers of xyloglucan, and they concluded that it was unlikely that accelerated rates of growth at pH 5 were due to accelerated rates of xyloglucan creep. By extrapolation, auxin may not affect in vivo hydrogen bonding between cellulose and xyloglucan. The second type of bond breakage which might allow loosening is a break in covalent bonds within the xyloglucan polymer, or between it and the pectic polymers. There is no evidence to show that the pH changes observed or applied during elongation could hydrolyze these bonds. The available data on enzymatic cleavage is equivocal in that the time course of increased activity is often too slow to account for elongation; assays usually measure exo-glycosidase activity when endo-glycosidase action is probably more important and the action of inhibitors have yielded conflicting conclusions. As mentioned at the beginning, the major block to progress in determining the action of auxin on wall loosening may be knowledge of wall structure. In the meantime, further work on endo-glycosidases seems warranted, and may help clarify the role of covalent bonding in maintaining cell wall structure and in cell elongation.

8.5.1.3. Turnover of wall components and synthesis of new wall material. Turnover of various wall components is a possible mechanism whereby bond breakage and therefore wall loosening might be presumed to occur. Ray (1969) reviewed the then existing data and concluded that the early turnover studies were not interpretable for technical reasons related to the presence of starch, differential solubility of various fractions and ill-defined nature of the wall components. In a recent study of the effect of auxin on turnover of wall polysaccharides in pea stem segments which had been prelabelled with [^{14}C] glucose, Labavitch and Ray (1974a) noted that there was no effect of the hormone on turnover of the majority of the wall carbohydrates, but they did note changes in a pectinase-degradable xyloglucan. They found that auxin induced the conversion of an insoluble wall xyloglucan into a water-soluble form. The release of the xyloglucan was measurable within 15 min of hormone treatment, and was dependent upon cellular metabolism but independent of elongation (Labavitch and Ray 1974b). Subsequently, Jacobs and Ray (1975 and 1976) showed that acidification could mimic the auxin effect and cause the release of a water-soluble xyloglucan from cell walls of *Pisum*. There was a release of non-cellulosic wall glucose from *Avena* sections treated with auxin (Loescher and Nevins 1973) and, although the time course of the response was very slow, it may represent a similar change in wall structure to that predicted by Labavitch and Ray (1974a). Other reports that auxin promoted turnover of wall polysaccharides (Matchett and Nance 1962; Katz and Ordin 1967; Wada et al. 1968) are even more difficult to interpret than the above because of the nature of the fractionation techniques. timing of exposure to hormone and the magnitude of the responses.

At present it is not possible to relate the data obtained from these studies on turnover or appearance of wall fragments to those in which bond breakage and enzymatic activity have been examined. It is to be expected that the appearance of soluble wall components during elongation is the result of enzymatic or other hydrolysis.

While it is usually considered that cell wall synthesis must occur during sustained growth and that auxin promotes that synthesis (Bonner 1934b), it is not clear what role synthesis plays in rapid growth responses. Ray (1962) has shown, by the use of inhibitors, that under certain conditions wall synthesis can be separated experimentally from elongation, in that some inhibitors, such as Ca^{2+}, Hg^{2+}, F^{+}, inhibit elongation much more than wall synthesis, whereas low temperature, dinitrophenol, mannitol and azide inhibit both to similar extents. Baker and Ray (1965a, b and c) showed that IAA promoted uptake and incorporation of [^{14}C] glucose into all cell wall fractions of *Avena* and the response was measurable by 4 hr. Similar results were obtained for pea epicotyl (Abdul-Baki and Ray 1971), except that the response could now be measured within 1–2 hr. In an extensive search for an effect of auxin on the activity of an enzyme involved in pea wall polysaccharide formation, it was shown that only one enzyme, β-glucan synthetase, one of the last enzymes in the biosynthetic pathway, showed any response (Abdul-Baki and Ray 1971). Thus, they concluded that the hormone did not act on a number of possible regulatory enzymes in wall synthesis. In most of the results presented above, the timing of observed changes in synthetic activity rule out the possibility that such changes are related to the rapid action of auxin; however, Ray (1973a and b) has shown that in pea epicotyl segments a two-fold increase in the activity of UDP-glucose-dependent β-glucan synthetase occurs under specialized conditions of additional sucrose and depleted levels of endogenous enzyme. The activity of the enzyme began to increase 10–15 min after exposure to the hormone, which brings the response into the range of the rapid effects observed on growth. The effect was independent of elongation as shown by inhibitor studies and by the fact that the enzyme of the subapical non-elongating regions also responded to the hormone. It was also shown to be a non-allosteric activation which was energy-dependent but independent of RNA and protein synthesis, hydroxyproline synthesis and Ca^{2+} ions (Ray 1973b). Ray (1967) had shown previously that normal (unstimulated) wall synthesis occurred in such a way (by apposition) that it could not be implicated in elongation, but that auxin did stimulate hemicellulose synthesis throughout the wall (intussusception), suggesting that such synthesis might be involved in promoting elongation. There is a single report of a slight effect of an auxin on the activity of a glucan synthetase from onion stem (VanDerWoude et al. 1972). These authors have shown that the activity of plasma membrane-associated glucan synthetase was stimulated in vitro, by the auxin 2,4-D, to the extent of 11% over the control. An 18-hr treatment of the tissue in vivo stimulated the activity by 24%. Analysis of the polysaccharides synthesized in vitro revealed that not all fractions were stimulated. The largest response was seen in the hot water-soluble fraction with some stimulation of the hot NaOH-insoluble fraction. However, it should be noted that this report could not be confirmed (Spencer et al. 1972).

The activity of wall synthesizing enzymes such as β-glucan synthetase (Ray 1973a) and the release of soluble xyloglucan fractions from auxin-stimulated pea epicotyls (Labavitch and Ray 1974b) may be reconciled if one proposes that wall elongation proceeds by a "break and repair" mechanism. Breaks might be induced by enzymatic or chemical means with concomitant release of chopped out sections. Turgor would cause separation between breaks and these could be repaired enzymatically.

The rapid responses to auxin and acid necessitate fine control over proteins and/or enzymes if they are involved in wall loosening. Wall enzymes are generally considered to be very stable and thus turnover of the enzymes would not facilitate regulation. Allosteric interaction seems unlikely from the available data. Based upon kinetics of elongation in the presence of cycloheximide, the existence of a group of growth-limiting proteins with a short half-life has been postulated (Cleland 1971; Penny 1971; see also Chapter 12, Volume II). There is little evidence to support the notion, except that they are conceptually attractive in that they would provide the necessary flexibility for control.

8.5.2. The effect of auxin on cell walls during cell expansion

An extensive investigation of the role of cellulase in auxin-induced radial expansion of pea epicotyls has shown a positive correlation between enzyme level and expansion activity (Fan and Maclachlan 1966 and 1967; Datko and Maclachlan 1968; Ridge and Osborne 1969; Ferrari and Arnison 1974). A role for cellulase in lateral expansion of cell is difficult to visualize, because changes in enzyme activity tend to follow rather than precede expansion. However, detailed analysis of the isozymes of cellulase present in the wall immediately after auxin application should be studied, since at least two forms have been detected in expanding pea epicotyls (Ferrari and Arnison 1974; Byrne et al. 1975) and seven isozymes have been found in abscission zones of *Citrus sinensis* (Goren and Huberman 1976). Byrne et al. (1975) found that both forms of the enzyme which they isolated responded to auxin, while Ferrari and Arnison (1974) showed that low concentrations of auxin promoted the activity of one form and ethylene promoted the other. Osborne et al. (1972) found that ethylene did not affect cellulase levels during radial cell expansion. The status of wall cellulase during expansion growth is not clear and, even if it were, the only role which it might play would be to provide primer ends for further cellulose deposition. Furthermore, based upon the discussion in section 8.5.1.2, it is to be expected that disruption of bonds other than those of cellulose might be expected during expansion as well as extension growth. The orientation of the cellulose microfibrils, and thus the microtubules, would also be expected to be altered (see section 8.4.1) from radial to longitudinal or random array (Veen 1970; Ridge 1973).

A number of other hydrolases (β-1,3-glucanase, pectinase) and synthetases (alkali-soluble and alkali-insoluble glucan synthetase, UDP-glucose and GDP-glucose synthetase) were also examined for changes in specific activity following auxin treatment, and while the levels of these were maintained by the hormone there were no overall increases in specific activity (Datko and Maclachlan 1968; Spencer et al. 1971). However, if endogenous levels were allowed to deplete following decapitation and auxin was added after three days, then particulate glucan synthetase levels were increased by what appeared to be new protein synthesis (i.e. inhibited by cycloheximide) (Spencer et al. 1972). These authors postulated an effect of auxin on turnover of the enzyme and also an effect on the activity of an activator of the enzyme. The supporting data for these conclusions is, so far, not very strong. A more promising approach has been that taken by Shore and Maclachlan (1973 and 1975), who have fractionated the β-1,4-glucan synthetase into four fractions and examined the fine detail of enhanced levels of each fraction in response to auxin. They obtained preliminary evidence to show that the activity of three fractions was controlled by auxin (Shore and Maclachlan 1973). Subsequently, they showed that the hormone may also influence the intracellular distribution of alkali-insoluble β-1,4-glucan synthetase activity (Shore and Maclachlan 1975). Improved assay conditions using tissue slices has shown that over 90% of the active cellulose-synthesizing enzyme was localized at the cell surface, and that intracellular forms of the enzyme were probably in transit from site of biosynthesis to site of activity (Shore et al. 1975). Cellulose synthetase level and cellulose deposition increased and paralleled one another and the changes occurred subsequent to growth. It is difficult to reconcile the high levels of cellulose hydrolyzing and synthesizing enzymes which are both present in the expanding hypocotyl, particularly since both appear to be stimulated by auxin with similar kinetics. It is necessary to invoke a mechanism of cell expansion which involves continual hydrolysis and addition of new polysaccharide units.

The involvement of ethylene and auxin and their possible interaction in cell expansion has merited considerable attention. During swelling of pea epicotyls it was shown that both ethylene and auxin stimulated cell expansion but, while auxin promoted cellulase activity, ethylene did not affect the enzyme (Ridge and Osborne 1969). Fractionation of cellulases into intracellular and wall forms did not reveal any selective effects of the gas on cellulases (Osborne et al. 1972), but these authors did show that ethylene increased wall peroxidase and enhanced the level of wall protein rich in hydroxyproline (Ridge and Osborne 1970 and 1971; Osborne et al. 1972). Ethylene altered the microfibrillar orientation in the cell walls of *Pisum* epicotyls induced to swell by the gas (Apelbaum and Burg 1971). This was accompanied by reduced incorporation of proline into wall protein relative to that of leucine, and conversion of proline to hydroxyproline was also supressed (Eisinger and Burg 1972). Thus, the quality of the wall protein in expanded cells may be considerably altered. However, Osborne et al. (1972) found that levels of hydroxyproline-rich protein were increased in ethylene-treated epicotyls. Difficulties in reconciling these sets of data are confounded, in that one study was short-term and consisted of labelling experiments (Eisinger and Burg 1972), while the other was long-term and total hydroxyproline was determined (Osborne et al. 1972). Neither result helps explain radial expansion of cells. Since ethylene was shown to enhance the level of one form of cellulase by Ferrari and Arnison (1974), and Osborne et al. (1972) found no change in enzyme levels, it is difficult to assign a role to the enzyme in expansion induced by the gas. In other long-term studies ethylene has been shown to increase pectic substances, and therefore it has been considered to inhibit extensibility (Mondal and Nance 1975). There is considerably less known about changes in cell walls associated with radial expansion induced by either auxin or ethylene than there is about the walls in auxin-induced cell elongation. This may in part be due to the long-term nature of the response and in part to the predominating interest in cell elongation.

Cell elongation and expansion in response to cytokinins and gibberellins has been studied, but data on structural changes in walls are relatively rare. GA_3 stimulated wall synthesis of *Avena* stem segments probably by increasing the synthesis of all wall components (Montague and Ikuma 1975), and this was accompanied by increased wall plasticity (Adams et al. 1975) during hormone-induced cell elongation. In longer-term studies on elongation in pea epicotyls, Mondal (1975) showed that GA_3 lowered the levels of pectic substances in the wall. Confirming the findings of Montague and Ikuma (1975) for *Avena,* Kawamura et al. (1976) showed that GA_3-dependent elongation in lettuce was accompanied by increased synthesis of cellulose, pectin and hemicellulose. These authors related their biochemical findings to changes in physical properties of the wall. However, in none of these studies is it possible to deduce the mechanism of alteration of wall structure.

Cytokinin-induced swelling of cells (Katsumi 1962; Victor and Vanderhoef 1975) is accompanied by increased levels of pectic substances (Nance 1973; Mondal and Nance 1975). Again, the way in which these changes occur is not understood.

8.5.3. The cell wall during abscission

The process of abscission has been dealt with in detail elsewhere (Chapter 7, Volume II), but one aspect is of further interest here. This involved the process whereby cells separate from one another and the question of how this is effected. An obvious candidate is an enzyme or enzyme complex which might degrade the middle lamellae or the total cell wall. However, a direct demonstration of a specific enzymatic event has been difficult to characterize. A distinguishing feature of cells of the abscission layers is that there is a general breakdown of the middle lamella involving the loss of pectic substances, magnesium and calcium followed by wall disruption (Bornman 1967; Webster 1968). This led to the proposal that pectinase (depolymerization and dissolution of pectins) activity was closely correlated with the onset of abscission in *Phaseolus vulgaris* (Morre 1968). Osborne (1958) and LaMotte et al. (1969) have shown that decreased pectin methylesterase activity was associated with the abscission zones of bean and *Coleus* leaves. This was later shown to be an artefact of extraction and no changes in pectin methyl esterases preceded abscission in members of several plant families (Moline et al. 1972). This was confirmed for *Citrus* by Ratner et al. (1969). A role for cellulase in abscission was first suggested by Horton and Osborne (1967), who demonstrated regulation of enzyme levels by auxin and ethylene in bean petioles. The gas was shown to promote abscission of explants and to raise the level of enzyme, while auxin had the opposite effect. Abeles (1969) also showed that cellulase levels appeared to be under at least partial control by ethylene, auxin and cytokinin, and that the level of enzyme was raised during abscission (Craker and Abeles 1969). Polygalacturonase activity was also associated with abscission in *Citrus* (Riov 1974), the enzyme was localized in the separation zone and displayed a time course of activity which implicated it in the separation process. Other hydrolytic enzymes, β-1,3-glucanase and chitinase, have been shown to increase in activity during ethylene-induced abscission of *Phaseolus* explants (Abeles et al. 1970), but their role in cell wall dissolution was not clarified. Early evidence that cellulase was responsible for abscission was strongly supported by Abeles and Leather (1971) and by Rasmussen (1973). They concluded that the gas promoted both enzyme synthesis and its secretion to the cell wall.

Two cellulase forms in *Phaseolus* were described by Lewis and Varner (1970), and they showed that, while the salt-soluble form increased with abscission, ethylene promoted the activity of the water-soluble form and gibberellic acid, which had no effect on abscission, also promoted the activity of the water-soluble form. The increase in cellulase paralleled but did not precede abscission, thus suggesting with all of the above data that changes in cellulase levels were only incidentally associated with abscission. In a subsequent study it was shown that salt solubility was probably an invalid criterion for the fractionation of cellulases, and that two forms of the enzyme could be separated based upon their isoelectric point (Reid et al. 1974). The basic form of the enzyme increased in response to ethylene and was associated with abscission zones. It was also thought to be extracellular in location, but when measured in relation to breakstrength of the abscission zones, it was shown that the level of enzyme did not rise until breakstrength dropped to zero (Reid et al. 1974). Thus, it was concluded again that the level of cellulase was only indirectly associated with abscission, probably with some of the late phases of cell separation and lysis. Huberman et al. (1975) concluded that pH was an important factor in cellulase extraction and that ethylene may further affect extractability. Several isozymes of cellulase have been separated by Goren and Huberman (1976) using electrophoresis, and they concluded that measurement of total enzyme activity may be misleading if abscission was to be related to levels of the enzyme. Variation in levels of various isozymes under different conditions led them to suggest, however, that cellulase could not be considered as the only factor in abscission because fluctuations in isozyme levels did not correlate well with abscission. Subtle changes in enzyme levels and the location of those enzymes are thus likely to be important factors in wall dissolution. It is to be expected that other enzymes of the wall may be involved in abscission, especially endo-glucanases, and as wall structure becomes understood it will be easier to predict the enzyme activity which might be most closely associated with abscission.

In summary, some of the longer-term responses to hormones in abscission may be partially mediated through hydrolytic enzymes, which are probably controlled at the level of synthesis of new enzyme or isozymes and the redistribution of existing enzyme to new sites of action. However, the early events in the process of abscission remain as a most potentially exciting area of investigation, and their elucidation will explain an interesting phenomenon in terminal differentiation.

8.5.4. Gibberellins and cell wall disruption

In mature barley aleurone tissue, GA_3 inhibits cell wall biosynthesis and simultaneously promotes cell wall degradation (Johnson and Chrispeels 1973). The rather irregular erosion of the wall was first described by Paleg and Hyde (1964), and has been described a number of times since (Jones 1969d; Taiz and Jones 1970; Ashford and Jacobsen 1974a). Similar patterns of wall erosion have been described in the aleurone of germinating barley (Van der Eb and Niewdorp 1967; Ashford and Jacobsen 1974b) and germinating wheat (Fulcher et al. 1972). Digestion began in the inner wall adjacent to the plasmalemma and also adjacent to the plasmodesmata, especially in the centre of the wall, apparently in the middle lamella (Taiz and Jones 1970; Ashford and Jacobsen 1974a). These areas enlarged and became continuous, with large areas at the cell corners which dissolved until after some time a major portion of the cell wall was digested. However there was a thin layer of wall adjacent to the plasmodesmata and the plasmalemma (apparently a continuous layer) which was resistent to hydrolysis (Taiz and Jones 1973; Ashford and Jacobsen 1974a). This layer of wall was characterized in barley by being silver-hexamine-positive and staining green with toluidine blue (Jacobsen et al. 1971; Taiz and Jones 1973). A similar layer was described in wheat where it was also resistant to digestion, and it was suggested that it contained ferulic acid (Fulcher et al. 1972). Thus, GA_3 leads to the dissolution of the cell walls, except for a thin layer which remained and acted as a skeleton and protective coating of the aleurone cytoplasm and plasmodesmata. There is evidence that this layer is largely cellulose in barley cell walls (Pyliotis and Jacobsen, unpublished).

A study designed to determine how enzyme molecules traverse the aleurone cell walls (Ashford and Jacobsen 1974a) found a correlation between areas of cell wall hydrolysis and the occurrence of acid phosphatase, part of which probably originated in the aleurone cytoplasm. During the period of GA_3-induced acid phosphatase release by the tissue, the enzyme occurred almost exclusively in pockets of hydrolyzed cell walls and not uniformly thoughout the wall. This led to the conclusion that the release of enzyme from the tissue, secreted through the plasmalemma, occurred only through digested wall areas. A similar parellelism between wall digestion and secreted enzymes has been found for peroxidase (Jacobsen and Knox 1972). Therefore, it would appear that one of the purposes of wall digestion in aleurone is to facilitate enzyme release.

The enzymes responsible for cell wall digestion are now becoming known. Based on cytochemical techniques, it was once thought that the walls of aleurone were of β-1,3-glucan (Taiz and Jones 1970), which seemed reasonable since β-glucanase is one of the enzymes released by aleurone cells after GA_3 treatment (MacLeod and Millar 1962; Briggs 1963; Taiz and Jones 1970). However, the specificity of the staining techniques is in doubt (Fulcher et al. 1972), and it now appears that aleurone walls are not largely β-glucan. Recent analysis shows that they are probably largely arabinoxylan (85%) with 8% cellulose and 6% protein (McNeil et al. 1975). The arabinoxylan has a β-1,4-xylan backbone substituted with arabinofuranosyl residues. GA_3 stimulated the activity of endo-β-1,4-xylanase and two glycosidases, β-xylopyranosidase and α-arabinofuranosidase in isolated aleurone layers (Taiz and Honigman 1976). It is likely that these enzymes are involved in digestion of aleurone cell walls, and that self-tunnelling of the wall may be necessary for release of these and other hydrolases from the tissue. Tunnelling of the wall, however, may not only involve enzymes secreted from the aleurone cytoplasm. The walls of mature dry aleurone contain 6% protein which is not rich in hydroxyproline (McNeil et al. 1975). At least some of the protein is enzymatic, because acid phosphatase and peroxidase have been localized in aleurone cell walls (Jacobsen and Knox 1972; Ashford and Jacobsen 1974a and b), and it is possible that these, especially the phosphatase, are involved with cell wall modification following GA_3 treatment.

Other examples of gibberellin-stimulated disruption of cell walls include wall lysis, which occurs in the abscission zones of *Coleus* and cotton *(Gossypium hirsutum)* after GA_3 treatment (Bornman 1967), although it is ethylene and not GA which is recognized as the prime regulator of abscission. It seems that modification involves removal of the pectic substances, because the middle lamella, which is very wide in these tissues, is fairly uniformly dissolved, leaving only narrow bands of primary cell wall around each cell. The end point of this is that cells separate from each other and permit abscission.

A similar situation seems to exist in celery *(Apium graveolens)* endosperm, where gibberellin causes the cells to separate and become individual spheroplasts (Jacobsen et al. 1976). The cell walls were originally thick but, following GA_3 treatment, they became much reduced in thickness. By plasmolyzing the spheroplasts, it could be shown that each cell was still surrounded by a thin layer (about 0.5 μm in width) of resistant cell wall, so it seems that the central part of the cell wall is dissolved. The nature of the wall components which are digested and the enzymes responsible are not yet known.

In all of these cases, there is a preference for hydrolysis to occur at least initially in the central wall area and probably the middle lamella. In cotton and *Coleus* abscission zones, there is direct evidence that the pectic substances of the middle lamella dissolve, thus causing cell separation (Bornman 1967), and there is evidence that the same is true for abscission in *Phaseolus* – in fact, changes in pectinase activity have been measured which correlate with abscission (Morré 1968). Therefore it would seem that, in the cases so far examined, the common denominator of GA_3-promoted cell wall degradation is probably at least in part lysis of the middle lamella by pectin hydrolyzing enzymes released from the cytoplasm. Depending on the cell wall type, other enzymes may also be required, e.g. cellulase and polygalacturonase, in abscission zones (Horton and Osborne 1967; Abeles 1968; Riov 1974) (see section 8.5.3).

8.6. Protoplasmic streaming

An excellent example of structural modification of cellular components by a plant hormone was provided by the demonstration that auxin stimulated protoplasmic streaming in *Avena* coleoptiles within one or two minutes of application (Sweeney and Thimann 1937 and 1942). Protoplasmic streaming was thought to be involved in distribution of cellular metabolites and organelles, and it was therefore postulated that streaming could mediate the action of auxin by directing ions, enzymes or other metabolites to a process which initiated cell elongation (Thimann 1969). While there is no direct evidence to suggest that streaming redistributes metabolites, this attractive hypothesis of auxin action was rendered unlikely by the finding that the fungal metabolite cytochalasin B, which inhibited streaming completely, did not entirely inhibit auxin-dependent elongation of oat and maize coleoptiles or pea stem segments (Cande et al. 1973). Thus, it was concluded that streaming was not essential but might be beneficial for cell elongation. Therefore, it seems that streaming, while rapidly responsive to physiologically active auxin, does not take part in the subsequent cell elongation. This suggests that, if auxin has a single primary action on these cells, protoplasmic streaming and cell elongation are two separate consequences of the phenomenon.

Growth and protoplasmic streaming in lettuce root-hair cells were inhibited by cytochalasin B (Sawhney and Srivastava 1974). Growth could be partially restored by 6-benzylaminopurine but protoplasmic streaming remained inhibited, again suggesting that streaming was unnecessary for growth.

8.7. Summary and conclusions

There is a great deal of evidence showing that phytohormones affect subcellular structure, both within the short- as well as the long-term. These range from the easily demonstrated changes in protoplasmic streaming to more subtle changes in cell wall extension, microtubule orientation and Golgi bodies. It is expected that, as techniques become refined, molecular changes such as those involved in receptor binding, membrane transport and regulation of transcription will be described. The vast majority of the evidence suggests that hormones bring about structural changes in cells, but there is very little indication, at present, of the importance of structure in hormone perception. It is clear that structural integrity of the cell is necessary for hormone action, which implies a requirement for the structural and biochemical continuum of the cell. No convincing data has been obtained to show that hormones will carry out substantial parts of their activities in cell-free systems. We conclude from the available data that hormone action in plant cells is a highly complex series of events which cannot easily be separated by conventional fractionation techniques, and that any approach to a description of their mechanism of action must consider organization at the cellular level.

References

Abdul-Baki, A.A. and Ray, P.M. (1971) Plant Physiol., 47, 537–544.
Abeles, F.B. (1968) Plant Physiol., 43, 1577–1586.
Abeles, F.B. (1969) Plant Physiol., 44, 447–452.
Abeles, F.B. (1972) Ann. Rev. Plant Physiol., 23, 259–292.
Abeles, F.B. and Leather, G.R. (1971) Planta, 97, 87–91.
Abeles, F.B., Boshart, R.P., Forrence, L.E. and Habig, W.H. (1970) Plant Physiol., 47, 129–134.
Adams, P.A., Montague, M.J., Tepfer, M., Rayle, D.L., Ikuma, H. and Kaufman, P. (1975) Plant Physiol., 56, 757–760.
Albersheim, P. (1965) in: Plant Biochemistry, J. Bonner and J.E. Varner (eds.), Academic, New York, pp. 151–186.
Albersheim, P. and Anderson-Prouty, A.J. (1975) Ann. Rev. Plant Physiol., 26, 31–52.
Altman, A. and Wareing, P.F. (1975) Physiol. Plant., 33, 32–38.
Apelbaum, A. and Burg, S.P. (1971) Plant Physiol., 48, 648–652.
Ashford, A.E. and Jacobsen, J.V. (1974a) Planta, 120, 81–105.
Ashford, A.E. and Jacobsen, J.V. (1974b) in: Mechanisms of Regulation of Plant Growth, R.L. Bieleski, A.R. Ferguson and M.M. Cresswell (eds.), The Royal Society of New Zealand, Wellington, pp. 591–599.
Atsmon, D. (1972) in: Plant Growth Substances 1970, D.J. Carr (ed.), Springer, Berlin, pp. 222–227.
Baker, D.B. and Ray, P.M. (1965a) Plant Physiol., 40, 345–352.
Baker, D.B. and Ray, P.M. (1965b) Plant Physiol., 40, 353–360.
Baker, D.B. and Ray, P.M. (1965c) Plant Physiol., 40, 360–368.
Bamberger, E.S. (1971) Phytochemistry, 10, 957–966.
Barendse, G.W.M. (1974) in: Plant Growth Substances 1973, Hirokawa, Tokyo, pp. 332–341.
Barkley, G.M. and Evans, M.L. (1970) Plant Physiol., 45, 143–147.
Batt, S. and Venis, M.A. (1976) Planta, 130, 15–21.
Batt, S., Wilkins, M.B. and Venis, M.A. (1976) Planta, 130, 7–13.
Bauer, W.D., Talmadge, K.W., Keegstra, K. and Albersheim, P. (1973) Plant Physiol., 51, 174–187.
Bendana, F.E., Galston, A.W., Kaur-Sawhney, R. and Penny, P.J. (1965) Plant Physiol., 40, 977–983.
Bennett, P. and Chrispeels, M.J. (1972) Plant Physiol., 49, 445–447.
Ben-Tal, Y. and Varner, J.E. (1974) Plant Physiol., 54, 813–816.
Berridge, M.V., Ralph, R.K. and Letham, D.S. (1970) Biochem. J., 119, 75–84.
Berridge, M.V., Ralph, R.K. and Letham, D.S. (1972) in: Plant Growth Substances 1970, D.J. Carr (ed.), Springer, Berlin, pp. 248–255.
Bonner, J. (1934a) Protoplasma, 21, 406–423.
Bonner, J. (1934b) Proc. nat. Acad. Sci. (Wash.), 20, 393–397.
Bonner, J. (1936) J. gen. Physiol., 20, 1–11.
Booth, A., Moorby, J., Davies, C.R., Jones, H. and Wareing, P.F. (1962) Nature (Lond.), 194, 204–205.
Bornman, C.H. (1967) Sth. Afr.. J. med. Sci., 63, 325–331.
Bos, C.J. and Emmelat, P. (1974) Chem. biol. Interact., 8, 349–361.
Bowen, M.R. and Wareing, P.F. (1971) Planta, 99, 120–132.
Brandes, H. and Kende, H. (1968) Plant Physiol., 43, 827–837.
Briggs, D.E. (1963) J. Inst. Brew. (Lond.), 69, 13–19.
Burke, D., Kaufman, P., McNeil, M. and Albersheim, P. (1974) Plant Physiol., 54, 109–115.
Byrne, H., Christou, N.V., Verma, D.P.S. and Maclachlan, G.A. (1975) J. biol. Chem., 250, 1012–1018.
Cande, W.Z., Goldsmith, M.H.M. and Ray, P.M. (1973) Planta, 111, 279–296.
Chambers, J.W., Georg, R.H. and Bass, A.D. (1965) Molec. Pharmacol., 1, 66–76.
Chambers, J.W., Georg, R.H. and Bass, A.D. (1968) Endocrinology, 83, 1185–1192.
Chambers, J.W., Georg, R.H. and Bass, A.D. (1970) Endocrinology, 87, 366–370.

Chang, K.-J., Marcus, N.A. and Cuatrecasas, P. (1974) J. biol. Chem., 249, 6854–6865.
Chen, K., Heller, J. and Canellakis, E.S. (1976) Biochem. biophys. Res. Commun., 68, 401–408.
Chrispeels, M.J. (1972) Planta, 108, 283–287.
Chrispeels, M.J. (1976) Ann. Rev. Plant Physiol., 27, 19–38.
Chrispeels, M.J. and Varner, J.E. (1967a) Plant Physiol., 42, 398–406.
Chrispeels, M.J. and Varner, J.E. (1967b) Plant Physiol., 42, 1008–1016.
Cleland, R. (1967) Planta, 74, 197–209.
Cleland, R.E. (1968) Science, 160, 194–195.
Cleland, R.E. (1971) Ann. Rev. Plant Physiol., 22, 197–222.
Cleland, R.E. (1973) Proc. nat. Acad. Sci. (Wash.), 70, 3092–3093.
Cleland, R.E. (1975) Planta, 127, 233–242.
Cleland, R.E. (1976) Biochem. biophys. Res. Commun., 69, 333–338.
Collins, G.G., Jenner, C.F. and Paleg, L.G. (1972) Plant Physiol., 49, 404–410.
Craker, L.E. and Abeles, F.B. (1969) Plant Physiol., 44, 1139–1143.
Cuatrecasas, P. (1974) Ann. Rev. Biochem., 43, 169–214.
Datko, A.H. and Maclachlan, G.A. (1968) Plant Physiol., 43, 735–742.
Dauwalder, M. and Whaley, W.G. (1974) J. Cell Sci., 4, 455–497.
Dauwalder, M., Whaley, W.G. and Kephart, J.E. (1972) Sub-Cell. Biochem., 1, 225–276.
Davies, C.R. and Wareing, P.F. (1965) Planta, 65, 139–156.
Davies, E. (1976) Plant Physiol., 57, 516–518.
Davies, E. and Larkins, B.A. (1973) Plant Physiol., 52, 339–345.
Dela Fuente, R.K. and Leopold, A.C. (1970) Plant Physiol., 46, 186–189.
Dela Fuente, R.K. and Leopold, A.C. (1973) Plant Physiol., 51, 845–847.
Dowler, M.J., Rayle, D.L., Cande, W.Z., Ray, P.M., Durand, A. and Zenk, M.H. (1974) Plant Physiol., 53, 229–232.
Eastwood, D. and Laidman, D.L. (1971) Phytochemistry, 10, 1275–1284.
Eisinger, W.R. and Burg, S.P. (1972) Plant Physiol., 50, 510–517.
Epstein, E. (1961) Plant Physiol., 36, 437–444.
Etherton, B. (1970) Plant Physiol., 45, 527–528.
Evans, M.L. (1974) Plant Physiol., 54, 213–215.
Evans, M.L. and Ray, P.M. (1969) J. gen. Physiol., 53, 1–20.
Evins, W.H. (1971) Biochemistry, 10, 4295–4303.
Evins, W.H. and Varner, J.E. (1971) Proc. nat. Acad. Sci. (Wash.), 68, 1631–1633.
Fan, D-F. and Maclachlan, G.A. (1966) Canad. J. Bot., 44, 1025–1034.
Fan, D-F. and Maclachlan, G.A. (1967) Canad. J. Bot., 45, 1837–1844.
Fellenberg, G. (1969) Planta, 84, 195–198.
Feng, K.A. (1973) Plant Physiol., 51, 868–870.
Ferrari, T.E. and Arnison, P.G. (1974) Plant Physiol., 54, 487–493.
Filner, P. and Varner, J.E. (1967) Proc. nat. Acad. Sci (Wash.), 58, 1520–1526.
Firn, R.D. (1975) Planta, 125, 227–233.
Firn, R.D. and Kende, H. (1974) Plant Physiol., 54, 911–915.
Fisher, M.L. and Albersheim, P. (1974) Plant Physiol., 53, 464–468.
Fowke, L.C. and Setterfield, G. (1969) Canad. J. Bot., 47, 1873–1877.
Fox, J.E. and Erion, J.L. (1975) Biochem. biophys. Res. Commun., 64, 694–700.
Fulcher, R.G. (1972) Observations on the Aleurone Layer with Emphasis on Wheat, Ph.D. Thesis, Monash University, Melbourne, 299 pp.
Fulcher, R.G., O'Brien, T.P. and Lee, J.W. (1972) Aust. J. biol. Sci., 25, 23–34.
Gardiner, M. and Chrispeels, M.J. (1975) Plant Physiol., 55, 536–541.
Gawlik, S.R. and Shen-Miller, J. (1974) Plant Physiol., 54, 217–221.
Gibson, R.A. and Paleg, L.G. (1972) Biochem. J., 128, 367–375.
Gibson, R.A. and Paleg, L.G. (1975) Aust. J. Plant Physiol., 2, 41–49.
Gibson, R.A. and Paleg, L.G. (1976) J. Cell Sci., 22, 413–425.

Glinka, Z. (1973) Plant Physiol., 51, 217–219.
Glinka, Z. and Reinhold, L. (1971) Plant Physiol., 48, 103–105.
Glinka, Z. and Reinhold, R. (1972) Plant Physiol., 49, 602–606.
Gordon, M.E. (1973) Cytokinins and the Expansion of Radish Cotyledons, Ph.D. Thesis, Australian National University, Canberra, 186 pp.
Goren, R. and Huberman, M. (1976) Physiol. Plant., 37, 123–130.
Gorski, J. and Gannon, F. (1976) Ann. Rev. Physiol., 38, 425–450.
Green, D.E., Blondin, G., Kessler, R. and Southard, J.H. (1975) Proc. nat. Acad. Sci. (Wash.), 72, 896–900.
Gwozdz, E., Wozny, A. and Szweykowska, A. (1974) Biochem. Physiol. Pflanzen, 165, 82–92.
Hager, A., Menzel, H. and Kraus, A. (1971) Planta, 100, 47–75.
Hallam, N.D. (1970) J. exp. Bot., 21, 1031–1038.
Halperin, W. and Jensen, W.A. (1967) J. ultrastruct. Res., 18, 428–443.
Hanson, A.D. and Kende, H. (1975) Plant Physiol., 55, 663–669.
Hardin, J.W., Cherry, J.H., Morré, D.J. and Lembi, C.A. (1972) Proc. nat. Acad. Sci. (Wash.), 69, 3146–3150.
Hardin, J.W., O'Brien, T.J. and Cherry, J.H. (1970) Biochim. biophys. Acta (Amst.), 224, 667–670.
Harmey, M.A. and Murray, A.M. (1968) Planta, 83, 387–389.
Hepler, P.K. and Palevitz, B.A. (1974) Ann. Rev. Plant Physiol., 25, 309–362.
Hertel, R., Thomson, K.-St. and Russo, V.E. (1972) Planta, 107, 325–340.
Heyn, A.N.J. (1940) Bot. Rev., 6, 515–574.
Heyn, A.N.J. (1970) Science, 167, 874–875.
Higinbotham, N. (1974) Plant Physiol., 54, 454–462.
Hodges, T.K., Leonard, R.T., Bracker, C.E. and Keenan, T.W. (1972) Proc. nat. Acad. Sci. (Wash.), 69, 3307–3311.
Hollenberg, M.D. and Cuatrecasas, P. (1975) Fed. Proc., 34, 1556–1563.
Holmes, P.L. and Speakman, P.T. (1973) Nature New Biol., 242, 190–192.
Horton, R.F. and Moran, L. (1972) Z. Pflanzenphysiol., 66, 193–196.
Horton, R.F. and Osborne, D.J. (1967) Nature (Lond.), 214, 1086–1088.
Huberman, M., Goren, R. and Birk, Y. (1975) Plant Physiol., 55, 941–945.
Ilan, I. (1971) Physiol. Plant., 25, 230–233.
Ilan, I. and Reinhold, L. (1963) Physiol. Plant., 16, 596–603.
Irvine, R.F. and Osborne, D.J. (1973) Biochem. J., 136, 1133–1135.
Iwata, T. and Stowe, B.B. (1973) Plant Physiol., 51, 691–701.
Jablanovic, M. and Nooden, L.D. (1974) Plant Cell Physiol., 15, 687–692.
Jacobs, M. and Ray, P.M. (1975) Plant Physiol., 56, 373–376.
Jacobs, M. and Ray, P.M. (1976) Plant Physiol., 58, 203–209.
Jacobsen, J.V. (1972) in: Plant Growth Substances 1970, D.J. Carr (ed.), Springer, Berlin, pp. 336–343.
Jacobsen, J.V. and Knox, R.B. (1972) in: Plant Growth Substances 1970, D.J. Carr (ed.), Springer, Berlin, pp. 344–351.
Jacobsen, J.V. and Knox, R.B. (1973) Planta, 112, 213–224.
Jacobsen, J.V., Knox, R.B. and Pyliotis, N.A. (1971) Planta, 101, 189–209.
Jacobsen, J.V. and Varner, J.E. (1967) Plant Physiol., 42, 1596–1600.
Jacobsen, J.V. and Zwar, J.A. (1974a) Aust. J. Plant Physiol., 1, 343–356.
Jacobsen, J.V. and Zwar, J.A. (1974b) Proc. nat. Acad. Sci. (Wash.), 71, 3290–3293.
Jacobsen, J.V., Pressman, E. and Pyliotis, N.A. (1976) Planta, 129, 113–122.
Jefferson, L.S., Schworer, C.M. and Tolman, E.L. (1975) J. biol. Chem., 250, 197–204.
Jensen, E.V. and DeSombre, E.R. (1972) Ann. Rev. Biochem., 41, 203–230.
Johnson, K.D. and Chrispeels, M.J. (1973) Planta, 111, 353–364.
Johnson, K.D. and Kende, H. (1971) Proc nat. Acad. Sci. (Wash.), 69, 2674–2677.
Johnson, K.D., Daniels, D., Dowler, M.J. and Rayle, D.L. (1974) Plant Physiol., 53, 224–228.
Johri, M.M. and Varner, J.E. (1968) Proc. nat. Acad. Sci. (Wash.), 59, 269–276.

Jones, R.J. and Mansfield, T.A. (1972) Physiol. Plant., 26, 321–327.
Jones, R.L. (1968) Plant Physiol., 43, 442–444.
Jones, R.L. (1969a) Planta, 85, 359–375.
Jones, R.L. (1969b) Planta, 87, 119–133.
Jones, R.L. (1969c) Planta, 88, 73–86.
Jones, R.L. (1969d) Plant Physiol., 44, 1428–1438.
Jones, R.L. (1971) Plant Physiol., 47, 412–416.
Jones, R.L. (1972) Planta, 103, 95–109.
Jones, R.L. (1973) Plant Physiol., 52, 303–308.
Jones, R.L. and Price, J.M. (1970) Planta, 94, 191–202.
Kang, B.G. and Burg, S.P. (1971) Proc. nat. Acad. Sci. (Wash.), 68, 1730–1733.
Kasamo, K. and Yamaki, T. (1974) Plant Cell Physiol., 15, 965–970.
Kasamo, K. and Yamaki, T. (1976) Plant Cell Physiol., 17, 149–164.
Katsumi, M. (1962) Physiol. Plant., 15, 115–121.
Katz, M. and Ordin, L. (1967) Biochim. biophys. Acta (Amst.), 141, 118–125.
Kawamura, H., Kamisaka, S. and Masuda, Y. (1976) Plant Cell Physiol., 17, 23–34.
Keegstra, K., Talmadge, K.W., Bauer, W.D. and Albersheim, P. (1973) Plant Physiol., 51, 188–196.
Kelley, S. (1947) Amer. J. Bot., 34, 521–526.
Kende, H. and Gardner, G. (1976) Ann. Rev. Plant Physiol., 27, 267–290.
Kennedy, C.D. and Harvey, J.M. (1972) Pestic. Sci., 3, 715–727.
Key, J.L. (1969) Ann. Rev. Plant Physiol., 20, 449–474.
Klämbt, D. (1976) Plant Cell Physiol., 17, 73–76.
Kobayashi, K. (1974) in: Plant Growth Substances 1973, Hirokawa, Tokyo, pp. 573–581.
Koehler, D.E. and Varner, J.E. (1973) Plant Physiol., 52, 208–214.
Kohler, D. (1956) Planta, 47, 159–164.
Labavitch, J.M. and Ray, P.M. (1974a) Plant Physiol., 53, 669–673.
Labavitch, J.M. and Ray, P.N. (1974b) Plant Physiol., 54, 499–502.
Lai, Y.F. and Thompson, J.E. (1972) Canad. J. Bot., 50, 327–332.
LaMotte, C.E., Gochnauer, C., LaMotte, L.R., Mathur, J.R. and Davies, L.L.R. (1969) Plant Physiol., 44, 21–26.
Lamport, D.T.A. (1970) Ann. Rev. Plant Physiol., 21, 235–270.
Lawson, V.R. and Weintraub, R.L. (1975) Plant Physiol., 55, 1062–1066.
LeJohn, H.B., Cameron, L.E., Stevenson, R.M. and Meuser, R.U. (1974) J. biol. Chem., 249, 4016–4020.
Leonard, R.T. and Hanson, J.B. (1972) Plant Physiol., 49, 430–435.
Leopold, A.C., Poovaiah, B.W., Dela Fuente, R.K. and Williams, R.J. (1974) in: Plant Growth Substances 1973, Hirokawa, Tokyo, pp. 780–788.
Lewis, L.N. and Varner, J.E. (1970) Plant Physiol., 46, 194–199.
Liao, S. and Hamilton, R.H. (1966) Science, 151, 822–824.
Lipetz, J. (1971) Experientia (Basel), 27, 933–934.
Lips, S.H. and Roth-Bejerano, N. (1974) in: Plant Growth Substances 1973, Hirokawa, Tokyo, pp. 719–724.
Livingston, J.N. and Lockwood, D.H. (1975) J. biol. Chem., 250, 8353–8360.
Loescher, W.H. and Nevins, D.J. (1973) Plant Physiol., 52, 248–251.
McCalla, D.R., Morré, D.J. and Osborne, D.J. (1962) Biochim. biophys. Acta (Amst.), 55, 522–528.
McCombs, P.J.A. and Ralph, R.K. (1972) Biochem. J., 129, 403–417.
McLeod, A.M. and Millar, A.S. (1962) J. Inst. Brew. (Lond.), 68, 322–332.
McNeil, M., Albersheim, P., Taiz, L. and Jones, R.L. (1975) Plant Physiol., 55, 64–68.
Mansfield, T.A. and Jones, R.J. (1971) Planta, 101, 147–158.
Marchesi, V.T., Furthmayer, H. and Tomita, M. (1976) Ann. Rev. Biochem., 45, 667–698.
Marré, E., Lado, P., Caldogno, F.R. and Colombo, R. (1973) Plant Sci. Lett., 1, 179–184.
Marré, E., Lado, P., Ferroni, A. and Ballarin Denti, A. (1974) Plant Sci. Lett., 2, 257–265.

Marré, E., Lado, P., Rasi-Caldogno, F., Colombo, R., Cocucci, R. and De Michelis, M.I. (1975) Physiol. Veg., 13, 797–811.
Masuda, Y. and Wada, S. (1967) Bot. Mag. (Tokyo), 80, 100–102.
Masuda, Y., Yamamoto, R., Kawamura, H. and Yamagata, Y. (1974) Plant Cell Physiol., 15, 1083–1092.
Matchett, W.H. and Nance, J.F. (1962) Amer. J. Bot., 49, 311–319.
Matthysse, A.G. and Abrams, M. (1970) Biochim. biophys. Acta (Amst.), 199, 511–518.
Matthysse, A.G. and Phillips, C. (1969) Proc. nat. Acad. Sci. (Wash.), 63, 897–903.
Mehard, C.W., Lyons, J.M. and Kumamoto, J. (1970) J. Membrane Biol., 3, 173–179.
Melcher, U. and Varner, J.E. (1971) J. Inst. Brew. (Lond.), 77, 456–461.
Milborrow, B.V. (1974) Ann. Rev. Plant Physiol., 25, 259–307.
Moline, H.E., LaMotte, C.E., Gochnauer, C. and McNamer, A. (1972) Plant Physiol., 50, 655–659.
Mondal, M.H. (1975) Plant Physiol., 56, 622–625.
Mondal, H., Mandal, R.K. and Biswas, B.B. (1972) Nature New Biol., 240, 111–114.
Mondal, M.H. and Nance, J.F. (1975) Plant Physiol., 55, 450–454.
Montague, M.J. and Ikuma, H. (1975) Plant Physiol., 55, 1043–1047.
Morgan, H.E. and Whitfeld, C.F. (1973) Curr. Top. Membrane Transport, 4, 255–303.
Morré, D.J. (1968) Plant Physiol., 43, 1545–1559.
Mothes, K. and Engelbrecht, L. (1961) Phytochemistry, 1, 58–62.
Muller, K. and Leopold, A.C. (1966) Planta, 68, 167–185.
Mullins, M.G. (1970) Ann. Bot. (Lond.), 34, 897–909.
Murray, A.K. and Bandurski, R.S. (1975) Plant Physiol., 56, 143–147.
Musgrave, A., Kays, S.E. and Kende, H. (1969) Planta, 89, 165–177.
Musgrave, A., Kays, S.E. and Kende, H. (1972) Planta, 102, 1–10.
Nagai, R. and Rebhun, L.I. (1966) J. ultrastruct. Res., 14, 571–589.
Nakamura, T. and Takahashi, N. (1974) in: Plant Growth Substances 1973, Hirokawa, Tokyo, pp. 559–566.
Nance, J.F. (1973) Plant Physiol., 51, 312–317.
Nevins, D.J. (1970) Plant Physiol., 46, 458–462.
Nevins, D.J. (1975) Plant Cell Physiol., 16, 347–356.
Nissl, D. and Zenk, M.H. (1969) Planta, 89, 323–341.
Northcote, D.H. (1972) Ann. Rev. Plant Physiol., 23, 113–132.
Obata, T. and Suzuki, H. (1976) Plant Cell Physiol., 17, 63–71.
O'Brien, T.P. and Thimann, K.V. (1966) Proc. nat. Acad. Sci. (Wash.), 56, 888–894.
O'Malley, B.W. and Means, A.R. (1974) Science, 183, 610–620.
Oostrom, H., Van Loopik-Detmers, M.A. and Libbenga, K.R. (1975) FEBS Lett., 59, 194–197.
Osborne, D.J. (1958) J. exp. Bot., 9, 446–457.
Osborne, D.J., Ridge, I. and Sargent, J.A. (1972) in: Plant Growth Substances 1970, D.J. Carr (ed.), Springer, Berlin, pp. 534–542.
Palade, G. (1975) Science, 189, 347–358.
Paleg, L.G. (1960) Plant Physiol., 35, 293–299.
Paleg, L.G. and Hyde, B. (1964) Plant Physiol., 39, 673–680.
Patrick, J.W. (1976) in: Transport and Transport Processes in Plants, I.F. Wardlaw and J.B. Passioura (eds.), Academic, New York, pp. 433–446.
Patrick, J.W. and Wareing, P.F. (1972) in: Plant Growth Substances 1970, D.J. Carr (ed.), Springer, Berlin, pp. 695–700.
Penny, P. (1969) N.Z. J. Bot., 7, 290–301.
Penny, P. (1971) Plant Physiol., 48, 720–723.
Penot, M. and Hourmant, A. (1969) C.R. Acad. Sci. (Paris), 269, 1046–1409.
Penot, M. and Hourmant, A. (1974) in: Plant Growth Substances 1973, Hirokawa, Tokyo, pp. 737–751.
Pollard, C.J. and Singh, B.N. (1968) Biochem. biophys. Res. Commun., 33, 321–326.
Poovaiah, B.W. and Leopold, A.C. (1973a) Plant Physiol., 51, 848–851.

Poovaiah, B.W. and Leopold, A.C. (1973b) Plant Physiol., 52, 236–239.
Preston, R.D. (1974) The Physical Biology of Plant Cell Walls, Chapman and Hall, London, 491 pp.
Probine, M.C. (1964) Proc. roy. Soc. B, 161, 526–537.
Rains, D.W. (1969) Plant Physiol., 44, 547–554.
Rains, D.W. and Floyd, R.A. (1970) Plant Physiol., 46, 93–98.
Rasmussen, G.K. (1973) Plant Physiol., 51, 626–628.
Ratner, A., Goren, R. and Monselise, S.P. (1969) Plant Physiol., 44, 1717–1723.
Ray, P.M. (1962) Amer. J. Bot., 49, 928–939.
Ray, P.M. (1967) J. Cell Biol., 35, 659–674.
Ray, P.M. (1969) in: Communication in Development, A. Lang (ed.), Academic, New York, pp. 172–205.
Ray, P.M. (1973a) Plant Physiol., 51, 601–608.
Ray, P.M. (1973b) Plant Physiol., 51, 609–614.
Ray, P.M. and Baker, D.B. (1962) Nature (Lond.), 195, 1322.
Ray, P.M. and Ruesink, A.W. (1962) Dev. Biol., 4, 377–397.
Ray, P.M., Shininger, T.L. and Ray, M.M. (1969) Proc. nat. Acad. Sci. (Wash.), 64, 605–612.
Rayle, D.L. (1973) Planta, 114, 63–73.
Rayle, D.L. and Cleland, R. (1970) Plant Physiol., 46, 250–253.
Rayle, D.L. and Cleland, R. (1972) Planta, 104, 282–296.
Rayle, D.L., Haughton, P.M. and Cleland, R.E. (1970) Proc. nat. Acad. Sci. (Wash.), 67, 1814–1817.
Reed, N.R. and Bonner, B.A. (1974) Planta, 116, 173–185.
Reid, P.D., Strong, H.G., Lew, F. and Lewis, L.N. (1974) Plant Physiol., 53, 732–737.
Ridge, I. (1973) Acta bot. neerl., 22, 144–158.
Ridge, I. and Osborne, D.J. (1969) Nature (Lond.), 223, 318–319.
Ridge, I. and Osborne, D.J. (1970) J. exp. Bot., 21, 720–734.
Ridge, I. and Osborne, D.J. (1971) Nature New Biol., 229, 205–208.
Rijven, A.H.G.C. (1976) Aust. J. Plant Physiol., 3, 567–574.
Riov, J. (1974) Plant Physiol., 53, 312–316.
Risueno, M.C., Diez, J.L., Gimenez-Martin, G. and De Le Torre, C. (1971) Protoplasma, 73, 323–328.
Ruesink, A.W. (1969) Planta, 89, 95–107.
Sargent, J.A. and Osborne, D.J. (1975) Planta, 124, 199–205.
Sarkissian, I.V. (1970) Biochem. biophys. Res. Commun., 40, 1385–1390.
Sawhney, V.K. and Srivastava, L.M. (1974) Canad. J. Bot., 52, 259–264.
Schenk, E. (1974) Acta bot. neerl., 23, 739–746.
Seth, A. and Wareing, P.F. (1967) J. exp. Bot., 18, 67–77.
Shaner, D.L., Mertz, S.M., Jr. and Arntzen, C.J. (1975) Planta, 122, 79–90.
Shaw, M. and Manocha, M.S. (1965) Canad. J. Bot., 43, 747–756.
Shibaoka, H. (1974) Plant Cell Physiol., 15, 255–263.
Shibaoka, H. and Thimann, K.V. (1970) Plant Physiol., 46, 212–220.
Shore, G. and Maclachlan, G.A. (1973) Biochim. biophys. Acta (Amst.), 329, 271–282.
Shore, G. and Maclachlan, G.A. (1975) J. Cell Biol., 64, 557–571.
Shore, G., Raymond, Y. and Maclachlan, G.A. (1975) Plant Physiol., 56, 34–38.
Short, K.C., Tepfer, D.C. and Fosket, D.E. (1974) J. Cell Sci., 15, 75–87.
Singer, S.J. (1974) Ann. Rev. Biochem., 43, 805–833.
Sloane, M. and Sadava, D. (1975) Nature (Lond.), 257, 68.
Snyder, J.A. and McIntosh, J.R. (1976) Ann. Rev. Biochem., 45, 699–720.
Spencer, F.S., Shore, G., Ziola, B. and Maclachlan, G.A. (1972) Arch. Biochem., 152, 311–317.
Spencer, F.S., Ziola, B. and Maclachlan, G.A. (1971) Canad. J. Biochem., 49, 1326–1332.
Srivastava, L.M., Sawhney, V.K. and Taylor, I.E.P. (1975) Proc. nat. Acad. Sci. (Wash.), 72, 1107–1111.
Stoddart, J., Breidenbach, W., Nadeau, R. and Rappaport, L. (1974) Proc. nat. Acad. Sci. (Wash.), 71, 3255–3259.
Stowe, B.B. and Dotts, M.A. (1971) Plant Physiol., 48, 559–563.

Sugiura, M. (1963) Bot. Mag. (Tokyo), 76, 359–362.
Sweeney, B.M. and Thimann, K.V. (1937) J. gen. Physiol., 21, 439–461.
Sweeney, B.M. and Thimann, K.V. (1942) J. gen. Physiol., 25, 841–854.
Taiz, L. and Honigman, W.A. (1976) Plant Physiol., 58, 380–386.
Taiz, L. and Jones, R.L. (1970) Planta, 92, 73–84.
Taiz, L. and Jones, R.L. (1973) Amer. J. Bot., 60, 67–75.
Talmadge, K.W., Keegstra, K., Bauer, W.D. and Albersheim, P. (1973) Plant Physiol., 51, 158–173.
Tanada, T. (1973) Plant Physiol., 51, 150–153.
Tanimoto, E. and Masuda, Y. (1968) Physiol. Plant., 21, 820–826.
Tata, J.R. (1970) in: Biochemical Actions of Hormones, G. Litwack (ed.), Academic, New York, pp. 89–133.
Tata, J.R. (1971) Sub-Cell Biochem., 1, 83–89.
Tautvydas, K.J. and Galston, A.W. (1972) in: Plant Growth Substances 1970 D.J. Carr (ed.), Springer, Berlin, pp. 256–264.
Tews, J.K., Woodcock, N.A. and Harper, A.E. (1970) J. biol. Chem., 245, 3026–3032.
Thimann, K.V. (1969) in: Plant Physiology: A Treatise, Vol. 6B, F.C. Steward (ed.), Academic, New York, pp. 237–272.
Travis, R.L., Anderson, J.M. and Key, J.L. (1973) Plant Physiol., 52, 608–612.
Trewavas, A.J. (1968) Phytochemistry, 7, 678–681.
Valdovinos, J.G., Jensen, T.E. and Sicko, L.M. (1971) Plant Physiol., 47, 162–163.
Valdovinos, J.G., Jensen, T.E. and Sicko, L.M. (1972) Planta, 102, 324–333.
Valent, B.S. and Albersheim, P. (1974) Plant Physiol., 54, 105–108.
Van der Eb, A.A. and Nieuwdorp, P.J. (1967) Acta bot. neerl., 15, 690–699.
Vanderhoef, L.N. and Stahl, C.A. (1975) Proc. nat. Acad. Sci. (Wash.), 72, 1823–1825.
VanDerWoude, W.J., Lembi, C.A. and Morré, D.J. (1972) Biochem. biophys. Res. Commun., 46, 245–253.
Varner, J.E. (1964) Plant Physiol., 39, 413–415.
Varner, J.E. and Mense, R. (1972) Plant Physiol., 49, 187–189.
Varner, J.E., Chandra, G.R. and Chrispeels, M.J. (1965) J. cell. comp. Physiol., 66, Suppl. 1, 55–68.
Varty, K. and Laidman, D.L. (1976) J. exp. Bot., 27, 748–758.
Veen, B.W. (1970) Proc. kon. ned. Akad. Wet., 73, 118–121.
Veen, H. (1966) Acta bot. neerl., 15, 419–433.
Venis, M.A. (1971) Proc. nat. Acad. Sci. (Wash.), 68, 1824–1827.
Verma, D.P.S., Maclachlan, G.A., Byrne, H. and Ewings, D. (1975) J. biol. Chem., 250, 1019–1026.
Victor, T.S. and Vanderhoef, L.N. (1975) Plant Physiol., 56, 845–846.
Vigil, E.L. and Ruddat, M. (1973) Plant Physiol., 51, 549–558.
Villiers, T.A. (1968) Planta, 82, 342–354.
Wada, S., Tanimoto, E. and Masuda, Y. (1968) Plant Cell Physiol., 9, 369–376.
Wardlaw, I.F. (1974) Ann. Rev. Plant Physiol., 25, 515–539.
Wardlaw, I.F. and Moncur, L. (1976) Planta, 128, 93–100.
Webster, B.D. (1968) Plant Physiol., 43, 1512–1544.
Wood, A. and Paleg, L.G. (1972) Plant Physiol., 50, 103–108.
Wood, A. and Paleg, L.G. (1974) Aust. J. Plant Physiol., 1, 31–40.
Wood, A., Paleg, L.G. and Sawhney, R. (1972) in: Plant Growth Substances 1970, D.J. Carr (ed.), Springer, Berlin, pp. 37–43.
Wooding, F.B.P. (1967) Protoplasma, 64, 315–324.
Yamagata, Y., Yamamoto, R. and Masuda, Y. (1974) Plant Cell Physiol., 15, 833–841.
Yamamoto, R., Maki, K., Yamagata, Y. and Masuda, Y. (1974) Plant Cell Physiol., 15, 823–831.
Yomo, H. (1960) Hakko Kyokaishi, 18, 600–602.
Yomo, H. (1961) Hakko Kyokaishi, 19, 284–285.
Zwar, J.A. and Jacobsen, J.V. (1972) Plant Physiol., 49, 1000–1006.

Phytohormones and Related Compounds – A Comprehensive Treatise, Volume I
Editors: Letham, Goodwin and Higgins

9

The Influence of Plant Hormones on Selected Aspects of Cellular Metabolism

T.J.V. HIGGINS and J.V. JACOBSEN

9.1. Introduction

Interest in the involvement of plant hormones in general cellular metabolism has fluctuated considerably over the past forty years. As more information became available on the various metabolic cycles and on organelles and their associated metabolic activity, there usually followed an intensive examination of the role plant hormones might play in regulating some aspect of metabolism in that system. There has been relatively little research activity in metabolic studies in recent years. It has been largely replaced by the current interest in aspects of transcription and translation and the synthesis of other macromolecules, such as those involved in cell walls.

A great variety of responses to phytohormones has been recorded, many of which are part of highly complex developmental processes, often requiring many days before manifestation. They are thus often poorly studied because of the difficulty of assigning a direct role to the hormone. A prevailing view of hormone action is that a single event or target is affected and that all measurable responses are derived from this initial interaction. This has led to an approach to hormone research dominated by short-term studies attempting to demonstrate mechanism of action. In the event that a phytohormone acts in a sequential manner in which the hormone is required for each step in the sequence, and each step is dependent upon those preceding, it becomes important to study hormone action in long-term complex developmental situations. Such opportunities are provided by the processes of photosynthesis, respiration and nitrogen metabolism.

Thus, we have attempted to extract from the literature some of the data obtained in studies of photosynthesis and associated carbohydrate metabolism, respiration, lipid and nitrogen metabolism. The relevant aspects of ripening and senescence will be covered in Volume II and have not been examined here, except to the extent that they complement the existing presentation. We have been particularly interested in obtaining evidence for a role for hormones in key reactions in each of the major metabolic systems, and we have examined a wide range of studies which might point in this direction. However, the majority of the research to date is descriptive, and it is generally not yet possible to categorically assign a definitive mediatory role to hormones in any individual process.

9.2. Photosynthesis and related carbohydrate metabolism

The following are photosynthetic processes which may be controlled by hormones:

(a) differentiation and maintenance of chloroplast membrane structures;

(b) development of the chloroplast enzyme complement and particularly the regulation of key enzymes, such as those involved in chlorophyll formation as well as ribulose diphosphate carboxylase (RuDP carboxylase) and the associated oxygenase;

(c) transport of chloroplast components including enzymes or enzyme subunits, membrane subunits and small metabolites, to and from the cytoplasm;

(d) electron flow and photophosphorylation;

(e) export of photosynthate from mesophyll cells into phloem;

(f) sucrose formation or any of the steps leading to the formation of 6-carbon sugars;

(g) the relative contribution to total photosynthesis made by C_3/C_4 pathways in relevant species which could be controlled by key enzymes, or by the intracellular transport of metabolites.

In the following sections, specific examples of modifications to photosynthesis by plant hormones are mentioned and are arbitrarily divided into effects on photochemical and enzymatic processes (both sugar synthesis and hydrolysis) and effects on organelle development. Some of these are, of course, interdependent and in some studies more than one aspect was examined.

9.2.1. Auxin

Assimilation of CO_2 has been shown to be stimulated rapidly by auxin application to leaves (Turner and Bidwell 1965; Bidwell and Turner 1966), to cells isolated from leaves (Kulandaivelu and Gnanam 1975) and to isolated chloroplasts (Tamas et al. 1972). It was initially proposed, without evidence (Turner and Bidwell 1965), that the hormone affected enzymatic processes rather than photochemical events, but subsequent analysis using isolated chloroplasts and chloroplast fragments showed that indoleacetic acid (IAA) (10^{-7} *M*) stimulated photophosphorylation within 5 min (Tamas et al. 1972), and that coupling of electron transport to ATP formation was enhanced, as measured by ATP/O_2 ratios (Tamas et al. 1974). Thus, some aspect of the photochemical process was stimulated and, although an effect on enzymatic processes was not ruled out, the photochemical response could account for the enhanced CO_2 fixation. It is possible that the efficiency of the energy-capturing process was increased by a modification of the organization of the chloroplast membrane system, although there is no data available which points to the primary action of the hormone. Specific effects on the reduction/oxidation of pyridine nucleotide have not been ruled out, and there is no definitive study on the effect of the hormone on early individual events of reducing power production or ATP synthesis.

Some support for the notion that IAA was not enhancing photosynthesis through an anti-oxidizing effect, as proposed by Ray (1974), is obtained from the data of Kulandaivelu and Gnanam (1975). They noted that CO_2 fixation by isolated leaf cells was enhanced by both IAA and 2,4-D. The synthetic auxin (2,4-D) does not have the anti-oxidizing capacity of IAA, and it is therefore possible that auxin stimulated CO_2 assimilation by means other than by its anti-oxidant properties. On the other hand, there are some disturbing features of this isolated cell and cell-free work in that photophosphorylation was uncoupled from ATP synthesis, which probably accounted for the observed enhanced electron transport. It may be possible to explain these anomalous results by the fact that high concentrations (10^{-4} *M*) of auxin were used in the study of the partial reactions.

There are other instances of the use of supraoptimal levels of auxin in the study of aspects of photosynthesis. One of these relates to the early suggestion of Turner and Bidwell (1965) that auxin might increase the activity of enzymes involved in CO_2 fixation. However, the opposite proved to be the case, since Feierabend (1970) showed that auxin at very high levels inhibited the development of carboxydismutase in light-grown rye *(Secale cereale)* seedlings, which was probably related to the inhibition of chloroplast development caused by the hormone. Neither the inhibition of the development of plastids nor the destruction of mature chloroplast membrane systems, e.g. in bean *(Phaseolus vulgaris)* chloroplasts (Hallam 1970), is understood at the molecular level, although the mechanism is of interest especially for the understanding of herbicidal action.

Development of the chloroplast as affected by exogenous auxin has been studied most intensively under tissue culture conditions. It has been found that growth-promoting concentrations of the hormone inhibit organelle development (Sunderland and Wells 1968; El Hinnaway 1974; Kaminek and Lustinec 1974), yet promote replication of mature plastids under cell expansion conditions (Boasson et al. 1972). These development-related differences in organelle biogenesis and proliferation as affected by auxin are not understood.

9.2.2. Gibberellin

Photosynthetic responses to gibberellin (GA) range from inhibition to stimulation of CO_2 assimilation and, in general, detailed studies have concentrated on enzyme activity, not on photochemical events as in auxin research. There are many long-term studies in which the response to GA has been measured, and it is difficult to assess the possible role of the hormone in regulating gas exchange, energy capture, production of reducing power or ATP synthesis.

The assimilation of CO_2 by leaves treated with GA_3 was unaffected in oat *(Avena sativa)* and pea *(Pisum sativum)* (Haber and Tolbert 1957), promoted in tomato *(Lycopersicon esculentum)*, maize *(Zea mays)*, red clover *(Trifolium praetense)* and bean (Coulombe and Paquin 1959; Treharne and Stoddart 1968; Wareing et al. 1968b; Treharne et al. 1970; Marcelle et al. 1974), and inhibited in *Pennisetum typhoides*, a plant utilizing the C_4 pathway of photosynthesis (Huber and Sankhla 1974a). It appears that the response to the hormone is complicated by a dependence upon the developmental stage of the tissue, e.g. bean (Treharne et al. 1970), or upon the environmental conditions, e.g. wheat *(Triticum aestivum)* (Treharne et al. 1972; Marcelle et al. 1974), thus making it difficult to interpret the significance of reports of promotion or inhibition of photosynthesis, if a variety of conditions and stages of tissue maturity were not employed.

The only enzyme which has been studied extensively in relation to regulation of photosynthesis by GA is RuDP carboxylase, but it is noteworthy that, in those cases presented above in which the enzyme was monitored, there was a parallel increase or decrease in the enzyme activity with photosynthesis (Treharne and Stoddart 1968; Treharne et al. 1970 and 1972; Huber and Sankhla 1974a; Marcelle et al. 1974). The activities of the C_4 enzymes, phosphoenolpyruvate carboxylase (PEP carboxylase) and malic enzyme were also inhibited during GA_3-induced inhibition of photosynthesis in *Pennisetum* (Huber and Sankhla 1974a). Furthermore, the pattern of the various photosynthetic products was altered suggesting that there were selective effects on certain enzymes; this potentially interesting finding does not appear to have been studied further. The question of specificity of the response to the hormone has rarely been resolved satisfactorily and it is disturbing to note that, when examined, non-chloroplast enzymes, such as acid phosphatase and glutamate-oxaloacetic transaminase, have been found to respond to GA_3 with the same kinetics as RuDP carboxylase (Treharne and Stoddart 1968). This may suggest an overall effect upon metabolism and growth without a specific effect on photosynthesis.

A similar diversity of responses to GA is evident in studies on chloroplast development. GA_3 inhibited the development of chloroplasts in tissue culture (Stetler and Laetsch 1965), but promoted organelle development and increased the activity of RuDP carboxylase in greening cereal leaves (Wellburn et al. 1973). Similarly, chlorophyll synthesis and degradation were unaffected in young wheat seedlings (Wolf and Haber 1960) while, in senescing tissues. GA_3 stimulated the disappearance of chlorophyll *a* in bean leaves (Szalai 1968), but maintained chlorophyll levels in *Rumex* leaves (Goldthwaite and Laetsch 1968). From these data it is impossible to draw conclusions which would support a unified hypothesis on GA-induced alterations in photosynthesis.

A direct role for gibberellin in the metabolism of starch and sucrose is well-documented. Starch hydrolysis in the endosperm of many Gramineae is thought to be controlled by gibberellin produced by the embryo and transported to the aleurone layer, where it induces the synthesis of a range of hydrolytic enzymes, including α-amylase (for review see Yomo and Varner 1971). The amylase is released from the aleurone layer into the starchy endosperm. The mechanism of action of GA_3 in controlling hydrolase synthesis, especially in barley *(Hordeum vulgare)* aleurone layers, has been extensively studied (see Chapter 10, section 10.3.2.4), but the primary response to the hormone has not been determined. Other instances of the promotion of starch catabolism by GA_3 have been described for wheat coleoptiles (Boothby and Wright 1962), bean hypocotyls (Clum 1967), tobacco *(Nicotiana tabacum)* leaves (Lee and Rosa 1969) and maize shoots (Katsumi and Fukahara 1969).

Invertase activity has been enhanced in GA_3-treated artichoke *(Helianthus tuberosus)* slices, sugar cane *(Saccharum officinarum)* tissue slices, oat internodes, etiolated lentil *(Lens culinaris)* epicotyls, and tobacco leaves (Edelman and Hall 1964; Gayler and Glasziou 1968; Kaufman et al. 1968; Seitz and Lang 1968; Lee and Rosa 1969). Many other examples are included in Table 9.1, but this short list serves to indicate that sucrose metabolism is also probably under GA control in a range of species and tissue types.

It is evident that gibberellin can regulate various aspects of photosynthesis including chloroplast development, activities of photosynthetic enzymes, starch hydrolysis and sucrose degradation, and it is known that gibberellin is produced in the chloroplast

(Stoddart 1968). However, there is no evidence for a single primary response in all tissues, nor is there good data on primary effects in any of the above processes, with the possible exception of α-amylase induction in barley aleurone layers (see Jacobsen 1977 for review).

9.2.3. Cytokinin

The slowing of senescence in many tissues by cytokinin has led to an extensive literature on the hormone-dependent preservation of various metabolic processes including photosynthesis (see Chapter 7, Volume II for details). Thus, the mechanism by which active growth is maintained has merited attention and the enhancement by cytokinin of the rate of photosynthesis in maize (Wareing et al. 1968b) and the maintenance of CO_2 assimilation by bean leaves (Adedipe et al. 1971) may help explain senescence processes. It was suggested that decreased mesophyll resistance to CO_2 diffusion was the cause of increased CO_2 fixation in bean (Fletcher and Adedipe 1972), but many other possibilities were not considered, such as increased levels of chlorophyll (Sugiura 1963), increased photosynthetic enzyme activity (Feierabend 1969; Harvey et al. 1974) or altered rates of photochemical events, for which there appears to be little published data except a preliminary report suggesting that kinetin partially prevented the decay of Hill activity in isolated spinach *(Spinacea oleracea)* chloroplasts (Banerji and Kumar 1975). The rate of chlorophyll synthesis or degradation is changed by cytokinins, depending upon the physiological state of the tissue. Early reports showed that in either senescing or detached leaves, cytokinin retarded chlorophyll degradation (Richmond and Lang 1957; Srivastava 1963; Fletcher 1969). But in young tissues, such as the developing cotyledons of *Cucumis sativus* and etiolated barley leaf, as well as tobacco callus cells and senescing *Nicotiana rustica* leaves, cytokinin was shown to increase chlorophyll levels (Kursanov et al. 1964; Fletcher and McCullagh 1971; Stobart et al. 1972; Kaul and Sabharwal 1974). The specific activity of the enzyme δ-aminolevulinic acid dehydrase, which catalyzes the formation of porphobilinogen from two molecules of δ-aminolevulinic acid, has been shown to be increased by kinetin in tobacco callus cells, suggesting a possible control over chlorophyll synthesis in this system (Kaul and Sabharwal 1974). While certain aspects of the pathway of chlorophyll biosynthesis are still controversial, there is a very thorough understanding of most of the critical steps (Rebeiz and Castelfranco 1973), and it is suggested that the cytokinin control of chlorophyll biosynthesis could be elucidated at the enzymological level using our current knowledge.

The activity of the photosynthetic enzymes RuDP carboxylase and NADP-glyceraldehyde-3-phosphate dehydrogenase was increased by cytokinin in rye seedlings and cucumber *(Cucumis sativus)* cotyledons (Feierabend 1969; Harvey et al. 1974). Although the hormone appeared to stimulate protein accumulation and non-photosynthetic enzyme activity in primary leaves, there was evidence that the developmental stage and the type of tissue was important, and that photosynthetic enzymes and those enzymes of the microbody most closely associated with photosynthesis, glycolate oxidase and hydroxypyruvate reductase, were selectively stimulated (Feierabend 1969; De Boer and Feierabend 1974). It is not clear whether synthesis, activation, or inhibition of degradation of the enzymes was the cause of enhanced activity, because the only available data is based upon equivocal inhibitor studies (e.g. Harvey et al. 1974), but it is evident that kinetin inhibited degradation of RuDP carboxylase in senescing barley leaves (Peterson and Huffaker 1975).

Chloroplast development was promoted in senescing *N. rustica* leaves and in tobacco tissue culture cells (Kursanov et al. 1964; Stetler and Laetsch 1965; Sveshnikova et al. 1966), but it was not affected in the young leaves of *Canavalia ensiformis* (Alberte and Naylor 1975). Thus, again it is obvious that the developmental state of the tissue determines the response to added hormone and may in part be a reflection of endogenous levels of cytokinins. The key events in chloroplast formation and maintenance which appear to be sensitive to cytokinins are not known.

Starch catabolism was promoted by kinetin in wheat endosperm (Boothby and Wright 1962), although GA_3 was vastly more effective. However, the reverse was the case in the control of α-amylase activity in bean hypocotyl. Clum (1967) showed that kinetin stimulated amylase activity, while GA_3 was relatively ineffective. Invertase activity was increased by kinetin in chicory *(Chicorium intybus)* roots (Rutherford and Bard 1971).

Thus, there is evidence that cytokinin is involved in chloroplast development or maintenance, in enhancement of the activity of some chloroplast and microbody enzymes and in control of carbohydrate metabolism. This variety of responses may reflect action of the hormone at a number of separate sites, or it may be the result of one or two key regulatory responses, such as maintenance of chloroplast membrane structure and/or control of mRNA levels for a key enzyme such as RuDP carboxylase or α-amylase. Further analyses of the separate processes controlling photosynthesis which might be regulated by cytokinin are required in order to determine the mechanism of action of the hormone in this major event in plant growth.

9.2.4. Abscisic acid

Abscisic acid (ABA) and the closely related compound phaseic acid are known to inhibit (Mittelheuser and Van Steveninck 1971; Poskuta et al.. 1972; Sankhla and Huber 1974 and 1975; Kriedemann et al. 1975 and 1976; Bauer et al. 1976; McLaren and Smith 1976) and promote (Hoffman and Kull 1974; McLaren and Smith 1976) the rate of photosynthesis in a range of plant species. The variation between studies again probably reflects differences in the stage of development of the experimental tissue, the time course of the experiments and other features such as concentration of the hormones. ABA very probably affects the diffusive resistance of stomata to CO_2. It was first shown by Mittelheuser and Van Steveninck (1969) that low concentrations of ABA closed stomata of wheat and barley leaves, and it was later shown to decrease CO_2 assimilation (Mittelheuser and Van Steveninck 1971). However, considerably higher levels (up to 1000-fold) of ABA were required to inhibit photosynthesis than were needed to close stomata (Poskuta et al. 1972). Non-stomatal inhibition of photosynthesis has been demonstrated in *Vitis vinifera* leaves and *Spinacea oleracea* leaf slices (Kriedemann et al. 1975 and 1976), and, furthermore, in *Lemna minor,* which contains open stomata only, ABA promoted CO_2 fixation for the first two days of the experiment (McLaren and Smith 1976), and caused a gradual decline thereafter (Bauer et al. 1976; McLaren and Smith 1976). Isolated protoplasts of *Petunia* are also free of stomata and, in short-term experiments, ABA promoted CO_2 fixation (Hoffman and Kull 1974). Thus, there is the possibility that ABA can regulate photosynthesis by controlling the supply of substrate CO_2, but there is good evidence that it can also act at some other level, e.g. on enzyme activity. Changes in the activity of photosynthetic enzymes in response to ABA have been reported. The activity of RuDP carboxylase was lowered in the leaves of cereals and *Pennisetum* (Wellburn et al. 1973; Sankhla and Huber 1974 and 1975) but, on the other hand, RuDP carboxylase activity was unaffected by phaseic acid in *Vitis vinifera* leaves, or by ABA in *Lemna minor* (Kriedemann et al. 1975; Bauer et al. 1976). The activity of several photosynthetic enzymes was increased by ABA in *Lemna* (PEP carboxylase, NAD-malate dehydrogenase, NADP-specific malic enzyme and NADP-malate dehydrogenase), and the activity of PEP carboxylase was increased in wheat leaves (Bauer et al. 1976). Thus, there is no simple relation between reduced rates of photosynthesis and the activity of photosynthetic enzymes. However, there is the potentially interesting observation that, in those cases where CO_2 fixation was reduced in response to ABA either by reduced RuDP carboxylase activity as in wheat (Sankhla and Huber 1974) or by some other means as in *Lemna minor* (Bauer et al. 1976), there was a slight compensatory increase in the activity of PEP carboxylase, an enzyme typical of the C_4 pathway of CO_2 fixation. This enzyme has a higher affinity for CO_2 than does RuDP carboxylase, and it is possible that the hormone controlled the level of the gas, thus permitting utilization of PEP carboxylase. However, the fact that the stomata of *Lemna* do not respond to ABA (McLaren and Smith 1976) makes it unlikely that CO_2 is limiting in this system, and it is therefore worth exploring the possibility that ABA may be able to regulate the ratio of C_3/C_4 photosynthesis.

Photochemical events may also be regulated by phaseic acid or ABA. Kriedemann et al. (1975) proposed that phaseic acid inhibited photosynthesis by lowering the rate of electron flow, while Bauer et al. (1976) suggested that ABA altered photosystem II activity, since the Hill reaction was inhibited by 50% by ABA. Thus, these preliminary observations on the effects of phaseic acid and ABA on the reduced capacity for electron flow during photosynthesis should be explored in some detail.

Effects on chloroplast development and stability are also possible ways in which ABA might regulate overall levels of photosynthesis, and, in seedlings of wheat and barley, Wellburn et al. (1973) showed that ABA inhibited plastid development both in vivo and in vitro during the early hours of greening. No adverse effects of ABA on the mature chloroplasts of *Lemna minor* were noted by electron microscopy (Bauer et al. 1976). However, some changes must have occurred, since the ratio of chlorophyll *a:b* was increased from 3.2 to 7.2, carotenoid levels were increased by 40% and starch grain size was increased (McLaren and Smith 1976), any or all of which might contribute to changes in the structural organization of the organelle, and therefore to changes in the rate of photosynthesis.

Starch hydrolysis in cereal endosperm, which is regulated in part by GA-induced α-amylase synthesis, is antagonized by ABA (Thomas et al. 1965; Chrispeels and Varner 1966). Since the way in which GA increases the level of α-amylase is not known, any suggestion as to how ABA antagonizes this response is speculation. It has been suggested

(Ho and Varner 1976) that ABA may inhibit the translation of mRNA for α-amylase, although there is some evidence that ABA may regulate the GA_3-induced increase in the level of mRNA for the enzyme (Higgins et al. 1976). Sucrose metabolism may be altered by ABA in carrot *(Daucus carota)* root tissue, sugar cane stalks and sugar beet *(Beta vulgaris)* roots through an effect on invertase levels or activity. It was shown that ABA could raise or lower the level of invertase activity in carrot, depending upon the hormone concentration (Saunders and Poulson 1968), and the level of invertase activity was increased slightly in sugar beet and sugar cane (Cherry 1968; Gayler and Glasziou 1969). Although it was suggested that ABA may have allosteric effects on invertase based upon the reaction of isolated fungal enzyme to the hormone (Saunders and Poulson 1968), this was not confirmed in the case of the higher plant enzyme (Gayler and Glasziou 1969), and it seems likely that the increased enzyme activity in sugar beet and sugar cane was due to increased synthesis of enzyme. No further analysis of the response has emerged, and it is not clear how enzyme synthesis was affected by the hormone.

As for the other hormones, ABA appears to be involved in a number of aspects of photosynthesis, including the development of plastids, photosynthetic enzyme activity, photochemical events and starch and sucrose hydrolysis. Not all processes are affected in all cases, and the stage of development of the cell seems to be an important factor in determining the nature of the response. In no case has the primary target of ABA in regulating photosynthesis been defined.

9.2.5. Ethylene

Ethylene is generally associated with ripening of fruit, senescence and abscission of fruit and leaves (Abeles 1973), and in this context has been shown to be involved in chloroplast disintegration and the disruption of photosynthesis. The chloroplast has been shown to be a possible site of ethylene production (Elstner and Konze 1974), and Alscher and Castelfranco (1972) observed that the gas stimulated chlorophyll synthesis in seedlings of *Cucumis sativus.* However, the precise role of the gas in development of the organelle or in the photochemical or biochemical reactions of photosynthesis has not been elucidated.

Invertase levels in a variety of tissues are increased by ethylene (see Abeles 1972 for review), but there is no information on how these levels were controlled. The level of α-amylase in barley aleurone was increased to a small extent by ethylene (Jacobsen 1972), but the response has not been further characterized.

9.2.6. Conclusions

There is ample evidence to suggest that the plant hormones regulate various aspects of photosynthesis and starch and sugar metabolism. There is now enough descriptive information available to enable a choice of system for further biochemical analysis along the lines suggested in section 9.2 of this chapter. So far, there has been no systematic analysis of the sequence of events which precede an observed effect on the photochemical or enzymatic processes of photosynthesis.

9.3. Influence of plant hormones on other pigment systems

The carotenoids are important pigments because of their proposed dual role as accessory light harvesters and protectants against photooxidation in chloroplasts. Their biosynthesis in many plants is considered to be controlled by another pigment, phytochrome (Cohen and Goodwin 1962), and there is some evidence to suggest that hormones may interact with phytochrome in controlling the level of carotenoids in peas (Kang and Burg 1972). In general GA_3, kinetin and ethylene have been shown to lower the final levels of carotenoids in different tissues of different species (Dostal and Leopold 1967; Khudairi and Arboleda 1971; Kang and Burg 1972), while ABA enhanced the level of carotenoids in developing tomato fruit both in the light and in darkness (Khudairi and Arboleda 1971). Thus, there is not a simple relationship between hormonal and phytochrome control of the level of these pigments in plant cells, nor is there any data on the relevant biosynthetic enzymes which might be regulated by phytohormones.

Another pigment system which may be under the control of phytochrome and/or phytohormones is the flavonoid complex. It is considered here mainly because of the great interest in the enzyme phenylalanine ammonialyase (PAL), which occupies an important position as a potential regulator of phenolic biosynthesis in higher plants (Zucker 1972; Camm and Towers 1973; Stafford 1974). Flavonoid pigments such as anthocyanins and betacyanins have been used as genetic, phylogenetic and taxonomic markers for a long time (see Harborne 1967), and details of hormonal interaction with them are of interest from this point of view. Furthermore, flavonoid conjugates themselves have been shown to possess some growth-regulatory activity (Galston 1969) and GA_3 inhibited the red light-induced increase in one of these conjugates, kaempferol triglucoside, in internode tissue of peas (Russell and Galston 1969). See also Chapter 7, section 7.7.

Anthocyanin levels were increased by cytokinin in *Impatiens balsamina* flowers (Klein and Hagen 1961) and in green olive *(Olea europaea)* fruit, while gibberellin, auxin and ethylene all appeared to lower the levels or to delay the appearance of anthocyanins in a range of tissues including cell cultures, hypocotyls and stem internodes (Vince 1968; Constabel et al. 1971; Craker et al. 1971; Kang and Burg 1973; Gregor 1974). Complicated responses to ethylene have been noted (Craker 1975) in which anthocyanin levels were raised if the gas was present during the lag phase before pigment appearance, but lowered if the gas was present during the phase of rapid pigment accumulation. The response to auxin in *Sorghum* and *Haplopappus* seems to be one of retardation of pigment appearance rather than complete inhibition of production (Vince 1968; Constabel et al. 1971).

The levels of betacyanins also appear to be raised by cytokinin in many tissues (Piatelli et al. 1971; Koehler 1972b; Kinsman et al. 1975; Singh and Chhabra 1976). Variable responses to the other hormones have been reported, ranging from no effect of auxin in *Amaranthus tricolor* and of GA_3 in *Celosia argenta* (Stobart et al. 1970; Singh and Chhabra 1976) to inhibition by auxin, gibberellin and ABA in other systems (Stobart et al. 1970; Kinsman et al. 1975) and promotion of betacyanin levels in *Celosia argenta* by auxin, ethylene, ABA and, as mentioned above, cytokinin (Singh and Chhabra 1976).

Most of this work is still in a descriptive phase and has not progressed past the stage where RNA and protein synthesis are being implicated in anthocyanin (Craker 1975) and betacyanin synthesis (Piatelli et al. 1971; Koehler 1972a). Recent progress in the elucidation of the pathway of flavonoid biosynthesis (Hahlbrock et al. 1971; Kreuzaler and Hahlbrock 1972; Grisebach and Hahlbrock 1974) should greatly aid future studies on the mechanism of hormone-induced or -repressed pigment levels.

A role for PAL in flavonoid biosynthesis, while still somewhat circumstantial, is becoming increasingly well-established (Zucker 1972; Camm and Towers 1973; Grisebach and Hahlbrock 1974). A major difficulty in accommodating all of the data on PAL and a group of other enzymes involved in phenylpropanoid metabolism is that they reach maximum activity long before a second group of enzymes involved in the formation of flavone glycosides (Hahlbrock et al. 1971). This, together with the fact that phenylpropanoids are precursors for lignins as well makes it very unlikely that PAL is the only regulatory enzyme in flavonoid biosynthesis. Given that the above limitation is accepted, there is extensive evidence that PAL levels are regulated by phytochrome (Smith and Attridge 1970) and by phytohormones (see also Tables 9.1 and 9.2). Inhibition of anthocyanin production by GA_3 in cell cultures of *Haplopappus gracilis* was correlated with reduced levels of PAL (Gregor 1974) but the retardation of pigment production by auxin in the same species was difficult to correlate with enzyme levels because of the 6–8-day lag in the appearance of the pigment, especially since the final level of the pigment was not altered by auxin (Constabel et al. 1971). Similarly, although ethylene stimulated both PAL activity and the level of isochlorogenic acid in carrot root tissue (Sarkar and Phan 1974a and b), there need not be a causal relationship between the two events.

The production of flavonoid pigments is complicated in that hormones and phytochrome may interact in controlling overall levels, while the enzymatic processes have only recently begun to be clarified, and the role of the best-characterized enzyme (PAL) as a regulator of pigment level is in some doubt. Nevertheless, there is now considerable scope for further work on the mechanism of hormone action in controlling these pigments. The large body of data on chemical characterization, genetics and phylogenetics of these compounds provides further incentive for detailed investigation of the regulation of flavonoid biosynthesis.

9.4. Respiration

The vital process of respiration in plants has long been known to be similar in most respects to that described for animals and microorganisms (Beevers 1961), but some of the unique regulatory mechanisms operating at the level of enzymes, substrates and energy production have yet to be elucidated in detail (Bonner 1973). Therefore, it becomes extremely difficult to interpret the fairly extensive data available on the effects of phytohormones on respiration, particularly when one considers that many of the studies have been long-term and that the variety of tissue types and developmental stages has been enormous, ranging from germinating seed to excised hypocotyl. Mitochondria have been studied in vitro in the absence of many of the presumed regulatory influences of the cytoplasm and other organelles, and even isolated enzymes have been examined for an in vitro response to phytohormones.

Taking a simplified view, one can envisage that hormones might regulate respiration by:

(a) a general maintenance of cell activity with respect to membrane, nucleic acid and protein biosynthesis, which in turn might ensure stability of relevant organelles, especially mitochondria and the glyoxysomes or equivalent microbodies in non-storage tissues;

(b) altering the level of activity of enzymes involved in one of the cycles of carbohydrate, fat or protein metabolism. A rate-limiting enzyme in a particular pathway should be the first choice for study;

(c) switching from the metabolism of one substrate (e.g. carbohydrate) to another (e.g. fat), a process which would also involve regulation of enzyme level or activity;

(d) regulating the rate of electron flow, in the case of aerobic respiration, and by altering the coupling between ATP formation and substrate oxidation;

(e) regulating the path of electron flow, e.g. diverting electrons from the normal cytochrome a,b,c system into the cyanide insensitive pathway.

These are possible points of control and there are, no doubt, many more sensitive controls which may operate but which may be difficult to monitor using current knowledge and techniques.

9.4.1. Auxin

Since Bonner (1933) first described a 20% increase in the respiration rate of *Avena* coleoptile sections in response to an auxin, there have been extensive studies on the effects of auxin on respiration in a variety of tissues. Early difficulties, related to reproducibility and magnitude of the response, have largely been resolved by attention to hormone concentration (root tissue required very low (10^{-10} *M*) levels to stimulate respiration (Audus and Garrard 1953)), and the use of defined tissue types. A causal relationship between enhanced growth and enhanced respiration was shown to be unlikely when non-growing tissues also demonstrated enhanced respiration (Michel 1951; Anker 1953), and subsequently the use of a range of metabolic inhibitors indicated that respiration and growth were not linked in a simple manner. Excellent reviews of the early work in this field have been presented by Cleland (1961) and Thimann (1972).

As discussed in Chapter 8, there is extensive evidence that auxin maintains the cellular structure of young and maturing tissues, and one of the ways in which this may be achieved is through maintenance of controlled substrate metabolism and the preservation of functional organelles. Even the protein synthesizing capacity of isolated mitochondria (from etiolated *Vigna sinensis)* was slightly stimulated by auxin (Bhattacharyya and Roy 1969), suggesting that structural or enzymatic protein of the organelle might be affected by the hormone. Mitochondria isolated from auxin-treated soybean *(Glycine max)* exhibited enhanced size and protein synthesizing capacity (Key et al. 1960; Baxter and Hanson 1968). However, there is no firm data on the control of the processes which regulate the structural and functional integrity of mitochondria, as is the case for other subcellular structures.

There are conflicting reports on whether auxin might selectively regulate a particular pathway of carbohydrate degradation or even a change from one predominant substrate (carbohydrate) to another (fat). Respiratory quotients (CO_2 evolution/O_2 consumption) were shown to be less than unity (the control value) in auxin-treated excised pea and kidney bean segments (Christiansen and Thimann 1950; Michel 1951), and these data were interpreted as indicating preferential fat metabolism in auxin-treated tissue. Some supporting data were provided by Christiansen and Thimann (1950) who noted that lipid levels were depleted in hormone-treated tissue. Enzymological analyses are required for characterization of the response, particularly in view of the ambiguity which is known to arise from determinations of respiratory quotient in whole tissues. Since total combustion of substrates to CO_2 and H_2O is relatively rare, and there is considerable utilization of intermediates as precursors for new synthesis, and since respiratory CO_2 is available for carboxylation, Bonner (1973) has warned against interpreting respiratory quotient data as reflecting the source of respiratory substrate without a more complete understanding of the metabolism of the cell type under study.

Glycolysis and the hexose monophosphate pentose pathway are two major routes of glucose degradation in plants. The respiration of glucose in *Avena* was increased by low levels of IAA, but both pathways seemed to be stimulated equally (Gibbs and Beevers 1955). Similarly, the pathway of glucose catabolism was not altered by physiological levels of exogenous IAA in seedlings of *Zea mays, Avena* and *Triticum aestivum* (Humphreys and Dugger 1957; Shaw et al. 1958), but herbicidal levels of auxin, which inhibited respiration, increased the activity of the hexose monophosphate-pentose pathway (Humphreys and Dugger 1957; Shaw et al. 1958). This intriguing observation may merely reflect the increased availability of glucose for this pathway when glycolysis was inhibited or it may reflect a requirement by the tissue for NADPH and pentoses, intermediates commonly associated with biosynthetic reactions in actively growing seedling tissue. Their role in the herbicidal activity of auxin may be worth exploring.

In contrast to the above data on monocotyledons, it has been reported that, in the dicotyledon, mung bean *(Phaseolus aureus),* auxin-stimulated respiration was mediated through stimulation of glycolysis and the Krebs cycle without altering the activity of the hexose monophosphate-pentose pathway (Carlier and van Hove 1964). From further analysis of various steps in glycolysis and the Krebs cycle using metabolic inhibitors, it was concluded that a step between pyruvic acid and oxalosuccinic acid was stimulated by the hormone (van Hove and Carlier 1968). It is unfortunate that the well-characterized enzymes of this region of the Krebs cycle were not studied to determine whether a rate-limiting enzyme was stimulated.

Selective increases or decreases in the activity of individual enzymes of the various pathways or selective effects on substrate levels might be expected to play a role in altered rates of respiration. Direct data on substrates are fairly rare; some have been obtained by interpolation from studies where the contribution of one pathway appeared to be changed, thus making available more glucose for another. Auxin-induced respiration increases in *Avena* were dependent upon the addition of 4-carbon organic acids such as malate and fumarate (Commoner and Thimann 1941), and it is known that the Krebs cycle in plants may be under complex control by malic acid levels (Bonner 1973). Whether this response reflected an acidification of the cellular environment which affected auxin uptake, or whether the effect was to release a rate-limiting step in oxidation, is not clear. Selective effects of auxin on the activity of various enzymes of glycolysis, the hexose monophosphate-pentose pathway and the Krebs cycle have been demonstrated in a variety of tissues. Alcohol dehydrogenase, malic dehydrogenase and fumarase were unaffected in *Avena* (Berger and Avery 1943), but, in tomato fruit, auxin stimulated malic and fumaric dehydrogenase and inhibited glutamic and succinic dehydrogenase (Teubner and Murneek 1952). Isocitrate lyase was shown to be increased in potato *(Solanum tuberosum)* tuber discs (Datta et al. 1965), and citrate synthetase was increased in bean shoots (Mitchell and Sarkissian 1966). Both enzymes occur in glyoxysomes (microbodies), and citrate synthetase is thought to be an important regulator of Krebs cycle activity. The glycolytic enzyme, aldolase, was increased in wheat seedlings as were the hexose monophosphate-pentose enzymes, glucose-6-phosphate dehydrogenase and 6-phosphogluconate dehydrogenase (Likholat and Morozova 1972; Likholat and Levina 1973). In photosynthetic tissue of rye, glucose-6-phosphate dehydrogenase activity was stimulated by auxin, but in cultured cells it was inhibited, as were the other pentose shunt enzymes, 6-phosphogluconate dehydrogenase and transketolase (Scott et al. 1964).

The altered activity of enzymes was considered in some cases to be mediated by an allosteric effect of auxin (Teubner and Murneek 1952; Mitchell and Sarkissian 1966; Likholat and Morozova 1972; Likholat and Levina 1973), the rationale being that, if activation of enzymes by the hormone could be demonstrated in vitro, or if increased enzyme activity was insensitive to nucleic acid and protein synthesis inhibitors, then an allosteric effect was indicated. The in vitro activation of enzymes by hormones has had a controversial history in most cases, and an early example serves to illustrate the point. In the cases of malic and fumaric dehydrogenases, the enzymes isolated from *Avena* were not affected by auxin in vitro (Berger and Avery 1943), but those isolated from tomato fruit did respond to the hormone (Teubner and Murneek 1952). In a more recent example which concerns citrate synthetase, the enzyme from beans was shown to be activated by auxin under extremely exacting conditions of hormone, substrate and protein concentrations, and it was claimed that the treated enzyme underwent an extensive "conformational change", which could be monitored by gel chromatography (Mitchell and Sarkissian 1966; Sarkissian and Schmalsteig 1969a and b; Sarkissian 1970 and 1972). However, the activation of the isolated bean enzyme could not be confirmed by Brock and Fletcher (1969), nor was purified pig heart citrate synthetase affected by auxin (Zenk and Nissl 1968). These workers may not have been able to reproduce the exacting (and sometimes changing) conditions specified by Sarkissian. It seems to be essential to confirm or convincingly refute the data on citrate synthetase, because the enzyme occupies an important position in cellular metabolism, and it is important to know whether auxin regulates its activity. A role for hormones as allosteric effectors of enzymes in plants has received relatively little support. The other important way in which enzyme activity is regulated is by controlling synthesis of new enzyme or degradation of existing enzyme. There is relatively little data on this aspect of respiratory enzyme control, but increased isocitrate lyase activity in potato tuber slices was dependent upon RNA and protein synthesis, suggesting that de novo synthesis was involved (Datta et al. 1965). There is clearly a need for detailed analyses of the processes involved in increasing or decreasing enzyme activity during respiration.

Isolated mitochondria provide an excellent opportunity to study some aspects of aerobic respiration, especially the Krebs cycle enzymes, pathways of electron transport and oxidative phosphorylation. The role of plant hormones in modifying some of these processes has been studied in mitochondria isolated from treated tissues and in isolated mitochondria treated with hormone in vitro. Respiratory activity of mitochondria isolated from auxin-treated soybean was reported to be greater than the control in one report (Key et al. 1960), but to be unaffected in another report from the same laboratory (Baxter and Hanson 1968). Parameters which were monitored included respiration, phosphorylation and coupling. Mitochondria from auxin-treated maize were shown to have enhanced respiratory activity only under conditions of carefully controlled concentrations of applied hormones, and only in selected varieties of *Zea mays* (McDaniel and Sarkissian 1966; Sarkissian and McDaniel 1966). Thus, mitochondria isolated from hormone-treated tissue displaying enhanced respiration measured at the tissue level do not appear to maintain the response in vitro, although the organelles do show evidence of increased integrity and ability to synthesize macromolecules (Key et al. 1960; Baxter and Hanson 1968). This may imply a requirement for other components derived from the intact cell, or it may reflect loss of structure and functional capacity due to improper isolation techniques. The latter point was stressed by Bonner (1973), who pointed out that many studies on respiratory activity of isolated mitochondria were not interpretable because of the poor quality of the isolated organelles.

Similar caution should probably be applied to those studies in which mitochondrial suspensions were treated with auxin and the respiratory response noted. Organelles isolated from *P. aureus* showed no change in the rate of respiration or in P/O ratio at low auxin concentrations (Poljakoff-Mayber 1955), whereas those isolated from soybean displayed decreased respiration over a wide range of concentrations, and the P/O ratio was only reduced at high concentrations (10^{-3} *M*) (Switzer 1957). On the other hand, there was evidence that auxin could cause a transient and small increase in ATP levels in pea stem segments (Marre and Forti 1958), which was thought to reflect enhanced oxidative phosphorylation and, in keeping with this notion, it was shown that mitochondria isolated from inbred maize responded to low concentrations of auxin by increasing phosphorylation under strictly defined conditions of incubation, which included optimal levels of exogenous acetyl-CoA (Sarkissian and McDaniel 1966).

The potential offered by isolating part of the respiratory process in the form of mitochondria does not appear to have been realized, mainly because the responses to auxin have been very small, and the exacting conditions necessary for that response may hamper the kind of systematic study which would be necessary for a convincing evaluation of the role of the hormone. It is still possible that improved isolation techniques may enable further study of the role of hormones in regulating certain aspects of mitochondrial activity.

9.4.2. Gibberellin and abscisic acid

The effect of these two phytohormones is often studied simultaneously because, shortly after the discovery of ABA, it was shown that many of the gibberellin-promoted responses were antagonized by ABA (Addicott and Lyon 1969). Neither hormone affected respiration in barley aleurone layers, despite their dramatic effects upon α-amylase activity and associated starch hydrolysis in barley endosperm tissue (Chrispeels and

Varner 1966). In contrast to the inhibitory influence of ABA upon starch hydrolysis in barley and *Pennisetum typhoides* (Huber and Sankhla 1974b), it appeared to enhance hydrolysis of the polymer in *Coleus blumei,* an event which was associated with increased levels of saturated lipids (Kull and Unger 1974). ABA treatment of *Pennisetum typhoides* and *Lemna minor* resulted in increased levels of carbohydrate biosynthesis and accumulation, while respiration, which decreased in *Lemna,* was not studied in *P. typhoides* (Huber and Sankhla 1974b; McLaren and Smith 1976). ABA has not been extensively studied in relation to respiration, but it was shown to stimulate the onset of oxygen consumption by senescing *Rumex* leaf discs and to antagonize the GA_3-inhibited oxygen consumption in the same tissue (Goldthwaite 1974). In contrast, oxygen consumption and glucose respiration were promoted by GA_3 while delaying senescence in Navel orange *(Citrus sinensis)* (Lewis et al. 1967). Thus, the way in which senescence is delayed by GA_3 in *Rumex* and *Citrus* in quite different if studied at the level of respiration.

Germination is another highly complicated process in which GA_3 has been studied and related to respiration. In general, no measurable increase in respiration was noted in GA_3-promoted germination of hazel *(Corylus avellana),* wheat or wild oat *(Avena fatua),* but in all cases there was evidence, some of it indirect, that the hormone somehow altered the pathway of glucose degradation from the glycolytic to the hexose monophosphate-pentose pathway (Pinfield 1968b; Simmonds and Simpkins 1971; Likholat and Morozova 1972; Likholat and Levina 1973). Pinfield (1968a) first showed that the glyoxylate cycle enzyme, isocitrate lyase, was increased during GA_3-promoted germination of hazel, and later showed that respiratory metabolites were being diverted from the Krebs cycle to biosynthetic reactions (Pinfield 1968b). Based upon analyses of C_6/C_1 ratios during the germination of wild oat, Simmonds and Simpkins (1971) concluded that GA_3 switched glucose metabolism from glycolysis to the pentose shunt. These conclusions were not tested by examining enzyme activities. However, in wheat coleoptiles undergoing cell division, the activity of aldolase, glucose-6-phosphate and 6-phosphogluconate dehydrogenases was increased by GA_3 and, unlike the auxin-stimulated increase of the same enzymes in elongating coleoptiles, actinomycin D inhibited the response, suggesting that new enzyme synthesis, rather than activation, was involved (Likholat and Morozova 1972; Likholat and Levina 1973).

The majority of the data relating GA_3 and ABA to changes in respiration suggest that oxygen consumption may not reveal the full complexity of the responses. Few and small changes in oxygen consumption have been detected, but some fairly extensive changes in respiratory enzyme activities have been noted, and thus it appears as if detailed analyses of selected aspects of the pentose shunt following hormone treatment may be worth exploring in greater detail. Judging by some of the data presented above, gluconeogenesis may be under some form of hormonal control, especially for GA, since Newman and Briggs (1976) have shown that GA_3 can stimulate sucrose formation from lipids. Furthermore, the direct effects of hormones, such as GA_3, on protein synthetic activity of isolated mitochondria (De and Roy 1974) merits further examination in relation to changes in respiration.

9.4.3. Cytokinin

Cytokinins delay senescence in many tissues and, because ageing is usually associated with increased rates of respiration (e.g. Tetley and Thimann 1974), it is not too surprising that there is a voluminous literature on the effects of cytokinin on respiratory activity, especially in tissues whose senescence has been slowed by the hormone (see Ch. 7 and 10, Vol. II). In most cases cytokinin inhibited the respiratory rise associated with ageing as, for example, in *Brassica oleracea,* tobacco, *Narcissus pseudonarcissus, Rumex* spp., oat and radish *(Raphanus sativus)* (Dedolph et al. 1962; Sugiura 1963; Ballantyne 1966; Goldthwaite 1974; Tetley and Thimann 1974; Gordon and Letham 1975). In other instances, such as those involving tissue culture, the hormone has stimulated respiration, e.g. in tobacco pith (Glasziou 1957) and soybean callus (Moore and Miller 1972), or at high concentrations it has inhibited respiration e.g. in tissue cultures of *Acer pseudoplatanus* (Simpkins and Street 1970) and soybean callus (Moore and Miller 1972). Thus, there are a number of different systems available for the study of cytokinin action in respiration.

Alterations in enzyme activity and the pathway of glucose utilization have been most commonly studied in the search for regulatory sites. The reduced respiration of cytokinin-treated *Brassica oleracea* was correlated with lowered kinase activities; hexokinase and pyruvic kinase were shown to be inhibited and other kinases were also thought to be affected (Tuli et al. 1964). Therefore, glycolyis was likely to be retarded. Other views, which are based on similarly indirect or even less convincing evidence, include the notion that cytokinin enhances coupling between respiration and phosphorylation in senescing oat leaves (Tetley and Thimann 1974), and that energy metabolism was slowed down in ageing *Rumex* leaf discs (Goldthwaite 1974).

In tissues other than ageing leaf segments, such as tissue culture or pea stem segments, progress has been made in attempting to determine the action of cytokinin on respiration. Glucose metabolism was altered in cultured carrot tissue and in pea stem segments, probably by suppressing the hexose monophosphate-pentose pathway and stimulating glycolysis (Lustinec et al. 1962; Kaminek and Stemberova 1967). While none of the relevant enzymes were studied, Simpkins and Street (1970) noted that phosphoglucose isomerase and glucose-6-phosphate dehydrogenase activities were decreased by high levels of kinetin in cultured *A. pseudoplatanus* cells. Thus, they provided evidence which might explain how both glycolysis and the pentose shunt might be inhibited. Further analysis led them to conclude, however, that the conversion of glucose to other sugars was the sensitive rate-limiting step which was affected by the hormone, rather than a particular step in glycolysis or the pentose shunt. On the other hand, Moore and Miller (1972) concluded that, in cultured soybean callus, cytokinin might regulate the entry of pyruvate into the Krebs cycle. This proposal was based upon the results of a series of experiments using selected metabolic inhibitors. The respiratory responses were small but merit further study at the enzymological level. No effect of cytokinin on ATP synthesis or oxidative phosphorylation could be demonstrated by Moore and Miller (1972), nor were isolated mitochondria from cytokinin-treated *A. pseudoplatanus* different from the control with respect to enzyme activity (Simpkins and Street 1970). However, it was shown that isolated mitochondria from etiolated *Vigna sinensis* responded to kinetin in vitro by increasing protein synthesis (Bhattacharyya and Roy 1969). This finding has not been analyzed further, but it might reflect enhanced biosynthetic activity of the organelle which would provide an environment suitable for stabilized respiratory activity.

9.4.4. Ethylene

The ripening of fruit has been associated with ethylene for a long time and early work showed that respiration was rapidly and reversibly stimulated by the gas in lemons *(Citrus limon)* and bananas *(Musa cavendishii)* (Denny 1924; Regeimbal et al. 1927) (see Ch. 10 Vol. II, for details of ethylene in relation to ripening). The concentrations of ethylene required to increase respiration were three orders of magnitude greater than those needed for ripening, suggesting that the two are separate phenomena. As with the other hormones a range of possible ways in which respiration might be increased has been examined, including a search for new protein subunits in ripening banana (Brady and O'Connell 1976). The techniques used by these workers could, at best, resolve 10^2 different polypeptides, while higher eucaryotic cells may be producing 1×10^4 to 3×10^4 different polypeptides at any given time (Bishop et al. 1974; Axel et al. 1976). Thus, the finding by Brady and O'Connell (1976), that there were no major qualitative changes in the polypeptide spectrum during ripening, does not rule out the possibility that there were extensive changes in minor species. Their approach, however, was consistent with early reports that enzyme levels were altered in response to ethylene treatment, and it was logical to determine whether these changes might be detected at the polypeptide level. It was suggested that glycolysis is enhanced in ripening fruit and in potato tubers in which respiration had been stimulated by ethylene. This conclusion was based upon changes in the level of fructose diphosphate, glucose-6-phosphate, 3-phosphoglyceric acid, malic enzyme, phosphoenolpyruvate and the rate of pyruvic decarboxylation (Baker and Solomos 1962; Chalmers and Rowan 1971, Hulme et al. 1971b; Solomos and Laties 1974 and 1975; Salminen and Young 1975). The suggestion that regulatory enzymes in glycolysis were affected by the gas followed from these data (e.g. Chalmers and Rowan 1971), and Salminen and Young (1975) provided evidence that there was a major increase in the activity of phosphofructokinase (PFK) in ripening banana fruit, whereas there were only slight changes in the activities of glucose-6-phosphate dehydrogenase and aldolase. However, Solomos and Laties (1974, 1975 and 1976), in studies employing cyanide and anaerobiosis, observed that the stimulation of glycolysis was unlikely to be due to enhanced enzyme levels alone, and they provided evidence that ethylene diverted electron transport from the usual cytochrome pathway, which is sensitive to cyanide, to an alternative cyanide-insensitive electron transport system. These results might account for the altered levels of glycolytic intermediates observed earlier, in that electron transport may be limiting the utilization of substrates, and the release of electron flow may cause the observed changes. There is no data available to indicate a mechanism whereby ethylene might activate the cyanide-insensitive path of electron flow, and there is relatively little known about the components of this alternative electron transport system (Bonner 1973).

Wade (1974) analyzed the respiratory response to ethylene in ripening bananas by Michaelis-Menten kinetics, and concluded that the hormone might be acting as a coupling-activator of respiration at specified concentrations of oxygen. This model has very little experimental support at present.

9.4.5. Conclusions

The most dramatic respiratory response to a plant hormone is that induced by ethylene. The recent proposal that the gas may be diverting electrons through an alternative electron transport system seems attractive, and offers two exciting opportunities for further study. The first involves the question of how these electrons might be diverted and the second, the possibility of studying and elucidating the mechanism of cyanide-insensitive electron transport in plants with the help of the regulator, ethylene.

The majority of the respiration-changes effected by the other hormones tend to be rather small, and thus may have contributed to the difficulty in finding sites of regulation. However, it has emerged that control commonly involves activity or level of a limited number of enzymes, but no particular enzyme functions as a regulator. Switching of respiratory pathways or from one source of substrate to another does not appear to be under strict hormonal control, although these features have not been studied in great depth.

The general maintenance or disruption of cellular and organellar integrity probably plays a major role in respiratory changes but, since the control of cellular organization is so complex, it is difficult to assess the usefulness of the structural approach alone as an aid to understanding changes in respiration. Finally, the process of respiration itself is so complex as to provide a formidable system in which to search for regulatory sites which might respond to plant hormones.

9.5. Lipid metabolism

Hormonal control of lipid metabolism in eucaryotic cells can be inferred from studies on diabetic and embryonic animals, and it has been shown that insulin, hydrocortisone, growth hormone, glucagon and cAMP can regulate the levels of fatty acid synthetase and acetyl-CoA carboxylase in avian and mammalian cells (Lakshmanan et al. 1972; Joshi and Sidbury 1976). It has been proposed that enzyme activity is regulated both at the level of activation or inactivation by such modifications as phosphorylation (Lee et al. 1973) and by modified rates of synthesis or degradation (Joshi and Sidbury 1976). The control of lipid metabolism in plant cells has not been analyzed at a comparable level, and this may in part be due to the fact that the pathways of biosynthesis and degradation and the subcellular location of these processes have only recently been elucidated, while details of some aspects of metabolism such as desaturation are still being worked out (Stumpf 1976).

Regulation of lipid metabolism is of interest from a number of viewpoints, but particularly because (a) lipids are involved in membrane biosynthesis and the role of membranes in cellular organization is extremely important, (b) they are involved in the storage of reserve material in many seeds, and (c) the degree of saturation of lipids may be important from a nutritional aspect. There is evidence that plant hormones play some part in lipid metabolism in all three cases mentioned above.

The gibberellin-induced changes in the organization and perhaps levels of the endoplasmic reticulum (ER) in barley aleurone layers, as described by Jones (1969) and Vigil and Ruddat (1973), and at the biochemical level (Evins and Varner 1971; Johnson and

Kende 1971; Ben-Tal and Varner 1974; Firn and Kende 1974) have been discussed in Chapters 8, 10 and 11. In brief, it appears as if the hormone induces changes in the assembly of the membranes from existing lipid material. The enzymatic studies to date would suggest that enzyme activation rather than de novo synthesis is involved (Ben-Tal and Varner 1974); however, progress in this direction has so far been limited.

It is known that GA_3 accelerates the disappearance of triglycerides from the aleurone layers of barley, and that gluconeogenesis is the probable means of utilization (Newman and Briggs 1976). While these authors did not examine lipase activity or study enzymes of β-oxidation, they did show that the activity of two enzymes of the glyoxalate cycle, isocitrate lyase and malate synthetase, increased in response to GA_3 within 20–40 hr. Thus, it appears that GA_3 may regulate gluconeogenesis in the barley endosperm, but the way in which lipid degradation occurs deserves further study.

The interaction of gibberellins with artificial membranes has been studied and is discussed in Chapters 8 (section 8.2.2.1) and 11 (section 11.3.1). This approach further highlights the importance which is attached to the regulation of membrane organization and, thus, lipid metabolism. Short-chain fatty acids have also been shown to interact with GA_3 in the barley aleurone layer during the induction of reducing sugar production (Buller et al. 1976). The nature of the interaction has not been studied further.

Changes in the organization of cellular membranes in response to other hormones have been demonstrated. Ethylene produced an effect in abscission zones of flower pedicels and in expanding pea epicotyl (Valdovinos et al. 1971; Sargent and Osborne 1975), while changes in membrane organization of pea epicotyl are implied by studies with auxin (Davies 1976), in which it was shown that the hormone caused increased attachment of ribosomes to the ER. The biochemical details of these changes are even less well-characterized than those caused by GA, and it is impossible to suggest a mechanism for hormone action.

Rapid responses to hormones attract attention because of the possibility that the observed response may be causally related to the overall hormone response. Thus, an observation that auxin stimulated the incorporation of [^{32}P]orthophosphate into an unidentified compound, possibly containing glycerol and phosphate, within 5 min in pea epicotyls (Neumann 1971) should be confirmed and examined further. Speculations suggesting a role in cell wall extension and intracellular location remain to be substantiated.

The storage reserves of many seeds are fats and metabolism of these to provide energy and biosynthetic precursors during germination is an important aspect of early plant development (Beevers 1961). The role of hormones in these processes has not been studied in great detail, although there are a few exceptions. Notable in this regard is the work of Tavener and Laidman (1972a and b). They have studied the distribution of fat reserves in the various tissues of the wheat grain and attempted to determine the controls operating in the mobilization of this fat in each tissue. They found that metabolism was independently controlled in each tissue, particularly by nitrogenous compounds such as glutamine and by cytokinin and IAA. Lipase activity, however, did not respond to hormones, especially in the starchy endosperm, and the enzyme appeared to be under complex control and often required a range of cellular metabolites with or without cytokinin and auxin to increase its activity. Thus, the role of the hormones, although apparently minor, seemed to be closely coordinated with the production of intermediate metabolites. However, their precise involvement in metabolism was not clarified. In contrast to the small effect of hormones on lipase activity discussed above, it was shown that GA_3 stimulated enzyme activity in half-seeds of cotton, although the time course of exposure to the hormone (3–5 days) would suggest that intermediate metabolites may also have been involved in the increased enzyme activity (Black and Altshul 1965). RNA synthesis was required for the increased enzyme activity, but the role of this RNA is difficult to assess from the available data.

The mobilization of stored fats in germinating castor bean *(Ricinus communis)* has been studied in depth (Beevers 1961), and more recently the regulation of lipid breakdown (Marriott and Northcote 1975a) and some of the enzymatic processes (Marriott and Northcote 1975b) have been examined. Two lipase activities (acid and alkaline) increased during the first few days of germination but did not respond to GA_3, even in the absence of the embryo. β-oxidation was monitored by studying 3-hydroxyacyl-CoA dehydrogenase and glyoxylate cycle by isocitrate lyase. In both cases, GA_3 stimulated total enzyme activity and increased the rate of appearance of the enzymes (Marriott and Northcote 1975b). The presence of the embryo was immaterial. The time course of appearance of enzyme activity was long, ranging over 3 days, and no other enzymes were investigated. However, the responses were marked, and the system should be studied further to determine how the hormone mediated these changes.

In another study, the effect of GA_3 on lipid metabolism was examined during the germination of *Corylus avellana* and it was shown that the incorporation of $[^{14}C]$acetate into lipids was reduced concomitant with an increased rate of lipid breakdown (Bradbeer and Pinfield 1967). The relationship between the two sets of data were not further clarified.

There are other instances of hormonal control of lipid metabolism, such as in senescing tissues or during release of lateral buds from apical dominance. Kinetin prevented senescence of detached wheat leaves and maintained levels of lipase, suggesting that lipid degradation was associated with maintenance of cellular and tissue integrity (Sodek and Wright 1969). The breaking of bud dormancy in tobacco by cytokinins was associated with enhanced methylation of neutral and polar lipids, especially phosphatidylcholine, as judged by the incorporation of methyl groups from labelled methionine (Schaeffer and Sharpe 1971; Schaeffer et al. 1972). The axillary buds of *Cicer arietinum* were also released from dormancy by cytokinin (Usciati et al. 1974), and in this instance there was a rapid increase in total lipid levels without any change in the proportion of the common fatty acids. In addition, it was shown that $[^{14}C]$acetate incorporation into lipids was stimulated within one hour, further emphasizing the rapidity of the response. Enzymes were not studied, and the mechanism of lipid accumulation (whether by increased synthesis or decreased degradation) is not understood.

The level and composition of lipids in leaves, buds, chloroplasts and tissue cultures are also known to be altered by plant hormones, but we are not aware of any reported changes in oil seeds. The latter are important sources of nutritional fats, and it would be interesting to know if composition of the lipids could be altered. It was noted that, when senescence of spinach leaves was retarded by kinetin, chloroplasts isolated from treated leaves contained an altered ratio of linolenic to stearic acid, kinetin causing the ratio to rise from 1.6 to 3.3 (Donaldson et al. 1969). The level of stearic acid was lowered, while the levels of oleic and linoleic acids were relatively unaffected. Furthermore, kinetin inhibited incorporation of radiolabelled acetate into all lipid fractions, suggesting that the high levels of linolenate were maintained by selective preservation from degradation.

As noted previously, cytokinin treatment released the axillary buds of *C. arietinum* from apical dominance and rapidly increased the total lipid levels (Usciati et al. 1974). However, no compositional change was found, since the proportions of the major fatty acids remained constant. In contrast, Kull and Buxenstein (1974) noted that cytokinin treatment of leaves of *Coleus, Impatiens* and *Populus* resulted in changes in the fatty acid composition of the lipids over a period of 4 weeks although not all genera responded in the same way. In *Impatiens,* the level of linolenic acid increased while that of palmitic acid dropped in response to cytokinin; the same trend was apparent in *Coleus,* although the changes were much smaller, and no changes were observed in *Populus.* High temperature conditions increased the unsaturated to saturated fatty acid ratio, but cytokinin lowered this to the levels found at low temperature (Kull and Buxenstein 1974). ABA treatment of the *Coleus* apex resulted in lower lipid levels with increased proportions of the saturated fatty acids, palmitic and stearic, while the proportion of linolenic acid was dramatically decreased (Kull and Unger 1974). There was little change in linoleic acid. Thus, there appears to be scope to study the mechanism of hormonal control of the saturation level of fatty acids in some leaves, although the long-term experiments mentioned above probably conceal an enormous complexity.

Another long-term study of the control of lipid level and composition by plant hormones was carried out in soybean suspension cultures (Stearns and Morton 1975). There was a general tendency toward higher levels of polyunsaturated fatty acids in hormone-treated cells. Auxin increased the level of saturated fatty acids, while GA_3 and cytokinin reduced them. Cytokinin together with GA_3 greatly reduced the levels of linolenic acid and increased the level of linoleic.

Thus, it appears to be possible to alter the composition of plant lipids with hormones, especially in leaves. The increasing knowledge of the pathways of lipid biosynthesis (Stumpf 1976) should allow detailed analysis of the ways in which hormones regulate lipid level and composition. The evidence presented above, although not very extensive, is sufficient to demonstrate that lipid metabolism can be regulated by plant hormones.

9.6. Nitrogen metabolism

There can be little doubt that hormones are involved in various aspects of nitrogen metabolism when one considers that much of Chapters 8–11 is concerned with hormone-induced changes in nucleic acids and proteins. The lists of enzymes in tables 9.1 and 9.2 provide some indication of the vast array of enzymes whose activity and/or synthesis may be controlled by hormones. Where there was sufficient data on the control of that activity, many of these enzymes have been further discussed under the appropriate heading. This section is limited to overall changes in cellular nitrogen including inorganic forms, and it concentrates mainly on those forms of nitrogen not covered elsewhere.

Table 9.1.

Examples of Increased Enzyme Level or Activity Following Hormone Treatment

Enzyme	Tissue	Hormone	Reference
Acid phosphatase	*Rhoeo discolor* leaf sections	ABA	De Leo and Sacher (1970)
	Jerusalem artichoke tubers	Auxin	Palmer (1970)
	Cotton leaf blade and pulvinus	Ethylene	Herrero and Hall (1960)
	Bean petioles	Ethylene	Pooviah and Rasmussen (1974)
Acid phosphatase (phosphomonoesterase and phosphodiesterase)	Barley aleurone	GA_3	Briggs (1963); McLeod et al. (1964); Pollard and Nelson (1971)
N-Acyl aspartate synthesis	Pea stem	Auxin	Sudi (1964, 1966)
ATPase	Mung bean hypocotyl	Auxin	Kasamo and Yamaki (1974)
	Tomato and sunflower plants	Ethylene	Madeikyte and Turkova (1965)
	Tomato leaf and stem	Ethylene	Turkova et al. (1965)
S-Adenosylmethionine:tyramine methyltransferase	Barley root tips	Cytokinin	Steinhart et al. (1964)
Aldolase	Banana fruit	Ethylene	Tager and Biale (1957)
	Wheat coleoptile	GA_3 and IAA	Likholat and Morozova (1972)
Alkaloid synthesis	*Ipomea violacea* L. seed	GA_3	Genest (1966)
Amaranthin synthesis	*Amaranthus caudatus* seedlings	Cytokinin	Kinsman et al. (1975)
α-Amylase	Bean cotyledons	Cytokinin (kinetin)	Gepstain and Ilan (1970)
	Bean hypocotyl discs	Cytokinin	Clum (1967)
	Wheat grain	Cytokinin (kinetin)	Boothby and Wright (1962)
	Cotton leaf blade and pulvinus	Ethylene	Herrero and Hall (1960)
	Tobacco leaves	Ethylene	Smirnov and Krainev (1940)
	Pea internodes	GA_3	Broughton and McComb (1971)
	Rice leaf sheaths	GA_3	Hayashi et al. (1956)
	Rice grain	GA_3	Matsunaka and Yarragisawa (1963)
	Maize leaf sheaths	GA_3	Katsumi and Fukuhara (1969)
	Lupin seed	GA_3	Mierzwinska (1973)
	Barley aleurone	GA_3	Yomo (1960a); Paleg (1960)
	Wheat grain	GA_3	Boothby and Wright (1962)

Benzoylaspartate synthetase	Pea epicotyls	Auxin	Venis (1964)
Benzoylmalate synthetase	Pea epicotyls	Auxin	Venis (1972)
Carboxydismutase	Rye seedling	Cytokinin	Feierabend (1969)
Carbohydrase	Jerusalem artichoke tuber	Auxin (2,4-D)	Wain et al. (1964)
Catalase	Mango *(Mangifera indica)* fruit	Ethylene	Bannerjee and Kar (1941)
	Mango fruit slices	Ethylene	Matoo and Modi (1969)
	Potato tuber tissues and sprouts	Ethylene	Elmer (1936)
	Cotton leaf blades and pulvinus	Ethylene	Herrero and Hall (1960)
	Pear, orange and cucumber fruit	Ethylene	Ivanov et al. (1931)
Cellobiase	Pea seedlings, etiolated	Auxin (IAA)	Ridge and Osborne (1969)
Cellulose synthetase	*Avena* coleoptiles	Auxin	Hall and Ordin (1968)
Cellulase	Pea epicotyl	Auxin (IAA)	Fan and Maclachlan (1966)
	Pea roots	Auxin (2,4-D)	Giovannozzi-Sermanni and Pietrosanti (1973)
	Pea, bean and maize seedlings	Auxin (IAA)	Maclachlan et al. (1968)
	Pea seedlings (etiolated)	Auxin (IAA)	Ridge and Osborne (1969)
	Red kidney bean seedlings	Auxin (IAA)	Lewis et al. (1974)
	Red kidney bean, cotton and *Coleus* petiole abscission zones	Ethylene	Abeles (1969)
	Bean petiole explants	Ethylene	Horton and Osborne (1967)
	Red kidney bean petiole abscission zone	Ethylene	Lewis and Varner (1970)
	Orange petiole explants	Ethylene	Ratner et al. (1969)
	Red kidney bean abscission zone	GA_3	Lewis and Varner (1970)
	Orange explants	GA_3	Ratner et al. (1969)
	Barley grain	GA_3	Sandegren and Beling (1958)
Chitinase	Red kidney bean leaves	Ethylene	Abeles et al. (1971)
Chlorogenic acid synthesis	Sweet potato root slices	Ethylene	Imaseki et al. (1968)
Chlorogenic acid (iso) synthesis	Carrot roots	Ethylene	Sarkar and Phan (1974a)
Cinnamic acid 4-hydroxylase	Detached pea epicotyl tissue	Ethylene	Hyodo and Yang (1971)
Daidzein glucoside synthesis	Soybean tissue culture	Cytokinin	Miller (1969)
Dehydrogenase (organic acid)	Bean petioles	Ethylene	Poovaiah and Rasmussen (1974)
Deoxyribonuclease	Detached barley leaf	ABA	Srivastava (1968b)
DNA polymerase (insoluble)	Soybean hypocotyl	Auxin	O'Brien et al. (1968)
Dipeptidase	Squash cotyledon	GA_3; Cytokinin	Tsay and Ashton (1974)

Endopentosanase	Barley aleurone	GA_3	MacLeod and Millar (1962)
Esterase	Wheat, detached leaves	Cytokinin (kinetin)	Sodek and Wright (1969)
Eugenin synthesis	Carrot root	Ethylene	Sarkar and Phan (1974a)
Fructose-1,6-diphosphatase	Castor bean endosperm	GA_3	Scala et al. (1969)
α-Galactosidase	Barley aleurone	GA_3	Briggs (1963)
β-Galactosidase	Barley aleurone	GA_3	Pollard and Nelson (1971)
β-1,3-Glucanase	Pea epicotyl	Auxin	Datko and Maclachlan (1968)
	Bean petiole explants	Ethylene	Abeles and Forrence (1970)
	Red kidney bean leaves	Ethylene	Abeles et al. (1971)
	Barley aleurone	GA_3	MacLeod and Millar (1962)
β-Glucan synthetase	Pea coleoptile and stem	Auxin	Ray (1973)
	Onion stem	Auxin	VanDerWoude et al. (1972)
Glucose-6-phosphate dehydrogenase	Wheat coleoptile	GA_3; Auxin	Likholat and Levina (1973)
α-Glucosidase	Barley aleurone	GA_3	Briggs (1963)
Glyceraldehyde phosphate dehydrogenase	Rye seedlings	Cytokinin (kinetin)	Feierabend (1969)
Hexosidase	Barley aleurone	GA_3	Pollard (1969)
IAA oxidase	Tobacco callus	Auxin (IAA)	Lee (1971)
	Cotton seedlings	Ethylene	Hall and Morgan (1964)
Invertase	Sugar beet	ABA	Cherry (1968)
	Sugar cane internodes	ABA	Gayler and Glasziou (1969)
		Auxin (NAA, IAA)	Sacher and Glasziou (1962)
	Chicory root discs	Auxin, GA_3	Rutherford and Bard (1971)
	Sugar cane storage parenchyma	Auxin (IAA)	Glasziou and Waldron (1964)
	Sugar cane internodes	Cytokinin	Gayler and Glasziou (1969)
	Pumpkin fruit	Ethylene	Ivanov (1932)
	Orange, lemon, mandarin, cucumber and tomato fruit	Ethylene	Kursanov and Krukova (1938)
	Pineapple fruit	Ethylene	Regeimbal and Harvey (1927)
	Tobacco leaves	Ethylene	Shcherbakov (1939)
Invertase	Jerusalem artichoke tuber	GA_3	Edelman and Hall (1964)
	Pea internodes	GA_3	Broughton and McComb (1971)
	Avena internodes	GA_3	Kaufman et al. (1968)

	Carrot root slices	GA_3	Ricardo (1976)
	Corn stamen filaments	GA_3	Schaeverbeke (1966)
	Lentil epicotyl	GA_3	Seitz and Lang (1968)
Isocitric lyase	Potato tuber	Auxin	Datta et al. (1965)
	Cotyledons of squash	Cytokinin	Penner and Ashton (1967)
	Barley endosperm	GA_3	Newman and Briggs (1976)
Kinase (protein)	Soya-bean callus	Cytokinin	Elliott and Murray (1974)
Lipase	Wheat aleurone	Auxin (IAA)	Tavener and Laidman (1972b)
	Wheat, detached leaves	Cytokinin (kinetin)	Sodek and Wright (1969)
	Cotton, cotyledons	GA_3	Black and Altschul (1965)
	Apple embryos	GA_3	Smolenska and Lewak (1974)
Malate dehydrogenase (NAD and NADP)	*Lemna minor*	ABA	Bauer et al. (1976)
Malate synthetase	Barley endosperm	GA_3	Newman and Briggs (1976)
Malic enzyme (NADP)	*Lemna minor*	ABA	Bauer et al. (1976)
Malic enzyme	Apple fruit	Ethylene	Hulme et al. (1971a)
Maltase	Wild oat grain	GA_3	Simpson and Naylor (1962)
MMHD (isocoumarin) formation	Carrot root disks	Ethylene	Chalutz et al. (1969)
Nitrate reductase	*Agrostemma githago* embryo	Cytokinin	Borriss (1967; Kende et al. (1971); Hirschberg et al. (1972)
	Tissue cultures of a variety of species	(kinetin)	Gunther (1974)
	Tobacco leaves	Cytokinin; GA_3	Roth-Bejerano and Lips (1970)
3′-Nucleotidase	Wheat embryo	GA_3	Shuster and Gifford (1962)
Oxidase activity	Potato tubers and sprouts	Ethylene	Elmer (1936)
Pectinase	Pea epicotyl	Auxin	Datko and Maclachlan (1968)
	Bean petiole explants	Ethylene	Morre (1968)
Pectin esterase	Pea epicotyl	Auxin	Datko and Maclachlan (1968)
	Cotton leaf blade and pulvinius	Ethylene	Herrero and Hall (1960)
Peroxidase	Tobacco pith explants	ABA	Leshem (1971)
	Lentil root	Auxin	Miassod et al. (1970)
	Lentil root and barley coleoptiles	Cytokinin	Darimont et al. (1971)
	Tobacco pith	Cytokinin	Galston et al. (1968)
	Sugar cane internode	Cytokinin	Gayler and Glasziou (1969)
	Tobacco pith	Ethylene	Adams and Galston (1974)

	Sweet potato root	Ethylene	Stahmann et al. (1966); Imaseki et al. (1968); Gahagan et al. (1968)
	Cotton leaf blade and pulvinus	Ethylene	Herrero and Hall (1960)
	Pumpkin fruit	Ethylene	Ivanov (1932)
	Cucumber, tomato, apple and mandarin fruit	Ethylene	Ivanov et al. (1931)
	Mango fruit slices	Ethylene	Matoo et al. (1968)
	Pea epicotyl	Ethylene	Ridge and Osborne (1970)
	Dwarf corn, expanding leaves	GA_3	McCune and Galston (1959)
	Barley aleurone	GA_3	Briggs (1963)
Peroxidase	Barley aleurone	GA_3	Duffus (1966); Harmey and Murray (1968)
PEP carboxylase	*Lemna minor*	ABA	Bauer et al. (1976)
	Wheat leaves	ABA	Bauer et al. (1976)
Phenylalanine ammonia-lyase	Gherkin hypocotyl	Ethylene	Engelsma and van Bruggen (1971)
	Pea, excised seedlings	Ethylene	Hyodo and Yang (1971)
	Sweet potato root slices	Ethylene	Imaseki et al. (1968)
	Grapefruit peel	Ethylene	Riov et al. (1969)
	Carrot root	Ethylene	Sarkar and Phan (1974b)
6-Phosphogluconate dehydrogenase	Wheat coleoptile	GA_3; Auxin	Likholat and Levina (1973)
Phosphorylcholine-cytidyl transferase	Barley aleurone	GA_3	Johnson and Kende (1971)
Phosphorylcholine-glyceride transferase	Barley aleurone	GA_3	Johnson and Kende (1971); Ben-Tal and Varner (1974)
Phytase	Barley aleurone	GA_3	Pollard and Singh (1968)
Polygalacturonase	Pea roots	Auxin (2,4-D)	Giovannozzi-Sermanni and Pietrosanti (1973)
Polyphenol-oxidase	Tobacco leaves	Ethylene	Shcherbakov (1939)
	Sweet potato roots, potato tubers and parsnip roots	Ethylene	Stahman et al. (1966)
Polyphenol-oxidase (catecholase)	Cotton leaf blade and pulvinus	Ethylene	Herrero and Hall (1960)
(catecholase) (chlorogenic acid	Sweet potato root slices	Ethylene	Imaseki et al. (1968)
Protease	Squash cotyledons	Cytokinin 6-benzylaminopure	Penner and Ashton (1966)

	Pineapple fruit	Ethylene	Regeimbal and Harvey (1927)
	Barley aleurone	GA_3	Yomo (1960b); MacLeod and Millar (1962); Jacobsen and Varner (1967)
Pyruvic decarboxylase	Apple fruit	Ethylene	Hulme et al. (1971a)
Ribonuclease	*Rhoeo discolor* leaf sections	ABA	De Leo and Sacher (1970)
	Barley, excised leaves	ABA	Srivastava (1968b)
	Oat leaf tissue	ABA	Wyen et al. (1972)
	Pea epicotyls	Auxin	Birmingham and Maclachlan (1972)
	Pharbitis nil flowers	Ethylene	Kende and Baumgartner (1974)
	Barley aleurone	GA_3	Chrispeels and Varner (1967)
Ribonuclease	Lupin seed	GA_3	Mierzwinska (1973)
RNA polymerase	Soybean hypocotyl	Auxin	O'Brien et al. (1968)
Ribulose diphosphate carboxylase	Rye seedlings	Cytokinin	Feierabend (1969)
	Cucumber cotyledons	Cytokinin	Harvey et al. (1974)
	Red clover	GA_3	Treharne and Stoddart (1968)
Sucrose synthetase	Sugar beet root	Auxin	Pavlinova and Prasolova (1970)
	Potato tuber	Auxin	Lavintman and Cardini (1968)
Tyramine methylpherase	Barley seedling root tips	Cytokinin (kinetin)	Mann et al. (1963)

Table 9.2.

Examples of Decreased Enzyme Level or Activity Following Hormone Treatment

Enzyme	Tissue	Hormone	Reference
Amaranthin synthesis (light-induced)	*Amaranthus caudalus* seedlings	GA_3	Kinsman et al. (1975)
α-Amylase	Barley aleurone	ABA	Cornforth et al. (1965); Chrispeels and Varner (1966)
Catalase	Tomato, cucumber, apple and mandarin fruit	Ethylene	Ivanov et al. (1931)
Cellobiase	Pea seedlings	Ethylene	Ridge and Osborne (1969)
Cellulase	Pea seedlings	Ethylene	Ridge and Osborne (1969)
Deoxyribonuclease	Detached barley leaves	Cytokinin (kinetin)	Srivastava and Ware (1965); Srivastava (1968b)
Dipeptidase	Squash cotyledon	ABA and IAA	Tsay and Ashton (1974)
Deoxyribonucleic acid polymerase	*Pisum sativum* seedling apices	Ethylene	Apelbaum et al. (1974)
Esterase	Detached tobacco leaves	Cytokinin (kinetin)	Balz (1966)
Gibberellin biosynthesis	Spinach plants and detached corn shoots	ABA	Wareing et al. (1968a)
β-1,3-Glucanase	*Phaseolus vulgaris* abscission zones	GA_3, IAA Cytokinin	Abeles and Forrence (1970)
β-1,3-Glucan synthetase	Bean callus	ABA	Haddon and Northcote (1976)
Invertase	Carrot root discs	ABA	Saunders and Poulson (1968)
		Kinetin	Ricardo (1976)
Invertase (insoluble)	Artichoke tuber	IAA	Edelman and Hall (1964)
	Sugar beet root	GA_3	Cherry (1968)
Isocitratase	Cotton cotyledons	ABA	Ihle and Dure (1972c)
Nitrate reductase	*Agrostemma githago* embryo	ABA	Kende et al. (1974)
	Mungbean root and hypocotyl	ABA	Higgins et al. (1974)
3′-Nucleotidase	Wheat embryos	IAA	Shuster and Gifford (1962)
Pectinase	Pea roots	Auxin (2,4-D)	Giovannozzi-Sermanni and Pietrosanti (1973)

Peroxidase	Tobacco pith	Auxin	Galston et al. (1968)
	Sugar cane	Auxin and GA_3	Glasziou et al. (1968)
	Tobacco pith	Ethylene	Adams and Galston (1974)
	Tobacco leaf discs	Ethylene	Nakaguchi et al. (1970)
	Green pea stem	Auxin	Ockerse et al. (1966)
	Dwarf corn and peas	GA_3	McCune and Galston (1959)
Peptidase (exo)	Cotton embryos	ABA	Ihle and Dure (1972c)
	Detached barley leaves	Cytokinin (kinetin)	Atkin and Srivastava (1969)
	Tobacco leaf discs	Cytokinin (kinetin)	Anderson and Rowan (1966)
Phenylalanine ammonia-lyase	Bean callus	ABA	Haddon and Northcote (1976)
Phosphatase (acid)	*Rhoeo discolor* leaf sections	Auxin	De Leo and Sacher (1970)
	Detached wheat leaves	Cytokinin (kinetin)	Sodek and Wright (1969)
	Detached tobacco leaves	Kinetin	Balz (1966)
Protease	Detached tobacco leaves	Cytokinin (kinetin)	Balz (1966)
	Tropaelum majus leaf discs	Cytokinin (kinetin)	Beevers (1968a)
Ribonuclease	Bean pulvinal tissue	Auxin	Abeles et al. (1968)
	Tomato leaflets	Auxin (IAA)	Dove (1971)
	Rhoeo discolor leaf sections	Auxin	De Leo and Sacher (1970)
	Bean endocarp	Auxin	Sacher (1968)
	Wheat coleoptile sections and green pea stem sections	Auxin (IAA)	Truelsen (1967)
	Tobacco leaves	Cytokinin	Bagi and Farkas (1967)
Ribonuclease (acid)	Detached tobacco leaves	Cytokinin (kinetin)	Balz (1966)
	Pea epicotyls	Cytokinin (6-benzylaminopurine)	Birmingham and Maclachlan (1972)
	Tomato leaves	Cytokinin (6-benzylaminopurine)	McHale and Dove (1969)
	Detached wheat leaves	Cytokinin (kinetin)	Sodek and Wright (1969)

	Detached barley leaves	Cytokinin (kinetin)	Srivastava (1968a)
	Oat leaves	Cytokinin	Udvardy et al. (1967)
Ribonucleic acid polymerase	Maize coleoptiles	ABA (kinetin)	Bex (1972)
	Radish hypocotyl homogenate	ABA	Pearson and Wareing (1969)
Ribulose diphosphate carboxylase	Cereal leaf	ABA	Wellburn et al. (1973)
	Pennisetum typhoides seedling	ABA	Sankhla and Huber (1974, 1975)
Transaminase	Detached lentil roots	ABA	Pilet (1971)

9.6.1. Nitrate reduction

The enzyme nitrate reductase (NR) is subject to substrate and hormonal control in a variety of plant tissues. Cytokinin was shown to increase NR activity in the embryos of *Agrostemma githago,* probably through a mechanism unrelated to nitrate accumulation and different from that involved in substrate induction (Kende et al. 1971; Hirschberg et al. 1972; Kende et al. 1974). These conclusions were based upon studies which showed that cytokinin did not affect nitrate levels and did not increase nitrite reductase, whereas the NO_3^--induced system included NR and nitrite reductase. Furthermore, the increase in NR by cytokinin was more sensitive to RNA and protein synthesis inhibitors than the substrate-induced enzyme, while the reverse proved to be the case when ABA and amino acids were used to inhibit enzyme induction (Kende et al. 1974). Cytokinin could not increase NR activity in tissue-cultured *A. githago,* while tissue cultures of many other plants responded strongly to the hormone (Gunther 1974), although it is possible that culture conditions and tissue age were not optimal in the case of *A. githago.* Therefore, there are developmental and physiological differences between the increase of NR by cytokinin and by the substrate, and these differences may allow further analysis of the mechanism of NR induction by either stimulus.

Cytokinin did not increase NR activity in mung bean hypocotyl or root (Higgins et al. 1974) and only slightly promoted enzyme activity in tobacco leaves (Roth-Bejerano and Lips 1970). However, GA_3 stimulated NR activity in tobacco leaves (Roth-Bejerano and Lips 1970) but had no effect in mung bean seedlings (Higgins et al. 1974). ABA inhibited the substrate-induced enzyme in *A. githago* and mung bean (Higgins et al. 1974; Kende et al. 1974) and, although an attempt was made to determine how ABA mediated the inhibition in mung bean, and it was concluded that the hormone was acting upon posttranscriptional events rather than upon the inhibition of mRNA synthesis (Higgins 1974), there are many limitations in the experimental approach which make such conclusions of doubtful validity (see Chapter 11, section 11.3.2 for general criticisms of this type of conclusion). Both the hormone-promoted and hormone-inhibited increases in NR are attractive responses for further study, especially if specific mRNA could be isolated. This requires details of the properties of the pure enzyme, which may soon follow, since purification seems to be quite possible (Solomonson 1975).

9.6.2. Nitrogen fixation

Nitrogen fixation is a fundamental aspect of the overall production of nitrogen for plant and animal use, and an understanding of the processes involved is extremely important (Bergersen 1971; Dilworth 1974). The complex interaction between the higher plant root and the procaryotic organisms required for fixation demand highly coordinated expression of genetic information and the production and redistribution of energy and intermediary metabolites. The nodules are highly organized and metabolically very active. It seems probable that the nodules and the process of nitrogen fixation are under control by plant hormones in order to maintain the degree of organization, gene expression, metabolism and cellular integrity needed to carry out the processes involved in nitrogen assimilation.

Correlative inhibition of nodule development was demonstrated by Nutman (1952), who showed that, by removing the root tip or nodules from *Trifolium praetense,* nodule number could be increased. This suggested that factors produced by the root meristem or existing nodules controlled the further development of nodules and therefore displayed characteristics commonly associated with plant hormones, such as in release of lateral buds from apical dominance. Subsequently, the level of hormones in nodules compared with surrounding tissue, the level of hormone degrading enzymes in nodules and the effect of exogenous hormones on nodule development were all studied to varying extents. The conclusions which can be drawn from the available data are not particularly satisfying, since no clear-cut role (as determined by our current views on how hormones might act) for the regulators in controlling any aspect of nitrogen fixation has emerged.

Higher levels of growth promoting substances were found in the nodules of *Pisum arvense, Ulex europaeus,* garden pea and bean than in the surrounding root tissue (Pate 1958; Radley 1961), while growth inhibitors could be extracted late in the development of the nodule (Pate 1958). Subsequently, when these substances were characterized, it was found that IAA occurred at high concentrations in the nodules of *Lupinus luteus* and the non-leguminous *Alnus glutinosa,* while the roots of the same plants contained low concentrations of IAA and indolecarboxylic acid (Dullaart 1970a and b). Radley (1961) had earlier identified gibberellin in the nodules of pea and bean. Furthermore, cytokinins are known to be produced by cultured *Rhizobium* spp. in sufficient amounts to induce polyploid mitoses typical of nodule primordia (Phillips and Torrey 1970 and 1972). Cytokinins are known to be present in the root nodules of bean and pea (Puppo et al. 1974; Syono and Torrey 1976). Thus, nodules are well-supplied with a range of the known hormones, but ethylene levels do not appear to have been studied.

At least two proposals have been made to account for the regulation of IAA levels in nodules. Mennes (1973) showed that nodules contained lower levels of IAA-oxidase than the surrounding root tissue, which might explain the differential distribution of the auxin. Ferrous leghaemoglobin promoted the oxidation of IAA in vitro (Puppo and Rigaud 1975) but the significance of this as an in vivo control of IAA levels is difficult to assess.

In an attempt to mimic the in vivo initiation and development of nodules, plant hormones have been applied to the leaves and roots of a variety of species. Spraying GA_3 onto the leaves of dwarf bean inhibited nodulation (Thurber et al. 1958), but the same treatment did not affect nodule initiation or development in *Trifolium repens* (Fletcher et al. 1958). Including gibberellin in the growing medium inhibited nodulation of *Trifolium subterraneum* (Kefford et al. 1960) and beans (Cartwright 1967), but concentration was important, since Kefford et al. (1960) found that low concentrations (10^{-8} *M*) had no effect on nodulation. Similarly, high concentrations of auxin and kinetin inhibited nodule development in *T. subterraneum* and bean while low concentrations had no effect, but the anti-auxin, α-(4-chlorophenoxy)isobutyric acid, increased nodule numbers on both species (Kefford et al. 1960; Cartwright 1967). A partially purified fraction from coconut *(Cocos nucifera)* water stimulated nodule formation on bean root segments and, although the component(s) were not identified, it was concluded from their behaviour that they did not resemble auxins or cytokinins (Schaeffer and Alexander 1967). This is one of the few leads on the nature of the "nodulation factor", and should be pursued. ABA inhibited nodulation of pea roots without having any effect on bacterial or infection thread growth (Phillips 1971) and, by elimination, it was concluded that the hormone inhibited early polyploid mitoses necessary for nodule initiation.

Despite the results obtained from exogenous hormone treatments, and probably because of determinations of endogenous hormone levels in nodules, it has been suggested that a finely controlled balance between auxin and cytokinin may regulate nodule initiation, and that some (if not all) of this information is transmitted by the infection thread (Phillips and Torrey 1970; Libbenga and Torrey 1973; Libbenga et al. 1973). The evidence for this proposition is not very strong, and experimental proof will be difficult to obtain. A possible approach would be to use mutant strains of *Rhizobium* incapable of producing the hormone(s), and supplementing them with a wide range of cytokinin/auxin ratios.

Although it may be possible to implicate hormones in nodule initiation through regulation at the level of mitosis, it is unclear why the nodules contain such high levels of growth-promoting substances. There does not appear to be any study of their role in leghaemoglobin or nitrogenase biosynthesis, nor is there any information on whether they are involved in maintaining the highly organized structural and metabolic integrity of the nodules. It is possible but improbable that they represent "secondary" metabolic products of the bacterial – plant interaction which have no role in the metabolism of the nodule.

9.6.3. The stored nitrogenous reserves of seeds

The deposition and mobilization of seed storage reserves are major metabolic events which may at least be partially under hormonal control. The current state of knowledge on reserve materials in seeds, especially for proteins, is given in reviews by Mayer and Shain (1974), Dure (1975), Millerd (1975) and Ashton (1976). It will be clear that details of the processes involved in the biosynthesis, deposition and mobilization of reserves are not well understood. In the case of the proteins, a major stumbling block is the lack of information about the structural characteristics of the macromolecules, which makes it difficult to study any of the processes involved in deposition and mobilization.

It is not known whether hormones affect the quantitative or qualitative character of deposited nitrogenous reserves, although some correlations can be made between the absolute levels of endogenous hormones and how they vary during seed development when reserves are being laid down (e.g. Wheeler 1972; Eeuwens and Schwabe 1975; King 1976). However, there is no information on the role these regulators play in deposition of reserves, except possibly from the data of Ihle and Dure (1972a, b and c). These workers suggested a model which might account for the preservation of reserves during seed development if one accepts their previous data (see Chapter 11, section 11.3.2), thought to show the existence of hydrolase mRNA which was prevented from being translated during development by ABA and released from hormonal control during germination. This model, although controversial, could help account for the maintenance of storage compounds, but provides no insight into the regulation of reserve biosynthesis. No change in the qualitative pattern of deposited proteins could be detected in cultured pea cotyledons in response to a range of hormones (Millerd et al. 1975), although the technique used would not detect changes of reasonable magnitude, and quantitative changes were not examined.

Considerably more is known about the effect of hormones on mobilization of reserves in germinating seeds (Mayer and Shain 1974; Ashton 1976) than is known about biosynthesis and deposition. Proteolytic enzymes have been purified and well-characterized in a number of cases (see Ashton 1976), and their role in protein degradation during germination is being elucidated. There are some major problems associated with assay procedures (artificial substrates are commonly used) and with distinctions between enzyme-type being assayed. These are being resolved, and it will be more feasible to study hormonal control when conditions are standardized and it is clear whether an endopeptidase or exopeptidase is being studied.

Degradation of stored protein in the cotyledons of germinating squash *(Cucurbita maxima)* may be effected by endopeptidases and dipeptidases, and at least some forms of both type of enzyme appear to be under hormonal control (Penner and Ashton 1966; Tsay and Ashton 1974). Cytokinin could replace the embryonic axis in enhancing the activity of an endopeptidase (Penner and Ashton 1966), while cytokinin, GA_3, auxin and ABA were all shown to have some effect on the appearance of dipeptidase activity (Tsay and Ashton 1974; Ashton 1976). Since the physical characteristics and the physiological development of the dipeptidase during germination have been carefully studied, this would seem to be a good system to use to further characterize the cytokinin response, since relatively little is known about how the hormone regulates the level of enzyme activity.

GA_3 stimulated the de novo synthesis of a protease in barley aleurone layers and the enzyme was released into the medium (Jacobsen and Varner 1967). It was speculated that endosperm proteins could be digested by such an enzyme, providing amino acids for the developing embryo. Four endopeptidases (Sundblom and Nikola 1972) were partially characterized within GA_3-treated aleurone layers. Thus it may be possible to analyze the role of peptidases in germination and the way in which GA_3 regulates their activity. In contrast to the foregoing, however, a peptidase of germinating barley did not respond to GA_3 (Burgher and Siegelman 1966), thus raising the questions of enzyme localization and the possibility of multiple forms of proteolytic enzymes, only some of which may respond to hormones. GA_3 also enhanced the degradation of storage protein in maize endosperm by an unspecified type of protease (Ingle and Hageman 1965), and Harvey and Oaks (1974) have purified an endopeptidase which showed increased activity during germination, and this increase was correlated with endosperm-protein breakdown. Whether gibberellin controls the enzyme activity, as suggested by Ingle and Hageman (1965), was not studied specifically, but ABA inhibited the increased activity while dwarf mutants known to be GA-deficient responded to exogenous GA_3. This data is, therefore, consistent with, but does not prove, the view that GA controls an endopeptidase in maize.

Most of the studies above describe the phenomenon of hormonal enhancement of proteolytic activity in the reserve tissues of seeds, but do not examine the possible ways by which enzyme activity might be regulated. In an extensive study (Ihle and Dure 1969 and 1972a, b and c), it was shown that an exopeptidase (carboxypeptidase) developed in the cotyledons of germinating cotton *(Gossypium hirsutum)* in the presence of actinomycin D (but not in cycloheximide), suggesting to the authors that the increased enzyme activity was independent of transcriptional control. They then studied the developing cotton seed for sensitivity to the drug in an attempt to determine time of transcription of the mRNA for the carboxypeptidase. They found that, about two-thirds of the way through development, the tissue became insensitive to the inhibitor, suggesting that the mRNA was transcribed during embryogenesis and not translated until germination. The presence of ABA was correlated with the "stored mRNA", and a role for the hormone in controlling translation was postulated. There was no examination of the control of transcription. Certain features of this elegant model have been questioned recently, especially the validity of the "stored mRNA" proposal (Smith et al. 1974; Radin and Trelease 1976), since the results with actinomycin D have not been confirmed, probably because of the mode of application (see Chapter 11, section 11.3.2). If the carboxypeptidase could be purified, then it may be possible to isolate the mRNA and thus resolve some of the controversy, and it would also be possible to study control of transcription and the lifetime of the template.

It has become apparent, from careful study of proteolytic enzyme levels in cotyledons of germinating pea and bean, that regulation of enzyme activity is a complex process. In peas, it appears to involve at least hormones, unknown factors from the embryonic axis, end-product inhibition and sink activity of the axis (Varner et al. 1963; Chin et al. 1972; Yomo and Varner 1973). It was suggested, therefore, that the regulation of carbohydrate and protein degradation is quite different, since α- and β-amylases but not protease could be controlled by ABA (Yomo and Varner 1973). In beans (Yomo and Srinivasan 1973) a slightly different situation was described, since here ABA inhibited the development of both amylase and protease. Some of the technical difficulties involved in protease assay, such as heterologous substrates and the fact that these proteolytic enzymes have not been characterized as to number of forms together with precise assay requirements, may make it difficult to monitor changes effected by hormones.

There seems to be scope for a thorough study of the role of hormones in the biosynthesis and deposition of reserves in seeds, since this is an important physiological event which might be expected to be subject to regulation by hormones. Certain features of the study of mobilization of reserves, such as increased knowledge of the nature of the proteolytic enzymes and new methods of assay, may allow further characterization of events regulated by hormones. The existing information on the control of carbohydrate metabolism by GA-induced α-amylase in barley is becoming voluminous and, although the mechanism of action of GA in this system is still not understood, a lot of progress has been made. Similar progress could be expected in the case of hormonal control of dipeptidase activity in germinating squash if this system received some attention. Similarly, the carboxypeptidase of cotton should be reexamined in order to understand the nature of the control being exerted by ABA.

9.6.4. General protein synthesis and amino acid metabolism

When hormone-induced changes in the level of macromolecules were first described, one of the general conclusions was that protein synthesis or amount was generally enhanced following auxin, GA or kinetin treatment (see Trewavas 1968; Key 1969). Two approaches yielded this type of picture. One was to measure total protein levels some time after exposure to the hormone, and the other involved measuring rates and level of accumulation of radioactive amino acids in protein by tissues treated or pretreated with the regulator. The data were not always consistent, and different tissues responded differently. For example, in auxin-treated *Avena* coleoptiles net protein content did not change (Boroughs and Bonner 1953), while net protein level was increased in auxin-treated expanding tuber tissues (Thimann and Loos 1957). Radioactive labelling experiments have been most commonly used, and a vast amount of data has been published on the stimulation of protein synthesis by hormones as measured by the incorporation of labelled amino acids. However, there are many ways in which radioactivity from amino acids might be preferentially detected in protein other than by a direct stimulation of protein biosynthesis. Some of these have been discussed by Kuraishi (1973), who has pointed out that alterations in specific activity of amino acid pools, differential incorporation of amino acids and differential metabolism of amino acids may all lead to false conclusions about the effect of hormones on protein synthesis. The fate of exogenously supplied amino acids, the preferential uptake of some amino acids and the specific activity of the charged tRNA species should also be considered in such studies.

The enhanced protein synthesis or incorporation of amino acids into protein by ethylene in ripening fruit (Hulme 1948; Richmond and Biale 1966; Frankel et al. 1968; Brady et al. 1970; Marei and Romani 1971) serves to illustrate some of the complexity involved in interpreting the various data. Many of the studies have used incorporation of labelled amino acids to monitor protein synthesis, while total protein has been measured in other cases (e.g. Hulme 1948; Wade et al. 1972). Changes in protein quantity and subunit composition have been carefully examined in banana, where it was shown that ethylene enhanced incorporation of amino acids into protein (Brady et al. 1970), but there were no changes in the amount of protein (Wade et al. 1972) nor in the subunit composition of that protein, as determined by sodium dodecyl sulphate-polyacrylamide gel electrophoresis (SDS-PAGE) and a dual labelling technique (Brady and O'Connell 1976). Thus, it has been assumed that ethylene must enhance the turnover of most fractions of the protein. It is possible, however, that other explanations, as outlined above, could account for the observations on radioactive amino acid incorporation.

Ethylene-enhanced amino acid incorporation during abscission (Abeles and Holm 1966) has not been well-documented, but an interesting feature was reported by Webster (1968) when she showed selective labelling (by autoradiography) of the proximal cortical cells of bean explants. Whether these results reflect net accumulation of protein or increased turnover is not clear, and the relationship between this labelling and the changes in enzyme levels considered to be associated with abscission (see Chapter 8, section 8.5.3) remains to be determined.

A major problem associated with some of the foregoing data is the low resolution of the protein fractions being studied in response to hormone. Two approaches have changed this, allowing alterations in minor or individual protein species to be studied. The first employs dual labelling of proteins, followed by separation of the proteins by gel filtration chromatography (Patterson and Trewavas 1967) or separation of the subunits by sodium dodecyl sulfate – polyacrylamide gel electrophoresis (Brady and O'Connell 1976). In brief, [^{3}H]amino acid(s) and the same amino acid(s) labelled with ^{14}C or ^{35}S may be used to label the tissue. In the control, a tissue sample is labelled with ^{3}H and another with ^{14}C, both in the absence of hormone. They are then mixed, extracted, analyzed by gel filtration or SDS-PAGE, and the ratio of ^{14}C to ^{3}H is determined. In this case (control) the ratio should be constant across the gel. In the treatment, a hormone-treated tissue sample labelled with ^{14}C is mixed with a non-treated sample labelled with ^{3}H, and they are extracted and analyzed as above. If the hormone treatment has affected incorporation of label into any specific fractions then, depending upon the length of the column or the gel, its resolving power and the size of the fractions or gel slices taken for radioactive determination, the ratio of ^{14}C to ^{3}H should be altered from the constant determined in the control in selected regions of the separation profile. Results obtained by Patterson and Trewavas (1967) using auxin treatment of pea epicotyls followed by gel filtration analysis of the labelled proteins indicated that there were differential effects on a number of unspecified proteins. The data of Brady and O'Connell (1976) have been discussed earlier. They found no changes in the pattern of labelling of protein subunits following treatment of banana slices with ethylene in the presence of labelled amino acids. In the event that differences are detected by this sensitive method, there still remains the problem of identifying the altered protein species. This could be overcome by purifying a selected enzyme or protein thought to be undergoing changes in turnover or rate of synthesis. Using the second approach to the study of protein metabolism, Peterson and Huffaker (1975) have studied the effect of auxin and kinetin on the rate of degradation of RuDP carboxylase in relation to other proteins in senescing barley leaves. Auxin did not affect degradation of the enzyme but kinetin slowed down the process, although other leaf proteins were more effectively protected from degradation than was RuDP carboxylase. This study was limited in that the effect of the hormones on rate of enzyme synthesis was not studied.

An extensive study of the rates of synthesis and degradation of total protein in *Lemna* by Trewavas (1972) showed that, under optimal growing conditions, ABA reduced the rate constant of protein synthesis and increased the rate constant of degradation. Benzylaminopurine, on the other hand, increased the rate constant of synthesis but did not alter degradation. When the duckweed was transferred to water, the growth rate was reduced, the rate constant of protein degradation increased and the rate constant of synthesis decreased. Under these conditions benzylaminopurine reduced the rate constant of degradation without changing the rate constant of synthesis. This elegant approach to the study of protein metabolism is to be commended, and the major limitation (that individual species were not examined) could be overcome by combining this technique with the approach of Peterson and Huffaker (1975) discussed above.

Ways in which protein metabolism might be altered by plant hormones at the level of posttranslational modification are discussed in Chapter 11, section 11.2.3. This type of change might be expected to alter protein activity, thus accounting for the other major way (other than synthesis or degradation) in which proteins might be regulated by hormones. Clearly, advances are being made in the study of overall protein metabolism in response to regulators, and it is to be hoped that individual protein species will be studied in more detail.

Amino acid metabolism following hormone treatment has received relatively little attention, but it might be expected to regulate various aspects of protein metabolism, especially biosynthesis. Alanine aminotransferase activity in the leaves of *Lolium temulentum* appeared to be selectively affected, in that GA_3 reduced and kinetin increased enzyme activity, while the activity of aspartate aminotransferase was unaltered (Hedley and Stoddart 1971). During germination of *Pennisetum typhoides,* both ABA and kinetin stimulated the activities of alanine and aspartate aminotransferases, while ABA inhibited the activity of glutamic dehydrogenase (GDH). Kinetin overcame the inhibition of GDH by ABA, again indicating complex control of amino acid metabolism by hormones in this system (Huber and Sankhla 1974c). Kuraishi (1973) presented some indirect evidence that auxin controlled amino acid biosynthesis from sugars and amino acid interconversions in mung bean hypocotyl. There is scope for further study of amino acid metabolism, especially upon the biosynthetic and interconverting enzymes and how changes in metabolism relate to protein biosynthesis and degradation.

9.7. Aspects of mineral metabolism

A relatively neglected area of hormone physiology is the interaction between minerals and growth regulators. As mentioned in Chapter 8, there is some fairly recent information on the role of hormones in mobilization and secretion of minerals, particularly from storage tissue. There is also some information about hormonal effects on accumulation of ions in treated tissues far removed from the source of the ion (see e.g. Seth and Wareing 1967; Patrick 1976). There may be some important nutritional implications in the fact that auxin caused the accumulation of N, P and K in the internodes of dwarf bean (Phillips 1968), or that presoaking seeds of maize in auxin solutions resulted in altered uptake of nutrients from soils of varying salinity: sodium in particular was excluded (Darra and Saxena 1973). Thus, the overall mineral composition of plants may be under partial hormonal control, and gradients of hormone concentration no doubt result in differential nutrient distribution within the plant, as shown by the work of Phillips (1968). The very selective and rapid effect of ABA on K^+ efflux from guard cells of stomata is an elegant example of an altered mineral status in response to a hormone (Mansfield and Jones 1971). In addition, mineral deficiency can result in altered levels of ABA, as shown for nutritionally stressed *Nicotiana rustica* (Mizrahi and Richmond 1972).

In the process of tumour and teratomata formation in disbudded tobacco plants, boron interacted with GA_3 in promoting intense xylem proliferation (Skok 1968). The GA_3-enhanced synthesis of α-amylase in barley aleurone was shown to have a requirement for ferric ions, and calcium ions were needed at a specified time during the lag period (Goodwin and Carr 1970 and 1972). Other examples of hormone interactions with minerals in a number of physiological responses have been presented by Leopold et al. (1974).

For the purposes of this chapter, it is sufficient to note that hormones alter mineral status and vice versa in plants. There are clear implications for control of enzyme activity, since various mineral elements are involved as cofactors in a wide range of enzymes. This possibility of regulation of metabolism by hormones should not be ruled out.

9.8. Summary and conclusions

Plant hormone research over the past 15 years has been concentrated upon aspects of gene expression and a few rapid responses. It is apparent that basic processes such as photosynthesis, nitrogen, carbohydrate and lipid metabolism and respiration have been generally neglected. There are some exceptions, however, the most notable being the extensive data accumulated on those aspects of nitrogen metabolism which involve protein biosynthesis, although protein turnover, which is relevant in this connection, has received very little attention. It is probably fair to say that considerably more is known about the details of the above processes than is known about gene expression in plants, and thus they represent more realistic targets for research with respect to plant hormones. However, most of the studies on respiration and hormones are quite old, except the work with ethylene. It is clear that there are some exciting prospects in the further elucidation of the alternative electron transport pathway which appears to be stimulated by the gas. A general revival of interest in intermediary metabolism might encourage further research on the role of hormones in this area, and it is to be hoped that a systematic study of the control of respiratory enzyme activity will be undertaken in a suitable system.

The current interest in plant membranes may result in the further study of electron transport in plant mitochondria and chloroplasts, since this might be expected to yield information about membrane organization and energy production. Electron transport, although not particularly well understood, is being intensively studied in mammalian cells, and our knowledge of this event in plant cells will no doubt benefit and provide the opportunity to study hormone action in this important process. Some recent studies on plant hormones and photosynthesis have indicated that electron transport may be a target for auxin and abscisic acid. The biogenesis of the mitochondrion and chloroplast and the control of enzyme activity are areas which would greatly benefit from further study using plant hormones and, again because of the interest in membrane organization, it is to be expected that there will be active research in this area.

Analysis of lipid metabolism has involved primarily the study of endoplasmic reticulum changes, especially during alterations in protein synthesis, and although these studies are still at a formative stage they indicate that hormones play a regulatory role in the organization of this membrane system. Whether hormones influence the lipid incorporation into other membranes such as the plasmalemma or those of the organelles does not appear to have been studied in any detail. Few aspects of lipid biosynthesis or degradation, with the exception of the enzymes of the glyoxalate cycle, have been studied in relation to hormonal control.

A feature of nitrogen metabolism which has merited some attention has been the cytokinin control of nitrate reduction and, thus, availability of ammonia for incorporation into amino acids and the regulation of proteolytic enzymes in germinating seeds. The picture emerging from these studies is one in which amino acid availability may be under cytokinin control in developing seedlings. There are data on these processes which should allow a fruitful study of the mechanisms involved in their control. A particularly neglected area is the study of the role of minerals in cellular metabolism and their interaction with hormones in controlling many plant processes. Recent work in this area has begun to highlight the importance of K^+, H^+ and Ca^{2+} in mediating hormonal responses, and it is to be expected that the list will be extended rapidly. In conclusion, it should be clear that, although many aspects of the basic metabolic processes have not been studied recently in conjunction with hormones, there is now great scope for further work, especially because of the relatively good state of our knowledge of some of those processes, and because preliminary experimental approaches have indicated that hormones may play a regulatory role.

References

Abeles, F.B. (1969) Plant Physiol., 44, 447–452.
Abeles, F.B. (1972) Ann. Rev. Plant Physiol., 23, 259–292.
Abeles, F.B. (1973) Ethylene in Plant Biology, Academic, New York.
Abeles, F.B. and Forrence, L.E. (1970) Plant Physiol., 45, 395–400.
Abeles, F.B. and Holm, R.E. (1966) Plant Physiol., 41, 1337–1342.
Abeles, F.B., Holm, R.E. and Gahagan, H.E. (1968) in: Biochemistry and Physiology of Plant Growth Substances, F. Wightman and G. Setterfield (eds.), Runge, Ottawa, pp. 1515–1523.
Abeles, F.B., Bosshart, R.P., Forrence, L.E. and Harbig, W.H. (1971) Plant Physiol., 47, 129–134.
Adams, W.R. and Galston, A.W. (1974) Plant Physiol., 53, 931–933.
Addicott, F.T. and Lyon, J.L. (1969) Ann. Rev. Plant Physiol., 20, 139–164.
Adedipe, N.O., Hunt, L.O. and Fletcher, R.A. (1971) Physiol. Plant., 25, 151–153.
Alberte, R.S. and Naylor, A.W. (1975) Plant Physiol., 55, 1079–1081.
Alscher, R.G. and Castelfranco, P.A. (1972) Plant Physiol., 50, 400–403.
Anderson, J.W. and Rowan, K.S. (1966) Biochem. J., 98, 401–404.
Anker, L. (1953) Acta bot. neerl., 2, 22–65.
Apelbaum, A., Sfakiotakis, E. and Dilley, D.R. (1974) Plant Physiol., 54, 125–128.
Ashton, F.M. (1976) Ann. Rev. Plant Physiol., 27, 95–117.
Atkin, R.K. and Srivastava, B.I.S. (1969) Plant Physiol., 22, 742–750.
Audus, L.J. and Garrard, A. (1953) J. exp. Bot., 4, 330–348.
Axel, R., Feigelson, P. and Schutz, G. (1976) Cell, 7, 247–254.
Bagi, G. and Farkas, G.L. (1967) Phytochemistry, 6, 161–169.
Ballantyne, D.J. (1966) Canad. J. Bot., 44, 117–119.
Balz, H.P. (1966) Planta, 70, 207–236.
Banerji, D. and Kumar, N. (1975) Biochem. biophys. Res. Commun., 65, 940–944.
Bannerjee, H.K. and Kar, B.K. (1941) Curr. Sci., 10, 289–290.
Barker, J. and Solomos, T. (1962) Nature (Lond.), 196, 189.
Bauer, R., Huber, W. and Sankhla, N. (1976) Z. Pflanzenphysiol., 77, 237–246.
Baxter, R. and Hanson, J.B. (1968) Planta, 82, 246–260.
Beevers, H. (1961) Respiratory Metabolism in Plants, Row Peterson, New York.
Beevers, L. (1968a) in: Biochemistry and Physiology of Plant Growth Substances, F. Wightman and G. Setterfield (eds.), Runge Press, Ottawa, pp. 1417–1435.
Beevers, L. (1968b) Phytochemistry, 7, 1837–1844.
Ben-Tal, Y. and Varner, J.E. (1974) Plant Physiol., 54, 813–816.
Berger, J. and Avery, G.S., Jr. (1943) Amer. J. Bot., 30, 297–302.
Bergerson, F.J. (1971) Ann. Rev. Plant Physiol., 22, 121–140.
Berridge, M.V. and Ralph, R.K. (1971) Plant Physiol., 47, 562–567.
Bex, J.H.M. (1972) Planta, 103, 11–17.
Bhattacharyya, J. and Roy, S.C. (1969) Biochem. biophys. Res. Commun., 35, 606–610.
Bidwell, R.G.S. and Turner, W.B. (1966) Plant Physiol., 41, 267–270.
Birmingham, B.C. and Maclachlan, G.A. (1972) Plant Physiol., 49, 371–375.
Bishop, J., Morton, J.G., Rosbash, M. and Richardson, M. (1974) Nature (Lond.), 250, 199–204.

Black, H.S. and Altschul, A.M. (1965) Biochem. biophys. Res. Commun., 19, 661–664.
Boasson, R., Bonner, J.J. and Laetsch, W.M. (1972) Plant Physiol., 49, 97–101.
Bonner, J. (1933) J. gen. Physiol., 17, 63–76.
Bonner, W.D., Jr. (1973) in: Phytochemistry, Vol. 3, L.P. Miller (ed.), Van Nostrand, Reinhold, New York, pp. 221–261.
Boothby, D. and Wright, S.T.C. (1962) Nature (Lond.), 196, 389–390.
Boroughs, H. and Bonner, J. (1953) Arch. Biochem., 46, 279–296.
Borriss, H. (1967) Wiss. Z. Univ. Rostock, 16, 629–639.
Bradbeer, J.W. and Pinfield, N.J. (1967) New Phytol., 66, 515–523.
Brady, C.J. and O'Connell, P.B.H. (1976) Aust. J. Plant Physiol., 3, 301–310.
Brady, C.J., Palmer, J.K., O'Connell, P.B.H. and Smillie, R.M. (1970) Phytochemistry, 9, 1037–1047.
Briggs, D.E. (1963) J. Inst. Brew.. (Lond.), 69, 13–19.
Brock, B.L.W. and Fletcher, R.A. (1969) Nature (Lond.), 224, 184–185.
Broughton, W.J. and McComb, A.J. (1971) Ann. Bot. (Lond.), 35, 213–228.
Buller, D.C., Parker, W. and Grant-Reid, J.S. (1976) Nature (Lond.), 260, 169–170.
Burgher, W.C. and Seigelman, H.W. (1966) Physiol. Plant., 19, 1089–1093.
Camm, E.L. and Towers, G.H.N. (1973) Phytochemistry, 12, 961–973.
Carlier, A. and Van Hove, C. (1964) Nature (Lond.), 201, 677–679.
Cartwright, P.M. (1967) Wiss. Z. Univ. Rostock, 16, 537–538.
Chalmers, D.J. and Rowan, K.S. (1971) Plant Physiol., 48, 235–240.
Chalutz, E., De Vay, J.E. and Maxie, E.C. (1969) Plant Physiol., 44, 235–241.
Cherry, J.H. (1968) in: Biochemistry and Physiology of Plant Growth Substances, F. Wightman and G. Setterfield (eds.), Runge, Ottawa, pp. 417–431.
Chin, T.Y., Poulson, R. and Beevers, L. (1972) Plant Physiol., 49, 482–489.
Chrispeels, M.J. and Varner, J.E. (1966) Nature (Lond.), 212, 1066–1067.
Chrispeels, M.J. and Varner, J.E. (1967) Plant Physiol., 42, 398–406.
Christiansen, G.S. and Thimann, K.V. (1950) Arch. Biochem., 26, 248–259.
Cleland, R.E. (1961) in: Handbuch der Pflanzenphysiologie, W. Ruhland (ed.), Springer, Berlin, pp. 754–783.
Clum, H.H. (1967) Plant Physiol., 42, 568–572.
Cohen, R.Z. and Goodwin, T.W. (1962) Phytochemistry, 1, 67–72.
Commoner, B. and Thimann, K.V. (1941) J. gen. Physiol., 24, 279–296.
Constabel, F., Shyluk, J.P. and Gamborg, O.L. (1971) Planta, 96, 306–316.
Cornforth, J.W., Milborrow, B.V., Ryback, G. and Wareing, P.F. (1965) Nature (Lond.), 205, 1269–1270.
Coulombe, L-J. and Paquin, R. (1959) Canad. J. Bot., 37, 897–901.
Craker, L.E. (1975) Phytochemistry, 14, 151–153.
Craker, L.E., Standley, L.A. and Starbuck, M.J. (1971) Plant Physiol., 48, 349–352.
Darimont, E., Gaspar, T. and Hofinger, M. (1971) Z. Pflanzenphysiol., 64, 232–240.
Darra, B.L. and Saxena, S.N. (1973) Plant Soil, 38, 657–661.
Datko, A.H. and Maclachlan, G.A. (1968) Plant Physiol., 43, 735–742.
Datta, A., Sen, S.P. and Datta, A.G. (1965) Biochim. Biophys. Acta (Amst.), 108, 147–149.
Davies, E. (1976) Plant Physiol., 57, 516–518.
De, D.K. and Roy, S.C. (1974) Physiol. Plant, 31, 51–54.
De Boer, J. and Feierabend, J. (1974) Z. Pflanzenphysiol., 71, 261–270.
Dedolph, R.R., Wittwer, S.H., Tuli, V. and Gilbert, D. (1962) Plant Physiol., 37, 509–512.
De Leo, P. and Sacher, J.A. (1970) Plant Physiol., 46, 806–811.
Denny, F.E. (1924) Bot. Gaz. (Chic.), 77, 322–329.
Dilworth, M.J. (1974) Ann. Rev. Plant Physiol., 25, 81–114.
Donaldson, R., Loescher, W. and Newman, D.W. (1969) Amer. J. Bot., 56, 1167–1172.
Dostal, M.C. and Leopold, A.C. (1967) Science, 158, 1579–1580.
Dove, L.D. (1971) New Phytol., 70, 397–401.

Duffus, J.H. (1966) J. Inst. Brew. (Lond.), 72, 569–573.
Dullaart, J. (1970a) Acta bot. neerl., 19, 573–615.
Dullaart, J. (1970b) J. exp. Bot., 21, 975–984.
Dure, L.S., III, (1975) Ann. Rev. Plant Physiol., 26, 259–278.
Edelman, J. and Hall, M.A (1964) Nature (Lond.), 201, 296–297.
Eeuwens, C.J. and Schwabe, W.W. (1975) J. exp. Bot., 26, 1–14.
El Hinnaway, E. (1974) Z. Pflanzenphysiol., 74, 95–105.
Elliot, D.C. and Murray, A.W. (1974) in: Plant Growth Substances 1973, Hirokawa, Tokyo, pp. 691–698.
Elmer, O.H. (1936) J. agr. Res., 52, 609–626.
Elstner, E.F. and Konze, J.R. (1974) FEBS Lett., 45, 18–21.
Engelsma, G. and Van Bruggen, J.M.H. (1971) Plant Physiol., 48, 94–96.
Evins, W.H. and Varner, J.E. (1971) Proc. nat. Acad. Sci. (Wash.), 68, 1631–1633.
Fan, D.F. and Maclachlan, G.A. (1966) Canad. J. Bot., 44, 1025–1034.
Feierabend, J. (1969) Planta, 84, 11–29.
Feierabend, J. (1970) Planta, 94, 1–15.
Firn, R.D. and Kende, H. (1974) Plant Physiol., 54, 911–915.
Fletcher, R.A. (1969) Planta, 89, 1–8.
Fletcher, R.A. and Adedipe, N.O. (1972) in: Plant Growth Substances 1970, D.J. Carr (ed.), Springer, Berlin, pp. 571–580.
Fletcher, R.A. and McCullagh, D. (1971) Planta, 101, 88–90.
Fletcher, W.W., Alcorn, J.W.S. and Raymond, J.C. (1958) Nature (Lond.), 182, 1319–1320.
Frenkel, C., Klein, I. and Dilley, D.R. (1968) Plant Physiol., 43, 1146–1153.
Gahagan, H.E., Holm, R.E. and Abeles, F.B. (1968) Physiol. Plant., 21, 1270–1279.
Galston, A.W. (1969) in: Perspectives in Phytochemistry, J.B. Harborne and T. Swain (eds.) Academic, London, pp. 193–204.
Galston, A.W., Lavee, S. and Siegel, B.Z. (1968) in: Biochemistry and Physiology of Plant Growth Substances, F. Wightman and G. Setterfield (eds.), Runge, Ottawa, pp. 455–472.
Gayler, K.R. and Glasziou, K.T. (1968) Phytochemistry, 7, 1247–1251.
Gayler, K.R. and Glasziou, K.T. (1969) Planta, 84, 185–194.
Genest, K. (1966) Experientia (Basel), 22, 681–682.
Gepstain, S. and Ilan, I. (1970) Plant Cell Physiol., 11, 819–822.
Gibbs, M. and Beevers, H. (1955) Plant Physiol., 30, 343–347.
Giovannozzi-Sermanni, G. and Pietrosanti, T. (1973) Agrochimica, 18, 46–52.
Glasziou, K.T. (1957) Nature (Lond.), 179, 1083–1084.
Glasziou, K.T. and Waldron, J.C. (1964) Nature (Lond.), 203, 541–542.
Glasziou, K.T., Gayler, K.R. and Waldron, J.C. (1968) in: Biochemistry and Physiology of Plant Growth Substances, F. Wightman and G. Setterfield (eds.), Runge, Ottawa, pp. 433–442.
Goldthwaite, J. (1974) Plant Physiol., 54, 399–403.
Goldthwaite, J.J. and Laetsch, W.M. (1968) Plant Physiol., 43, 1855–1858.
Goodwin, P.B. and Carr, D.J. (1970) Cytobios, 2, 165–174.
Goodwin, P.B. and Carr, D.J. (1972) in: Plant Growth Substances 1970, D.J. Carr (ed.), Springer, Berlin, pp. 378–387.
Gordon, M.E. and Letham, D.S. (1975) Aust. J. Plant Physiol., 2, 129–154.
Gregor, H.D. (1974) Protoplasma, 80, 273–277.
Grisebach, H. and Hahlbrock, K. (1974) in: Recent Advances in Phytochemistry, Vol. 8, V.C. Runeckles and E.E. Conn (eds.), Academic, New York, pp. 22–52.
Gunther, G. (1974) in: Plant Growth Substances 1973, Hirokawa, Tokyo, pp. 671–674.
Haber, A.H. and Tolbert, N.E. (1957) Plant Physiol., 32, 152–153.
Haddon, L. and Northcote, D.H. (1976) J. Cell Sci., 20, 47–55.
Hahlbrock, K., Ebel, J., Ortman, R., Sutter, A., Willmann, E. and Grisebach, H. (1971) Biochim. biophys. Acta (Amst.), 244, 7–15.

Hall, M.A. and Ordin, L. (1968) in: Biochemistry and Physiology of Plant Growth Substances, F. Wightman and G. Setterfield (eds.), Runge, Ottawa, pp. 659–671.
Hall, W.C. et Morgan, P.W. (1964) in: Regulateurs naturels de la croissance vegetale, J.P. Nitsch (ed.), C.N.R.S., Paris, pp. 727–745.
Hallam, N.D. (1970) J. exp. Bot., 21, 1031–1038.
Harborne, J.B. (1967) Comparative Biochemistry of the Flavonoids, Academic, London.
Harmey, M.A. and Murray, A.M. (1968) Planta, 83, 387–389.
Harvey, B.M.R. and Oaks, A. (1974) Plant Physiol., 53, 453–457.
Harvey, B.M.R., Lu, B.C. and Fletcher, R.A. (1974) Canad. J. Bot., 52, 2581–2586.
Hayashi, T., Murakami, Y. and Matsunaka, S. (1956) Bull. agr. Chem. Soc. (Japan), 20, 159–164.
Hedley, C.L. and Stoddart, J.L. (1971) J. exp. Bot., 22, 249–261.
Herrero, F. and Hall, W.C. (1960) Physiol. Plant., 13, 736–750.
Higgins, T.J.V. (1974) in: Plant Growth Substances 1973, Hirokawa, Tokyo, pp. 599–607.
Higgins, T.J.V., Goodwin, P.B. and Carr, D.J. (1974) Aust. J. Plant Physiol., 1, 1–8.
Higgins, T.J.V., Zwar, J.A. and Jacobsen, J.V., in: Acides Nucleiques et Sythese des Proteines chez les Vegetaux 261, C.N.R.S., Paris, pp. 481–486.
Hirschberg, K., Hubner, G. and Borriss, H. (1972) Planta, 108, 333–337.
Ho, D.T.H. and Varner, J.E. (1976) Plant Physiol., 57, 175–178.
Hoffman, F. and Kull, U. (1974) Experientia (Basel), 30, 746–747.
Horton, R.F. and Osborne, D.J. (1967) Nature (Lond.), 214, 1086–1088.
Huber, W. and Sankhla, N. (1974a) Planta, 116, 55–64.
Huber, W. and Sankhla, N. (1974b) Z. Pflanzenphysiol., 71, 275–280.
Huber, W. and Sankhla, N. (1974c) Z. Pflanzenphysiol., 73, 160–166.
Hulme, A.C. (1948) Biochem. J., 43, 343–349.
Hulme, A.C., Rhodes, M.J.C. and Wooltorton, L.S.C. (1971a) Phytochemistry, 10, 749–756.
Hulme, A.C., Rhodes, M.J.C. and Wooltorton, L.S.C. (1971b) Phytochemistry, 10, 1315–1323.
Humphreys, T.E. and Dugger, W.H. (1957) Plant Physiol., 32, 136–140.
Hyodo, H. and Yang, S.F. (1971) Plant Physiol., 47, 765–770.
Ihle, J.N. and Dure, L., III (1969) Biochem. biophys. Res. Commun., 36, 705–710.
Ihle, J.N. and Dure, L., III (1972a) J. biol. Chem., 247, 5034–5040.
Ihle, J.N. and Dure, L., III (1972b) J. biol. Chem., 247, 5041–5047.
Ihle, J.N. and Dure, L., III (1972c) J. biol. Chem., 247, 5048–5055.
Imaseki, H., Uchiyama, M. and Uritani, I. (1968) Agr. biol. Chem., 32, 387–389.
Ingle, J. and Hageman, R.H. (1965) Plant Physiol., 40, 672–675.
Ivanov, N.N. (1932) Biochem. Z., 254, 71–87.
Ivanov, N.N., Prokoshev, S.M. and Gabunya, M.K. (1931) Bull. appl. Bot. Genet. Plant Breed., 25, 262–278.
Jacobsen, J.V. (1972) in: Plant Growth Substances 1970, D.J. Carr (ed.), Springer, Berlin, pp. 336–343.
Jacobsen, J.V. (1977) Ann. Rev. Plant Physiol., 28, 537–564.
Jacobsen, J.V. and Varner, J.E. (1967) Plant Physiol., 42, 1596–1600.
Johnson, K.D. and Kende, H. (1971) Proc. nat. Acad. Sci. (Wash.), 68, 2674–2677.
Jones, R.L. (1969) Planta, 87, 119–133.
Joshi, V.C. and Sidbury, J.B. (1976) Arch. Biochem., 173, 403–414.
Kaminek, M. and Lustinec, J. (1974) Z. Pflanzenphysiol., 73, 65–73.
Kaminek, M. and Stemberova, A. (1967) Biol. Plant., 9, 142–148.
Kang, B.G. and Burg, S.P. (1972) Plant Physiol., 49, 631–633.
Kang, B.G. and Burg, S.P. (1973) Planta, 110, 227–235.
Kasamo, K. and Yamaki, T. (1974) Plant Cell Physiol., 15, 965–970.
Katsumi, M. and Fukahara, M. (1969) Physiol. Plant., 22, 68–75.
Kaufman, P.B., Ghosheh, N.S. and Ikuma, H. (1968) Plant Physiol., 43, 29–34.
Kaul, K. and Sabharwal, P.S. (1974) Plant Physiol., 54, 644–648.

Pinfield, N.J. (1968b) J. exp. Bot., 19, 452–459.
Poljakoff-Mayber, A. (1955) J. exp. Bot., 6, 321–327.
Pollard, C.J. (1969) Plant Physiol., 44, 1227–1232.
Pollard, C.J. and Nelson, D.O. (1971) Biochim. biophys. Acta (Amst.), 244, 372–376.
Pollard, C.J. and Singh, B.N. (1968) Biochem. biophys. Res. Commun., 33, 321–326.
Pooviah, B.W. and Rasmussen, H.P. (1974) Amer. J. Bot., 61, 68–73.
Poskuta, J., Antoszewski, R. and Faltynowicz, M. (1972) Photosynthetica, 6, 370–374.
Puppo, A. and Rigaud, J. (1975) Physiol. Plant., 35, 181–185.
Puppo, A. Rigaud, J. and Barthe, P. (1974) C.R. Acad. Sci. (Paris), 279, 2029–2032.
Radin, J.W. and Trelease, R.N. (1976) Plant Physiol., 57, 902–905.
Radley, M.E. (1961) Nature (Lond.), 191, 684–685.
Ratner, A., Goren, R. and Monselise, S.P. (1969) Plant Physiol., 44, 1717–1723.
Ray, P.M. (1973) Plant Physiol., 51, 601–608.
Ray, P.M. (1974) Rec. Advanc. Phytochem., 7, 93–122.
Rebeiz, C.A. and Castelfranco, P.A. (1973) Ann. Rev. Plant Physiol., 24, 129–172.
Regeimbal, L.O. and Harvey, R.B. (1927) J. Amer. Chem. Soc., 49, 1117–1118.
Regeimbal, L.O., Vacha, G.A. and Harvey, R.B. (1927) Plant Physiol., 2, 357–359.
Ricardo, C.P.P. (1976) Phytochemistry, 15, 615–617.
Richmond, A. and Biale, J.B. (1966) Plant Physiol., 41, 1247–1253.
Richmond, A.E. and Lang, A. (1957) Science, 125, 650–651.
Ridge, I. and Osborne, D.J. (1969) Nature (Lond.), 223, 318–319.
Ridge, I. and Osborne, D.J. (1970) J. exp. Bot., 21, 843–856.
Riov, J., Monselise, S.P. and Kahan, R.S. (1969) Plant Physiol., 44, 631–635.
Roth-Bejerano, N. and Lips, S.H. (1970) New Phytol., 69, 165–169.
Russell, D.W. and Galston, A.W. (1969) Plant Physiol., 44, 1211–1216.
Rutherford, P.P. and Bard, D.R. (1971) Phytochemistry, 10, 1635–1638.
Sacher, J.A. (1968) Plant Physiol., 44, 313–314.
Sacher, J.A. and Glasziou, K.T. (1962) Biochem. biophys. Res. Commun., 8, 280–282.
Salminen, S.O. and Young, R.E. (1975) Plant Physiol., 55, 45–50.
Sandegren, E. and Belling, H. (1958) Die Brauerei, 11, 231–238.
Sankhla, N. and Huber, W. (1974) Physiol. Plant., 30, 291–294.
Sankhla, N. and Huber, W. (1975) Z. Pflanzenphysiol., 74, 267–271.
Sargent, J.A. and Osborne, D.J. (1975) Planta, 124, 199–205.
Sarkar, S.K. and Phan, C.T. (1974a) Physiol. Plant., 30, 72–76.
Sarkar, S.K. and Phan, C.T. (1974b) Physiol. Plant., 32, 318–321.
Sarkissian, I.V. (1970) Biochem. biophys. Res. Commun., 40, 1385–1390.
Sarkissian, I.V. (1972) in: Plant Growth Substances 1970, D.J. Carr (ed.), Springer, Berlin, pp. 265–271.
Sarkissian, I.V. and McDaniel, R.G. (1966) Biochim. biophys. Acta (Amst.), 128, 413–418.
Sarkissian, I.V. and Schmalsteig, F.C. (1969a) Life Sci., 8, 933–938.
Sarkissian, I.V. and Schmalsteig, F.C. (1969b) Naturwissenschaften, 56, 284.
Saunders, P.F. and Poulson, R.H. (1968) in: Biochemistry and Physiology of Plant Growth Substances, F. Wightman and G. Setterfield (eds.), Runge, Ottawa, pp. 1581–1591.
Scala, J., Patrick, C. and Macbeth, G. (1969) Phytochemistry, 8, 37–44.
Schaeffer, A.G. and Alexander, M. (1967) Plant Physiol., 42, 563–567.
Schaeffer, G.W. and Sharpe, F.T., Jr. (1971) Physiol. Plant., 25, 456–460.
Schaeffer, G.W., St. John, J.B. and Sharpe, F.T., Jr. (1972) Biochim. biophys. Acta (Amst.), 261, 38–43.
Schaeverbeeke, J. (1966) C.R. Acad. Sci. (Paris) Ser. D, 262, 277–280.
Scott, K.J., Daly, J. and Smith, H.H. (1964) Plant Physiol., 39, 709–711.
Seitz, K. and Lang, A. (1968) Plant Physiol., 43, 1075–1082.
Seth, A.K. and Wareing, P.F. (1967) J. exp. Bot., 18, 65–77.

Shaw, M., Samborski, D.J. and Oaks, A. (1958) Canad. J. Bot., 36, 233–237.
Shcherbakov, A.P. (1939) Bull. Acad. Sci. URSS Ser. Biol., 6, 975–988.
Shuster, L. and Gifford, R.H. (1962) Arch. Biochem., 96, 534–540.
Simmonds, J.A. and Simpson, G.M. (1971) Canad. J. Bot., 49, 1833–1840.
Simpkins, J. and Street, H.E. (1970) J. exp. Bot., 21, 170–185.
Simpson, G.M. and Naylor, J.M. (1962) Canad. J. Bot., 40, 1659–1673.
Singh, O.S. and Chhabra, N. (1976) Z. Pflanzenphysiol., 78, 189–196.
Skok, J. (1968) Plant Physiol., 43, 384–388.
Smirnov, A.I. and Krainev, S.I. (1940) Bull. Acad. Sci. URSS Ser. Biol., 4, 577–588.
Smith, H. and Attridge, T.H. (1970) Phytochemistry, 9, 487–495.
Smith, R.H., Schubert, A.M. and Benedict, C.R. (1974) Plant Physiol., 54, 197–200.
Smolenska, G. and Lewak, S. (1974) Planta, 116, 361–370.
Sodek, L. and Wright, S.T.C. (1969) Phytochemistry, 8, 1629–1640.
Solomos, T. and Laties, G.G. (1974) Plant Physiol., 54, 506–511.
Solomos, T. and Laties, G.G. (1975) Plant Physiol., 55, 73–78.
Solomos, T. and Laties, G.G. (1976) Biochem. biophys. Res. Commun., 70, 663–671.
Solomonson, L. (1975) Plant Physiol., 56, 853–855.
Srivastava, B.I.S. (1963) Arch. Biochem., 103, 200–205.
Srivastava, B.I.S. (1965) Arch. Biochem., 110, 97–103.
Srivastava, B.I.S. (1968a) Biochem. biophys. Res. Commun., 32, 533–538.
Srivastava, B.I.S. (1968b) Biochim. biophys. Acta (Amst.), 169, 534–536.
Srivastava, B.I.S. and Ware, G. (1965) Plant Physiol., 40, 62–64.
Stafford, H.A. (1974) Ann. Rev. Plant Physiol., 25, 459–486.
Stahmann, M.A., Clare, B.G. and Woodbury, W. (1966) Plant Physiol., 41, 1505–1512.
Stearns, E.M. Jr. and Morton, W.T. (1975) Phytochemistry, 14, 619–622.
Steinhart, C.E., Mann, J.D. and Mudd, S.H. (1964) Plant Physiol., 39, 1030–1038.
Stetler, D.A. and Laetsch, W.M. (1965) Science, 149, 1387–1388.
Stewart, G.R. and Smith, H. (1972) J. exp. Bot., 23, 875–885.
Stobart, A.K., Pinfield, N.J. and Kinsman, L.T. (1970) Planta, 94, 152–155.
Stobart, A.K., Shewry, P.R. and Thomas, D.R. (1972) Phytochemistry, 11, 571–577.
Stoddart, J.L. (1968) Planta, 81, 106–112.
Stumpf, P.K. (1976) in: Plant Biochemistry, J. Bonner and J.E. Varner (eds.), Academic, New York, pp. 428–461.
Sudi, J. (1964) Nature (Lond.), 201, 1009–1010.
Sudi, J. (1966) New Phytol., 65, 9–21.
Sugiura, M. (1963) Bot. Mag. (Tokyo), 76, 309–310.
Sundblom, N. and Nikola, J. (1972) Physiol. Plant., 27, 281–284.
Sunderland, N. and Wells, B. (1968) Ann. Bot. (Lond.), 32, 327–346.
Sveshnikova, I.N., Kulaeva, O.N. and Bolyakina, Y.P. (1966) Fiziol. Rast., 13, 769–774.
Switzer, C.M. (1957) Plant Physiol., 32, 42–44.
Syono, K. and Torrey, J.G. (1976) Plant Physiol., 57, 602–606.
Szalai, I. (1968) Planta, 83, 161–165.
Tager, J.M. and Biale, J.B. (1957) Physiol. Plant., 10, 79–85.
Tamas, I.A., Atkins, B.D., Ware, S.M. and Bidwell, R.G.S. (1972) Canad. J. Bot., 50, 1523–1527.
Tamas, I..A., Schwartz, J.W., Breithaupt, B.J., Hagin, J.M. and Arnold, P.H. (1974) in: Plant Growth Substances 1973, Hirokawa, Tokyo, pp. 1159–1168.
Tavener, R.J.A. and Laidman, D.L. (1972a) Phytochemistry, 11, 981–987.
Tavener, R.J.A. and Laidman, D.L. (1972b) Phytochemistry, 11, 989–997.
Tetley, R.M. and Thimann, K.V. (1974) Plant Physiol., 54, 294–303.
Teubner, F.G. and Murneek, A.E. (1952) Science, 116, 39–41.
Thimann, K.V. (1972) in: Plant Physiology. A Treatise, Vol. 6b, F.C. Steward (ed.), Academic, New York, pp. 1–365.

Thimann, K.V. and Loos, G.M. (1957) Plant Physiol., 32, 274–279.
Thomas, T.H., Wareing, P.F. and Robinson, P.M. (1965) Nature (Lond.), 205, 1269–1272.
Thurber, G.A., Douglas, J.R. and Galston, A.W. (1958) Nature (Lond.), 181, 1082–1083.
Treharne, K.J. and Stoddart, J.L. (1968) Nature (Lond.), 220, 457–458.
Treharne, K.J., Stoddart, J.L. and Headley, C.L. (1972) in: Second International Conference on Photosynthesis Research, Stressa, 1971, G. Forti, M. Avron and A. Melandri (eds.), Dr. W. Junk, The Hague, pp. 2497–2509.
Treharne, K.J., Stoddart, J.L., Pughe, J., Paranjothy, K. and Wareing, P.F. (1970) Nature (Lond), 228, 129–131.
Trewavas, A. (1968) in: Progress in Phytochemistry, Vol. 1, L. Reinhold and Y. Liwschitz (eds.) Interscience, London, pp. 113–160.
Trewavas, A.J. (1972) Plant Physiol., 49, 47–51.
Truelsen, T.A. (1967) Physiol. Plant., 20, 1112–1119.
Tsay, R. and Ashton, F.M. (1974) Phytochemistry, 13, 1759–1763.
Tuli, V., Dilley, D.R. and Wittwer, S.H. (1964) Science, 146, 1477–1479.
Turkova, N.S., Vasileva, L.N., Cheremukhina, L.F. (1965) Sov. Plant Physiol., 12, 721–726.
Turner, W.B. and Bidwell, R.G.S. (1965) Plant Physiol., 40, 446–451.
Udvardy, J., Farkas, G.L., Marre, E. and Forti, G. (1967) Physiol. Plant., 20, 781–788.
Usciati, M., Codaccioni, M., Mazliak, P. and Guern, J. (1974) Plant Sci. Lett., 2, 295–301.
Valdovinos, J.G., Jensen, T.E. and Sicko, L.M. (1971) Plant Physiol., 47, 162–163.
VanDerWoude, W.J., Lembi, C.A. and Morré, D.J. (1972) Biochem. biophys. Res. Commun., 46, 245–253.
Van Hove, C. and Carlier, A. (1968) Z. Pflanzenphysiol., 58, 395–401.
Varner, J.E., Balce, L.V. and Huang, R.C. (1963) Plant Physiol., 38, 89–92.
Venis, M.A. (1964) Nature (Lond.), 202, 900–901.
Venis, M.A. (1972) Plant Physiol., 49, 24–27.
Vigil, E.L. and Ruddat, M. (1973) Plant Physiol., 51, 549–558.
Vince, D. (1968) Planta, 82, 261–279.
Wade, N.L. (1974) J. exp. Bot., 25, 955–964.
Wade, N.L., O'Connell, P.B.H. and Brady, C.J. (1972) Phytochemistry, 11, 975–979.
Wain, R.L., Rutherford, D.P., Weston, E.W. and Griffiths, C.M. (1964) Nature (Lond.), 203, 504–506.
Walton, D.C. and Sondheimer, E. (1968) Plant Physiol., 43, 467–469.
Wareing, P.F., Good, J. and Manuel, J. (1968a) in: Biochemistry and Physiology of Plant Growth Substances, F. Wightman and G. Setterfield (eds.), Runge, Ottawa, pp. 1561–1579.
Wareing, P.F., Khalifa, M.M. and Treharne, K.J. (1968b) Nature (Lond.), 220, 453–457.
Webster, B.D. (1968) Plant Physiol., 43, 1512–1544.
Wellburn, F.A.M., Wellburn, A.R., Stoddart, J.L. and Treharne, K.J. (1973) Planta, 111, 337–346.
Wheeler, A.W. (1972) Ann. appl. Biol., 72, 327–334.
Wolf, F.T. and Haber, A.H. (1960) Nature (Lond.), 186, 217–218.
Wood, A. and Paleg, L.G. (1974) Aust. J. Plant Physiol., 1, 31–40.
Wood, A., Paleg, L.G. and Spotswood, T.M. (1974) Aust. J. Plant Physiol., 1, 167–169.
Wyen, N.V., Erdei, S., Udvardy, J., Bagi, G. and Farkas, G.L. (1972) J. exp. Bot., 23, 37–44.
Yomo, H. and Srinivasan, K. (1973) Plant Physiol., 52, 671–673.
Yomo, H. and Varner, J.E. (1971) Curr. Top. Develop. Biol., 6, 111–144.
Yomo, H. and Varner, J.E. (1973) Plant Physiol., 51, 708–713.
Yomo, J. (1960a) Hakko Kyokaishi, 18, 494–499.
Yomo, J. (1960b) Hakko Kyokaishi, 18, 600–602.
Zenk, M.H. and Nissl, D. (1968) Naturwissenschaften, 55, 84–85.
Zucker, M. (1972) Ann. Rev. Platn Physiol., 23, 133–156.

Phytohormones and Related Compounds – A Comprehensive Treatise, Volume I
Editors: Letham, Goodwin and Higgins

10

The Influence of Phytohormones on Replication and Transcription

JOHN V. JACOBSEN and T.J.V. HIGGINS

10.1. Introduction

The exogenous application of growth regulators to plant tissues often causes rather dramatic changes in growth and development. Such changes involve cell division, cell elongation and expansion and cell differentiation. Because cell division involves DNA synthesis and because cell elongation, expansion and differentiation involve changes in the protein complement of cells, and since protein synthesis is the end result of a series of events involving DNA-dependent RNA synthesis, the possibilities that hormones act by regulating DNA replication and its transcription to mRNA have received considerable attention.

There is a large body of literature describing hormone-evoked changes in nucleic acids in a wide variety of plant systems and tissues. The approach commonly used has been to measure quantitative and qualitative changes in DNA and RNA following application of plant hormones using either analytical methods or incorporation of radioactive precursors. Although this has produced a wealth of knowledge about cell division, cell differentiation and the mode of action of hormones in various systems, the interpretation of many of the data is complicated and consequently progress towards understanding mechanisms of hormone action is slow.

A unifying hypothesis regarding hormonal control of replication, transcription or translation has not emerged from the data so far. This is due in part to the complexity of living systems, especially the eucaryotes in which any result obtained must be interpreted within the framework of current knowledge which is constantly being updated, and thus any particular interpretation is only one of many possible interpretations. Similarly, interpretation of in vivo radioactive tracer experiments is very difficult because of artefacts due to imprecise knowledge of nucleic acid metabolism, (especially turnover rates) and the nature of precursor pool location, size and turnover. Furthermore, experiments have been conducted over relatively long times thus making it difficult to decide whether a modification of nucleic acid metabolism is the result or the cause of changes in protein synthesis, growth or morphogenesis. There have been relatively few studies of very short term effects of hormones so far although such studies have been hampered, inter alia, by lack of technology sufficiently sensitive to detect small changes in nucleic acid levels or labelling, as well as by problems of fractionating and identifying both major and minor nucleic acid species. Until the late 1960's, there was no technology available for routine examination of the minor RNA fractions and so research on RNA metabolism and hormones was confined mainly to the major species. Recently, great progress in the knowledge of messenger RNA (mRNA) structure has led to the development of methods which allow the isolation and purification of mRNA to be accomplished rather easily and furthermore, methods are now available for the in vitro translation of such mRNA with a high degree of fidelity. Similarly, recent advances in elucidation of chromosome structure and control of DNA transcription should greatly accelerate the understanding of eucaryotic DNA. Because of these developments it seems likely that a new level of nucleic acid and hormone research has been reached.

The aim of this chapter is to determine what effects hormones have on nucleic acid metabolism and to assess whether any changes are primary responses to the hormone. In doing this, it may be of value to establish a framework within which the genome and its expression might be modified by plant hormones. The following are possibilities.

10.1.1. DNA synthesis

(1) Replication of the whole or much of the nuclear genome leading to polyploidy or polyteny or perhaps to cell division and the formation of clones of functionally similar cells.

(2) Replication of certain parts of the nuclear genome leading to amplification of certain genes or other parts of the DNA.

(3) Replication of all or part of the non-nuclear organellar DNA.

10.1.2. RNA synthesis

(1) Regulation of transcription quantitatively by increasing or decreasing the numbers of existing transcripts. This could be done by: (a) regulation of the number of gene copies per cell as in 1 or 2 above; (b) regulating RNA polymerase level or activity; (c) regulation of the rate of initiation, termination and release of RNA molecules; (d) RNA-dependent RNA polymerase (see Duda 1976) which could amplify transcripts by making negative strands and then positive copies from these; (e) stabilization or promotion of turnover of transcripts.

(2) Regulation of transcription qualitatively by causing the initiation of new transcription or the suppression of existing transcription by: (a) controlling template availability, perhaps by modification of histones and other chromosomal proteins or by the operation of regulator genes; (b) altering affinity of polymerase for certain template sites, perhaps by regulation of the formation of suitable initiator complexes involving the hormone, receptor molecules and non-histone nuclear proteins.

(3) Regulating post-transcriptional cleavage, e.g. processing of ribosomal RNA (rRNA) precursors and heterogeneous nuclear RNA (HnRNA).

10.2. Hormones and DNA metabolism

10.2.1. Auxin

In the first of its type, the study of Silberger and Skoog (1953) showed that low concentrations of indoleacetic acid (IAA) increased cell division and DNA content of cultured tobacco *(Nicotiana tabacum)* pith. However, although increase in DNA content occurred at higher auxin concentrations than were optimal for the stimulation of cell enlargement and increase in RNA content, there was no cell division. It was concluded that cytokinesis did not necessarily follow increase in DNA. Later a report from the same laboratory (Patau et al. 1957) examined the relationship between DNA content, cell division, auxin and kinetin, and concluded that auxin was necessary for DNA increase and mitosis but that mitosis and cytokinesis were under the control of kinetin. This study of total nucleic acid levels demonstrated new relationships between plant hormones and apparent nucleic acid synthesis and, in particular, it showed that auxin could increase DNA content with and without subsequent cytokinesis and that auxin-induced cell enlargement was associated with an increase in DNA as well as RNA.

The relationship between auxin, nucleic acid and cell enlargement has been examined in a number of systems. In lentil *(Lens culinaris)* epicotyls, there was a correlation between auxin-induced cell elongation and DNA synthesis without cell division (Nitsan and Lang 1965, 1966) but subsequently it was shown that there was sufficient cell division occurring in this system to account for the rise in DNA (Atsmon 1972) and that this DNA synthesis was not required for cell elongation to occur (Lang and Nitsan 1967). Similarly, auxin-induced growth in corn *(Zea mays)* coleoptiles and artichoke *(Helianthus tuberosus)* tuber sections was not sensitive to fluoro-deoxyuridine (FUDR) (Maheshwari and Nooden 1971) which is an inhibitor of DNA synthesis. These and similar findings in oat *(Avena sativa)* coleoptiles, pea *(Pisum sativum)* stems and soybean *(Glycine max)* hypocotyls indicated that auxin-induced increases in DNA were unrelated to cell elongation. However, in cultured tobacco pith where high levels of auxin caused an increase in nuclear DNA and cell expansion without cell division, data from inhibitor studies indicated that this DNA synthesis was probably necessary for the induction of cell enlargement but not to sustain growth (Maheshwari and Nooden 1971). This seems to be the only case known where DNA synthesis has been shown to be necessary for auxin action and it is only based upon the use of FUDR as an inhibitor of DNA synthesis. To explain this result, a number of suggestions have been advanced. Firstly, the requirement for DNA synthesis may mean that the number of copies of certain genes must be increased. Such genes could be ribosomal since there is a large increase in ribosomal RNA associated with auxin-induced cell expansion. However, it remains to be shown that the increase in rRNA is necessary for cell expansion since it is not required for cell enlargement in other systems (Key 1969). Secondly, because auxin increased respiration in tobacco pith before it affected growth (Newcombe 1960), the additional DNA could be associated with increased numbers of mitochondria. However in artichoke tuber discs where auxin also increased respiration, the increase in respiration could be prevented but growth continued (Nooden 1968), and in any case there was no evidence for an increase in mitochondrial numbers. Thirdly, perhaps the genes necessary for auxin action are transcribed only during their replication, as in *Chlorella* where it has been found that nitrite metabolizing enzymes are most effectively induced (by nitrite) during the period of DNA synthesis (Knutsen 1965) although there is no corresponding example in higher plants. Fourthly, larger cells may require more copies of all genes to service their larger protoplasts. The above are merely possibilities and clearly there is no evidence to indicate what role DNA synthesis may play in auxin control of cell enlargement. However, in many cases it is not required at all.

There are two reports describing promotive effects of auxin on incorporation of labelled precursors into DNA. Both report an enhancement of thymidine incorporation into DNA, one into isolated coconut *(Cocos nucifera)* nuclei (Roychoudhury and Sen 1964a), although the authenticity of the isolated nuclei is questionable (see section 10.3.1.3), and another into stem segments of flowering tobacco plants occurring within an hour (Wardell 1975). Thus, it seems that auxin, at least in tobacco stem tissue, can very rapidly enhance DNA synthesis.

On the other hand, auxins may have no effect on, or may even inhibit, DNA synthesis. In cultured pea root cortex cells, there was enhanced DNA synthesis in the presence of auxin and kinetin but not when auxin was present alone (Libbenga and Torrey 1973). Growth-promotive auxin concentrations inhibited DNA synthesis in internodes of vegetative tobacco plants (as opposed to flowering plants described above) (Wardell and Skoog 1973) and had both inhibitory and slightly promotive effects in pea epicotyls depending on the tissue (Sherwin and Gordon 1974). High concentrations of auxins inhibited DNA synthesis in vivo in peanut *(Arachis hypogea)* cotyledons (Carpenter and Cherry 1966b) and in vitro using *E. coli* polymerase and plant template (Schwimmer 1968). Thus, the effects of auxins on DNA synthesis depend on the physiological state of the tissue and on whether the concentration of the auxin promotes or inhibits growth.

There is some evidence that enhanced DNA synthesis may result from regulation of the activity of DNA polymerase by auxin. The polymerase of soybean hypocotyls is chromatin-bound and the activity of this polymerase was increased by 30% after 12 hr treatment of hypocotyls with 2,4-dichlorophenoxyacetic acid (2,4-D) (Leffler et al. 1971). This corresponded with the onset of mitosis which occurred after 12–14 hr in auxin-treated hypocotyls. Whether the prime effect of auxin was on polymerase or on cell division (or perhaps both) was not determined. Both could be results of 2,4-D-induced changes in transcription which occurred in this tissue within 4 hr of hormone treatment (O'Brien et al. 1968b).

10.2.2. Ethylene

There is relatively little information on effects of ethylene on DNA replication but that which is available comes from relatively well-defined systems. Ethylene did not affect [^{14}C] thymidine incorporation into DNA of explants of bean *(Phaseolus vulgaris)* petioles during abscission (Holm and Abeles 1967) but in growing tissues, where ethylene generally reduces growth and causes lateral expansion of cells rather than longitudinal expansion, DNA synthesis has been affected in the following ways. In the apical dividing zone of soybean hypocotyls, ethylene inhibited DNA accumulation (in parallel with RNA accumulation) (Holm et al. 1970) and likewise in the meristematic zone of etiolated

Alaska pea seedlings, ethylene (as well as high 2,4-D levels) inhibited DNA synthesis (measured by radioactive precursor incorporation) and cell division in parallel (Apelbaum and Burg 1972). In the sub-hook region of pea seedlings where cell division does not occur, ethylene also inhibited DNA synthesis, thus demonstrating that ethylene did not act on DNA synthesis via cell division. Likewise, in the subapical tissues of the soybean hypocotyl (both basal and elongating regions) where there was little or no cell division, ethylene also affected DNA levels (Holm and Abeles 1968; Holm et al. 1970), although the level was increased (within 24 hr) rather than decreased. Thus, although extractable DNA has been studied in some cases and precursor incorporation in others, the effects of ethylene on DNA seem to vary from tissue to tissue as do those of auxin. The effect of ethylene in dividing tissues, however, seems to be consistent in that DNA synthesis is inhibited. This, in addition to the fact that ethylene stops cell division before prophase seems to indicate that ethylene prevents cell division by regulating DNA synthesis, and subsequent studies have shown that treatment of intact pea plants (but not the polymerase reaction mixture) for 24 hr with ethylene (50 μl/l) reduced the soluble polymerase (or its activity) obtained from the apical portions (plumule plus plumular hook) by about two-thirds (Apelbaum et al. 1974). There was some indication that this was the result of the appearance of a non-competitive inhibitor of DNA polymerase activity. An attractive feature of this work was the use of solubilized DNA-dependent DNA polymerase thus eliminating the possibility of measuring effects on template availability, but a major question regarding the total level of polymerase remained unanswered. Ethylene may have altered the distribution between the soluble and chromatin-bound fractions (Robinson and Bryant 1975 and references therein). It should also be noted that the inhibition of cell division by ethylene requires very high ethylene concentrations (50 μl/l) and the inhibition of DNA synthesis may have nothing to do with responeses of plant tissues to more physiological ethylene levels of two orders of magnitude less. Where ethylene has promoted DNA synthesis, there is as yet no demonstration that it enhanced DNA polymerase activity.

10.2.3. Gibberellin

It was first noticed in lettuce *(Latuca sativa)* seedlings and lentil epicotyls that both natural and GA_3-promoted elongation was prevented by inhibitors of DNA synthesis (Nitsan and Lang 1965). In particular, it was FUDR-sensitive and this was reversed completely by thymidine but not by uridine. Growth in lentil epicotyls apparently occurred by cell elongation with very little or no cell division (Nitsan and Lang 1966). It was also shown that GA_3 increased DNA content. Evidence that GA-dependent cell elongation required DNA synthesis was also obtained from other systems. In mustard *(Sinapis alba)* hypocotyls, stems of *Kalanchoe* plantlets and sections of cucumber *(Cucumis sativus)* hypocotyls, GA_3-induced elongation is also FUDR-sensitive. Only in some cases has the essential control experiment been carried out, that is, an experiment to investigate whether the inhibition was reversible by thymidine but not by uridine, as shown for cucumber (Lang 1966).

Whether DNA synthesis occurs in the elongating cells and whether elongation growth somehow involves cell division, thus requiring DNA synthesis has not yet been satisfactorily resolved although there is some evidence for DNA synthesis in elongating cells. In dark-grown wheat *(Triticum aestivum)* coleoptiles, early growth, which occurs by cell enlargement alone, was promoted by GA_3 and the increase in growth was prevented by FUDR and restored by thymidine (Lang 1966) indicating a requirement for DNA synthesis within the enlarging cells.

However, more recent studies have indicated that increased DNA may not be confined to elongating cells and even if it is, it may not be all nuclear DNA. Whereas lentil epicotyls were originally thought to be relatively free of cell division, Atsmon (1972) found that, in fact, there was sufficient meristematic activity around and within the vascular bundles to account for the additional DNA. However, in cucumber hypocotyls, where growth is FUDR-sensitive, there was also an increase in cell number but it was insufficient to account for the large increase in GA_3-promoted DNA synthesis (Degani and Atsmon 1970b). This DNA, however, seems to be largely organellar. The major increase in DNA synthesis as measured by [^{3}H] thymidine incorporation occurred in the DNA-containing organelles such as chloroplasts and mitochondria (Degani and Atsmon 1970a), although without data on thymidylate pool sizes in the different organelles and without specific activitites of the nuclear and non-nuclear DNAs, it is difficult to be sure of the conclusion. Nevertheless, the result did correlate with increased numbers of chloroplasts in the elongating zone after GA_3 treatment (Kadouri and Atsmon 1974). The DNA of hypocotyls has been examined in some detail recently (Kadouri et al. 1975) and has been found to contain two satellite components comprising 41% of the total DNA. Large proportions (1/3) of these satellites were organellar DNA (chloroplast DNA in one and mitochondrial DNA in the other). The syntheses of both main band and satellite DNAs were enhanced in the elongating zone by 100% and in the meristematic zone by 27% in response to GA_3. However, this increase in DNA synthesis was not reflected by an increase in total DNA and a hormonal effect on DNA turnover or on nucleotide pools was possible.

The question arises as to whether the DNA synthesis is a prerequisite or a result of growth. In fact, chloroplast proliferation and cytoplasmic DNA synthesis can be prevented by chloramphenicol but GA_3-induced cell elongation cannot, so it seems that the cytoplasmic DNA synthesis is not a pre-requisite for growth. If so, what DNA synthesis is required? Atsmon's group has shown that colchicine, which inhibits cell division, but not DNA synthesis, also inhibited GA_3-induced cell elongation indicating that it is cell division which is required and not DNA synthesis per se. How cell division is involved is not clear but it seems likely that it contributes to GA_3-induced elongation by providing more cells to undergo elongation. Nevertheless it is still possible that a small amount of nuclear DNA synthesis is required for GA-induced elongation, although the original concept that DNA synthesis is of prime importance in an elongating cell has little support from studies on stem tissues. For pea epicotyl tissue, in which GA_3 also promotes cell elongation, there is almost complete agreement that DNA synthesis is not required. In mature internode tissue, GA_3 was shown to promote DNA synthesis but only 24–48 hr after it promoted elongation and RNA synthesis, and the DNA synthesis was associated with enhanced cell division (Broughton 1968, 1969). Furthermore, inhibition of DNA synthesis with FUDR did not inhibit GA_3-induced cell elongation. These results are in agreement with other studies showing that after GA_3 treatment of mature pea internodes, there was no change in the amount of DNA per cell (Gamburg et al. 1965). Similarly, studies on less mature pea epicotyl tissue also showed that GA_3 promoted elongation of larger differentiated cells unaccompanied by DNA synthesis although it also appeared to promote the progression of smaller cells from the G1 phase to the S phase and thus promoted cell division (Nakamura and Takahashi 1973). Therefore, in this tissue, elongation was apparently a product of cell division as well as cell elongation and the enhancement of DNA levels seemed to be a result of the cell division and not associated directly with elongating cells.

From the above, it is clear that interpretation of data from systems in which any amount of cell division is occurring is complicated and perhaps misleading even although the data from pea seedlings seems convincing. Use of tissues in which there is no cell division is preferable. Such studies have been made and the data support the contention that GA action does not involve DNA synthesis. In gamma-irradiated wheat plantlets there was no DNA synthesis or mitosis (Haber and Foard 1964) but GA_3 stimulated growth of the coleoptile and the first leaf while inhibitors of DNA synthesis did not interfere with the response (Rose and Adamson 1969). Similarly, GA_3 promoted germination of gamma-irradiated lettuce seeds indicating again that DNA synthesis was not required. Supporting evidence comes from work on soybean hypocotyl. The apical section of the hypocotyl elongates by cell division (and DNA synthesis) and cell elongation, and GA_3-promoted elongation of this section is FUDR-sensitive. However, the basal section of the hypocotyl elongates only by cell elongation and inhibitors of DNA synthesis do not interfere with GA_3-promotion of this elongation (Holm and Key 1969). Therefore, it seems that GA_3 can promote cell elongation in tissues where there is no DNA synthesis or cell division. Furthermore, in general, colchicine inhibits GA_3 responses which are inhibited by FUDR; this reinforces the idea that where DNA synthesis is required for GA_3-action, the requirement seems to be for meristematic activity and at least in some systems, meristematic activity may be required for hormone-induced enlargement of adjacent cells (Atsmon 1972).

10.2.4. Cytokinins

Historically, cytokinins have been considered to be promoters of cell-division and indeed derived their name from this function, but surprisingly, although they have been implicated, little work has been done on their involvement with DNA synthesis and the mitotic cycle. In general, recent studies have supported the idea that cytokinins are involved in the latter and have tended to dissociate cytokinins from direct effects on DNA synthesis. After the initial demonstration by Miller and co-workers that kinetin promoted cell multiplication in tobacco callus (Miller et al. 1955), the first evidence that it affected the mitotic cycle was provided by the cytological studies of Guttman in which kinetin was shown to cause polyploidy in meristematic cells of onion *(Allium cepa)* roots, to reduce the time for interphase and to affect coiling and uncoiling of chromosomes (Guttman 1956). Subsequently, there has been an effort to determine how cytokinins promote cell division; whether by promoting DNA synthesis, by controlling mitosis itself or both.

There is some support for the idea that cytokinins affect DNA synthesis. In tobacco pith explants, kinetin was shown to promote DNA synthesis in one case after 1 day (Patau et al. 1957) and in another after 2 days (Zwar 1973). In both cases it occurred before the beginning of cell division, which implied that DNA synthesis was the first process affected although the initial biochemical events of cytokinesis probably occurred well before cell division could be observed, and could precede the onset of DNA synthesis. Another implication was that cytokinin alone could cause the effect when in fact auxin must also be present. Using aged tobacco pith blocks, Simard (1971) found that kinetin alone did not enhance precursor incorporation into DNA, although auxin or auxin plus kinetin did promote incorporation. The aging process may have served to deplete the tissue of other hormones and if so, the results would indicate that auxin has a permissive role in DNA synthesis and kinetin a stimulatory role only. This may well substantiate the suggestion of Cline (1976) that the previous hormone history and hormone status of a tissue is of paramount importance when attempting to interpret results of hormone experiments using excised tissues.

It is also possible for kinetin to induce cytokinesis without DNA synthesis as in tobacco pith explants (Patau and Das 1961; Patau et al. 1957). In support of this, Jouanneau and colleagues, using hormone-starved tobacco cell suspension cultures, were able to induce partially synchronous mitosis by the addition of cytokinin (Jouanneau 1971), thus producing additional evidence that cytokinins could control cytokinesis directly. Subsequent studies by Jouanneau and co-workers showed that DNA replication was not controlled by cytokinins and that cytokinins could promote mitosis in the absence of DNA synthesis although protein synthesis was required (Jouanneau and Tandeau de Marsac 1973; Jouanneau 1975), thus supporting the hypothesis that cytokinins regulate mitosis by controlling synthesis of mitosis-specific proteins.

In summary, although there were early indications that cytokinins could affect the entire cell cycle including DNA synthesis in interphase, the evidence that cytokinins have no direct effect on DNA synthesis is now very strong. Indeed, at present it seems that cytokinins act primarily as promoters of cytokinesis. Reports such as that of Holm and Key (1969) in which cytokinin restored DNA synthesis in the apical section of soybean hypocotyls deprived of their roots, while perhaps requiring further analysis, could well be explained by hormonal control of cytokinesis alone.

10.2.5. Abscisic acid

The first report on the effects of abscisic acid (ABA) on nucleic acid synthesis provided evidence that perhaps ABA inhibited growth of duckweed *(Lemna)* by inhibiting DNA synthesis (Van Overbeek et al. 1967). The incorporation of [^{32}P]orthophosphate into all nucleic acid fractions (discernible by methylated albumin kieselguhr (MAK) chromatography) during a 2–16 hr pulse was less after 24 hr pre-treatment with ABA than in controls, but the data indicated that suppression of incorporation into the DNA fraction occurred before inhibition of incorporation into RNA fractions. Incorporation into DNA was also the first to recover on addition of cytokinin, which reversed the effects of ABA on growth of the duckweed. This led the authors to conclude that control of DNA synthesis was probably the prime effect of ABA and they speculated that DNA polymerase was an allosteric enzyme and that ABA and cytokinins could act as allosteric effectors. Presumably, inhibition of DNA synthesis would lead ultimately to a reduction in RNA synthesis and thus explain the sequentiality of the responses. Nevertheless, the [^{32}P] orthophosphate incorporation in the vicinity of the DNA peak was never shown to be in plant nuclear DNA. Indeed, as the authors acknowledged, this peak may contain some RNA, as a DNA-RNA hybrid, which becomes rapidly labelled (Cherry 1964; Chroboczek and Cherry 1965; Ingle and Key 1965; Ingle et al. 1965; Cherry and Chroboczek 1966; Chroboczek and Cherry 1966; Ewing and Cherry 1967) and incorporation of [^{32}P] orthophosphate into such RNA could account for the bulk of the "DNA" labelling. However, there is now rather compelling evidence that DNA which is rapidly labelled with [^{32}P] orthophosphate is bacterial DNA (see section 10.2.6) and that phytohormones can affect the rate of growth of bacteria on plant surfaces.

Similar criticisms apply to studies implicating ABA in the control of elongating embryonic bean axes. The hormone inhibited the elongation of the axes and simultaneously inhibited [^{32}P] orthophosphate incorporation into DNA (as well as other nucleic acid fractions), but again the radioactivity was associated with the DNA peak in MAK column fractionation and was not shown to be primarily in plant nuclear DNA (Walton et al. 1970). Furthermore, 5-fluorouracil (5-FU), while not affecting axis elongation, inhibited label incorporation into the DNA fraction and zeatin, which reversed the inhibition of growth by ABA, did not restore label incorporation into the DNA; thus there was no apparent correlation between "DNA" synthesis and growth. From these data, it seems that even if the [^{32}P] orthophosphate incorporation was into plant DNA, the effect of ABA had little to do with its inhibition of growth. This is supported by the fact that the inhibitors of DNA synthesis, phleomycin and FUDR, did not inhibit axis elongation. Better evidence, at least technically, that DNA synthesis is involved in ABA action comes from a more recent study of *Lemna*. Soon after ABA application (1–3 hr), [^{3}H] thymidine incorporation into DNA was strongly inhibited and it was thought that this preceded inhibition of RNA synthesis (measured by [^{32}P] orthophosphate and [^{14}C] orotate incorporation) which occurred from 3–6 hr after ABA application (Steward and Smith 1972). This evidence was claimed to be technically sound because the DNA was isolated on 2.2% polyacrylamide gels which were believed to separate DNA from any rapidly labelled RNA species, and the label used was a specific precursor for DNA, maybe even specific for plant (as opposed to bacterial) DNA (Julien and Delseny 1972; Sarrouy-Balat et al. 1973). Therefore, it seems likely that the fraction in question was DNA but whether it was all plant nuclear DNA was not clear since bacterial DNA would probably have similar electrophoretic mobility. Furthermore, in this study it was shown that FUDR inhibited growth, indicating that DNA synthesis was required and thus interference with its synthesis by ABA could have inhibitied growth. If the above results are confirmed and the DNA verified as being of plant origin, then the *Lemna* study (Stewart and Smith 1972) provides good evidence that ABA interferes with DNA synthesis. Whether this precedes the effect on growth remains to be determined.

There is good evidence that ABA inhibits growth and DNA synthesis in radish *(Raphanus raphanistrum)* seedlings (Sarrouy-Balat et al. 1973). The DNA was thymidine-labelled and identified as main band plant DNA on cesium chloride gradients, so the results probably do demonstrate inhibition of DNA synthesis although inhibition of DNA turnover is still a possibility (see metabolic DNA in Section 10.2.6).

The studies on *Lemna,* bean axes and radish seedlings described above have been examined primarily on the basis of whether the ABA-inhibited DNA synthesis was authentic plant DNA synthesis. Where it does seem to involve plant DNA, the question of whether the inhibition of synthesis occurs before or after growth inhibition has not been unequivocally answered and thus whether ABA effects on DNA are primary or secondary also remains unanswered.

There is evidence against a role for DNA synthesis in ABA action and although it comes from systems in which there is no cell division, it too has some inadequacies. In lettuce seed germination, the usual effects of GA (promotion) and ABA (inhibition) were found to occur even in gamma-irradiated seed in which there was no nuclear DNA synthesis detectable by Feulgen microspectrophotometry and [^{3}H] thymidine incorporation and autoradiography (Haber et al. 1969). In addition, using the same methods, it was not possible to detect any DNA synthesis in mature barley *(Hordeum vulgare)* aleurone tissue. These results led to the argument that ABA-inhibition of lettuce seed germination and of GA_3-induced amylase synthesis in barley aleurone (Thomas et al. 1965; Addicott et al. 1966; Chrispeels and Varner 1966) could not involve DNA synthesis. Furthermore, FUDR did not inhibit amylase induction by GA, thus making it very unlikely that any inhibitory effect of ABA on DNA synthesis would interfere with amylase induction. We are unaware of any similar data for lettuce seed. However, more recently, DNA synthesis (apparently nuclear) has been detected by [^{3}H] thymidine incorporation in aleurone (Taiz 1975). This incorporation was unaffected by GA_3 and it is clearly not important in amylase induction, but it does raise questions about the limits of detection of DNA synthesis used in the lettuce seed studies (Haber et al. 1969). In studies on isolated radish leaf discs where ABA enhances senescence, involvement of DNA synthesis was also found to be unlikely since again no DNA synthesis could be detected (Wareing et al. 1968b; Paranjothy and Wareing 1971).

To date, the major part of the evidence suggests that ABA does not directly affect DNA synthesis. However, it would seem that ABA does or can inhibit DNA synthesis (or turnover) but what sort of DNA is affected and what relevance this has to ABA action is not at all clear. Perhaps it is significant that in discs of mature leaves, germinating lettuce seed, aleurone layers and elongating bean axes where it seems that DNA synthesis is not involved with ABA action, there is little or no cell division, but in *Lemna* and perhaps radish seedlings where it seems that DNA synthesis is associated with ABA action, cell division is probably a component of growth. In such systems, it could be that the primary effect of ABA is on cell division which leads to inhibition of DNA synthesis rather than on DNA synthesis itself. It has also been pointed out that in radish seedlings, ABA may also "retard" DNA synthesis by delaying replication (Sarrouy-Balat et al. 1973). There is a partial synchrony of DNA synthesis in young seedlings and if replication was delayed by ABA, labelling during a period of slow DNA synthesis would give a very low and misleading rate of incorporation. Therefore, in such studies it is necessary to pay attention to the timing of successive S phases of cell division.

10.2.6. Problems of DNA labelling experiments

The interpretability of DNA-labelling studies depends on the specificity of labelling and on the ability to distinguish DNA from contaminants, to fractionate DNA itself and particularly to distinguish plant from bacterial DNA. In many studies of plant hormones and DNA metabolism, the labelled compound used has not been a specific DNA precursor and the "DNA" fraction obtained by MAK column chromatography, the most commonly used fractionation technique, is a complex mixture of nucleic acids as described below.

In the previous section (section 10.2.5), it was mentioned that early work by Cherry (1964) indicated that labelling of the DNA-RNA fraction from MAK chromatography may have involved RNA rather than DNA. Indeed the DNA-RNA from peas has been separated into two fractions, one fraction being designated DNA I or metabolic DNA, which turned over rapidly, and the other DNA II, which was chromosomal DNA (Sampson and Davies 1966 and references therein). However, subsequently, DNA I has been shown to be either all or in part RNA (Bryant and Wildon 1971). RNA, both free and bound, has also been found in the DNA-RNA from radish seedlings (Julien and Guitton 1972). Therefore, it seems that label in the DNA peak could, at least in part, be associated with RNA.

The complexity of the DNA composition of the DNA-RNA fraction is evident from studies of the complex from radish seedlings (Julien and Guitton 1972). The DNA-RNA contained not only hybrid DNA-RNA, but also free DNA and RNA. The hybrid DNA was very high in cytidine (C) and the hybrid RNA correspondingly high in guanine (G). The DNA sedimented in cesium chloride gradients not as main band DNA but as a more dense satellite which contained 59% G and C compared to 38% for main-band DNA. It labelled rapidly and had characteristics similar to those of the crab satellite consisting of a $3' \rightarrow 5'$ copolymer of deoxyribonucleotides of adenosine and thymidine (poly d(AT)) (Skinner and Triplett 1967) indicating that it has a very peculiar structure. This DNA was synthesized in the absence of cell division, it was rapidly labelled and was labile and therefore qualified as metabolic DNA (Pelc 1968). It appears to be the same as, or similar to, labile DNA discussed by Stern (1964). Similar satellites have been found in cultured tobacco cells (Bendich and Filner 1971) and roots of broad bean *(Vicia faba)* and corn (Julien and Delseny 1972). This satellite (buoyant density 1.722 $g.ml^{-1}$) was called stress-induced satellite because it appeared under stress conditions and, from hybridization experiments, it appeared to contain rRNA genes (Guillé et al. 1968; Quétier et al. 1968). However, in cold stored (stressed) cucurbit seeds it was found to be due to bacterial DNA (Pearson and Ingle 1972).

There is now strong evidence that in many cases, incorporation of [^{32}P]orthophosphate into nucleic acids of the DNA peak is due to bacterial contamination (Lonberg-Holm 1967; Tester and Dure 1967; Sabota et al. 1968). Labelling of the DNA peak was strongly and preferentially inhibited if antibiotics such as penicillin, streptomycin, gramicidin and chloramphenicol were present. Furthermore, bacterial DNA would label rapidly and is G + C-rich like the rapidly labelled material associated with the DNA fraction from radish seedlings. In addition to the heavy "stress-induced" DNA satellite described above, a number of other satellites in DNA isolated from pea, oat and rye *(Secale cereale)* seedlings have also been found to be bacterial DNA (Bendich 1972). It has been argued that the satellite DNA could not be bacterial because it was affected by plant hormones. However it has been shown in radish seedlings, where labelling of the satellite and main band DNA with [^{32}P] orthophosphate and thymidine respectively was enhanced by GA_3 and inhibited by ABA and where the satellite DNA corresponded in buoyant density to bacterial DNA, that the numbers of bacteria in treatments varied in parallel with satellite labelling, and surface-sterilization of the tissue eliminated satellite labelling (Sarrouy-Balat et al. 1973). However, both GA_3 and ABA enhanced growth of cultured bacteria so that at least for ABA the effect on bacterial growth is indirect. Somehow then, plant hormones affect plant tissues in such a way as to regulate the numbers of contaminating micro-organisms.

Finally, it should be noted that whereas rapidly turning over (metabolic) DNA may or may not exist in plants as satellite DNA, it does seem to exist in main band DNA. Metabolically labile main band DNA (thymidine labelled) has been shown to occur in aseptically grown pea seedlings monitored for sterility (Bryant 1974) and the existence of metabolic DNA is supported by in vivo thymidine labelling studies using autoradiography on root tip cells of *Pisum sativum* and *Vicia faba* (Wildon and Wong 1971; MacLeod 1972).

In summary, it seems that the rapidly [^{32}P]orthophophate-labelled DNA-RNA is probably bacterial and the effects of plant hormones on this reflect some plant–bacterial relationship. Since this DNA does not become labelled with thymidine and since main band DNA does (Julien and Delseny 1972), studies on plant hormones using thymidine as a label probably do measure plant DNA synthesis, but they may also reflect DNA turnover and not net synthesis since metabolically labile DNA does appear to exist in plant tissues. Thus, it is clear that any study of plant DNA using labelled precursors must pay particular attention to microbial contamination and, if possible, should involve organisms which can be cultured under sterile conditions (e.g. *Lemna* and *Spirodela*). It must be shown that labelled DNA is in fact plant DNA. It should be determined what sort of plant DNA is labelled (organellar, nuclear, main band, satellite) and whether it represents net synthesis or turnover. Because such information is not available for the early studies on plant hormones and nucleic acid synthesis, conclusions based upon incorporation of label into "DNA" may require reevaluation.

10.2.7. Effects of hormones on DNA composition

In general, DNA from higher plants and animals consists of three base sequence classes; highly repeated, moderately repeated and unique, the latter probably giving rise to mRNA. These fractions can be detected, characterized and quantitated by studying re-annealing of denatured sheared DNA. Highly repeated sequences reassociate rapidly at low C_0t values (C_0t = concentration of DNA in mols nucleotide per litre × time in seconds) and unique sequences reassociate slowly at high C_0t values. Very few studies of this type have been done with plant hormones and plant DNA but those which have been reported testify to rather dramatic changes in DNA composition.

Studies of DNA from untreated and auxin-treated artichoke tuber showed that auxin caused the apparent loss of a rapidly reannealing fraction and caused changes in the C_0t values for other fractions (Kovoor and Melet 1972). A similar study of DNA from cucumber by Kessler (1973) showed that while GA_3 had no effect on DNA from roots, it changed the DNA from shoots and leaves by increasing the amount of intermediate reassociating (moderately repeated) DNA. This corresponded to about a 9-fold increase in average sequence reiteration. This was in contrast to the effects of cyclic 3,5-adenosine monophosphate (cAMP) which caused major changes in the average reiteration of rapidly reassociating root DNA and slowly reassociating leaf and shoot DNA but had no effect on intermediate reassociating DNA in either roots or shoots and leaves. In *Cymbidium* protocorm tissue, auxin, cytokinin and GA_3 all causes major (and different) changes in the thermal denaturation profiles of DNA, auxin causing increased AT-rich DNA and GA_3 causing increased GC-rich DNA (Nagl and Rucker 1976). Similar cases of differential DNA replication are known to occur during normal development of wheat embryos (Kern and Grubler 1976), pea epicotyls (Van Oostweldt and Van Parijs 1976) and in many other tissues (Nagl 1975).

The hormone-induced differential DNA replication appears to involve major changes in the DNA composition, Questions relating to the significance of the changes and to whether they are primary or secondary effects are yet to be answered. The intermediate reassociating or moderately reiterated sequences are thought by Britten and Davidson (1969) to play a regulatory role but why GA would lead to amplification of regulator sequences is difficult to answer at present although it may have something to do with amplification of sensor genes with which the hormone might bind. The results of the work described above, if confirmed, suggest that hormones might effect major changes in DNA. This is of great interest because the concept of modulation of DNA itself by hormones is not generally considered likely since it would involve very mobile "metabolic" DNA, the existence of which would undermine the concept of basic stability of the cell's information. However, as demonstrated in the previous section (section 10.2.6), the criticism is not necessarily a valid one. The proposed system would involve the production of some DNA and the destruction of other DNA with each stimulus. That such changes would occur seems to be supported by the fact that the nature of the DNA from cucumber depends on the organ from which it was obtained and the age of the tissue (Kessler 1973).

10.2.8. Conclusions

The relationship between hormones and DNA metabolism is a badly neglected field of physiology. This is, in part, due to our limited knowledge of the organization of eucaryotic DNA and the role of the various abundance classes. Much of the evidence for involvement of DNA synthesis in growth processes comes from the use of FUDR. However, while the ability of thymidine, but not uridine, to reverse the effects of FUDR is accepted as good evidence for a requirement for DNA synthesis, these reversibility experiments are not always done, and furthermore, as with all inhibitor studies, it is still possible that such results are misleading. For example, a "no-effect" result must always be a "non-result". Clearly there is a case for judicious selection of the experimental plant system to be used and careful analysis of the quantitative and qualitative aspects of DNA metabolism. Much of the information available now involves measurement of overall rates of synthesis and total contents of DNA and in none of the studies so far has it been possible to eliminate the possibility that small changes in DNA synthesis and turnover are involved in early responses to the hormone. Indeed, overall rates of synthesis are usually estimated from rates of incorporation of radioactive precursors which may not reflect the true rate of synthesis. Attempts to improve measurements of the rates of synthesis by simultaneously measuring the specific activity of the precursor pool, the pool size and its turnover (Katterman and Clay 1975) are rarely, if ever, made. Rapidly turning over DNA, termed metabolic DNA, seems to exist in plants and warrants further investigation in relation to hormone action. In general, there is some evidence that hormones alter DNA metabolism and probably DNA polymerase activity but in most cases it is not clear whether or not it reflects a primary response to the hormone.

10.3. Hormones and RNA metabolism

10.3.1. Auxin

10.3.1.1. Perception; receptors and mediators. Studies of how plant cells perceive auxin outnumber those concerning other plant hormones, but still the picture is very unclear (see also Chapter 8, section 8.2.1.3). In general, it is currently thought that there may be one or two cellular sites at which auxin might interact with a macromolecule(s). One such site has been thought to be located in plasma-membranes and could be related inter alia to auxin transport and regulation of activity and function of certain membrane-bound proteins perhaps concerned with proton transport (this is discussed further in Chapter 12, Volume II and Chapter 8), and the other could involve interaction with non-plasma-membrane factors originating in the nuclei or perhaps the cytoplasm. As far as transcription is concerned, auxin could be perceived in both ways. It could cause release and/or activation of membrane-bound as well as perhaps non-membrane-bound factors which could act as mediators and regulate transcription alone, or it could complex with nuclear and perhaps cytoplasmic factors and the complex could alter transcription. The detection and identification of auxin-perceiving sites or macromolecules has been rather fragmented so far. Different workers have described sites or factors apparently recognizing auxin but there has been no attempt to unify the approach and to correlate the results of different groups. Furthermore, there has been little effort to characterize recognition factors in any detail or to determine if there is a multiplicity of factors in any cell or tissue. The composition of reconstituted systems used to assay for biological activity of auxin-recognition factors has not been well defined thus increasing the chance of artefact, and much of the work is unconfirmed. The rest of this section documents and attempts to unify the research where possible but for a more comprehensive assessment of the field of auxin binding, the reader is referred to a review by Kende and Gardner (1976).

There are two reported cases in plants where auxin, in combination with a presumed macromolecular receptor has influenced transcription in cell-free systems, although the systems concerned were poorly characterized. In the first case, auxin (2,4-D) stimulation of RNA synthesis by nuclei isolated from tobacco and soybean tissue culture depended on the isolation occurring in the presence of the hormone (Matthyse and Phillips 1969).

Nuclei isolated in its absence would not respond to auxin but responsiveness was restored by adding supernatant from nuclei centrifuged in the absence of auxin. An auxin-receptor complex was therefore envisaged and was considered to be bound in the nuclei. The factor was apparently a protein, being heat labile and ribonuclease (RNase) stable, and was resolved into two fractions with molecular weights of 20,000 and 200,000. It could be obtained from tobacco, soybean or pea bud nuclei. The second case involved the occurrence of an apparently similar factor in coconut nuclei. A transcribing system consisting of α-amanitin-sensitive RNA polymerase (and therefore presumably polymerase II), native DNA and an initiation factor, did not respond to auxin alone but when added with a nucleoplasmic factor, IAA stimulated RNA synthesis 2–3 fold (Mondal et al. 1970a, 1972a, b). Evidence to show that the IAA–factor complex reacted directly with the DNA was presented although the data are not convincing and in particular did not demonstrate unequivocally that IAA was involved in the binding. The factor was probably protein, had a molecular weight of 100,000, was a monomer, and bound auxin non-covalently in the ratio of 2:1 (receptor: auxin) (Ganguly et al. 1973; Biswas et al. 1975). While these cases describe receptor molecules which have not been well characterized or isolated and the nature of the auxin binding is unclear, they nevertheless provide preliminary support for the existence of auxin-receptor proteins. It should also be noted that whereas it is convenient to call the factors described above acceptor, receptor or hormone-active molecules, it is yet to be demonstrated that any of them are really auxin receptor molecules.

There is another group of hormone-associated transcription factors which do not act in association with hormone molecules but which are thought to be activated or mobilized in some way by auxin and which apparently regulate transcription in the absence of hormone. Such factors will be referred to as mediator molecules (Venis 1973) rather than as receptors. One of these factors has been isolated from pea and corn shoots and from soybean hypocotyls by the use of affinity chromatography involving bound 2,4-D (Venis 1971). The factor was soluble although it could perhaps have arisen from anywhere in the cell. It seemed to be a very high molecular weight protein which, although isolated by 2,4-D affinity chromatography, did not bind auxin in equilibrium dialysis experiments. It stimulated cell-free transcription of corn or pea DNA using *E. coli* polymerase. Further characterization of the factor and its involvement in transcription seems worthwhile. It seems to be very much like the soybean cotyledon factor described by Hardin and colleagues. This (soybean) factor stimulated chromatin-bound polymerase isolated from untreated tissue but not from chromatin of auxin-treated tissue (Hardin et al. 1970). It was claimed that this factor was bound to plasma membrane fragments isolated from soybean hypocotyl and onion stems and that auxin caused its release (Hardin et al. 1972). The factor stimulated polymerase activity in vitro and the stimulation was α-amanitin-sensitive indicating that it was preferentially regulating the nucleoplasmic enzyme (RNA polymerase II). The association of the putative transcription factor with the plasmalemma was not convincingly demonstrated, but the stimulation of in vitro transcription was encouraging. However, the data are preliminary and the transcription products should be characterized.

A number of aspects of auxin perception by the cell remain to be clarified. There is now good evidence for binding of auxin to membranes (for reviews see Kende and Gardner 1976; Venis 1973 Venis and Batt 1975) and it is of particular interest to determine if there is any relation between these membrane binding sites and the transcription-regulating factors described above, particularly the soybean factor which was thought to be a membrane component. The mediator factors are of particular interest in that they can tentatively be considered in terms of auxin perception since they appear to be hormone-activated, but they seem to be able to regulate transcription without direct participation of the hormone. Activation could involve, for example, relocation of the factor to a responsive site or a hormone-induced configurational change in the mediator molecule. Finally, all of the factors described above seem to affect polymerase II-dependent transcription but verification of this and determination of whether polymerase I is also affected can only come from the use of well-defined transcription assay systems where transcription products have been characterized.

10.3.1.2. Interaction with DNA and chromatin proteins. The interaction of an auxin-receptor complex with DNA as claimed by Mondal and colleagues is consistent with the possibility that auxin, perhaps as part of a complex, may have some direct effect on DNA and such effects have been reported (Fellenberg 1969a, b, c, 1971b). The absorption maximum of DNA from pea in alkaline solution showed a bathochromic effect in the presence of IAA (and in the absence of any other factors) which was reversed by urea but not salt. IAA also reduced the melting point of DNA. These results indicated that auxin bound to DNA by weak hydrogen bonding and that this weakened the helical structure. Similar results were obtained by Bamberger (1971) who demonstrated the effects at pH 7.0. These results led Fellenberg to regard auxin as a true initiator of RNA synthesis but it is disturbing that auxin effects occurred also with calf thymus DNA (Bamberger 1971), indicating a lack of base sequence specificity, and that they occurred apparently preferably in neutral to alkaline conditions although the nuclear environment is probably slightly acid. In this regard a more recent study by Penner and Early (1972) seems to be particularly relevant. They showed that the concentrations of growth regulators used in the earlier DNA melting studies were sufficiently high to exceed the capacity of the buffers used and that the pH changes were sufficient to affect the melting point of DNA. Prevention of pH changes negated the effects of hormones on DNA and both animal and plant DNA was affected similarly. As noted by Bamberger (1971), reduction in melting point of DNA requires extensive (and probably unspecific) modification of helical structure and while it is perhaps still possible that plant hormones have some non-specific effects on DNA structure and hence transcription, it is intuitively more appealing that regulation of transcription would involve more subtle and less extensive changes. One further point worth considering is that whereas the DNA preparations used in these studies were purified and attempts were made to eliminate protein and RNA, there is no guarantee that the DNA was in fact RNA- and protein- as well as perhaps carbohydrate-free and no analyses were performed to assess purity. If the DNA was not pure, the effects of hormones on DNA structure do not necessarily involve direct hormone-base interaction.

The studies described above have not yielded positive results although such studies should perhaps be encouraged. It seems possible that hormone-receptor complexes may recognize DNA sequences and in this way lead to some initiation complex with polymerase and other factors. Therefore it seems equally acceptable that some of the binding specificity of the complex resides in the hormone itself but clearly more work is needed to prove or disprove this. There is a tendency among physiologists to discount such studies as meaningless because many compounds such as dyes, antibiotics and carcinogenic hydrocarbons will bind to DNA, but in fact the biological effects of these compounds can often be associated with their interactions with nucleic acids (Levine et al. 1963; Arya and Yang 1975). Thus, there would seem to be some correlation between ability to bind to nucleic acids and biological activity. That there may also be nuclear environmental factors which confer specificity upon hormone-DNA interactions is indicated by studies of steroid hormones and animal DNA. The results were initially similar to those from plants (Goldberg and Atchley 1966) in that the structure of native DNA was found to become unstable, but later studies found that steroids bound only to denatured DNA, not native DNA (Cohen et al. 1969; Cohen and Kidson 1969; Kidson et al. 1970). Such inconsistency may be explained by the recent finding that preference for native or denatured DNA depends on temperature and electrolyte concentration (Arya and Yang 1975). The highest affinity was achieved when the DNA structure was intermediate between a compact double helix and complete disorder and the overall effect was to stabilize the DNA structure. These results are described to show that hormone-DNA interaction, if it occurs, may be very complex and until the nature of the nuclear environment is known the results of experiments such as those described above will be very difficult to interpret.

The stability of chromatin is also affected by auxin. At pH 8.0, the hormone bound to pea epicotyl chromatin and the melting point of the chromatin was reduced indicating a loosening of histone-DNA binding (Fellenberg 1969b, c). This was supported by studies of artificially reconstructed nucleoproteins in which it was shown that IAA and α-naphthaleneacetic acid (NAA), but not β-NAA or tryptophan, reduced the thermal stability of the chromatin without causing loss of protein (Fellenberg 1969b, 1971a). These protein studies have been examined more closely and generally confirmed in subsequent studies which also found that auxin could only induce instability in the decondensed regions of chromatin (Fellenberg and Schomer 1975), that the effects of auxin (and GA_3) were strongly dependent on pH and hormone concentration (Grieshaber-Scheubel and Fellenberg 1972) and that the effect was accentuated (although still small) in nuclear membrane-free chromatin (Fellenberg and Voges 1975). Furthermore, chromatin isolated from IAA, NAA and 2,4-D-treated pea cotyledons contained auxin activity, thus indicating a non-covalent association between hormone and chromatin in vivo (Fellenberg 1970), although the possibility that the auxin-chromatin association occurred during tissue homogenization is difficult to eliminate. These studies have led Fellenberg and colleagues to the general conclusion that auxin and other hormones may have a selective histone-loosening effect in vivo and that this may be an integral part of transcription regulation. There are various aspects of this work which remain to be clarified, especially those relating to specificity of the effect, and until this is done the claim that the effect is relevant in vivo must be regarded as speculative. The dependence of the effect on pH may indicate that it is partly due to an effect of hormones on pH of the medium such as that described by Penner and Early (1972). Spiker and Chalkley (1972) were unable to detect an effect of hormones on stability of chromatin from peas.

Covalent interaction of auxin with chromatin has been demonstrated although this involves basic proteins. During aging of excised pieces of tuber tissues, basic proteins decreased and acidic proteins increased (Kamisaka and Masuda 1968). 2,4-D promoted similar changes during callus formation in pea epicotyls (Yasuda and Yamada 1970) and the loss involved primarily lysine-rich proteins. In the latter case, there appeared to be a direct hormone-chromatin interaction since radioactivity from 2,4-D was incorporated into all histones but especially into lysine-rich ones, and the radioactivity was recoverable as 2,4-D. At the same time, radioactivity in acidic proteins increased indicating that formation of histone-2,4-D complexes caused a reduction in basicity of very basic (lysine-rich) histones, which were then recoverable as acidic proteins. Such a change in basicity would cause a loosening of their attachment to DNA. Indeed, lysine-2,4-D compounds were demonstrated to occur in histone from auxin-treated tissues (Kamisaka and Masuda 1968; Yamada and Yasuda 1971; Yamada et al. 1971). Similar compounds have been found in Jerusalem artichoke tuber tissue (Yamada et al. 1973). Such changes in histone-DNA affinity could cause or assist in derepression of genetic information. However, these results should be regarded with caution since firstly, the studies were done with high concentrations of 2,4-D (10^{-5} M or so) which, although callus-inducing, are nevertheless growth inhibitory concentrations and secondly, the specificity of the effect has not been established in that it has not been shown to occur only in responsive tissues following application of active auxins.

Chromatin consists of DNA, protein and RNA in a ratio of about 1:1.5–2.5:0.05 and the protein can be further subdivided into histone protein, which is very high in arginine, histidine and lysine, and non-histone protein (NHP), which is rich in acidic amino acids. The ratio of histone to NHP is 1:0.5–1.5. This and other basic information used below is discussed more fully in a number of review articles (Bonner et al. 1968; Elgin and Bonner 1972; Olson and Busch 1974; Stein et al. 1974b; Elgin and Weintraub 1975; Stein et al. 1975).

The histones commonly consist of five major classes separable on acrylamide gel and the uniformity of these proteins from species to species, from tissue to tissue and one physiological state to another, indicates that they are not involved in fine control of transcription, especially since they change qualitatively very little, if at all, even though major changes in transcription may have occurred. They seem to have a major structural role in DNA organization and chromosomal structure and thus the early postulate that histones were responsible for repression and de-repression of transcription may be correct only in a general and not a specific way. Plant histones have been studied very little but the level of transcription of pea chromatin has been shown to be inversely related to the quantity of histones associated with the DNA and although the same histones occur in different tissues, quantitative differences between histones have been found to occur in different organs of pea plants and at different stages of development of pea cotyledons (Bonner et al. 1968; Fambrough et al. 1968). In addition, only one histone was found in barley endosperm nuclei (Duffus 1972) and quantitative differences between histones occurred with vernalization in wheat embryos (Teraoka 1967, 1968), thus providing further evidence of histone mobility. However, balanced against these earlier reports of histone changes are more recent studies showing that histone variations can be artefacts due to contamination by basic cytoplasmic proteins and to changes in existing histones during preparation (Panyim et al. 1968). Re-examination of the work of Teraoka in wheat using techniques which overcame these problems (Spiker and Chalkley 1971) showed that there were no differences in histones with development of wheat seedlings (Spiker and Krishnaswamy 1973). Thus, there is very little to encourage a search for hormone-induced changes in histones at present but nevertheless they may still occur. In plants nothing is known about phosphorylation, acetylation and methylation of histones although such modifications are known to occur in animal histones.

Currently, the NHP seem to be likely candidates as regulators of transcription, and thus hormone-NHP interactions in plants should receive much more attention than they have been given in the past. In animal cells, the non-histone proteins are very heterogeneous, consisting of a number of enzymes and structural proteins, they are species-specific, tissue-specific and they change with the pattern of gene activity and are often thought to be responsible for such differences. Indeed cross-reconstitution experiments involving NHP and DNA from different tissues of the same animal, have shown that the expression of the globin gene is directed by the non-histone proteins (see reviews listed above). The NHP are characteristically extensively phosphorylated (mainly on serine) accounting for most (90%) of the protein-bound phosphorus and the phosphorylation pattern changes with different states of gene activity and in some cases after hormone (steroid) treatment (Stein et al. 1974b; Cohen and Kleinsmith 1976; Kleinsmith et al. 1976). Furthermore, it seems to be a very specific process in that only certain proteins are phosphorylated. The nature of the interaction of these proteins with DNA is still rather obscure but it is suggested by the in vitro reconstitution experiments that the NHP do interact specifically with certain regions of the DNA. The precise mechanism of the interaction is unknown and how the NHP regulate transcription is also obscure. However, there is evidence that they may regulate histone-DNA binding at specific loci (Stein et al. 1974a). Another aspect of the mobility of constitution of NHP probably involves the import and export of newly synthesized or relocated regulatory factors such as the mediators and receptors (of hormones and other external stimuli) which would direct and modulate polymerase activity. There is also evidence for the existence among the NHP of acceptor proteins for hormone-receptor complexes. The presence of acceptor proteins has been shown to be necessary for binding of progesterone-receptor complex to chromatin and the effect is specific to NHP of target cells (Stein et al. 1974b; Spelsberg et al. 1976).

In plants, it is known that NHP are zone-specific in bean hypocotyl (Spelsberg and Sarkissian 1968), that they are phosphorylated in *Lemna* and barley embryos by nuclear kinases and that the phosphorylation and the protein patterns vary during germination of barley embryos and during light-dark transitions in *Lemna* (Chapman et al. 1975; Van Loon et al. 1975; Trewavas 1976). Although there is presently no good evidence to suggest that auxin acts through the NHP, we have presented the above discussion in the belief that further study of the role of NHP is likely to lead to elucidation of some events in the regulation of transcription and thus plant hormones should be considered as potentially having effects on these proteins. Changes in NHP, perhaps through modification of existing proteins or through addition or loss of enzymes, structural proteins, hormone-receptor complexes or mediator molecules, acceptor proteins and polymerase cofactors, all either pre-existing or synthesized elsewhere in the cell, would seem to be likely ways of altering gene expression.

10.3.1.3. Transcription in vitro following in vivo hormone treatment; nuclei, chromatin and DNA. The aim of this section is to assess whether auxin may regulate gene expression as determined by assays based upon cell-free transcription of DNA or chromatin (and nuclei) isolated from hormone-treated tissue. There are two major ways of judging whether there have been transcriptional changes: the first entails a search for new gene transcripts and the second relies upon a change, either in level or specificity of RNA polymerase activity. When studying isolated nuclei or chromatin, it is very difficult to distinguish between changes in template availability and changes in polymerase. One initial approach to the problem has involved the use of *E. coli* RNA polymerase core enzyme which lacks specificity and which probably transcribes all available template. In systems saturated with this enzyme, maintenance of a difference in level of RNA synthesis between nuclei or chromatin from treated and untreated tissue would signify an effect of the hormone on template availability, while elimination of a difference would indicate an effect of the hormone on polymerase level or activity. However, it is pertinent to the following discussion that since hormone-induced changes in template in most tissues might be expected to be minor, the *E. coli* polymerase saturation technique is probably not sufficiently sensitive to give a reliable estimate of whether or not template availability has been affected.

In general, chromatin and nuclei isolated from auxin-treated tissues have shown a greater capacity for RNA synthesis and a greater RNA content than chromatin and nuclei isolated from untreated tissues. Nuclei isolated from soybean seedlings treated with 2,4-D for 24 hr contained the same amount of DNA but the amounts of RNA and protein had doubled (Chen et al. 1975). Since the nucleoli had also enlarged, the increased RNA was probably mostly rRNA. By using inhibitors it was shown that the levels of polymerase I- and II-dependent transcription in isolated nuclei had risen; that of polymerase I being much greater than that of polymerase II. The RNA synthesized by nuclei of treated seedlings was not characterized but in nuclei of untreated seedlings, 35% of the RNA synthesized was apparently rRNA, dependent upon polymerase I, and the rest was synthesized primarily by polymerase II (Gurley et al. 1976). Thus in treated seedlings the RNA synthesized was presumably largely rRNA. Results obtained from studies of isolated chromatin have been similar. Chromatin from soybean seedlings which had been treated with auxin for 4 hr or more, showed increased ability to synthesize RNA. Addition of saturating levels of *E. coli* polymerase core enzyme increased greatly (11-fold) the transcription of both treated and control chromatin, leaving only a very small difference due to auxin (O'Brien et al. 1968a, b). This was interpreted to mean that auxin did not increase the total available template, but in fact the actual difference in native chromatin was maintained in polymerase-saturated conditions indicating that template availability had been increased. Analysis of the RNA synthesized using MAK chromatography showed that control chromatin made RNA which was released (soluble), consisting of tenaciously-bound RNA (TB-RNA) and heterodisperse RNA, and RNA which was not released (bound) from the chromatin consisting of 4S – 5 S RNA. Auxin caused a major enhancement of synthesis of TB-RNA and several fractions of heterodisperse RNA, and a small enhancement of 4S – 5S RNA synthesis. Furthermore, the transcription products supported amino acid incorporation in a cell-free system and auxin caused a doubling of the effect. Unfortunately, the cell-free translation products were not characterized. This linked transcription – translation system should be further explored and characterized. Subsequently, the RNA transcribed from chromatin of treated seedlings was found to have different nearest neighbour frequencies to that transcribed from chromatin of control tissue, demonstrating that auxin had caused qualitative as well as quantitative changes in transcription (Holm et al. 1970). Almost everything about the work on soybean points to the major effect of auxin being on mRNA and/or HnRNA synthesis or at least polymerase II activity. Nevertheless, it would seem that polymerase I is somehow involved since in one study (Holm et al. 1970) the effects of auxin on chromatin from apical, elongating and basal sections of soybean hypocotyl were eliminated by saturating amounts of *E. coli* polymerase, and the effects of auxin on chromatin from each section paralleled the ability of auxin to promote or retard RNA accumulation in the section. The RNA was undoubtedly largely rRNA and transfer RNA (tRNA) (or perhaps pre-tRNA) which are the products of polymerase I and polymerase III respectively (Weinman and Roeder 1974; Weil and Blatti 1976). Thus, it would seem that in soybean hypocotyls, auxin treatment caused changes in transcription which were dependent on all polymerases although differences between studies are confusing. Nevertheless, these studies did provide evidence that in vivo regulation of RNA levels could well have its origin in chromatin activity rather than for example in RNA conservation.

Experiments involving other plant tissues have also produced evidence that auxin can influence polymerase activity and/or template availability. In a similar study of cucumber embryonic axes, chromatin isolated from auxin-treated seedlings synthesized more RNA than chromatin from untreated seedlings, but while for the first 6 hr of treatment *E. coli* polymerase eliminated the difference, thereafter it did not (Johnson and Purves 1970), indicating that early effects of the hormone involved enhancement of polymerase activity while later effects involved transcription of additional template. The first effect could involve any polymerase but the template effect probably involved polymerase II-dependent transcription.

Washing of excised pieces of storage tissues usually leads to changes in respiration and changes in enzyme activities which depend on RNA and protein synthesis. In sugar beet *(Beta vulgaris)* slices, invertase is synthesized and this is enhanced by GA_3 and retarded by auxin. The transcriptional activity of chromatin isolated from washed tissue had increased 7-fold after 25 hr of washing and there was also a 3-fold increase in template availability (Duda and Cherry 1971), indicating that both the polymerase activity and template had increased. The RNA synthesized by "washed" chromatin was different to that synthesized by control chromatin on the basis of nearest neighbour frequencies, both in the presence and absence of *E. coli* polymerase, thus suggesting that new DNA was available for transcription. "Unwashed" chromatin was not affected by in vitro application of hormones but it was changed by including hormones in the washing solution. IAA in the solution did not affect the endogenous transcription level of isolated chromatin but it did increase the total template available as determined in *E. coli* polymerase-saturated conditions.

In auxin-treated lentil roots, Teissere and co-workers claimed that the first response was a change in template (Teissere et al. 1973). Chromatin from treated roots (1.5 hr) transcribed 25% more RNA than chromatin from untreated roots and there was no change in activity of any of the isolated polymerases thus leaving the alternative that template had changed. If this were so, removing histones from the chromatin should have eliminated the difference as well as increased template availability. Trypsin treatment did in fact increase total template and the authors claimed a significant reduction in the auxin effect thus confirming their hypothesis that only template availability was involved. However, in fact, the effect of auxin was still an increase of between 10% and 20% depending on how much trypsin was used, indicating not only that differences in polymerase activity still existed (apparently the polymerases were trypsin-stable), but that such differences could account for between 40% and 80% of the initial auxin-induced difference in chromatin activity. It should perhaps be asserted here that changes in chromatin-bound polymerases do not necessarily survive isolation (Guilfoyle et al. 1975; Lin et al. 1977). Modulation of chromatin-bound activity could occur which would involve no increase in polymerase molecules and which would occur only on chromatin. Nevertheless, after 14 hr of auxin treatment, there was a mjaor increase in nucleolar polymerase I activity but still no change in polymerase II.

These studies support the idea that auxin application to responsive tissues leads to transcription of additional template dependent on polymerase II as well as to increased overall polymerase activity. Analysis of transcripts has, in general, supported changes in template, showing that the auxin-induced changes have involved rather heterodisperse transcripts and TB-RNA which could have been mRNA. Certainly there is no evidence from these studies that synthesis of major species of RNA is affected. However it is difficult to have any confidence in such an analysis because there has been so little characterization of the RNA synthesized by isolated chromatin, and little can be achieved until the products are characterized or chromatin can be made to produce more discrete and recognizable products. The results of the lentil root studies led Teissere and colleagues to hypothesize that there was a sequence of responses involving, firstly, an increase in template availability (presumably involving synthesis of new HnRNA and mRNA) followed by an increase in the polymerase I-dependent species (rRNA). Their studies of intact roots, to be discussed later, provide evidence for such a sequence but they have received little support from studies on chromatin isolated from lentil roots or any other tissue. Indeed, the cucumber hypocotyl studies indicate a reversal of this sequence. A polymerase II – polymerase I-dependent sequence of changes in transcription has also been proposed for estrogen action in uterine tissue (Glasser et al. 1972; Hardin et al. 1976; Raynaud-Jammet et al. 1972) although here too good evidence seems to be lacking.

10.3.1.4. Polymerases. Although increases in template availability are thought to involve the derepression of DNA by removal or addition of proteins or some other regulatory factor such as a hormone or hormone-complex, they could involve changes in the affinity of mRNA-transcribing polymerases for certain sites on the chromatin. Such considerations, as well as changes in total polymerase activity measured by the *E. coli* polymerase saturation technique have led, in several cases, to measurement of the effects of auxin on the activity or amount of individual polymerases.

The increased protein and RNA contents and enlarged nucleoli of nuclei isolated from auxin-treated soybean seedlings were accompanied by increases in activity of polymerase isolated from both nuclei and chromatin (Chen et al. 1975; Guilfoyle et al. 1975). After 24 hr treatment with auxin, nuclei contained 25% more polymerase II activity but 2.5-fold more polymerase I activity. The enzymes were distinguished on the basis of differential α-amanitin sensitivity.These increases were paralleled by the ability of the nuclei to synthesize polymerase I- and II-dependent RNA indicating that the auxin-induced changes in polymerases were responsible for increased RNA synthesis. The RNA was not characterized other than to show that in control nuclei, 35% of the RNA synthesized was rRNA, dependent on polymerase I, and the remainder was dependent on polymerase II (Gurley et al. 1976), indicating that auxin-treated nuclei were synthesizing polymerase I- and polymerase II-dependent RNA in about equal proportions. Whereas the auxin-induced changes in polymerase I have been examined further (see below), the changes in polymerase II have received little attention. Indeed, in this study the effect of auxin on polymerase II activity was described as insignificant. Clearly it is important and reflects either transcription of new template, in which case it is consistent with the effects of auxin on increased template availability described earlier, or it reflects increased transcription of existing template or perhaps both. Either way it is worthy of further attention especially when mRNA seems to be the only RNA whose synthesis is required for auxin action (see later).

With regard to changes in polymerase activity, it is of interest to determine whether or not the changes are the result of increased numbers of molecules of the same or a variant type, or whether they result from changes in activity of existing molecules. For polymerase I, studies from different laboratories are not entirely in agreement. In one case, increase in activity of polymerase bound to chromatin isolated from auxin-treated soybean was maintained in solubilized enzyme (Hardin and Cherry 1972), indicating that there was a real increase in the amount of enzyme although this is not necessarily the correct interpretation. The chromatin-bound polymerase was a mixture of polymerases but it was evident (from their table V) that auxin caused an increase in polymerase I-dependent and not polymerase II-dependent transcription, although apparently polymerase II constituted only a very small fraction of the original total polymerase II, most of which was lost from the chromatin during certain procedures of chromatin purification (Rizzo et al. 1974). In the other case, in studies in Key's laboratory, auxin-induced increases in chromatin-bound polymerase have been confirmed and the results are generally similar to those of Rizzo et al. (1974), but the analysis of the effect has been more extensive giving rise to somewhat different conclusions (Guilfoyle and Hanson 1973; Guilfoyle and Hanson 1974; Lin et al. 1974). These studies led to the development of a chromatin fractionation procedure which produced chromatin containing almost exclusively polymerase I (α-amanitin-insensitive). The enhancement of polymerase after

24 hr of in vivo auxin treatment (here measured as 3–5-fold) was due to the same amount of polymerase molecules becoming more active and producing much longer chains per unit time, although there was a small (20–50%) increase in the number of chains. Thus, it seemed that auxin primarily caused activation of already existing polymerase molecules. However, as in the first case, the activity increase (in the basal hypocotyl section) was maintained in solubilized enzyme (Guilfoyle et al. 1975), demonstrating that the modification of the polymerase, whatever it was, did not require the environment of the chromatin for stability. It is perhaps significant that the increase in polymerase II transcription did not survive solubilization, perhaps indicating that it was the result of increased template rather than a change in the enzyme. It is appealing to think that perhaps the increased polymerase I activity leads to transcription of more contiguous ribosomal genes or at least transcription of a large part of a pre-rRNA transcript, but in fact the transcripts with or without auxin were considerably smaller than light ribosomal RNA (1rRNA). Therefore, while the effect is an interesting one, it is difficult to interpret in terms of synthesis of additional, entire rRNA. A more recent study from Key's laboratory has given support to the concept of modulation of activity of existing enzyme, but the results seem to be at some variance with the earlier reports. Differences existing between polymerases I and II in nuclei from auxin-treated apical and basal sections of hypocotyl tissue were maintained only in part, if at all, in solubilized enzyme (Lin et al. 1977). In particular, in the meristematic apical region, auxin suppressed polymerase I activity (as well as in vivo RNA level) but the level of polymerase, as judged from solubilized preparations, was almost unaffected, Whereas this indeed indicates modulation of polymerase enzyme within the nucleus, it seems to be at odds with a previous report which claimed that increased polymerase activity was maintained in solubilized enzyme (Guilfoyle et al. 1975). There may be a difference in stability of the effect on soluble enzyme (perhaps indicating a different mechanism) depending on whether it is a modulation, up or down, of enzyme activity.

Studies essentially similar to those on soybean hypocotyl have been carried out on lentil root and the conclusions are very similar (Teissere et al. 1973). After a short auxin treatment of a few hours there was no change in any of the solubilized polymerases but after 14 hr treatment, the level of solubilized polymerase I had risen by about 100% while the levels of polymerases II and III were unchanged. The change appeared to involve altered activity of existing enzyme molecules and not synthesis of new ones.

Much less attention has been given to effects of auxin on polymerase II but nevertheless there is a report of in vivo auxin treatment of corn seedlings causing increased activity of soluble polymerase II (Arens and Stout 1972). However, the enhancement could be demonstrated only in dialyzed preparations. Whether the change involved increased enzyme or increased activity was not determined and whether the inhibitor has any in vivo significance was not clear.

In summary, there would seem to be agreement that auxin treatment somehow leads to changes in activity of polymerase I. The changes can be either increase or decrease of enzyme activity. Neither seem to involve major changes in the amount of polymerase enzyme and it seems that the effect of auxin is to modulate enzyme activity within the nucleus. The situation with respect to polymerase II is very unclear. Polymerase II-dependent transcription clearly increases in soybean in response to auxin and the increased activity is not maintained by the solubilized enzyme, thus indicating that the amount of enzyme is not increased by auxin. However in corn, there is an increase in soluble polymerase II activity but only after removal of an inhibitory, dialyzable component. Perhaps a similar inhibitor remained in the soybean polymerase II preparations. Further work is required to clarify this situation although auxin regulation of both enzyme activity and available template seems to be possible. Purification and characterization of the various polymerases will be essential if an analysis of regulatory factors which alter activity is to be possible. Characterization of transcripts will also be necessary.

10.3.1.5. Effect of auxin and receptors or mediators on RNA synthesis in vitro. We have discussed some of the evidence suggesting that auxin treatment leads to increased template availability and increased polymerase activity. In an attempt to elucidate how auxin might mediate changes in transcription, we will consider studies in which auxin has been added to cell-free transcribing systems, namely isolated nuclei and isolated chromatin and DNA. This approach has some attractive features in that various components of the transcribing system can be isolated and reconstituted and that hormone complexes, if they occur, might be traced to their point of origin and their site of action.

The first demonstration of an in vitro effect of any plant hormone on nucleic acid metabolism was that of auxin on the nucleic acids of coconut nuclei, although as a result of recent work, this must perhaps be reassessed. In a series of papers in 1964 and 1965, Roychoudhury and Sen showed that, in a preparation described as isolated nuclei resuspended in coconut-milk supernatant, auxin enhanced incorporation of labelled thymine into DNA, uracil into RNA and [^{32}P]orthophosphate into both (Roychoudhury and Sen 1964a, b, c; Roychoudhury et al. 1965). In general, the effects were rapid, being evident well within an hour and they were long lasting. Similar results were obtained with kinetin and GA_3. The RNA synthesis was actinomycin D-sensitive. There was very little effort to characterize the RNA synthesized except to show that incorporation of label into buffer-soluble RNA was still increasing with IAA levels up to 10^{-3} *M* while incorporation into phenol-soluble RNA had an optimum at 10^{-5} *M* IAA, thus more closely resembling effects of auxin on growth. There are many aspects of this work which require clarification. It is surprising that these nuclei seemed to respond for long periods of time to all hormones when one might expect to see some limit to their competence. No effort was made to keep the system sterile, but even more confusing is the fact that, more recently, coconut-water has been found to contain no free nuclei (Mondal et al. 1970c). All of the nucleic acids in coconut water are soluble and not associated with any organelle (Mondal et al. 1970b). Thus, it would seem that the studies of Roychoudhury and Sen did not involve isolated nuclei but perhaps whole cells dislodged from the semisolid endosperm. Other early reports of auxin enhancement of RNA synthesis by isolated nuclei involved nuclei isolated from soybean (Cherry 1967; O'Brien 1967), pea and mung bean *(Phaseolus aureus)* seedlings and mature onion bulb (Maheshwari et al. 1966), tobacco and soybean tissue culture and pea buds (Matthyse and Phillips 1969) and sugar beet root (Duda and Cherry 1971). Centrifugation of nuclei from the original homogenates of pea, soybean, mungbean, onion and sugar beet did not eliminate activity although addition of supernatant back to resuspended nuclei of pea, mung bean and onion (Maheshwari et al. 1966) did enhance incorporation. Thus, if co-factors for auxin action were required, they were largely retained in the isolated nuclei. The first evidence that co-factors were required came from the studies of tobacco and soybean tissue cultures where it was found that the auxin-stimulation of RNA synthesis was dependent on isolation of nuclei occurring in the presence of the hormone. Furthermore, the macromolecular factor (see section 10.3.1.1) could be obtained not only from tobacco and soybean tissue culture nuclei but also from pea buds, indicating that while pea and probably other nuclei contained and probably required the factor, they simply did not lose it during isolation.

While in most of the studies on isolated nuclei described above, there was some evidence that the labelled RNA precursors were indeed incorporated into RNA, the RNA was not characterized except for one case involving sugar beet nuclei (Duda and Cherry 1971). Firstly, the MAK column profile of RNA synthesized was very similar to that expected from in vivo experiments, and secondly, auxin had no effect on soluble, 4S and 5S RNA (sRNA), DNA-RNA hybrid or heavy ribosomal plus "messenger" (hr + "m") RNA. The so-called mRNA of this study is usually refered to as DNA-like RNA (D-RNA). Auxin promoted (by 109%) the incorporation of precursors into lrRNA. TB-RNA was not studied. The effect on lrRNA indicated an effect on polymerase I-dependent transcription but since lrRNA and hrRNA are probably transcribed as one long transcript or pre-rRNA molecule, as in animal cells (see for example Lewin 1976), which is then processed to give 18 and 28S rRNA, it seems unlikely that lrRNA synthesis would be enhanced preferentially. The identification of this RNA as lrRNA is also perhaps doubtful since a fraction of RNA synthesized by pea nuclei appeared to be rRNA according to MAK chromatography but sucrose gradient analysis showed it to have an S value of only 6–10 (Johri and Varner 1968). Thus, it would seem that the identity of the lrRNA cannot be taken for granted. Therefore, probably all that the isolated nuclei studies tell us is that auxin may have associated with at least one macromolecular nuclear co-factor, perhaps a receptor, to cause transcription to be modified, but there is no further information concerning the sequence of events in the process.

On the other hand, studies on isolated chromatin, although few in number, have been somewhat more informative. In general, both auxin (plus co-factors) and mediator molecules added in vitro have been shown to increase chromatin transcription. Studies on chromatin from tobacco and soybean tissue culture (Matthyse and Phillips 1969) showed that auxin and co-factor together (but not alone) caused a 30–150% increase in RNA synthesis. The effect still occurred in *E. coli* polymerase-saturating conditions indicating that, at least in part if not completely, it was due to an increase in template availability and involved polymerase II-dependent transcription. However, using purified pea DNA, there was no effect of auxin plus co-factor, indicating that the auxin-co-factor complex did not recognize new DNA sequences in the absence of chromatin proteins. In assessing these results, it should be noted that the DNA studies did not involve plant polymerase but *E. coli* polymerase, which may not have had the capacity to mediate an auxin response. However, results of studies on coconut DNA transcription by Mondal and colleagues lead to a different conclusion (Mondal et al. 1970a, 1972a, b). In this case, a system consisting of native DNA, polymerase sensitive to α-amanitin and initiation factors, all purified from coconut chromatin, synthesized RNA 2–3 times faster in response to auxin plus a partially purified nucleoplasmic factor. The factor, apparently an IAA-receptor protein, did not interact with the polymerase or the initiation factors but it did interact with the DNA although the nature of the interaction is still unclear. Hybridization studies indicated that new classes of RNA were synthesized in the presence of IAA, and polyacrylamide gel fractionation of products showed that, whereas in the absence of IAA, RNA synthesized migrated primarily as 20S RNA, in its presence 12S, 25S as well as 20S RNAs were made. The polymerase used here was α-amanitin-sensitive and was presumably polymerase II. Thus, it would seem that here auxin plus the factor led to the transcription of new template dependent on polymerase II. Evidence was presented to show that base-sequence recognition was involved in the mechanism although the evidence was deficient in some respects (see Kende and Gardner 1976) and there was no analysis to show that the DNA was indeed free of protein and other contaminants. Both of these studies indicate that auxin plus cofactors can regulate polymerase II transcription at the level of chromatin or DNA and while there is some evidence for direct interaction with DNA, it requires support from other systems.

An entirely different mechanism of transcription regulation is indicated by the mediator molecule studies. A factor from soybean cotyledons and from membrane fragments isolated from soybean hypocotyl and onion stems, stimulated RNA synthesis by chromatin isolated from untreated soybean tissue but not from auxin-treated tissue (Hardin et al. 1970, 1972; Cherry 1974). The stimulation was α-amanitin-sensitive, indicating that the factor was preferentially stimulating polymerase II-dependent transcription. A factor from pea and corn shoots (Venis 1971) isolated by affinity chromatography stimulated transcription in chromatin as well as in a system containing pure homologous DNA and *E. coli* polymerase. The additional RNA synthesis seemed to involve enhanced chain initiation at new sites although the possibility of more rapid initiation at existing ones could not be ruled out. It is unfortunate that the relationship between the above "factor" described by Venis was not related to the recent thorough study of an auxin-binding component found in maize membranes (Batt and Venis 1976; Batt et al. 1976).

Both of the above studies seem to suggest that regulation of polymerase II-dependent transcription involves an unknown mediator molecule with the properties of a transcription initiation factor. Although these studies seem to demonstrate a separate transcription-regulating mechanism where a hormone is not directly involved, it is possible that all of the macromolecular cofactors are similar and really require only contact with the hormone for activation as suggested by Venis. This could be the case for the receptors studied by Matthyse and Phillips and by Mondal and colleagues since, although the hormone and cofactors obviously needed each other for activity, it was not shown that they are both required at the site of transcription. In support of these suggestions, all of the factors were thought to regulate polymerase II-dependent transcription. Thus, perhaps for the sake of a working hypothesis, it is preferable to regard the factors as all being similar and being macromolecules which somehow interact with polymerase and/or other chromatin components including DNA to specify initiation sites.

Examination of the limited data on polymerase I leads to a similar conclusion. It has been described above how the auxin-induced rise in polymerase I activity seems to result not from enzyme synthesis but from modification of the properties of already existing enzyme, and studies on lentil roots have produced a multiplicity of polymerase I-regulating factors among acidic chromatin proteins (Teissere et al. 1975). Four factors, termed alpha, beta, gamma and delta, were fractionated from the stimulatory proteins, one of which (delta) stimulated polymerase I activity by an enormous 700%. The levels of factors alpha, beta and delta were unaffected by in vivo auxin treatment but the activity of gamma was doubled. Factor gamma regulated only polymerase I while delta affected both polymerases I and II. Thus the activity of existing polymerase I could well be regulated by the appearance of factor gamma at the transcription site. There was evidence that the factors were involved in chain initiation and that they facilitated opening of the DNA helix and the recognition of "promoter-like" regions. They had no effect on chain elongation (Teissere et al. 1976), which seems to be at odds with the reported effect of auxin on soybean polymerase activity which involved primarily chain elongation rather than chain initiation (Guilfoyle and Hanson 1974). Protein factors regulating polymerase II activity have also been reported in animal tissues such as calf thymus (Stein and Hansen 1970), rat liver (Seifart 1970) and Novikoff ascites tumour cells (Lee and Dahmus 1973; Dahmus 1976). All of these factors are specific for polymerase II, having no effect on bacterial polymerase and at best, the ascites factor is specific for native DNA (having no effect on denatured DNA), causing a change in base composition of RNA synthesized. There seems to be a multiplicity of such factors in a single tissue, as appears to be the case in plants, but as yet they have not been shown to be related to hormone action.

In summary, the principle seems to be emerging that the activities of both polymerase I and polymerase II can be affected by macromolecules, probably proteins, acting with or without auxin, which in the main seem to be involved in chain initiation, controlling either rate or specificity of initiation or perhaps both. Such factors, therefore, appear to be similar to and have a similar function to the polymerase-regulating factors, such as sigma factor, in bacteria. Purification of the enzymes and these postulated factors will have to be achieved before a final assessment can be made. From the limited definitive data, it also seems that auxin can cause changes in levels of such factors in chromatin, involving either synthesis or transport from some other site, at present tenuously thought to be a membrane. Hormonal control of such factors and the way in which the factors regulate polymerase activity and specificity would now seem to be a very promising field of research and central to the mechanism of auxin regulation of transcription. Perhaps higher-resolution techniques will demonstrate a multiplicity of transcription factors in single tissues, the factors perhaps showing different initiation specificities.

In general, there are many apparent similarities between the scheme outlined above for auxin and steroid hormone action which has received a lot of attention and which has been thoroughly reviewed recently (Means and O'Malley 1972; Jensen and DeSombre 1973; Stein et al. 1974b, 1975). For example, estradiol in uterus tissue enters the cell and combines with a protein receptor which occurs either as an 8S molecule of molecular weight of 200,000 or as its 4S subunits, each with a molecular weight of about 75,000. The complex is non-covalent and stable. It then undergoes a temperature-dependent transformation to a 5S complex which migrates to the nucleus where it somehow regulates transcription. The action at least partly involves regulation of polymerase I activity rather than level. The hormone alone and the untransformed complex are ineffective in regulating transcription. Thus, for steroids a protein transformation is involved, as could be the case for auxin receptor, but there is also a hormone-receptor complex which is not yet established for auxin. The regulation of polymerase activity is common to both steroids and auxin. However, what the steroid-receptor complex binds to in the nucleus (DNA, protein or both), and whether steroids act by promoting recognition of specific gene sequences or by being involved in complex protein-DNA interactions is still almost as obscure in animal cells (Yamamoto and Alberts 1974) as it is for auxin action in plants.

10.3.1.6. RNA metabolism in intact tissues. The previous sections have dealt with the effects of auxin on cell-free transcription and in general suggest that auxin can lead to modification of RNA synthesis in isolated nuclei and chromatin. The purpose of this section is to assess whether these effects are reflected in RNA metabolism in vivo and whether they are directly related to the effects of auxin on growth, development and protein synthesis.

It is of historical interest that the first connection between plant hormones and nucleic acid metabolism was the demonstration by Silberger and Skoog (1953) that auxin increased the RNA (and DNA) contents of cultured tobacco pith tissue. This was associated with very little cell division but extensive cell enlargement (Skoog 1954). The increase in nucleic acids occurred before cell enlargement but subsequently there was a proportional increase in cell size and RNA content. The implication of this work was that cell enlargement was dependent on changes in nucleic acid metabolism; however, the first experimental evidence for this is commonly attributed to Masuda (1959) who showed that treatment of oat coleoptiles with ribonuclease prevented auxin-promoted cell elongation. Whether the ribonuclease was indeed destroying intracellular RNA is perhaps doubtful, but at least the study led to the finding that inhibitors of RNA synthesis, like actinomycin D, as well as inhibitors of protein synthesis, prevented growth in the presence and absence of hormone (Nooden and Thimann 1963; Key 1964). Today the specificity of actinomycin D as an inhibitor of transcription is in doubt, but at least the results of these early experiments, whether they were right or wrong, provided an hypothesis for the mechanism of growth regulator action which has been and still is being examined in some depth.

Since these early experiments, there have been numerous claims that auxin regulates RNA metabolism and content in a wide variety of plant systems, and since many of these have been reviewed by Key (1969) and by Trewavas (1968c), only a brief outline of this work will be presented here. In intact plants, auxins generally cause cessation of cell division and cell enlargement in apical tissues, while in the more mature tissues RNA metabolism is stimulated and cells proliferate, RNA accumulation (largely ribosomal) preceding cell division. In excised plant parts incubated in the absence of auxin, there is usually a loss of RNA and although auxin affects this, the way in which it is affected depends on the system. In corn mesocotyl, physiological concentrations of auxin enhance the rate of loss of RNA, apparently by enhancing breakdown with no effect on synthesis. In keeping with this, auxin has been shown to enhance RNase activity in this tissue. In the upper regions of soybean hypocotyls there is also a small loss of RNA during incubation, but auxin prevents this, apparently by stimulating synthesis. In this tissue there is no detectable regulation of RNase activity. However, in the maturing zone of the hypocotyls, auxin causes a large increase in RNA content while in the absence of auxin there is no change during incubation, which makes this an ideal system to work with and most of the studies of Key and Cherry on auxin and RNA metabolism have been on this tissue. Labelled precursor incorporation studies showed that the rate of RNA synthesis was doubled after 2 to 3 hr but even during the lag phase, there was a real increase in RNA synthesis. There are now many such cases in which auxin apparently increases RNA synthesis, some occurring within minutes and others in a matter of hours. It is unlikely that such auxin-induced increases in RNA are all the result of cell growth because in oat coleoptiles, Masuda and colleagues have shown that auxin can increase RNA even when cell enlargement is prevented with an osmoticum and in the basal non-elongating mature section of soybean hypocotyl, RNA level increases in response to auxin in the absence of growth. There are also a number of reports describing increases in ribosomes in auxin-treated tissues (Key and Hanson 1961; Chrispeels and Hanson 1962; Fowke and Setterfield 1968; Fites et al. 1969).

In general, it seems that in many instances and especially in the long term, auxin enhances RNA synthesis whereupon the question of whether the increase is necessary for growth arises. Auxin-induced growth in oat coleoptiles and soybean hypocotyls is inhibited by actinomycin D and there are now many similar reports involving other systems and other inhibitors of RNA synthesis like the purine base analogs, 8-azaguanine, 6-methylpurine and cordycepin. It is interesting to note however, that pyrimidine base analogs, like 5-FU do not prevent cell enlargement in soybean hypocotyl and artichoke tuber (Key 1964; Key and Ingle 1964; Nooden 1968). Therefore, not only does auxin increase RNA synthesis but at least some of the RNA synthesis seems to be required for auxin action.

A clue as to which species of RNA are required and perhaps those which are not required is obtained from consideration of inhibitor specifities. 5-FU usually inhibits RNA synthesis but not cell enlargement and since it inhibits rRNA synthesis preferentially and tRNA to some extent, it seems that rRNA synthesis and perhaps tRNA are not necessary for auxin promotion of cell elongation. Actinomycin D combines with guanine residues in DNA and inhibits probably all DNA-dependent RNA synthesis. Therefore, accepting the limitations of inhibitor studies, it seems that the required RNA is neither ribosomal nor tRNA. It follows that the most likely candidates are the minor RNA species. Key and colleagues (Key and Ingle 1968; Vanderhoef and Key 1968) were the first to describe how the level of the AMP-rich TB-RNA was related to auxin-induced cell growth making it a likely candidate for the required RNA species. TB-RNA has since been shown to be poly(A)-containing RNA or mRNA (Key and Silflow 1975; Schmid et al. 1975). TB-RNA having 40% AMP was first reported in pea roots in 1967 (Ewing and Cherry 1967) and later in soybean hypocotyls and oat coleoptiles (Tester and Dure 1967).

The RNA essential for cell elongation has a fairly short life. In soybean hypocotyls and other systems, antinomycin D stopped elongation after 1–2 hr (Key et al. 1967) and the lag for RNA synthesis was somewhat shorter so that the life of the required RNA is probably considerably less than 2 hr which is very short compared to say plant rRNA which has a half life of 4–5 days (Trewavas 1970). Perhaps in keeping with this, auxin must be present the whole time for elongation to continue, presumably, to ensure a supply of the essential RNA. However, a role for the hormone in other processes has not been ruled out. Therefore, evidence from inhibitor studies indicates that auxin-promoted cell elongation requires the synthesis of a short-lived RNA which may be AMP-rich but that it does not require synthesis of rRNA or tRNA.

What evidence is there that auxin stimulates RNA synthesis and is there any evidence for preferential stimulation of an RNA fraction which qualifies as the essential RNA? Beginning in 1964 with the work of Key and colleagues, there have been a number of studies comparing labelled precursor incorporation into RNA in the presence and absence of auxin which have indicated that the hormone can stimulate synthesis of all species of RNA discernible by MAK column fractionation, provided the incubation period was long enough. However, synthesis of rRNA as determined by isotope incorporation, as for example in pea stems (Trewavas 1968a) and in soybean hypocotyl (Key and Ingle 1968), led to the proposal that auxin caused a preferential increase in rRNA synthesis although inhibitor studies had indicated that this was not an essential RNA. Furthermore, the enhancement of rRNA synthesis usually occurred well after auxin had affected growth. Label accumulation into D-RNA and TB-RNA was also increased by auxin and although specific activities of individual RNA fractions were not always measured, in some cases, as in the studies of Masuda and colleagues and Tester and Dure with *Avena* coleoptiles and pea internode sections (Masuda and Tanimoto 1967; Masuda et al. 1967) and in soybean hypocotyls (Fites et al. 1969), auxin stimulated label accumulation into the AMP-rich species more or earlier than into other species. Indeed, the synthesis of rapidly labelled polydisperse, AMP-rich RNA in *Avena* coleoptiles was reported to be increased within 1 hr of auxin treatment (Cline and Rehm 1974). Nevertheless, until the late 1960's, it was common to find effects of auxin on synthesis (or accumulation) of all RNA species although the RNA which was affected most seemed to be variable and so the interpretation of the results was, in general, very difficult. Changes in the minor species of RNA were rarely measured.

In the general context of labelling experiments, evidence that auxin does exercise some control over mRNA synthesis comes from studies of lentil root where auxin stimulated de novo synthesis of ribosome-bound peroxidase (Miassod et al. 1970; Penon 1972). A large part of the rapidly labelled RNA (30 min with [^{32}P]orthophosphate) was tenaciously bound to MAK columns, and was therefore TB-RNA and it could be fractionated into two parts. One was uridine-rich, heterodisperse, of very high molecular weight, it had a half-life of less than 1 hr and was primarily of nuclear origin. The other was adenine-rich, sedimented at 12S, had a half-life of greater than 2 hr and was largely polysomal in origin although some could be found in the nucleus. Short-term (45 min) auxin treatment enhanced label accumulation into the nuclear fraction preferentially whereas longer treatment (5 hr) enhanced incorporation into both fractions. The authors concluded that the nuclear fraction was HnRNA and that the polysomal material was mRNA and if so, this is the first study of plant tissues distinguishing between the two species. There is evidence for a sequence of responses which is consistent with the current hypothesis that HnRNA is processed to give mRNA. In addition, the authors have evidence for competition between these two fractions in binding to purified non-repetitive DNA which indicates similarities in base sequences (Teissere et al. 1972). If the HnRNA is mRNA precursor, it is not only an excellent system in which to study the short-term effects of auxin but it is also the first good evidence from a whole plant system for auxin-directed mRNA synthesis. After longer periods of auxin treatment (15 hr), ribosome numbers increased so that the sequence of events would seem to involve changes in synthesis of HnRNA, then mRNA accumulation followed by changes in rRNA synthesis.

In general most of the in vivo labelling experiments performed are difficult to assess and interpret for a number of reasons. Unless a comprehensive time sequence study of changes in RNA labelling was made, the data throw little light on sequences of effects. Also, in general there is a scarcity of specific activity data and of identification of minor RNA fractions. There are also the effects of hormones on label uptake and processing to be considered. However, perhaps most important of all, Trewavas has recently pointed out that in interpreting labelling experiments, consideration of the turnover rates of individual RNA species is of paramount importance (Trewavas 1977). If this is done, it can be shown that differential increases in specific activities of different RNA species are to be expected from an overall increase in RNA synthesis. An example cited is that the increases in specific activity of 47%, 38% and 24% for rRNA, D-RNA and TB-RNA respectively after a 2 hr auxin treatment of soybean hypocotyls (Key and Ingle 1968) could well result from an overall increase of 50% in RNA synthesis rather than from a preferential increase in rRNA synthesis. Furthermore, it can also be shown that after a shorter auxin treatment period, like 30 min, a similar overall increase in RNA synthesis would cause the increases in specific activity to be reversed compared to the order given above. Thus, as perhaps in the lentil root studies, one would expect that the measured short-term effect of an overall increase in RNA synthesis would be a preferential increase in specific activity of mRNA and the later effect to be a preferential increase in specific activity of rRNA leading to the conclusion that there is a sequence of responses in individual RNA species. In addition, the specific activity of mRNA is expected to rise very rapidly compared to rRNA. Thus, it is clear that as a result of these considerations as well as many other problems concerning label uptake and conversion, it is extremely difficult to interpret in vivo labelling experiments with any degree of certainty and probably many have been misinterpreted. Nevertheless, these studies did indicate that auxin could affect RNA metabolism, probably synthesis, and they drew attention to the possible importance of AMP-rich RNA in hormone action.

What other evidence is there that auxin regulates levels of mRNA in vivo? The amount of polysomes in a tissue is probably an estimate of the rate of protein being synthesized in a tissue as well as a crude estimate of the level of translatable mRNA. Auxin has been shown to increase the formation of polysomes from monosomes usually within a few hours which is at least consistent with enhanced mRNA levels (Key and Ingle 1964; Trewavas 1968b; Fites et al. 1969; Davies and Larkins 1973; Travis et al. 1973; Davies 1976). The increases in polysomes usually occurred before growth began and before any major increase in RNA levels indicating a sequence in the increase of mRNA and rRNA levels. Furthermore, it was actinomycin D-sensitive but unaffected by 5-FU indicating that synthesis of new ribosomes was unnecessary for polysome formation but another species, presumably mRNA was required. However, increase in polysomes does not necessarily reflect increased mRNA levels and actinomycin D sensitivity does not necessarily reflect increased mRNA synthesis. For example, polysome formation could be dependent on some ribosome modification with a requirement for continued synthesis of already existing mRNA. Indeed, changes in ribosome efficiency in translation have been measured in response to light (Williams and Novelli 1968; Travis et al. 1972), with age of the tissue (Lin et al. 1973) and following auxin treatment (Key and Ingle 1964; Travis et al. 1973). Thus, formation of polysomes from monosomes seems to be dependent on activation of ribosomes as well as on a continuous requirement for RNA synthesis which is not tRNA or rRNA synthesis but which is responsible for increase in levels of mRNA species, some perhaps not pre-existing. Thus, whereas increases in polysomes have been taken to be indicative of new or enhanced mRNA synthesis, clearly this cannot be done with any degree of confidence.

The only reported effort to detect qualitative changes in RNA of auxin-treated tissue by RNA-DNA hybridization techniques was performed on auxin-treated pea stem tissue but no changes could be detected (Thompson and Cleland 1971). At face value, this would argue against auxin initiating synthesis of new mRNA species. However, this study was subject to a number of limitations, one of which is that the experiments were performed in such a way that only mRNA species complementary to intermediate or highly repeated genes would have been detected and changes in mRNA species present in relatively small numbers transcribed from unique sequence DNA, as one might expect cellulase to be (see below), would not have been detected.

To date, the only good evidence for an auxin-induced change in mRNA complement stems from the work of Maclachlan and colleagues on cellulase induction in pea epicotyls. The work is the result of a search for an enzyme mediating cell-wall loosening accompanying auxin-induced cell expansion. About 6 hr after application, IAA causes an increase in cellulase activity in young pea epicotyls which is dependent on continued RNA and protein synthesis (Fan and Maclachlan 1966, 1967a, b; Datko and Maclachlan 1968; Davies and Maclachlan 1968) and microsomes isolated from IAA-treated epicotyls were found to be capable of synthesizing cellulase although this was probably run-off of pre-initiated polypeptides rather than initiation and synthesis of complete molecules (Davies and Maclachlan 1969). Nevertheless, it demonstrated that IAA enhanced the level of membrane-bound polysomes capable of synthesizing cellulase. This was accompanied by an overall increase in RNA and protein as well as a maintenance of the levels of polysomes in IAA-treated tissue (Birmingham and Maclachlan 1972) whereas in control epicotyl segments, the polysome level declined. Evidence that polysome production was associated with enhanced mRNA levels has recently been obtained (Verma et al. 1975). The poly(A)RNA was isolated and translated in a cell-free protein synthesizing system from wheat germ. Cellulase was identified by immunoprecipitation with monospecific antibodies and by its electrophoretic mobility. The control tissue contained no translatable cellulase mRNA whereas in treated tissue, the level increased rapidly without appreciable lag. This meant that the level of cellulase mRNA was not correlated with the rate of cellulase synthesis since the latter began only after a 24 hr lag. Therefore, a secondary control over translation was invoked but there are alternative explanations. The implication is that translatable cellulase mRNA exists within the cell but is not translated. One alternative is that the mRNA exists in the cell in the latter stages of processing and that during isolation or incubation with the wheat germ system, the final change(s) allowing translatability occur. Another alternative is that cellulase itself is produced in an inactive form requiring time dependent activation. The point to be made is that much more work is needed to prove the existence of control over the translation of fully processed mRNA. Nevertheless, in developing cotton embryos there is some evidence for control over translation of pre-existing mRNA, probably by ABA, (Ihle and Dure 1969, 1970, 1972) but there is no evidence that the mRNA exists in a fully processed translatable form (see also Chapter 11, section 11.3.2). It has been suggested that some processing step is required for activation of mRNA (Walbot et al. 1974) although proof of this requires more direct experimentation. Nevertheless, the study of cellulase induction has demonstrated an auxin-promoted level of translatable cellulase mRNA and was the first to do so for any plant hormone. The problem now is to determine how the increase in mRNA occurs. The studies of in vitro transcription provide grounds for thinking that new synthesis is involved but there is some evidence that the auxin-promoted level of invertase in sugar cane *(Saccharum officinarum)* involved stabilization of mRNA (Gayler and Glasziou 1969). Such a stablization could conceivably be associated with the suppression of RNase by auxin found to occur in wheat coleoptiles (Truelsen 1967), some leaf sections (De Leo and Sacher 1970) and root tips (Pilet 1969; Pilet 1970; Pilet and Braun 1970; Pilet 1972) although the stability of any RNA species has never been shown to be dictated by ribonuclease levels (see Chapter 11, section 11.2.1).

10.3.1.7. Summary and conclusions. Treatment of intact tissues with auxin usually leads to regulation of total RNA levels. The levels can either rise or fall depending on the tissue. Such changes, however, reflect changes in the major species which appear to be unnecessary for auxin action. Auxin also seems to enhance the level of mRNA, causing preferential increases of some species, and these changes appear to be essential for auxin action.

Although it is possible that such changes in RNA levels could be the result of auxin effects on RNA stability and/or processing, the effects of auxin on cell-free transcription provide preliminary evidence that the in vivo effects may in part result from regulation of transcription. Modulation of polymerase activity would seem to account for increases and decreases in ribosomal gene transcription and hence rRNA accumulation and, in part, loss. They are well correlated in various parts of the soybean hypocotyl. However, while increase in polymerase activity is claimed to be due to auxin-promoted production of macromolecular modulating (probably initiation) factors which would enhance initiation rate or indeed promote transcription of previously untranscribed "reserve" ribosomal genes, this is still unproven. There is no provision for auxin-dependent decreases in polymerase activity within the current context of thought. Available data suggests that auxin has little effect on the number of polymerase I molecules. How polymerase I modulation occurs in terms of macromolecular interactions and polymerase structure will probably be studied intensively.

There is also some evidence that auxin induces changes in mRNA in vivo and that this may be due to changes in polymerase II-dependent transcription. Consideration of the evidence for auxin-induced changes in in vitro transcription products, in template availability and the α-amanitin sensitivity of these changes, suggests that auxin-protein complexes or perhaps even just the protein itself can affect polymerase II-dependent transcription. There is no evidence that auxin causes any changes in the level of polymerase II enzyme although changes in activity of isolated enzyme have been demonstrated. Changes in RNA synthesis indicate that changes in polymerase II-dependent transcription involves transcription of new template as well as perhaps more rapid transcription of existing template. However, whether the new transcription really reflects synthesis of new mRNA species must await the application of technology permitting the identification of new mRNA species, either by translation into recognizable proteins or by hybridization to copy DNA (DNA complementary to authentic, purified mRNA preparations). The former approach depends on the development of in vitro systems from plants which synthesize in vivo-like RNA although this has so far received little attention. Copy DNA has been used with considerable success by O'Malley and colleagues studying the induction of ovalbumin synthesis in chick oviduct by oestrogen (Harris et al. 1976). They have evidence that the level of mRNA for ovalbumin is increased by the hormone in association with a receptor. This seems to be the result of mRNA synthesis which is regulated by affinity of some factor(s) for a site on the DNA which controls the rate at which polymerase II transcribes the gene. Whether auxin regulates changes in polymerase II activity in plants and thus alters the rate of mRNA synthesis or type of DNA which is transcribed is not clearly proven as yet but there is evidence consistent with the idea that auxin, in conjunction with tissue-specific factors, specifies new transcription sites or alters the rate at which existing sites are transcribed. Thus, tissue competence could depend on the presence of gene-specific auxin-reactive macromolecules. However, while there is good evidence that auxin interacts with some components of the cell (Kende and Gardner 1976), there is no convincing evidence that it ever interacts directly with specific genes. In fact, the evidence suggests that polymerase II activity can be regulated by activated mediator molecules in the absence of the hormone. Whether changes in histones and non-histone chromatin proteins are also involved in auxin control of transcription is still an open question. Although there is currently a pre-occupation with auxin-induced increases in mRNA levels, it is likely that auxin-controlled decreases in certain mRNA species may also occur. Little attention has been given to this or to auxin-promoted decreases in polymerase I activity.

Consideration of the importance of turnover rates raises the possibility that whereas differential changes in mRNA species may indeed involve selective transcription, by analogy with an alternative explanation developed by Trewavas for differential effects of auxin on protein and RNA synthesis (Trewavas 1977), the following alternative is presented to explain the differential effects of auxin on mRNA levels. The half-lives of very few plant mRNA species are known but it seems very likely that each has its characteristic life. If mRNA synthesis was raised overall, for example, by activating polymerase II, there would be rapid and large increases in species with very short half-lives and slower less dramatic increases in species with long half-lives giving the impression of a sequence of responses. This derives partly from the fact that species with short lives would exist in very small quantities while those with long lives would exist in large quantities. This presents the possibility that auxin may cause differential responses in levels of all RNA species by raising the overall level of transcription (involving all polymerases) because different RNAs have different half-lives. Similarly, differential rises in levels of different mRNA species (depending on their half-lives) could also be accomplished by increasing the overall activity of polymerase II. Given that there were no restrictions on translation, the levels of individual mRNA species would then dictate the rates of synthesis of the proteins they code for. There is essentially no data available to test such a scheme, that is, whether rapidly responding mRNA species have short half-lives, but it is interesting that the predicted shape of the curve of increase in activity which rises and then plateaus, is the sort of curve which has been obtained for the rise in levels of the two hormone-regulated mRNA species to have been measured in plants, those coding for cellulase in peas and amylase in barley (see section 10.3.2.4). Furthermore, one may well ask why different RNA species or any other molecules have different half-lives.

The proposal that there is a sequence of responses by different RNA species to auxin, mRNA preceding rRNA, and that new mRNA may be required for synthesis of polymerase I-regulating factors (Teissere 1975) has little convincing support. Evidence from in vivo radioactive tracer experiments (section 10.3.1.6) is tempting but not yet convincing because of the difficulties involved in interpreting such experiments. Similarly, the evidence for early increases in template availability (section 10.3.1.3) have not been unequivocally shown to occur before any changes in polymerase I-dependent transcription. Increases in mRNA and polysome formation thought to occur before major increases in rRNA and ribosome levels are probably the result of prolonged stimulation of polymerase I activity. Early changes in polymerase I activity would be unlikely to be detected in terms of rRNA levels since so much rRNA already exists. The conclusion is that convincing evidence for a sequence of responses is yet to be produced.

Finally, there is a paradox in the results of studies of auxin and RNA synthesis. There is evidence that auxin enhances the activity of polymerase I and increases the rate of rRNA synthesis although rRNA synthesis can be abolished with 5-FU with no effect on growth. These results indicate that rRNA synthesis is unnecessary for auxin action and unnecessary for growth although paradoxically its synthesis is enhanced by auxin. In the long term, clearly growth cannot proceed without rRNA synthesis but most experiments have been conducted over relatively short terms and the number of ribosomes may be adequate for the level of protein synthesis required. Thus it seems (although not proven) that rRNA synthesis is not required initially. However, while it may be reasonable that rRNA synthesis is enhanced by auxin but not required for auxin action, a reappraisal of the RNA labelling and 5-FU experiments may be required.

10.3.2. Gibberellin

10.3.2.1. Perception. There is very little known about perception of GA by plant cells and the evidence that GA receptor molecules exist is not yet convincing (Chapter 8, section 8.2.1.1). Responsive cells in pea stems accumulated more GA than unresponsive cells (Musgrave et al. 1969) and the hormone-responsive region of the stem of *Pharbitis nil* also accumulated more GA than the unresponsive region (Barendse 1974) both indicating the existence of receptor sites. In addition, two non-covalent GA-protein complexes have been isolated from dwarf pea seedlings with radioactive GA_3 (Stoddart et al. 1974) but that the proteins act in vivo as GA receptors remains to be shown. Attempts to localize binding sites in cells by microautoradiography and cell fractionation studies have also had very limited success (for review see Kende and Gardner 1976).

The only indication that GA may interact with a macromolecule of some sort and alter transcription comes from a study by Johri and Varner (1968) who found that RNA synthesis by isolated pea nuclei did not respond to GA_3 but if the nuclei were isolated in the presence of GA_3, transcription was affected. This suggested that a factor, possibly a receptor, was lost from nuclei during their isolation. The phenomenon is similar to that observed for the auxin receptor of Matthyse and Phillips but by no means proves a role for GA receptor molecules in transcription.

10.3.2.2. Interaction with chromatin proteins and DNA. The studies of the effects of GA on chromatin have mostly been done in conjunction with other homones and the results for GA are very similar to those for auxin. The studies by Fellenberg showed that GA_3, like auxin, affected the physical characteristics of pea chromatin (Fellenberg 1969a, b, c, 1971b). It bound to the chromatin at pH 8 and reduced its melting point. This effect was also shown to occur with artificially reconstituted nucleoprotein and it occurred without loss of protein (Fellenberg 1971a). Thus, GA_3 caused a loosening of the chromatin-protein binding. Similarly, GA_3 caused the absorption maximum of purified pea DNA in alkaline solution to show a bathochromic effect which was reversed by urea but not salt indicating that the hormone loosened the DNA helical structure as well and that it involved hydrogen bonding.

A similar reduction of melting point of pea seedling DNA by GA_3 has been measured by Bamberger (1971) although fairly high concentrations (10^{-3}–10^{-4} *M*) were used. It had the same effect on calf thymus DNA demonstrating lack of specificty for base sequences although the DNA purity was unproven. Thus, as for auxin, GA_3 seemed to loosen the whole chromatin structure indicating to Fellenberg that it too was involved in transcription initiation and could alter template availability. However, the criticisms of this work raised previously in relation to auxin (section 10.3.1.2) also apply here. Particularly, the effects of GA_3 on DNA melting point may be due to a reduction of pH caused by high hormone concentrations and not to any specific GA_3-DNA interaction (Penner and Early 1972). The studies of Kessler and colleagues have described similar GA-DNA interactions but the effect seemed to involve a stabilization rather than a loosening of the DNA duplex (Kessler and Snir 1969; Kessler 1971a, b). GA_7 (but not GA_3) interacted with DNA from a number of plant sources as well as some phage DNA but it did not interact with DNA from animal and bacterial sources. The effect was well established by measurements of thermal denaturation, optical rotatory dispersion, circular dichroism and changes in density in cesium chloride gradients. The change occurred at physiological levels of GA_7 (about 10^{-6} *M*), it required Mg^{2+} and the DNA was purified to the extent that RNA and protein were below the levels of detection of the orcinol and Folin – Ciocalteau reagents respectively but there was no apparent attempt to ensure freedom from carbohydrate. In general, GA_7 did not interact with DNA with GC content more than about 35% but by using d(AT) and d(GC) polymers, it was found that the interaction was dependent on high AT rather than low GC content. In addition, in the presence of DNA ligase, GA_7 caused the formation of internal, convalently linked, loops in cucumber DNA while cytokinins and auxin had no effect. These characteristics of the interaction indicate specificity and it was proposed that GA_7 was intercalated in the DNA duplex at AT-rich sites. Kessler (1973) has argued that the data are in line with suggestions that highly reiterated nucleotide sequences are concerned with regulation of gene activity (Britten and Davidson 1969) and that such sequences should be able to bind specific inducing agents such as hormones, in this case GA_7. Recently it was found that at low concentrations ethidium bromide, which is an intercalating agent, acted synergistically with $GA_{4/7}$ and GA_3 in promoting elongation of cucumber hypocotyls (Snir and Kessler 1975), perhaps by exposing binding sites in the DNA and thus making GA more efficient. Whatever the mechanism, the results were taken to indicate that GA could regulate transcription by interacting directly with DNA.

The conclusions which may be drawn from the above are probably limited to the fact that chromatin can interact with small molecules including GA, other plant hormones and hydrogen ions. Whether the effects of hormones are specific or non-specific or a mixture of the two remains to be seen. In particular, it is difficult to understand how GA caused DNA structure to weaken in one study and to tighten in another and why GA_3, which is commonly a very active GA, did not interact with DNA while GA_7 did. In general, there is little indication that these effects of GA on DNA have any relevance in vivo; however, as in other cases, the whole phenomenon is difficult to dismiss, since the hormone does have an effect at physiological concentrations and there is no reason to believe that these effects do not occur in vivo. Indeed, purified DNA from the elongating zone of pea epicotyls contained small quantities of GA-like substances but DNA from the same tissue of GA_3-treated seedlings contained more hormone (Nakamura and Takahashi 1973). Such results have many alternative explanations but at least they are consistent with the concept of in vivo GA-DNA interactions. Biochemical assays for hormone-DNA interaction are required to supplement the physical data and to authenticate the claims.

10.3.2.3. Transcription in vitro. Chromatin isolated from GA-treated tissues has been studied in cell-free transcription systems in order to determine whether there was enhanced transcriptional activity over that found in untreated tissue. Results have varied widely and are difficult to interpret because polymerase activity and template changes were often not separable or even examined. Also, it is not clear whether hormone treatment altered the population of chromatin components being studied by affecting the extractability of the tissue.

Chromatin from pea seedlings treated in vivo for 6 hr had increased transcription capacity and the effect increased with time matching the kinetics for GA_3-enhancement of internode growth (McComb et al. 1970). This was determined to be due to increased polymerase activity with no change in template availability, although these interpretations are questionable because of the sensitivity of the assay. Similar results were obtained with chromatin from cucumber seedlings treated for 12 hr in vivo (Johnson and Purves 1970) although by 24 hr, increased template had developed. On the other hand, in two other systems the results have been quite different. In dormant hazel nuts *(Corylus avellana),* GA_3 stimulated germination by substituting for an after-ripening period and in doing so, it stimulated RNA synthesis long before any change in fresh or dry weight of the embryo. The transcription capacity of the chromatin rose dramatically after about 10 hr which coincided with the rise in RNA synthesis; however, the available template increased very rapidly after GA_3 treatment and had reached a plateau by the time the transcription capacity of the chromatin began to rise. Thus, the initial response involved template availability and the later response overall chromatin-bound polymerase level. The effect of GA_3 on washed sugar beet tissue also apparently involved, at least initially, changes in template availability (Duda and Cherry 1971). The change of 28% in transcription capacity of chromatin from treated tissue was perhaps largely accounted for by a doubling of template availability, although it is not clear whether total polymerase activity increased as well. In none of the studies described above were the polymerases or the products of transcription characterized, so that it is probably premature to make any conclusions except that the major changes in transcription capacity could account for overall rises in tissue RNA (probably rRNA). It seems that GA can affect both template and polymerase activity in some cases but what happens depends on the tissue and there is no general rule. It may be pertinent that the tissues found to respond to GA with template changes, sugar beet and hazel nut, were non-growing tissues, while responses involving no template changes but altered polymerase activity were obtained from stem tissues that were already growing, so that perhaps the response depends on the type of tissue and its history.

The results of adding GA to cell-free transcription systems are only slightly more informative than the studies described above and they are still few in number and involve only isolated nuclei and not chromatin or DNA. Roychoudhury and colleagues were the first to report an effect of GA_3 on RNA (and DNA) synthesis in coconut nuclei measured as $[^{32}P]$orthophosphate and $[^{14}C]$uracil incorporation. The effect was rapid, occurring within 15 min, and was actinomycin D-sensitive. GA_3 also promoted release of RNA from the nuclei (Roychoudhury and Sen 1964b; Roychoudhury et al. 1965). However, the response was not characterized more than this and the criticisms offered earlier (see section 10.3.1.5) relating to contamination and to the authenticity of the nuclei are pertinent here also. The first well-established case of regulation of cell-free transcription by GA came from the studies of Johri and Varner (1968) on isolated pea nuclei. GA_3 added during homogenization of the tissue caused nuclei to synthesize RNA faster than those from control tissue, and the RNA was different as determined by nearest neighbour frequencies. There was no response to GA_8, which is inactive in dwarf pea. MAK fractionation of control RNA showed that most of the RNA migrated as rRNA but that some was associated with DNA (4–10%) and some was TB-RNA (10–15%). Only specific activities of the minor fractions were enhanced by GA_3. The identity of all these fractions is unknown. The "rRNA" had an S value of only 6–10 in sucrose gradient analysis and was clearly not intact rRNA. The "DNA" could have been DNA or perhaps RNA associated with DNA. The TB-RNA could have been poly(A) RNA. The results suggest that GA_3 affected mRNA synthesis and it would perhaps be worthwhile re-examining this system with a view to characterizing the TB-RNA. It is encouraging that the nearest neighbour frequencies indicated a tendency towards higher AMP content in RNA-synthesized by GA_3-treated nuclei. However, it is disturbing that these nuclei did not synthesize either 4–5S RNA or intact rRNA. Nevertheless the results suggest that GA_3 caused the nuclei to transcribe new DNA and possibly new genes. Results similar in some respects were obtained by Duda and Cherry (1971) from sugar beet nuclei. GA_3 added in vitro to a nuclei-rich fraction did not affect the synthesis of S-RNA or DNA-RNA, but lrRNA and (hr + "m") RNA were enhanced to a large extent. The RNA fractions were not characterized further and TB-RNA was not eluted from the column, but nevertheless, if the "m" RNA was Hn RNA (or pre-mRNA) as could be the case, the results would have something in common with those of Johri and Varner (1968). Of course the effect on putative rRNA is not consistent with the pea nuclei work, but again this may be a function of tissue competence or the method of isolation of the nuclei. These systems are very promising but should be re-examined with the aid of new technology to further clarify the response.

In summary, although only crude systems have been examined and any conclusions can only be tentative, it appears that GA treatment can cause alterations in the transcriptional capacity of chromatin and that these changes can involve transcription of additional template as well as changes in polymerase activity. However, there has been no direct study of polymerases, and there is no compelling evidence that GA_3 causes transcription of new mRNA genes, so that while GA_3 can apparently affect transcription, the mechanism involved is obscure.

10.3.2.4. Regulation of RNA metabolism in vivo. That GA can cause changes in RNA levels in many intact tissues has been demonstrated in a variety of ways. For example, the total RNA content has been found to increase in GA_3-treated barley leaf segments, clover *(Trifolium hirtum)* seedlings, pea seedlings, lentil epicotyls, senescing leaf tissues and in potato *(Solanum tuberosum)* buds (Broughton 1968; Rappaport and Wolf 1968; Broughton 1969; Key 1969; Poulson and Beevers 1970; Shih and Rappaport 1970; McComb and Broughton 1972), and the incorporation of labelled precursors into all major RNA species has been enhanced by GA_3 in pea epicotyls (Nakamura et al. 1970), corn embryos (Wielgat et al. 1974), barley leaves (Poulson and Beevers 1970) and protoplasts made from corn seedlings (Wasilewska and Kleczkowski 1974). Differential effects on RNA labelling have been reported in corn scutellum where rRNA and D-RNA but not tRNA were enhanced (Jankowski et al. 1975) and in corn seedling protoplasts where there was a preferential effect on heterodisperse RNA, HnRNA and poly(A) RNA (Wasilewska and Kleczkowski 1974). GA_3 has also been reported to cause enlarged nucleoli in lettuce cotyledons and in discs of Jerusalem artichoke and carrot *(Daucus carota)* (Chapman and Jordan 1971; Das 1973) and to cause increased labelling of *Avena fatua* aleurone nuclei (Naylor 1966). Increases in polysomes have been described in barley leaf segments (Poulson and Beevers 1970), lettuce hypocotyls (Srivastava et al. 1975) and barley embryos (Rao and Khan 1975). Without going into these reports in detail, it is clear that they constitute strong evidence that GA_3 can affect the levels of different RNA species, depending on the tissue, although the labelling studies should be examined taking into account RNA turnover as described in section 10.3.1.6.

Much of the work on regulation of RNA metabolism by GA has been done on barley aleurone and this system has now provided strong evidence for a selective effect of GA on at least one particular mRNA species. In this tissue, Varner and Chandra (1964) found that the GA_3-dependent production and release of amylase was inhibited by RNA synthesis inhibitors, in particular actinomycin D, indicating that DNA-dependent synthesis of RNA was necessary for the production of amylase, and this led to the hypothesis that GA_3 controlled the synthesis of amylase by causing the production of specific messengers (Varner 1964; Varner and Chandra 1964). Although actinomycin D prevented α-amylase synthesis if it was added at the same time as (or before) GA_3, there was no effect on amylase synthesis if it was added after 8 hr even though RNA synthesis was inhibited by 65% over the period 8–12 hr after the beginning of hormone treatment (Chrispeels and Varner 1967). Added at 4 hr, it was less effective than if added at the same time as GA_3 but more effective than at 8 hr, so there was a gradual escape from actinomycin D inhibition indicating an escape from dependence on RNA synthesis. Similar results have been obtained using cordycepin as the RNA synthesis inhibitor (Ho and Varner 1974). In a more detailed study, Goodwin and Carr (1972) determined that the period most sensitive to actinomycin D was late in the lag phase, so that GA_3 action would appear to require RNA synthesis beginning just before amylase synthesis commences and not during the very early events of the lag phase. Although the actinomycin D data seem to limit the requirement for RNA synthesis to the lag phase, amylase synthesis (and RNA synthesis) is sensitive to other RNA synthesis inhibitors, 6-methylpurine and 8-azaguanine, added after the lag phase. Clearly, there is a problem of inhibitor specificities and whether there is a requirement for RNA synthesis after the lag phase is still unresolved.

As with auxin and ethylene, experiments with 5-FU led to the conclusion that the RNA required for GA_3 action was not rRNA or tRNA. Extremely high concentrations of 5-FU added at the same time as GA_3 did not inhibit amylase induction but almost eliminated rRNA synthesis and severely inhibited tRNA synthesis (Jacobsen and Zwar 1974a). However, the synthesis of polydisperse RNA was relatively unaffected, and thus it was a candidate for the required RNA fraction. Since there seems to be a requirement for RNA synthesis during GA_3-induced amylase synthesis, attempts have been made to determine which fractions of RNA are affected by GA_3. The first of these was by Varner, Chandra and Chrispeels in 1965 who found that GA_3 stimulated accumulation of labelled precursors (uridine, adenine and orthophosphate) into salt-soluble RNA, which was a large part of the RNA of the cell and which presumably reflected changes primarily in rRNA (Varner et al. 1965). The effect was evident between 6 and 8 hr after the beginning of incubation and it was not the result of increased label uptake. The RNA synthesized (labelled) in the presence of GA_3 had a different base composition to the RNA synthesized in its absence, suggesting that GA_3 caused both quantitative and qualitative changes. Most of these experiments involved the application of treatments (GA_3 and labelled precursors) to dry half-seeds (de-embryonated seeds) and the results could be confounded by an effect of GA_3 on imbibition of water and associated changes which would then affect RNA synthesis, as has been pointed out by Chrispeels and Varner (1967). In this study, it is worth noting that although labelling studies indicated that GA_3 stimulated RNA synthesis to a large extent, there was no increase in the total RNA during the period of GA_3 treatment. Presumably the amount of additional RNA synthesized was very small compared to the amount of RNA already present.

These data, at least in principle, seemed to support a subsequent study by Evins (1971) on the effects of GA_3 on ribosomes in aleurone. He found that GA_3 caused an increase in the numbers of ribosomes recovered from aleurone beginning about 4 hr after commencement of treatment. After 15 hr the ribosome numbers had doubled. Simultaneously, there was a major increase in the fraction of the ribosomes appearing as polysomes. After 10 hr incubation, between 40% and 50% of the ribosomes in control tissue were in polysomes and in GA_3-treated tissue, the figure was 76%. Holmes and Speakman (1973) have obtained data which are in general agreement with this. They found that in barley half seeds, the ribosome numbers decreased for the first 20 hr incubation, both in the presence and absence of GA_3, but after 20 hr in its presence, there was a large increase in ribosomes. This was associated with an increase in accumulation of [^{3}H]UMP into ribosomes. Nevertheless, these experiments involving ribosome isolation should be assessed in the light of a number of problems of ribosome isolation now known to give misleading results. If polysomes were prepared by the usual methods involving centrifugation through a dense sucrose cushion, a significant proportion of the ribosomes were not recovered and there was a preferential recovery of large polysomes (Leaver and Dyer 1974; Larkins and Davies 1975) unless centrifugation continued for a number of hours. This would mean that more ribosomes could be automatically recovered from tissue with bigger or more polysomes if the centrifugation period was only 2 hr or so as in Evins' work. Differential extractability of ribosomes from cytokinin-treated radish cotyledons has been described recently although the hormone did not cause any increase in rRNA (Gordon et al. 1975). There may well be such a problem with aleurone tissue, since GA_3-treatment causes breakdown of aleurone cell walls (Jones 1969; Ashford and Jacobsen 1974) which could be more easily ruptured during homogenization. Such a problem would probably be accentuated where a very gentle grinding technique is used, as in polysome isolation. Thus, again recovery of more polysomes and ribosomes from GA_3-treated aleurone could be artificial. We have also come across a problem involving variations in quantities of RNA, recovered from control and GA_3-treated barley aleurone, which depend on the RNA isolation technique (and not on cell breakage) (Zwar, J.A. and Jacobsen, J.V. unpublished data) but we do not know whether this is relevant to ribosome recovery.

In spite of the above arguments against an increase in ribosomes, there is ultrastructural evidence to support the case for an increase in the organelles. Electron micrographs usually show more ribosomes and more polysomes in GA_3-treated tissue than in untreated tissue (Van der Eb and Niewdorp 1967; Jones 1969; Vigil and Ruddat 1973). In untreated tissue the ribosomes are spread out and seem to be few in number whereas in treated tissue, they are concentrated on stacks of endoplasmic reticulum and certainly give the impression of more ribosomes; however, no ribosome counts from micrographs have been published.

What is the evidence that major RNA species are unaffected by GA? Increased numbers of ribosomes would presumably be reflected in increased RNA content of GA_3-treated tissue but independent measurements have shown that this does not occur (Chandra and Varner 1965; Jacobsen and Zwar 1974a). Therefore, any increase in label accumulation into RNA caused by GA_3 would have to result from increased turnover, not net synthesis. However, in experiments designed to circumvent the problems of doing radioactive tracer experiments with aleurone, we found that GA_3 had no effect on incorporation of precursors into total RNA or the major species (Zwar and Jacobsen 1972) and that the specific activity of total RNA was the same in control and GA_3-treated tissue (Zwar and Jacobsen, unpublished data). Whether GA_3 enhances ribosome and/or rRNA levels in aleurone is still not entirely clear, but there is increasing evidence that GA_3 has no effect on the levels of major RNA species. However, from the point of view of amylase induction, the outcome is of little consequence since the synthesis of major RNA species is not required for normal promotion of enzyme synthesis (Jacobsen and Zwar 1974a).

Some interesting observations are derived from the above. Since amylase induction occurs in the absence of rRNA synthesis and assuming that there is no stockpile of rRNA which can form into new ribosomes, it follows that GA_3-induced increase in polysome size or number involves pre-existing ribosomes. This should perhaps be accepted with some reservations because although an increase in polysome size and/or numbers seems to be an integral part of the response to GA_3, it has not been shown to be necessarily associated with amylase synthesis, nor has it been shown to occur in the absence of rRNA synthesis. However, it would seem that the GA_3-induced changes in polysomes could involve either enhanced level of translatable mRNA or activation of ribosomes (or perhaps both) as described for auxin (section 10.3.1.6). GA_3 has been shown to enhance methylation of purine residues in rRNA (Chandra and Duynstee 1971) which could perhaps have something to do with ribosome function but changes, for example in ribosomal proteins (see Chapter 11, section 11.3.3) as described for auxin, have not been reported for GA.

However, clearly some sort of RNA synthesis, presumably of a minor species, is required for amylase induction, and there is now evidence that GA_3 regulates metabolism of mRNA and that this may account for the requirement for RNA synthesis. Early indications of this came from the studies by Chandra and Duynstee (1968), in which it was shown that GA_3 stimulated label incorporation into RNA migrating on MAK columns as a hrRNA shoulder or D-RNA, and from our studies, which produced evidence that GA_3 did not affect label incorporation into 4S, 5S and rRNA but did enhance incorporation into a minor fraction of RNA which was polydisperse on acrylamide gels (Zwar and Jacobsen 1972). Furthermore, appearance of the effect was correlated with amylase synthesis. Neither of these minor GA_3-promoted fractions were characterized further but subsequent studies showed that GA_3 promoted labelling of poly(A) RNA by about 70% (Ho and Varner 1974; Jacobsen and Zwar 1974a, b) which could account for both of the early reports. Although previous studies on plant hormones had dealt with the AMP-rich species, D-RNA and TB-RNA, both of which had been referred to as mRNA at one time or another without much supporting evidence, these studies on poly(A) RNA were the first to have detected a hormone-promoted accumulation of label into RNA which had many of the characteristics of mRNA. Furthermore, it appears to be a specific effect and not the result of an overall regulation of RNA metabolism. The evidence indicated that the enhancement involved a number of poly(A) RNA species and that synthesis of both the heteropolymeric (presumably template) RNA and the poly(A) tracts were enhanced equally (Jacobsen and Zwar 1974b). It also indicated that GA_3 regulated the levels of complete adenylated mRNA molecules and not the adenylation of preexisting unadenylated RNA, although it should be remembered that such overall data do not necessarily imply a similar response in all mRNA species. Nevertheless, the effects of GA_3 on total poly(A) RNA were correlated with amylase synthesis (Ho and Varner 1974; Jacobsen and Zwar 1974a), the lag period for enhanced poly(A) RNA labelling being 3–4 hr which correlated well with the lag period for amylase synthesis (see later). In addition, 5-FU, which does not inhibit amylase synthesis, had no effect on poly(A) RNA synthesis while strongly inhibiting synthesis of major RNA species (Jacobsen and Zwar, unpublished data). These data provided some reason to think that GA_3 caused changes in mRNA levels which were both correlated with, and necessary for, amylase induction and there is now some evidence which is consistent with this view.

In a study involving translation of aleurone poly(A) RNA in a cell-free system from wheat embryos, mRNA for amylase was detected and translated into recognizable amylase protein (Higgins et al. 1976). The amount of translatable amylase mRNA began to increase between 2 and 3 hr after GA_3 treatment and the level of mRNA continued to increase up to 12 hr and correlated well with the in vivo rate of amylase synthesis. There was no lag in translation of the mRNA as in the case of auxin-controlled cellulase synthesis (Verma et al. 1975). Critical measurement of the lag phase for amylase synthesis showed it to be of 2–3 hr duration and not in excess of 6 hr as often quoted. Therefore, the GA_3-promotion of the level of translatable amylase mRNA could account for control of amylase synthesis although this does not eliminate other possibilities. The problem now is to determine if the rise in level of amylase mRNA is due to increased synthesis, activation of pre-existing mRNA or a decrease in the rate of breakdown, or some combination of these. Because GA_3 increased the labelling of total mRNA, the first or last alternatives seem most likely and consideration of the effects of GA_3 on in vitro transcription (section 10.3.2.3) perhaps makes the first alternative most likely of all.

Direct proof is needed to show that regulation of transcription is involved in amylase mRNA production, since there is one report that the mRNA may exist, at least in part, before GA_3 application. By using 5-FU which was presumably incorporated into mRNA causing miscoding, and by detecting aberrant amylase molecules, Carlson (1972) obtained evidence for pre-existing amylase mRNA indicating that GA_3 regulated amylase synthesis by a post-transcriptional change. If this is so, and it is considered unlikely (Rodaway and Kende 1978) the mRNA does not exist in a translatable form and the change would presumably involve some sort of activation of the gene product (Chandra 1974). Adenylation has been proposed as the processing step necessary for translation of mRNA for enzymes involved in germination of cotton seed (Walbot et al. 1974), although in animal cells poly(A) is certainly not absolutely required for translation of mRNA

(Schochetman and Perry 1972; Bard et al. 1974; Fromson and Verma 1976). Poly(A) is more likely to be involved with translation life time or stability of the mRNA (Mendecki et al. 1972; Sheiness and Darnell 1973; Huez et al. 1974). Effects of hormones on poly(A) polymerase have not yet been reported although the enzyme has been demonstrated in plant tissues associated with chromatin in maize (Benson and Mans 1972) and wheat (Sasaki and Tazawa 1973), and in chloroplasts (Burkard and Keller 1974). There is also indirect evidence that GA_3 treatment may result in increased levels of invertase in sugar cane by causing stabilization of mRNA for invertase (Gayler and Glasziou 1969). Therefore, although there are problems in interpreting some of these studies (see Chapter 11, section 11.3.1), it is still possible that GA acts by regulating translatability of preexisting amylase mRNA.

10.3.2.5. Summary, and conclusion. Some progress has been made towards describing and understanding the effects of GA on RNA levels in vivo. There are some good correlations between altered RNA metabolism and subsequent changes in protein synthesis but clearly a larger body of evidence is necessary to verify what is known from a very limited number of systems. Thus, care must be taken not to over-interpret the limited data available at present.

GA does appear to affect the transcriptional capacity of extractable chromatin and of nuclei and the changes indicate regulation of polymerase II-dependent transcription. Whether the enzyme activity is somehow altered either by increased affinity for particular initiation sites or by changes in the initiation sites themselves has not been examined so far. Such areas as the development of reconstituted RNA-synthesizing systems and identification of components, particularly regulatory factors (if they exist), close examination of systems where GA enhances rRNA levels, and analysis of polymerase activity and of the enzymes themselves, plus further characterization of the extent and nature of the effect of GA on the mRNA complement as a whole (including suppression as well as enhancement) are all aspects of GA and RNA metabolism awaiting attention. There are some apparent similarities (although few) between effects of auxin and GA on cell-free and in vivo RNA metabolism. However, if auxin research is directed mostly towards polymerase I action and the effort in GA research towards polymerase II action, as is the case now, the development of information enabling comparisons to be made will be a slow process.

In conclusion, it can be said that the studies described above provide a basis on which to propose that regulation of RNA levels in vivo by GA could involve effects on transcription, but at the present time, posttranscriptional effects of the hormone (Chapter 11) are equally likely.

10.3.3. Cytokinin

10.3.3.1. Perception. There are several reports that cytokinins bind to components of plant cells but there is little known about the binding sites themselves except perhaps in the case of ribosomes where the binding seems to be to a ribosomal protein (Berridge et al. 1970, 1972; Fox and Erion 1975). Binding of cytokinins in pea bud nuclei has also been detected (Berridge et al. 1970) but in this case, as well as in others, there was no evidence that the binding was connected with regulation of transcription. There is only one case known where cytokinin, in combination with a macromolecule, has been linked to transcription, but even so, cytokinin binding to the macromolecule has not been demonstrated and thus it does not really qualify as a receptor (Matthyse and Abrams 1970). In this case a transcribing system, composed of isolated pea bud or pea root chromatin saturated with *E. coli* polymerase, did not respond to kinetin alone but it did respond to kinetin plus a partially characterized factor isolated from crude pea chromatin. Presumably the factor was lost during chromatin purification. The factor appeared to be a heat-stable protein with a monomer molecular weight of 10,000 although polymers (aggregates) of this unit appeared to exist. The kinetin–factor combination also stimulated transcription in a system containing pure DNA and *E. coli* polymerase and the effect was specific for homologous DNA. Thus the kinetin-factor combination alone seemed quite capable of regulating transcription. Although this study is encouraging, some time has passed since it was reported and McCombs and Ralph (1972) have been unable to confirm the results.

In summary, macromolecular co-factors appear to be required for cytokinin action as judged by these preliminary data but there is no evidence for a cytokinin receptor molecule as such. The general criticisms of the auxin and GA-perception studies apply to this work also, although the cytokinin data are even more incomplete.

10.3.3.2. Interaction with chromatin proteins and DNA. As for GA and auxin, kinetin has also been found to bind to DNA and chromatin causing physical changes in the molecules, but the kinetin effect shows an interesting difference. Whereas kinetin caused a urea-reversible bathochromic change in the absorption maximum of DNA in alkaline solution, indicating hydrogen bonding as for the other hormones, it caused an increase in the melting point of DNA and not a decrease as for auxin and GA (Fellenberg 1969a, b, c, 1971b). That this may not be a specific effect, however, is indicated by the study of Penner and Early (1972) who showed that the effects of hormones on DNA melting point might be related to their effects on pH of the solution. Whereas, GA and auxin caused a reduction in pH and thus a reduction in DNA melting point, kinetin caused an increase in pH which could explain its opposite effect on melting point.

The effect of kinetin in reducing the melting point of isolated chromatin and artificially reconstituted nucleoproteins (Fellenberg 1971a) was the same as for the other hormones. Reduction of the melting point of chromatin has also been demonstrated by cytochemical means in nuclei isolated from kinetin-treated onion root tips (Piesco and Alvarez 1972), indicating that the effect is probably not an artefact caused by isolation of chromatin and that the characteristics of chromatin are changed in vivo by cytokinins. Thus, whereas kinetin decreased the binding of histones to DNA, it somehow stabilized the DNA duplex structure. Stabilization of DNA structure has been reported for 6-benzylaminopurine, as well as kinetin, when both are applied at high concentrations (10^{-4} *M*) (Bamberger 1971). While the loosening of DNA structure led Fellenberg to think of auxin and GA as promoters of the initiation of transcription, it is difficult to ascribe a function to cytokinin-promoted DNA stabilization unless it reflects a suppression of DNA transcription but there is no support for this from either in vitro or in vivo studies of transcription. Because of this, the explanation of the kinetin effect on DNA provided by Penner and Early (1972) appears attractive. The effect of cytokinin on chromatin-protein binding may well have significance in transcription in vivo (see section 10.3.3.3) and apart from a direct in vitro effect on protein-binding, there may also be less direct effects in vivo since it was shown cytochemically that kinetin induced an increase in the arginine to lysine ratio in histones bound to DNA in cytokinin-treated onion root tips (Piesco and Alvarez 1972), which presumably reflected a decrease in histone basicity and a reduction in binding strength. These studies are subject to the same criticisms as similar studies on other hormones.

10.3.3.3. Transcription in vitro. There are few informative studies on the effects of kinetin, added either in vivo or in vitro, on transcription by isolated nuclei, chromatin or DNA, but there are indications that template availability can be altered indicating that polymerase II-dependent transcription might be involved. However, no studies discriminating between the polymerases have been reported.

There are two studies on transcription by in vitro systems isolated from cytokinin-treated tissues. Chromatin from three-day-old cucumber seedlings which had been treated with kinetin for 12 hr exhibited increased endogenous polymerase and increased template activity (Johnson and Purves 1970) but no attempt was made to examine short-term effects. Chromatin from kinetin-treated dormant pear *(Pyrus communis)* embryos also synthesized more RNA but there was no further examination of the effect (Khan 1972). These unconfirmed and preliminary studies indicate that kinetin acts by causing alterations in transcriptional activity and the results of studies by Matthyse and Abrams (1970) on chromatin isolated from pea tissues indicate that the effect may be a result of direct kinetin-chromatin interaction. RNA synthesis by isolated pea bud or pea root chromatin under *E. coli* polymerase-saturated conditions was enhanced by the addition of kinetin plus a macromolecular protein isolated from chromatin (Matthyse and Abrams 1970). The hormone–protein combination also stimulated transcription of pure DNA (plus *E. coli* polymerase) and the effect was specific for homologous DNA. Again there was no guarantee of the purity of the DNA and other factors may have been present, but these data must serve as an indication that the kinetin – protein combination may have recognized certain DNA sequences and specified transcription of new DNA. However, in this study there seemed to be no requirement for homologous polymerase for the effect in a pure DNA system. None of the transcription products were examined in this study, but this, plus further examination of the whole phenomenon, would seem to be very worthwhile.

In general, studies on isolated nuclei have not produced any further insights. In coconut nuclei, as for other hormones, kinetin enhanced [^{32}P] orthophosphate incorporation into DNA and RNA and thymine and uracil incorporation into DNA and RNA in as little as 15 min and it promoted release of RNA from the nuclei (Roychoudhury and Sen 1964b; Roychoudhury et al. 1965) although there are reservations about this work relating to contamination and authenticity of the isolated nuclei preparations which have already been expressed (section 10.3.1.3). Similarly, zeatin was shown to enhance

[^{32}P]orthophosphate labelling of onion, mung bean and pea nuclei over a 4 hr period (Maheshwari et al. 1966). In both of these studies, there was probably little doubt that the label was, at least in part, incorporated into nucleic acids but which nucleic acids and whether they were made in the nuclei or by microorganisms is not quite clear. In relation to contamination, it is unfortunate that [^{32}P]orthophosphate was used as a label, since bacteria incorporate phosphate very rapidly, at least into DNA. Nevertheless, a more recent examination of the kinetin effect demonstrated that the hormone (like auxin) increased [^{14}C]adenosine triphosphate labelling of RNA (TCA-precipitable and KOH-digestible) in nuclei-rich preparations from soybean and tobacco callus and from pea buds (Matthyse and Phillips 1969), and this seems to validate the earlier reports. Notably in this study, the effects were detected within 5–10 min and at their maximum were 40–100% over control nuclei. It is also notable that the crude nuclei preparations seemed to retain whatever cofactors were required for kinetin action.

Thus, it is evident that studies on the effect of kinetin on cell-free transcription, although getting off to an apparently good start, have languished. Although it seems that, as for other hormones, kinetin plus cofactor(s) can affect transcription, not only of isolated chromatin but also of DNA, strongly indicating a role for the hormone in vivo at this level, neither the transcription products nor the polymerase(s) involved have ever been examined. These aspects plus the possibility of the occurrence of DNA of base sequence-dependent hormone-protein recognition sites seem to be worthy of attention.

10.3.3.4. RNA metabolism in intact tissues. Kinetin has been shown to increase RNA content of a number of plant systems. The first demonstration of this was by Guttman (1957) who showed cytochemically that kinetin increased the nuclear RNA content of onion bulb tissue within 24 hr. Later it was shown by autoradiographic techniques that there was in fact a sequence of kinetin effects. After 1 hr, kinetin increased label incorporation ([^{14}C]adenine) into nucleoli. After 3 hr, the increase had spread to the whole nucleus and after 24 hr, the cytoplasmic labelling was enhanced (Olszewka 1959), thus indicating an effect of kinetin on transcription. There has been a similar demonstration (using cytochemical techniques) of an effect of kinetin on RNA content of onion root tip cells within 30 min of the commencement of treatment (Jensen et al. 1964). Although these studies have since been re-examined and evidence produced that they may have been misleading in some respects (Chaley and Setterfield 1972), they supported the concept that cytokinin enhanced levels of RNA perhaps by promoting synthesis. All of these studies served to illustrate the usefulness of the cytochemical approach which has been rarely applied in recent times.

In 1962, Osborne suggested that kinetin might act by regulating the synthesis of a particular RNA fraction (Osborne 1962) but evidence for this has not been easy to obtain. The early studies indicated changes in a multiplicity of RNA species. In very early studies on effects of cytokinins on whole fruit and vegetables, it seemed that cytokinins might regulate synthesis of ribosomal and soluble RNA (Salunkhe et al. 1962; Sugiura et al. 1962) and in a more detailed study involving MAK fractionation of RNA, it was found that after 2 hr, 6-benzylaminopurine promoted [^{32}P]orthophosphate accumulation into all RNA fractions of peanut cotyledons (Carpenter and Cherry 1966a). Because all species were affected, the authors suggested that cytokinins may act at the nucleotide precursor level but no evidence for this was presented. In *Lemna*, where growth is promoted by cytokinins, both RNA and protein content increased within 1 hr of kinetin treatment although the effects were transient and by 8 hr, the RNA and protein contents were back to control levels; nevertheless, the changes were rapid and large (Erisman and Farkhauser 1967). In a more detailed study of the same system, Van Overbeek and colleagues (1967) showed that 6-benzylaminopurine enhanced [^{32}P] orthophosphate labelling of all nucleic acid species fractionated by MAK column chromatography, indicating again that cytokinins cause an overall and not a selective increase in nucleic acids. The system was not analyzed further with respect to regulation of transcription. Similar results have come from studies of dark grown (dormant) *Spirodela* in which kinetin enhanced RNA, protein and DNA synthesis within 1 hr (McCombs and Ralph 1972). The synthesis of the major (rRNA and tRNA) RNA species were increased to similar extents but the minor species were not examined.

Overall changes in RNA metabolism have also been found in cytokinin-promoted proliferation of tobacco pith explants (Zwar 1973). Previous to this study, all that was known of the response was that DNA content increased before cell division began (Das et al. 1956). Zwar found that after 2 days incubation with kinetin, there was no clear effect on RNA synthesis but after 5 days there was a large increase in total RNA and a stimulation of label accumulation into hrRNA (69%), lrRNA (100%) and tRNA (23%) as well as in high molecular weight species which were polydisperse on polyacrylamide gels. This occurred before any major cell division. However, these effects are very long-term and there was no evidence that a change in any particular species was an early event following cytokinin treatment although the author concedes that the technology was such that even major early changes in mRNA synthesis probably would not have been detected. There was certainly no evidence for an early preferential effect on tRNA synthesis but again the sensitivity of the technology leaves the question open. Studies on growth-regulator-starved tobacco pith tissue, however, have given some evidence for more rapid control of nucleic acid synthesis by hormones in this tissue (Klämbt 1974). Within 3 hr of addition, kinetin and auxin together enhanced [^{32}P] orthophosphate incorporation into polysomal RNA, however, whereas auxin had some effect on its own, kinetin alone had none indicating that it had no major rapid effects on RNA metabolism itself but that it somehow enhanced auxin action. It has recently been shown to enhance protein synthesis directly in vitro (Klämbt 1976) (see Chapter 11, section 11.3.4) which may be related to its auxin-enhancing effects. Alternatively, it may promote RNA conservation by suppressing RNase in the way described by Birmingham and Maclachlan (1972). The latter would classify as a posttranscriptional effect but nevertheless may require RNA synthesis especially if regulation of RNase level is involved.

A great deal of the effort in cytokinin research has been expended on tRNA biochemistry and physiology because of the occurrence of cytokinins in these species (see Chapter 11, section 11.3.4). This has occurred at the expense of the other RNA species. Because of this, it is perhaps not surprising to find that the only really differential effects of cytokinins on RNA levels involve tRNA. Increases in specific isoaccepting leucyl-tRNA species have been detected in cotyledons and hypocotyl of soybeans following cytokinin treatment (Anderson and Cherry 1969; Bick et al. 1970; Venkataraman et al. 1970) but whereas this could be a result of directed synthesis, it could also be a result of posttranscriptional changes like hormone-induced methylation (Venkataraman et al. 1970) or some other aspect of pre-tRNA processing or perhaps by artefact involving production of tRNA fragments from pre-existing species which are still capable of being acylated.

Cytokinins promote expansion of cotyledons and whereas nucleic acid levels have been shown to increase in fenugreek *(Trigonella foenum-graecum)* (Rijven and Parkash 1970) and radish cotyledons (Gordon and Letham 1975), these increases occurred after expansion had begun. Growth was inhibited by 8-azaguanine and actinomycin D but in fenugreek, 5-FU only prevented the increase in RNA and did not affect expansion. Therefore, synthesis of some minor RNA species seems to be necessary for expansion to occur. Subsequently in fenugreek, kinetin was found to cause increased microsomal effectiveness in amino acid incorporation but not, incidentally, in components of the post-ribosomal supernatant including tRNA activity. It was considered that the factor limiting microsomal function could have been mRNA (Rijven and Parkash 1971) although there was no direct evidence for this. On the contrary, more recent evidence indicated that the effect on microsomes as well as the enhancement of protein synthesis, at least in part, was due to increased ribosomal efficiency. Kinetin seemed to enhance the initiating capacity of ribosomes from fenugreek cotyledons (Rijven 1974), and in radish cotyledons, 6-benzylaminopurine stimulated protein synthesis when transcription was severely inhibited (Gordon and Letham 1975). Indeed, cytokinin has recently been shown to enhance protein synthesis slightly in vitro (Klämbt 1976) and so there is a variety of data accumulating that suggest that cytokinin may regulate protein synthesis directly (see Chapter 11 for more detailed discussion.) However, RNA synthesis is still necessary for cotyledon expansion and therefore, although the effect on translation has a tendency to direct attention away from transcription, it is still quite likely that the critical effect of cytokinin may be to regulate mRNA levels, and the effect on protein synthesis may be secondary.

Much of the effort in cytokinin research has been expended on determining how cytokinins prevent senescence of excised plant tissues and in general, here also, RNA levels appear to be important. There are a number of changes associated with senescence of leaf tissue which include loss of chlorophyll and other pigments, starch, lipid, sugar, protein and nucleic acids, both RNA and DNA, a loss of ability to synthesize protein and RNA (Shaw et al. 1965; Srivastava 1967) and a loss of polysomes (Srivastava and Arglebe 1967) and in some cases kinetin (and GA) retard these changes (Richmond and Lang 1957; Brian et al. 1959). With a view to explaining this in terms of nucleic acid metabolism, there have been a number of reports which show that kinetin increases (or at least maintains) levels of both RNA and DNA in excised leaves of tobacco (Sugiura et al. 1962), wheat (Shaw et al. 1965), cockleburr *(Xanthium pennsylvanicum)* (Osborne 1962) and barley (Srivastava and Ware 1965). In most of these tissues as well as in radish leaves (Burdett and Wareing 1966), cytokinins have caused enhanced labelling of RNA with radioactive precursors. However, perhaps there is a case for re-examining some of this work since recently it was shown that the loss of RNA occurring in excised tobacco leaf discs was accounted for, at least over the first few days, by loss of chloroplast rRNA (Takegami 1975) rather than cytoplasmic rRNA. The loss of chloroplast rRNA was arrested by 6-benzylaminopurine but the level of cytoplasmic rRNA was unaffected. Examination of kinetin-treated wheat leaf discs by electron microscopy showed that they contained more ribosomes and more endoplasmic reticulum than untreated discs (Shaw and Manocha 1965) and using differential centrifugation, it was shown that the effect in tobacco involved increases in both the microsomal and the supernatant RNA (Sugiura et al. 1962). In barley leaves incubated with kinetin for 4 days (Srivastava 1965), there was evidence for promotion of rRNA accumulation but there were small decreases in other fractions, apparently tRNA. In the same study, no changes in base ratios or melting profiles of nucleic acids from treated leaf discs were noted. All of these studies agree in general that at least one of the long term effects of cytokinins in senescing leaf tissues is a stimulation of rRNA and ribosome accumulation compared to controls. In a more detailed study of the effect of cytokinin on senescence in radish leaf discs (Burdett and Wareing 1966), it was shown that after 21 hr of incubation in kinetin, there were only small but real increases in accumulation of labelled orotic acid into lr and hrRNA and tRNA but there was a large difference (36%) in material remaining in the phenol phase during the initial phenol extraction but extractable from the phenol with 4-amino salicylate which extracts high molecular weight polydisperse RNA which could include mRNA (Kidson and Kirby 1964). This would seem to be the only evidence available that cytokinins have a preferential effect on mRNA-like RNA synthesis in senescing leaf tissues. All other studies have found a uniform effect of cytokinins on RNA synthesis. In general, it would seem that maintenance of chlorophyll, protein and RNA levels is associated with stimulation of RNA synthesis but there is no good evidence available that it involves any particular sort of RNA.

Enhancement of label accumulation, as in the experiments described above, is commonly taken to mean increased synthesis but in the case of senescing leaf tissue, there is evidence that it may involve changes in RNA breakdown. In barley leaf discs, kinetin depresses the activities of both deoxyribonuclease and ribonuclease which normally increase in senescing tissues (Srivastava and Ware 1965) and similar results have been obtained for other leaf tissues (Anderson and Rowan 1966; Bagi and Farkas 1967; McHale and Dove 1969; Sodek and Wright 1969; Udvardy et al. 1969). In particular, in *Avena* leaf sections, it has been shown that the effect of kinetin is not on all nucleases present but that it regulated one in particular (Wyen et al. 1972). The same nuclease was the only one whose activity was regulated by ABA. Such regulation of nuclease activity by cytokinins could account for the maintenance of nucleic acid levels in kinetin-treated tissues but no causal relationship has yet been established (see also Chapter 11, section 11.2.1).

The cases described above involve promotional effects of cytokinins but cases of inhibition of growth by cytokinins have also been examined. Kinetin inhibited both auxin-induced accumulation of ribosomes as well as the auxin-induced growth in the elongating zone of soybean hypocotyls (Shannon and Key 1963). A detailed analysis of the effects of kinetin on RNA metabolism (Vanderhoef and Key 1968) showed that kinetin did inhibit auxin-enhanced accumulation of label ([^{14}C]uridine) into RNA but that it acted in a differential way. After a 6 hr incubation (auxin and kinetin together), kinetin at 1.2×10^{-4} *M* inhibited label accumulation into both lr and hr-RNA almost completely (90%) and TB-RNA was affected to some extent (30–50%) although D-RNA was not affected. The authors made the point that the effects of kinetin on RNA synthesis were very similar to those of 5-FU with the exception that kinetin inhibits growth as well as TB-RNA synthesis whereas 5-FU does not, which implies that growth, whether auxin-promoted or kinetin-inhibited, is dependent on TB-RNA supply but not D-RNA.

10.3.3.5. Summary and conclusions. The effort expended on the relation between cytokinins and RNA metabolism is enormous and rivals that expended on auxin research but, as is obvious from the preceding sections, there is very little strong support for any selective effect on a particular species. It is becoming clear that levels of RNA change after cytokinin treatment, that RNA synthesis, especially of minor species, is required for cytokinin effects on growth, and that cytokinins can enhance label incorporation into various species of RNA; whereas the cell-free transcription studies offer a good opportunity for studying these changes in depth, this potential has not been realized so far. Regulation of RNA stability through nucleases should perhaps receive more attention. Detection of early effects of cytokinins on RNA metabolism, which seems to have been the primary aim in many studies of other hormones, has been relatively neglected in the case of cytokinins. The field would seem to offer the possibility of great rewards from the application of modern techniques of molecular biology to the study of both in vitro and in vivo RNA metabolism. The occurrence of cytokinins in tRNA has not been discussed here because it has been reviewed at length by other authors (see Chapter 11, section 11.3.4). However, in general it seems that the occurrence of cytokinins in tRNA is no longer viewed as being an integral part of cytokinin action and regulation of nucleic acid levels. (Kende 1971; Hall 1973; Skoog 1973; Burrows 1975). Perhaps this, as well as the recent progress made in elucidating modes of action of other hormones, might lead to a redirection of interest and resources into the metabolism of RNA species other than tRNA in cytokinin-treated tissues in the hope that older studies will be updated and clarified.

10.3.4. Ethylene

10.3.4.1. Perception. Essentially nothing is known about how plant cells perceive ethylene. The structure of the molecule and some of its physiological effects suggest that it may affect the structure and function of membranes and thus their permeability, particularly that of the tonoplast (Burg 1968; Abeles 1972; Hanson and Kende 1975), and the interaction of ethylene with carbon monoxide indicated that it could bind to a metallo-receptor (Burg and Burg 1967; Burg 1973), although attempts to characterize the receptor have been unsuccessful (Abeles et al. 1972; Beyer 1972). Ethylene has been found to affect the surface tension of artificial lipid and protein films (Mehard et al. 1970), although non-specifically, and to be incorporated to a small extent into various cellular components of pea seedlings (Beyer 1975), but whether these effects have anything to do with ethylene perception is unknown. As a result, there is no evidence that any perception mechanism might in some way be linked to transcription.

10.3.4.2. Transcription in vitro. Cell-free transcription of chromatin isolated from ethylene–treated tissue has yielded encouraging results in a single study. Ethylene and auxin had very similar effects on RNA metabolism in soybean hypocotyls and there were parallels between their abilities to enhance or retard RNA levels in the apical, elongating and basal sections of soybean hypocotyl and their effects on chromatin-directed RNA synthesis (Holm et al. 1970). Accumulation (basal sections) and loss (apical sections) of RNA were associated with enhanced and retarded chromatin activity, respectively, indicating not only that the quantitative effects of auxin and ethylene were similar, but that the regulation of in vivo RNA synthesis may have its origin in chromatin activity, although this is not meant to imply that the mechanisms of activation of the chromatin by the two hormones are identical. The ethylene effect on chromatin was maximal after 3 hr and it caused qualitative changes in the RNA synthesized as evidenced by changes in nearest neighbour frequencies, and these differences were maintained in *E. coli* polymerase-saturating conditions indicating that new template was being transcribed. The changes in template caused by ethylene were different to those caused by auxin. Thus, there is evidence that within 3 hr, ethylene can affect polymerase II-dependent transcription. Clearly this work holds some promise for further analysis of the role of regulatory factors, polymerase activity or altered template activity.

10.3.4.3. Regulation of RNA metabolism in vivo. The first link between ethylene and RNA metabolism appeared in the mid 1960's when it was shown that, during ethylene-induced epinasty, RNA levels in tomato *(Lycopersicon esculentum)* leaves and stems were enhanced (Turkova et al. 1965). Subsequent to this, most of the work has involved the process of abscission in which ethylene is believed to play a major role and which involves synthesis and secretion of cellulase and perhaps other enzymes including pectinase into the cell walls (Horton and Osborne 1967; Morré 1969) (see Chapter 8, section 8.5.3). Ethylene-promoted abscission in cotton *(Gossypium hirsutum), Coleus* and bean was found to be actinomycin D-sensitive indicating a requirement for RNA synthesis, and in support of this, ethylene was found to enhance [^{32}P] orthophosphate incorporation into RNA of bean petiole explants 1 hr after the commencement of treatment (Abeles and Holm 1966, 1967). The effect was mainly in the whole abscission zone although the authors referred to it as the separation layer. The effect on label incorporation as well as promotion of abscission by ethylene is dependent on an ageing process in the explants about which very little seems to be known. Holm and Abeles (1967) subsequently examined ethylene-regulated RNA synthesis in more detail by MAK chromatography and differential extraction techniques, and found that after a 6 hr incubation of aged bean explants with ethylene, accumulation of label in all fractions of RNA was enhanced although the enhancement was greatest in rRNA and "m"RNA. After 3 hr, an enhancement was apparent in rRNA, "m"RNA and DNA-RNA but not in tRNA. However, part of this enhancement was probably not necessary for ethylene to cause abscission, because 5-FU while having no effect on abscission, inhibited the ethylene effect on synthesis of rRNA, tRNA, DNA-RNA and "m"RNA by 53%, 49%, 46% and 35% respectively. Actinomycin D inhibited synthesis of all RNA species as well as abscission. Therefore, although the data are by no means clear-cut, it would seem that synthesis of only some species of RNA is required for abscission to occur. The authors suggest that species inhibited most by 5-FU are not required for abscission but that those inhibited least, mRNA in particular, may be required. As pointed out earlier, what the authors referred to as mRNA is more commonly called D-RNA and this may be HnRNA and not mRNA. In any case, there is some indication that ethylene stimulated RNA synthesis in petioles and although inhibitor studies indicate a requirement for concurrent RNA synthesis, the required RNA species has not been identified. Whatever species is required, its synthesis is not required throughout the whole induction and abscission period because ethylene-induced abscission escapes from actinomycin D control after about 6 hr. Presumably within this time, sufficient of the required RNA has been synthesized (Abeles and Holm 1966).

In general, similar results have been obtained from studies of ethylene-promoted fruit ripening. Ethylene promotes ripening of apples *(Pyrus malus)* and it has been shown that in peel discs, ethylene stimulated RNA synthesis (as well as protein synthesis) just before the respiration climacteric (Hulme et al. 1971) although the particular species of RNA (if selective) was not determined. Ethylene also hastens ripening of figs *(Ficus carica)* and a more detailed study of RNA synthesis using this system has shown that after a 6 hr treatment of tissue slices with ethylene, the sedimentation profiles of ribosomes and MAK column profiles of nucleic acids were the same as controls but incorporation of [5-^{3}H]uridine into ribosomes and tRNA had been enhanced to a large extent (Marei and Romani 1971). Thus, there was apparently an overall stimulation of RNA synthesis but whether this involved the minor RNA species was not determined. Effects of ethylene on tRNA metabolism have been examined further in tomato fruit and it was found that 10 hr after ethylene treatment, changes occurred in the relative amounts of iso-accepting species of leucyl and lysyl-tRNA and methionyl and tyrosinyl-tRNA, with changes in the last two being less extensive and taking longer to occur (Mettler and Romani 1976). These were taken to be quantitative changes in iso-accepting tRNA species related to ageing of the tissue, but whether they were direct or indirect consequences of ethylene action was not evident. There were overall decreases in chargeability of most tRNA species with ageing, so that while the changes in relative amounts of iso-accepting species could well have been quantitative changes, they could also have reflected differences in chargeability.

10.3.4.4. Summary and conclusions. Ethylene treatment does affect RNA levels in vivo and in one case appeared to cause a change in polymerase II-dependent transcription measured in vitro. In addition, ethylene control of abscission appears to have a requirement for synthesis of minor RNA. These results provide an encouraging basis for further work.

10.3.5. Abscisic acid

10.3.5.1. Transcription of isolated DNA and chromatin. There have been two studies of the effects of ABA on cell-free transcription and both detected inhibitory responses. The first involved radish hypocotyl tissue in which ABA inhibits growth (Pearson and Wareing 1969). Adding ABA directly to the purified chromatin assay system had no effect on [^{3}H]uridine triphosphate incorporation into RNA but if it was added to the tissue-homogenizing medium, incorporation of UTP (in vitro) was inhibited. There was no consistent quantitative effect on the subsequent activity of chromatin isolated from hypocotyls pretreated with ABA although qualitative differences were not studied. Therefore, it seems that ABA may react with some cytoplasmic factor in order to regulate chromatin activity, but if this is so, it is difficult to understand why chromatin from pretreated hypocotyls was unaffected unless a high concentration of ABA is necessary to maintain the structure of some ABA-factor-chromatin complex. Whether the reduced transcription was due to decreased polymerase activity or decreased template availability was not determined. In a second study it was demonstrated that ABA treatment of nondormant pear embryos caused a reduction in transcriptional capacity of isolated chromatin as well as inhibition of the increase in transcriptional capacity resulting from kinetin treatment (Khan 1972). However, no further analysis of the ABA effect was undertaken. The effect of in vivo application of ABA on in vitro transcription is not consistent with the studies on radish hypocotyls but at this stage it is clear that ABA can reduce the capacity of chromatin to direct RNA synthesis and it would seem that, as for kinetin, auxin and gibberellin, there is a requirement for a cofactor.

10.3.5.2. Polymerase. There has been no systematic analysis of the polymerases affected by ABA either in vivo or in isolated nuclei or chromatin, but the results of a single study on polymerase from maize give some indication that ABA may modulate polymerase activity. In maize coleoptiles, ABA was found to inhibit RNA synthesis and to cause a reduction in the activity of a crude preparation of RNA polymerase, which in these coleoptiles was apparently all soluble, and not bound to chromatin (Bex 1972a, b). Unfortunately, the enzyme was not identified although in other studies, soluble polymerase isolated from maize seedlings has been found to be primarily polymerase II (Benson 1971; Strain et al. 1971). The interesting aspect of this work is that ABA inhibited polymerase activity when it was added to the grinding medium and also when it was added to a polymerase-calf-thymus-DNA system indicating that the effect was on the polymerase or some associated co-factor which modulated enzyme activity or altered DNA-binding activity. Nevertheless, it seems likely that modulation of polymerase activity rather than reduction of enzyme level was involved. There is very little definitive data on the in vitro transcripts, the polymerases responsible and the timing in relation to in vivo responses, but it nevertheless suggests that ABA does regulate transcription.

10.3.5.3. RNA metabolism in vivo. The role of ABA is generally considered to be that of an inhibitor. It prevents or retards growth and morphogenesis and either maintains a status quo situation or alternatively leads to senescence (Milborrow 1974). There are also innumerable cases where it antagonizes the responses of plant tissues to auxin, gibberellin and cytokinin. Since the promotive hormones have been thought to act via RNA metabolism, the obvious possibility is that ABA may antagonize these responses at the level of RNA metabolism. Thus, ABA research has consisted primarily of attempts to show that it reduces RNA levels and inhibits RNA synthesis, and support for this has come from a number of plant types and tissues including leaves, coleoptiles, embryonic axes, barley aleurone and *Lemna*.

The common result of studies on ABA promotion of leaf senescence is an apparent overall inhibition of RNA synthesis, although minor species have not been measured directly. Analysis of RNA from radish leaf discs, which had been incubated with ABA for 5.5–16 hr and then pulsed with [^{32}P]orthophosphate for 3 hr, by sucrose density gradients, MAK chromatography and polyacrylamide gel electrophoresis showed that precursor incorporation into all RNA fractions was inhibited by the hormone although the effect on hrRNA and lrRNA was greater than on tRNA (Wareing et al. 1968a, b). Such overall inhibition of RNA synthesis occurred not only in radish leaf discs but also in leaves of *Taraxacum, Nasturtium, Solanum tuberosum* and barley (Poulson and Beevers 1970). However, the effects of ABA on the major species appeared to be of little consequence to the action of ABA as shown by the fact that 5-FU did not enhance ABA-promoted senescence although ribosome synthesis was apparently completely suppressed (Paranjothy and Wareing 1971). The implication, as for all other hormones, was that some minor species was important and indeed when rRNA synthesis was suppressed, it could be seen that synthesis of polydisperse RNA migrating on polyacrylamide gels in the rRNA region was inhibited by ABA and promoted by kinetin. Judging by its mobility characteristics, this RNA was probably a mixture of mRNA and HnRNA.

Studies on coleoptiles have yielded similar results. In maize coleoptiles, after 3 hr, ABA began to inhibit incorporation of labelled cytidine especially into rRNA and the inhibition became progressively greater with time (Bex 1972a). The effect was greater for rRNA than for tRNA. ABA inhibited cytidine uptake into coleoptiles as previously reported (Pearson 1969) but the effect was not obvious until 3 hr after ABA had begun to inhibit incorporation of label into RNA. Nevertheless, it would seem that part of the inhibition of incorporation into RNA could be accounted for by inhibition of label uptake. Even faster effects of ABA have been found in *Avena* coleoptiles; after only 30 min, incorporation of [^{32}P]orthophosphate into total RNA was depressed (Cline and Rehm 1974). The nucleic acids examined here were rapidly labelled and polydisperse on acrylamide gels. No qualitative differences due to ABA were observed and the effect appeared to be transient in that after 1 hr incubation, no qualitative or quantitative differences in RNA synthesis could be detected. Also there was no increase in UMP content of rapidly labelled RNA after ABA treatment as has been reported for pear embryos and lentil roots (see later).

In embryonic bean axis, ABA (after 10 hr incubation) inhibited cell elongation and incorporation of [^{32}P]orthophosphate into rRNA, tRNA and DNA but not into TB-RNA (as for 5-FU), but the effect on RNA was probably not mediated by an effect on DNA synthesis (Walton et al. 1970). Also, it was shown that growth during the test period was insensitive to 5-FU and therefore probably not dependent on rRNA or tRNA synthesis. Therefore, although ABA reduced the incorporation of label (and apparent synthesis) into the major species, it would seem that it did not affect growth in this way. Neither 5-FU nor ABA inhibited incorporation into TB-RNA or D-RNA (hr-RNA shoulder), although since actinomycin D inhibited growth and label incorporation into these peaks (Walton and Soofi 1969), it seems that one or the other or both of these RNA fractions contain species essential for growth. The notable result then is that ABA did not appear to inhibit label incorporation into these species while still inhibiting growth. Nevertheless, the possibility of small undetected changes in these fractions was acknowledged. Studies on germinating pear embryos (pre-chilled) produced very similar results (Khan and Heit 1969). ABA inhibited [^{32}P]orthophosphate incorporation into tRNA, DNA-RNA and lrRNA within 1–4 hr after the beginning of hormone treatment but appeared to have very little effect on hrRNA and mRNA (D-RNA), although hrRNA and mRNA were not well separated. TB-RNA was not examined in this study but nevertheless, the rapidly labelled species in embryos, in general, seem to be insensitive to ABA while the major species are sensitive. Similarly, associated with ABA-inhibited germination of *Phaseolus vulgaris*, there was inhibition of label incorporation into all major species. Minor species were not examined (Walbot et al. 1975). It was also shown that these effects occurred in the absence of any major effects on respiration, adenosine uptake, phosphorylation to ATP or on the ATP pool-size, indicating that ABA did not act by limiting supply of energy or nucleic acid precursors. Studies on *Fraxinus* embryos are consistent with the above data in that ABA inhibited labelled RNA precursor incorporation (measured autoradiographically), but studies of polysomes produced an interesting variation. ABA inhibited the formation of polysomes but not protein synthesis (Villiers 1968). On the premise that availability of mRNA limits polysome formation, these data were interpreted to mean that ABA prevented the formation of messenger species essential for germination. However, since protein synthesis was unaffected, the synthesis of "essential" proteins would constitute only a small part of the total protein synthesis. Therefore, the question arises as to why ABA caused such a large decrease in polysome formation while having no effect on protein synthesis. Considering all of these studies together, it seems that in germinating embryos, ABA inhibits the synthesis of the major species of RNA preferentially although it is possible that this is a result rather than a cause of growth inhibition. Again, synthesis of major species is not necessary for short-term growth and so the effects of ABA on growth are unlikely to involve the major species. Clearly there is a need to examine the early effects of ABA on mRNA synthesis in embryos more closely because it would seem that the synthesis of a minor RNA fraction, perhaps mRNA, is critical in ABA action.

In the general context of studies of embryonic systems, base composition changes in RNA following ABA treatment have been reported. In the rapidly labelled TB-RNA, DNA-RNA and the (D-RNA + rRNA) fraction, (called (mRNA + rRNA) in this study) of pear embryos, ABA caused a major increase in UMP and a major decrease in guanosine monophosphate (GMP) while cytidine monophosphate (CMP) and adenosine monophosphate (AMP) remained unchanged (Khan and Anojulu 1970). Similar experiments on lentil roots produced similar results (Khan et al. 1970). The values for UMP were in the 30–40% range indicating that the newly synthesized RNA could have contained reasonably long poly(U) tracts. This is interesting, especially in the light of recent reports that RNA controlling translation has a very high (50%) UMP content (see Chapter 11, section 11.2.2.1). However, since the effect of ABA is to inhibit RNA synthesis and since the RNA synthesized early in the germination of some seeds has also been shown to contain very high levels of UMP (40–50%) (Watanabe et al. 1973; Dobrzanska and Buchowicz 1976), the possibility exists that ABA may have inhibited the synthesis of most species except UMP-rich RNA.

Evidence for an overall effect of ABA came also from studies on *Lemna* where ABA causes the cessation of growth and the production of dormant fronds (turions) in 3–5 days. Van Overbeek and colleagues measured an overall decrease in [^{32}P]orthophosphate incorporation into all nucleic acids after 1 day of exposure to ABA (uptake was not taken into account), although DNA was affected most (Van Overbeek et al. 1967). These effects could be measured much earlier, however, as shown by a more detailed study of the same system by Steward and Smith (1972). [^{3}H]Thymidine incorporation into DNA was depressed within 1–3 hr after the beginning of ABA treatment, and orotate and [^{32}P]-orthophosphate incorporation into RNA within 3–6 hr. The effects were seen in rRNA and polydisperse RNA. Comparisons were made between specific activities of RNA fractions and corrections made for uptake differences, and so this constitutes good evidence that an overall effect of ABA on RNA synthesis occurred; however, this could have resulted from an initial inhibition of DNA synthesis.

Studies on barley aleurone tissue do not conform to the general picture of overall inhibition of RNA synthesis which has emerged from other systems. Although ABA is a potent inhibitor of GA_3-induced amylase synthesis in aleurone, in one study there was no inhibition of RNA synthesis as measured by [^{14}C]uridine incorporation (Chrispeels and Varner 1966) and in another, ABA inhibited incorporation of [^{3}H]uridine and [^{3}H]-adenosine into all major species of RNA only slightly but eliminated the apparently large stimulation of label incorporation into polydisperse minor species caused by GA_3 (Zwar and Jacobsen 1972). Some of these minor species could be mRNA. In keeping with this result, ABA was shown to antagonize the GA_3-induced stimulation of poly(A) RNA synthesis (Ho and Varner 1974). Therefore, in aleurone, it seems that there is no major overall inhibition of RNA synthesis associated with inhibition of amylase synthesis but that there is a reversal of GA_3-induced changes in minor species. Similarly, in mung bean roots, where ABA prevented nitrate reductase induction by nitrate, there was no detectable inhibition of rRNA or mRNA (poly(A) RNA) synthesis, although small changes in the latter may not have been detected (Higgins 1974).

It is abundantly clear that a major effort in this field has produced little evidence of a clear-cut effect of ABA on mRNA levels. As for other hormones, such direct evidence has been unavailable because of the inability to measure the levels and rates of synthesis of individual mRNA species. However, recent progress in the assay for specific mRNAs has allowed us to measure the effect of ABA on the level of translatable mRNA for α-amylase (Higgins et al. 1977). ABA added at the same time as GA_3 inhibited amylase synthesis and there was a concomitant reduction in the level of translatable amylase mRNA indicating that ABA controlled amylase synthesis by regulating the level of its mRNA. It is yet to be determined whether this effect involves inhibition of amylase mRNA synthesis, inactivation, or breakdown of the mRNA. This result does indicate, however, that ABA probably does not act only by inhibiting translation of amylase mRNA and suggests strongly that the hormone may (also) act by affecting biosynthesis, degradation or activity of amylase mRNA.

10.3.5.4. Summary and conclusions. The effects of ABA on labelling of RNA in intact tissues indicates that in most cases, it causes an inhibition of synthesis of the major species. However, as in some embryos and leaf tissues, inhibition of synthesis of the major RNA species does not enhance or duplicate the effects of ABA. The fact that ABA does not inhibit synthesis of the major species in aleurone while strongly inhibiting synthesis of some enzymes also argues for the contention that while ABA may directly inhibit synthesis of major RNA species, depending on the competence of the tissue or perhaps the requirements of the tissue response, this is not the RNA of importance to its action, at least initially.

There is every indication that synthesis of minor species, probably mRNA, is involved in ABA action and that ABA can rapidly affect the metabolism of this RNA. In tissues of bean and pear embryos where ABA is physiologically effective but has no apparent effects on mRNA-like RNA, it must be considered very likely that the numbers of mRNA species affected could be very small in relation to unaffected species and that techniques used would not have detected such changes. The inhibition of polysome formation or promotion of disaggregation in several cases is consistent with effects on mRNA levels although there is some indication that such changes may not be entirely dependent on mRNA levels.

The lack of effect of ABA on putative mRNA in embryos implies that ABA inhibits RNA species differentially. Indeed, there is some indication that ABA does not inhibit transcription which is not hormone-promoted. In aleurone, ABA does not affect poly(A) RNA synthesis occurring in the absence of GA_3 (Zwar and Jacobsen, unpublished data), but it does appear to inhibit GA_3-induced changes in poly(A) RNA. Therefore, it is tempting to interpret an "overall" inhibition of RNA synthesis as inhibition of transcription which is occurring only in response to a promotive hormone. Indeed, ABA treatment at levels which are strongly inhibitory to various processes never completely eliminates RNA synthesis.

The depression of translatable amylase mRNA by ABA complements a body of evidence which points to ABA acting by regulating levels of individual mRNA species. That such changes are the result of changes in transcription is strongly suggested by the effects of ABA on in vitro transcription and polymerase activity. Modulation of polymerase activity of ABA seems to be a real possibility, especially in the light of similar effects by auxin (see section 11.3.1.4). Although the qualitative effects of ABA on transcription and polymerases are yet to be examined, it seems very likely that the inhibitory effects of ABA on RNA levels and labelling in vivo can be accounted for in the main by its effects on transcription. Nevertheless, the situation is not quite that simple because there is indirect evidence firstly that ABA can control the translation of pre-existing mRNA and secondly that in some systems it is physiologically active when transcription is not required for normal metabolism (see Chapter 11, section 11.3.2). Thus, while in some cases the inhibitory effects of ABA may involve inhibition of transcription of certain mRNA species, there are other instances where ABA apparently has a posttranscriptional action, but even here there is evidence that transcription is required (see Chapter 11, section 11.3.2).

References

Abeles, F.B. (1972) Ann. Rev. Plant Physiol., 23, 259–292.
Abeles, F.B. and Holm, R.E. (1966) Plant Physiol., 41, 1337–1342.
Abeles, F.B. and Holm, R.E. (1967) Ann. N.Y. Acad. Sci., 144, 367–373.
Abeles, F.B., Ruth, J.M., Forrence, L.E. and Leather, G.R. (1972) Plant Physiol., 49, 669–671.
Addicott, F.T., Ohkuma, K., Smith, O.E. and Thiessen, W.E. (1966) Advanc. Chem., 53, 47.
Anderson, J.W. and Rowan, K.S. (1966) Biochem. J., 101, 15–18.
Anderson, M.B. and Cherry, J.H. (1969) Proc. nat. Acad. Sci. (Wash.), 62, 202–209.
Apelbaum, A. and Burg, S.P. (1972) Plant Physiol., 50, 117–124.
Apelbaum, A., Sfakiotakis, E. and Dilley, D.R. (1974) Plant Physiol., 54, 125–128.
Arens, M.Q. and Stout, E.R. (1972) Plant Physiol., 50, 640–641.
Arya, S.K. and Yang, J.T. (1975) Biochemistry, 14, 963–969.
Ashford, A.E. and Jacobsen, J.V. (1974) Planta, 120, 81–105.
Atsmon, D., (1972) in: Plant Growth Substances 1970, D.J. Carr (ed.), Springer, Berlin, pp. 222–227.
Bagi, G. and Farkas, G.L. (1967) Phytochemistry, 6, 161–169.
Bamberger, E.S. (1971) Phytochemistry, 10, 957–966.
Bard, E., Efron, D., Marcus, A. and Perry, R.P. (1974) Cell, 1, 101–106.
Barendse, G.W.M. (1974) in: Plant Growth Substances 1973, Hirokawa, Tokyo, pp. 332–341.
Batt, S. and Venis, M.A. (1976) Planta, 130, 15–21.
Batt, S., Wilkins, M.B. and Venis, M.A. (1976) Planta, 130, 7–13.
Bendich, A.J. (1972) Biochim. biophys. Acta (Amst.), 272, 494–503.
Bendich, A.J. and Filner, P. (1971) Mut. Res., 13, 199–214.
Benson, R.H. (1971) Plant Physiol., 47, 5–36.
Benson, R.H. and Mans, R.J. (1972) FASEB Abst., 31, 1179.
Berridge, M.V., Ralph, R.K. and Letham, D.S. (1970) Biochem. J., 119, 75–84.
Berridge, M.V., Ralph, R.K. and Letham, D.S. (1972) in: Plant Growth Substances 1970, D.J. Carr (ed.), Springer, Berlin, pp. 248–255.
Bex, J.H.M. (1972a) Planta, 103, 1–10.
Bex, J.H.M. (1972b) Planta, 103, 11–17.
Beyer, E.M. (1972) Plant Physiol., 49, 672–675.
Beyer, E.M. (1975) Nature (Lond.), 255, 144–146.
Bick, M.D., Liebke, H., Cherry, J.H. and Strehler, B.Z. (1970) Biochim biophys. Acta (Amst.), 204, 175–182.
Birmingham, B.C. and Maclachlan, G.A. (1972) Plant Physiol., 49, 371–375.
Biswas, B.B., Ganguly, A., Das, A. and Roy, P. (1975) in: Regulation of Growth and Differentiation Function in Eukaryotic Cells, G.P. Talwar (ed.), Raven Press, New York, pp. 461–477.
Bonner, J.M., Dahmus, M.E., Fambrough, D., Huang, R.C., Murashige, K. and Tuan, D.Y.H. (1968) Science, 159, 47–56.
Brian, P.W., Petty, J.H.P. and Richmond, P.T. (1959) Nature (Lond.), 183, 58–59.
Britten, R.J. and Davidson, E.H. (1969) Science, 165, 345–357.
Broughton, W.J. (1968) Biochim. biophys. Acta (Amst.), 155, 308–310.
Broughton, W.J. (1969) Ann. Bot. (Lond.), 33, 227–243.
Bryant, J.A. (1974) Planta, 118, 17–24.
Bryant, J.A. and Wildon, D.C. (1971) Biochim. biophys. Acta (Amst.), 232, 624–629.
Burdett, N. and Wareing, P.F. (1966) Planta, 71, 20–26.
Burg, S.P. (1968) Plant Physiol., 45, 1503–1511.
Burg, S.P. (1973) Proc. nat. Acad. Sci. (Wash.), 70, 591–597.
Burg, S.P. and Burg, E.A. (1967) Plant Physiol., 42, 144–152.
Burkard, G. and Keller, E.R. (1974) Proc. nat. Acad. Sci. (Wash.), 71, 389–393.
Burrows, W.J. (1975) Curr. Advanc. Plant Sci., No. 21, 837–847.

Carlson, P.S. (1972) Nature New Biology, 237, 39–41.
Carpenter, W.J.G. and Cherry, J.H. (1966a) Biochim. biophys. Acta (Amst.), 114, 640–642.
Carpenter, W.J.G. and Cherry, J.H. (1966b) Plant Physiol., 41, 919–922.
Chaley, N. and Setterfield, G. (1972) Planta, 108, 363–368.
Chandra, G.R. (1974) Nature (Lond.), 248, 161–162.
Chandra, G.R. and Duynstee, E.E. (1968) in: Biochemistry and Physiology of Plant Growth Substances 1967, F. Wightman and G. Setterfield (eds.) Runge Press, Ottawa, pp. 723–745.
Chandra, G.R. and Duynstee, E.E. (1971) Biochim. biophys. Acta (Amst.), 232, 514–523.
Chandra, G.R. and Varner, J.E. (1965) Biochim. biophys. Acta (Amst.), 108, 583–592.
Chapman, J.M. and Jordan, E.G. (1971) J. exp. Bot., 22, 620–626.
Chapman, K.S.R., Trewavas, A. and Van Loon, L.C. (1975) Plant Physiol., 55, 293–296.
Chen, Y., Lin, C., Chang, H., Guilfoyle, T.J. and Key, J.L. (1975) Plant Physiol., 56, 78–82.
Cherry, J.H. (1964) Science, 146, 1066–1069.
Cherry, J.H. (1967) Ann. N.Y. Acad. Sci., 144, 154–168.
Cherry, J.H. (1974) In: Plant Growth Substances 1973, Hirokawa, Tokyo, pp. 752–759.
Cherry, J.H. and Chroboczek, H. (1966) Phytochemistry, 5, 411–422.
Chrispeels, M.J. and Hanson, J.B. (1962) Weeds, 10, 123–125.
Chrispeels, M.J. and Varner, J.E. (1966) Nature (Lond.), 212, 1066–1067.
Chrispeels, M.J. and Varner, J.E. (1967) Plant Physiol., 42, 1008–1016.
Chrispeels, M.J., Jacobsen, J.V. and Varner, J.E. (1966) Plant Physiol., Suppl. 41, IXXVII.
Chroboczek, H. and Cherry, J.H. (1965) Biochem. biophys. Res. Commun. 20, 774–779.
Chroboczek, H. and Cherry, J.H. (1966) J. mol. Biol., 19, 28–37.
Cline, M.G. (1976) Plant Sci. Lett., 6, 181–184.
Cline, M.G. and Rehm, M.M. (1974) in: Plant Growth Substances 1973, Hirokawa, Tokyo, pp. 582–591.
Cohen, M.E. and Kleinsmith, L.J. (1976) Biochim. biophys. Acta (Amst.), 435, 159–166.
Cohen, P., Chin, R.C. and Kidson, C. (1969) Biochemistry, 8, 3603–3609.
Cohen, P. and Kidson, C. (1969) Proc. nat. Acad. Sci. (Wash.) 63, 458–464.
Dahmus, M.E. (1976) FEBS. Lett., 61, 227–230.
Das, G. (1973) Ann. Bot. (Lond.), 37, 999–1004.
Das, N.K., Patau, K. and Skoog, F. (1956) Physiol. Plant., 9, 640–651.
Datko, A.H. and Maclachlan, G.A. (1968) Plant Physiol., 43, 735–742.
Davies, E. (1976) Plant Physiol., 57, 516–518.
Davies, E. and Larkins, B.A. (1973) Plant Physiol., 52, 339–345.
Davies, E. and Maclachlan, G.A. (1968) Arch. Biochem., 128, 595–600.
Davies, E. and Maclachlan, G.A. (1969) Arch. Biochem., 129, 581–587.
Degani, Y. and Atsmon, D. (1970a) Exp. Cell Res., 61, 226–229.
Degani, Y. and Atsmon, D. (1970b) Nature (Lond.), 228, 554–555.
De Leo, P. and Sacher, J.A. (1970) Plant Physiol., 46, 806–811.
Dobrzanska, M. and Buchowicz, J. (1976) Biochim. biophys. Acta (Amst.), 432, 73–79.
Duda, C.T. (1976) Ann. Rev. Plant Physiol., 27 119–132.
Duda, C.T. and Cherry, J.H. (1971) Plant Physiol., 47, 262–268.
Duffus, C.M. (1972) Symp. Biol. Hung., 13, 273–275.
Edmonds, M. and Caramela, M.G. (1969) J. biol. Chem., 244, 1314–1324.
Elgin, S.C.R. and Bonner, J. (1972) in: The Biochemistry of Gene Expression in Higher Organisms J.K. Pollock and J.W. Lee (eds.), Australia and New Zealand Book Co., Sydney, pp. 142–163.
Elgin, S.C.R. and Weintraub, H. (1975) Ann. Rev. Biochem., 44, 725–774.
Erismann, K.H. and Farkhauser, M. (1967) Experientia (Basel) 23, 621–622.
Evins, W.H. (1971) Biochemistry, 10, 4295–4303.
Ewing, E.E. and Cherry, J.H. (1967) Phytochemistry, 6, 1319–1328.
Fambrough, D.M., Fujimura, F. and Bonner, J. (1968) Biochemistry, 7, 575–585.
Fan, D.F. and Maclachlan, G.A. (1966) Canad. J. Bot., 44, 1025–1034.

Fan, D.F. and Maclachlan, G.A. (1967a) Canad. J. Bot., 45, 1837–1844.
Fan, D.F. and Maclachlan, G.A. (1967b) Plant Physiol., 42, 1114–1122.
Fellenberg, G. (1969a) Planta, 84, 195–198.
Fellenberg, G. (1969b) Planta, 84, 324–338.
Fellenberg, G. (1969c) Z. Pflanzenphysiol., 60, 457–466.
Fellenberg, G. (1970) Planta, 95, 359–361.
Fellenberg, G. (1971a) Planta, 100, 347–356.
Fellenberg, G. (1971b) Z. Naturforsch., 26b, 607–612.
Fellenberg, G. and Schomer, V. (1975) Z. Pflanzenphysiol., 75, 449–456.
Fellenberg, G. and Voges, B. (1975) Z. Pflanzenphysiol., 76, 456–461.
Fites, R.C., Hanson, J.B. and Slife, F.W. (1969) Bot. Gaz. (Chic.), 130, 118–126.
Fowke, L.C. and Setterfield, G. (1968) in: Biochemistry and Physiology of Plant Growth Substances, F. Wightman and G. Setterfield (eds.), Runge Press, Ottawa, pp. 581–602.
Fox, J.E. and Erion, J.L. (1975) Biochem. biophys. Res. Commun., 64, 694–700.
Fromson, D. and Verma, D.P.S. (1976) Proc. nat. Acad. Sci. (Wash.), 73, 148–151.
Gamburg, K.Z., Mal'tseva, V.N. and Kobyl'skii, G.I. (1965) Soviet Plant Physiol., 12, 120–124.
Ganguly, A., Das, A., Mondal, H., Mandal, R.K. and Biswas, B.B. (1973) FEBS Lett., 34, 27–30.
Gayler, K.R. and Glasziou, K.T. (1969) Planta, 84, 185–194.
Glasser, S.R., Chytel, F. and Spelsberg, T.C. (1972) Biochem. J., 130, 947–957.
Goldberg, M.L. and Atchley, W.A. (1966) Proc. nat. Acad. Sci. (Wash.), 55, 989–996.
Goodwin, P.B. and Carr, D.J. (1972) Planta, 106, 1–12.
Gordon, M.E. and Letham, D.S. (1975) Aust. J. Plant Physiol., 2, 129–154.
Gordon, M.E., Letham, D.S. and Beaver, J.E. (1975) Physiol. Plant., 35, 27–33.
Grieshaber-Scheubel, D. and Fellenberg, G. (1972) Z. Pflanzenphysiol., 66, 106–112.
Guilfoyle, T.J. and Hanson, J.B. (1973) Plant Physiol., 51, 1022–1025.
Guilfoyle, T.J. and Hanson, J.B. (1974) Plant Physiol., 53, 110–113.
Guilfoyle, T.J., Chen. Y.M., Nagao, R.T. and Key, J.L. (1974) Biochem. biophys. Res. Commun., 60, 498–505.
Guilfoyle, T.J., Lin, C.Y., Chen, Y.M., Nagao, R.T. and Key, J.L. (1975) Proc. nat. Acad. Sci. (Wash.), 72, 69–72.
Guillé, E., Quétier, F. and Huguet, T. (1968) C.R. Acad. Sci. (Paris), Sér. D, 266, 836–838.
Gurley, W.B., Lin, C., Guilfoyle, T.J., Nagao, R.T. and Key, J.L. (1976) Biochim. biophys. Acta (Amst.), 425, 168–174.
Guttman, R. (1956) Chromosoma, 8, 341–350.
Guttman, R. (1957) Biophys. biochem. Cytol., 3, 129–131.
Haber, A. and Foard, D.E. (1964) Amer. J. Bot., 51, 151–159.
Haber, A.H., Foard, D.E. and Perdue, S.W. (1969) Plant Physiol., 44, 463–467.
Hall, R.H. (1973) Ann. Rev. Plant Physiol., 24, 415–444.
Hanson, A.D. and Kende, H. (1975) Plant Physiol., 55, 663–669.
Hardin, J.W. and Cherry, J.H. (1972) Biochem. biophys. Res. Commun., 48, 299–306.
Hardin, J.W., Cherry, J.H., Morré, D.J. and Lembi, C.A. (1972) Proc. nat. Acad. Sci. (Wash.), 69, 3146–3150.
Hardin, J.W., Clark, J.H., Glasser, S.R. and Peck, E.J. (1976) Biochemistry, 15, 1370–1374.
Hardin, J.W., O'Brien, T.J. and Cherry, J.H. (1970) Biochim. biophys. Acta (Amst.), 224, 667–670.
Harris, S.E., Schwartz, R.J., Tsai, M. and O'Malley, B.W. (1976) J. biol. Chem., 251, 524–529.
Higgins, T.J.V. (1974) in: Plant Growth Substances 1973, Hirokawa, Tokyo, pp. 599–607.
Higgins, T.J.V., Zwar, J.A. and Jacobsen, J.V. (1976) Nature (Lond.), 260, 166–169.
Higgins, T.J.V., Zwar, J.A. and Jacobsen, J.V. (1977) in: Acides Nucleiques et Synthese des proteines chez les Vegetaux 261, C.N.R.S. Paris, pp. 481–486.
Ho, D.T.H. and Varner, J.E. (1974) Proc. nat. Acad. Sci. (Wash.), 71, 4783–4786.
Holm, R.E. and Abeles, F.B. (1967) Plant Physiol., 42, 1094–1102.
Holm, R.E. and Abeles, F.B. (1968) Planta, 78, 293–304.

Holm, R.E. and Key, J.L. (1969) Plant Physiol., 44, 1295–1302.
Holm, R.E., O'Brien, T.J., Key, J.L. and Cherry, J.H. (1970) Plant Physiol., 45, 41–45.
Holmes, P.L. and Speakman, P.T. (1973) Nature New Biology 242, 190–192.
Horton, R.F. and Osborne, D.J. (1967) Nature (Lond.), 214, 1086–1089.
Huez, G., Marbaix, G., Hubert, G., Leclercq, M., Nudel, U., Soreq, H., Salomon, R., Lebleu, B., Revel, M. and Littauer, U.Z. (1974) Proc. nat. Acad. Sci. (Wash.), 71, 3143–3146.
Hulme, A.C., Rhodes, M.J.C. and Wooltorton, L.S.C. (1971) Phytochemistry, 10, 749–756.
Ihle, J.N. and Dure, L.S. III (1969) Biochem. biophys. Res. Commun., 36, 705–710.
Ihle, J.N. and Dure, L.S. III (1970) Biochem. biophys. Res. Commun., 38, 995–1001.
Ihle, J.N. and Dure, L.S. III (1972) J. biol. Chem., 247, 5048–5055.
Ingle, J. and Key, J.L. (1965) Plant Physiol., 40, 1212–1219.
Ingle, J.L., Key, J.L. and Holm, R.E. (1965) J. molec. Biol., 11, 730–746.
Jacobsen, J.V. and Zwar, J.A. (1974a) Aust. J. Plant Physiol., 1, 343–356.
Jacobsen, J.V. and Zwar, J.A. (1974b) Proc. nat. Acad. Sci. (Wash.), 71, 3290–3293.
Jankowski, J., Wielgat, B. and Kleczkowski, K. (1975) Plant. Sci. Lett., 347–350.
Jensen, E.V. and DeSombre, E.R. (1973) Science, 182, 126–134.
Jensen, W.A., Pollock, E.G., Healey, P. and Ashton, H. (1964) Exp. Cell Res., 33, 523–530.
Johnson, K.D. and Purves, W.K. (1970) Plant Physiol., 46, 581–585.
Johri, M.M. and Varner, J.E. (1968) Proc. nat. Acad. Sci. (Wash.), 59, 269–276.
Jones, R.L. (1969) Planta, 88, 73–86.
Jouanneau, J.P. (1971) Exp. Cell Res., 67, 329–337.
Jouanneau, J.P. (1975) Exp. Cell Res., 91, 184–190.
Jouanneau, J.P. and Tandeau de Marsac, N. (1973) Exp. Cell Res., 77, 167–174.
Julien, R. and Delseny, M. (1972) Biochem. biophys. Res. Commun., 48, 578–584.
Julien, R. and Guitton, Y. (1972) Symp. Biol. Hung., 13, 33–41.
Kadouri, A. and Atsmon, D. (1974) in: Mechanisms of Regulation of Plant Growth, R.L. Bieleski, A.R. Ferguson and M.M. Creswell (eds.), Roy. Soc. New Zealand, Wellington, pp. 339–343.
Kadouri, A., Atsmon, D. and Edelman, M. (1975) Proc. nat. Acad. Sci. (Wash.), 72, 2260–2264.
Kamisaka, S. and Masuda, Y. (1968) Plant Cell Physiol., 9, 61–67.
Katterman, F.R.H. and Clay, W.F. (1975) Plant Physiol., 56, 523–525.
Kende, H. (1971) Int. Rev. Cytol., 31, 301–338.
Kende, H. and Gardner, G. (1976) Ann. Rev. Plant Physiol., 27, 267–290.
Kern, H. and Grubler, J. (1976) Z. Naturforsch., 31c, 104–107.
Kessler, B. (1971a) Biochim. biophys. Acta (Amst.), 232, 611–613.
Kessler, B. (1971b) Biochim. biophys. Acta (Amst.), 240, 330–343.
Kessler, B. (1973) in: The Biochemistry of Gene Expression in Higher Organisms, J.K. Pollack and J.W. Lee (eds.), Australia and New Zealand Book Co. Sydney, pp. 333–356.
Kessler, B. and Snir, I. (1969) Biochim. biophys. Acta (Amst.), 195, 207–218.
Key, J.L. (1964) Plant Physiol., 39, 365–370.
Key, J.L. (1969) Ann. Rev. Plant Physiol., 20, 449–474.
Key, J.L. and Hanson, J.B. (1961) Plant Physiol., 36, 145–152.
Key, J.L. and Ingle, J. (1964) Proc. nat. Acad. Sci. (Wash.), 52, 1382–1388.
Key, J.L. and Ingle, J. (1968) in: Biochemistry and Physiology of Plant Growth Substances, F. Wightman and G. Setterfield (eds.), Runge Press, Ottawa, pp. 711–722.
Key, J.L. and Silfow, C. (1975) Plant Physiol., 56, 364–369.
Key, J.L., Barnett, N.M. and Lin, C.Y. (1967) Ann. N.Y. Acad. Sci., 144, 49–62.
Khan, A.A. and Anojulu, C.C. (1970) Biochem. biophys. Res. Commun., 38, 1069–1075.
Khan, A.A. and Heit, C.E. (1969) Biochem. J., 113, 707–712.
Khan, A.A., Anderson, L. and Gaspar, T. (1970) Plant Physiol., 46, 494–495.
Kidson, C. and Kirby, K.S. (1964) J. molec. Biol., 10, 187–198.
Kidson, C., Thomas, A. and Cohen, P. (1970) Biochemistry, 9, 1571–1576.
Klämbt, D. (1974) Planta, 118, 7–16.

Klämbt, D. (1976) Plant Cell Physiol., 17, 73–76.
Kleinsmith, L.J., Stein, J. and Stein, G. (1976) Proc. nat. Acad. Sci. (Wash.), 75, 1174–1178.
Knutsen, G. (1965) Biochim. biophys. Acta (Amst.), 103, 495–502.
Kovoor, A. and Melet, D. (1972) Symp. Biol. Hung., 13, 29–32.
Lang, A. (1966) 25th Symp. Soc. Dev. Biol., 251–287.
Lang, A. and Nitsan, J. (1967) Ann. N.Y. Acad. Sci., 144, 180–190.
Larkins, B.A. and Davies, E. (1975) Plant Physiol., 55, 749–756.
Leaver, C.J. and Dyer, J.A. (1974) Biochem. J., 144, 165–167.
Lee, S. and Dahmus, M.C. (1973) Proc. nat. Acad. Sci. (Wash.), 70, 1383–1387.
Leffler, H.R., O'Brien, T.J., Glover, D.V. and Cherry, J.H. (1971) Plant Physiol., 48, 43–45.
Levine, L., Gordon, L.A. and Jencks, W.P. (1963) Biochemistry, 2, 168–175.
Lewin, B. (1976) Nature (Lond.), 260, 574–576.
Libbenga, K.R. and Torrey, J.G. (1973) Amer. J. Bot., 60, 293–299.
Lin, C.Y., Travis, R.L., Chia, L.S.Y. and Key, J.L. (1973) Phytochemistry, 12, 515–522.
Lin, C.Y., Guilfoyle, T.J., Chen, Y.M., Nagao, R.T. and Key, J.L. (1974) Plant Physiol., 53, 110–113.
Lin, C.Y., Chen, Y.M., Guilfoyle, T.J. and Key, J.L. (1977) Plant Physiol.
Lonberg-Holm, K.K. (1967) Nature (Lond.), 213, 454–457.
McComb, A.J. and Broughton, W.J. (1972) in: Plant Growth Substances 1970, D.J. Carr (ed.), Springer, Berlin pp. 407–413.
McComb, A.J., McComb, J.A. and Duda, C.T. (1970) Plant Physiol., 46, 221–223.
McCombs, P.J.A. and Ralph, R.K. (1972) Biochem. J., 129, 403–414.
McHale, J.S. and Dove, L.D. (1969) New Phytol., 68, 79–91.
MacLeod, R.D. (1972) Z. Pflanzenphysiol., 68, 379–387.
Maheshwari, S.C., Guha, S. and Gupta, S. (1966) Biochim. biophys. Acta (Amst.), 117, 470–472.
Maheshwari, M.G. and Nooden, L.D. (1971) Physiol. Plant., 24, 282–287.
Marei, N. and Romani, R.J. (1971) Plant Physiol., 48, 806–808.
Masuda, Y. (1959) Physiol. Plant., 12, 324–335.
Masuda, Y. and Tanimoto, E. (1967) Plant Cell Physiol., 8, 459–465.
Masuda, Y., Tanimoto, E. and Wada, S. (1967) Physiol. Plant., 20, 713–719.
Matthyse, A.G. and Abrams, M. (1970) Biochim. biophys. Acta (Amst.), 190, 511–518.
Matthyse, A.G. and Phillips, C. (1969) Proc. nat. Acad. Sci. (Wash.), 63, 897–903.
Means, A.R. and O'Malley, B.W. (1972) Metabolism, 21, 357–370.
Mehard, C.W., Lyons, J.M. and Kumamoto, J. (1970) J. Membr. Biol., 3, 173–179.
Mendecki, J., Lee, S.Y. and Brawerman, G. (1972) Biochemistry, 11, 792–798.
Mettler, I.J. and Romani, R.J. (1976) Phytochemistry, 15, 25–28.
Miassod, R., Penon, P,, Teissere, M., Ricard, J. and Cecchini, J. (1970) Biochim. biophys. Acta (Amst.), 224, 423–440.
Milborrow, B.V. (1974) Ann. Rev. Plant. Physiol., 25, 259–307.
Miller, C.O., Skoog, F., Von Saltza, M.H. and Strong, F.M. (1955) J. Amer. Chem. Soc., 77, 1392.
Mondal, H., Ganguly, A., Das, A., Mandal, R.K. and Biswas, B.B. (1972) Europ. J. Biochem., 28, 143–150.
Mondal, H., Mandal, R.K. and Biswas, B.B. (1970a) Biochem. biophys. Res. Commun., 40, 1194–1200.
Mondal, H., Mandal, R.K. and Biswas, B.B. (1970b) Plant Cell Physiol., 11, 177–181.
Mondal, H., Mandal, R.K. and Biswas, B.B. (1970c) The Nucleus XIII, 10–18.
Mondal, H., Mandal, R.K. and Biswas, B.B. (1972a) Europ. J. Biochem., 25, 463–470.
Mondal, H., Mandal, R.K. and Biswas, B.B. (1972b) Nature New Biology, 240, 111–114.
Morré, D.J. (1968) Plant Physiol., 43, 1545–1559.
Musgrave, A., Kays, S.E. and Kende, H. (1969) Planta, 89, 165–177.
Nagl, W. (1975) Progr. Bot., 37, 186–210.
Nagl, W. and Rucker, W. (1976) Nucleic Acid Res., 3, 2033–2039.
Nakamura, T. and Takahashi, N. (1973) In: Plant Growth Substances 1973, Hirokawa, Tokyo, pp. 559–466.

Nakamura, T., Yabuuchi, Y. and Takahashi, N. (1970) Develop. Growth Diff., 12, 189–197.
Naylor, J.M. (1966) Canad. J. Bot., 44, 19–32.
Newcombe, E.N. (1960) Physiol. Plant., 13, 459–467.
Nitsan, J. and Lang, A. (1965) Develop. Biol., 12, 358–376.
Nitsan, J. and Lang, A. (1966) Plant Physiol., 41, 965–970.
Nooden, L.D. (1968) Plant Physiol., 43, 140–150.
Nooden, L.D. and Thimann, K.V. (1963) Proc. nat. Acad. Sci. (Wash.), 50, 194–200.
O'Brien, T.J. (1967) Induction of Ribonucleic acid polymerase activity in soybean hypocotyls with 2,4-Dichlorophenoxyacetic Acid. PhD Thesis, University of Illinois.
O'Brien, T.J., Jarvis, B.C., Cherry, J.H. and Hanson, J.B. (1968a) in: Biochemistry and Physiology of Plant Growth Substances, F. Wightman and G. Setterfield (eds.), Runge Press, Ottawa, pp. 747–759.
O'Brien, T.J., Jarvis, B.C., Cherry, J.H. and Hanson, J.B. (1968b) Biochim. biophys. Acta (Amst.), 169, 35–43.
Olson, J.O.J. and Busch, H. (1974) in: The Cell Nucleus, Vol III, H. Busch (ed.), Academic Press, New York, pp. 211–268.
Olszewka, M.J. (1959) Exp. Cell Res., 16, 193–201.
Osborne, D.J. (1962) Plant Physiol., 37, 595–602.
Panyim, S., Jensen, R.H. and Chalkley, G. (1968) Biochim. biophys. Acta (Amst.), 160, 252–255.
Paranjothy, K. and Wareing, P.F. (1971) Planta, 99, 112–119.
Patau, K. and Das, N.K. (1961) Chromosoma, 11, 553–572.
Patau, K., Das, N.K. and Skoog, F. (1957) Physiol. Plant., 10, 949–966.
Pearson, G.G. and Ingle, J. (1972) Cell Diff., 1, 43–51.
Pearson, J.A. (1969) Affects of abscisic acid on nucleic acid metabolism in plants, Ph.D. Thesis, University of Wales, Aberystwyth.
Pearson, J.A. and Wareing, P.F. (1969) Nature (Lond.), 221, 672–673.
Pelc, S.R. (1968) Nature (Lond.), 219, 162–163.
Penner, D. and Early, R.W. (1972) Phytochemistry, 11, 3135–3138.
Penon, P. (1972) Symp. Biol. Hung., 13, 325–334.
Piesco, N.P. and Alvarez, M.R. (1972) Exp. Cell Res., 73, 129–139.
Pilet, P.E. (1969) C.R. Acad. Sci. (Paris), Ser. D, 268, 930–932.
Pilet, P.E. (1970) J. exp. Bot., 21, 446–451.
Pilet, P.E. (1972) Symp. Biol. Hung., 13, 313–323.
Pilet, P.E. and Braun, R. (1970) Physiol. Plant., 23, 245–250.
Poulson, A. and Beevers, L. (1970) Plant Physiol., 46, 782–785.
Quétier, F., Guillé, E. and Vedel, F. (1968) C.R. Acad. Sci. (Paris), Sér. D, 266, 735–738.
Rao, V.S. and Khan, A.A. (1975) Biochem. biophys. Res. Commun., 62, 25–30.
Rappaport, L. and Wolf, N. (1968) in: Proceedings of International Symposium on Plant Biochemical Regulation in Diseased Plants or Injury, Phytopathological Society of Japan, Tokyo, pp. 203–211.
Raynaud-Jammet, C., Cattelli, M.G. and Baulieu, E.E. (1972) FEBS Lett., 22, 93–96.
Richmond, A.E. and Lang, A. (1957) Science, 125, 650–651.
Rijven, A.H.G.C. (1974) Nature (Lond.), 252, 257–259.
Rijven, A.H.G.C. and Parkash, V. (1970) Plant Physiol., 46, 638–640.
Rijven, A.H.G.C. and Parkash, V. (1971) Plant Physiol., 47, 59–64.
Rizzo, P.J., Cherry, J.H., Pedersen, K. and Durham, V.L. (1974) Plant Physiol., 54, 349–355.
Robinson, N.E. and Bryant, J.A. (1975) Planta, 127, 69–75.
Rodaway, S. and Kende, H. (1978) Plant Physiol., in press.
Rose, R.J. and Adamson, D. (1969) Planta, 88, 274–281.
Roychoudhury, R. and Sen, S.P. (1964a) Biochem. biophys. Res. Commun. 14, 7–11.
Roychoudhury, R. and Sen, S.P. (1964b) Bull. Bot. Soc. (Bengal), Suppl. 18, 191–198.
Roychoudhury, R. and Sen, S.P. (1964c) Physiol. Plant., 17, 352–362.

Roychoudhury, R., Datta, A. and Sen, S.P. (1965) Biochim. biophys. Acta (Amst.), 107, 346–351.
Sabota, A.E., Leaver, C.J. and Key, J.L. (1968) Plant Physiol., 43, 907–913.
Salunkhe, D.K., Dhaliwol, A.S. and Boe, A.A. (1962) Nature (Lond.), 195, 724–725.
Sampson, M. and Davies, D.D. (1966) Exp. Cell Res., 43, 669–673.
Sarrouy-Balat, H., Delseny, M. and Julien, R. (1973) Plant Sci. Lett., 1, 287–292.
Sasaki, K. and Tazawa, T. (1973) Biochem. biophys. Res. Commun., 52, 1440–1449.
Schmid, B.D., Siegel, N.R. and Vanderhoef, L.N. (1975) Plant Physiol., 55, 277–281.
Schochetman, G. and Perry, R.P. (1972) J. molec. Biol., 63, 591–596.
Schwimmer, S. (1968) Plant Physiol., 43, 1008–1010.
Seifart, K.H. (1970) Cold Spr. Harb. Symp. quant. Biol., 35, 719–725.
Shannon, J. and Key, J. (1963) Plant Physiol., 38, Suppl. XIV.
Shaw, M., Bhattacharya, P.K. and Quick, W.A. (1965) Canad. J. Bot., 43, 739–746.
Shaw, M. and Manocha, M.S. (1965) Canad. J. Bot., 43, 747–755.
Sheiness, O. and Darnell, J.E. (1973) Nature New Biology, 241, 265–268.
Sherwin, J.E. and Gordon, S.A. (1974) Planta, 116, 65–72.
Shih, C.Y. and Rappaport, (1970) Plant Physiol., 45, 33–36.
Silberger, J. and Skoog, F. (1953) Science, 118, 443–444.
Simard, A. (1971) Canad. J. Bot., 49, 1541–1549.
Skinner, D.M. and Triplett, L.L. (1967) Biochem. biophys. Res. Commun., 28, 892–897.
Skoog, F. (1954) Brookhaven Symp. Biol., No. 6, 1–21.
Skoog, F. (1973) Biochem. Soc. Symp., 38, 195–215.
Snir, I. and Kessler, B. (1975) Physiol. Plant., 35, 191–193.
Sodek, L. and Wright, S.T.C. (1969) Phytochemistry, 8, 1629–1630.
Spelsberg, T.C. and Sarkissian, I.V. (1968) Phytochemistry, 7, 2083–2088.
Spelsberg, T.C., Webster, R.A. and Pikler, G.M. (1976) Nature (Lond.), 262, 65–67.
Spiker, S. and Chalkley, R. (1971) Plant Physiol., 47, 342–345.
Spiker, S. and Chalkley, R. (1972) Planta, 102, 362–367.
Spiker, S. and Krishnaswamy, L. (1973) Planta, 110, 71–76.
Srivastava, B.I.S. (1965) Arch. Biochem., 110, 97–103.
Srivastava, B.I.S. (1967) Int. Rev. Cytol., 22, 349–387.
Srivastava, B.I.S. and Arglebe, E. (1967) Plant Physiol., 42, 1497–1503.
Srivastava, B.I.S. and Ware, G. (1965) Plant Physiol., 40, 62–64.
Srivastava, L.M., Sawhney, V.K. and Taylor, I.E.P. (1975) Proc. nat. Acad. Sci. (Wash.), 72, 1107–1111.
Stein, G.S., Hunter, G. and Lavie, L. (1974a) Biochem. J., 139, 71–76.
Stein, G.S., Spelsberg, T.C. and Kleinsmith, L.J. (1974b) Science, 183, 817–824.
Stein, G.S., Stein, J.S. and Kleinsmith, L.J. (1975) Scientific Amer., 232, 46–57.
Stein, H. and Hansen, P. (1970) Europ. J. Biochem., 14, 270–277.
Stern, H. (1964) in: Regulateurs Naturels de la Croissance Vegetale, J.P. Nitsch (ed.), C.N.R.S., Paris, pp. 19–31.
Stewart, G.R. and Smith, H. (1972) J. exp. Bot., 23, 875–885.
Stoddart, J., Breidenbach, W., Nadeau, R. and Rappaport, L. (1974) Proc. nat. Acad. Sci. (Wash.), 71, 3255–3259.
Strain, G.C. Mullinix, K.R. and Bogorad, L. (1971) Proc. nat. Acad. Sci. (Wash.), 68, 2647–2651.
Sugiura, M., Umemura, K. and Oota, Y. (1962) Physiol. Plant., 15, 457–464.
Taiz, L. (1975) Plant Physiol. 56, Suppl. 45.
Takegami, T. (1975) Plant Cell Physiol., 16, 417–425.
Teissere, M., Penon, P., Azou, Y. and Ricard, J. (1976) Plant Sci. Lett., 6, 49–55.
Teissere, M., Penon, P. and Ricard, J. (1973) FEBS Lett., 30, 65–70.
Teissere, M., Penon, P., Ricard, J. and Ratle, G. (1972) FEBS Lett., 23, 65–68.
Teissere, M., Penon, P., Van Huystee, R.B., Azou, Y. and Ricard, J. (1975) Biochim. biophys. Acta (Amst.), 402, 391–402.
Teraoka, H. (1967) Plant Cell Physiol., 8, 87–95.

Teraoka, H. (1968) Plant Cell Physiol., 9, 819–823.
Tester, C.F. and Dure, L.S. III (1967) Biochemistry, 6, 2532–2537.
Thomas, T.H., Wareing, P.F. and Robinson, P.M. (1965) Nature (Lond.), 205, 1270–1272.
Thompson, W.F. and Cleland, R. (1971) Plant Physiol., 48, 663–670.
Travis, R.L., Anderson, J.M. and Key, J.L. (1973) Plant Physiol., 52, 608–612.
Travis, R.L., Lin, C.Y. and Key, J.L. (1962) Biochim. biophys. Acta (Amst.), 277, 606–614.
Trewavas, A.J. (1968a) Arch. Biochem. Biophys., 123, 324–335.
Trewavas, A. (1968b) Phytochemistry, 7, 673–681.
Trewavas, A. (1968c) Prog. Phytochem., 1, 113–160.
Trewavas, A. (1970) Plant Physiol., 45, 742–751.
Trewavas, A. (1976) Phytochemistry, 15, 363–366.
Trewavas, A.J. (1977) in: Molecular Aspects of Gene Expression in Plants, J.A. Bryant (ed.), Academic Press, London, pp. 249–298.
Truelson, T.A. (1967) Physiol. Plant., 20, 1112–1119.
Turkova, N.S., Vasileva, L.N. and Cheremukhina, L.F. (1965) Soviet Plant Physiol., 12, 721–726.
Udvardy, J., Farkas, G.L. and Marre, E. (1969) Plant Cell Physiol., 10, 375–386.
Van der Eb, A.A. and Niewdorp, P.J. (1967) Acta bot. neer., 15, 690–699.
Vanderhoef, L.N. and Key, J.L. (1968) Plant Cell Physiol., 9, 343–351.
Van Loon, L.C. Trewavas, A. and Chapman, K.S.R. (1975) Plant Physiol., 55, 288–292.
Van Oostveldt, P. and Van Parijs, R. (1976) Exp. Cell Res., 98, 210–221.
Van Overbeek, J., Loeffler, J.E. and Mason, M.I.R. (1967) Science, 156, 1497–1499.
Varner, J.E. (1964) Plant Physiol., 39, 413–415.
Varner, J.E. and Chandra, G. (1964) Proc. nat. Acad. Sci. (Wash.), 52, 100–106.
Varner, J.E. Chandra, G.R. and Chrispeels, M.J. (1965) J. Cell Comp. Physiol., Suppl. 66, 55–68.
Venis, M.A. (1971) Proc. nat. Acad. Sci. (Wash.), 68, 1824–1827.
Venis, M.A. (1973) Curr. Advanc. Plant Sci. 3, 21–28.
Venis, M.A. and Batt, S. (1975) Biochem. Soc. Trans., 3, 1148–1151.
Venkataraman, R., De Leo, P., Anderson, M.B. and Cherry, J.H. (1970) Plant Physiol., Suppl. 45, 9.
Verma, D.P.S., Machlachlan, G.A., Byrne, H. and Ewings, D. (1975) J. biol. Chem., 250, 1019–1026.
Vigil, E. and Ruddat, M. (1973) Plant Physiol., 51, 549–558.
Villiers, T.A. (1968) Planta, 82, 342–354.
Walbot, V., Capdevila, A. and Dure, L.S. III (1974) Biochem. biophys. Res. Commun., 60, 103–110.
Walbot, V., Clutter, M. and Sussex, I. (1975) Plant Physiol., 56, 570–574.
Walton, D.C. and Soofi, G.S. (1969) Plant Cell Physiol., 10, 307–315.
Walton, D.C., Soofi, G.S. and Sondheimer, E. (1970) Plant Physiol., 45, 37–40.
Wardell, W.H. (1975) Plant Physiol., 56, 171–176.
Wardell, W.H. and Skoog, F. (1973) Plant Physiol., 52, 215–220.
Wareing, P.F., Good, J. and Manuel, J. (1968a) in: Biochemistry and Physiology of Plant Growth Substances 1967, F. Wightman and G. Setterfield (eds.), Runge Press, Ottawa, pp. 1561–1579.
Wareing, P.F., Good, J., Potter, H. and Pearson, A., (1968b) Soc. Chem. Ind. Monogr., 31, 191–207.
Wasilewska, L.D. and Kleczkowski, K. (1974) FEBS Lett., 44, 164–168.
Watanabe, A., Nitta, T. and Shiroya, T. (1973) Plant Cell Physiol., 14, 29–38.
Weil, P.A. and Blatti, S.P. (1976) Biochemistry, 15, 1500–1509.
Weinmann, R. and Roeder, R.G. (1974) Proc. nat. Acad. Sci. (Wash.), 71, 1790–1794.
Wielgat, B., Wasilowska, L.D. and Kleczkowski, K. (1974) in: Plant Growth Substances 1973, Hirokawa, Tokyo, pp. 573–598.
Wildon, D.C. and Wong, D. (1971) Biochem. J., 124, 12P.
Williams, G.R. and Novelli, G.D. (1968) Biochim. biophys. Acta (Amst.), 155, 183–192.
Wyen, N.V., Udvardy, J., Erdei, S. and Farkas, G.L. (1972) Symp. Biol. Hung., 13, 293–297.
Yamada, Y. and Yasuda, T. (1971) Biochem. biophys. Res. Commun., 43, 488–493.
Yamada, Y., Yasuda, T., Kaga, M. and Sekiya, J. (1971) Agric. Biol. Chem., 35, 99–104.
Yamada, Y., Yasuda, T. and Yajima, Y. (1973) in: Plant Growth Substances 1973, Hirokawa, Tokyo, pp. 549–558.

Yamamoto, K.R. and Alberts, B. (1974) J. biol. Chem., 249, 7076–7086.
Yasuda, T. and Yamada, Y. (1970) Biochem. biophys. Res. Commun., 40, 649–653.
Yasuda, T. and Yamada, Y. (1971) FEBS Lett., 18, 115–117.
Zwar, J.A. (1973) J. exp. Bot., 24, 701–710.
Zwar, J.A. and Jacobsen, J.V. (1972) Plant Physiol., 49, 1000–1006.

Phytohormones and Related Compounds – A Comprehensive Treatise, Volume I
Editors: Letham, Goodwin and Higgins

11

Posttranscriptional, Translational and Posttranslational Effects of Plant Hormones

J.V. JACOBSEN and T.J.V. HIGGINS

11.1. Introduction

It is appealing to think of hormones as having a single primary site of action. While this may be true or at least a useful basis on which to approach the problem of hormonal control, clearly one cannot overlook the possibility that it is an oversimplification and that hormones have multiple sites of action, even within the same cell. In the latter case the response would be a concomitant complex of events coordinated in such a way as to ensure normal development. Although this view may not be strongly defensible at present, there is an indication that some processes are affected by hormones in several ways. While this may indicate some sort of feedback or other secondary control, the possibility that it reflects multiple independent primary control points is an important consideration. Thus, this chapter is not meant to infer that non-transcriptional (posttranscriptional, translational and posttranslational) controls are alternatives to transcriptional control. At this time it should be considered that any control may operate alone or in conjunction with others.

Responses are often referred to as being direct or indirect but in the context of this chapter these terms require reconsideration. As a result of common usage, a direct effect on translation, for example, would be an effect which did not operate via transcription. However, since plant hormones may have a variety of initial points of action and since hormone action may not be simply a matter of transcription versus translation, the terms direct and indirect assume new meanings. A direct effect then is one where the site of response is the primary site of action. Thus, modification of membrane function may hypothetically lead to a change in translation but not transcription, in which case the effect on translation would in the past have been referred to as a direct effect, meaning that it did not involve transcription, whereas clearly it would more appropriately be called an indirect or secondary effect because the translation process was not the primary site of hormone action. Similarly, changes in polypeptide synthesis may result from changes in various components of the polysomal complex, including not only mRNA level but also other components such as initiation, elongation, termination or interference factors and any such changes may result from initial regulation of transcription. Thus, it is most important to determine primary sites of action and to determine whether hormonal effects are primary, in that they follow hormone perception immediately, or whether they are secondary.

This chapter begins with an examination of the possible non-transcriptional sites of hormone action and then with this background and using the principles developed above, continues with an assessment of the effects of phytohormones on these processes.

11.2. Possible control points

Current knowledge of the metabolism of nucleic acids after transcription and the events associated with the translation processes as well as the posttranslational modification of proteins provides the opportunity to study the effect of hormones on a great number of potential control points in cellular metabolism. Although rapid advances are being made in the understanding of most of the above processes in plants, there is still a great deal to be learned about the features of posttranscriptional events. There appears to be considerable similarity between nucleic acid processing, protein biosynthesis and processing in animals and in plants (in those cases where the same process has been studied). Thus, we have used data obtained from both sources to outline those steps in non-transcriptional events which might be considered to offer control sites for hormone action.

11.2.1. Processing and turnover of nucleic acids

11.2.1.1. Messenger RNA. Currently, there are two major models for the origin of messenger RNA (mRNA). One suggests that the mRNA is transcribed from DNA, a unit similar to or only very slightly larger in size than that found in the polysomes (for review see Perry 1976). Direct evidence to support this model has been obtained for mRNAs for globin (MacNaughton et al. 1974), ovalbumin (McKnight and Schimke 1974) and silk fibroin (Lizardi 1976) by hybridization analysis using DNA complementary to the specific mRNA sequences and various sized fractions of the nuclear RNA. The other model envisages mRNA as arising from precursor mRNA (pre-mRNA) which is part of the heterogeneous nuclear RNA (HnRNA) and which is generally considered to be very much larger than the mRNA found on polysomes (Perry 1976). The evidence for this model is derived largely from analysis of the kinetics of labelling of nuclear and cytoplasmic RNA and is thus of an indirect nature. There is, however, evidence obtained by in vitro translation of HnRNA from avian erythrocytes that globin sequences were present in the very large (greater than 45S) RNA transcripts (Ruiz-Carrillo et al. 1973) and hybridization data (Spohr et al. 1974) also suggested that the HnRNA of avian erythroblasts contained precursors to globin mRNA with a wide range of sizes.

Both models have attracted criticism. With regard to the first model, it can be argued that degradation of the HnRNA was not adequately controlled and with regard to the second, that aggregation of the RNA occurred during its isolation. It should also be noted that the mRNAs which have been shown to be present in the HnRNA as molecules of similar size to those found in the cytoplasm, are also present as an abundant class of mRNA and may be subject to different processing mechanisms to the majority of mRNAs which are present to the extent of only a few copies per cell.

Studies on the origin of mRNA have been further complicated by the findings that both HnRNA and mRNA are polyadenylated on the 3′ end and "capped" on the 5′ end of the chain by an inverted guanosine triphosphate which becomes methylated (for review see Perry 1976). Thus there does not appear to be a simple relationship between HnRNA and mRNA since it now seems possible that pre-mRNA contains two or more mRNA molecules, one at the 5′ end and one at the 3′ end or that capping and polyadenylation of pre-mRNA and mRNA are independent events. Therefore, the questions of where the mRNA exists in the pre-mRNA and whether it is cleaved from the pre-mRNA remain unsolved. Thus, it is not yet possible to assess to what extent, if at all, pre-mRNA processing presents possibilities for control of mRNA levels. However, the post-transcriptional addition of poly(A) and capping are both clearly worthy of attention as possible control points whether or not pre-mRNA cleavage occurs. Another characteristic of pre-mRNA with a possible control function involves the occurrence of double stranded loops formed by base pairing of complementary sequences (Perry 1976).

From the little that is known, it seems that pre-mRNA and mRNA of plants is similar to that of mammalian cells. Plant nuclei contain large polyadenylated polynucleotides which seem to be pre-mRNA (Penon 1972; Schmid et al. 1975; Grierson and Covey 1976). It appears that these putative precursors are not much larger (perhaps only about 50%) than the derived mRNA (Schmid et al. 1975; Grierson and Covey 1976).

The addition of poly(A) and the cap to the nucleotide chain would seem to be of particular interest from a control point of view. The poly(A) consists of 50–200 nucleotides added to the 3′ end of the HnRNA and mRNA chains by poly(A) polymerase (Brawerman 1974). Whether HnRNA and mRNA poly(A) are the same sequences or whether the poly(A) turns over is not yet clear. The role of poly(A) is still being studied but suggestions include roles in HnRNA processing and export to the cytoplasm and regulation of the stability and thus life-time of mRNA (Perry 1976). Although poly(A) is not necessary for translation of mRNA as in the case of histones in vivo (Adesnik et al. 1972) and globin in vitro (Huez et al. 1974), it may affect translational efficiency of mRNA (Doel and Carey 1976). Capping occurs at the 5′ end of the chain and involves the addition of 7-methyl guanosine which is linked by a 5′-5′-di-or triphosphate bridge to a 2′ O methylated nucleoside (Ro-Choi et al. 1974; Perry et al. 1975; Wei and Moss 1975). Again, whether the HnRNA and mRNA caps are the same is unknown. The function of the cap is unknown but it has been thought to be involved in transcription (Furuichi 1974), in pre-mRNA processing (Rattman et al. 1974; Furuichi et al. 1975b; Rose and Lodish 1976), and in translation (Both et al. 1975; Furuichi et al. 1975a) although, whereas it is required for translation of some mRNAs (Muthukrishnan et al. 1975) it is not absolutely required for others (Wimmer 1972; Rose and Lodish 1976). At present, it seems most likely that the cap and other internal methylated bases at the 5′ end of the chain facilitate translation of some but not all mRNAs. Poly(A) RNA has been detected in higher plants (Higgins et al. 1973; Manahan et al. 1973; Van de Walle 1973; Harris and Dure 1974; Ho and Varner 1974; Jacobsen and Zwar 1974; Sagher et al. 1974; Stegman and Hahn 1974; Yoshida 1974; Esnault et al. 1975; Hammett and Katterman 1975; Key and Silflow 1975; Schmid et al. 1975; Tobin and Klein 1975; Covey and Grierson 1976). The poly(A) is also at the 3′ end of the chain (Ho and Varner 1974; Esnault et al. 1975) and the poly(A) RNA is, at least in part, mRNA (Verma et al. 1975; Higgins et al. 1976). There is also preliminary evidence that plant mRNA is capped (Kennedy and Lane 1975) and a capping enzyme(s) has been found in wheat germ (Muthukrishnan et al. 1975).

mRNA is longer than necessary for the proteins for which it codes (Brawerman 1974) demonstrating that in addition to poly(A), there are other sequences of nucleotides which are not translated. Untranslated sequences exist at both ends of the mRNA molecule (Proudfoot and Brownlee 1976). At the 5′ end of the chain, some of the non-coding sequences as well as the cap are apparently involved in the initiation complex. Although it has not yet been demonstrated in higher organisms, in bacteria, there is a ribosome binding site about 10 nucleotides long which base pairs with the sequences at the 3′ end of the 16S ribosomal RNA (Shine and Dalgarno 1974, 1975; Steitz and Jakes 1975). However, the role of the major part of the 5′ non-coding sequences is unknown. The 3′ non-coding region is longer than the 5′ region and although sequences adjacent to poly(A) have been determined in a number of eucaryotic mRNAs and a short homologous sequence has been found (Proudfoot and Brownlee 1976), essentially nothing is known of their function. However, the majority of the untranslated sequences in various mRNAs is largely different in each case indicating the possibility that they are messenger-specific and are perhaps involved with processing, longevity or translation. The same may well apply to the 5′ non-coding sequences.

Proteins associated with RNA, both in the nucleus and in the cytoplasm, may play a crucial role in cleavage, transport, stability and translatability. The proteins associated with sub-ribosomal mRNA particles, polysomal mRNA and the poly(A) fragment are being characterized (Barrieux et al. 1976; Fukami and Itano 1976) and although little messenger specificity has been demonstrated so far, there are claims that the isolated proteins can influence the efficiency of translation of the two globin mRNAs in a cell-free system (Fukami and Itano 1976).

Probably before leaving the nucleus, mRNA or some mRNA precursor combines with protein to form messenger ribonucleoprotein (mRNP) particles. In this form it can either be active mRNA or inactive stored mRNA existing as RNP particles originally known as informosomes (Spirin 1969; Georgiev and Samarina 1971; Spirin 1972). The process of activation of stored mRNP is currently receiving attention in many laboratories but no clear-cut mechanism has emerged. Informosomes have been found in a number of embryonic plant tissues (Ajtkhozhin et al. 1973; Ajtkhozhin and Akhanov 1974; Ajtkhozhin et al. 1975) and there is evidence that the storage mRNA in plant tissues (Marcus and Feeley 1964; Dure and Waters 1965; Chen et al. 1968; Weeks and Marcus 1971) also exists as mRNP or informosomes (Walbot 1974; Ajtkhozhin et al. 1976). In mammalian informosomes, the mRNA in mRNP is polyadenylated (Bester et al. 1975) although in cotton mRNP particles, if the RNA was mRNA, it was not adenylated (Walbot 1974). RNA presumably occurring as RNP has recently been isolated from the seeds of a number of species and it has been shown to contain poly(A) RNA and to have template activity (Gordon and Payne 1976).

Stability of mRNA sequences is another factor to be considered in control mechanisms. Once in the cytoplasm, mRNA has a finite life. There are relatively few estimates of mRNA stability in eucaryotes and those that have been made have involved very selected mRNA fractions and are also subject to limitations of methodology (for review see Sensky et al. 1975). However, in general, eucaryotic mRNAs are considered to be relatively stable. RNases of various kinds have been implicated in mRNA breakdown, but relatively little is known of the degradation of specific mRNA molecules.

In this sequence of events, there are many potential control points. Because of the possibility that they are involved in mRNA translation and stability, capping (including methylation) and adenylation are two of the most obvious control points for regulation of active mRNA levels. Regulation of pre-mRNA processing, proteinization and translocation out of the nucleus as well as activation and inactivation and degradation of mRNA all appear to be likely points at which mRNA translatability might be controlled. Likewise, it is possible that untranslated base sequences in mRNA could be recognition sites for repressor or activator molecules which would determine translatability, as proposed for the mechanism of action of translation-control RNA in controlling synthesis of myoglobin and myosin (Heywood et al. 1974).

The relatively high stability of eucaryotic mRNA has led to the thought that control of stability is unlikely to be involved in regulation of gene expression (Murphy and Attardi 1973; Singer and Penman 1973) although the possibility that control of stability of specific inducible mRNA species is important cannot be eliminated at present.

11.2.1.2. Ribosomal RNA. There are many ribosomal genes (160–28,000) in each diploid cell (Birnstiel et al. 1971) and they are clustered in the nucleolus. Each ribosomal RNA (rRNA) cistron contains the genes for both heavy (hrRNA) and light (lrRNA) cytoplasmic rRNA plus other sequences. The transcript or pre-rRNA is initially one very large molecule consisting of ribosomal and spacer RNA, which becomes associated with protein (pre-rRNP) and is methylated and processed in the nucleolus to give the 40S and 60S ribosomal subunits containing lrRNA and hrRNA respectively (Warner 1974; Perry 1976). In an actively growing tissue, most of the pre-rRNA is processed to maturity but in a resting tissue much of it is broken down (Liau et al. 1976). Thus, the efficiency of processing determines the rate at which ribosomes are produced and the efficiency of processing depends primarily on the synthesis of ribosomal proteins both in animals (Cooper and Gibson 1971; Pederson and Kumar 1971; Cooper 1973) and in plants (Gebauer et al. 1975) and on the methylation of pre-rRNA (Vaughan et al. 1967; Ascione and VanDerWoude 1969; Maden and Salina 1974). There is evidence that the activity of rRNA methylases can change and that the level of rRNA methylation can vary accordingly (Liau et al. 1976).

Protein constitutes about 60% of animal and plant ribosomes (Bayley 1964; Hsiao 1964; Berridge et al. 1970; Penon et al. 1970). In mammalian cells this protein is heterogeneous and consists of about 70 different polypeptides, 30 occurring in the small subunit and 40 in the large (Sherton and Wool 1974; Wool and Stoffler 1974; Howard et al. 1975; Martini and Gould 1975). A similar number of proteins occurs in the ribosomes of plants (Jones et al. 1972; Nagabhushan et al. 1974). The constitution of the protein is not constant and has been shown to change for example, during rRNA maturation (Kumar and Subramanian 1975) and during induction of cold hardiness in plant seedlings (Bixby and Brown 1975). Some of these proteins are phosphorylated (Traught and Porter 1976 and references therein; Trewavas 1976). Although changes in ribosomal proteins or their phosphorylation are yet to be closely correlated with changes in ribosomal function, such changes do occur. Alterations in phosphorylation of animal ribosomal proteins occur after hormone and other treatments (Blat and Loeb 1971; Barden and Labrie 1973). rRNA in *Lemna* has been estimated to have a half-life of about 4–5 days (Trewavas 1970). Presumably nucleases are responsible for breakdown and ribosome-bound nucleases are usually implicated. However, it now seems that nucleases bound to isolated ribosomes (Wilson 1971; Acton 1972; Gagnon and de Lamirande 1972) are probably adsorbed onto ribosomes during isolation and are not ribosome constituents in vivo (Dyer and Payne 1974). It is more likely that ribosomes are separated from nucleases in vivo (Payne and Boulter 1974). The enzymes probably occur mainly in vacuoles and lysozomes (Matile 1968; Pitt and Galpin 1973). Breakdown of ribosomes probably occurs inside these organelles (Dyer and Payne 1974). Thus, rRNA may have to be engulfed by lytic vacuoles before it can be degraded. Regulation of ribosome longevity by regulation of nuclease levels does not seem likely although the rate of movement of ribosomes into digestive organelles may be a potential control point.

11.2.1.3. Transfer RNA. The formation of transfer RNA (tRNA) is similar to that of other RNA species in that the initial transcript (pre-tRNA) is slightly larger than the final product (Burdon 1971; Choe and Taylor 1972; Chen and Siddiqui 1975). The pre-tRNA is cleaved by enzymes which can be isolated from ribosomes but little is known of this process except that there are multiple precursors (for a recent review of pre-tRNA formation and processing see Perry 1976).

A number of substitutional changes occur during maturation of the pre-tRNA. tRNA of plants contains cytokinins which normally appear to be formed by addition of side-chains to preformed pre-tRNA (Kende 1971; Hall 1973; Skoog 1973; Burrows 1975). tRNA is also methylated and thiolated (for reviews see Borek and Srinivasan 1966; Sueoka and Kano-Sueoka 1970). Methylation of both bases and sugars occurs and is accomplished by multiple specific methylases which use *S*-adenosylmethionine as the methyl donor. tRNA methylases and thiolases have been demonstrated in plant tissue (Abeels et al. 1972).

There are a number of steps in tRNA formation which could limit the rate of tRNA production and activation. The extent of methylation of pre-tRNA apparently does not limit processing (Chen and Siddiqui 1975) and is unlikely to be a control point, but methylation and reversible oxidation and reduction of thiol groups do affect tRNA acceptor ability, the ability to recognize the right codon and the ability to bind to ribosomes (Borek and Srinivasan 1966; Sueoka and Kano-Sueoka 1970). Changes in methylase activity and methylation have been demonstrated in animals accompanying changes in physiological state and hormonal treatment (Sharma and Borek 1974). The occurrence of multiple iso-accepting tRNA species and their possible significance is discussed in section 11.2.2.2.

All active tRNA molecules have a C-C-A terminal nucleotide sequence at the 3′-OH end of the chain. The sequence is added by a cytidine triphosphate (CTP), adenosine triphosphate (ATP)-terminal nucleotide transferase. Terminal C-C-A turns over during, and is required for, tRNA function. There is some evidence that it may control overall protein synthesis in fertilized sea urchin eggs (Glisin and Glisin 1964; Gross et al. 1965).

The rate of degradation of tRNA can also be envisaged as a control point in protein synthesis but very little is known about this process.

11.2.2. Translation

11.2.2.1. Chain initiation, elongation and termination and the ribosome cycle. The earliest event in the translational process involves the formation of a complex between initiator methionyl-tRNA (met-tRNA), the 40S ribosomal subunit, guanosine triphosphate (GTP) and the 5′ end of the mRNA chain, probably involving the cap and various other base sequences, and proteinaceous initiation factors (Lucas-Lenard and Lippman 1971; Anderson et al. 1974; Weissbach and Ochoa 1976). A similar kind of complex seems to form in plants (Seal et al. 1972; Weeks et al. 1972; Wells and Beevers 1973, 1974, 1975; Roman et al. 1976), although there is some controversy over its precise constitution and the order of its construction in plants as well as in other eucaryotes (Anderson et al. 1974).

Although some initiating factors may show no specificity for mRNA, there is evidence that some do. Until 1970, the occurrence of tissue or species-specific initiation factors was very doubtful, but then it was shown by Heywood that translation of muscle mRNA in vitro required initiation factors from muscle ribosomes (Heywood 1970; Rourke and Heywood 1972). This led to the description of proteinaceous initiation factors which specify translation of certain messenger species (Rourke and Heywood 1972; Nudel et al. 1973; Heywood et al. 1974). In one case, two proteins with translational-control capacity were found, one specifying myoglobin synthesis and the other specifying myosin synthesis (Heywood et al. 1974). These factors, isolated from the initiation factor 3 complex of chick red muscle, are envisaged as sequestering the non-specific components of the initiation complex and conferring upon them the specificity for translation of specific messengers. However, it was recently pointed out (Lodish 1976) that some of these studies have alternative explanations and that the case for mRNA-specific proteinaceous translation factors is still not strong.

Messenger-specific RNA components of initiation factor complexes have also been found in mammalian cells and these have been found to have both positive and negative control. A low molecular weight RNA has been isolated from the salt wash of human and rabbit reticulocyte ribosomes and it preferentially stimulates synthesis of one of the globin polypeptide chains (Fuhr and Natta 1972; Bogdanovsky et al. 1973; Fuhr and Overton 1975). It appeared that control was exerted on peptide chain initiation although it is as yet unproven. A search for the mechanism by which inactive mRNA, stored as free or ribosome-bound mRNP (see section 11.2.1.1), is prevented from being translated led to the discovery of translational-control RNA (tcRNA). It has been isolated from initiation factor preparations of polysomes and from mRNP particles (Heywood et al. 1974; Kennedy et al. 1974; Bester et al. 1975). tcRNA from the two sources is different in that tcRNA from mRNP inhibits translation of mRNP-mRNA, and has little effect on translation of polysomal mRNA, and tcRNA from polysomes has no effect on translation of mRNP-mRNA while it stimulates translation of polysomal mRNA. The factor from mRNP has a molecular weight of only 10,000, is rich in uridine and appears to prevent translation in part by base pairing with poly(A) (Heywood et al. 1975; Heywood and Kennedy 1976). Thus, there seems to be a multiplicity of factors, both proteins and nucleic acids, which can affect translational specificity in vitro; however, the significance of these factors in vivo remains to be established.

Peptide chain elongation proceeds when the 60S ribosomal subunit combines with the mRNA-40S subunit complex. The series of steps occurring in animal cells has been recently reviewed by Weissbach and Ochoa (1976) and although it is not yet entirely clear a similar sequence of events appears to occur in plants (Legocki and Marcus 1970; Yarwood et al. 1971; Seal et al. 1972; Lin and Key 1973; Lanzani et al. 1975; Sim and Klämbt 1976). How chain termination occurs in eucaryotes and what happens to the ribosomes before they can be re-used is currently being studied. However, it is known that termination involves release factors which recognize the chain termination codons (UAA, UGA or UAG) and other factors which facilitate cleavage of the C-terminal amino acid from its tRNA thus freeing the newly synthesized protein (Tate and Caskey 1974; Weissbach and Ochoa 1976). Ribosomes appear to dissociate before being re-used as in bacteria but what controls their dissociation is still unclear.

The intracellular arrangement of membranes may play a critical role in the control of translation in eucaryotic cells. Tata (1973) has proposed that ribosome-membrane interaction controls the segregation of different populations of ribosomes involved in synthesizing different classes of protein. A model proposed for the assembly of membrane-bound polysomes (Sabatini et al. 1972) assumes that the mRNA contains the information for binding of polysomes to specific membrane sites. If this is true, hormonal involvement in membrane-polysome associations would seem to be relegated to creating active sites, activating potential binding sites or to constructing membrane with active sites. There is evidence for either of the first two possibilities in that steroid hormones have been shown to be required for ribosome binding to smooth isolated endoplasmic reticulum (ER) (Sunshine et al. 1971). A refinement of the above proposal suggests that the nascent protein recognizes the membrane and controls binding of the mRNA-ribosome-nascent protein complex to the ER by virtue of the sequence of amino acids at the NH_2-terminal end of the nascent polypeptide (Blobel and Dobberstein 1975). This is known as the signal hypothesis and it attempts to explain the mechanism of synthesis of proteins destined for transport.

A model of the overall translation process has been developed by Vassart and colleagues (Vassart et al. 1971) which, although limited in some aspects, probably goes a long way towards understanding the kinetics of polypeptide synthesis and polysome formation. The model is based on the amount of mRNA present, the level of ribosomes and ribosome subunits, the rate of binding of subunits to mRNA (initiation), the rate of progression of ribosomes along mRNA (chain elongation) and the rate of ribosome release (chain termination). By computerizing the model it was possible to "construct" polysome profiles and to determine rates of protein synthesis, the distribution of polysomes along the mRNA, and the mean transit time of a ribosome along the message and to assess the effects of varying each parameter on these values. In general, it was found that ribosome release was probably not limiting but that chain initiation and elongation could limit translation. The picture which emerged was that:

(1) An increase or decrease in mRNA supply causes an increase or decrease in total polysomes with no change in the polysome profile.

(2) Increased initiation causes essentially no change in transit time and therefore more (and closer) ribosomes on each mRNA leading to more protein synthesis and a change in the polysome profile towards larger polysomes.

(3) Reduced initiation is again associated with essentially no change in transit time and therefore fewer ribosomes evenly distributed along each mRNA leading to less protein synthesis and a shift in polysome profile towards smaller polysomes.

(4) An enhanced rate of chain elongation would lead to smaller polysomes, a shorter mean transit time, enhanced protein synthesis and perhaps queuing of ribosomes at the 3′ end of the mRNA if ribosome release becomes limiting.

(5) A slower rate of chain elongation leads to larger polysomes, an increased mean transit time and decreased protein synthesis.

As a result of these considerations, the nature of any change occurring in polysome profiles is potentially useful in interpreting the cause. However, it should be noted that the interpretation is only as good as the quality of the isolated polysomes. They must not be broken down and must be representative of the in vivo polysome population (see Chapter 10, section 10.3.2.4 for problems in polysome isolation). It is extremely difficult to fulfill these criteria with any certainty.

The general conclusion of the polysome modelling study (Vassart et al. 1971) was that for linear regulation of protein synthesis, the most effective parameter to vary is the rate of initiation. There is evidence that in the same cell the rates of elongation and termination of all mRNAs seem to be the same but the rates of initiation of different mRNAs differ (Lodish 1976). The regulation of levels of any of the components of the initiation complex could control polypeptide synthesis. However, it is difficult to see how specificity for initiation of translation of specific messengers could involve the rRNA, although pairing of base sequences between the 3′ terminus of lrRNA and untranslated sequences at the 5′ end of the message (see section 11.2.1.1), if it occurs in eucaryotes, offers potential for specificity. It is also unlikely to involve met-tRNA if the initiator codon on all messengers is AUG; however, perhaps messengers for inducible enzymes have alternative initiator codons. It is much easier to imagine specificity built into the initiation factors and perhaps other ribosomal components in their interaction with particular mRNA species. Initiation specificity has been discussed in some detail recently by Lodish (1976).

Regulation of chain elongation is probably an unlikely point of translational control. Changes in the levels of elongation factors could conceivably change the overall rate of polypeptide synthesis but there would seem to be little advantage in this unless it was coordinated with changes in initiation. Ribosome dissociation and the supply of ribosomal subunits would seem to offer a very effective means of control of protein synthesis although it would seem to provide little specificity.

11.2.2.2. A regulatory role for tRNA. The possibility that protein synthesis might be regulated by tRNA metabolism was expressed many years ago by Itano (1963) during studies of haemoglobin synthesis. Ames and Hartman in 1963 and Stent in 1964 used this concept plus information on the genetic code to propose a scheme which could not only account for regulation of overall protein synthesis but also for the synthesis of individual proteins or groups of proteins (Ames and Hartman 1963; Stent 1964). The scheme is based on the premises that there are multiple iso-accepting species of tRNA (with different anticodons) for all amino acids and that the genetic code is degenerate. Because of this, different triplets, while coding for the same amino acid, could code for different tRNA iso-accepting species. If the amounts of iso-accepting tRNA species are widely different and there are major and minor (or limiting) species, selection for the limiting species by a particular triplet in the mRNA could limit the elongation of the nascent polypeptide to a rate in accordance with the occurrence (or activity) of the limiting tRNA. In this way, the placement of "limiting" codons into single mRNA molecules could account for regulation of synthesis of individual proteins and use of such codons in groups of mRNA molecules could account for regulation of the synthesis of functional groups of proteins. There is evidence that the levels of iso-accepting species of various tRNAs can be controlled by estrogens in avian liver and oviduct (Illinger et al. 1974) and that there is some tissue specificity displayed by tRNA from mammalian (Gilbert and Anderson 1970) and avian tissues (Gerlinger et al. 1975).

It is well established that iso-accepting species of a number of different tRNAs exist in plants (Bergquist and Robertson 1965; Legocki et al. 1968; Anderson and Cherry 1969; Burkard et al. 1970; Cherry and Osborne 1970; Vanderhoef and Key 1970; Merrick and Dure 1972). Some species may be artefacts caused by RNase action as described by Babcock and Morris (1973), by alteration of tertiary structure, by aggregation of tRNA and by changes in methylation, and some may be organellar species (Dure and Merrick 1971; Merrick and Dure 1972, 1973) although there still seems to be a good case for multiple cytoplasmic iso-accepting species. There is also some evidence that in plants, the different iso-accepting species show different codon recognition properties (Hague and Kofoid 1971) and that they show changes in relative abundance at different stages of development (Vold and Sypherd 1968a, b; Bick et al. 1970; Legocki and Wojciechowska 1970; Shugart 1972; Venkataraman and De Leo 1972; Patel and Pillay 1976). However, there are cases where development has not been accompanied by changes in iso-accepting tRNA (Vanderhoef and Key 1970; Merrick and Dure 1972). Nevertheless, in general, there is a case to be made for the operation of a tRNA-modulating scheme of translation control in plants. Whether changes in levels of iso-accepting species in plants involve changes in tRNA synthesis, processing or some reversible modification is unknown. However, it is perhaps worth noting that even though the occurrence of cytokinins in tRNA does not seem to be related to the mechanism of action of free cytokinin (section 11.3.4), the fact that the cytokinins in tRNA are placed rather strategically adjacent to the anticodon in some tRNA species may indicate a role for them in regulating tRNA activity, perhaps even the activity of individual iso-accepting species.

Active tRNA molecules are charged with the appropriate amino acid by multiple aminoacyl-tRNA synthetases which show specificity for both the amino acid and individual iso-accepting tRNA species (Anderson and Cherry 1969; Williams and Williams 1970; Kanabus and Cherry 1971; Lea and Norris 1972; Merrick and Dure 1973; Jeannin et al. 1976). Changes in the level of all or some aminoacyl synthetases could control the function of all or some tRNA species. Evidence for multiple synthetases and changes in synthetases accompanying development of plants has been obtained (Anderson and Rowan 1966; Anderson and Fowden 1969; Pollack and Filner 1970; Bick and Strehler 1971; Nathan and Richmond 1974; Guillemaut and Weil 1975; Patel and Pillay 1976).

11.2.3. Polypeptide processing and transport and protein turnover

Following the synthesis of polypeptide chains, a number of processes occur leading to the formation of fully functional proteins. Amino acids in the chain may be modified chemically in a variety of ways including phosphorylation, methylation and glycosylation (Holzer and Duntze 1971; Cantoni 1975; Rubin and Rosen 1975; Ruiz-Carrillo et al. 1975). Such changes may be necessary for protein function (and include the formation of allosteric sites), or for subsequent processing steps such as cleavage of the peptide chain (Hershko and Fry 1975) and the formation of secondary or tertiary structure. The post-translational glycosylation of precursor polypeptides was shown to be required for subsequent cleavage of eucaryotic viral proteins (Kaluza 1976). Formation of polymeric proteins requires the coordinated supply and correct assembly of subunits and many proteins require the addition of cofactors. Some proteins are made in inactive forms (zymogens) which can be stored and activated when required. Activation can involve proteolysis (for review see Neurath and Walsh 1976).

Completed proteins may be transported either to their sites of function within the cell or they may be transported out of the cell (for review see Chrispeels 1976). These events are presumably the result of characteristics of the protein, its messenger, its site of synthesis or some aspect of cellular arrangement. Having been located once, it is conceivable that proteins may be relocated depending on the requirements of the cell.

Somewhere in this sequence of events proteins are eventually broken down and so the level of a protein in a tissue is the net result of synthesis and degradation or turnover. There is a large body of evidence in general support of such a process in plants (Huffaker and Peterson 1974) but most of the information has been derived from studies of total protein rather than individual proteins. Nevertheless, there is circumstantial evidence for the turnover of a number of individual proteins, in particular, nitrate reductase (Travis et al. 1969), phenylalanine ammonia lyase (PAL) (Zucker 1971, 1972) and ribulose 1,5-diphosphate carboxylase (Peterson and Huffaker 1975). Given such a dynamic state, the possibility exists for control of protein levels by regulation of degradation or by inactivation of a protein by an inhibitor. Both seem to occur, even involving similar enzymes, since control of invertase levels in sugar cane *(Saccharum officinarum)* seems to involve enzyme degradation (Glasziou et al. 1966) but in corn *(Zea mays),* potatoes *(Solanum tuberosum)* and beets *(Beta vulgaris),* it seems to involve inhibitors (Pressey and Shaw 1966; Pressy 1968; Jaynes and Nelson 1971). Cases are described in section 11.3.4 where changes in proteolytic activity seem to account for protein degradation. Thus, regulation of the protease level or activity or the levels of enzyme inhibitors (a number of which are known to exist, for example for α-amylase (Marshall 1975), for carboxypeptidase (Ryan et al. 1974; Hass et al. 1975) and for many other proteinases (Vogel et al. 1968; Ryan 1973) would lead to changes in levels of active proteins.

In general very little is known about enzyme breakdown by proteolysis except that since some proteins may disappear while others remain, proteolysis would seem to be quite specific although there is very little information on protease specificity and heterogeneity in plants. There is circumstantial evidence for proteases specific for PAL, nitrate reductase, UDP-galactose polysaccharide transferase and fatty acid synthetase (Huffaker and Peterson 1974) so that regulation of such protease levels could control protein degradation, although it is probably difficult to decide whether the specificity resides in the protease or in the susceptibility of the substrate to proteolysis (through cofactor or configurational changes).

In principle, protein breakdown could be initiated or its rate regulated by: (1) changes in the substrate susceptibility by changes in configuration or its removal from a protective site perhaps on a membrane; (2) changes in protease involving (a) synthesis of protease, (b) activation of a zymogen, (c) changes in protease inhibitor, (d) changes in cofactors for proteolysis, (e) liberation from a bound (e.g. lysozomal) form or transport of substrate to the bound protease (e.g. into the lysozome) and (f) control of turnover of the protease itself.

11.2.4. Conclusions

The previous discussion indicates clearly that there is much to be learnt about non-transcriptional processes in animal cells and although it is possible to define some potential points of control, in most cases such definition can only be of a general nature due to the lack of basic knowledge. In plants, the lack of knowledge is particularly severe and the number of workers interested in expanding this knowledge is very small. Until the insight into plant processes is deepened and the possibilities for control become better defined, plant scientists must rely on the animal literature and on techniques designed for animal systems. In particular, there would seem to be a great need for information on pre-mRNA, its processing, the structure of mRNA and the nature of polypeptide chain initiation (or perhaps suppression) factors. The need for information on these aspects in particular will become clear from the early indications of non-transcriptional control in plants described in the next section.

11.3. Evidence for non-transcriptional control in plants

11.3.1. Gibberellin

Evidence for non-transcriptional effects of gibberellins (GA) has been provided by study of a variety of systems, but especially the barley *(Hordeum vulgare)* aleurone layer. In this tissue GA_3 promotes, inter alia, the de novo synthesis of a number of enzymes including amylase, protease, ribonuclease (Rnase) and β-1,3-glucanase (Filner and Varner 1967; Jacobsen and Varner 1967; Bennett and Chrispeels 1972). Synthesis of the minor RNA species is required for this but a requirement for the synthesis of rRNA and tRNA could not be demonstrated (see Chapter 10, section 10.3.2.4). However, although GA_3 caused a major change in poly(A) RNA metabolism and increased the level of translatable mRNA for α-amylase (see Chapter 10, section 10.3.2.4), there is still no direct evidence that the process limiting hydrolase synthesis and mRNA accumulation is mRNA synthesis, at least early in the induction period.

The stabilization or activation of mRNA has been suggested as a process by which GA_3 regulates hydrolase synthesis in barley aleurone and sugar cane. 5-Fluorouracil (5-FU) can be incorporated into mRNA and cause misreading of codons during translation (Mandel 1969) resulting in proteins with changed physical characteristics. Carlson has used this principle in an attempt to distinguish between the possibilities that amylase mRNA is made before or after addition of GA_3 (or both) (Carlson 1972). Aleurone was incubated with 5-FU for varying times before and during GA_3 treatment and amylase was examined for changes in thermolability. Among other things, it was found that tissue "labelled" with 5-FU before hormone treatment and washed and incubated in uracil to get rid of excess 5-FU, synthesized some thermolabile amylase in response to GA_3 indicating that the mRNA for this amylase had been made before GA_3 was added and GA_3 somehow regulated its translatability rather than its synthesis. However, interpretation of the results is complicated, firstly, by the possibility that 5-FU may be incorporated into tRNA as well as mRNA and this may cause misreading of the mRNA especially if the 5-FU was in the anticodon (and hence increased thermolability of the protein); secondly, that amylase made by aleurone consists of two quite distinct proteins with different characteristics (Jacobsen et al. 1970; Jacobsen and Knox 1973) and in such a study it is desirable to distinguish between these two amylases; and thirdly, it was not demonstrated that 5-FU in fact became incorporated into either mRNA or tRNA of barley. Furthermore, Carlson's results could not be repeated in a comprehensive study which examined purified, as opposed to crude, amylase (Rodaway 1976). Thus, Carlson's experiments may not constitute good evidence for non-transcriptional control by GA_3. A different approach was taken in the study of GA_3 (and auxin) in relation to invertase levels in sugar cane (Sacher and Glasziou 1962; Sacher et al. 1963). By using inhibitors to block RNA and protein synthesis separately and at various times, a system was developed in which the effects of various hormones on different processes could be examined. By applying this technique, it was found that GA_3 (and auxin) was only effective during a stage when RNA levels were limiting (Glasziou et al. 1966; Glasziou et al. 1968) indicating an effect on mRNA level, and it was an effect upon mRNA stability rather than upon synthesis (Gayler and Glasziou 1969). This approach is an interesting and useful one but it is limited by the use of inhibitors with doubtful specificities.

Chemical modification of RNA in response to GA has been studied as a possible regulatory event. GA_3-induced changes in RNA methylation have been detected indicating the possibility of hormonal control of RNA processing (Chandra and Duynstee 1971). The changes involved increased methylation of purine residues in hrRNA and tRNA over a 30 hr period. Our studies over a 16 hr period demonstrated a strong promotion of methylation (but not synthesis, see Chapter 10, section 10.3.2.4) of 4S RNA but no effect on either lrRNA or hrRNA (Zwar, J.A. and Jacobsen, J.V., unpublished data). Even so, some or all of the tRNA labelling could have been due to amino-acylation of tRNA with labelled methionine. Thus, in the short term, it seems unlikely that any GA_3-induced changes in ribosomes could be due to methylation but there may be changes in tRNA function due to GA_3-promoted methylation. Large (5–8 fold) GA_3-promoted increases in aminoacyl tRNA synthetases have been found in pea *(Pisum sativum)* seedlings (Henshall and Goodwin 1964) which could account for enhanced tRNA charging capacity, but whether such an effect is general or whether it is a primary or secondary effect is unknown. Whether GA_3 affected methylation of mRNA was not determined in any of the studies.

In such experiments as those described above, it is important to determine whether the additional methylation occurs on newly synthesized or on preexisting RNA, since only in this way can effects of hormones on processing be distinguished from effects on polynucleotide synthesis.

Increased polysome size may reflect posttranscriptional alterations in protein synthesis (see section 11.2.2.1). GA_3 appears to cause increases in the number or size of polysomes in many tissues (Villiers 1968; Poulson and Beevers 1970; Evins 1971; Evins and Varner 1972) although there are some tissues in which it has no effect on polysome profiles (Chen and Osborne 1970; Pearson and Wareing 1970; Bonnafous et al. 1973). Changes in polysome number probably reflect variation in the level of mRNA due to changes in mRNA synthesis, activation or stabilization but increases in polysome size probably reflect changes in translation. GA_3-induced changes in translation or ribosome characteristics do not seem to have been reported, however, such research would seem to be worthwhile because of the instances of GA_3-induced changes in polysome profiles and because auxin-induced changes in ribosome efficiency have been described (section 11.3.3).

Since intracellular membranes play such an important role in controlling protein biosynthesis in eucaryotic cells (Tata 1973; Palade 1975), it is worth considering whether GA might affect translational processes by altering the membrane arrangement in responsive cells (see Chapter 8, section 8.4.2). Electron microscope data indicated that ER proliferation followed GA_3 treatment of barley aleurone layers (Jones 1969a, b; Vigil and Ruddat 1973). However, using density gradient fractionation of cellular components coupled with assay of cytochrome *c* reductase, which is an ER marker enzyme, Gibson (1974) was unable to detect any increase in ER following GA_3 treatment of wheat *(Triticum aestivum)* aleurone. Similar cell fractionation studies have not been repeated on barley although it must be remembered that claims for proliferating ER in barley aleurone from electron microscope studies have not been supported by quantitative data. The biochemical data relating to membrane changes are similarly equivocal. Four to five hours after the beginning of GA_3 treatment, there was enhanced choline incorporation into membranes (Evins and Varner 1971) and enhanced [^{32}P] orthophosphate incorporation into phospholipids (Koehler and Varner 1973), but there was no effect on acetate or glycerol incorporation into lipids or on types of lipids being synthesized (Koehler and Varner 1973; Firn and Kende 1974). Thus, GA_3 apparently stimulated lipid turnover and not de novo synthesis but whether this reflects net increase in ER or rearrangement and reconstruction of existing ER is not clear. Membrane construction or rearrangement in aleurone or any other secreting tissue could be of particular importance, since many of the GA_3-induced enzymes are secreted and secreted proteins are considered to be synthesized on rough ER (Tata 1968; Palade 1975). Such considerations have already led to the proposal that construction of membranes with appropriate sites for translation of proteins destined for secretion may limit the translation of hydrolase mRNA (Johnson and Kende 1971). In addition, there is evidence that control of membrane lipid supply (either by synthesis or rearrangement of preexisting lipids), and thus membrane construction, is not dependent on RNA or protein synthesis. Activities of the particulate enzymes of the cytidine diphosphate-choline pathway, which are involved in lecithin synthesis, were enhanced by GA_3 (Johnson and Kende 1971) and the effect seemed to begin immediately after GA_3 addition. Although these responses were sensitive to cycloheximide and actinomycin D, the phosphorylcholine glyceride transferase increase was subsequently shown to be insensitive to a range of amino acid analogues as well as cordycepin (Ben-Tal and Varner 1974), indicating that neither functional proteins nor RNA synthesis were required for membrane formation. These data are convincing because inhibition of nitrate reductase was used as an internal control to demonstrate effectiveness of the inhibitors, but it still remains to be explained why cycloheximide and actinomycin D inhibited the response. In any case, it seems possible that lipid synthesis and/or turnover might be controlled by GA_3 at some non-transcriptional and non-translational step and that this may hypothetically give rise to newly constructed membrane with sites which would permit translation of preexisting or new hydrolase mRNA.

Although there are data suggesting that GA may affect membrane permeability (see Chapter 8, section 8.2.2.1), there is little evidence that membranes are primary sites of GA action. Probably the only demonstration of a direct effect of GA on a biological membrane system relates to the GA_3-promoted retention of enzymes such as glycolic oxidase in microbodies during isolation (Lips and Roth-Bejerano 1974), although it was not clear whether this was an effect on the membrane or the enzyme and specificity of the effect was not examined in any detail. However, there is considerable indirect evidence that GA can affect membranes. Wood, Paleg and others have examined the effects of GA_3 on artificially constituted membranes. Phospholipid (lecithin) and sterol spherules or micelles called liposomes were made by dispersing soybean *(Glycine max)* phospholipids and β-sitosterol in aqueous media. The liposomes thus formed showed many of the properties of natural membranes (for review see Wood and Paleg 1972). If they were constructed so as to contain uncharged (glucose and sucrose) or charged (bromate) particles, GA_3 enhanced leakage of the particles from the liposomes (Wood and Paleg 1972; Wood et al. 1972; Wood and Paleg 1974a, b) and this occurred at very low hormone to lipid ratios. The effect was temperature-dependent and the dependence varied with the hormone concentration and the constitution of the membrane. The permeability of such membranes exhibited thermal transitions presumably reflecting rather abrupt changes in fluidity of the membrane and it seemed that GA_3 acted by lowering the temperature at which thermal transition occurred. Direct interaction between GA_3 and auxin (which apparently also changes liposome permeability) with phosphatidyl-choline, a normal membrane constituent, has been detected by NMR (Paleg et al. 1974), although the interactions were detected in organic medium where the phospholipid is disorganized in contrast to the aqueous in vivo situation where lipid is organized. The GA_3-lipid interaction involved the carboxyl group of GA_3 in a 1:1 association with lipid but the IAA interaction probably did not involve the carboxyl group and occurred in a 2:1 IAA: lipid ratio. These studies indicate that GA_3 might cause permeability changes in plant tissues and the physical changes of the sort caused by GA_3 in liposomes can also change activities of membrane-bound enzymes (Lyons 1973), although there is still no direct evidence that GA_3 can act this way. However, it seems possible to repeat the permeability experiments using biological membranes which can be isolated and partly fractionated (Hartman et al. 1975).

11.3.2. Abscisic acid

Until about 1969, the data on the mechanism of action of abscisic acid (ABA) was interpreted as indicating that ABA acted on transcription rather than at a posttranscriptional site since, in general, ABA had been found to have large effects on nucleic acid synthesis but not on protein synthesis (see Chapter 10, section 10.3.5.3). Therefore, since the main effect of ABA seemed to be on nucleic acid synthesis, this was the likely site of action and therefore it was unlikely to be acting on translation. To explain this, it was proposed that ABA was an allosteric effector of both DNA and RNA polymerases (Van Overbeek et al. 1967, 1968; Pearson and Wareing 1969) and the effects of ABA on enzymes like invertase (Saunders and Poulson 1968) and α-amylase (Hemberg 1967; Chabaud et al. 1969; Mousseron-Canet et al. 1970; Hemberg 1975) seemed to support the case that ABA was an allosteric effector even though the invertase was from yeast and the amylases were of fungal origin. Thus, what was considered to be a primary effect of ABA on transcription was really a case of non-transcriptional control of transcription.

In the late 1960's, an argument in favour of non-transcriptional effects of ABA was developed. We have chosen 6 examples to illustrate the type of evidence presented which was considered, at the time, to implicate ABA in the control of posttranscriptional events rather than in the synthesis of mRNA. The examples include studies on the early events in the germination of wheat and cotton *(Gossypium hirsutum)*, the growth of oat *(Avena sativa)* coleoptile and bean *(Phaseolus vulgaris)* embryo, and enzyme induction in barley aleurone and sugar cane discs.

The early work on expansion of wheat embryos indicated that during the first 12 hr of germination, there was no mRNA transcribed and furthermore, that GA_3 and ABA had no effect on RNA synthesis during this time, although the possibility that very small changes in RNA synthesis would not have been detected was recognized (Chen and Osborne 1970). Using hybridization techniques, no differences in total hybridizable (to DNA) RNA could be found between control, GA_3-treated and ABA-treated embryos although again the limitations of the technique were recognized. There was no evidence for effects on tRNA synthesis or charging and there were no differences in polysome profiles. However, the cell-free protein synthesizing capacity of polysomes isolated from GA_3-treated tissue was greater than that of polysomes from control tissue which in turn was greater than that of polysomes from ABA-treated tissue. It was shown that both GA_3 and ABA could promote and inhibit, respectively, germination as well as total protein synthesis well within the first 12 hr of inbibition, although the effects of GA_3 and ABA were not reflected in cell-free ribosome activity using poly(U) as messenger. Thus, although there was no strong evidence for effects of GA and ABA on translation, there was apparently good reason to think that transcription was not involved and the study concluded that whatever effects the hormones had on transcription, they were secondary to control at the level of translation and, in general, it seemed that both GA_3

and ABA regulated the translation or the translatability of stored mRNA. These conclusions seemed reasonable until it was found that there was apparently mRNA (polydisperse and polysomal) synthesis going on within the first 12 hr of germination (Dobrzanska et al. 1973; Rejman and Buchowitz 1973; Doshchanov et al. 1975), although demonstration of early poly(A) RNA synthesis was only presented in one case (Spiegel et al. 1975). GA and ABA have not been shown to regulate this early mRNA synthesis, but if they do, this could account for their effects on growth and protein synthesis and this is contrary to the conclusions of Chen and Osborne (1970).

An argument similar to that used by Chen and Osborne, was used by Bonnafous and colleagues (1973) to support the conclusion that ABA prevented oat coleoptile growth by some translational (non-transcriptional) step. ABA inhibited growth and protein synthesis but not total RNA synthesis, although the criticism that crucial but small and undetected changes in RNA metabolism could have occurred applies here also. These authors pointed out as further evidence for their conclusion that changes in protein synthesis probably occurred before ABA affected RNA polymerase (Bex 1972b).

Results of studies of the effects of ABA on nucleic acid metabolism in the embryonic axis of bean seed are not entirely consistent with those from wheat embryos, but in principle they are similar. In bean embryos, ABA inhibited cell elongation but it was associated with inhibition of [^{32}P] orthophosphate incorporation into rRNA, tRNA and DNA, indicating that (unlike wheat), synthesis of these species was retarded by ABA while there was no effect on incorporation into tenaciously-bound RNA (TB-RNA) (mRNA) (Walton et al. 1970). These effects are similar to those produced by 5-FU which does not inhibit growth of bean axes. This, together with the fact that ABA rapidly stopped elongation of already elongating axes with kinetics more like those of cycloheximide than 8-azadenine, led the authors to speculate that ABA may be acting on some translational step. The possibility that ABA was inhibiting the production of a few limiting TB-RNA species cannot be eliminated.

The first evidence for effects of ABA on posttranscriptional events came from the studies of Dure and colleagues on cotton embryos, where the appearance of certain enzyme activities, including isocitrate lyase, during germination seemed to be dependent on stored messengers which were thought to be synthesized during embryogenesis (Ihle and Dure 1969). These authors produced correlative evidence that (in vivo) ABA prevented the translation of these messengers in the later stages of embryogenesis. By washing developing embryos, the translational constraint could be removed and by adding ABA, it could be re-imposed (Ihle and Dure 1970), thus indicating that ABA was acting at some transcriptional site. In addition, it was found that actinomycin D prevented the inhibitory effect of ABA which led to the hypothesis that ABA controlled messenger translation by a mechanism involving transcription. However, two subsequent studies have raised questions concerning the early observations of Dure and colleagues. The evidence that isocitrate lyase depended on preformed mRNA involved the demonstration that enzymes developed in the presence of actinomycin D during precocious embryo germination or normal seed germination (Ihle and Dure 1969). However, recently it was shown that if actinomycin D was given adequate time to enter the tissue, the production of isocitrate lyase was blocked (Smith et al. 1974; Radin and Trelease 1976) thus providing evidence that the enzyme was not translated from preformed mRNA which is more in line with isocitrate lyase development in other fatty seeds (Hock and Beevers 1966; Gientka-Rychter and Cherry 1968). While the data presented by Ihle and Dure seemed to provide a convincing case for translational control by ABA, the objections raised by Smith et al. (1974) and Radin and Trelease (1976) should be regarded as reopening the question. However, the controversy will not be resolved by using actinomycin D or other inhibitors, and quantification of the mRNA for the enzyme at various stages of seed development will be necessary.

Indications of non-transcriptional control by ABA were obtained in the studies of Gayler and Glasziou on sugar cane discs where ABA (as well as NAA and GA_3) stimulated invertase synthesis (Gayler and Glasziou 1969). However, unlike NAA and GA_3, ABA was effective during the time when RNA level was non-limiting, that is, when inhibitors of transcription were not effective, and therefore its site of action was considered to be at some unspecified posttranscriptional point although it had no posttranslational effects on invertase synthesis (Gayler and Glasziou 1969). This system should also be reexamined using more direct methods of mRNA isolation to determine whether the effect of ABA was strictly posttranscriptional.

One of the best cases for non-transcriptional control by ABA has come from study of its action in barley aleurone. GA_3-promoted amylase synthesis was found to be inhibited by addition of transcription inhibitors at the same time as the hormone, but enzyme synthesis became insensitive to these inhibitors (actinomycin D and cordycepin) after about 8–10 hr even though RNA synthesis was still inhibited (Chrispeels and Varner 1967; Goodwin and Carr 1972; Ho and Varner 1974) thus indicating that RNA synthesis was no longer limiting. ABA added after 10 hr (midcourse) could still inhibit amylase synthesis indicating that it did not act on RNA synthesis by, for example, inhibiting amylase mRNA synthesis, but that it acted at some translational (non-transcriptional) site (Chrispeels and Varner 1967; Ho and Varner 1974). So far, the argument is convincing; however, it relies on evidence of an indirect nature and on the assumption that the inhibitors of RNA synthesis do in fact inhibit mRNA synthesis added midcourse although this has not been demonstrated. The data require re-examination by direct measurement of amylase mRNA levels and turnover. Also at this stage, the possibility that ABA has both transcriptional and non-transcriptional effects cannot be eliminated. More recently, the inhibitory effect of ABA on midcourse amylase synthesis was found to be prevented by cordycepin (Ho and Varner 1976), leading to a conclusion similar to that reached in studies on ABA in cotton embryo development (Ihle and Dure 1970), namely that RNA synthesis was required for ABA to prevent amylase synthesis. The ABA-controlled RNA synthesis would presumably give rise to something like tcRNA or a translation product of some ABA-promoted RNA such as a tc-protein (see section 11.2.2.1) or an mRNA-specific nuclease whereby translation would be controlled in a way similar to that described by Tomkins and colleagues (Tomkins et al. 1969, 1972). However, again the evidence is indirect, relying on inhibitor studies and apart from needing verification, it requires support by direct examination of amylase mRNA. It should also be noted that a transcription requirement for ABA action does not necessarily indicate that ABA directs RNA synthesis. It is equally consistent with a mechanism involving the continuous synthesis and turnover of a regulatory RNA or protein which can repress certain mRNAs only in conjunction with ABA.

The data of Ho and Varner (1974, 1976) are consistent with the idea that ABA may control the translation of α-amylase mRNA during the later stages of enzyme synthesis. However, by direct measurement of amylase mRNA, it has been demonstrated that ABA, added at the same time as GA_3, inhibits the development of translatable (in vitro) amylase mRNA (Higgins et al. 1977). Whereas this may reflect posttranscriptional inactivation of otherwise complete mRNA molecules or increased degradation of newly produced or activated mRNA for amylase (or a combination of some of these), it could also reflect an effect on transcription of α-amylase mRNA.

ABA may control stability and lifetime of RNA and nucleases have been examined in this regard. As a result of studies involving the application of exogenous RNase and rifampicin (to inhibit endogenous RNase) to barley endosperm, Leshem (1971) suggested that ABA exerted non-transcriptional control over reducing-sugar production in this system. RNase, like ABA, inhibited GA_3-induced reducing-sugar production and rifampicin had no effect. The deductions that polymerase activity was unnecessary for amylase synthesis and that ABA acted by regulating nucleases are of doubtful validity because the significance of the RNase data cannot be assessed and it was not shown that rifampicin, which is considered to be primarily an inhibitor of procaryotic polymerases (Wehrli and Staehelin 1970) can affect polymerase II activity. Furthermore, in barley aleurone, there is evidence that ABA inhibits rather than promotes nuclease activity (Chrispeels and Varner 1967). Nevertheless, concomitant with its promotion of senescence, ABA does enhance nuclease levels in many leaf tissues. For example, in excised barley leaves ABA enhanced the natural increase in chromatin-bound RNase and deoxyribonuclease (DNase) (Srivastava 1968b) and in excised *Avena* leaves it acted specifically on one (out of four) particular nuclease isozyme (Wyen et al. 1972a, b). The nuclease was the only one which was RNA-specific having purine-specific endonuclease activity giving 2′:3′-cyclic phosphates as products. However, perhaps such results should be regarded with caution since it has now been shown that ABA can have positive or negative effects on RNase activity depending on the water status of the tissue (Arad et al. 1973). While ABA can (directly or indirectly) cause changes in nuclease levels, there is no direct evidence that these nucleases are responsible for breakdown of any particular RNA species and reduction of nucleic acid levels does not always parallel increase in nuclease. In lentil *(Lens culinaris)* roots, ABA inhibited growth and reduced RNA level and there was a concomitant rise in RNase (Pilet 1970). However, in maize coleoptiles ABA inhibited radioactive precursor accumulation into RNA within 3 hr whereas RNase level was unaffected for about 8 hr (Bex 1972a) and so could not have been involved. Also in radish *(Raphanus raphanistrum)* hypocotyls, ABA inhibited growth but had no perceptible effect on chromatin-bound RNase (Pearson and Wareing 1969). In senescing *Phaseolus* leaves, nucleic acid and RNase levels declined in parallel (Phillips et al. 1969). Thus, although important but minor changes in nuclease and RNA levels may have remained undetected in these and other studies, there is at present no evidence for an obligate relationship between RNase and RNA levels following ABA treatment and models based on such a relationship such as the one described above for aleurone are little more than speculation.

In the cotton and aleurone studies, there is indirect evidence that ABA does not exercise control over mRNA breakdown. In aleurone, midcourse inhibition of amylase synthesis by ABA was almost instantaneously reversed by addition of cordycepin which presumably acted by inhibiting transcription (Ho and Varner 1976). A depleted amylase mRNA level is unlikely to be replenished so quickly under conditions of inhibited transcription. Similar arguments apply to cotton embryos where actinomycin D prevented ABA from inhibiting the development of certain enzyme activities (Ihle and Dure 1970). However, some less destructive process, like for example, removal of some nucleotide sequence or removal or modification of the mRNA cap structure could have an effect on the translatability of the mRNA and such modifications could be reversible.

A systematic approach to the study of the stability and lifetime of RNA in *Lemna* was undertaken by Trewavas (1970). The study was limited to the major species of RNA and would be extremely useful if it could now be extended to include individual species of mRNA. In *Lemna,* the rates of synthesis and degradation of salt-soluble RNA (probably largely tRNA) were reduced by ABA but only the synthesis of rRNA was inhibited (Trewavas 1970) while the degradation was unaffected. Further analysis of the difference between the species is required. It should be noted that the results are contrary to the contention that ABA enhances nucleic acid breakdown through RNase.

Chemical modification of RNA has not been studied with respect to ABA although adenylation has been proposed as a control point for the translatability of mRNA in cotton embryos (Walbot et al. 1974; Hammett and Katterman 1975). Even though there seems to be some role for cytoplasmic adenylation in mRNA function in some systems such as sea urchin eggs (Slater and Slater 1974), it is not required for translation of the sea urchin mRNA (Mescher and Humphries 1974; Fromson and Duchastel 1975) or for many others (see section 11.2.1.1). Since mRNA of many seeds appears to be adenylated (Gordon and Payne 1976) and the basic question about the validity of stored mRNA in cotton is still open, the role of ABA in adenylation of cotton mRNA remains to be elucidated. However, methylation, capping and adenylation all remain as possible ways in which mRNA stability or translatability might be controlled by hormones, including ABA.

The translational process involving initiation, initiation factors, elongation, termination, polysome formation and ribosomal or tRNA activity has received little definitive analysis. Polysome levels have been studied but in many instances the profiles were so poor as to indicate considerable degradation during extraction, thus making interpretation of the data very equivocal. Perhaps this is reflected in the rather inconsistent reports on the effects of ABA on polysome profiles. The evidence available indicates that while ABA can have rather drastic inhibitory effects on growth and development processes, it does not necessarily have major discernible effects on polysomes. It prevented germination of wheat embryos and inhibited protein synthesis (Chen and Osborne 1970) but did not affect polysome distribution or level. ABA inhibited light-promoted unrolling and greening of etiolated wheat leaves but it did not prevent light-induced polysome formation (Pearson and Wareing 1970). On the other hand, in etiolated barley seedling leaves, ABA not only reduced polysomes in the dark but it inhibited the light-induced increase in polysomes as well as leaf unrolling (Poulson and Beevers 1970). It is difficult to reconcile these conflicting results from what appear to be very similar systems but at least it seems that ABA can prevent leaf unrolling while not necessarily having major effects on polysome levels and distribution. It is also of interest to note that although effects on polysomes may accompany hormone-induced changes in growth or form, it is possible to increase polysomes without causing other changes, for example polysomes were increased by light in ABA-inhibited etiolated wheat leaves as described above and by sucrose in etiolated pea epicotyls (Trewavas 1968) where sucrose had no effects on growth or form. ABA also inhibited growth of barley coleoptiles over a 5 hr period without affecting polysome profiles or the amount of ribosomes but it did inhibit protein synthesis (Bonnafous et al. 1973). The hormone inhibited polysome formation in barley aleurone tissue but these polysomes were GA_3-induced (Evins and Varner 1972). The polysome profiles in some of these studies are not typical and problems of polysome isolation described in Chapter 10, section 10.3.2.4 may be applicable in some cases; however, there appears to be no consistent relationship between ABA application and polysome numbers. This would seem to indicate that ABA probably acts by selectively regulating the levels or translatability of varying numbers of mRNA species rather than having an overall effect on translation level. Inhibition of GA_3-promoted mRNA levels (Ho and Varner 1974; Higgins et al. 1977) is consistent with this. Thus, examination of the effects of ABA on polysomes gives no indication that ABA may act on anything other than levels of active mRNA but apart from changes in phosphorylation of ribosomal proteins, there seems to have been no attempt to look for ABA-induced changes in ribosomal characteristics including overall efficiency.

In *Lemna*, there is a ribosomal (small subunit) protein of molecular weight 42,000 which is phosphorylated, and ABA inhibited both growth and the phosphorylation of this protein (Trewavas 1973). The effect on phosphorylation began after about 5 hr and the effect on growth after about 12–16 hr so that they may be causally related. Protein kinases have been shown to be associated with ribosomes of pea shoots and *Lemna* and they can phosphorylate ribosomal proteins although they lack the specificity observed in in vivo phosphorylation (Keates and Trewavas 1974). The implication of such results is that ABA-induced changes in ribosomal protein phosphorylation may regulate ribosomal efficiency, although at present the evidence is, at best, correlative. The possible significance of these and other cases of protein phosphorylation have been discussed in a recent review by Trewavas (1976).

Finally, in one of the few well documented cases, ABA was shown to regulate protein levels in *Lemna* by affecting protein degradation as well as synthesis (Trewavas 1972), but there was no indication of how either was controlled.

In summary, there seems to be a case for ABA acting at some non-transcriptional site, perhaps in addition to effects on transcription, although even the non-transcriptional effect might be mediated by transcription. It may involve the production or activation of factors which would specifically prevent the translation of certain mRNAs, modify the translatability of the mRNA or promote its degradation. Of these the first two possibilities seem most likely but even so, the proposal that ABA acts non-transcriptionally in vivo still requires definite demonstration.

11.3.3. Auxin

It has usually been found that auxin effects are dependent upon RNA synthesis, particularly on RNA thought to be messenger-like, and the hormone has been shown to regulate poly(A) RNA levels (see Chapter 10, section 10.3.1.6). Thus, while it has been concluded that auxin can control mRNA levels, there is also evidence that it may regulate non-transcriptional events.

In pea epicotyls, the indoleacetic acid (IAA) stimulated cellulase activity has been shown to be accompanied by a rise in translatable cellulase mRNA (Verma et al. 1975). Because the process required RNA synthesis, the authors noted that while the data were consistent with de novo synthesis of cellulase mRNA, they could not rule out the possibility that preexisting (pre-) mRNA was activated in some way which required RNA synthesis. An unexpected feature of this work was that the increase in translatable cellulase mRNA preceded the accumulation of cellulase. The authors should perhaps have been comparing level of cellulase mRNA with the rate of cellulase synthesis but even so, the mRNA still appeared to increase prior to enzyme accumulation. This was accompanied by a temporary reduction in the ability of supernatant factors from auxin-treated tissue to support cell-free translation, which could not only explain the lag between appearance of mRNA and its translation but also indicates that auxin may have non-transcriptional as well as transcriptional sites of action. However, the possibility that the data have alternative explanations should also be considered (see chapter 10, section 10.3.1.6).

Examination of the effects of auxin on polysome formation and ribosome characteristics has produced the best evidence available for hormone-directed changes in ribosomal components as well as changes in mRNA levels. Auxin has been shown to promote polysome formation in pea epicotyls (Trewavas 1968; Birmingham and Maclachlan 1972; Davies and Larkins 1973) and in the elongating and basal (non-elongating) regions of soybean hypocotyls (Travis et al. 1973). Studies on both species have provided evidence for transcriptional and non-transcriptional effects of auxin.

The studies of Davies and Larkins showed that decapitation of pea epicotyls caused a shift from polysomes to monosomes evident within 5 hr and that auxin (and GA_3) prevented this (Davies and Larkins 1973). In this study, the polysome distributions (with and without auxin) were very similar and although it was shown later that a non-representative sample of polysomes had been examined (Larkins and Davies 1975), the polysome profiles indicated that it was mRNA supply and not initiation which limited polysome formation. A decrease in initiation should have led to a decrease in large polysomes and an increase in small ones (section 11.2.2.1). In aged (2 days) decapitated epicotyls, where the proportion of monosomes was very high, IAA treatment led to the formation of more and bigger polysomes as well as more ribosomes. Analysis of the profiles after various times of hormone treatment, assuming that one mRNA molecule was associated with one polysome, showed that after 3 hr of IAA treatment, there was an increase in mRNA associated with increased polysomes, whereas there was no increase in ribosomes. Nevertheless, it would seem that the change in polysome distribution could also indicate a change in initiation rate. Between 3 and 6 hr, there were increases in polysomes, mRNA and ribosomes, but the increase in mRNA was much greater than the increase in ribosomes again indicating that the increase in polysomes was due primarily to more mRNA as well as perhaps more initiation. Most ribosome synthesis occurred between 6 and 10 hr. A recent reexamination of some of this work using techniques designed to examine all polysomes, confirmed the results that in aged pea stems, auxin gave rise to more polysomes as well as a greater proportion of big polysomes (Davies 1976), again indicating higher mRNA levels and greater initiation rate. Therefore, the sequence of events accompanying auxin-promoted polysome increases seemed to be: increased mRNA then possibly increased initiation, then ribosome production and then cell expansion. This study is of great interest in that it provides some indication that both mRNA supply and initiation changed in the early hours of auxin action, although it should be remembered that the bigger polysomes were primarily membrane-bound (Davies 1976) and may be translating different newly synthesized mRNA molecules with inherently greater rates of initiation than those mRNA molecules which preexisted.

Results indicating a similar early dual effect of auxin on mRNA supply as well as initiation have been obtained from studies on soybean hypocotyls. In the basal non-growing segments of soybean hypocotyls, auxin caused an increase in polysomes after 1–2 hr (which was obviously not related to growth) and this occurred in the absence (5-FU-inhibited) of rRNA synthesis (Travis et al. 1973). Therefore, the transition from monosomes to polysomes was not directed by the level of rRNA. In general, it seemed to be primarily dependent on and perhaps a result of auxin-enhanced synthesis of mRNA-like RNA (Key 1972) although there were some changes in ribosomes which accompanied the formation of polysomes (Travis et al. 1973). The cell-free protein synthesizing activity of isolated 80S ribosomes (poly(U)-directed phenylalanine incorporation) was enhanced by auxin and the increase was biphasic in parallel with polysome formation although only the second phase was auxin-dependent. The increase in ribosome efficiency was associated with an increase in the level of peptidyl tRNA isolated with the ribosomes, indicating that the change in ribosomes may have involved enhanced initiation capacity. Thus, it was proposed that formation of polysomes was dependent on ribosome activation as well as mRNA supply. The obvious possibility that auxin caused changes in initiation factors has not yet been tested although auxin-induced changes in ribosomal protein synthesis associated with increased activity of (pre-existing) ribosomes has recently been demonstrated (Travis and Key 1976). Whether such changes regulate ribosome activity or are the result of engagement of ribosomes in protein synthesis due to additional mRNA synthesis remains to be shown. Dilution of stationary phase cell cultures of peanut *(Arachis hypogea)* into fresh medium also caused an increase in ribosome efficiency which was associated with increased polysomes and protein synthesis (Verma and Marcus 1974). This tends to indicate that, as a general rule, where overall increases in rate of protein synthesis occur, increases in ribosome activity may be an obligate adjunct to enhanced mRNA levels rather than being a hormone-directed event.

Although the evidence is rather indirect, there is a case to be made for an effect of auxin on mRNA levels by regulation of mRNA stability. The production of invertase by sugar cane internode slices was enhanced by auxin (as well as GA_3) which acted only when mRNA was limiting. These and other data (Sacher and Glasziou 1962; Sacher et al. 1963) indicated that the hormone regulated mRNA supply by an effect on its stability rather than on its synthesis (see also section 11.3.1). There are a number of cases where a high auxin level has been related to suppressed RNase activity and increased RNA level (Truelson 1967; Pilet 1969, 1970, 1972; Pilet and Braun 1970) although again, no direct relationship between nuclease levels and RNA stability has been established. On the other hand, in pea epicotyls IAA caused a large increase in membrane-bound (microsomal) RNase which was suppressed by kinetin (Birmingham and Maclachlan 1972) and similarly, intensive growth in callus cultures from wheat root induced by auxins was positively correlated with some nucleases (Udvardy et al. 1976). Thus, auxin does not always suppress RNase and in some cases it may be necessary not only to create new polysomes, as in pea epicotyls, but also, to control their turnover (Birmingham and Maclachlan 1972).

There were early indications that auxin may act by complexing in some way with tRNA. There was evidence that complexes could form in vitro (Kefford et al. 1963; Galston et al. 1964) and in vivo (Bendaña et al. 1965). Such studies led to the hypothesis that auxin could act by regulating tRNA function in a manner similar to cytokinins (Armstrong 1966). However, the proposal has lost most of its support since labelled auxin has been shown to be incorporated into all RNA species, not only tRNA (Key and Ingle 1968), probably by use of auxin breakdown-products, and that most of the radioactive auxin associated with tRNA could be separated from it by further fractionation (Davies 1971; Davies and Galston 1971). Reasons why more recent claims of auxin-tRNA complexes formed in vivo (Kobayashi and Yamaki 1972; Kobayashi 1974) may not be supported by the data have recently been described by Kende and Gardner (1976). Thus, there appears to be no good evidence that auxin has any direct connection with tRNA function and so it is difficult to understand why RNA from auxin-treated tissues should be more RNase-resistant than RNA from untreated tissues (Bendaña and Galston 1965).

Finally, there are reports of direct effects of auxin on enzymes and apparently direct effects of auxin on membrane function, particularly in relation to promotion of cell elongation, both of which have been described elsewhere (see Chapters 12, Volume II, and Chapter 10). The nature of these interactions has not been examined in any detail and they will not be discussed further here.

11.3.4. Cytokinin

Evidence that cytokinins might regulate plant growth through effects on posttranscriptional events has accumulated from a variety of studies but the idea probably obtained its initial impetus from the fact that cytokinins were found in certain tRNAs (see Kende 1971). Other findings, such as cytokinin-induced changes in ribosomal activity and levels of tRNA iso-acceptors, enhanced protein synthesis in the absence of new transcription, cytokinin binding to ribosomes and effects of cytokinins on cell-free protein synthesis, have all contributed to the notion of posttranscriptional control by cytokinins.

The occurrence of cytokinins adjacent to the anticodon in some tRNA species reading codons beginning with U (for reviews see Skoog and Armstrong 1970; Kende 1971) presented the possibility that this hormone may exert some sort of control over translation by regulating levels or activity of the tRNA molecules. Changes in alanyl- and isoleucyl-tRNA have occurred following 6-benzylaminopurine treatment of lupin *(Lupinus luteus)* embryos (Legocki and Wojciechowska 1970) and increases in acceptor capacity of specific iso-accepting leucyl-tRNA species have occurred in soybean tissues (Anderson and Cherry 1969; Bick et al. 1970; Wright et al. 1972/3; Pillay and Cherry 1974), but whether this involves increased levels or increased activity of tRNA molecules is not known. In any case, it now seems that occurrence of cytokinins in tRNA and cytokinin-induced changes in certain species may not be related, since there is now strong evidence that there is no relationship between the action of free cytokinin and its occurrence in tRNA (for reviews see Burrows 1975; Kende and Gardner 1976). A recent study by Burrows (1976) concluded that there was no evidence for an association between the action of the non-purine cytokinin, diphenylurea, and the occurrence of cytokinins in tRNA. Thus, the regulation of tRNA levels by cytokinins described above, although still perhaps reflecting some sort of translational control by modulation of tRNA levels by synthesis or activation, is probably unrelated to the occurrence of cytokinins in tRNA. Thus, Kende and Gardner (1976) concluded that "the combined evidence justifies shifting the search for the site of action of cytokinins from tRNA to a hormonal receptor with which the hormone is expected to interact non-covalently".

Nevertheless, it is clear that cytokinins can affect tRNA activity and the question is whether this is caused by changes in tRNA synthesis or in some other way. There is some evidence to suggest that changes may be associated with changes in methylation (Venkataraman et al. 1970). tRNA methylation in tobacco was found to change with development, and the naturally-occurring cytokinin N^6-(Δ^2-isopentenyl)adenine enhanced methylation when added to an in vitro methylase-tRNA system (Abeels et al. 1972). However, on the other hand four adenosine analog cytokinins were found to inhibit methylation of tRNA in vitro (Wainfau and Landsberg 1971) although the methylases were from animal and bacterial cells and the concentrations of cytokinin needed to be effective were very high; 3 m*M* cytokinin was required to inhibit methylation by 30%. It seems unlikely that this effect is important in cytokinin action in plants. The possibility that methylation is important in vivo has not been examined further although it would perhaps seem worthwhile to do so. Changes in tRNA levels may also involve regulation of breakdown, although the only evidence for this comes from a study of *Lemna* in which it was found that the rate of degradation as well as the rate of synthesis of salt-soluble RNA (probably largely tRNA) was enhanced by 6-benzylaminopurine (Trewavas 1970). So far, the possibility that changes in amino-acyl synthetase activity may be involved in cytokinin action has not received much support. Treatment of senescing tobacco leaf discs with kinetin did cause an increase in synthetase activity but the effect was too slow to account for the senescence-inhibiting effects of the hormone (Anderson and Rowan 1966). Furthermore in vitro studies of acylation using tRNA and synthetases isolated from different tissues of 6-benzylaminopurine-treated soybean seedlings indicated that the hormone affected the tRNA level but had no effect on the synthetases (Anderson and Cherry 1969). mRNA synthesis did not appear to be required for the early cytokinin-induced protein synthesis in radish cotyledons (Gordon and Letham 1975). These data were based upon a relatively insensitive measure of mRNA synthesis (fractionation of labelled RNA by gel electrophoresis) and upon the action of actinomycin D, neither of which could be relied upon to detect changes in the metabolism of minor RNA species. Similarly, a case was made for non-transcriptional control in soybean cells where cytokinin stimulated polysome formation in the presence of actinomycin D (Short et al. 1974). These data indicate, but do not prove, that mRNA synthesis was not occurring and further analysis is warranted before it can be firmly concluded that transcription did not play a role in the responses.

Polysome formation in tissues treated with cytokinins has not received the same attention as those treated with auxin but as mentioned earlier, the appearance of polysomes in soybean cells was stimulated by cytokinins, apparently in the absence of mRNA synthesis (Short et al. 1974). Cytokinin-induced changes in polysome profiles of expanding radish cotyledons appeared to be due to enhanced extractability of the hormone-treated tissue (Gordon 1973). In several studies, kinetin was found to augment the effects of auxin in promoting polysome formation but did not promote polysome formation itself (Klämbt 1974; Birmingham and Maclachlan 1972). This has been interpreted as indicating that kinetin can exert control over translation but not over RNA synthesis or levels.

Evidence suggesting that cytokinins might act at the level of translation was provided by the findings that ribosomes could display fairly selective binding of cytokinins (Berridge et al. 1970, 1972; Fox and Erion 1975). Cell-free systems have been developed to which exogenous cytokinins have been added, but no quantitative changes in protein synthesis could be demonstrated (Berridge et al. 1970, 1972; Yakovleva et al. 1976). In contrast, Klämbt (1976) showed that N^6-(Δ^2-isopentenyl)adenine did stimulate cell-free protein synthesis slightly in systems derived from tobacco *(Nicotiana tabacum)* pith and from corn using special conditions of dialysis of the extracts. In the above cell-free systems, it is likely that initiation of polypeptide synthesis was occurring at very low rates and that elongation was the predominant event being measured. Using two cell-free systems, one consisting of pea polysomes, which was synthesizing protein mainly by chain elongation, and the other derived from wheat germ which was both initiating and elongating polypeptide chains using pea cotyledon mRNA, a range of cytokinins and cytokinin metabolites has been tested for effects on the quantitative and qualitative translation products. It was found that neither system responded quantitatively nor qualitatively to any of the cytokinins tested (Higgins and Letham, unpublished). Further analysis is clearly required and the biological significance of cytokinin binding to ribosomes remains to be resolved.

The effects of cytokinins on organellar "cell-free" protein synthesis may support the case for non-transcriptional control. The incorporation of amino acids into isolated plastids and mitochondria was promoted very rapidly, within 5 min, by cytokinin (Davies and Cocking 1967; Bhattacharya and Roy 1969) and this has been interpreted as an effect on translation although it is still possible that it involved transcription of the organellar genome or perhaps changes in membrane permeability.

Changes in ribosome function have been sought by treating tissue for varying periods of time and then isolating the particle and testing its capacity for protein synthesis. The cytokinin-promoted expansion of fenugreek *(Trigonella foenum-graecum)* cotyledons was accompanied by an increase in the activity of the ribosomes in cell-free protein synthesis which was accounted for, at least in part, by an increase in initiation activity (Rijven and Parkash 1971), although it may have been non-specific initiation since poly(U) and not natural mRNA was the template. The activity was reduced by washing the ribosomes in KCl which was interpreted to mean that changes in levels (or activities) of initiating factors were responsible for the effect although there are alternative interpretations and initiating factor activity was not assayed directly. Further analysis, using an inhibitor of initiation, showed that initiating activity of the ribosomes was stimulated (Rijven 1974). This is a promising approach and may provide an account of cytokinin action during this important step in translation if further studies, using natural mRNA, purified initiation components and partial reactions of the initiation sequence are carried out.

Modification of ribosomal proteins has also been studied as a hormone-responsive phenomenon related to protein synthesis. Cytokinins have been found to affect protein phosphorylation and ribosomal proteins have been shown to be phosphorylated in several plant tissues by kinases attached to the ribosomes. There are two cases where cytokinins have been found to inhibit the activity of the ribosomal protein kinases in vitro (Ralph et al. 1972; Keates and Trewavas 1974) indicating a possible way for cytokinins to regulate ribosome activity. However, the levels of cytokinins required for the effect were very high, high enough to inhibit the growth of *Lemna,* and it has been suggested that the in vitro effect may be due to cytokinin competing with ATP in the kinase reaction (Keates and Trewavas 1974) rather than any specific effect. Nevertheless, transfer of a cytokinin-requiring soybean callus to a cytokinin-deficient medium caused a loss of polysomes and an increase in phosphorylation of only a few ribosomal proteins (Tepfer and Fosket 1975), although it was not shown that reapplication of cytokinin enhanced polysomes and depressed phosphorylation. Thus, although there is no good evidence that cytokinin reduces phosphorylation and enhances translation in vivo, at least the results are in keeping with studies in animal cells where phosphorylation of ribosomal proteins is associated with inactivity of ribosomes (Kabat 1970; Monier et al. 1972).

Cytokinin depression of protein phosphorylation in isolated membranes by membrane-bound kinases has recently been correlated with cytokinin-promotion of expansion in Chinese cabbage *(Brassica pekinensis)* leaf discs (Ralph et al. 1976). The results were interpreted in terms of control of membrane permeability by phosphorylation but direct evidence for this was not presented. The arguments for specificity of cytokinin action on certain kinases were compelling but still more evidence is required. On the other hand in soybean callus, cytokinins were shown to enhance phosphorylation of a variety of particulate and non-particulate acidic and basic proteins (Elliott and Murray 1973). Clearly much more work on protein phosphorylation particularly in relation to the rapidity and specificity of the effect, must be done before it can be seriously considered as a control mechanism in plant tissues, but, as recently pointed out (Trewavas 1976), the importance of protein phosphorylation in animal cells is good reason to believe that research on protein phosphorylation in plants is likely to be very rewarding.

There is little evidence that cytokinins can affect rRNA metabolism non-transcriptionally except for one case involving *Lemna.* It was found that whereas 6-benzylaminopurine applied under optimal growth conditions had no effects on rRNA metabolism, under suboptimal conditions (growing on water only) it caused increases in the rates of both synthesis and degradation of rRNA (Trewavas 1970). The physiological significance of this effect is unknown.

The possibility that cytokinins regulate nuclease levels and thus control nucleic acid levels has been studied. In senescing leaf tissues there are usually losses of RNA and protein, and cytokinins prevent these losses from occurring (see Chapter 10, section 10.3.3.4). Increases in RNase and DNase occur at the same time as nucleic acid losses, and the increases in nucleases are suppressed by cytokinins. Consequently, it seemed that cytokinins might cause changes in nucleic acid levels by regulating nuclease level or activity (Srivastava and Ware 1965; Srivastava 1968a, b). As described previously for ABA (see section 11.3.2), there is no good evidence for a direct relationship between nucleic acid levels and nuclease activity and the same applies to cytokinins. However, at least for cytokinins, there is evidence for control of a particular nuclease which indicates there may be some specificity. In *Avena* leaves it was found that the rise in RNase occurring after leaf excision occurred in both soluble and particulate RNase and that kinetin suppressed the rise in both fractions (Udvardy et al. 1967). Similar results were reported for wheat leaves (Sodek and Wright 1969). However, the *Avena* leaf nucleases were fractionated into component enzymes and specificity of kinetin action was demonstrated. The effects of kinetin and ABA (but not GA_3 or IAA) on RNase levels were due primarily to rapid changes in one nuclease (Wyen et al. 1972a, b). It was one of four well-characterized nucleases and the only one which was RNA specific, having endo-RNase activity (purine specific) giving 2′:3′-cyclic phosphates as products. Cytokinins do not always suppress nucleases since in barley leaves, kinetin can have both positive and negative effects on RNase activity. Leaves grown in the absence of water stress responded to kinetin by suppressing RNase but in water-stressed tissue, kinetin enhanced RNase activity (Arad et al. 1973). The control may operate indirectly through effects on tissue water level but in the case of ABA (see section 11.3.2) the relationship between hormones, nucleases and nucleic acids is a complex one.

Finally, there is some evidence that cytokinins can cause changes in protein degradation. It is well known that they retard the loss of protein which usually occurs in senescing leaf tissues and there is evidence that this is due to reduced protein degradation rather than, or in addition to, increase in synthesis. Kinetin retarded the loss of radioactive label from prelabelled protein in leaves of *Brassica rapa* L. (Kuraishi 1968) and 6-benzylaminopurine had a similar effect in corn leaves (Tavares and Kende 1970). In the second case, it was clear that there was no major effect on protein synthesis so that the protein-conserving effect of cytokinin was due largely but not entirely to effects on protein degradation. Similarly, inhibitor studies on tobacco leaf discs led to the conclusion that 6-benzylaminopurine retarded protein loss both by promoting synthesis and by inhibiting protein degradation (Takegami 1975). Since cytokinins have been shown to suppress the increase in protease which normally occurs in senescing leaves (Balz 1966; Atkin and Srivastava 1969), it would seem that cytokinins could regulate overall protein level by regulation of proteolysis. The recently described conservation of ribulose-1,5-diphosphate carboxylase in senescing wheat leaves by kinetin application may well be such a case since the effect was correlated with suppression of proteolytic activity (Peterson and Huffaker 1975). Since cycloheximide prevented the increase in protease activity, it was suggested that the enzyme was being synthesized during senescence. In *Lemna*, cytokinins as well as ABA regulated protein levels but the effects were on both protein degradation and synthesis (Trewavas 1972). Under optimal growth conditions, 6-benzylaminopurine enhanced protein levels by enhancing synthesis, but in plants grown on water only, the hormone reduced protein degradation. Why the mechanisms differed remains obscure. Whereas an effect of cytokinin on protein stability (mediated by protease) can be considered as a non-transcriptional control, changes in the activity of protease could well involve transcriptional control.

11.3.5. Ethylene

There is very little evidence to indicate a primary non-transcriptional role for ethylene. However, as the gibberellin-controlled membrane construction in barley aleurone cells may offer a plausible basis for non-transcriptional control of protein synthesis and distribution (see section 11.3.1), the ethylene-controlled increases in length of endoplasmic reticulum profiles in pedicel abscission zone cells of tobacco (Valdovinos et al. 1971, 1972), and in the hook region of etiolated *Pisum sativum* seedlings (Osborne et al. 1972; Sargent et al. 1973; Sargent and Osborne 1975) might also represent a way of altering protein synthesis and the pattern of cell growth. In *Pisum,* very long profiles of both rough and smooth ER appeared within 6 hr of ethylene application and since phospholipid levels declined (Irvine and Osborne 1973), it appeared that the ER changes resulted from membrane reorganization rather than proliferation, or perhaps from membrane synthesis from reserve lipid as appears to occur in barley aleurone (Koehler and Varner 1973; Firn and Kende 1974).

Ethylene has been found to affect some properties of artificial lipid and protein membranes although in a non-specific way (Mehard et al. 1970). This, in addition to the long-recognized lipophilic nature of ethylene, suggested the possibility that ethylene may act in vivo by directly affecting membrane characteristics. However, at present there is little support for this idea. The ethylene-controlled aging of morning glory *(Ipomoea tricolor)* flowers has been associated with changes in tonoplast permeability (Kende and Baumgartner 1974; Hanson and Kende 1975) but although this was an early result of ethylene action, it took 2 hr to appear, making it seem unlikely that the membrane was the primary site of ethylene action.

Thus, there is very little data at present to suggest a non-transcriptional site(s) of action for ethylene. However, ethylene can act very rapidly as in pea seedlings where it depressed growth rate with a lag of only about 10 min (Warner and Leopold 1971), and so the possibility of non-transcriptional control by ethylene certainly cannot be ruled out. The timing of many responses in other systems has also often been used to distinguish between transcriptional and non-transcriptional effects. The validity of this approach is highly dubious and until the rates of transcription and translation in a variety of conditions are known it is dangerous to interpret rapid responses as being unrelated to transcription and translation.

11.4. Discussion and conclusions

From the preceding discussion it is clear that there is very little convincing support for non-transcriptional control of metabolism by plant hormones. Except in a few instances,

References

Abeels, M., Digneffe, C. and Dubois, E. (1972) Symp. Biol. Hung., 13, 69–74.
Acton, G.J. (1972) Nature New Biology, 236, 255–256.
Adesnik, M., Salditt, M., Thomas, W. and Darnell, J.E. (1972) J. molec. Biol., 71, 21–30.
Ajtkhozhin, M.A. and Akhanov, A.U. (1974) FEBS Lett., 41, 275–279.
Ajtkhozhin, M.A., Akhanov, A.U. and Doschanov, Kh.J. (1973) FEBS Lett., 31, 104–106.
Ajtkhozhin, M.A., Polimbetova, N.S. and Akhanov, A.U. (1975) FEBS Lett., 54, 212–216.
Ajtkhozhin, M.A., Doschanov, Kh.J. and Akhanov, A.U. (1976) FEBS Lett., 66, 124–126.
Ames, B.N. and Hartman, P. (1963) Cold Spr. Harb. Symp. quant. Biol., 28, 349–356.
Anderson, J.W. and Fowden, L. (1969) Plant Physiol., 44, 60–68.
Anderson, J.W. and Rowan, K.S. (1966) Biochem. J., 101, 15–18.
Anderson, M.B. and Cherry, J.H. (1969) Proc. nat. Acad Sci. (Wash.), 62, 202–209.
Anderson, W.F., Bosch, L., Gros, F., Grunberg-Manago, M., Ochoa, S., Rich, A. and Staehelin, T. (1974) FEBS Lett., 48, 1–6.
Arad, S., Mizrahi, Y. and Richmond, A.E. (1973) Plant Physiol., 52, 510–512.
Armstrong, D.J. (1966) Proc. nat. Acad. Sci. (Wash.), 56, 64–66.
Ascione, R. and VanDerWoude, G.P. (1969) J. Virol., 4, 727–737.
Atkin, R.K. and Srivastava, B.I.S. (1969) Physiol. Plant., 22, 742–750.
Babcock, D.F. and Morris, R.O. (1973) Plant Physiol., 52, 292–297.
Balz, H.D. (1966) Planta, 70, 207–236.
Barden, N. and Labrie, F. (1973) Biochemistry, 12, 3096–3102.
Barrieux, A., Ingraham, H.A., Nystul, S. and Rosenfield, G. (1976) Biochemistry, 15, 3523–3528.
Bayley, S.T. (1964) J. molec. Biol., 8, 231–238.
Bendaña, F.E. and Galston, A.W. (1965) Science, 150, 69–70.
Bendaña, F.E., Galston, A.W., Kaur-Sawhney, R. and Penny, P.J. (1965) Plant Physiol., 40, 977–983.
Bennett, P.A. and Chrispeels, M.J. (1972) Plant Physiol., 49, 445–447.
Ben-Tal, Y. and Varner, J.E. (1974) Plant Physiol., 54, 813–816.
Bergquist, P.L. and Robertson, J.M. (1965) Biochim. biophys. Acta (Amst.), 95, 357–359.
Berridge, M.V., Ralph, R.K. and Letham, D.S. (1970) Biochem. J., 119, 75–84.
Berridge, M.V., Ralph, R. and Letham, D. (1972) in: Plant Growth Substances 1970. D.J. Carr ed. Springer, Berlin, pp. 248–255.
Bester, A.J., Kennedy, D.S. and Heywood, S.M. (1975) Proc. nat. Acad. Sci. (Wash.), 72, 1523–1527.
Bex, J.H.M. (1972a) Planta, 103, 1–10.
Bex, J.H.M. (1972b) Planta, 103, 11–17.
Bhattacharya, J. and Roy, S.C. (1969) Biochem. biophys. Res. Commun., 35, 606–610.
Bick, M.D., Liebke, H., Cherry, J.H. and Strehler, B.L. (1970) Biochim. biophys. Acta (Amst.), 204, 172–182.
Bick, M.P. and Strehler, B.L. (1971) Proc. nat. Acad. Sci. (Wash.), 68, 224–228.
Birmingham, B.C. and Maclachlan, G.A. (1972) Plant Physiol., 49, 371–375.
Birnstiel, M.L., Chipchase, M. and Speirs, J. (1971) Prog. Nucleic Acid Res. Molecular Biol., 11, 351–389.
Bixby, J.A. and Brown, G.N. (1975) Plant Physiol., 56, 617–621.
Blat, C. and Loeb, J.E. (1971) FEBS Lett., 18, 124–126.
Blobel, G. and Dobberstein, B. (1975) J. Cell Biol., 67, 835–851.
Bogdanovsky, D., Herman, W. and Schapira, G. (1973) Biochem. biophys. Res. Commun., 54, 25–32.
Bonnafous, J.C., Mousseron-Canet, M. and Olivé, J.L. (1973) Biochim. biophys. Acta (Amst.), 312, 165–171.
Borek, E. and Srinivasan, P.R. (1966) Ann. Rev. Biochem., 35, 275–298.
Both, G.W., Banerjee, A.K. and Shatkin, A.J. (1975) Proc. nat. Acad. Sci. (Wash.), 72, 1189–1193.
Brawerman, G. (1974) Ann. Rev. Biochem., 43, 621–642.
Burdon, R.H. (1971) Prog. Nucleic Acid Res. Molecular Biol., 11, 33–72.

Burkard, G., Guillemaut, P. and Weil, J.H. (1970). C.R. Soc. Biol. (Paris), 163, 2731–2736.
Burrows, W.J. (1975) Curr. Advanc. Plant. Sci., No. 21, 837–847.
Burrows, W.J. (1976) Planta, 130, 313–316.
Cantoni, G.L. (1975) Ann. Rev. Biochem., 44, 335–451.
Carlson, P.S. (1972) Nature New Biol., 237, 39–41.
Chabaud, J., Mousseron-Canet, M. and Durand, B. (1969) C.R. Acad. Sci. (Paris) Ser. C., 269, 865–868.
Chandra, G.R. and Duynstee, E.E. (1971) Biochim. biophys. Acta (Amst.), 232, 514–523.
Chen, D. and Osborne, D.J. (1970), Nature (Lond.), 226, 1157–1160.
Chen, D., Sarid, S. and Katchalski, E. (1968) Proc. nat. Acad. Sci. (Wash.), 60, 902–909.
Chen, G.S. and Siddiqui, M.A.Q. (1975) J. Molec. Biol., 96, 153–170.
Cherry, J.H. and Anderson, M.B. (1972) in: Plant Growth Substances 1970, D.J. Carr, (ed.), Springer, Berlin, pp. 181–189.
Cherry, J.H. and Osborne, D.J. (1970) Biochem. biophys. Res. Commun., 40, 763–769.
Choe, B.K. and Taylor, M.W. (1972) Biochim. biophys. Acta (Amst.), 272, 275–287.
Chrispeels, M.J. and Varner, J.E. (1967) Plant Physiol., 42, 1008–1016.
Chrispeels, M.J. (1976) Ann. Rev. Plant Physiol., 27, 19–38.
Cooper, H.L. (1973) J. Cell Biol., 59, 250–254.
Cooper, H.L. and Gibson, E.M. (1971) J. biol. Chem., 246, 5059–5066.
Covey, S.N. and Grierson, D. (1976) Europ. J. Biochem., 63, 599–606.
Davies, E. (1976) Plant Physiol., 57, 516–518.
Davies, E. and Larkins, B.A. (1973) Plant Physiol., 52, 339–345.
Davies, J.W. and Cocking, E.C. (1967) Biochem. J., 104, 23–33.
Davies, P.J. (1971) Plant Cell Physiol., 12, 785–789.
Davies, P.J. and Galston, A.W. (1971) Plant Physiol., 47, 435–441.
Dobrzanska, M., Tomaszewski, M., Grzelczak, Z., Rejman, E. and Buchowicz, J. (1973) Nature (Lond.), 244, 507–508.
Doel, M.T. and Carey, N.H. (1976) Cell, 8, 51–58.
Doshchanov, K.I., Ajtkhozhin, M.A. and Darkanbaev, T.B. (1975) Fiziol. Rast., 22, 368–375.
Dure, L. III and Merrick, W.C. (1971) in: Autonomy and Biogenesis of Mitochondria and Chloroplasts, N.K. Boardman, A.W. Linnane and R.M. Smillie (eds.), North Holland, Amsterdam, pp. 413–421.
Dure, L. III and Waters, L.C. (1965) Science, 147, 410–412.
Dyer, T.A. and Payne, P.I. (1974) Planta, 117, 259–268.
Elliott, D.C. and Murray, A.W. (1973) in: Plant Growth Substances 1973, Hirokawa, Tokyo, pp. 691–698.
Esnault, R., Trafy, S. and Van Huystee, R.B. (1975) Europ. J. Biochem., 60, 83–89.
Evins, W.H. (1971) Biochemistry, 10, 4295–4303.
Evins, W.H. and Varner, J.E. (1971) Proc. nat. Acad. Sci. (Wash.), 68, 1631–1633.
Evins, W.H. and Varner, J.E. (1972) Plant Physiol., 49, 348–352.
Filner, P. and Varner, J.E. (1967) Proc. nat. Acad. Sci. (Wash.), 58, 1520–1526.
Firn, R.D. and Kende, H. (1974) Plant Physiol., 54, 911–915.
Fox, J.E. and Erion, J.L. (1975) Biochem. biophys. Res. Commun., 64, 694–700.
Fromson, D. and Duchastel, A. (1975) Biochim. biophys. Acta (Amst.), 378, 394–404.
Fuhr, J.E. and Natta, C. (1972) Nature New Biology, 240, 274–276.
Fuhr, J.E. and Overton, M. (1975) Biochem. biophys. Res. Commun., 63, 742–747.
Fukami, H. and Itano, H. (1976) Biochemistry, 15, 3529–3535.
Furuichi, Y. (1974) Nucleic Acid Res., 1, 809–822.
Furuichi, Y., Morgan, M., Muthukrishnan, S. and Shatkin, A.J. (1975a) Proc. nat. Acad. Sci. (Wash.), 72, 362–365.
Furuichi, Y., Morgan, M., Shatkin, A.J., Jelinek, W., Salditt-Georgieff, M. and Darnell, J.E. (1975b) Proc. nat. Acad. Sci. (Wash.), 72, 1904–1908.

Gagnon, C. and deLamirande, G. (1972) Arch. Biochem. Biophys., 150, 573–577.

Galston, A.W., Jackson, P., Kaur-Sawhney, R., Kefford, N.P. et Meudt, W.J. (1964) in: Regulateurs Naturels de la Croissance Vegetale, J.P. Nitsch (ed.), Centre National de la Recherche Scientifique, Paris, pp. 251–264.

Gayler, K.R. and Glasziou, K.T. (1969) Planta, 84, 185–194.

Gebauer, H., Seitz, U. and Seitz, U. (1975) Z. Naturforsch., 30, 213–218.

Georgiev, G.P. and Samarina, O.P. (1971) Advanc. Cell Biol., 2, 47–110.

Gerlinger, P., Le Meur, M.A. and Ebel, J.P. (1975) FEBS Lett., 49, 376–379.

Gibson, R.A. (1974) Ph.D. Thesis, The Lysosomal Nature of Hormonally Induced Enzymes in Wheat Aleurone Cells, University of Adelaide, 277 pp.

Gientka-Rychter, A. and Cherry, J.H. (1968) Plant Physiol., 43, 653–659.

Gilbert, J.M. and Anderson, W.F. (1970) J. biol. Chem., 245, 2342–2349.

Glasziou, K.T., Waldron, J.C. and Bull, T.A. (1966) Plant Physiol., 41, 282–288.

Glasziou, K.T., Gayler, K.R. and Waldron, J.C. (1968) in: Biochemistry and Physiology of Plant Growth Substances, F. Wightman and G. Setterfield (eds.) Runge Press, Ottawa, pp. 433–442.

Glisin, V.R. and Glisin, M.V. (1964) Proc. nat. Acad. Sci. (Wash.), 52, 1548–1553.

Goodwin, P.B. and Carr, D.J. (1972) Planta, 106, 1–12.

Gordon, M.E. (1973) Ph.D. Thesis, Cytokinins and the Expansion of Radish Cotyledons, Aust. Nat. Univ. Canberra., 186 pp.

Gordon, M.E. and Letham, D.S. (1975) Aust. J. Plant Physiol., 2, 129–154.

Gordon, M.E. and Payne, P.I. (1976) Planta 130, 269–273.

Grierson, D. and Covey, S. (1976) Planta, 130, 317–321.

Gross, P.R., Kraemer, K. and Malkin, L.I. (1965) Biochem. biophys. Res. Commun., 18, 569–575.

Guillemaut, P. and Weil, J.W. (1975) Biochim. biophys. Acta (Amst.), 407, 240–248.

Hague, D.R. and Kofoid, E.C. (1971) Plant Physiol., 48, 305–311.

Hall, R.H. (1973) Ann. Rev. Plant Physiol., 24, 415–444.

Hammett, J.R. and Katterman, F.R. (1975) Biochemistry, 14, 4375–4379.

Hanson, A.D. and Kende, H. (1975) Plant Physiol., 55, 663–669.

Harris, B. and Dure, L. III (1974) Biochemistry, 13, 5463–5467.

Hartman, M.A., Normand, G. and Benveniste, P. (1975) Plant. Sci. Lett., 5, 287–292.

Hass, G.M., Nau, H., Biemann, K., Grahn, D.T., Ericsson, L.H. and Neurath, H. (1975) Biochemistry, 14, 1334–1342.

Hemberg, T. (1967) Acta Chem. Scand., 21, 1665–1666.

Hemberg, T. (1975) Physiol. Plant., 35, 11–15.

Henshall, J.D. and Goodwin, T.W. (1964) Phytochemistry, 3, 677–691.

Hershko, A. and Fry, M. (1975) Ann. Rev. Biochem., 44, 775–797.

Heywood, S.M. (1970) Proc. nat. Acad. Sci. (Wash.), 67, 1782–1788.

Heywood, S.M. and Kennedy, D.S. (1976) Biochemistry, 15, 3314–3319.

Heywood, S.M., Kennedy, D.S. and Bester, A.J. (1974) Proc. nat. Acad. Sci. (Wash.), 71, 2428–2431.

Heywood, S.M., Kennedy, D.S. and Bester, A.J. (1975) Europ. J. Biochem., 58, 587–593.

Higgins, T.J.V., Mercer, J.F.B. and Goodwin, P.B. (1973) Nature New Biology, 246, 68–70.

Higgins, T.J.V., Zwar, J.A. and Jacobsen, J.V. (1976) Nature (Lond.), 260, 166–169.

Higgins, T.J.V., Zwar, J.A. and Jacobsen, J.V. (1977) in: Acides Nucléiques et Synthèse des Protéines chez les Végétaux, 261, C.N.R.S., Paris, pp. 481–486.

Ho, D.T. and Varner, J.E. (1974) Proc. nat. Acad. Sci. (Wash.), 71, 4783–4786.

Ho, D.T. and Varner, J.E. (1976) Plant Physiol., 57, 175–178.

Hock, B. and Beevers, H. (1966) Z. Pflanzenphysiol., 55, 405–414.

Holzer, H. and Duntze, W. (1971) Ann. Rev. Biochem., 40, 345–374.

Howard, G.A., Traugh, J.A., Crosser, E.A. and Traut, R.R. (1975) J. molec. Biol., 93, 391–404.

Hsiao, T.C. (1964) Biochim. biophys. Acta (Amst.), 91, 598–605.

Huez, G., Marbaix, G., Hubert, E., Leclercq, M., Nudel, U., Soreq, H., Salamon, R., Lebleu, B., Revel, M. and Littauer, U.Z. (1974) Proc. nat. Acad. Sci. (Wash.), 71, 3143–3146.

Huffaker, R.C. and Peterson, L.W. (1974) Ann. Rev. Plant Physiol., 25, 363–392.
Ihle, J.N. and Dure, L. III, (1969) Biochem. biophys. Res. Commun., 36, 705–710.
Ihle, J.N. and Dure, L. III, (1970) Biochem. biophys. Res. Commun., 38, 995–1001.
Illinger, D., Le Meur, M.A., Gerlinger, P. and Ebel, J.P. (1974) Biochimie, 56, 529–536.
Irvine, R.F. and Osborne, D.J. (1973) Biochem. J., 136, 1133–1135.
Itano, H. (1963) Proc. Symp. Abnormal Haemoglobins. Ibadan. Nigeria.
Jacobsen, J.V. and Knox, R.B. (1973) Planta, 112, 213–224.
Jacobsen, J.V. and Varner, J.E. (1967) Plant Physiol., 42, 1596–1600.
Jacobsen, J.V. and Zwar, J.A. (1974) Proc. nat. Acad. Sci. (Wash.), 71, 3290–3293.
Jacobsen, J.V., Scandalios, J.G. and Varner, J.E. (1970) Plant Physiol., 45, 367–371.
Jaynes, T.A. and Nelson, O.E. (1971) Plant Physiol., 47, 629–634.
Jeannin, G., Burkard, G. and Weil, J.H. (1976) Biochim. biophys. Acta (Amst.), 442, 24–31.
Johnson, K.D. and Kende, H. (1971) Proc. nat. Acad. Sci. (Wash.), 68, 2674–2677.
Jones, B.L., Nagabhushan, N., Gulyas, A. and Zalik, S. (1972) FEBS Lett., 23, 167–170.
Jones, R.L. (1969a) Planta, 87, 119–133.
Jones, R.L. (1969b) Planta, 88, 73–86.
Kabat, D. (1970) Biochemistry, 9, 4160–4174.
Kaluza, E. (1976) J. Virol., 19, 1–12.
Kanabus, J. and Cherry, J.H. (1971) Proc. nat. Acad. Sci. (Wash.), 68, 873–876.
Keates, R.A.B. and Trewavas, A.J. (1974) Plant Physiol., 54, 95–99.
Kefford, N.P., Kaur-Sawhney, R. and Galston, A.W. (1963) Acta chem. scand., 17, S 313–318.
Kende, H. (1971) Int. Rev. Cytol., 31, 301–338.
Kende, H. and Baumgartner, B. (1974) Planta, 116, 279–289.
Kende, H. and Gardner, G. (1976) Ann. Rev. Plant Physiol., 27, 267–290.
Kennedy, D.S., Bester, A.J. and Heywood, S.M. (1974) Biochem. biophys. Res. Commun., 61, 415–423.
Kennedy, T.D. and Lane, B.G. (1975) Canad. J. Biochem., 53, 1346–1348.
Key, J.L. (1972) Symp. Biol. Hung., 13, 15–28.
Key, J.L. and Ingle, J. (1968) in: Biochemistry and Physiology of Plant Growth Substances, F. Wightman and G. Setterfield (eds.), Runge Press, Ottawa, pp. 711–722.
Key, J.L. and Silflow, C. (1975) Plant Physiol., 56, 364–369.
Klämbt, D. (1974) Planta, 118, 7–16.
Klämbt, D. (1976) Plant Cell Physiol., 17, 73–76.
Kobayashi, K. (1974) in: Plant Growth Substances 1973, Hirokawa, Tokyo, pp. 573–581.
Kobayashi, K. and Yamaki, T. (1972) Plant Cell Physiol., 13, 49–65.
Koehler, D.E. and Varner, J.E. (1973) Plant Physiol., 52, 208–214.
Kumar, A. and Subramanian, A.R. (1975) J. molec. Biol., 94, 409–423.
Kuraishi, S. (1968) Physiol. Plant., 21, 78–83.
Lanzani, G.A., Caldiroli, E., Zocchi, G., Manzocchi, H., Bollini, R. and Alberti, L. (1975) Biochim. biophys. Acta (Amst.), 407, 449–458.
Larkins, B.A. and Davies, E. (1975) Plant Physiol., 55, 749–756.
Lea, P.J. and Norris, R.D. (1972) Phytochemistry, 11, 2897–2920.
Legocki, A.B. and Marcus, A. (1970) J. biol. Chem., 245, 2814–2818.
Legocki, A.B. and Wojciechowska, K. (1970) Bull Acad. Pol. Sci. Ser. Sci. Biol., 18, 7–13.
Legocki, A.B., Szymkowiak, A., Hierowski, M. and Pawelkiewicz, J. (1968) Acta biochim. polon., 15, 197–200.
Leshem, Y. (1971) Physiol. Plant., 24, 85–89.
Liau, M.C., Hunt, M.E. and Hurlbert, R.B. (1976) Biochemistry, 15, 3158–3164.
Lin, C.Y. and Key, J.L. (1973) Phytochemistry, 12, 43–53.
Lips, S.H. and Roth-Bejerano, N. (1974) in: Plant Growth Substances 1973, Hirokawa, Tokyo, pp. 719–724.
Lizardi, P.M. (1976) Cell, 7, 239–245.

Lodish, H.F. (1976) Ann. Rev. Biochem., 45, 39–72.
Lucas-Lenard, J. and Lipmann, F. (1971) Ann. Rev. Biochem., 40, 409–448.
Lyons, J.M. (1973) Ann. Rev. Plant Physiol., 24, 445–466.
McKnight, G.S. and Schimke, R.T. (1974) Proc. nat. Acad. Sci. (Wash.), 71, 4327–4331.
MacNaughton, M., Freeman, K.B. and Bishop, J.O. (1974) Cell, 1, 117–125.
Maden, B.E.H. and Salina, M. (1974) J. molec. Biol., 88,133–164.
Mandel, H.G. (1969) Prog. Molec. Subcell. Biol., 1, 82–135.
Manahan, C.O., App, A.A. and Still, C.C. (1973) Biochem. biophys. Res. Commun., 53, 588–595.
Marcus, A. and Feeley, J. (1964) Proc. nat. Acad. Sci. (Wash.), 51, 1075–1079.
Marshall, J.J. (1975) Ann.. Chem. Soc. Symp. Ser., 15, 244–266.
Martini, O.H.W. and Gould, H.J. (1975) Molec. Gen. Genet., 142, 299–316.
Matile, Ph. (1968) Z. Pflanzenphysiol., 58, 365–368.
Mehard, C.W., Lyons, J.M. and Kumamoto, J. (1970) J. Membr. Biol., 3, 173–179.
Merrick, W.C. and Dure, L. III (1972) J. biol. Chem., 247, 7988–7999.
Merrick, W.C. and Dure, L. III (1973) Biochemistry, 12, 629–635.
Mescher, A. and Humphries, T. (1974) Nature (Lond.), 249, 138–139.
Monier, D., Santhanam, K. and Wagle, S.R. (1972) Biochem. biophys. Res. Commun., 46, 1881–1886.
Mousseron-Canet, M., Mani, J., Durand, B., Nitsch, J., Dornand, J. and Bonnafous, J. (1970) C.R. Acad. Sci. (Paris), Ser.. D., 270, 1936–1939.
Murphy, W. and Attardi, G. (1973) Proc. nat. Acad. Sci. (Wash.), 70, 115–119.
Muthukrishnan, S., Both, G.W., Furuichi, Y. and Shatkin, A.J. (1975) Nature (Lond.), 255, 33–37.
Nagabhushan, N., Gulyas, A. and Zalik, S. (1974) Plant Physiol., 53, 516–518.
Nathan, I. and Richmond, A. (1974) Biochem. J., 140, 169–173.
Neurath, H. and Walsh, K.A. (1976) Proc. nat. Acad. Sci. (Wash.), 73, 3825–3832.
Nudel, V., Lebleu, B. and Revel, M. (1973) Proc. nat. Acad. Sci. (Wash.), 70, 2139–2144.
Osborne, D.J., Ridge, I. and Sargent, J.A. (1972) in: Plant Growth Substances 1970, D.J. Carr, (ed.), Springer, Berlin, pp. 534–542.
Palade, G. (1975) Science, 189, 347–358.
Paleg, L.G., Wood, A. and Spotswood, T.M. (1974) in: Plant Growth Substances 1973, Hirokawa, Tokyo, pp. 732–736.
Patel, H.V. and Pillay, D.T.N. (1976) Phytochemistry, 15, 401–405.
Payne, P.I. and Boulter, D. (1974) Planta, 117, 251-258.
Pearson, J.A. and Wareing, P.F. (1969) Nature (Lond.), 221, 672–673.
Pearson, J.A. and Wareing, P.F. (1970) Planta, 93, 309–313.
Pederson, T. and Kumar, A. (1971) J. molec. Biol., 61, 655–668.
Penon, P. (1972) Symp. Biol. Hung., 13, 325–334.
Penon, P., Cecchini, J., Miassod, R., Ricard, J., Teissere, M. and Pinna, M. (1970) Phytochemistry, 9, 73–86.
Perry, R.P. (1976) Ann. Rev. Biochem., 45, 605–692.
Perry, R.P., Kelley, D.E., Friderici, K. and Rottman, F. (1975) Cell, 4, 387–394.
Peterson, L.W. and Huffaker, R.C. (1975) Plant Physiol., 55, 1009–1015.
Phillips, D.R., Horton, R.F. and Fletcher, R.A. (1969) Physiol. Plant., 22, 1050–1054.
Pilet, P.E. (1969) C.R. Acad. Sci. (Paris) Ser. D, 268, 930–932.
Pilet, P.E. (1970) J. exp. Bot., 21, 446–451.
Pilet, P.E. (1972) Symp. Biol. Hung., 13, 313–323.
Pilet, P.E. and Braun, R. (1970) Physiol. Plant., 23, 245–250.
Pillay, D.T.N. and Cherry, J.H. (1974) Canad. J. Bot., 52, 2499–2504.
Pitt, D. and Galpin, M. (1973) Planta, 109, 233–258.
Pollack, B.M. and Filner, P. (1970) Plant Physiol., 46, Suppl, 30.
Poulson, R. and Beevers, L. (1970) Plant Physiol., 46, 782–785.
Pressey, R. (1968) Plant Physiol., 43, 1430–1434.
Pressey, R. and Shaw, R. (1966) Plant Physiol., 41, 1657–1661.

Proudfoot, N.J. and Brownlee, G.G. (1976) Nature (Lond.), 263, 211–214.
Radin, J.W. and Trelease, R.N. (1976) Plant Physiol., 57, 902–905.
Ralph, R.K., McCombs, P.J.A., Tener, G. and Wojcik, S.J. (1972) Biochem. J., 130, 901–911.
Ralph, R.K., Bullivant, S. and Wojcik, S.J. (1976) Biochim. biophys. Acta (Amst.), 421, 319–327.
Rattman, F., Shatkin, A.J. and Perry, R.R. (1974) Cell, 3, 197–199.
Rejman, E. and Buchowicz, J. (1973) Phytochemistry, 12, 271–276.
Rijven, A.H.G.C. (1974) Nature (Lond.), 252, 257–259.
Rijven, A.H.G.C. and Parkash, V. (1971) Plant Physiol., 47, 59–64.
Ro-Choi, T.S., Reddy, R., Choi Y-C., Raj, N.B. and Henning, D. (1974) Fed. Proc., 33, 1548.
Rodaway, S. (1976) Plant Physiol., 57, Suppl. 101.
Roman, R., Brooker, J.D., Seal, S.N. and Marcus, A. (1976) Nature (Lond.), 260, 359–360.
Rose, J. and Lodish, H.F. (1976) Nature (Lond.), 262, 32–37.
Rourke, A.W. and Heywood, M. (1972) Biochemistry, 11, 2061–2066.
Rubin, C.S. and Rosen, O.M. (1975) Ann. Rev. Biochem., 44, 831–887.
Ruiz-Carrillo, A., Beato, M., Schutz, G., Feigelson, P. and Allfrey, V.G. (1973) Proc. nat. Acad. Sci. (Wash.), 70, 3641–3645.
Ruiz-Carrillo, A., Wangh, L.J. and Allfrey, V.G. (1975) Science, 190, 117–128.
Ryan, C.A. (1973) Ann. Rev. Plant Physiol., 24, 173–196.
Ryan, C.A., Hass, G.M. and Kuhn, R.W. (1974) J. biol. Chem., 249, 5495–5499.
Sabatini, D., Borgese, N., Adelman, M., Kreibich, G. and Blobel, G. (1972) in: FEBS Symposia Series v. 27. RNA viruses: Replication and structure (and) Ribosome structure, Function and Biogenesis, H. Bleomendal, E.M.J. Jaspars, A. van Kammen and R.J. Planta (eds.), North Holland, Amsterdam, pp. 147–171.
Sacher, J.A. and Glasziou, K.T. (1962) Biochem. biophys. Res. Commun., 8, 280–282.
Sacher, J.A., Hatch, M.D. and Glasziou, K.T. (1963) Physiol. Plant., 16, 836–842.
Sagher, D., Edelman, M. and Jakob, K.M. (1974) Biochim. biophys. Acta (Aust.), 349, 32–38.
Sargent, J.A. and Osborne, D.J. (1975) Planta, 124, 199–205.
Sargent, J.A., Atack, A.V. and Osborne, D.J. (1973) Planta, 109, 185–192.
Saunders, P.F. and Poulson, R.H. (1968) in: Biochemistry and Physiology of Plant Growth Substances, F. Wightman and G. Setterfield (eds.), Runge Press, Ottawa, pp. 1581–1592.
Schmid, B.D., Siegel, N.R. and Vanderhoef, L.N. (1975) Plant Physiol., 55, 277–281.
Seal, S.N., Bewley, J.D. and Marcus, A. (1972) J. biol. Chem., 247, 2592–2597.
Sensky, T.E., Haines, M.E. and Rees, K.R. (1975) Biochim. biophys. Acta (Amst.), 407, 430–438.
Sharma, O.K. and Borek, E. (1974) Advanc. Enzyme Regulation, 12, 85–102.
Sherton, C.C. and Wool, I.G. (1974) J. biol. Chem., 249, 2258–2267.
Shine, J. and Dalgarno, L. (1974) Proc. nat. Acad. Sci. (Wash.), 71, 1342–1346.
Shine, J. and Dalgarno, L. (1975) Nature (Lond.), 254, 34–38.
Short, K.C., Tepfer, D.A. and Foskett, D.E. (1974) J. Cell Sci., 15, 75–87.
Shugart, L. (1972) Exp. Gerontol., 7, 251–262.
Sim, W. and Klämbt, D. (1976) Planta, 131, 47–50.
Singer, R.H. and Penman, S. (1973) J. molec. Biol., 73, 321–334.
Skoog, F. (1973) Biochem. Soc. Symp., 38, 195–215.
Skoog, F. and Armstrong, D.J. (1970) Ann. Rev. Plant. Physiol., 21, 359–384.
Slater, I. and Slater, D.W. (1974) Proc. nat. Acad. Sci. (Wash.), 71, 1103–1107.
Smith, R.H., Schubert, A.M. and Benedict, C.R. (1974) Plant Physiol., 54, 197–200.
Sodek, L. and Wright, S.T.C. (1969) Phytochemistry, 8, 1629–1640.
Spiegel, S., Obendorf, R. and Marcus, A. (1975) Plant Physiol., 56, 502–507.
Spirin, A.S. (1969) Europ. J. Biochem., 10, 20–35.
Spirin, A.S. (1972) Frontiers Biol., 27, 515–538.
Spohr, G., Imazumi, T. and Scherrer, K. (1974) Proc. nat. Acad. Sci. (Wash.), 71, 5009–5013.
Srivastava, B.I.S. (1968a) Biochem. biophys. Res. Commun., 32, 533–538.
Srivastava, B.I.S. (1968b) Biochim. biophys. Acta (Amst.), 169, 534–536.

Srivastava, B.I.S. and Ware, G. (1965) Plant Physiol., 40, 62–64.
Stegmann, J. and Hahn, H. (1974) Z. Pflanzenphysiol., 74, 143–151.
Steitz, J.A. and Jakes, K. (1975) Proc. nat. Acad. Sci. (Wash.), 72, 4734–4738.
Stent, G.S. (1964) Science, 144, 816–820.
Sueoka, N. and Kano-Sueoka, T. (1970). Prog. Nucleic Acid Res. Molec. Biol., 10, 23–55.
Sunshine, G.H., Williams, D.J. and Rabin, B.R. (1971) Nature New Biology, 230, 133–136.
Takegami, T. (1975) Plant Cell Physiol., 16, 407–416.
Tata, J.R. (1968) Nature (Lond), 219, 331–337.
Tata, J.R. (1973) in: Protein Synthesis in Reproductive Tissue, E. Diczfalusy (ed.), Karolinska Institute, Stockholm, pp. 192–224.
Tate, W.P. and Caskey, C.T. (1974) Molec. Cell Biochem., 5, 115–126.
Tavares, J. and Kende, H. (1970) Phytochemistry, 9, 1763–1770.
Tepfer, D.A. and Fosket, D.E. (1975) Phytochemistry, 14, 1161–1165.
Tobin, E.M. and Klein, A.O. (1975) Plant Physiol., 56, 88–92.
Tomkins, G.M., Thomas, D.G., Granner, D., Martin, D. Jr., Samuels, H.H. and Thompson, E.B. (1969) Science, 166, 1474–1480.
Tomkins, G.M., Levinson, B.B., Baxter, J.D. and Dethlefsen, L. (1972) Nature New Biology, 239, 9–14.
Traught, J.A. and Porter, G.G. (1976) Biochemistry, 15, 610–616.
Travis, R.L. and Key, J.L. (1976) Plant Physiol., 57, 936–938.
Travis, R.L., Jordan, W.R. and Huffaker, R.C. (1969) Plant Physiol., 44, 1150–1156.
Travis, R.L., Anderson, J.M. and Key, J.L. (1973) Plant Physiol., 52, 608–612.
Trewavas, A. (1968) Phytochemistry, 7, 673–681.
Trewavas, A. (1970) Plant Physiol., 45, 742–751.
Trewavas, A. (1972) Plant Physiol., 49, 47–51.
Trewavas, A. (1973) Plant Physiol., 51, 760–767.
Trewavas, A. (1976) Annu. Rev. Plant Physiol., 27, 349–374.
Truelson, T.A. (1967) Physiol. Plant., 20, 1112–1119.
Udvardy, J., Farkas, G.L., Marré, E. and Forti, G. (1967) Physiol. Plant., 20, 781–788.
Udvardy, J., Sivok, B. and Nemet, G. (1976) Z. Pflanzenphysiol., 78, 33–40.
Valdovinos, J.G., Jensen, T.E. and Sicko, L.M. (1971) Plant Physiol., 47, 162–163.
Valdovinos, J.G., Jensen, T.E. and Sicko, L.M. (1972) Planta, 102, 324–333.
Vanderhoef,L.W. and Key, J.L. (1970) Plant Physiol., 46, 294–298.
Van der Walle, C. (1973) FEBS Lett., 34, 31–34.
Van Overbeek, J., Loeffler, J.E. and Mason, M.I.R. (1967) Science, 156, 1497–1499.
Van Overbeek, J., Loeffler, J.E. and Mason, M.I.R. (1968) in: Biochemistry and Physiology of Plant Growth Substances, F. Wightman and G. Setterfield (eds.), Runge Press, Ottawa, pp. 1593–1607.
Vassart, G., Dumont, J.E. and Cantraine, F.R.L. (1971) Biochim. biophys. Acta (Amst.), 247, 471–485.
Vaughan, M.H., Soeira, R., Warner, J.R. and Darnell, J.E. (1967) Proc. nat. Acad. Sci. (Wash.), 58, 1527–1534.
Venkataraman, R. and De Leo, P. (1972) Phytochemistry, 11, 923–927.
Venkataraman, R., De Leo, P., Anderson, M.B. and Cherry, J.H. (1970) Plant Physiol., 45, Suppl, 9.
Verma, D.P.S., Maclachlan, G.A., Byrne, H. and Ewings, D. (1975) J. biol. Chem., 250, 1019–1026.
Verma, D.P..S. and Marcus, A. (1974) Plant Physiol., 53, 83–87.
Vigil, E.L. and Ruddat, M. (1973) Plant Physiol., 51, 549–558.
Villiers, T.A. (1968) Planta, 82, 342–354.
Vogel, R., Trautschold, I. and Werle, E. (1968) Natural Proteinase Inhibitors, Academic Press, New York.
Vold, B.S. and Sypherd, P.S. (1968a) Plant Physiol., 43, 1221–1226.
Vold, B.S. and Sypherd, P.S. (1968b) Proc. nat. Acad. Sci. (Wash.), 59, 453–458.
Wainfau, E. and Landsberg, B. (1971) FEBS Lett., 19, 144–148.

Walbot, V. (1974) Plant Physiol., 54, Suppl. 8.
Walbot, V., Capdevila, A. and Dure, L. III (1974) Biochem. biophys. Res. Commun., 60, 103–110.
Walton, D.C., Soofi, G.S. and Sondheimer, E. (1970) Plant Physiol., 45, 37–40.
Warner, H.L. and Leopold, A.C. (1971) Biochem. biophys. Res. Commun., 44, 989–994.
Warner, J.R. (1974) in: Ribosomes, M. Nomura, A. Tissieres and P. Lengyel (eds.), Cold Spring Harbour, New York, pp. 461–488.
Weeks, D.P. and Marcus, A. (1971) Biochim. biophys. Acta (Amst.), 232, 671–684.
Weeks, D.P., Verma, D.P.S., Seal, S.N. and Marcus, A. (1972) Nature (Lond.), 236, 167–168.
Wehrli, W. and Staehelin, M. (1970) in: RNA Polymerase and Transcription, L. Silvestri (ed.), North Holland, Amsterdam, pp. 65–70.
Wei, C.M. and Moss, B. (1975) Proc. nat. Acad. Sci. (Wash.), 72, 318–322.
Weissbach, H. and Ochoa, S. (1976) Ann. Rev. Biochem., 45, 191–216.
Wells, G.N. and Beevers, L. (1973) Plant Sci. Lett., 1, 281–286.
Wells, G.N. and Beevers, L. (1974) Biochem. J., 139, 61–69.
Wells, G.N. and Beevers, L. (1975) Arch. Biochem. Biophys., 170, 384–391.
Williams, G.R. and Williams, A.S. (1970) Biochem. biophys. Res. Commun., 39, 858–863.
Wilson, C.M. (1971) Plant Physiol., 48, 64–68.
Wimmer, E. (1972) J. Molec. Biol., 68, 537–540.
Wood, A. and Paleg, L.G. (1972) Plant Physiol., 50, 103–108.
Wood, A. and Paleg, L.G. (1974a) Aust. J. Plant Physiol., 1, 31–40.
Wood, A. and Paleg, L.G. (1974b) in: Plant Growth Substances 1973, Hirokawa, Tokyo, pp. 725–731.
Wood, A., Paleg, L.G. and Sawhey, R. (1972) in: Plant Growth Substances 1970, D.J. Carr (ed.), Springer-Verlag, Berlin, pp. 37–43.
Wool, I.G. and Stoffler, G. (1974) in: Ribosomes, M. Nomura, A. Tissieres and P. Lengyel (eds.), Cold Spring Harbour, New York, pp. 417–460.
Wright, R.D., Pillay, D.T.N. and Cherry, J.H. (1972/73) Mech. Age. Dev., 1, 403–412.
Wyen, N.V., Erdei, S., Udvardy, J., Bagi, G. and Farkas, G.L. (1972a) J. exp. Bot., 23, 37–44.
Wyen, N.V., Udvardy, J., Erdei, S. and Farkas, G.L. (1972b) Symp.. Biol. Hung., 13, 293–297.
Yakovleva, L.A., Klyachko, N.L. and Kulaeva, O.N. (1976) Soviet Plant Physiol., 22, 750–752.
Yarwood, A., Payne, E.S., Yarwood, J.N. and Boulter, D. (1971) Phytochemistry, 10, 2305–2311.
Yoshida, K. (1974) Plant Cell Physiol., 15, 441–456.
Zucker, M. (1971) Plant Physiol., 47, 442–444.
Zucker, M. (1972) Ann. Rev. Plant Physiol., 23, 133–156.

Subject Index